Science of Microscopy

Science of Microscopy

Edited by

Peter W. Hawkes
John C.H. Spence

Volume II

 Springer

Peter W. Hawkes
CEMES-LOE du CNRS
Toulouse
France

John C.H. Spence
Department of Physics
Arizona State University
Tempe, AZ
and
Lawrence Berkeley Laboratory
Berkeley, CA
USA

Library of Congress Control Number: 2005927385

ISBN 10: 0-387-25296-7
ISBN 13: 978-0387-25296-4

Printed on acid-free paper.

9 8 7 6 5 4 3 2 1

springer.com

Preface

In these two volumes, we have asked many of the leaders in the field of modern microscopy to summarize the latest approaches to the imaging of atoms or molecular structures, and, more especially, the way in which this aids our understanding of atomic processes and interactions in the organic and inorganic worlds.

Man's curiosity to examine the nanoworld is as at least as old as the Greeks. Aristophanes, in a fourth-century BC play, refers to a burning glass; the Roman rhetorician Seneca describes hollow spheres of glass filled with water being used as magnifiers, while Marco Polo in the thirteenth century remarks on the Chinese habit of wearing spectacles. Throughout this time it would have been common knowledge that a drop of water over a particle on glass will provide a magnified image, while a droplet within a small hole does even better as a biconvex lens. By the sixteenth century magnifying glasses were common in Europe, but it was Anthony van Leeuwenhoek (1632–1723) who first succeeded in grinding lenses accurately enough to produce a better image with his single-lens instrument than with the primitive compound microscopes then available. His 112 papers, published in *Philosophical Transactions of the Royal Society*, brought the microworld to the general scientific community for the first time, covering everything from sperm to the internal structure of the flea. Robert Hooke (1635–1703) developed the compound microscope, publishing his results in careful drawings of what he saw in his *Micrographia* (1665). The copy of this book in the University of Bristol library shows remarkable sketches of faceted crystallites, below which he has drawn piles of cannon balls, whose faces make corresponding angles. This strongly suggests that Hooke believed that matter consists of atoms and had made this discovery long before its official rediscovery by the first modern chemists, notably Dalton in 1803. (Greeks such as Leucippus (450 BC) had long before convinced themselves that a stone, cut repeatedly, would eventually lead to "a smallest fragment" or fundamental particle; Democritus once said that "nothing exists except atoms and empty space. All else is opinion" (!))

This atomic hypothesis itself has a fascinating history, and is intimately connected with the history of microscopy. It was Brown's observation in 1827 of the motion of pollen in water by optical microscopy which laid the basis for the modern theory of matter based on atoms. As late as 1900 many chemists and physicists did not believe in atoms, despite the many independent estimates that could be made of their size. These were summarized by Kelvin and Tait in an appendix to their *Treatise on Natural Philosophy*, together with an erroneous and rather superficial estimate of the age of the earth, to be used against Darwin. (This text was the standard English language physics text of the late nineteenth century, despite its failure to cover much of Maxwell's work.) Einstein's 1905 theory of Brownian motion, and Perrin's (1909) more accurate repetition of Brown's experiment, using microscope observations to estimate Avogadro's number, finally settled the matter regarding the existence of atoms. Einstein does not reference Brown's paper, but indicates that he had been told about it. But as Archie Howie has commented, it is interesting to speculate how different the history of science would be if Maxwell had read the Brown paper and applied his early statistical mechanics to it. By the time of Perrin's paper, Bohr, Thomson, Rutherford and others were well committed to atomic and even subatomic physics.

In biology, the optical microscope remained an indispensable tool from van Leeuwenhoek's time with many incremental improvements, able to identify bacteria and their role in disease, but not viruses, which were first seen with the transmission electron microscope (TEM) in 1938. With Zernike's phase contrast theory in the thirties a major step forward was taken, but the really dramatic and spectacular modern advances had to await the widespread use of the TEM, the invention of the laser and the CCD detector, the introduction of the scanning mode, computer control and data acquisition, and the production of fluorescent proteins.

The importance of this early history should not be underestimated—in the words of Feynman "If in some cataclysm, all scientific knowledge were to be destroyed and only one sentence passed on to the next generation of creatures, what statement would contain the most information in the fewest words? I believe it is the atomic hypothesis—that all things are made of atoms."

Images of individual atoms were first provided by Muller's field-ion microscope in the early 1950s, soon to be followed by Albert Crewe's Scanning Transmission Electron Microscope (STEM) images of heavy atoms on thin-film surfaces in 1970. With its subångström resolution, the modern transmission electron microscopes (TEM) can now routinely image atomic columns in thin crystals. For favorable surface structures, the scanning tunneling microscope has provided us with images of individual surface atoms since its invention in 1982, and resulted in a rich spin-off of related techniques.

Probes of condensed and biological matter must possess a long lifetime if they are to be used as free-particle beams. For the most part this has limited investigators to the use of light, X-rays, neutrons and electrons. The major techniques can then often be classified as imaging, diffraction, and spectroscopy. These may be used in both the transmission and reflection geometries, giving bulk and surface information respectively. Chapter 8 (Bauer) reviews both the low-energy electron microscope (LEEM) and spin-polarized LEEM methods which, using reflected electrons, have recently revolutionized surface science and thin-film magnetism. Here the high cross-section allows movies to made of surface processes at submicrometer resolution, while Auger electron spectroscopy is conveniently incorporated. Chapter 9 (Feng and Scholl) deals with the closely related photoelectron microscopy, where a LEEM instrument is used to image the photoelectrons excited by a synchrotron beam. Here the superb energy selectivity of optical excitation can be used to great advantage. Chapter 3 (Reichelt) describes advances in scanning electron microscope (SEM) research, where the lower-energy secondary electrons provide images with large depth of focus for the most versatile of all electron-optical instruments. The numerous modes of operation include X-ray analysis, cathodolumines-cence, low-voltage modes for insulators and the controlled-atmosphere environmental SEM (ESEM). Turning now to the transmission geo-metry, we review the latest work in atomic-resolution TEM in Chapter 1 (Kirkland and Hutchison), the technique which has transformed our understanding of defect processes in the bulk of solids such as oxides, and the STEM mode in Chapter 2 (Nellist). STEM provides an addi-tional powerful analytical capability, which, like the STM, can provide spectroscopy with atomic-scale spatial resolution. An entire chapter (Chapter 4, Botton) is then devoted to analytical TEM (AEM), with a detailed analysis of the physics and performance of its two main detec-tors, for characteristic X-ray emission and energy-loss spectroscopy. The remarkable recent achievements of *in-situ* TEM are surveyed in Chapter 6 (Ross), including transmission imaging of liquid cell elec-trolysis, observations of the earliest stages of crystal and nanotube growth, phase transitions and catalysts, superconductors, magnetic and ferroelectric domains and plastic deformation in thin films, all at nanometer resolution or better. Again, the large scattering cross-section of electron probes provides plenty of signal even from individual atoms, so that movies can be made. Chapter 5 (King, Armstrong, Bostanjoglo and Reed) summarizes the dramatic recent revival of time-resolved electron microscope imaging, which uses laser-pulses to excite pro-cesses in a sample. The excited state may then be imaged by passing the delayed pulse to the photocathode of the TEM in this "pump-probe" mode. Single-shot transmission electron diffraction patterns have now been obtained using electron pulses as short as a picosecond. Most of these techniques are undergoing a quiet revolution as electron-optical aberration corrector devices are being fitted to microscopes. The dramatic discovery, that, after 60 years of effort, aberration correction is now a reality, was made about ten years ago, and we review the rel-evant electron-optical theoretical background in Chapter 10 (Hawkes). Finally, in biology, potentially the largest scientific payoff of all is occur-ring in the field of cryo-electron microscopy, where single-particle images of macromolecules embedded in thin films of ice are imaged, and a three-dimensional reconstruction is made. While the structure of the ribosome and purple membrane protein (among many others) have already been determined in this way, the grand challenge of locating every protein and molecular machine in a single cell remains to be completed. We summarize this exciting field in Chapter 7 (Plitzko and Baumeister).

Electrons, with the largest cross-section and a source brighter than current generation synchrotrons, provide the strongest signal and hence the best resolution. They do this in a manner that can conveniently be combined with spectroscopy, and we now have aberration-corrected lenses for them. But multiple scattering and inelastic background scattering often complicate interpretation. X-ray imaging of nanostructures, even at synchrotrons, involves much longer data acquisition times but the absence of background and multiple scattering effects greatly improves quantification of data, and thicker samples can be examined. (It is easy to show that the small magnitude of the fine-structure constant will almost certainly never permit imaging of individual atoms using X-rays. We should also recall that in protein crystallography, about 98% of the X-ray beam hits the beam-stop and does not interact with the sample. Of the remaining 2%, 84% is annihilated in production of photoelectrons, and 8% in Compton scattering, while only the remainder produces Bragg diffraction. For light elements the inelastic cross-section for kilovolt electrons is about three times the elastic.) Success with X-rays has thus come mainly through the use of crystallographic redundancy to reduce radiation damage in protein crystallography. However, soft X-ray imaging with zone-plate lenses provides about 30 nm resolution in the "water window" with the advantages of thick samples and an aqueous environment. Applications have also been found in environmental science, materials science and magnetic materials. In addition, the equivalent of the STEM has been developed for soft X-rays: the scanning transmission x-ray microscope (STXM), which uses a zone-plate to focus X-rays onto a sample that can be translated by piezo motors. This arrangement can then provide spatially-resolved X-ray absorption spectroscopy. This work is reviewed in Chapter 13 (Howells, Jacobsen and Warwick).

Both X-ray and electron-beam imaging methods are limited in biology by the radiation damage they create, unlike microscopy with visible light, which also allows observations in the natural state. Optical microscopy is undergoing a revolution, with the development of super-resolution, two-photon, fluorescent labeling and scanning confocal methods. These methods are reviewed in Chapters 11 and 12. Chapter 11 (Diaspro, Schneider, Bianchini, Caorsi, Mazza, Pesce, Testa, Vicidomini, and Usai) discusses two-photon confocal microscopy, in which the spot-scanning mode is adopted, and a symmetrical lens beyond the sample collects light predominantly from the excitation region, thereby eliminating most of the "out-of-focus" background produced in the normal full-field "optical sectioning" mode. Three-dimensional image reconstruction is then possible. Two-photon microscopy combines this with a fluorescence process in which two low-energy incident photons are required to excite a detectable photon emitted at the sum of their energies. This has several advantages, by reducing radiation damage and background, and allowing observation of thicker samples. The method can also be used to initiate photochemical reactions for study. Chapter 12 (Hell and Schönle) describes the latest super-resolution schemes for optical microscopy, which have now brought the lateral resolution to about 28 nm and, by the symmetrical lens arrangement (4-π confocal), increased resolution measured along the optic axis by a factor of up to seven. The lateral resolution can be improved by modulating the illumination field or by using the stimulated emission depletion microscopy mode (STED), in which saturated excitation of a fluorophor produces nonlinear effects allowing the diffraction barrier to resolution to be broken.

For the scanning near-field probes new possibilities arise. Although restricted to the surface (the site of most chemical activity) and requiring in some cases complex image interpretation, damage is reduced, while the subångström resolution normal to the surface is unparalleled. The method is also conveniently combined with spectroscopy. Early work was challenged by problems of reproducibility and tip artifacts, but Chapters 14–17 in this book show the truly remarkable recent progress in surface science, materials science and biology. Chapter 14 (Nikiforov and Bonnell) describes the various modes of atomic force microscopy which can be used to extract atomic-scale information from the surfaces of modern materials, including oxides and semiconductors. Work-functions can be mapped out (a Kelvin probe with good spatial resolution) and a variety of useful signals obtained by modulation spectroscopy methods. In this way maps of magnetic force, local dopant density, resistivity, contact potential and topography may be obtained. Chapter 15 (Sutter) describes applications of the scanning tunneling microscope (STM) in materials science, including inelastic tunneling, surface structure analysis in surface science, the information on electronic structure which may be extracted, atomic manipulation, quantum size and subsurface effects, and high temperature imaging. Weierstall, in Chapter 16, reviews STM research at low temperatures, including a thorough analysis of instrumental design considerations and applications. These include measurements of local density-of-states oscillations, energy dispersion measurements, electron confinement, lifetime measurements, the Stark and Kondo effects, atomic manipulation, local inelastic tunneling spectroscopy, photon emission, superconductivity and spin-polarized tunneling microscopy. Finally, in Chapter 16, Amrein reviews the special problems that arise when the atomic force microscope (AFM) is applied to the imaging of biomolecules; much practical information on instrumentation and sample preparation is provided, and many striking examples of cell and macromolecule images are shown.

We include two chapters on unconventional "lensless" imaging methods—Chapter 18 (Dunin-Borkowski, Kasama, McCartney and Smith) deals with electron holography and Chapter 19 with diffractive imaging. Gabor's original 1948 proposal for holography was intended to improve the resolution of electron microscopes, and only recently have his plans been realized. Meanwhile, electron holography using Möllenstedt's biprism and the Lorentz mode has proved an extremely powerful method of imaging the magnetic and electrostatic fields within matter. Dramatic examples have included TEM movies of superconducting vortices as temperature and applied fields are varied, and ferroelectric and magnetic domain images, all within thin self-supporting films. Chapter 19 (Spence) describes the recent develop-

A projection from a three-dimensional image of a carbon nanotube with gold clusters attached. This was reconstructed by taking a series of projected STEM-ADF images at different tilt angles. A faceted gold cluster is shown in the inset. Electron tomography makes it possible to study the three-dimensional nanotube–metal contact geometry which determines the electrical contact resistance to the nanotubes (courtesy of J. Cha, M. Weyland, and D. Muller, 2006).

ment of new iterative solutions to the non-crystallographic phase problem, which now allows diffraction-limited images to be reconstructed from the far-field scattered intensity distribution. This has produced lensless atomic-resolution images of carbon nanotubes (reconstructed from electron microdiffraction patterns) and phase contrast images from both neutron and soft X-ray Fraunhofer diffraction patterns of isolated, non-periodic objects. In this work, lenses are replaced by computers, so that images may now be formed with any radiation for which no lens exists, free of aberrations. Our volumes end with a chapter by van Aert, den Dekker, van Dyck and van den Bos on the definition of resolution in all its forms.

Coverage has been limited to high-resolution methods, with the result that some important new microscopies have been omitted (such as magnetic resonance imaging (MRI), projection X-ray tomography, acoustic imaging etc.). Field-ion microscopy and near-field optical microscopy are also absent. Conversely, although there is no chapter on tomography in materials science, we must mention the rapid progress of these techniques, which has culminated in a remarkable near-atomic reconstruction by J. Cha, M. Weyland and D. Muller of a carbon nanotube to which gold clusters are attached (see figure). For further information on this branch of tomography, see Midgley and Weyland (2003), Midgley (2005), Weyland et al. (2006) and Kawase et al. (2006).

The ingenuity and creativity of the microscopy community as recorded in these pages are remarkable, as is the spectacular nature of the images presented. Neither shows any signs of abating. As in the past, we fully expect major advances in science to continue to result from breakthroughs in the development of new microscopies.

Peter W. Hawkes
John C.H. Spence

References

Kawase, N., Kato, M., Nishioka, H. and Jinnai, H. (2006). Transmission electron microtomography without the "missing wedge" for quantitative structural analysis. *Ultramicroscopy* (2006) forthcoming.

Midgley, P.A. (2005). Tomography using the transmission electron microscope. In *Handbook of Microscopy for Nanotechnology* (Yao, N. and Wang, Z.L., Eds) 601–627 (Kluwer, Boston).

Midgley, P.A. and Weyland, M. (2003). 3D electron microscopy in the physical sciences: the development of Z-contrast and EFTEM tomography. *Ultramicroscopy* **96**, 413–431.

Weyland, M., Yates, T.J.V., Dunin-Borkowski, R.E., Laffont, L. and Midgley, P.A. (2006). Nanoscale analysis of three-dimensional structures by electron tomography. *Scripta Mater.* **55**, 29–33.

Contents

VOLUME I

PART I IMAGING WITH ELECTRONS   1

Contributors

Matthias Amrein
Microscopy and Imaging Facility, Faculty of Medicine, Department
of Cell Biology and Anatomy, University of Calgary, Calgary, Canada

Michael R. Armstrong
University of California Chemistry and Materials Science Directorate,
Lawrence Livermore National Laboratory, Livermore, CA, USA

Ernst Bauer
Department of Physics, Arizona State University, Tempe, AZ, USA

Wolfgang Baumeister
Max Planck Institut of Biochemistry, Martinsried, Germany

Paolo Bianchini
Department of Physics, LAMBS-IFOM MicroScoBIO Research Centre,
University of Genoa, Genoa, Italy

Dawn A. Bonnell
Department of Materials Science and Engineering, Nano-Bio
Interface Center, University of Pennsylvania, Philadelphia, PA, USA

Oleg Bostanjoglo
Optisches Institut, Sekr. P1-1, Technische Universität Berlin, Berlin,
Germany

Gianluigi Botton
Department of Materials Science and Engineering, McMaster
University, Hamilton, Canada

Arnold J. den Dekker
Delft Centre for Systems and Control, Delft University of Technology,
Delft, The Netherlands

Rafal E. Dunin Borkowski
Department of Materials Science and Metallurgy, University of
Cambridge, Cambridge, UK

Valentinea Caorsi
LAMBS-IFOM MicroScoBIO Research Centre, Department of Physics,
University of Genoa, Genoa, Italy

Alberto Diaspro
Department of Physics, LAMBS-IFOM MicroScoBIO Research Centre,
University of Genoa, Genoa, Italy

Jun Feng
Lawrence Berkeley National Laboratory, Berkely, CA, USA

Peter W. Hawkes
CEMES CNRS, Toulouse, France

Stefan W. Hell
Department for NanoBiophotonics, Max Planck Insitute of
Biophysical Chemistry, Göttingen, Germany

Malcolm Howells
Advanced Light Source, Lawrence Livermore National Laboratory,
Livermore, CA, USA

John L. Hutchison
Department of Materials, University of Oxford, Oxford, UK

Christopher Jacobsen
Department of Physics and Astronomy, Stony Brook University, Stony
Brook, NY, USA

Takeshi Kasama
Frontier Research System, Institute of Physical and Chemical
Reasearch, Hatoyama, Japan, and Department of Materials
Science and Metallurgy, University of Cambridge, Cambridge,
UK

Wayne King
Chemistry and Materials Science Directorate, Lawrence Livermore
National Laboratory, Livermoe, CA, USA

Angus I. Kirkland
Department of Materials, University of Oxford, Oxford, UK

Davide Mazza
Department of Physics, LAMBS-IFOM MicroScoBIO Research Centre,
University of Genoa, Genoa, Italy

Martha R. McCartney
Department of Physics and Astronomy and Center for Solid-State
Science, Arizona State University, Tempe, AZ, USA

Peter D. Nellist
Department of Physics, Trinity College, Dublin, Ireland

Maxim P. Nikiforov
Nano-Bio Interface Center, University of Pennsylvania, Philadelphia,
PA, USA

Mattia Pesce
Department of Physics, LAMBS-IFOM MicroScoBIO Research Centre,
University of Genoa, Genoa, Italy

Juergen Plitzko
Max Planck Institute of Biochemistry, Martinsried, Germany

Bryan W. Reed
Lawrence Livermore National Laboratory, Livermore, CA, USA

Rudolf Reichelt
Institut für Medizinische Physik und Biophysik, Westfälische
Wilhelms-Universität, Münster, Germany

Frances M. Ross
IBM Research Division T. J. Watson Research Center, Yorktown
Heights, NY, USA

Marc Schneider
Biopharmaceutics and Pharmaceutical Technology, Saarbrücken,
Germany

Andreas Scholl
Lawrence Berkeley National Laboratory, Berkeley, CA, USA

Andreas Schönle
Department of NanoBiophotonics, Max Planck Institute of
Biophysical Chemistry, Göttingen, Germany

David J. Smith
Department of Physics and Astronomy and Center for Solid-State
Science, Arizona State University, Tempe, AZ, USA

John C.H. Spence
Department of Physics, Arizona State University Tempe, AZ, *and*
Lawrence Berkeley Laboratory, Berkeley, CA, USA

Peter Sutter
Center for Functional Nanomaterials, Brookhaven National
Laboratory, Upton, NY, USA

Ilaria Testa
Department of Physics, LAMBS-IFOM MicroScoBIO Research Centre,
University of Genoa, Genoa, Italy

Cesare Usai
National Research Council Institute of Biophysics, Genoa, Italy

Sandra Van Aert
University of Antwerp, Antwerp, Belgium

A. Van den Bos
Faculty of Applied Sciences, Delft University of Technology, Delft,
The Netherlands

D. Van Dyck
University of Antwerp, Antwerp, Belgium

Giuseppe Vicidomini
Department of Physics, LAMBS-IFOM MicroScoBIO Research Centre,
University of Genoa, Genoa, Italy

Tony Warwick
Advanced Light Source, Lawrence Berkeley National Laboratory,
Berkeley, CA, USA

Uwe Weierstall
Department of Physics, Arizona State University, Tempe, AZ, USA

Part II
IMAGING WITH PHOTONS

11

Two-Photon Excitation Fluorescence Microscopy

Alberto Diaspro, Marc Schneider, Paolo Bianchini,
Valentina Caorsi, Davide Mazza, Mattia Pesce,
Ilaria Testa, Giuseppe Vicidomini, and Cesare Usai

1 Introduction

Two-photon excitation (TPE) fluorescence microscopy (Denk et al., 1990; Pennisi, 1997; Esposito et al., 2004) can be considered an important example of the continuing growth of interest in optical microscopy (Diaspro, 1996; Koster and Klumperman, 2003). In spite of its low spatial resolution compared to other modern imaging techniques, such as scanning near-field microscopy (Dürig and Pohl, 1986), scanning probe microscopy (Binnig et al., 1986), or electron microscopy (Ruska and Knoll, 1931), light microscopy techniques, including TPE microscopy, have unique capabilities in the investigation of biological structures in a hydrated state, in living specimens, or at least under conditions that are close to physiological states (Pawley, 1995; Periasamy, 2001; Diaspro, 2002). This fact, coupled with advances in fluorescence labeling, permits the study of the complex and delicate relationships existing between structure and function in the four-dimension (x–y–z–t) biological systems domain (Arndt-Jovin et al., 1985; Beltrame et al., 1985; Wang and Herman, 1996; Herman and Tanke, 1998). As well, the advances achieved in the field of biological markers, especially the design of application-suited chromophores, the development of the so-called quantum dots (Jaiswal et al., 2004), visible fluorescent proteins (VFPs) from the green fluorescent protein (GFP) and its natural homologues to specifically engineered variants of these molecules (Patterson and Lippincott-Schwarz, 2002; Wiedenmann et al., 2004), and the improvements in resolution by means of special optical schemes (Egner et al., 2004; Gugel et al., 2004), are enabling TPE to move from microscopy to nanoscopy (Hell, 2003; Bastiaens and Hell, 2004).

There is also ongoing research to use TPE in new fields where its special features can be advantageously applied to improve and to optimize existing schemes (McConnell and Riis, 2004). This covers new online detection systems such as endoscopic imaging based on gradient refractive index fibers (Jung et al., 2004), the development of new substrates with higher fluorescence output (Kappel et al., 2004), as well as the use of TPE to systematically cross-link protein matrices and control the diffusion (Basu and Campagnola, 2004). Furthermore, the combination of TPE applications with other techniques has great potential use (Bird et al., 2004; Nemet et al., 2004; Periasamy and Diaspro, 2004).

TPE microscopy belongs to the category of three-dimensional (3D) optical microscopy methods, which have been widespread since the 1970s (Weinstein and Castleman, 1971; Agard et al., 1989; Bianco and Diaspro, 1989; Diaspro et al., 1990; Brakenhoff et al., 1979; Sheppard and Wilson, 1980; Wilson and Sheppard, 1984; Carlsson et al., 1985; Shotton, 1993). During the past 10 years, confocal microscopes have proved to be extremely useful research tools, notably in the life sciences. The evolution has also brought optical microscopy from 3D (x–y–z) to 5D (x–y–z–t–λ) analysis allowing researchers to probe even deeper into the intricate mechanisms of living systems (Cheng, 1994; Pawley, 1995; Masters, 1996; Sheppard and Shotton, 1997; Periasamy, 2001; Diaspro, 2002). Here, TPE microscopy (Denk et al., 1990; Diaspro, 1999a–c), or more generally multiphoton excitation (MPE) microscopy (König, 2000; Gratton et al., 2001), can probably be considered the most relevant advance in fluorescence optical microscopy since the introduction of confocal imaging in the 1980s (Wilson and Sheppard, 1984; White et al., 1987; Pawley, 1995; Webb, 1996; Sheppard and Shotton, 1997; Diaspro, 2002; Amos, 2000).

TPE microscopy couples a 3D intrinsic ability, shared with confocal microscopy, with almost five other interesting properties. First, TPE greatly reduces photo interactions and allows imaging of living specimens over long time periods. Second, it allows operation in a high-sensitivity background-free acquisition scheme. Third, TPE microscopy can penetrate turbid and thick specimens down to a depth of a few hundred micrometers. Fourth, due to the distinct character of the multiphoton absorption spectra of many of the fluorophores, TPE allows simultaneous excitation of different fluorescent molecules reducing colocalization errors. Fifth, TPE can prime photochemical reactions within a subfemtoliter volume inside solutions, cells, and tissues.

Furthermore, this form of nonlinear microscopy also favored the deveopment and application of several investigative techniques starting from TPE microscopy (Denk et al., 1990), namely, three-photon excited fluorescence (Hell et al., 1996; Maiti et al., 1997), second harmonic generation (Gannaway and Sheppard, 1978; Gauderon et al., 1999; Campagnola et al., 1999; Zoumi et al., 2002), third-harmonic generation (Mueller et al., 1998; Squier et al., 1998), fluorescence correlation spectroscopy (Berland et al., 1995; Schwille et al., 1999, 2000; Schwille, 2001; Heinze et al., 2004; Ruan et al., 2004), image correlation spectroscopy (Wiseman et al., 2000, 2002), lifetime imaging (König et al., 1996; French et al., 1997; Sytsma et al., 1998; Straub and Hell, 1998), single molecule detection schemes (Mertz et al., 1995; Xie and Lu, 1999; Sonnleitner et al., 1999, 2000; Cannone et al., 2003b; Chirico et al., 2001, 2003b), photodynamic therapies (Bhawalkar et al., 1997), two-photon photoactivation and photoswitching of visible fluorescent proteins (Chirico et al., 2004, 2005; Post et al., 2005; Schneider et al., 2005), and others (White and Errington, 2000; Masters, 2002; Periasamy,

2001; Periasamy and Diaspro, 2003; Cahalan et al., 2003; Miller et al., 2003; Diaspro, 1998, 2004; Piston, 2005, Cruz and Luscher, 2005).

2 Brief Chronological Notes

The important work done by Abbe indicated how to optimize the optical microscope. In 2005 the "Focus on Microscopy" conference, held in Jena, was dedicated to Abbe's work 100 hundred years after his death (www.focusonmicroscopy.org). Abbe's approach in defining factors influencing the microscope's resolution was fundamental. Confocal and TPE microscopy show how to extend optical parameters to obtain a better resolution. In 2000 the Optical Society of America honored Paul Davidovits, M. David Egger, and Marvin Minsky with the R.W. Wood Prize for "seminal contributions to confocal microscopy." The fact is that in 1975 Minsky invented a confocal microscope identical in the concept to the one developed by Egger and Davidovits at Yale (Davidovits and Egger, 1969, 1971), by Sheppard and Wilson at Oxford (Sheppard and Choudhury, 1977; Sheppard and Wilson, 1980; Wilson and Sheppard, 1984), and by Brakenhoff and colleagues in Amsterdam (Brakenhoff et al., 1979, 1989). As reported by Minsky (1988), the circumstances are also remarkable in that Minsky only published his invention as a patent (Figure 11–1). In addition, the idea for a confocal microscope was previously presented by Naora, who built an optical setup based upon a concept of Koana, as recently indicated by Guy Cox. It was not until the end of the 1970s, with the advent of affordable computers and lasers, and the development of digital image processing software, that the first single-beam confocal laser scanning microscopes became available in a number of laboratories and were applied to biological and materials specimens. A new revolution was developing (Sheppard and Kompfner, 1978): TPE second harmonic and fluorescence microscopy. The TPE story dates back to 1931, having its roots in the theory originally developed by Maria Göppert-Mayer (1931). The keystone of TPE theory lies in the prediction that one atom or molecule can simultaneously absorb two photons in the same quantum event.

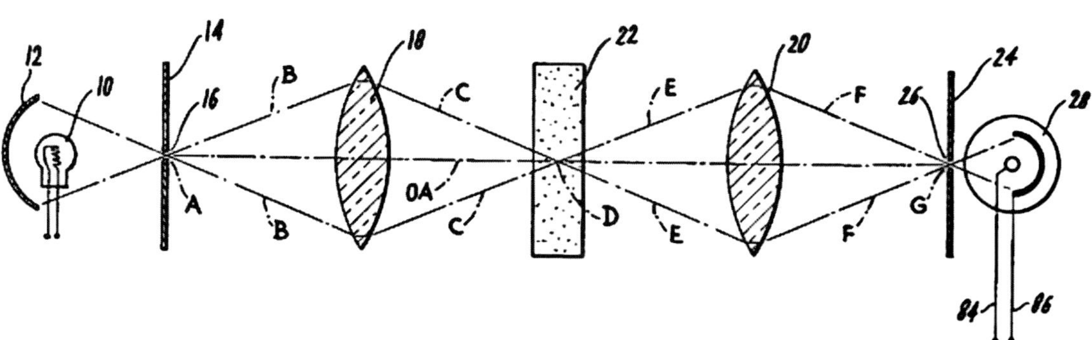

Figure 11–1. Historical sketch of the confocal setup as reported by Marvin Minsky in his patent (U.S. patent 3013467: Microscopy Apparatus, filed 7 November 1957).

Now, to indicate the rarity of the event, consider that "simultaneously" here implies "within a temporal window of 10^{-16}–10^{-15}s": in bright daylight a good one- or two-photon excitable fluorescent molecule absorbs a photon through a one-photon interaction about once a second and a photon pair by two-photon simultaneous interaction every 10 million years (Denk and Svoboda, 1997). To increase the probability of the event, a very high density of photons is needed, i.e., a laser source. As in confocal microscopy, the laser is the key to the development and dissemination of the technique. In fact, it was only in the 1960s, after the development of the first laser sources (Svelto, 1998; Wise, 1999), that it was possible to find experimental evidence for Maria Göppert-Mayer's prediction. Kaiser and Garret (1961) reported TPE of fluorescence in $CaF_2:Eu^{2+}$ and Singh and Bradley (1964) were able to estimate the three-photon absorption cross-section for naphthalene crystals. These two results consolidated other related experimental achievements obtained by Franken et al. (1961) of second harmonic generation in a quartz crystal using a ruby laser. Later, Rentzepis and colleagues (1970) observed three-photon excited fluorescence from organic dyes, and Hellwarth and Christensen (1974) collected second-harmonic generation signals from ZnSe polycrystals at a microscopic level. In 1976, Berns reported a probable two-photon effect as a result of focusing an intense pulsed laser beam onto chromosomes of living cells, and such interactions form the basis of modern nanosurgery (König et al., 1999). However, the original idea of generating 3D microscopic images by means of such nonlinear optical effects was first suggested and attempted in the 1970s by Sheppard, Kompfner, Gannaway, and Choudhury of the Oxford group (Sheppard et al., 1977; Gannaway and Sheppard, 1978; Sheppard and Kompfner, 1978). It should also be emphasized that for many years the application of two-photon absorption was mainly related to spectroscopic studies (Friedrich and McClain, 1980; Friedrich, 1982; Birge, 1986; Callis, 1997). The real "TPE boom" took place at the beginning of the 1990s at the W.W. Webb laboratories (Cornell University, Ithaca, NY). However, it was the excellent and effective work done by Winfried Denk and colleagues (Denk et al., 1990) that was responsible for spreading the technique and that revolutionized fluorescence microscopy imaging.

3 Basic Principles on Confocal and Two-Photon Excitation of Fluorescent Molecules

3.1 Fluorescence

Fluorescence optical microscopy is very popular for imaging in biology since fluorescence is highly specific either as exogenous labeling or endogenous autofluorescence (Beltrame et al., 1985; Arndt-Jovin et al., 1985; Periasamy, 2001). Fluorescent molecules allow both spatial and functional information to be obtained through specific absorption, emission, lifetime, anisotropy, photodecay, diffusion, and other contrast mechanisms (Diaspro, 2002; Zoumi et al., 2002). This means that it is possible to efficiently study, for example, the distribution and dynamics of proteins, DNA, and chromatin as well as ion concentration, voltage, and temperature within living cells (Chance, 1989; Tsien, 1995; Robinson, 2001). TPE of fluorescent molecules is a nonlinear process related to the simultaneous absorption of two photons whose total energy equals the energy required for conventional, one-photon excitation (Birks, 1970; Denk et al., 1995; Callis, 1997). In any case the energy required to prime fluorescence is the energy sufficient to produce a molecular transition to an excited electronic state. The excited fluorescent molecules then decay to an intermediate state giving off a photon of light having an energy lower than needed to prime excitation. This means that the energy E provided by photons should equal the molecule energy gap ΔE_g, and, considering the relationship between photon energy E and radiation wavelength λ, it follows that

$$\Delta E_g = E = \frac{hc}{\lambda} \tag{1}$$

where $h = 6.6 \times 10^{-34}\,\mathrm{J \cdot s}$ is Planck's constant and $c = 3 \times 10^{8}\,\mathrm{m\,s^{-1}}$ is the value of the speed of light (considered in vacuum and to a reasonable approximation). Conventional techniques for fluorescence excitation use ultraviolet (UV) or visible radiation and excitation occurs when the absorbed photons are able to match the energy gap to the ground from the excited state. Due to energetic aspects, the fluorescence emission is shifted toward a wavelength longer than the one used for excitation. This shift typically ranges from 50 to 200 nm (Birks, 1970; Cantor and Schimmel, 1980). For example, a fluorescent molecule that absorbs one photon at 340 nm, in the ultraviolet region, exhibits fluorescence at 420 nm in the blue region.

A 3D reconstruction of the distribution of fluorescence within a three-dimensional object such as a living cell starting with the acquisition of the two-dimensional distribution of specific intensive properties is one of the most powerful properties of the optical microscope. In fact, this allows complete morphological analysis through volume rendering procedures (Kriete, 1992; Robinson, 2001) of living biological specimens, where the opportunity of optical slicing allows information to be obtained from different planes of the specimen without being invasive, thus preserving the structures and functionality of the different parts (Weinstein and Castleman, 1971; Agard, 1984; Agard et al., 1989; Diaspro et al., 1990).

3.2 Confocal Principle and Laser Scanning Microscopy

Conventional wide-field microscopes involve a specimen entirely bathed in the radiation from the light source, viewed directly by eye or through any capture device [charge coupled device (CCD) camera, for instance or photosensitive film]. As reported in the paper by Minsky (1961), an ideal microscope would examine each point of the specimen and measure the amount of light scattered, absorbed, or emitted by that point, excluding contributions from another part of the sample from the actual or from adjacent planes (Figure 11–2). Unfortunately, in trying to obtain images by making many such measurements at the same time, every focal image point will be clouded by aberrant rays of scattered light deflected from points of the specimen that are not the point of interest. This means that samples undergo continuous full excitation, leading to in- and out-of-focus light points and contributing to overlapping and worsening axial resolution and producing that typical hazing in the collected images that, together with light-diffraction effects, limits the performance of the instrument. Most of those extra rays would be absent if we could illuminate only one specimen point at a time. There is no way of eliminating every undesired ray, because of multiple scattering, but it is comparatively straightforward to remove all rays not initially aimed at the focal point by using a sort of second microscope (instead of a condenser lens) to image a pinhole aperture (a small aperture in an opaque screen) on a single point of the specimen. This reduces the amount of light in the specimen by orders of magnitude without reducing the focal brightness. Even then, some of the initially focused light will be scattered by out-of-focus specimen points onto other points in the image plane affecting the clarity of the final acquisition, i.e., of the observed image o. But it is possible to reject undesired rays as well, by placing a second pinhole aperture in the image plane that lies beyond the exit side of the objective lens. We end up with an elegant, symmetrical geometry: a pinhole and an objective lens on each side of the specimen. This leads to the use of two lenses on both the excitation and detection sides of the microscope, combining the two lenses for a unique effect (Figure 11–3).

To acquire an image the excitation light has to be fully delivered to each point of the sample and the emission signal collected and displayed. This is usually accomplished by either of two possible but different strategies.

The first one is based upon scanning the sample in a raster pattern such that over every fixed period of time, the necessary amount of information from the focal plane is collected and the emitted light signal, usually detected through a photomultiplier tube (PMT), is displayed by a mapping of each single point light emission. Sometimes, the use of a slit moving in one direction (rather than a single point) is

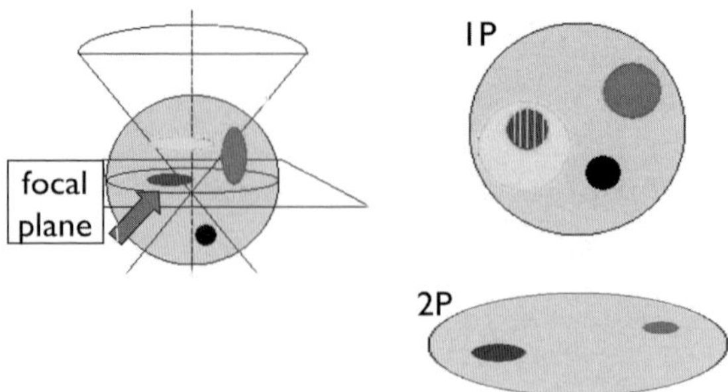

Figure 11–2. Comparison between conventional (1P) and two-photon (2P) excitation with respect to image formation. When focusing on the actual focal plane under 1P, a contribution from adjacent planes that are physically excluded in the 2P process is obtained, as happens in a confocal setup. (From Giuseppe Vicidomini, LAMBS, MicroScoBio, University of Genoa.) (See color plate.)

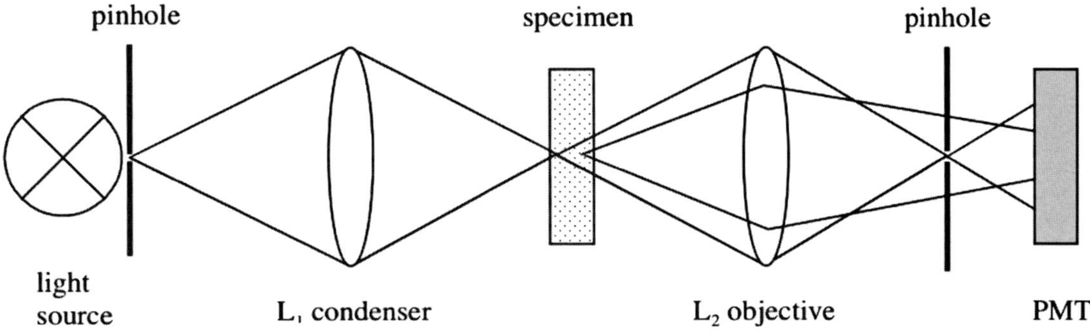

Figure 11–3. Equivalent optical configuration for a confocal setup (see text).

preferred for speeding up the scanning rate, although this leads to an evident worsening of the spatial resolution and of the three-dimensional imaging capability.

A second possible approach to form confocal images consists of employing a multipinhole Nipkow spinning disk (Petran et al., 1968; Kino and Corle, 1989). This is a disk containing multiple sets of spirally arranged holes placed in the image plane of the objective lens. A large parallel beam of light is then pointed at a particular region of the disk and the light passing through the illuminated pinholes is focused by the objective lens straight onto the specimen. When spinning the disk at a rapid rate, the sample may undergo excitation several hundred times per second: emitted light is collected and imaged typically by a high-resolution and high-quantum-efficiency CCD camera. Concerning optical sectioning, every architecture is built such that the sample is placed along the light path at a conjugate focal plane and the movements along the optical axis keep the focus at a fixed distance from the objective, making it possible to effectively scan different fields of view through the specimen (due to a step-by-step motor device attached to the fine focus) and collect a series of in-focus optical slices for 3D reconstruction.

The degree of confocality is a function of the pinhole size: the use of smaller pinholes improves the discrimination of focused light from stray light, thus involving a thinner plane in the image formation process and improving resolution, at the cost of lower light throughput, which makes things complicated when dealing with particularly dim samples.

In these architectures, z-resolution and optical sectioning thickness (which are basically the parameters involved in every optical sectioning process) depend on a number of factors such as the numerical aperture (NA) of the objective lens, the wavelength of the excitation/emission light, the pinhole size, the refractive index of components along the light path, and finally the overall alignment of the instrument.

3.3 Theoretical Analysis

The development of an effective theoretical model for describing the properties of an optical system needs some preliminary, *realistic* assumptions to be made to simplify calculations.

From this point of view, a linear space invariant (LSI) model is a good choice, pliable enough to obtain important insights and develop suitable mathematical tools for the analysis of most concrete situations. Let us consider the situation sketched in Figure 11–3, showing a typical confocal setup.

A monochromatic point light source is focused onto a sample in the focal plane through a lens L_1 (condenser) and the emitted radiation from the sample (which is also supposed to be monochromatic) is collected through a second lens L_2 (objective) by a point detector. Let h_{ex} and h_{em} be, respectively, the impulse response of L_1 and L_2, i.e., the lens response to a point-like light source. Under this hypothesis it can be written that $U_{ex}(x) = (h_{ex} \otimes \delta_s)(x) = h_{ex}(x)$, where the excitation light source is modeled by a Dirac impulse.

It can be shown that with U_{ex} being the signal reaching the sample, the emitted signal scales with the fluorescent dye density D according to $U_{em} = D \cdot U_{ex}$. The emitted radiation is then focused on the point detector through L_2.

This leads to $U_{det}(x) = [(h_{em} \otimes U_{em}) \cdot \delta_d](x)$ where the point detector function is assumed to be a Dirac impulse. The overall signal collected by the detector is

$$I_{tot} = \int U_{det}(x)dx = \int dx \delta_d(x)(h_{em} \otimes U_{em})(x)$$
$$= \int dx \delta_d(x) \int dy h_{em}(x-y)D(y)h_{ex}(y)$$
$$= \int dy D(y)h_{ex}(y) \int dx \delta_d(x)h_{em}(x-y)$$
$$= \int dy D(y)h_{ex}(y)h_{em}(-y)$$

If we now limit ourselves to a point-like sample:

$$\int U_{det}(x)dx = \int dy \delta(y)h_{ex}(y)h_{em}(-y) = h_{ex}(0)h_{em}(0)$$

where $h_{ex} = h_{em}$ under the hypothesis of $L_1 = L_2$ and $\lambda_{ex} = \lambda_{em}$.[1]

Since an x–y–z scanning process is generally coupled to the imaging one, it is natural to write, for a general point $P(x,y,z)$: $I_{tot} = h^2(x,y,z)$, which is the general expression for the point spread function (PSF), that is, the system impulse response. A mathematical expression for $h(x,y,z)$ can be obtained through the scalar electromagnetic waves theory.

The formulation, based on Fraunhofer diffraction, leads to

$$h(u,v) \propto \left| \int_0^1 J_0(v\rho)e^{-0.5iu\rho^2}\rho d\rho \right|^2 \tag{2}$$

Where u and v are suitable dimensionless variables defined according to the following:

$$u \propto z$$

$$v \propto \sqrt{x^2 + y^2}$$

---

[1] The equivalence of $h_{ex} = h_{em}$ is valid only for pinhole sizes ≤ 0.25 AU. AU is the so-called Airy unit, which represents the diameter of the Airy disk.

By limiting the discussion to the points along the optical axis and in the focal plane, we have:

$$h(0,v) \propto \left[\frac{2J_1(v)}{v}\right]^2 \qquad h(u,0) \propto \left[\frac{\sin(u/4)}{u/4}\right]^2$$

that is

$$I_{\text{tot}}(0,v) \propto \left[\frac{2J_1(v)}{v}\right]^4 \qquad I_{\text{tot}}(u,0) = \left[\frac{\sin(u/4)}{u/4}\right]^4$$

Compared to a conventional microscope, where $I_{\text{tot}} \approx h(u,v)$, the calculation of the fall width at half-maximum (FWHM), representing the system resolution, leads to an improvement in resolution by a factor 1.4 (Brakenhoff et al., 1979; Wilson and Sheppard, 1984; Diaspro et al., 1999a; Jonkman and Stelzer, 2002; Torok and Sheppard, 2002).

3.4 Remarks and Comments

Comparisons between the ideal PSF in the case of strict confocality with that of conventional microscopes account for improvements in resolution. However, despite the use of this mathematical formalism for concrete situations, further drawbacks need to be highlighted. First, there is a natural relation between the pinhole size and the PSF: the more the pinhole size is increased, the more the confocal microscope response tends to fit conventional responses.

This means that in the case of dim or highly photosensitive specimens some compromise has to be found between the resolution and the amount of the collected signals, according to the kind of analysis to be performed (whether a morphometric one or intensity one).

Second, the PSF is obviously dependent on many physical parameters, inter alia the refractive index of the sample, immersion medium, its turbidity, the degree of homogeneity of the sample, and the photochemical properties of the dyes used. For these reasons the development of complicated computations often leads to poorly applicable results in practice, since conditions often change dramatically for the different measurements.

One of the most meaningful factors on which PSF depends is the refractive index mismatch between the objective immersion medium and that of the sample solution. This results in a loss of axial resolution and a corruption of the shape (Diaspro et al., 2002a).

Table 11–1 reports the value of theoretical and experimental FWHM of confocal PSFs using different pinhole sizes. A sample of subresolu-

Table 11–1. FWHM of confocal PSFs for different pinhole sizes.

	Oil ($n = 1.5$)			
	Lateral (nm)		Axial (nm)	
	Pinhole 20 μm	Pinhole 50 μm	Pinhole 20 μm	Pinhole 50 μm
Experimental	186 ± 6	215 ± 5	489 ± 6	596 ± 4
Theoretical	180	210	480	560

Table 11–2. Variation of lateral FWHM with focusing depth.

Depth (µm)	Air		Glycerol		Oil	
	Lateral (nm)	Axial (nm)	Lateral (nm)	Axial (nm)	Lateral (nm)	Axial (nm)
0	187 ± 8	484 ± 24	183 ± 14	495 ± 29	186 ± 6	489 ± 6
30	244 ± 10	623 ± 9	221 ± 5	545 ± 12	197 ± 10	497 ± 21
60	269 ± 11	798 ± 10	252 ± 7	628 ± 9	186 ± 12	496 ± 19
90	277 ± 5	1063 ± 24	268 ± 8	797 ± 26	191 ± 9	484 ± 12

tion beads [Polyscience, diameter $= (0.064 \pm 0.009)$ µm] has been imaged by means of a 100× NIKON oil-immersion objective (NA = 1.3; WD = 0.20 mm) under argon laser excitation (λ = 488 nm). Theoretical values are those expected (in the absence of mismatch) and are calculated by means of web-based deconvolution software (http://www.powermicroscope.com).

As can be seen from the reported values, the system resolution is worse along the optical axis and is strictly dependent on the pinhole size: these results are in accordance with what is expected from the above theory. Asymmetry of the plots in the real case is typical and becomes even more evident when focusing through different stratified media (Diaspro et al., 2002a).

The theory, developed within the context of electromagnetic waves focusing across stratified media, suggests a progressive broadening of the PSF with respect to the focusing depth, becoming even more noticeable under refractive-index mismatch conditions. Furthermore, on a higher level of complexity, the coexistence of different refractive indices within the sample and the resulting artifacts can be considered.

As a consequence of this, the largest percentage of variation of the lateral FWHM, with respect to the focusing depth, goes from 6% (oil-immersed PSF), to 48% (air-immersed PSF), whereas the axial FWHM varies up to 130% (air-immersed PSF) (see Table 11–2).

This phenomenon is related to a subsequent weakening of the signal with respect to the focusing depth, which turns out to be more evident in the case of refractive index mismatch. Table 11–3 gives typical observed values of the percentage of variation of the PSF intensity peak under different mismatch conditions and at different focusing depths (referred to the coverslip).

Table 11–3. Variations in PSF intensity with mismatch conditions and focusing depths.

Medium	% at 30 µm depth	% at 60 µm depth	% at 90 µm depth
Oil	3	6	7
Glycerol	17	27	34
Air	44	51	60

3.5 Resolution and Three-Dimensional Optical Sectioning

Three-dimensional reconstruction of an object starting from the acquisition of 2D confocal slices is one of the most powerful procedures for morphological analysis and volume rendering, especially within biological sciences, where optical slicing allows information to be obtained from different planes of the specimen without being invasive, thus preserving the structure and functionality of the different parts.

Three-dimensional optical sectioning is intrinsically coupled with the axial resolution of the confocal microscope. For pinhole diameters smaller than 1 AU the approximation of a point-like pinhole is used. For this case the FWHM of the total PSF in the z direction can be expressed as

$$r_z = \frac{0.64\bar{\lambda}}{n - \sqrt{n^2 - NA^2}} \tag{3}$$

This expression describes the axial resolution and the effective optical slice thickness for the sectioning of the specimen. Here $\bar{\lambda}$ is $\bar{\lambda} \approx \sqrt{2}\left(\lambda_{em}\lambda_{ex}/\sqrt{\lambda_{ex}^2 + \lambda_{em}^2}\right)$: a mean wavelength. This technique is essentially based on an automatic fine z stepping either of the objective or of the sample stage, coupled with the usual x–y point-to-point scanning of the focal plane and image capturing.

The synchronous x–y–z scanning allows the collection of a set of in-focus 2D images, which are less affected by signal cross-talk from other planes of the sample as more strictly confocal conditions are respected.

This means that when a set of 2D images is acquired at various focus positions and under certain conditions, in principle the 3D shape of the object can be recovered. However, the observed image $o(x,y,z)$, produced by the true intensity distribution $i(x,y,z)$, is corrupted by the characteristic PSF of the image formation system $s(x,y,z)$, by noise stemming from different sources $n(x,y,z)$, and by cross information coming from different planes rather than from the focus one.

At a certain plane of focus z_0 within the sample or, at discrete planes along the z axis, the simplest way to describe such a process for the jth plane can be regarded as the following:

$$o_j = i_j \otimes s_0 + i_{j-1} \otimes s_{j-1} + i_{j+1} \otimes s_{j+1} + \text{(other plane contributions}$$
$$\text{if relevant)} + n \tag{4}$$

where the subscripts on i and o refer to the z plane numbers, while the subscripts on s refer to the number of interplane spacings z away from the "in-focus" position at the actual jth plane.

This relationship is usually transferred to the Fourier frequency domain, where the convolution operator becomes an algebraic multiplication (Diaspro et al., 1990). Image restoration algorithms (deconvolution) aim to invert such equations in order to extract the true measured quantity $i(x,y,z)$. Thus a 3D sample reconstruction is possible directly by piling up 2D images. Therefore further scale corrections are performed accounting for axial distortion phenomena linked to the refractive index mismatch.

The generic considerations for this method evidently demonstrate the crucial importance of getting to know the system performance under different working conditions, by means of its PSF. This knowledge will extend the possibilities of applying the confocal technique to a wide range of studies.

In practice, one wants to find the best estimate, accordingly to some criterion, of $i(x,y.z)$ through the knowledge of the observed images, the distortion or PSF of the image formation system, and the additive noise within a restoration scheme classical for space invariant linear systems (Diaspro et al., 1990; Bertero and Boccacci, 1998; Boccacci and Bertero, 2002). Figure 11–4 shows an example of digital restoration of microscopic data obtained after solving the appropriate set of equations. So far, this can be computationally done starting from any data set of optical slices. Recently a WWW service, named Power-Up-Your-Microscope (Diaspro et al., 2002c; Bonetto et al., 2004), has become available that produces the best estimate of $i(x,y,z)$ accordingly to the acquired data set of optical slices. Interested readers can find information and check the service through the webpage http://www.power-microscope.com for free (Figure 11–4).

Image restoration is needed only to correct PSF distortions that are less than in the conventional case. Unfortunately, a drawback occurs. In fact, during the excitation process of the fluorescent molecules the whole thickness of the specimen is harmed by every scan, within an hourglass-shaped region (Bianco and Diaspro, 1989). This means that even though out-of-focus fluorescence is not detected, it is generated, with the negative effect of potential induction of those photobleaching and phototoxicity phenomena previously mentioned. The situation becomes particularly serious when there is the need for 3D and temporal imaging coupled to the use of fluorochromes that require excita-

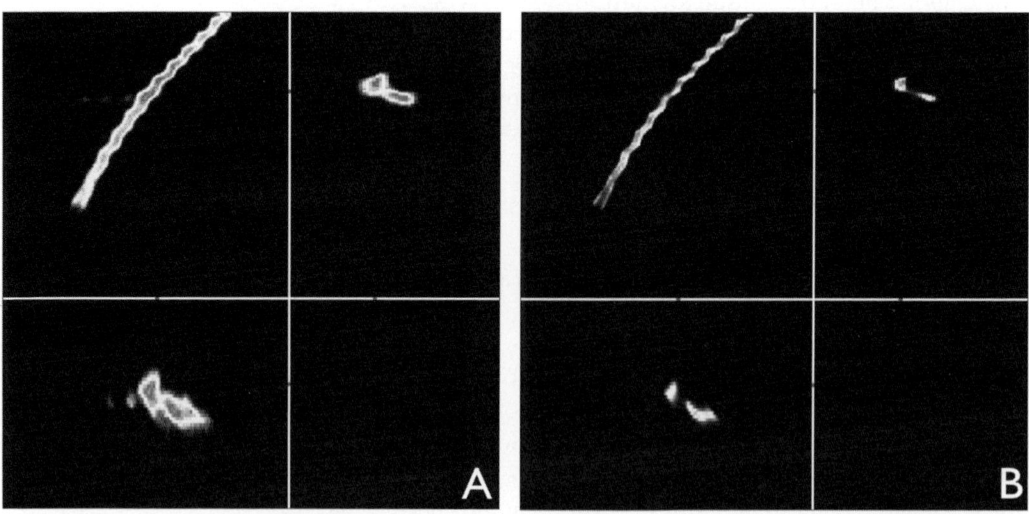

Figure 11–4. Comparison between 3D views of a helicoidal biological sample before (left) and after (right) image processing utilizing 3D deconvolution strategies as implemented at http://www.power-microscope.com as the web-based computational facility (Difato et al., 2004).

tion in the ultraviolet regime (Stelzer et al., 1994; Denk, 1996). As earlier reported by König and colleagues (1996a,b), even using UVA (320–400 nm) photons may modify the activity of the biological system. DNA breaks, giant cell production, and cell death can be induced at radiant exposures of the order of magnitude of a few J/cm^2, accumulated during 10 scans with a 5-μW laser scanning beam at approximately 340 nm and a 50-μs pixel dwell time. In this context, TPE of fluorescent molecules provides an immediate practical advantage over confocal microscopy (Denk et al., 1990; Potter, 1996; Centonze and White, 1998; Gu and Sheppard, 1995; Diaspro, 1998; Piston, 1999; Squirrel et al., 1999; Diaspro and Robello, 2000; So et al., 2000; Wilson, 2002). In fact, reduced overall photobleaching and photodamage are generally acknowledged as major advantages of TPE in laser scanning microscopy of biological specimens (Brakenhoff et al., 1996; Denk and Svoboda, 1997; Patterson and Piston, 2000) even though photobleaching in the focal plane can be accelerated (Patterson and Piston, 2000). However, the excitation intensity has to be kept low considering a regime under 10 mW of average power as a normal operation mode. When laser power is increased above 10 mW some nonlinear effects might arise, evidenced through abrupt rising of the signals (Hopt and Neher, 2001). Moreover photothermal effects should be induced, especially when focusing on single molecule detection schemes (Chirico et al., 2003a).

4 Two-Photon Excitation

Let us now move from conventional excitation of fluorescence as used in computational optical sectioning and confocal microscopy to a special case of multiphoton excitation, i.e., TPE. All considerations can be easily extended to the TPE. The physical suppression of contributions from adjacent planes is realized in a completely different way, thus moving again to 3D optical sectioning ability.

In TPE, two low-energy photons are involved in the interaction with absorbing molecules. The excitation process of a fluorescent molecule can take place only if two low-energy photons are able to interact simultaneously with the very same fluorophore. As mentioned in the introduction, the time scale for simultaneity is the time scale of molecular energy fluctuations at photon energy scales, as determined by the Heisenberg uncertainty principle, i.e., 10^{-16}–10^{-15} s (Louisell, 1973). These two photons do not necessarily have to be identical, but their wavelengths, λ_1 and λ_2, have to be such that

$$\lambda_{1P} \cong \left(\frac{1}{\lambda_1} + \frac{1}{\lambda_2} \right)^{-1} \tag{5}$$

where λ_{1P} is the wavelength needed to prime fluorescence emission in a conventional one-photon absorption process according to the energy relation given in Eq. (1). This situation, compared to the conventional one-photon excitation process shown in Figure 11–5, is illustrated using

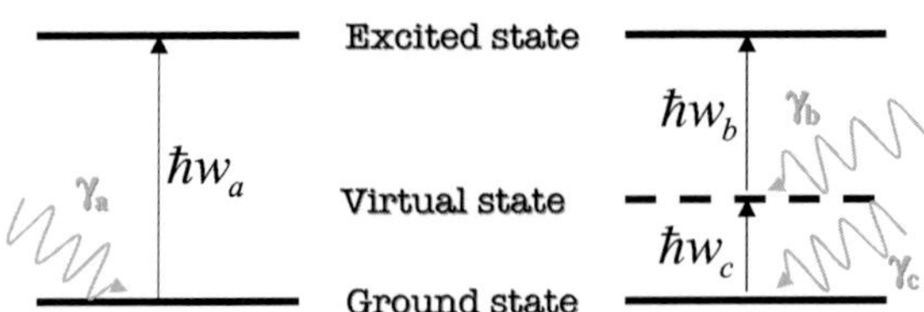

Figure 11–5. Perrin–Jablonsky-like diagram illustrating the difference between conventional (left) and two-photon (right) excitation (Esposito et al., 2004). In the second case, photons delivering one-half the energy conventionally needed for bringing the fluorescent molecule to an excited state are used (see text).

a Perrin–Jablonski-like diagram. It is worth noting that for practical reasons the experimental choice is usually such that (Denk et al., 1990; Diaspro, 2001; Girkin and Wokosin, 2002)

$$\lambda_{2P} = 2\lambda_{1P} \quad \lambda_1 = \lambda_2 \approx 2\lambda_{1P} \tag{6}$$

and

$$\Delta E_g = \frac{2hc}{\lambda_{1P}} \tag{7}$$

Considering this as a nonresonant process and assuming the existence of a virtual intermediate state, the resident time, τ_{virt}, in this intermediate state should be calculated using the time–energy uncertainty consideration for TPE:

$$\Delta E_g \cdot \tau_{virt} \cong h/2 \tag{8}$$

where $\hbar = h/2\pi$. It follows that

$$\tau_{virt} \cong 10^{-15}\text{--}10^{-16}\,\text{s} \tag{9}$$

This is the temporal window available to two photons to coincide in the virtual state.

So far, in a TPE process it is hence crucial to combine sharp spatial focusing with temporal confinement of the excitation beam.

The process can be extended to n-photons requiring higher photon densities temporally and spatially confined. Thus, near infrared (circa 680–1100 nm) photons can be used to excite UV and visible electronic transitions producing fluorescence. The typical photon flux densities are of the order of more than 10^{24} photons $\text{cm}^{-2}\,\text{s}^{-1}$, which implies intensities around MW–TW cm^{-2} (Goppert Mayer, 1931). An elegant treatment in terms of quantum theory for two-photon transition has been proposed by Nakamura (1999) using perturbation. He clearly described the process by a time-dependent Schrödinger equation, where the Hamiltonian contains electric dipole interaction terms. Using a perturbation expansion, the first-order solution is found to be related to one-photon excitation while higher order solutions are related to n-photon ones (Esposito et al., 2004).

The dependence of TPE on I^2 should be evident and is demonstrated by using simple arguments (Diaspro and Sheppard, 2002).

The fluorescence intensity per molecule, $I_f(t)$, can be considered to be proportional to the molecular cross-section $\delta_2(\lambda)$ and to the square of $I(t)$ as follows:

$$I_f(t) \propto \delta_2 \cdot I(t)^2 \propto \delta_2 \cdot P(t)^2 \left[\pi \frac{(NA)^2}{hc\lambda} \right]^2 \qquad (10)$$

where $P(t)$ is the laser power and (NA) is the numerical aperture of the focusing objective lens. The last term of Eq. (10) simply takes care of the distribution in time and space of the photons by using the paraxial approximation in an ideal optical system (Born and Wolf, 1980).

It follows that the time-averaged two-photon fluorescence intensity per molecule within an arbitrary time interval T, $\langle I_f(t)\rangle$, can be written as

$$\langle I_f(t)\rangle = \frac{1}{T}\int_0^T I_f(t)dt \propto \delta_2 \left[\pi \frac{(NA)^2}{hc\lambda} \right]^2 \frac{1}{T}\int_0^T P(t)^2\,dt \qquad (11)$$

in the case of continuous wave (CW) laser excitation.

Now, because the present experimental situation for TPE is related to the use of ultrafast lasers, we consider that for a pulsed laser $T = 1/f_P$, where f_P is the pulse repetition rate. This implies that a CW laser beam, where $P(t) = P_{ave}$, allows transformation of Eq. (11) into

$$\langle I_{f,cw}(t)\rangle \propto \delta_2 \cdot P_{ave}^2 \left[\pi \frac{(NA)^2}{hc\lambda} \right]^2 \qquad (12)$$

For a pulsed laser beam with pulse width, τ_p, repetition rate, f_p, and average power

$$P_{ave} = D \cdot P_{peak}(t)$$

where $D = \tau_p \cdot f_p$, the approximated $P(t)$ profile can be described as

$$\begin{aligned} P(t) &= P_{ave}/D &\quad \text{for } 0 < t < \tau_p \\ P(t) &= 0 &\quad \text{for } \tau_p < t < (1/f_p) \end{aligned} \qquad (13)$$

We can write Eq. (12) as:

$$\langle I_{f,p}(t)\rangle \propto \delta_2 \frac{P_{ave}^2}{\tau_p^2 f_P^2} \left[\pi \frac{(NA)^2}{hc\lambda} \right]^2 \frac{1}{T}\int_0^{\tau_P} dt = \delta_2 \frac{P_{ave}^2}{\tau_p f_P} \left[\pi \frac{(NA)^2}{hc\lambda} \right]^2 \qquad (14)$$

The conclusion here is that CW and pulsed lasers operate at the very same excitation efficiency, i.e., fluorescence intensity per molecule, if the average power of the CW laser is kept higher by a factor of $1/\sqrt{\tau \cdot f_P}$. This means that 10 W delivered by a CW laser, allowing the same efficiency of conventional excitation performed at approximately 10^{-1} mW, is nearly equivalent to 30 mW for a pulsed laser (Diaspro and Chirico, 2002).

5 Fluorescent Molecules under TPE Regime

The above steps lead to the most popular relationship reported below, which is related to the practical situation of a train of beam pulses focused through a high numerical aperture objective, with a duration τ_p and repetition rate f_p. In this case, the probability, n_a, that a certain fluorophore simultaneously absorbs two photons during a single pulse, in the paraxial approximation, is (Denk et al., 1990)

$$n_a \propto \frac{\delta_2 \cdot P_{ave}^2}{\tau_p f_p^2} \left(\frac{NA^2}{2hc\lambda} \right)^2 \tag{15}$$

where P_{ave} is the time-averaged power of the beam and λ is the excitation wavelength. Introducing 1 GM (Goppert-Mayer) = 10^{-58} (m^4·s), for a δ_2 of approximately 10 GM per photon (Denk et al., 1990; Xu, 2002), focusing through an objective of $NA > 1$, an average incident laser power of $\approx$1–50 mW, and operating at a wavelength ranging from 680 to 1100 nm with 80–150 fs pulsewidth and 80–100 MHz repetition rate, would saturate the fluorescence output as for one-photon excitation. This suggests that for optimal fluorescence generation, the desirable repetition time of pulses should be on the order of a typical excited-state lifetime, which is a few nanoseconds for commonly used fluorescent molecules. For this reason the typical repetition rate is around 100 MHz. A further condition that makes Eq. (15) valid is that the probability that each fluorophore will be excited during a single pulse has to be smaller than one. The reason lies in the observation that during the pulse time (10^{-13} s of duration and a typical excited-state lifetime in the 10^{-9} s range) the molecule has insufficient time to relax to the ground state. This can be considered a prerequisite for absorption of another photon pair. Therefore, whenever n_a approaches unity saturation effects start to occur. The use of Eq. (15) makes it possible to choose optical and laser parameters that maximize excitation efficiency without saturation. In case of saturation the resolution declines and the image becomes worse (Cianci et al., 2004). It is also evident that the optical parameter for enhancing the process in the focal plane is the lens numerical aperture, NA, even if the total fluorescence emitted is independent of this parameter as shown by Xu (2002). This is usually confined around 1.3–1.4 as the maximum value. Now, it is possible to estimate n_a for a common fluorescent molecule like fluorescein that possesses a two-photon cross-section of 38 GM at 780 nm (Diaspro and Chirico, 2003).

To this end, we can use $NA = 1.4$, a repetition rate at 100 MHz, and a pulse width of 100 fs within a range of P_{ave} values of 1, 10, 20, and 50 mW, and substituting the proper values in Eq. (15) we get $n_a \cong$ $5930 * P^2_{ave}$. This result for $P_{ave} = 1$, 20, as a function of 1, 10, 20, and 50 mW, gives values of 5.93×10^{-3}, 5.93×10^{-1}, 1.86, and 2.965, respectively. It is evident that saturation begins to occur at 10 mW (Diaspro and Sheppard, 2002).

The related rate of photon emission per molecule, at a nonsaturation excitation level, in the absence of photobleaching (Patterson and Piston, 2000; So et al., 2001), is given by n_a multiplied by the repetition rate of the pulses. This means approximately 5×10^7 photons s^{-1} in both cases. It is worth noting that, when considering the effective fluorescence emission, a further factor given by the so-called quantum efficiency of the fluorescent molecules should also be considered. At present, the quantum efficiency value is usually known from conventional one-photon excitation data (Diaspro, 2002).

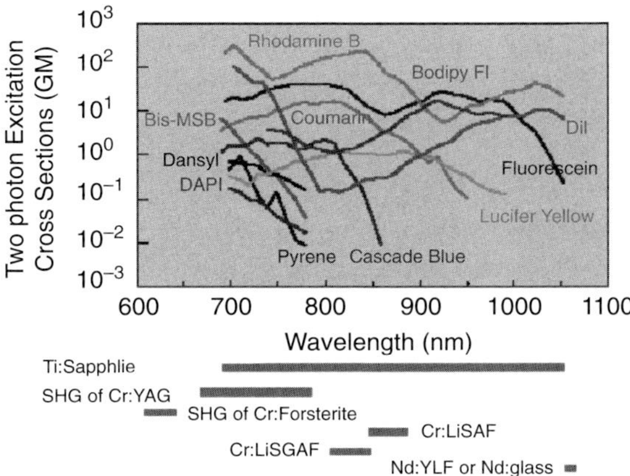

Figure 11–6. Two-photon cross-sections for popular fluorescent molecules as a function of the excitation wavelength. Red bars indicate the emission range of some common laser sources utilized in TPE microscopy and spectroscopy. (See color plate.)

Now, even if the quantum-mechanical selection rules for TPE differ from those for one-photon excitation, several common fluorescent molecules can be used. Unfortunately, knowing the one-photon cross-section for a specific fluorescent molecule does not allow any quantitative prediction of the two-photon trend, except for a sort of "rule of thumb." This simple rule states that, in general, a TPE cross-section may be expected to peak at double the wavelength needed for one-photon excitation. However, the cross-section parameter has been measured for a wide range of dyes (Xu et al., 1995; Albota et al., 1998b; Diaspro and Chirico, 2003). It is worth noting that due to the increasing dissemination of TPE microscopy, new "ad hoc" organic molecules, endowed with large engineered two-photon absorption cross-sections, have recently been developed (Albota et al., 1998; Abbotto et al., 2005). Figure 11–6 summarizes the properties of some commonly used fluorescent molecules under two-photon excitation (Xu et al., 1995; So et al., 2000). TPE fluorescence from NAD(P)H, flavoproteins (Piston et al., 1995; So et al., 2000), tryptophan, and tyrosine in proteins (Lakowicz and Gryczynski, 1992) has been measured. In addition, the autofluorescent biological proteins such as the GFP and its molecular variants are important molecular markers (Chalfie et al., 1994; Chalfie and Kain, 1998; Potter, 1996; Zimmer, 2002). Their TPE cross-sections are between 6 and 40 GM (Blab et al., 2001). As a comparison consider that the cross-section for NADH, at the excitation maximum of 700 nm, is approximately 0.02 GM (So et al., 2000). Moving to quantum dots there is an increase of cross-section up to 2000 GM.

6 Optical Consequences of TPE

In terms of optical consequences the two-photon effect limits the excitation region to within a subfemtoliter volume. The 3D confinement of the TPE volume can be understood with the aid of optical diffraction theory

(Born and Wolf, 1980). Using excitation light with wavelength λ, the intensity distribution at the focal region of an objective with numerical aperture $NA = \sin(\alpha)$ is described [see also Eq. (2)] in the paraxial regime (Born and Wolf, 1980; Sheppard and Gu, 1990) by

$$I(u,v) = \left| 2\int_0^1 J_0(v\rho)e^{-(i/2)u\rho^2}\rho\,d\rho \right|^2 \tag{16}$$

where rho is a dimensionless radial, J_o is the zeroth-order Bessel function, ρ is a radial coordinate in the pupil plane, and $u = 8\pi\sin^2(\alpha/2)z/\lambda$ and $v = 2\pi\sin(\alpha)r/\lambda$ are dimensionless axial and radial coordinates, respectively, normalized to the wavelength (Wilson and Sheppard, 1984). Now, the intensity of fluorescence distribution within the focal region has an $I(u, v)$ behavior for the one-photon case and $I^2(u/2, v/2)$ for the TPE case as demonstrated above. The arguments of $I^2(u/2, v/2)$ take into proper account the fact that in the latter case wavelengths are utilized that are approximatively twice those used for one-photon excitation. As compared with the one-photon case, the TPE intensity distribution is axially confined (Nakamura, 1993; Gu and Sheppard, 1995; Jonkman and Stelzer, 2002). In fact, considering the integral over v, keeping u constant, its behavior is constant along z for one-photon and has a half-bell shape for TPE. This behavior, better discussed in Wilson (2002), Torok and Sheppard (2002), and Jonkman and Stelzer (2002), explains the three-dimensional discrimination property in TPE.

Now, the most interesting aspect is that the excitation power falls off as the square of the distance from the lens focal point, within the approximation of a conical illumination geometry. In practice this means that the quadratic relationship between the excitation power and the fluorescence intensity results in the fact that TPE falls off as the fourth power of distance from the focal point of the objective. This fact implies that those regions away from the focal volume of the objective lens, directly related to the numerical aperture of the objective itself, therefore do not suffer photobleaching or phototoxicity effects and do not contribute to the signal detected when a TPE scheme is used. Because they are simply not involved in the excitation process, a confocal-like effect is obtained without the necessity of a confocal pinhole. It is also immediately evident that in this case an optical sectioning effect is obtained. In fact, the observed image $o(x,y,z)$ at a plane j, produced by the true fluorescence distribution $i(x,y,z)$ at plane j, distorted by the microscope through s, plus noise n, again corresponds to the confocal ideal situation where contributions from adjacent k planes can be set to zero as in the confocal situation: $o_j = i_j * s_j + n$.

This means that TPE microscopy is intrinsically three dimensional. It is worth noting that the optical sectioning effect is obtained in a very different way with respect to the confocal solution. No fluorescence has to be removed from the detection pathway. In this case it should be possible to collect as much fluorescence is possible. In fact fluorescence can come only and exclusively from the small focal volume traced in Figure 11–7, which also shows a comparison with the confocal mode, that is of the order of a fraction of a femtoliter.

In TPE over 80% of the total intensity of fluorescence comes from a 700- to 1000-nm-thick region about the focal point for objectives with numerical apertures in the range of 1.2–1.4 (Brakenhoff et al., 1979;

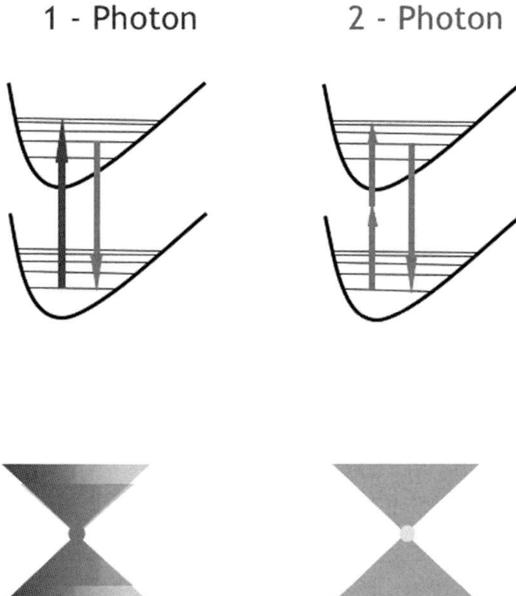

Figure 11–7. Illustration of the two different modalities for selecting 3D information under a confocal (left) and TPE regime (right). In the confocal case the selection is realized during the emission process. The different case of two-photon excitation shows how the 3D selection can be realized during the fluorescence excitation process.

Wilson and Sheppard, 1984; Wilson, 2002; Jonkman and Stelzer, 2002; Torok and Sheppard, 2002). This fact implies a reduction in background that allows compensation of the poorer spatial resolution compared to the single-photon confocal mode due to the longer wavelength utilized. However, the utilization of an infrared wavelength instead of UV-visible ones also allows deeper penetration than in the conventional case (So et al., 2001; Periasamy et al., 2002; König and Tirlapur, 2002). The long wavelengths used in TPE, or in general in multiphoton excitation, will be scattered less than the ultraviolet–visible wavelengths used for conventional excitation (de Grauw and Gerritsen, 2001). Hence deeper targets within a thick sample can be reached. Of course, for fluorescence light, scattering on the way back can be overcome by acquiring the emitted fluorescence using a large area detector and collecting not only ballistic photons (Soeller and Cannel, 1999; Bauhler et al., 1999; Girkin and Wokosin, 2002).

7 The Optical Setup

A TPE architecture including confocal modality includes the following: a high peak-power laser delivering moderate average power (femtosecond or picosecond pulsed at a relatively high repetition rate) emitting infrared or near-infrared wavelengths (650–1100 nm), CW laser sources for confocal modes, a laser beam scanning system or a confocal laser scanning head, high numerical aperture objectives (>1), a high-throughput microscope pathway, and a high-sensitivity detection system (Denk et al., 1995; So et al., 1996; Soeller and Cannell, 1996; Wokosin and White, 1997; Centonze and White, 1998; Potter et al., 1996; Wolleschensky et al., 1998; Diaspro et al., 1999a,b; Wier et al., 2000; Soeller and Cannell, 1999; Tan et al., 1999; Mainen et al., 1999; Majewska et al., 2000; Diaspro, 2002; Girkin and Wokosin, 2002; Iyer et al., 2002). Figure 11–8 shows a general scheme for a TPE microscope incorporating a confocal mode.

In typical TPE or confocal microscopes, images are built by raster scanning the x–y mirrors of a galvanometrically driven mechanical scanner (Webb, 1996). This fact implies that image formation speed is mainly determined by the mechanical properties of the scanner, i.e., for single line scanning it is of the order of milliseconds. Faster beam-scanning schemes can be realized, even if the "eternal triangle of compromise" should be considered for sensitivity, spatial resolution, and temporal resolution. While the x–y scanners provide lateral focal-point scanning, axial scanning can be achieved by means of different positioning devices, the most popular being a belt-driven system using a DC motor and a single objective piezo nanopositioner, such as the PIFOC (Physik Instrumente, Germany). Usually, it is possible to switch between confocal and TPE modes retaining x–y–z positioning on the sample being imaged (Diaspro, 2001; Diaspro and Chirico, 2003). Acquisition and visualization are generally completely computer controlled by dedicated software. Figure 11–9 shows a TPE microscope.

Let us now consider two popular approaches that can be used to perform TPE microscopy, namely, the descanned and nondescanned

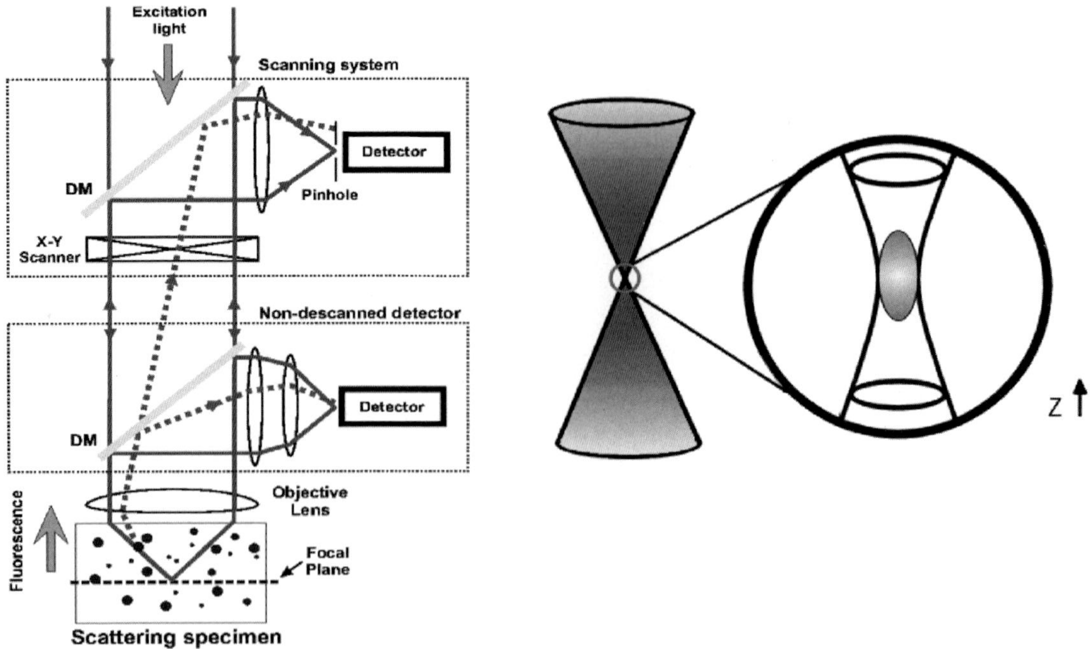

Figure 11–8. Optical configuration for a TPE microscope operating in a descanned (upper inset box) and nondescanned (lower inset box) mode; see text. (Courtesy of M. Cannel and C. Soeller.)

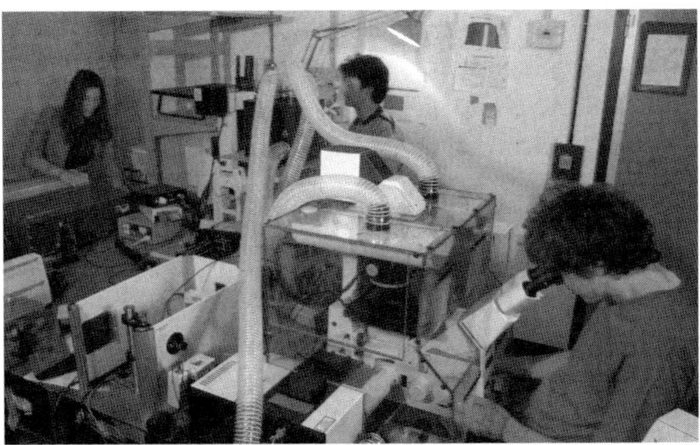

Figure 11–9. The TPE setup at LAMBS, MicroScoBio Research Center of the University of Genoa (from left to right: Ilaria Testa, Paolo Bianchini, and Davide Mazza).

modes. They are skectched in Figure 11–8. The former uses the very same optical pathway and mechanism employed in confocal laser scanning microscopy. The latter mainly optimizes the optical pathway by minimizing the number of optical elements encountered on the way from the sample to detectors, and increases the detector area. The TPE nondescanned mode provides very good performances giving a superior signal-to-noise ratio inside strongly scattering samples (Masters et al., 1997; Daria et al., 1998; Centonze and White, 1998; So et al., 2000).

In the descanned approach pinholes are removed or set to their maximum aperture and the emission signal is captured using an excitation scanning device on the back pathway. For this reason it is called the descanned mode. In the latter, the confocal architecture has to be modified in order to increase the collection efficiency: pinholes are removed and the emitted radiation is collected using dichroic mirrors on the emission path or external detectors without passing through the galvanometric scanning mirrors. A high-sensitivity detection system is another critical issue (Wokosin et al., 1998; So et al., 2000; Girkin and Wokosin, 2002).

The fluorescence emitted is collected by the objective and transferred to the detection system through a dichroic mirror along the emission path (Figure 11–8). Due to the high excitation intensity, an additional barrier filter is needed to avoid mixing the excitation and emission light at the detection system that is differently placed depending on the acquisition scheme being used. Photodetectors that can be used include photomultiplier tubes, avalanche photodiodes, and CCD cameras (Denk et al., 1995; Murphy, 2001). Photomultiplier tubes are the most commonly used. This is due to their low cost, good sensitivity in the blue-green spectral region, high dynamic range, large size of the sensitive area, and single-photon counting mode availability (Hamamatsu Photonics, 1999). They have a quantum efficiency around 20–40% in the blue-green spectral region that drops down to <1% moving to the red region. This is a good condition, especially in MPE mode, because it is desirable to reject as much as possible wavelengths above 680 nm that are mainly used for excitation. Another advantage is that the large size of the sensitive area of photomultiplier tubes allows efficient collection of signal in the nondescanned mode within a dynamic range of the order of 10^8. Avalanche photodiodes are excellent in terms of sensitivity exhibiting quantum efficiency close to 70–80% in the visible spectral range. Unfortunately they are high in cost and the small active photosensitive area, <1 mm size, could introduce drawbacks in the detection scheme and requires special descanning optics (Farrer et al., 1999). CCD cameras are used in video rate multifocal imaging (Fuijta and Takamatsu, 2002; Girkin and Wokosin, 2002). However, once the best quality image possible has been obtained then image restoration algorithms can be applied to enhance the features of interest to the biological researcher and to improve the quality of data to be used for three-dimensional modeling, such as those used for single-photon optical sectioning microscopy, available at http://www. powermicroscope.com (van der Voort et al., 1995; Shotton, 1995; Diaspro et al., 1990, 2000; Boccacci and Bertero, 2002; Carrington, 2002; Difato et al., 2004; Bonetto et al., 2004).

Laser sources, as often happened in optical microscopy, represent an important resource, especially in fluorescence microscopy (Gratton and van de Ven, 1995; Svelto, 1998). For nonresonant TPE, owing to the comparatively low TPE cross-sections of fluorophores, high photon flux densities are required, $>10^{24}$ photons cm^{-2} s^{-1} (König, 2000). Using radiation in the spectral range of 600–1100 nm for TPE, excitation intensities in the MW–GW cm^{-2} range are required. This high energy can be obtained by the combined use of focusing lens objectives and CW (Hanninen and Hell, 1994; König et al., 1995) or pulsed (Denk et al., 1990) laser radiation of 50 mW mean power or less (Girkin and Wokosin, 2002; Diaspro and Sheppard, 2002). TPE microscopes have been realized using CW, femtosecond, and picosecond laser sources (Periasamy, 2001; Diaspro, 2001, 2002; Masters, 2002). Since the original successful experiments in TPE microscopy, advances have been made in the technological field of ultrashort pulsed lasers. Today laser sources suitable for TPE can be described as "turnkey" compact systems (Fisher et al., 1997; Wokosin et al., 1996; Diaspro, 2001).

Figure 11–10 shows a new generation ultrafast Ti:sapphire laser source. The emission range between 700 and 1050 nm of the Ti:sapphire laser allows a large number of commonly used fluorescent molecules to be excited. Other laser sources used for TPE are Cr-LiSAF, pulse-compressed Nd-YLF in the femtosecond regime, and mode-locked Nd-YAG and picosecond Ti-sapphire lasers in the picosecond regime (Gratton and Van de Ven, 1995; Wokosin et al., 1996). Most of the laser sources used for TPE operate in a mode-locking mode. This endows the laser with the ability to generate a train of very short pulses by modulating the gain or excitation of a laser at a frequency with a period equal to the roundtrip time of a photon within the laser cavity (Fisher et al., 1997; Svelto, 1998) (Figure 11–11). The resulting pulsewidth is in the 50–150 fs regime. The parameters that are more relevant in the selection of the laser source are average power, pulsewidth and repetition rate, and wavelength also according to Eq. (15). The most

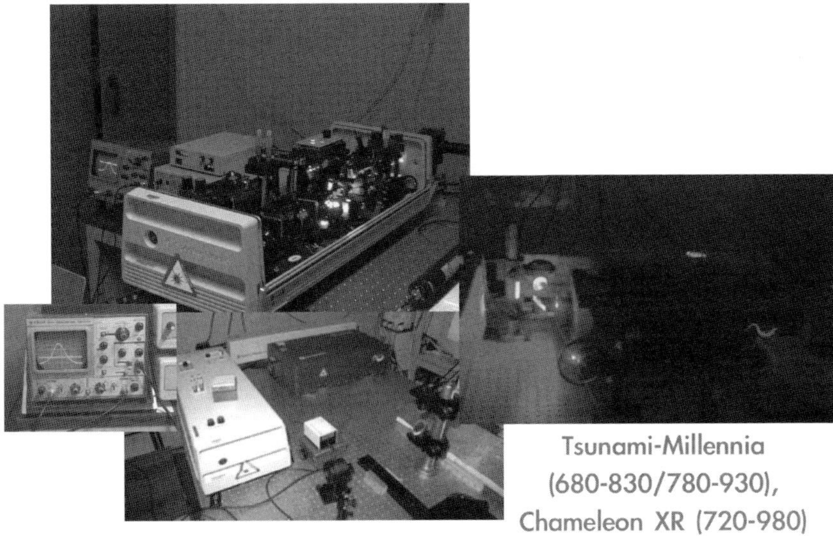

Figure 11–10. Typical laser sources in use for TPE microscopy.

popular features for an infrared pulsed laser are 700 mW–1 W average power, 80–100 MHz repetition rate, and 100–150 fs pulse width.

At present, the use of short pulses and small duty cycles are mandatory to allow image acquisition in a reasonable time while using power levels that are biologically tolerable (Denk et al., 1994; Denk, 1996; Koester et al., 1999; König et al., 1996, 1998; König, 2000; König and Tirlapur, 2002). To minimize pulse width dispersion problems König (2000) suggested working with pulses around 150–200 nm, and this constitutes a very good compromise both for pulse stretching and sample viability. It should always be remembered that a shorter pulse broadens more than a longer one. Pulse width measurement is a very delicate issue. In fact, because it is not very easy to measure it at the focal volume within the sample, little can be definitely said about it (Hanninen and Hell, 1994; Guild et al., 1997; Wolleschensky et al., 2002). Although users do not perform measurement of the pulse width at the

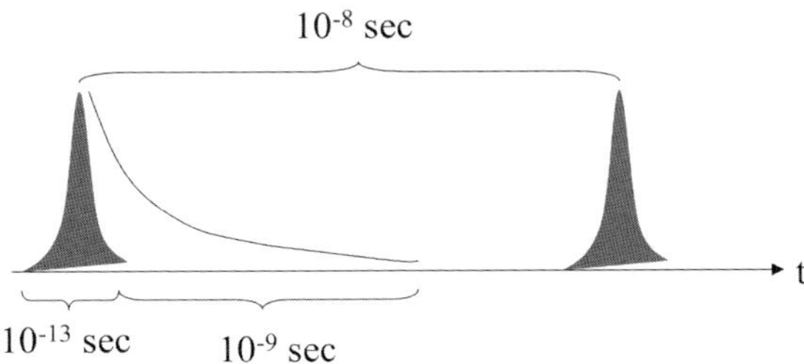

Figure 11–11. Laser emission time scale for TPE excitation: a short pulse at high photon density is released for approximately 100 fs; this laser shot is able to prime fluorescence without damaging the sample so fluorescence occurs in the next few nanoseconds. The laser is silent for 10 ns and then delivers a new high-density photon pulse. This modality allows TPE to be experienced at tolerable time-averaged power (see text).

sample when they use two-photon microscopy, which would require a specific procedure that, even if not too complex for a researcher in the field, could be irksome for the majority of users, it is a reasonable approximation to assume that at the focal volume, 1.5–2. times temporal pulse broadening occurs using high-quality optics (Wolleschensky, 2002; Girkin and Wokosin, 2002). As an example, for a measured laser pulse width of about 100 fs, an estimate at the sample is about 150–180 fs under favorable experimental conditions, sample characteristics included. Sample properties are mentioned because for thick samples the role played by thickness, also in terms of pulse width broadening, is not so obvious (de Grauw and Gerritsen, 2002; So et al., 2001; Gu et al., 2000; Saloma et al., 1998).

8 Conclusion

Confocal microscopy, in the authors' opinion, constitutes one of the most significant advances in optical microscopy within the past decades, and has become a powerful investigative tool for the molecular, cellular, and developmental biologist, the materials scientist, the biophysicist, and the electronic engineer. It is entirely compatible with the range of "classical" light microscopic techniques, and, at least in scanned beam instruments, can be applied to the same specimens on the same optical microscope stage. Its peculiar advantages result in its ability to generate multidimensional (x–y–z–t) images by noninvasive optical sectioning with a virtual absence of out-of-focus blur, its capacity for multiparametric imaging of multiply labeled samples, and its property of investigating at microscopic resolution large objects as a result of the rejection of scattered light. So far, the advent of confocal microscopy in the mid-1980s favored the rapid spreading of two- and multiphoton excitation microscopy, since Denk's report at the beginning of the 1990s, bringing dramatic changes in designing experiments that utilize fluorescent molecules and, more specifically, in fluorescence 3D optical microscopy.

While confocal microscopy is moving to spectral and fast-scanning architectures in terms of acquisition, it is mainly two-photon microscopy that occupies the scene of advances in fluorescence optical microscopy. TPE microscopy, with its intrinsic three-dimensional resolution, the absence of background fluorescence, and the attractive possibility of exciting UV excitable fluorescent molecules, thus increasing sample penetration, constitutes significant progress in science. In fact, in a TPE scheme two 720-nm photons combine to produce the very same fluorescence conventionally primed at ~360 nm, and to be utilized in a classical confocal microscope using conventional excitation of fluorescent molecules. The excitation of the fluorescent molecules bound to the specific components of the biological systems being studied mainly takes place (80%) in an excitation volume of the order of magnitude of 0.1 fl. This results in an intrinsic 3D optical sectioning effect. What is invaluable for cell imaging and, in particular, for live-cell imaging is the fact that weak endogenous one-photon absorption and

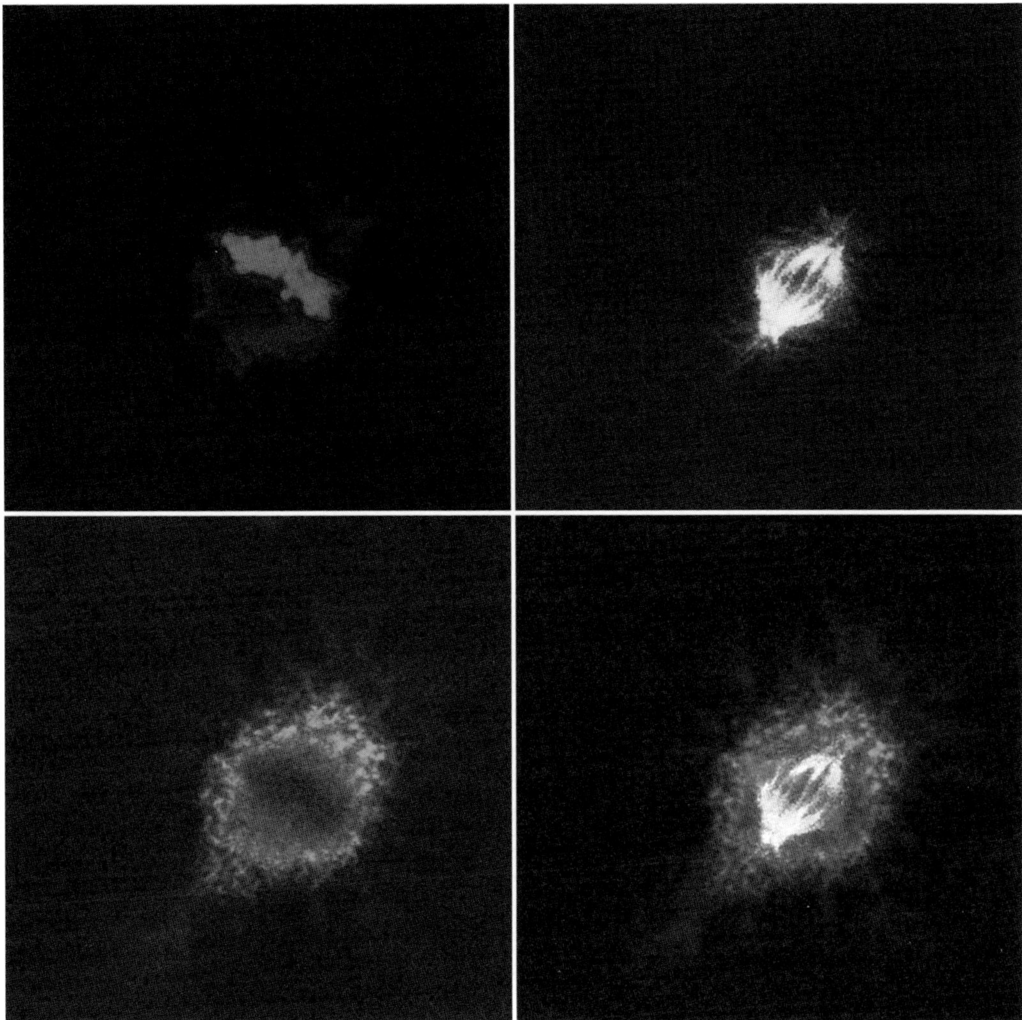

Figure 11–12. Multiple excitation of three fluorescent dyes using 740 nm under a TPE regime. The conventional excitation would have required the utilization of 360 or 405 nm, 488 nm, and 543 nm laser lines. The final image (lower right quadrant) is realized by merging the three subsets. (This image has been acquired by students of the Biotechnology School during the course of Advanced Microscopy Techniques activated at the University of Genoa, academic year 2005. Advisors: Grazia Tagliafierro and Alberto Diaspro.) (See color plate.)

highly localized spatial confinement of the TPE process dramatically reduce phototoxicity stress. To summarize the unique characteristics and advantages of TPE we recall the following properties:

1. Spatially confined fluorescence excitation in the focal plane of the specimen can be considered the key feature of TPE microscopy. It is one of the advantages over confocal microscopy, where fluorescence emission occurs across the entire thickness of the sample being excited by the scanning laser beam. A strong implication is that there is no photon signal from sources out of the geometric position of the optical focus within the sample. Therefore, the signal-to-noise ratio increases, photo-

degradation effects decrease, and optical sectioning is immediately available without the need for pinhole or deconvolution algorithms. In addition, very efficient acquisition schemes can be implemented such as the nondescanned one operating at an excellent signal-to-noise ratio.

2. The use of near-infrared (IR)/IR wavelengths permits examination of thick specimens in depth. This is due to the fact that, apart from some cases such as pigmented samples and portions of the absorption spectral window of water, cells and tissues absorb poorly in the near-IR/IR region. Cellular damage is globally minimized, thus allowing cell viability to be prolonged with long-term 3D sessions. Moreover, scattering is reduced and deeper targets can be reached with fewer problems than in one-photon excitation. The depth of penetration can be up to 0.5 mm. In addition, whereas in one-photon excitation, the emission wavelength is comparatively close to the excitation one (about 50–200 nm longer), in TPE the fluorescence emission occurs at a wavelength substantially shorter and at a larger spectral distance than in one-photon excitation. Thus separation of the excitation light and the emitted light can be easily performed.

Continuing research in this field is focused on very intriguing problems (www.focusonmicroscopy.org offers a complete scenario of the evolution of three-dimensional microscopy in the past 5 years) such as local heating from absorption of IR light by water at high laser power (Schonle and Hell, 1998) and photothermal effects on fluorescent molecules (Chirico et al., 2003a), phototoxicity from long wavelength IR excitation and short wavelength fluorescence emission (König et al., 1996c; Tyrrel and Keyse, 1990; König, 2000; Hopt and Neher, 2001; König and Tirlapur, 2002), photoactivation and photocycling of visible fluorescent proteins (Post et al., 2004; Chirico et al., 2004; Schnedier et al., 2005), development of new fluorochromes better suited for TPE and multiphoton excitation (Albota et al., 1998a; Abbotto et al., 2005), and the investigation of the cross-sections of uncharacterized molecules (Gostkowski et al., 2004; Wokosin et al., 2004).

One of the major benefits in setting up an MPE microscope is the flexibility in choosing the measurement modality favored by the simplification of the optical design. In fact, a TPE microscope offers a greater variety of measurement options without changing any optics or hardware. This means that during the very same experiments real multimodal information can be obtained from the specimen being studied (Zoumi et al., 2002; Wang et al., 2004). Moreover, the usefulness of the TPE scheme for spectroscopic and lifetime studies (So et al., 1996; Sytsma et al., 1998; Schwille et al., 2000; Diaspro et al., 2001; Wiseman et al., 2002), for optical data storage and microfabrication (Cumpston et al., 1999; Kawata et al., 2001), and for single molecule detection (Mertz et al., 1995; Farrer et al., 1999; So et al., 2000; Chirico et al., 2001; Cannone et al., 2003b) has been well documented. Other very interesting applications involve the study of impurities affecting the growth of protein crystals (Caylor et al., 1999), TPE imaging in the field of plant biology (Tirlapur and König, 2002), and measurements in living systems (Squirrel et al., 1999; Yoder and Kleinfeld, 2002; Diaspro et al., 2002b,d; Post et al., 2004). Here the combination of MPE and second-harmonic generation offers the opportunity to investigate the morphometric properties on the basis of the microstructure of blood cells (Zoumi et al., 2004). Another promising field is the investigation of complex formation where the TPE properties will improve the information accessible (Heinze et al., 2004). Another, more indirect usage that provides a look at the sample with nanometer resolution is the excitation of an evanescent wave at a metal surface (Novotny et al., 1998). For microscopic purposes the evanescent wave needs to be localized at a nanoparticle or a fine metal tip (Sánchez et al., 1999; Gerton et al., 2004). The MPE microscope can also be used as an active device, with increasing applications related to nanosurgery (König, 2000), selective uncaging of caged compounds (Diaspro et al., 2003), and photodynamic therapy (Bhalwalkar et al., 1997; So et al., 2000). Recently TPE microscopy, even if in an evanescent-field-induced configuration, has been extended to large area structures of the order of square centimeters (Duveneck et al., 2001). This has application in the realization of biosensing platforms such as genomic and proteomic microarrays based upon large planar waveguides. It is easy to perceive that the range of applicability of MPE microscopes is rapidly increasing in the biomedical, biotechnological, and biophysical sciences and is expanding to clinical applications (Diaspro, 2002; Masters, 2002; Periasamy and Diaspro, 2003).

Acknowledgments. The first Italian TPE architecture realized at LAMBS has been supported by INFM grants. LAMBS-MicroScoBio is currently funded by IFOM (Istituto FIRC di Oncologia Molecolare, FIRC Institute of Molecular Oncology, Milano). This chapter is dedicated to the memory of Osamu Nakamura, who passed away January 23, 2005 at Handai Hospital.

References

Abbe, E. (1910). In: Die Lehre von der Bildentstehung in Mikroskop (O. Lummer, Ed.). F. Reiche, Braunschweig.

Abbotto, A., Baldini, G., Beverina, L., Chirico, G., Collini, M., D'alfonso, L., Diaspro, A., Magrassi, R., Nardo, L. and Pagani, G.A. (2005). Dimethyl-Pepep: A Dna probe in two-photon excitation cellular imaging. *Biophys. Chemi.* **114**(1), 35–41.

Agard, D.A. (1984). Optical sectioning microscopy: Cellular architecture in three dimensions. *Annu. Rev. Biophys.* **13**, 191–219.

Agard, D.A., Hiraoka, Y., Shaw, P.J. and Sedat, J.W. (1989). Fluorescence microscopy in three-dimensions. *Methods Cell. Biol.* **30**, 353–378.

Albota, M., Beljonne, D., Bredas, J.L., Ehrlich, J.E., Fu, J.Y., Heikal, A.A., Hess, S.E., Kogej, T., Levin, M.D., Marder, S.R. and others. (1998a). Design of organic molecules with large two-photon absorption cross sections. *Science* **281**(5383), 1653–1656.

Albota, M.A., Xu, C. and Webb, W.W. (1998b). Two-photon fluorescence excitation cross sections of biomolecular probes from 690 to 960 nm. *Appl. Opt.* **37**, 7352–7356.

Amos, B. (2000). Lessons from the history of light microscopy. *Nature Cell Biol.* **2**, E151–E152.

Andrews, D.L. (1985). A simple statistical treatment of multiphoton absorption. *Am. J. Phys.* **53**, 1001–1002.

Arndt-Jovin, D.J., Nicoud, R.M., Kaufmann, J. and Jovin, T.M. (1985). Fluorescence digital-imaging microscopy in cell biology. *Science* **230**, 1333–1335.

Bastiaens, P.I. and Hell, S.W. (Eds.) (2004). Recent advances in light microscopy. *J. Struct. Biol.* **147**, 1–89.

Basu, S. and Campagnola, P.J. (2004). Properties of crosslinked protein matrices for tissue engineering applications synthesized by multiphoton excitation. *J. Biomed. Mat. Res.* Part A **71**A(2), 359–368.

Beltrame, F., Bianco, B., Castellaro, G. and Diaspro, A. (1985). Fluorescence, absorption, phase-contrast, holographic and acoustical cytometries of living cells. In: *Interactions between Electromagnetic Fields and Cells* (A. Chiabrera and H.P. Schwan, Eds.). NATO ASI Series, Vol. 97, pp. 483–498. (Plenum Press, New York).

Benedetti, P. (1998). From the histophotometer to the confocal microscope: The evolution of analytical microscopy. *Eur. J. Histochem.* **42**, 11–17.

Benham, G.S. and Schwartz, S. (2002). Suitable microscope objectives for multiphoton digital imaging. In: *Multiphoton Microscopy in the Biomedical Sciences II* (A. Periasamy and P.T.C. So, Eds.). Proc. SPIE 4620, 36–47.

Berland, K. (2002). Basics of fluorescence. In: *Methods in Cellular Imaging* (A. Periasamy, Ed.), pp. 5–19. (Oxford University Press, New York).

Berland, K.M., So, P.T.C. and Gratton, E. (1995). Two-photon fluorescence correlation spectroscopy: Method and application to the intracellular environment. *Biophys. J.* **68**, 694–701.

Berns, M.W. (1976). A possible two-photon effect in vitro using a focused laser beam. *Biophys. J.* **16**, 973–977.

Bertero, M. and Boccacci, P. (1998). Introduction to Inverse Problems in Imaging. (IOP Publishing, Bristol).

Bhawalkar, J.D., Kumar, N.D., Zhao, C.F. and Prasad, P.N. (1997). Two-photon photodynamic therapy. *J. Clin. Laser Med. Surg.* **15**, 201–204.

Bianco, B. and Diaspro, A. (1989). Analysis of the three dimensional cell imaging obtained with optical microscopy techniques based on defocusing. *Cell Biophys.* **15**(3), 189–200.

Binnig, G., Quate, C.F. and Gerber, C. (1986). Atomic force microscope. *Phys. Rev. Lett.* **56**(9), 930–933.

Bird, D.K., Eliceiri, K.W., Fan, C.H. and White, J.G. (2004). Simultaneous two-photon spectral and lifetime fluorescence microscopy. *Appl. Opt.* **43**(27), 5173–5182.

Birge, R.R. (1986). Two-photon spectroscopy of protein-bound fluorophores. *Acc. Chem. Res.* **19**, 138–146.

Birks, J.B. (1970). Photophysics of Aromatic Molecules. (Wiley Interscience, London).

Blab, G.A., Lommerse, P.H.M., Cognet, L., Harms, G.S. and Schmidt, T. (2001). Two-photon excitation action cross-sections of the autofluorescent proteins. *Chem. Phys. Lett.* **350**(1–2), 71–77.

Boccacci, P. and Bertero, M. (2002). Image restoration methods: Basics and algorithms. In: *Confocal and Two-Photon Microscopy: Foundations, Applications and Advances* (A. Diaspro, Ed.), pp. 253–270. (Wiley-Liss, New York).

Bonetto, P., Boccacci, P., Scarito, M., Davolio, M., Epifani, M., Vicidomini, G., Tacchetti, C., Ramoino, P., Usai, C. and Diaspro, A. (2004). Three-dimensional microscopy migrates to the Web with "PowerUp Your Microscope." *Microsc. Res. Tech.* **64**(2), 196–203.

Born, M. and Wolf, E. (1980). *Principles of Optics, 6th ed.* (Cambridge University Press, Cambridge, UK).

Brakenhoff, G.J., Blom, P. and Barends, P. (1979). Confocal scanning light microscopy with high aperture immersion lenses. *J. Microsc.* **117**, 219–232.

Brakenhoff, G.J., van Spronsen, E.A., van der Voort, H.T. and Nanninga, N. (1989). Three-dimensional confocal fluorescence microscopy. *Method. Cell. Biol.* **30**, 379–398.

Brakenhoff, G.J., Muller, M. and Ghauharali, R.I. (1996). Analysis of efficiency of two-photon versus single-photon absorption for fluorescence generation in biological objects. *J. Microsc.* **183**, 140–144.

Cahalan, M.D., Parker, I., Wei, S.H. and Miller, M.J. (2002). Two-photon tissue imaging: Seeing the immune system in a fresh light. *Nat. Rev. Immunol.* **2**(11), 872–880.

Callis, P.R. (1997). Two-photon-induced fluorescence. *Annu. Rev. Phys. Chem.* **48**, 271–297.

Campagnola, P., Mei-de Wei, Lewis, A. and Loew, L. (1999). High-resolution nonlinear optical imaging of live cells by second harmonic generation. *Biophys. J.* **77**, 3341–3351.

Cannell, M.B. and Soeller, C. (1997). High resolution imaging using confocal and two-photon molecular excitation microscopy. *Proc. R. Microsc. Soc.* **32**, 3–8.

Cannone, F., Chirico, G., Baldini, G. and Diaspro, A. (2003a). Measurement of the laser pulse width on the microscope objective plane by modulated auto-correlation method. *J. Microsc. (O_{XF})*, **210**(2), 149–157.

Cannone, F., Chirico, G. and Diaspro, A. (2003b). Two-photon interactions at single fluorescent molecule level. *J. Biomed. Opt.* **8**(3), 391–395.

Cantor, C.R. and Schimmel, P.R. (1980). *Biophysical Chemistry. Part II: Techniques for the Study of Biological Structure and Function.* (Freeman and Co., New York).

Carlsson, K., Danielsson, P.E., Lenz, R., Liljeborg, A., Majlof, L. and Aslund, N. (1985). Three-dimensional microscopy using a confocal laser scanning microscope. *Opt. Lett.* **10**, 53–55.

Carrington, W. (2002). Imaging live cells in 3-d using wide field microscopy with image restoration. In: *Confocal and Two-Photon Microscopy: Foundations, Applications and Advances* (A. Diaspro, Ed.), 33–346. (Wiley-Liss, New York).

Caylor, C.L., Dobrianov, I., Kimmer, C., Thorne, R.E., Zipfel, W. and Webb, W.W. (1999). Two-photon fluorescence imaging of impurity distributions in protein crystals. *Phys. Rev. E* **59**, 3831–3834.

Centonze, V.E. and White, J.G. (1998). Multiphoton excitation provides optical sections from deeper within scattering specimens than confocal imaging. *Biophys. J.* **75**, 2015–2024.

Chalfie, M. and Kain, S. (Eds.) (1998). *Green Fluorescent Protein. Properties, Applications and Protocols.* (Wiley-Liss, New York).

Chalfie, M., Tu, Y., Euskirchen, G., Ward, W.W. and Prasher, D.C. (1994). Green fluorescent protein as a marker for gene expression. *Science* **263**, 802–805.

Chance, B. (1989). *Cell Structure and Function by Microspectrofluorometry.* (Academic Press, New York).

Cheng, P.C. (Ed.) (1994). *Computer Assisted Multidimensional Microscopies.* (Springer-Verlag, New York).

Chirico, G., Cannone, F., Beretta, S., Baldini, G. and Diaspro, A. (2001). Single molecule studies by means of the two-photon fluorescence distribution. *Microsc. Res. Tech.* **55**, 359–364.

Chirico, G., Cannone, F., Baldini, G. and Diaspro, A. (2003a). Two-photon thermal bleaching of single fluorescent molecules. *Biophys. J.* **84**, 588–598.

Chirico, G., Cannone, F. and Diaspro, A. (2003b). Single molecule photodynamics by means of one- and two-photon approach. *J. Phys. D: Appl. Phys.* **36**, 1–7.

Chirico, G., Cannone, F., Diaspro, A., Bologna, S., Pellegrini, V., Nifosì, R. and Beltram, F. (2004). Multiphoton switching dynamics of single green fluorescent proteins. *Phys. Rev. E* **70**, 030901.

Chirico, G., Diaspro, A., Cannone, F., Collini, M., Bologna, S., Pellegrini, V. and Beltram, F. (2005). Selective fluorescence recovery after bleaching of single E2gfp proteins induced by two-photons excitation. *Chemphyschem* **6**(2), 328–335.

Cianci, G.C., Wu, J. and Berland, K. (2004). Saturation modified point spread functions in two-photon microscopy. *Microsc. Res. Tech.* **64**(2), 135–141.

Cox, G. (2002). Biological confocal microscopy. *Materials Today* **5**, No. 3, 34–41.

Cruz, H.G. and Luscher, C. (2005). Applications of two-photon microscopy in the neurosciences. *Front Biosci.* **10**, 2263–2278.

Cumpston, B.H., Ananthavel, S.P., Barlow, S., Dyer, D.L., Ehrlich, J.E., Erskine, L.L., Heikal, A.A., Kuebler, S.M., Lee, I.Y.S., McCord-Maughon, D. and others. (1999). Two-photon polymerization initiators for three-dimensional optical data storage and microfabrication. *Nature* **398**(6722), 51–54.

Daria, V., Blanca, C.M., Nakamura, O., Kawata, S. and Saloma, C. (1998). Image contrast enhancement for two-photon fluorescence microscopy in a turbid medium. *Appl. Opt.* **37**, 7960–7967.

Davidovits, P.D. and Egger, M.D. (1969). Scanning laser microscope. *Nature* **223**, 831.

Davidovits, P.D. and Egger, M.D. (1971). Scanning laser microscope for biological investigations. *Appl. Opt.* **10**, 1615–1619.

de Grauw, K. and Gerritsen, H. (2002). Aberrations and penetration depth in confocal and two-photon microscopy. In: *Confocal and Two-Photon Microscopy: Foundations, Applications and Advances* (A. Diaspro, Ed.), 153–170. (Wiley-Liss, New York).

Denk, W. (1996). Two-photon excitation in functional biological imaging. *J. Biomed. Opt.* **1**, 296–304.

Denk, W. and Svoboda, K. (1997). Photon upmanship: Why multiphoton imaging is more than a gimmick. *Neuron* **18**, 351–357.

Denk, W., Strickler, J.H. and Webb, W.W. (1990). Two-photon laser scanning fluorescence microscopy. *Science* **248**, 73–76.

Denk, W., Delaney, K.R., Gelperin, A., Kleinfeld, D., Strowbridge, B.W., Tank, D.W. and Yuste, R. (1994). Anatomical and functional imaging of neurons using two-photon laser scanning microscopy. *J. Neurosci. Methods* **54**, 151–162.

Denk, W., Piston, D. and Webb, W.W. (1995). Two-photon molecular excitation in laser scanning microscopy. In: *Handbook of Confocal Microscopy* (J.B. Pawley, Ed.), 445–457. (Plenum, New York).

Diaspro, A. (guest editor) (1996). New world microscopy. *IEEE Eng. Med. Biol. Mag.* **15**(1), 29–100.

Diaspro, A. (1998). Two-photon fluorescence excitation. A new potential perspective in flow cytometry. *Minerva Biotechnol.* **11**(2), 87–92.

Diaspro, A. (guest editor) (1999a). Two-photon microscopy. *Microsc. Res. Tech.* **47**, 163–212.

Diaspro, A. (guest editor) (1999b). Two-photon excitation microscopy. *IEEE Eng. Med. Biol. Mag.* **18**(5), 16–99.

Diaspro, A. (1999c). Two-photon excitation of fluorescence in three-dimensional microscopy. *Eur. J. Histochem.* **43**, 169–178.

Diaspro, A. (2001). Building a two-photon microscope using a laser scanning confocal architecture. In: *Methods in Cellular Imaging* (A. Periasamy, Ed.), 162–179. (Oxford University Press, New York).

Diaspro, A. (Ed.) (2002). *Confocal and Two-Photon Microscopy: Foundations, Applications, and Advances.* (Wiley-Liss, New York).

Diaspro, A. (2004). Rapid dissemination of two-photon excitation microscopy prompts new applications. *Microsc. Res. Tech.* **63**(1), 1–2.

Diaspro, A. and Chirico, G. (2003). Two-photon excitation microscopy. *Adv. Imaging Elect. Phys.* **126**, 195–286.

Diaspro, A. and Robello, M. (2000). Two-photon excitation of fluorescence for three-dimensional optical imaging of biological structures. *J. Photochem. Photobiol. B* **55**, 1–8.

Diaspro, A. and Sheppard, C.J.R. (2002). Two-photon excitation microscopy: Basic principles and architectures. In: *Confocal and Two-Photon Microscopy: Foundations, Applications, and Advances* (A. Diaspro, Ed.), 39–74. (Wiley-Liss, New York).

Diaspro, A., Sartore, M. and Nicolini, C. (1990). Three-dimensional representation of biostructures imaged with an optical microscope: I. Digital optical sectioning. *Image Vision Comp.* **8**, 130–141.

Diaspro, A., Beltrame, F., Fato, M., Palmeri, A. and Ramoino, P. (1997). Studies on the structure of sperm heads of eledone cirrhosa by means of CLSM linked to bioimage-oriented devices. *Microsc. Res. Tech.* **36**, 159–164.

Diaspro, A., Annunziata, S., Raimondo, M. and Robello, M. (1999a). Three-dimensional optical behaviour of a confocal microscope with single illumination and detection pinhole through imaging of subresolution beads. *Microsc. Res. Tech.* **45**(2), 130–131.

Diaspro, A., Corosu, M., Ramoino, P. and Robello, M. (1999b). Adapting a compact confocal microscope system to a two-photon excitation fluorescence imaging architecture. *Microsc. Res. Tech.* **47**, 196–205.

Diaspro, A., Annunziata, S. and Robello, M. (2000). Single-pinhole confocal imaging of sub-resolution sparse objects using experimental point spread function and image restoration. *Microsc. Res. Tech.* **51**, 464–468.

Diaspro, A., Chirico, G., Federici, F., Cannone, F., Beretta, S. and Robello, M. (2001). Two-photon microscopy and spectroscopy based on a compact confocal scanning head. *J. Biomed. Opt.* **6**, 300–310.

Diaspro, A., Federici, F. and Robello, M. (2002a). Influence of refractive-index mismatch in high-resolution three-dimensional confocal microscopy. *Appl. Opt.-OT* **41**, 685–690.

Diaspro, A., Silvano, D., Krol, S., Cavalleri, O. and Gliozzi, A. (2002b). Single living cell encapsulation in nano-organized polyelectrolyte shells. *Langmuir* **18**, 5047–5050.

Diaspro, A., Boccacci, P., Bonetto, P., Scarito, M., Davolio, M. and Epifani, M. (2002c). "Power-up your Microscope." www.powermicroscope.com.

Diaspro, A., Fronte, P., Raimondo, M., Fato, M., De Leo, G., Beltrame, F., Cannone, F., Chirico, G. and Ramoino, P. (2002d). Functional imaging of living paramecium by means of confocal and two-photon excitation fluorescence microscopy. In: *Functional Imaging* (D. Farkas, Ed.). Proc. SPIE 4622, 47–53.

Diaspro, A., Federici, F., Viappiani, C., Krol, S., Pisciotta, M., Chirico, G., Cannone, F. and Gliozzi, A. (2003). Two-photon photolysis of 2-nitrobenzaldehyde monitored by fluorescent-labeled nanocapsules. *J. Phy. Chem. B* **107**(40), 11008–11012.

Difato, F., Mazzone, F., Scaglione, S., Fato, M., Beltrame, F., Kubinova, L., Janacek, J., Ramoino, P., Vicidomini, G. and Diaspro, A. (2004). Improvement in volume estimation from confocal sections after image deconvolution. *Microsc. Res. Tech.* **64**(2), 151–155.

Dürig, U. and Pohl, D.W. (1986). Near-field optical-scanning microscopy. *J. Appl. Phys.* **59**(10), 3318–3327.

Duveneck, G.L., Bopp, M.A., Ehrat, M., Haiml, M., Keller, U., Bader, M.A., Marowsky, G. and Soria, S. (2001). Evanescent-field-induced two-photon fluorescence: Excitation of macroscopic areas of planar waveguides. *Appl. Phys.* B **73**, 869–871.

Esposito, A., Federici, F., Usai, C., Cannone, F., Chirico, G., Collini, M. and Diaspro, A. (2004). Notes on theory and experimental conditions behind two-photon excitation microscopy. *Microsc. Res. Tech.* **63**, 12–17.

Faisal, F.H.M. (1987). *Theory of Multiphoton Processes.* (Plenum Press, New York).

Farrer, R.A., Previte, M.J.R., Olson, C.E., Peyser, L.A., Fourkas, J.T. and So, P. T.C. (1999). Single molecule detection with a two-photon fluorescence microscope with fast scanning capabilities and polarization sensitivity. *Opt. Lett.* **24**, 1832–1834.

Fay, F.S., Carrington, W. and Fogarty, K.E. (1989). Three-dimensional molecular distribution in single cells analyzed using the digital imaging microscope. *J. Microsc.* **153**, 133–149.

Feynman, R.P. (1985). *QED: The Strange Theory of Light and Matter.* (Princeton University Press, Princeton, NJ).

Fisher, W.G., Watcher, E.A., Armas, M. and Seaton, C. (1997). Titanium: sapphire laser as an excitation source in two-photon spectroscopy. *Appl. Spectrosc.* **51**, 218–226.

Ford, B.J. (1991). *The Leeuwenhoek Legacy.* (Biopress and Parrand, Bristol).

Franken, P.A., Hill, A.E., Peters, C.W. and Weinreich, G. (1961). Generation of optical harmonics. *Phys. Rev. Lett.* **7**, 118–119.

French, T., So, P.T.C., Weaver, D.J., Coelho-Sampaio, T. and Gratton, E. (1997). Two-photon fluorescence lifetime imaging microscopy of macrophage-mediated antigen processing. *J. Microsc.* **185**, 339–353.

Friedrich, D.M. (1982). Two-photon molecular spectroscopy. *J. Chem. Educ.* **59**, 472–483.

Friedrich, D.M. and McClain, W.M. (1980). Two-photon molecular electronic spectroscopy. *Annu. Rev. Phys. Chem.* **31**, 559–577.

Gannaway, J.N. and Sheppard, C.J.R. (1978). Second harmonic imaging in the scanning optical microscope. *Opt. Quant. Electron.* **10**, 435–439.

Gauderon, R., Lukins, R.B. and Sheppard, C.J.R. (1999). Effects of a confocal pinhole in two-photon microscopy. *Microsc. Res. Tech.* **47**, 210–214.

Gerton, J.M., Wade, L.A., Lessard, G.A., Ma, Z. and Quake, S.R. (2004). Tip-enhanced fluorescence microscopy at 10 nanometer resolution. *Phy. Rev. Lett.* **93**(18), 180801.

Girkin, J. and Wokosin, D. (2002). Practical multiphoton microscopy. In: *Confocal and Two-Photon Microscopy: Foundations, Applications and Advances* (A. Diaspro, Ed.), 207–236. (Wiley-Liss, New York).

Göppert-Mayer, M. (1931). Über Elementarakte mit zwei Quantensprüngen. *Ann. Phys.* **9**, 273–295.

Gosnell, T.R. and Taylor, A.J. (Eds.) (1991). Selected Papers on Ultrafast Laser Technology. SPIE Milestone Series. (SPIE Press, Bellingham, WA).

Gostkowski, M.L., Allen, R., Plenert, M.L., Okerberg, E., Gordon, M.J. and Shear, J.B. (2004). Multiphoton-excited serotonin photochemistry. *Biophys. J.* **86**(5), 3223–3229.

Gratton, E. and van de Ven, M.J. (1995). Laser sources for confocal microscopy. In: *Handbook of Confocal Microscopy* (J.B. Pawley, Ed.), 69–97. (Plenum, New York).

Gratton, E., Barry, N.P., Beretta, S. and Celli, A. (2001). Multiphoton fluorescence microscopy. *Methods* **25**, 103–110.

Gu, M. and Sheppard, C.J.R. (1995). Comparison of three-dimensional imaging properties between two-photon and single-photon fluorescence microscopy. *J. Microsc.* **177**, 128–137.

Gu, M., Gan, X., Kisteman, A. and Xu, M.G. (2000). Comparison of penetration depth between two-photon excitation and single-photon excitation in imaging thorugh turbid tissue media. *Appl. Phys. Lett.* **77**(10), 1551–1553.

Guild, J.B., Xu, C. and Webb, W.W. (1997). Measurement of group delay dispersion of high numerical aperture objective lenses using two-photon excited fluorescence. *Appl. Opt.* **36**, 397–401.

Hamamatsu Photonics K.K. (1999). *Photomultiplier Tubes: Basics and Applications*, 2nd ed. (Hamamatsu Photonics K.K., Japan).

Hanninen, P.E. and Hell, S.W. (1994). Femtosecond pulse broadening in the focal region of a two-photon fluorescence microscope. *Bioimaging* **2**, 117–121.

Harper, I.S. (2002). Fluorophores and their labeling procedures for monitoring various biological signals. In: *Methods in Cellular Imaging* (A. Periasamy, Ed.), 20–39. (Oxford University Press, New York).

Haughland, P.R. (Ed.) (2002). *Handbook of Fluorescent Probes and Research Chemicals*. (Molecular Probes, Eugene, OR).

Heinze, K.G., Jahnz, M. and Schwille, P. (2004). Triple-color coincidence analysis: One step further in following higher order molecular complex formation. *Biophys. J.* **86**(1), 506–516.

Hell, S.W. (guest editor) (1996). Nonlinear optical microscopy. *Bioimaging* **4**, 121–172.

Hell. S.W. (2003). Toward fluorescence nanoscopy. *Nat. Biotechnol.* **21**, 1347–1355.

Hell, S.W., Bahlmann, K., Schrader, M., Soini, A., Malak, H., Gryczynski, I. and Lakowicz, J.R. (1996). Three-photon excitation in fluorescence microscopy. *J. Biomed. Opt.* **1**, 71–74.

Hellwarth, R. and Chistensen, P. (1974). Nonlinear optical microscopic examination of structures in polycrystalline ZnSe. *Opt. Commun.* **12**, 318–322.

Herman, B. and Tanke, H.J. (1998). *Fluorescence Microscopy.* (Springer-Verlag, New York).

Hooke, R. (1961). *Micrographia* (facsimile). (Dover, New York).

Hopt, A. and Neher, E. (2001). Highly nonlinear photodamage in two-photon fluorescence microscopy. *Biophys. J.* **80**, 2029–2036.

Iyer, V., Hoogland, T.M., Losavio, B.E., McQuiston, A.R. and Saggau, P. (2002). Compact two-photon laser scanning microscope made from minimally modified commercial components. In: *Multiphoton Microscopy in the Biomedical Sciences II* (A. Periasamy, P.T.C. So, Eds.). Proc. SPIE 4620, 274–280.

Jaiswal, J.K., Goldman, E.R., Mattoussi, H. and Simon, S.M. (2004). Use of quantum dots for live cell imaging. *Nat. Methods* **1**(1), 73–78.

Jonkman, J. and Stelzer, E. (2002). Resolution and contrast in confocal and two-photon microscopy. In: *Confocal and Two-Photon Microscopy: Foundations, Applications and Advances* (A. Diaspro, Ed.), 101–126. (Wiley-Liss, New York).

Jung, J.C., Mehta, A.D., Aksay, E., Stepnoski, R. and Schnitzer, M.J. (2004). In vivo mammalian brain imaging using one- and two-photon fluorescence microendoscopy. *J. Neurophysiol.* **92**(5), 3121–3133.

Kaiser, W. and Garrett, C.G.B. (1961). Two-photon excitation in CaF2:Eu2+. *Phys. Rev. Lett.* **7**, 229–231.

Kappel, C., Selle, A., Fricke-Begemann, T., Bader, M.A. and Marowsky, G. (2004). Giant enhancement of two-photon fluorescence induced by resonant double grating waveguide structures. *Appl. Phys. B–Lasers Opt.* **79**(5), 531–534.

Kawata, S., Sun, H.-B., Tanaka, T. and Takada, K. (2001). Finer features for functional microdevices. *Nature* **412**, 697–698.

Kino, G.S. and Corle, T.R. (1989). Confocal scanning optical microscopy. *Phys. Today* **42**, 55–62.

Koana, Z. (1943). *J. Illumination Eng. Inst.* **26**, 371.

Koester, H.J., Baur, D., Uhl, R. and Hell, S.W. (1999). Ca2+ fluorescence imaging with pico- and femtosecond two-photon excitation: Signal and photodamage. *Biophys. J.* **77**, 2226–2236.

König, K. (2000). Multiphoton microscopy in life sciences. *J. Microsc.* **200**, 83–104.

König, K. and Tirlapur, U.K. (2002). Cellular and subcellular perturbations during multiphoton microscopy. In: *Confocal and Two-Photon Microscopy: Foundations, Applications and Advances* (A. Diaspro, Ed.), 191–206. (Wiley-Liss, New York).

König, K., Liang, H., Berns, M.W. and Tromberg, B.J. (1995). Cell damage by near-IR microbeams. *Nature* **377**, 20–21.

König, K., Krasieva, T., Bauer, E., Fiedler, U., Berns, M.W., Tromberg, B.J. and Greulich, K.O. (1996a). Cell damage by UVA radiation of a mercury microscopy lamp probed by autofluorescence modifications, cloning assay and comet assay. *J. Biomed. Opt.* **1**, 217–222.

König, K., So, P.T.C., Mantulin, W.W., Tromberg, B.J. and Gratton, E. (1996b). Two-photon excited lifetime imaging of autofluorescence in cells during UVA and NIR photostress. *J. Microsc.* **183**, 197–204.

König, K., Liang, H., Berns, M.W. and Tromberg, B.J. (1996c). Cell damage in near infrared multimode optical traps as a result of multiphoton absorption. *Opt. Lett.* **21**, 1090–1092.

König, K., Boehme, S., Leclerc, N. and Ahuja, R. (1998). Time-gated autofluorescence microscopy of motile green microalga in an optical trap. *Cell. Mol. Biol.* **44**, 763–770.

König, K., Riemann, I., Fischer, P. and Halbhuber, K.J. (1999). Intracellular nanosurgery with near infrared femtosecond laser pulses. *Cell. Mol. Biol.* **45**, 195–201.

Kriete, A. (Ed.) (1992). *Visualization in Biomedical Microscopies.* (VCH, Weinheim).

Lakowicz, J.R. and Gryczynski, I. (1992). Tryptophan fluorescence intensity and anisotropy decays of human serum albumin resulting from one-photon and two-photon excitation. *Biophys. Chem.* **45**, 1–6.

Loudon, R. (1983). *The Quantum Theory of Light.* (Oxford University Press, London).

Louisell, W.H. (1973). *Quantum Statistical Properties of Radiation.* (Wiley, New York).

Mainen, Z.F., Malectic-Savic, M., Shi, S.H., Hayashi, Y., Malinow, R. and Svoboda, K. (1999). Two-photon imaging in living brain slices. *Methods* **18**, 231–239.

Maiti, S., Shear, J.B., Williams, R.M., Zipfel, W.R. and Webb, W.W. (1997). Measuring serotonin distribution in live cells with three-photon excitation. *Science* **275**, 530–532.

Majewska, A., Yiu, G. and Yuste, R. (2000). A custom-made two-photon microscope and deconvolution system. *Pflugers Arch. Eur. J. Physiol.* **441**(2/3), 398–408.

Masters, B.R. (1996). *Selected Papers on Confocal Microscopy.* SPIE Milestone Series. (SPIE Press, Bellingham, WA).

Masters, B.R. (2002). Selected Papers on Multi-photon Excitation Microscopy. SPIE Milestone Series. (SPIE Press, Bellingham, WA).

Masters, B.R., So, P.T. and Gratton, E. (1997). Multiphoton excitation fluorescence microscopy and spectroscopy of in vivo human skin. *Biophys J.* **72**, 2405–2412.

McConnell, G. and Riis, E. (2004). Two-photon laser scanning fluorescence microscopy using photonic crystal fiber. *J. Biomed. Opt.* **9**(5), 922–927.

Mertz, J., Xu, C. and Webb, W.W. (1995). Single molecule detection by two-photon excited fluorescence. *Opt. Lett.* **20**, 2532–2534.

Miller, M.J., Wei, S.H., Cahalan, M.D. and Parker, I. (2003). Autonomous T cell trafficking examined in vivo with intravital two-photon microscopy. *Proc. Natl. Acad. Sci. USA* **100**(5), 2604–2609.

Minsky, M. (1988). Memoir of inventing the confocal scanning microscope. *Scanning* **10**, 128–138.

Moscatelli, F.A. (1986). A simple conceptual model for two-photon absorption. *Am. J. Phys.* **54**, 52–54.

Mueller, M., Squier, J., Wilson, K.R. and Brakenhoff, G.J. (1998). 3D microscopy of transparent objects using third-harmonic generation. *J. Microsc.* **191**, 266–274.

Murphy, D.B. (2001). *Fundamentals of Light Microscopy and Electronic Imaging.* (Wiley-Liss, New York).

Nakamura, O. (1993). Three-dimensional imaging characteristics of laser scan fluorescence microscopy: Two-photon excitation vs. single-photon excitation. *Optik* **93**, 39–42.

Nakamura, O. (1999). Fundamentals of two-photon microscopy. *Microsc. Res. Tech.* **47**, 165–171.

Naora, H. (1951). Microspectrophotometry and cytochemical analysis of nucleic acids. *Science* **114**, 279–280.

Nemet, B.A., Nikolenko, V. and Yuste, R. (2004). Second harmonic imaging of membrane potential of neurons with retinal. *J. Biomed. Opt.* **9**(5), 873–881.

Novotny, L., Sanchez, E.J. and Xie, X.S. (1998). Near-field optical imaging using metal tips illuminated by higher-order Hermite-Gaussian beams. *Ultramicroscopy* **71**(1–4), 21–29.

Patterson, G.H. and Lippincott-Schwarz, J.A. (2002). Photoactivatable GFP for selective photolabeling of proteins and cells. *Science* **297**, 1873.

Patterson, G.H. and Piston, D.W. (2000). Photobleaching in two-photon excitation microscopy. *Biophys. J.* **78**, 2159–2162.

Pawley, J.B. (Ed.) (1995). *Handbook of Biological Confocal Microscopy.* (Plenum Press, New York).

Pennisi, E. (1997). Biochemistry: Photons add up to better microscopy. *Science* **275**, 480–481.

Periasamy, A. (Ed.) (2001). *Methods in Cellular Imaging.* (Oxford University Press, New York).

Periasamy, A. and Diaspro, A. (2003). Multiphoton microscopy. *J. Biomed. Opt.* **8**(3), 327–328.

Periasamy, A., Skoglund, P., Noakes, C. and Keller, R. (1999). An evaluation of two-photon excitation versus confocal and digital deconvolution fluorescence microscopy imaging in Xenopus morphogenesis. *Microsc. Res. Tech.* **47**, 172–181.

Periasamy, A., Noakes, C., Skoglund, P., Keller, R. and Sutherland, A.E. (2002). Two-photon excitation fluorescence microscopy imaging in Xenopus and transgenic mouse embryos. In: *Confocal and Two-Photon Microscopy: Foundations, Applications and Advances* (A. Diaspro, Ed.), 271–284. (Wiley-Liss, New York).

Petran, M., Hadravsky, M., Egger, M.D. and Galambos, R. (1968). Tandem-scanning reflected-light microscope. *J. Opt. Soc. Am.* **58**, 661–664.

Piston, D.W. (1999). Imaging living cells and tissues by two-photon excitation microscopy. *Trends Cell Biol.* **9**, 66–69.

Piston, D.W. (2005). When two is better than one: Elements of intravital microscopy. *PLoS Biol.* **3**(6), e207. Epub June 14, 2005.

Piston, D.W., Masters, B.R. and Webb, W.W. (1995). Three-dimensionally resolved NAD(P)H cellular metabolic redox imaging of the *in situ* cornea with two-photon excitation laser scanning microscopy. *J. Microsc.* **178**, 20–27.

Post, J.N., Lidke, K.A., Rieger, B. and Arndt-Jovin, D.J. (2005). One- and two-photon photoactivation of a paGFP-fusion protein in live Drosophila embryos. *FEBS Lett.* **579**(2), 325–330.

Potter, S.M. (1996). Vital imaging: Two-photons are better than one. *Curr. Biol.* **6**, 1596–1598.

Potter, S.M., Wang, C.M., Garrity, P.A. and Fraser, S.E. (1996). Intravital imaging of green fluorescent protein using two-photon laser-scanning microscopy. *Gene* **173**, 25–31.

Rentzepis, P.M., Mitschele, C.J. and Saxman, A.C. (1970). Measurement of ultrashort laser pulses by three-photon fluorescence. *Appl. Phys. Lett.* **17**, 122–124.

Robinson, J.P. (2001). *Current Protocols in Cytometry.* (John Wiley, New York).

Ruan, Q., Cheng, M.A., Levi, M., Gratton, E. and Mantulin, W.W. (2004). Spatial-temporal studies of membrane dynamics: Scanning fluorescence correlation spectroscopy (SFCS). *Biophys. J.* **87**(2), 1260–1267.

Rudolf, R., Mongillo, M., Magalhaes, P.J. and Pozzan, T. (2004). In vivo monitoring of Ca2+ uptake into mitochondria of mouse skeletal muscle during contraction. *J. Cell Biol.* **166**(4), 527–536.

Ruska, E. and Knoll, M. (1931). Die magnetische Sammelspule fuer schnelle Elektronenstrahlen. *Z. Tech. Phys.* **12**(389–400 and 488).

Saloma, C., Saloma-Palmes, C. and Kondoh, H. (1998). Site-specific confocal fluorescence imaging of biological microstructures in a turbid medium. *Phys. Med. Biol.* **43**, 1741.

Sánchez, E.J., Novotny, L. and Xie, X.S. (1999). Near-field fluorescence microscopy based on two-photon excitation with metal tips. *Phys. Rev. Lett.* **82**(20), 4014–4017.

Schneider, M., Barozzi, S., Testa, I., Faretta, M. and Diaspro, A. (2005). Two-photon activation and excitation properties of Pa-Gfp in the 720–920 nm region. *Biophys. J.* **89**(2), 1346–1352.

Schonle, A. and Hell, S.W. (1998). Heating by absorption in the focus of an objective lens. *Opt. Lett.* **23**, 325–327.

Schwille, P. (2001). Fluorescence correlation spectroscopy and its potential for intracellular applications. *Cell Biochem. Biophys.* **34**, 383–405.

Schwille, P., Haupts, U., Maiti, S. and Webb, W.W. (1999). Molecular dynamics in living cells observed by fluorescence correlation spectroscopy with one- and two-photon excitation. *Biophys. J.* **77**, 2251–2265.

Schwille, P., Kummer, S., Heikal, A.A., Moerner, W.E. and Webb, W.W. (2000). Fluorescence correlation spectroscopy reveals fast optical excitation-driven intramolecular dynamics of yellow fluorescent proteins. *Proc. Natl. Acad. Sci. USA* **97**, 151–156.

Sheppard, C.J.R. (2002). The generalized microscope. In: *Confocal and Two-Photon Microscopy: Foundations, Applications and Advances* (A. Diaspro, Ed.), 1–18. (Wiley-Liss, New York).

Sheppard, C.J.R. and Choudhury, A. (1977). Image formation in the scanning microscope. *Opt. Acta* **24**, 1051–1073.

Sheppard, C.J.R. and Gu, M. (1990). Image formation in two-photon fluorescence microscopy. *Optik* **86**, 104–106.

Sheppard, C.J.R. and Kompfner, R. (1978). Resonant scanning optical microscope. *Appl. Opt.* **17**, 2879–2882.

Sheppard, C.J.R. and Shotton, D.M. (1997). *Confocal Laser Scanning Microscopy.* (BIOS, Oxford, UK).

Sheppard, C.J.R. and Wilson, T. (1980). Image formation in confocal scanning microscopes. *Optik* **55**, 331–342.

Sheppard, C.J.R., Kompfner, R., Gannaway, J. and Walsh, D. (1977). The scanning harmonic optical microscope. *IEEE/OSA Conf. Laser Eng. Appl.*, (Washington, DC).

Shih, Y.H., Strekalov, D.V., Pittman, T.D. and Rubin, M.H. (1998). Why two-photon but not two photons? *Fortschr. Phys.* **46**, 627–641.

Shotton, D.M. (Ed.) (1993). *Electronic Light Microscopy: Techniques in Modern Biomedical Microscopy, Vol. 1.* (Wiley-Liss, New York).

Shotton, D.M. (1995). Electronic light microscopy—present capabilities and future prospects. *Histochem. Cell Biol.* **104**, 97–137.

Singh, S. and Bradley, L.T. (1964). Three-photon absorption in naphthalene crystals by laser excitation. *Phys. Rev. Lett.* **12**, 162–164.

So, P.T.C., Berland, K.M., French, T., Dong, C.Y. and Gratton, E. (1966). Two-photon fluorescence microscopy: Time resolved and intensity imaging. In: *Fluorescence Imaging Spectroscopy and Microscopy* (X.F. Wang, B. Herman, Eds.). Chemical Analysis Series, Vol. 137, 351–373. (John Wiley, New York).

So, P.T.C., Dong, C.Y., Masters, B.R. and Berland, K.M. (2000). Two-photon excitation fluorescence microscopy. *Annu. Rev. Biomed. Eng.* **2**, 399–429.

So, P.T.C., Kim, K.H., Buehler, C., Masters, B.R., Hsu, L. and Dong, C.Y. (2001). Basic principles of multi-photon excitation microscopy. In: *Methods in Cellular Imaging* (A. Periasamy, Ed.), 152–161. (Oxford University Press, New York).

Soeller, C. and Cannell, M.B. (1996). Construction of a two-photon microscope and optimisation of illumination pulse duration. *Pfluegers Arch. Eur. J. Physiol.* **432**, 555–561.

Soeller, C. and Cannell, M.B. (1999). Two-photon microscopy: Imaging in scattering samples and three-dimensionally resolved flash photolysis. *Microsc. Res. Tech.* **47**, 182–195.

Sonnleitner, M., Schutz, G.J. and Schmidt, T. (1999). Imaging individual molecules by two-photon excitation. *Chem. Phys. Lett.* **300**, 221–226.

Sonnleitner, M., Schutz, G., Kada, G. and Schindler, H. (2000). Imaging single lipid molecules in living cells using two-photon excitation. *Single Mol.* **1**, 182–183.

Spence, D.E., Kean, P.N. and Sibbett, W. (1991). 60-fsec pulse generation from a self-mode-locked Ti:sapphire laser. *Opt. Lett.* **16**, 42–45.

Squier, J.A., Muller, M., Brakenhoff, G.J. and Wilson, K.R. (1998). Third harmonic generation microscopy. *Opt. Express* **3**, 315–324.

Stelzer, E.H.K., Hell, S., Lindek, S., Pick, R., Storz, C., Stricker, R., Ritter, G. and Salmon, N. (1994). Non-linear absorption extends confocal fluorescence microscopy into the ultra-violet regime and confines the illumination volume. *Opt. Commun.* **104**, 223–228.

Straub, M. and Hell, S.W. (1998). Fluorescence lifetime three-dimensional microscopy with picosecond precision using a multifocal multiphoton microscope. *App. Phys. Lett.* **73**, 1769–1771.

Svelto, O. (1998). *Principles of Lasers*, 4th ed. (Plenum, New York).

Sytsma, J., Vroom, J.M., De Grauw, C.J. and Gerritsen, H.C. (1998). Time-gated fluorescence lifetime imaging and microvolume spectroscopy using two-photon excitation. *J. Microsc.* **191**, 39–51.

Tan, Y.P., Llano, I., Hopt, A., Wurriehausen, F. and Neher, E. (1999). Fast scanning and efficient photodetection in a simple two-photon microscope. *J. Neurosci. Methods* **92**, 123–135.

Tirlapur, U.K. and König, K. (2002). Two-photon near infrared femtosecond laser scanning microscopy in plant biology. In: *Confocal and Two-Photon Microscopy: Foundations, Applications and Advances* (A. Diaspro, Ed.), 449–468. (Wiley-Liss, New York).

Torok, P. and Sheppard, C.J.R. (2002). The role of pinhole size in high aperture two and three-photon microscopy. In: *Confocal and Two-Photon Microscopy: Foundations, Applications and Advances* (A. Diaspro, Ed.), 127–152. (Wiley-Liss, New York).

Tsien, R.Y. (1995). Design of Molecules to Probe Living Cells. Abstracts of Papers of the American Chemical Society 209:4-SOCED.

Tyrrell, R.M. and Keyse, S.M. (1990). The interaction of UVA radiation with cultured cells. *J. Photochem. Photobiol.* B **4**, 349–361.

Volkmann, J. (1996). Ernst Abbe and his work. *Appl. Opt.* **5**, 1720–1731.

Wang, X.F. and Herman, B. (1996). *Fluorescence Imaging Spectroscopy and Microscopy.* (Wiley-Liss, New York).

Wang, Y., Wang, X.F., Wang, C. and Ma, H. (2004). Simultaneously multiparameter determination of hematonosis cell apoptosis by two-photon and confocal laser scanning microscopy. *J. Clin. Lab. Anal.* **18**(5), 271–275.

Webb, R.H. (1996). Confocal optical microscopy. *Rep. Progr. Phys.* **59**, 427–471.

Weinstein, M. and Castleman, K.R. (1971). Reconstructing 3-D specimens from 2-D section images. *Proc. SPIE* **26**, 131–138.

White, J.G., Amos, W.B. and Fordham, M. (1987). An evaluation of confocal versus conventional imaging of biological structures by fluorescence light microscopy. *J. Cell Biol.* **105**, 41–48.

White, N.S. and Errington, R.J. (2000). Improved laser scanning fluorescence microscopy by multiphoton excitation. *Adv. Imag. Elect. Phys.* **113**, 249–277.

Wiedenmann, J., Ivanchenko, S., Oswald, F., Schmitt, F., Rocker, C., Salih, A., Spindler, K.-D. and Nienhaus, G.U. (2004). EosFP, a fluorescent marker protein with UV-inducible green-to-red fluorescence conversion. *Proc. Natl. Acad. Sci. USA* **101**(45), 15905–15910.

Wier, W.G., Balke, C.W., Michael, J.A. and Mauban, J.R. (2000). A custom confocal and two-photon digital laser scanning microscope. *Am. J. Physiol.* **278**, H2150–H2156.

Wilson, T. (2002). Confocal microscopy: Basic principles and architectures. In: *Confocal and Two-Photon Microscopy: Foundations, Applications and Advances* (A. Diaspro, Ed.), 19–38. (Wiley-Liss, New York).

Wilson, T. and Sheppard, C.J.R. (1984). *Theory and Practice of Scanning Optical Microscopy.* (Academic Press, London).

Wise, F. (1999). Lasers for two-photon microscopy. In: *Imaging: A Laboratory Manual* (R., Yuste, F., Lanni, A. Konnerth, Eds.), 18.1–9. (Cold Spring Harbor Press, Cold Spring Harbor, NY).

Wiseman, P.W., Squier, J.A., Ellisman, M.H. and Wilson, K.R. (2000). Two-photon image correlation spectroscopy and image cross-correlation spectroscopy. *J. Microsc.* **200**, 14–25.

Wiseman, P.W., Capani, F., Squier J.A. and Martone, M.E. (2002). Counting dendritic spines in brain tissue slices by image correlation spectroscopy analysis. *J. Microsc. (Oxf)* **205**, 177–186.

Wokosin, D.L. and White, J.G. (1997). Optimization of the design of a multiple-photon excitation laser scanning fluorescence imaging system. In: *Three-Dimensional Microscopy: Image, Acquisition and Processing* IV. Proc. SPIE 2984, 25–29.

Wokosin, D.L., Centonze, V.E., White, J., Armstrong, D., Robertson, G. and Ferguson, A.I. (1996). All-solid-state ultrafast lasers facilitate multiphoton excitation fluorescence imaging. *IEEE J. Selected Top. Quant. Electron.* **2**, 1051–1065.

Wokosin, D.L., Amos, W.B. and White, J.G. (1998). Detection sensitivity enhancements for fluorescence imaging with multiphoton excitation microscopy. *Proc. IEEE Eng. Med. Biol. Soc.* **20**, 1707–1714.

Wokosin, D.L., Loughrey, C.M. and Smith, G.L. (2004). Characterization of a range of Fura dyes with two-photon excitation. *Biophys. J.* **86**(3), 1726–1738.

Wolleschensky, R., Feurer, T., Sauerbrey, R. and Simon, U. (1998). Characterization and optimization of a laser scanning microscope in the femtosecond regime. *Appl. Phys.* B **67**, 87–94.

Wolleschensky, R., Dickinson, M. and Fraser, S.E. (2002). Group velocity dispersion and fiber delivery in multiphoton laser scanning microscopy. In: *Confocal and Two-Photon Microscopy: Foundations, Applications and Advances* (A. Diaspro, Ed.), 171–190. (Wiley-Liss, New York).

Xie, X.S. and Lu, H.P. (1999). Single molecule enzymology. *J. Biol. Chem.* **274**, 15967–15970.

Xu, C. (2002). Cross-sections of fluorescence molecules used in multiphoton microscopy. In: *Confocal and Two-Photon Microscopy: Foundations, Applications and Advances* (A. Diaspro, Ed.), 75–100. (Wiley-Liss, New York).

Xu, C., Guild J., Webb, W.W. and Denk, W. (1995). Determination of absolute two-photon excitation cross sections by *in situ* second-order autocorrelation. *Opt. Lett.* **20**, 2372–2374.

Yoder, E.J. and Kleinfeld, D. (2002). Cortical imaging through the intact mouse skull using two-photon excitation laser scanning microscopy. *Microsc. Res. Tech.* **56**(4), 304–305.

Zimmer, M. (2002). Green fluorescence protein (GFP): Applications, structure, and related photophysical behavior. *Chem. Rev.* **102**(3), 759–781.

Zoumi, A., Yeh, A. and Tromberg, B.J. (2002). Imaging cells and extracellular matrix in vivo by using second-harmonic generation and two-photon excited fluorescence. *Proc. Natl. Acad. Sci. USA* **99**(17), 11014–11019.

Zoumi, A., Lu, X., Kassab, G.S. and Tromberg, B.J. (2004). Imaging coronary artery microstructure using second-harmonic and two-photon fluorescence microscopy. *Biophys. J.* **87**(4), 2778–2786.

12

Nanoscale Resolution in Far-Field Fluorescence Microscopy

Stefan W. Hell and Andreas Schönle

1 Introduction

The discovery of the diffraction barrier in 1873 by Ernst Abbe (1873) has shown that, relying on propagating light waves and regular lenses, the traditional light microscope cannot resolve spatial structures that are smaller than about half the wavelength of the focused light. This physical insight has been accepted as an unalterable limitation of focusing light microscopy and has consequently triggered the invention of nonoptical imaging techniques, such as electron and scanning probe microscopy. In spite of the tremendous improvement in resolution brought about by these methods, light microscopy has maintained its importance in many fields of science. The reasons are mainly a number of exclusive advantages, the most prominent of which is the ability to noninvasively image (living) specimens. Light microscopy also entails the possibility of using fluorescence as a highly specific signature of the specimen features of interest. Fluorescence is particularly attractive when provided by endogenous fluorescence markers, i.e., proteins in physiologically intact cells. Mapped with a confocal or multiphoton excitation microscope (Sheppard and Kompfner, 1978; Wilson and Sheppard, 1984; Denk et al., 1990), fluorescence emission readily yields protein three-dimensional (3D) distributions, or that of other fluorescently labeled molecules from the strongly convoluted inside of biological specimens.

Abandoning the concept of focusing light altogether, near-field optical microscopy is the earliest practically relevant attempt to overcome the diffraction barrier with visible light (Pohl and Courjon, 1993). To this end, scanning near-field optical microscopes employ ultrasharp tips or tiny apertures to confine the interaction of the light field with the object to subdiffraction dimensions. However, applying this technique to soft biological matter has not been very successful to date and it has to be applied carefully to avoid imaging artifacts (Hecht et al., 1997). In any case, a near-field optical microscope is confined to imaging surfaces, again underscoring the importance of improving the resolution in an optical microscope that still preserves focusing. For convenience we refer to the traditional light microscope as a "far-field" instrument.

This formidable problem has been approached by many scientists (Toraldo di Francia, 1952; Lukosz, 1966), but only in recent years have methods been characterized (Hell and Kroug, 1995; Hell, 1997, 2003) that effectively broke Abbe's diffraction barrier, implying the potential to attain molecular resolution with regular lenses and visible light (Hell et al., 2003; Hell, 2003; Westphal and Hell, 2005).

2 The Resolution Limit

The limited resolution of a focusing light microscope is readily described by the form and extent of the effective focal spot, commonly referred to as the (effective) point spread function (PSF). The PSF is the image that an infinitely small object would create. For an incoherent imaging mode such as fluorescence, the image is therefore given by the convolution of the object with the PSF:

$$I = h \otimes G \tag{1}$$

Here I, h, and G denote the image, the PSF, and the object, respectively, with the object consisting of, e.g., a distribution of fluorophores in space. The convolution actually means that the object is "smeared out" by the focal spot. While a careful analysis of the imaging properties requires a detailed analysis of the PSF, for PSFs with a single peak, assessing the full width half maximum (FWHM) usually gives a good estimate of the microscope's resolution. If identical molecules are within the FWHM distance, the molecules cannot be separated in the image. Therefore, it becomes evident that improving the resolution is largely equivalent to narrowing the PSF of the microscope.

In a conventional fluorescence microscope, the FWHM of the PSF is about $\lambda/(2n\sin\alpha) = \lambda/(2\,\mathrm{NA})$, with λ denoting the wavelength, n the refractive index, α the semiaperture angle, and NA the numerical aperture of the lens. Without changes to the principal method, the spot size can only be decreased by using shorter wavelengths or larger aperture angles (Abbe, 1873; Born and Wolf, 1993). However, the lens half-aperture is technically limited to ~70° and the wavelength λ cannot be reduced below 350 nm because shorter wavelengths are not compatible with live cell imaging. In the best case, established far-field microscopes resolve 180 nm in the focal plane (x,y) and merely 500–800 nm along the optic axis (z) (See Figure 2–1a) (Pawley, 1995). All attempts to improve the resolution by a mere improvement of the optical components remain limited by diffraction. This is arguably better understood in the frequency domain. Here, the resolving power is described by the optical transfer function (OTF) giving the strength with which these spatial frequencies are transferred from the object to the image. The OTF is readily computed as the Fourier transform of the PSF (Wilson and Sheppard, 1984; Goodman, 1968). Due to the convolution theorem, Eq. (1) becomes

$$\hat{I} = \hat{h}\hat{G} = o\hat{G} \tag{2}$$

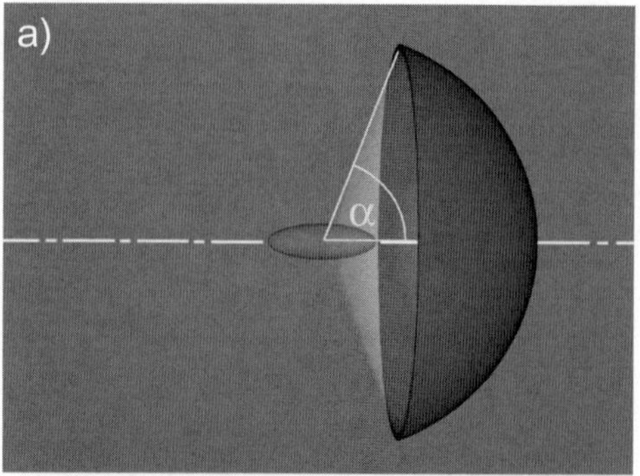

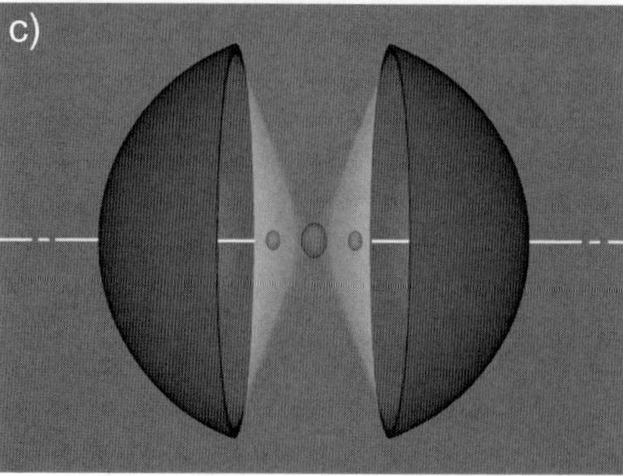

Figure 12–1. (a) The wavefront created by a single lens is a spherical cap. For the highest available semiaperture angle (~70°), diffraction theory dictates an elongated spot along the optic axis (z) that is approximately three times longer than it is wide (xy). The elongation can also be explained by the fact that the wavefront is only a cap rather than a sphere, bringing about an asymmetry of the focusing process. (b) Coherent addition of two spherical wavefronts created by opposing lenses completes a major part of the missing wavefront toward a more spherical one, which is the tenet of the "4Pi concept." (c) The central focal spot, i.e., the main diffraction maximum, becomes more spherical as a result. However, because the wavefront is not approaching a sphere side-maxima appear along the optic axis.

where the hat signifies a 3D Fourier transform and o is the OTF. Due to the multiplication on the right hand side it is evident that even under optimal imaging conditions, object frequencies are lost where the OTF is zero. This loss is physically irrecoverable. Hence, the ultimate resolution limit is given by the highest frequency where the OTF is nonzero, i.e., the extent of the support of the OTF.

The highest spatial frequency produced by a (quasi)monochromatic wave passing through the setup is $k = 2\pi n/\lambda$. For focused light, the Fourier transform of its electric field, $C(k)$, is therefore given by a spherical cap with radius k, as seen in Figure 12–2. The excitation probability is well approximated by the absolute square of the electric field, corresponding to an autocorrelation in frequency space. This signifies that frequencies of up to $2k$ are contained in the excitation OTF. However, resolving optics can be employed both in the illumination and the detection path. This introduces a spatially varying detection probability for the emitted fluorescence. Since the phase of the illuminating light is lost during excitation, fluorescence is emitted incoherently. Therefore the effective PSF is given by the multiplication of the excitation PSF with its normalized detection counterpart, i.e., the detection PSF (Wilson and Sheppard, 1984), which is equivalent to a convolution in frequency space. Thus the achievable cutoff is at $2k_{ex} + 2k_{det}$ where k_{ex} and k_{det} are the frequencies corresponding to the excitation and detection wavelengths in the medium, respectively.

These considerations are illustrated in Figure 12–2. As an example, if fluorescence is excited at 488 nm and detected at around 530 nm, the cutoff is given by $(2n/488\,\text{nm} + 2n/530\,\text{nm})^{-1} \cong 127\,\text{nm}/n$,

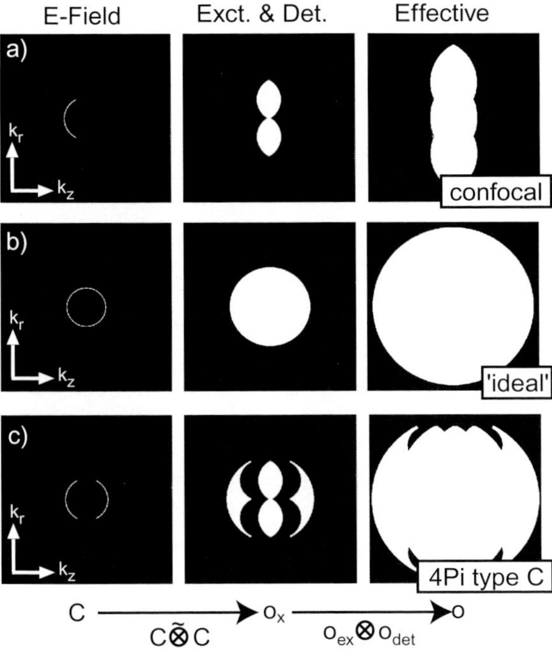

Figure 12–2. OTF supports in frequency space. (a) For a single lens, the Fourier transform of the electric field, $C(k)$, is a spherical cap with a cutoff angle determined by the aperture angle of the objective lens, which has been technically limited to <70°. Excitation and detection (intensity) PSFs depend on the absolute square of the electric field; hence, the respective OTFs are given by the autocorrelation of the cap. Finally, in a confocal fluorescence microscope, the effective PSF is given by the product of the excitation and detection (intensity) PSF. Its OTF is therefore the convolution of the corresponding excitation and detection OTF, resulting in a squeezed OTF along the z direction accounting for the degraded axial resolution. (b) With a scalar field a truly spherical wavefront could ideally be generated. The autocorrelation of such a complete spherical shell is nonzero within a sphere of radius $2k$. Convolution of the excitation and detection OTFs again results in a spherical OTF support of radius $2k_{ex} + 2k_{det}$, which is tantamount to isotropic resolution. This radius defines the theoretical resolution limit of all microscopes in which the resolution is solely created by the optical arrangement and in which the phase of light is lost at the sample, as in the excitation–emission process. (c) The 4Pi microscope of type C with coherent excitation and detection through both lenses with large aperture angles generates the largest wavefront currently achievable in practice. For the highest available cutoff angles of approximately 70° this results in an almost spherical effective OTF. The resulting resolution is very close to the possible limit. While a semiaperture angle somewhat greater than 70° would still give a better coverage, the study also shows that the true solid angle of 4π is actually not required to gain the desired improvement in axial resolution. The putative use of aperture angles approaching 80° and higher is also complicated by the transverse nature of electromagnetic waves.

corresponding to approximately 83 nm for oil and 96 nm for water immersion. Therefore, all far-field light microscopes relying solely on improvements of the instrument have an ultimate resolution limit of fust below 100 nm in all directions, even when all conceivable spatial frequencies are transmitted and restored in the image. This theoretical benchmark is indeed almost achieved by the 4Pi microscope described in more detail later (Egner et al., 2002; Hell et al., 1997).

To overcome this fundamental limit, higher frequencies need to be generated. Obviously this can be achieved by introducing nonlinearities in the interaction of the excitation light with a dye: For example, when using multiphoton (m-photon) excitation, the probability of producing a fluorescence photon depends on the mth order of the illumination intensity (Sheppard and Kampfner, 1978; Göpper-Mayer, 1931; Bloembergen, 1965). The original excitation OTF will be convolved m times with itself, thus extending the support region to m times higher frequencies. Therefore, it has sometimes been argued that superresolution can be attained by multiphoton excitation. It is, however readily understood why the resolution limit cannot really be pushed (Denk et al., 1990). While it is true that the excitation volume is narrower than the focal spot, which is due to the nonlinear dependence, using m-photon excitation to excite the same dye also means that the excitation energy has to be split between the photons (Denk et al., 1990). Consequently the photons have an m times lower amount of energy, thus an m times longer wavelength $m\lambda_{ex}$, which results in m times larger focal spots to begin with. Consequently, the theoretical cutoff remains equal at $2m(k_{ex}/m) + 2k_{det} = 2k_{ex} + 2k_{det}$. The multiplication by m stems from the m correlations or convolutions in frequency space and the division from the longer excitation wavelength. Obviously, this equally holds for excitation with several photons of different energies such as two-color two-photon excitation (Lakowicz et al., 1996). However, in the case of multiphoton excitation, higher frequencies within the support are usually damped, making the resolution poorer as compared to the 1-photon case. Multiphoton excitation is therefore not a viable option for significantly increasing the resolution. In addition, multiphoton excitation, especially for $m > 3$, requires very high intensities (Xu et al., 1996) leading to damage to the sample. Bleaching is largely restricted to the focal plane, but there it can be much stronger than for one-photon excitation. For some dyes, however, the excitation wavelength can be chosen to be shorter than $m\lambda_{ex}$ and when imaging deep into scattering samples such as brain slices, multiphoton imaging can feature a superior signal-to-noise ratio (SNR). In these cases, multiphoton excitation is a valuable method for reasons unrelated to the classical resolution issue.

Recognizing that the energy subdivision prevents any resolution increase, "multiphoton" concepts have been proposed, where the detection of a photon occurs only after the consecutive absorption of multiple excitation photons. Since the photons induce a linear optical transition in the dye, high intensities are not required and photon-energy subdivision does not occur (Hänninen et al., 1996; Schönle et al., 1999; Schönle and Hell, 1999). However, these concepts require specific conditions or complicated dye systems, also complicating their practical realization.

A radically different concept for improving the far-field fluorescence microscopy resolution appeared in the mid-1990s in the form of *s*timulated *e*mission *d*epletion (STED) and *g*round *s*tate *d*epletion (GSD) microscopy (Hell and Wichmann, 1994; Hell and Kroug, 1995). Both concepts were based on one common principle: The required nonlinearities in the relationship between the irradiating light and the signal detected from the dye are provided by a spatially modulated *re*versible *s*aturable *o*ptically *l*inear (*f*luorescence) *t*ransition between two molecular states (RESOLFT). Since there is no physical limit to the degree of saturation, there is no longer a theoretical limit to the resolution. The limit is largely the size of the marker molecule itself. Therefore, concepts based on the RESOLFT principle, such as STED or GSD microscopy, truly break the diffraction barrier. Due to their principles of operation they can achieve nanoscale resolution in *all* directions (Hell, 1997; Hell et al., 2003; Heintzmann et al., 2002).

Fundamentally departing from earlier ap-proaches to resolution increase, an introduction to the general idea will follow in the second part of this chapter, and we shall outline different possibilities for realizing the RESOLFT concept. We will see that initial applications of RESOLFT-based superresolution microscopy have recently yielded encouraging results, such as the first demonstration of a spatial resolution of $\lambda/25$ with focused light using regular lenses (Westphal et al., 2003). A potential road map toward imaging with nanometer resolution in live cells will be discussed, opening the prospect of bridging the gap between electron and current light microscopy, and providing a "nanoscope" working with focused light.

3 Axial Resolution Improvement by Aperture Enlargement: 4Pi Microscopy and Related Approaches

Figure 12–2 illustrates that the suboptimal axial resolution of a far-field light microscope can readily be motivated by the fact that the focusing angle of the objective lens is far from covering a full solid angle of 4π. The axially elongated PSF evidently also implies an axially compressed OTF. If the focused wavefronts were almost spherical, so would the focal spot, as well as the OTF of the system. In this case, the resolution would be largely isotropic. Therefore, an obvious way to achieve optimal optical resolution is to synthesize a focusing angle that comes closer to the solid angle of 4π. Since the focusing angle of a lens is limited by technical constraints, a larger wavefront can be obtained only by pasting together two or more wavefronts. A wavefront synthesis sharpening the focus along the optic axis is attained by employing two opposing lenses (Figure 12–1) that either coherently illuminate the sample from both sides or that detect the light through both lenses in a coherent manner (Hell, 1990; Hell and Stelzer, 1992a; Gustafsson et al., 1995). The resultant support of the OTF is displayed in Figure 12–2c. In the spatial domain the effect can be pictured as two counterpropagating focused beams interfering either at the focal spot or at a common point at the detector, for the illumination and the detection, respectively. For the common high-angle lenses with a semiaperture $64° < \alpha < 72°$, constructive interference produces a maximum with an approximately three to four times narrower axial FWHM. The basic idea of the 4Pi microscope is to produce and take advantage of this axially narrowed PSF. Due to the fact that the semiaperture angle α is considerably smaller than $90°$, the main maximum is also accompanied by one or two axial satellite lobes originating from the "missing lateral part" of a potentially full solid angle (Figure 12–1c).

Importantly, the narrowed main focal maximum is not $\lambda/(4n)$, as would be anticipated for a flat standing wave, but somewhat larger. By the same token, the side maxima are not located at $m\lambda/(2n)$, ($m = 0, 1, 2, 3\ldots$), but slightly farther away from the focal point, constantly decreasing in height with increasing order. As we shall see later, this difference is essential to the axial resolution improvement brought about by the coherent use of two opposing lenses and the actual reason why it is not possible to improve the axial resolution just by the inter-ference of counterpropagating plane waves (Lanni, 1986; Bavley et al., 1993). The latter is the tenet of the standing wave microscope (SWM), which utilizes a *flat* standing wave of laser light and thus a set of exci-tation nodal planes in the sample along the optic axis. The SWM uses wide-field detection on a camera, like any conventional fluorescence microscope; the flat standing wave of excitation light is produced either with a mirror beneath the sample or by adding counterpropagating waves from two lenses. A substantial increase in axial resolution has initially been claimed by SWM. However, the resulting multiple thin interference layers in a sample (Lanni, 1986) do not unambiguously resolve features extending over a minimal axial range that is larger than half the wavelength (0.2–0.4 µm) (Krishnamurthi et al., 1996; Freimann et al., 1997). While SWM might prove useful for specialized applications, it fails in delivering axial images of arbitrary objects (Nagorni and Hell, 2001a, 2001b). An explanation for this is that the interference excitation maxima have almost equal intensity, leading to strongly blurred "ghost images." The deeper-rooted physical reason is that SWM does not increase the aperture of the system. This can be readily observed when analyzing the SWM in the frequency space (Figure 12–3). The excitation OTF has only three delta peaks located at $-k$, 0, and k on the inverse optical axis. When convolved with the detec-tion OTF, large gaps remain in the support of the effective OTF repre-senting potential object frequencies that are not transferred from the object to the image.

It is a major physical insight (into the problem of axial resolution improvement with coherently used opposing lenses) that these gaps can be closed only when the light is *focused*, that is with spherical wave-fronts. Focusing to the same point requires the accurate alignment of the two lenses, but the real physical challenge is to identify feasible mechanisms helping to further reduce the fluorescence contributed by the side-maxima still present when focusing at $64° < \alpha < 72°$. Reduction of the contribution from the side-maxima is equivalent to completing the support of the OTF and eventually to avoiding imaging artifacts. When spot-scanning confocal, nonconfocal 4Pi microscopy (Hell, 1990;

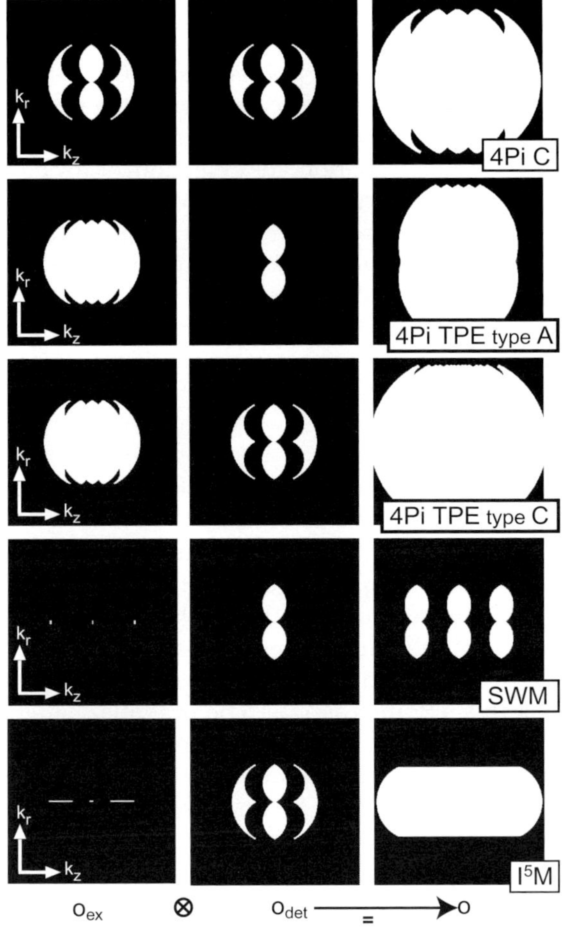

Figure 12–3. Comparison of the supports of excitation (left column), detection (middle column), and effective OTFs for techniques using two opposing lenses coherently. The effective OTF is the convolution of the excitation and detection OTFs. For two-photon excitation in the 4Pi mode, the Fourier transform of the excitation intensity is given by the excitation OTF of the 4Pi type C microscope but scaled by 488 nm/800 nm = 0.61. The excitation OTF is therefore very similar to the effective OTF of 4Pi type C. Note the gaps in the support of the SWM microscope's effective OTF.

Hell and Stelzer, 1992b) (Figure 12–3) and wide-field I^5M microscopy (Gustafsson et al., 1995) finally demonstrated the improvement of axial resolution (Hell et al., 1997; Schrader and Hell, 1996; Gustafsson, 1999; Gustafsson et al., 1999), each method used lenses with high numerical aperture and relied at least on one of the following three lobe-reducing mechanisms: (1) Using both focused excitation *and* confocal detection (Hell, 1990; Hell and Stelzer, 1992a) suppresses fluorescence from higher order side lobes of the excitation PSF; (2) two-photon excitation (TPE) (Hell and Stelzer, 1992b) emphasizes differences in the intensities of the main and the side-maxima due to its quadratic dependence on the excitation intensity; and (3) finally, spatial disparities between the maxima of excitation and fluorescence wavelengths can be used (Hell and Stelzer, 1992b; Gustafsson et al., 1995) if light is detected coherently through both lenses. That is, the intensity maxima for excitation and detection are located at different points in space.

Three major types of 4 Pi microscopy have been reported (Hell and Stelzer, 1992a). They differ on whether the spherical wavefronts are coherently added for illumination, for detection, or for both simultaneously; they are referred to as type A, B, and C, respectively. Usually the detection has been confocalized, but in conjunction with TPE successful axial separation with nonconfocal detection has also been reported. Here we will concentrate on the TPE 4 Pi (type A), the 4 Pi type C, and the TPE 4 Pi type C confocal microscopes. Of these three, the TPE 4 Pi confocal microscope has been applied to the largest number of imaging problems. It uses the very effective lobe-reducing measure of TPE combined with "point-like" detection. In reality the size of the "point-like" detector amounts to about the size of the main maximum of the diffraction-limited fluorescence spot (Airydisk), when imaged into the focal plane of the objective lens.

Clearly, nonconfocal wide-field detection and regular illumination would make 4 Pi microscopy more versatile. Therefore, the related approach of I^5M (Gustafsson et al., 1995, 1996, 1999; Gustafsson, 1999) confines itself to using the simultaneous interference of both the excitation and the (Stokes-shifted) fluorescence wavefront pairs; the latter are spherical as in a 4 Pi microscope. The potential benefits of I^5M are readily stated: single-photon excitation with arguably less photobleaching, an additional 20–50% gain in fluorescence signal, and lower cost. This method has so far yielded 3D images of actin filaments with an axial resolution slightly better than 100 nm in fixed cells (Gustafsson et al., 1999). To remove the side-lobe artifacts, I^5M-recorded data are deconvolved offline. While the consideration of the OTF support in Figure 12–3 suggests that this single mechanism is indeed sufficient, it turns out that the relaxation of the side-lobe suppression comes at the expense of an increased vulnerability to sample-induced aberrations, especially with nonsparse objects (Nagorni and Hell, 2001a, 2001b). Thus I^5M imaging, which has so far relied on oil immersion lenses, has required mounting the cell in a medium with $n = 1.5$ (Gustafsson et al., 1999). Live cells inevitably necessitate aqueous media ($n = 1.34$). Moreover, water immersion lenses have an inferior focusing angle and therefore larger lobes to begin with (Bahlmann et al., 2001). Potential strategies for improving the tolerance of I^5M are the implementation of a nonlinear excitation mode and its combination with pseudoconfocal or patterned illumination (Gustafsson, 2000). While these measures again add physical complexity, they may have the potential to render I^5M more suitable for live cells.

However, at this stage, the implementation of at least two of the mechanisms above proved more reliable: After initial demonstration of TPE 4 Pi confocal microscopy (Schrader and Hell, 1996), superresolved axial separation was applied to fixed cells (Hell et al., 1997). The image quality could be improved further by applying nonlinear restoration (Holmes, 1988; Carrington et al., 1995; Holmes et al., 1995). Under biological imaging conditions, this typically improves the resolution up to a factor of two in both the transverse and the axial direction.

Therefore, in combination with image restoration, TPE 4 Pi confocal microscopy has resulted in a resolution of ~100 nm in all directions, as first witnessed by the imaging of filamentous actin[19] and immunofluorescently labeled microtubules (Nagorni and Hell, 1998; Hell and Nagorni, 1998) in mouse fibroblasts.

While a very useful and explanative comparison of the OTF *supports* has been published (Gustafsson, 1999) it is obvious from the above that consideration of the supports alone is not sufficient to understand the respective benefits and limitations of SWM, I⁵M, and 4 Pi microscopy. Rather than comparing their technical implementation (Gustafsson et al., 1999), a quantitative analysis of their PSFs and OTFs is needed to clarify under which conditions these microscopes will be able to deliver 3D-resolved images with superior resolution. The success of increasing the axial resolution with coherently used opposing lenses not only depends on the achievable bandwidth, but also on the strength with which the respective systems transfer the spatial frequencies *within* this bandwidth. Gaps and weak parts that occur in some systems (Gustafsson et al., 1995; Krishnamurthi et al., 1996) must be quantified for a particular optical setting because they may render the removal of artifacts impossible, thus precluding an increase in axial resolution. Below we demonstrate that these gaps are intimately connected with the optical arrangement and therefore are inherent to some of the methods described.

3.1 The Optical Transfer Function of 4 Pi-Microscopy and Related Systems

Before expanding on the analysis of PSFs and OTFs, we quickly review their theoretical derivation. The excitation PSF of the confocal microscope, as well as the *detection* PSF of the wide-field, confocal, SWM, and 4 Pi-type microscopes, are regular intensity PSFs. Assuming that excitation and emission involve a dipole transition of the dye, the excitation and emission PSF of a single lens is well approximated by the absolute square of the electric field in the focal region:

$$h = |\mathbf{E}_1(z, r, \phi)|^2 \tag{3}$$

If depends on the axial and radial distance to the geometric focus (z and r, respectively) and the polar angle ϕ. The field is usually calculated using the vectorial theory of Richards and Wolf (1959). In our case, we assumed circular polarization of the light. In frequency space the Fourier transform of the electrical field is then simply given by a spherical cap of radius k_{ex} (Figure 12–2). This is equivalent to approximating the spherical wavefronts emerging at the exit pupil as plane waves close to the focal spot. The absolute square in Eq. (3) corresponds to an autocorrelation in frequency space. When calculating the OTF, this convolution can be carried out directly (Schönle and Hell, 2002; Sheppard et al., 1993) or the OTF can be determined by Fourier transformation of Eq. (3).

For the excitation PSF of the 4 Pi confocal microscopes (type A and C) and for the detection PSF of the I⁵M and the 4 Pi confocal (type B and C), a calculation of constructive interference between the two spherical wavefronts is required. The PSF is therefore given by the coherent addition of two beams

$$h = |\mathbf{E}_1(z, r, \phi) + \mathbf{E}_2(z, r, \phi)|^2 \tag{4}$$

The field of the opposing lens is given by

$$\mathbf{E}_2(\mathbf{r}) = \mathbf{M}\mathbf{E}_1(\mathbf{M}^{-1}\mathbf{r}) \tag{5}$$

$\mathbf{M}$ is the coordinate transform from the system of lens number 2 to lens number 1 and is a diagonal matrix inverting the z-components for a triangular cavity and the y- and z-components for a rectangular cavity (Bahlmann and Hell, 2000). In frequency space, the Fourier transform of the electric field is now given by two caps corresponding to the two focused wavefronts. Consequently, the OTF consists of an autocorrelation part equivalent to that for single lens excitation or detection and a cross-correlation of two opposite spherical caps represented by the outer "brackets" in Figures 12–2, 12–3, and 12–4. For TPE the excitation PSF is simply given by squaring the one-photon PSF scaled to the

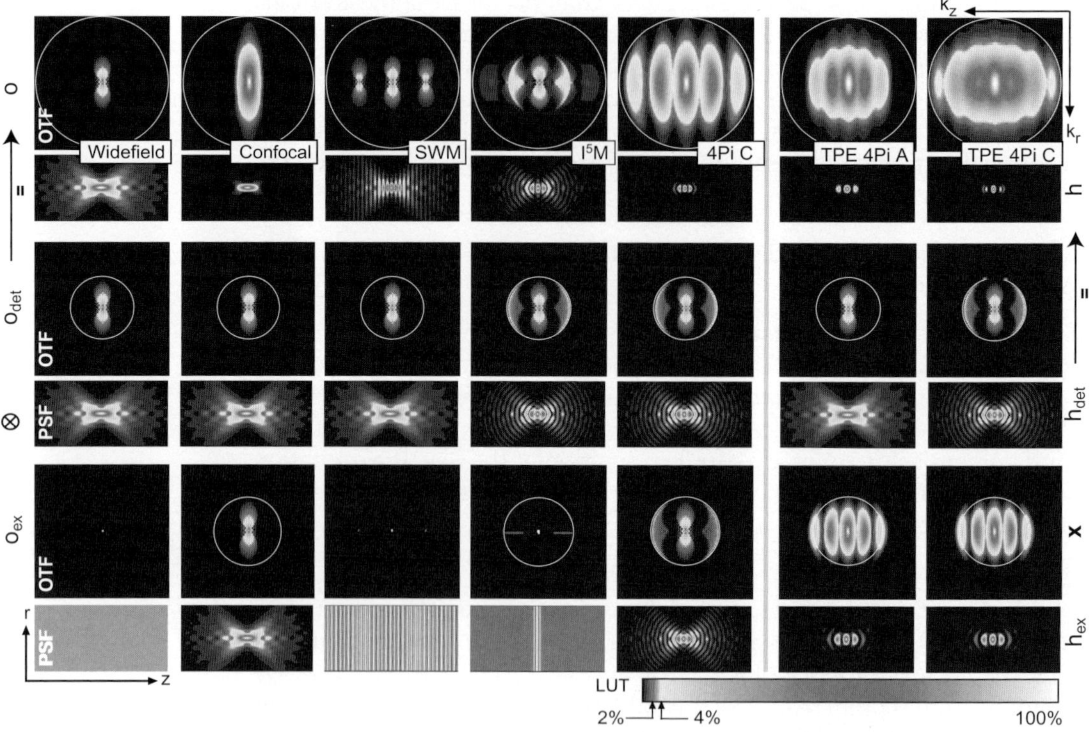

Figure 12–4. Overview of the excitation (bottom), detection (center), and effective (top) OTFs' modulus and the corresponding PSFs of the wide-field, confocal, standing wave (SWM), I^5M, 4Pi confocal type C, TPE 4Pi confocal type A, and TPE 4Pi confocal type C microscopes. The color look-up table (LUT) has been designed to emphasize the important weak OTF regions. The OTFs are shown in the squares above the corresponding PSFs; the zero frequency point is in the center and the largest frequency displayed is $2\pi/80\,\mathrm{nm}^{-1}$. The circles represent the maximum possible carrier as explained in Figure 12–3. For TPE, the excitation OTFs slightly extend over these circles because the excitation wavelength of 800 nm is less than double the one-photon excitation wavelength of 488 nm. While all these methods extend the OTF along the axial direction, they fundamentally differ in contiguity and absolute strength within the support region. For example, there are pronounced frequency gaps for the SWM and depressions for the I^5M. The rectangular images of the PSFs represent a region of $5 \times 2.5\,\mu m$ with the geometric focus in the center. (See color plate.)

appropriate TPE wavelength and similarly the OTF is the autoconvolution of the scaled one-photon OTF (Denk et al., 1990).

The excitation intensity of the SWM is given by a plane standing wave along the optic axis,

$$h = I_0 \cos^2(k_{ex})$$ (6)

where I_0 denotes a constant, k is the wave-number; and we assumed constructive interference to occur at the common geometric focus. The excitation OTF is its Fourier transform and given by

$$o = I_0[\delta(k) + \delta(k - 2k_{ex})/2 + \delta(k + 2k_{ex})/2]/2$$ (7)

Again, this is the result of auto-correlating the electric field's Fourier transform that consists of delta peaks at $\pm 2k_{ex}$.

While the 4 Pi microscope uses a spatially coherent point-like laser illumination, in the I^5M microscope wide-field illumination is used, normally in the Köhler mode, either with a lamp or a laser. The physical consequences of this difference are best explained as follows. If the two spherical wavefronts of the 4 Pi illumination are decomposed into plane waves incident from different angles and corresponding to different points of the illumination apertures, *all* plane waves of the aperture interfere with each other in the focal region. In the I^5M the illumination light is not coherent throughout the aperture. Therefore only *pairs* of plane waves originating from corresponding points (mirror images about the focal plane) of the illumination apertures are mutually coherent, forming a standing wave in the focal region. The period of these standing waves scales with the cosine of the azimuth angle θ. The excitation PSF of the I^5M can then be calculated by adding the intensity of these plane standing waves. Assuming uniform intensity throughout the exit pupil of the lens, the PSF is given by

$$h(z) = I_0 \int d\phi \int_0^\alpha d\theta \sin\theta \cos^2(k_{ex} z \cos\theta)$$
$$= 2\pi I_0 \int_0^\alpha d\theta \sin\theta \cos^2(k_{ex} z \cos\theta)$$ (8)

The support of the excitation OTF is readily inferred: For each θ the integrand in Eq. (8) is the excitation PSF of the SWM and thus the total OTF consists of a superposition of expressions of Eq. (6) for wave vectors ranging from $k_{ex} \cos(\alpha)$ to k_{ex}. Loosely speaking, this incoherent superposition smears out the delta peaks at the sides, forming the lines in Figures 12–3 and 12–4. This excitation mode contributes to avoiding the gaps that remain in the SWM's support after convolution with the detection OTF. If, on the other hand, the incoherent light source is imaged into the focal plane of the lens, i.e., critical illumination, mutually coherent points form wavefronts focused onto and interfering at the conjugate point in the image of the light source. The individual 4 Pi PSFs produced by each point of the light source as a result are incoherently summed up, giving an integral of the 4 Pi excitation PSF over the field of view in the focal plane. The OTF becomes nonzero exclusively on the inverse optical axis where it is given by the values of the 4 Pi OTF, altogether leading to a result not very much different from that predicted by Eq. (8). However, critical illumination is problematic due to potential nonuniformities in the light source and will be omitted in our analysis.

In any case, once the fluorescence light is generated in the sample, the I⁵M collects the spherical fluorescent wavefronts just as in a 4Pi microscope of type B or C. The two counterpropagating spherical wavefronts of fluorescence collected by each lens interfere constructively in a common point on the camera. Disturbance of the interference pattern of neighboring points of fluorescence emission is damped by the spatial incoherence in the focal plane: the radius of spatial coherence is largely given by the Airy disk associated with the fluorescence light at the aperture in use. Thus the I⁵M implements the highest possible degree of parallelization of 4Pi detection.

Figure 12–4 shows the numerically calculated excitation, detection, and effective PSFs/OTFs of the 4Pi, I⁵M, and SWM setups, along with those of the conventional epifluorescence and confocal microscope. The epifluorescence microscope features uniform illumination intensity throughout the sample volume; its OTF is a single deltapeak at the origin. To obtain a practically relevant comparison, we assumed a numerical aperture of $NA = 1.35$ and oil immersion with a refractive index $n = 1.51$. In the case of single-photon excitation of the dye, an excitation and detection (i.e., central fluorescence) wavelength of 488 nm and 530 nm, respectively, is assumed. For TPE, an excitation wavelength of 800 nm was chosen. Finite-sized pinholes can be taken into account by convolving the detection PSF with the image of the pinholes in the focal plane: a disk of radius r_{PH}. In frequency space this corresponds to a multiplication with the disk's Fourier transform given by

$$\hat{h}_{PH} = J_1(kr_{PH})/k \tag{9}$$

The first root of the Bessel function is at ~3.83 and thus the detection OTF becomes zero at $k = 3.83/r_{PH}$. However, the largest frequency present in the detection OTF is given by half the wavelength and therefore its support is unaltered if the pinhole radius is smaller than $3.83\lambda/(4\pi NA) \cong 0.3\lambda/NA$, which corresponds to half the Airy disk radius. Pinholes smaller than this can virtually be neglected in the computation, while for sizes around this value and larger, the PSF will widen laterally, suppressing the OTF at higher lateral frequencies. The effect of the pinhole size on axial resolution remains small as long as it does not exceed that of the Airy disk. We will therefore neglect the pinhole in our further analysis. All PSFs were numerically computed in a volume of $128 \times 128 \times 512$ pixels in x-, y-, and z-directions, respectively, for cubic pixels with 20 nm length. The OTFs were calculated by Fourier transformation; of the 512 pixels in the z-direction only data based on the central 256 pixels are shown in Figure 12–4. The color look-up-table (LUT) has been chosen so that the regions of weak signal are emphasized for both PSF and OTF. This reveals important differences between the systems. Areas of low but nonnegligible intensity are important since they cover a large volume and substantially contribute to the image formation.

Let us first consider the PSFs (Figure 12–4, narrow columns). Immediately, some differences between the various approaches become apparent. While the excitation modes of the SWM and I^{5}M are similar, the local minima are not zero for the latter due to the incoherent addition of the standing wave spectrum. The most important difference is observed when comparing the 4 Pi microscopes. As a result of focusing, their PSFs are confined in the lateral direction so that contributions from the outer lateral parts of the focal region are reduced. The confinement has important consequences. Whereas the 4 Pi confocal microscope, especially its two-photon version, exhibits only two pronounced but rather low lobes, the I^{5}M and even more so the SWM feature a multitude of lobes and fringes on either side of the focal plane, despite the fact that all of them rely on the same aperture. The second consequence is that due to its quadratic or cubic dependence on the excitation distributions, the 4 Pi confocal PSFs can be separated into an axial and a radial function in good approximation (Hell et al., 1995; Schrader et al., 1998):

$$h(r, z) \cong c(r)h_1(z) \tag{10}$$

Separability is a particular feature of the 4 Pi confocal and multiphoton arrangements and we shall see later that it is the prerequisite for simple online removal of sidelobe effects in the image. TPE leads to a further suppression of the outer parts of the excitation focus and thus of the side lobes of the 4 Pi illumination mode. Figures 12–3 and 12–4 also reveal that in conjunction with coherent detection (type C), TPE 4 Pi confocal microscopy features an almost lobe-free PSF. Its OTF almost fills the maximum support region. In the SWM and the I^{5}M, the number and relative heights of the lobes increase dramatically when moving away from the focal point because an effective suppression mechanism is missing. In the SWM the lobes become even higher than the central peak itself. In the I^{5}M the secondary maxima are as high as the first maxima of the single-photon 4 Pi confocal microscope of type C.

3.2 Removing Periodic Artifacts through Deconvolution

Even if they are small, side lobes and resulting ghost images in the raw data remain a common feature of all methods employing two lenses coherently. Next, we turn to the OTF to understand the circumstances under which image processing can effectively be used to remove the artifacts induced by the lobes. It is obvious from Eq. (2) that if the OTF were nonzero everywhere, we could divide the Fourier transform of the image by it. Subsequent Fourier back-transformation of the data would render the object. In practice, the OTF is limited in bandwidth and has weak regions. As division outside the OTF support is impossible, these frequencies are lost. But even in regions where the OTF is small, division strongly amplifies noise-producing artifacts.

Image restoration techniques, whether linear or nonlinear, aim at restoring as many frequencies as possible while trying to avoid this effect. Linear deconvolution is based on the division approach but introduces a special treatment for frequencies not transmitted by the OTF. It is capable of restoring frequencies only where the OTF is above the noise level. If frequencies are missing, a correct representation of the object in the image can be given only if these frequencies are extracted from a priori knowledge of the object. This extraction is mathematically more complex and often not viable. Linear deconvolution, on the other hand, is computationally facile and fast. Speed is of particular importance because the interference artifacts are ideally removed online making the final image readily accessible. Therefore, one of the most prominent advantages of a uniformly strong OTF is the possibility of applying a linear deconvolution.

The comparison of the effective OTFs in Figure 12–4 highlights the severe gaps in the SWM, making deconvolution impossible. Linear deconvolution is reportedly possible in the I^{5}M (Gustafsson et al., 1999). However, as the gaps in the I^{5}M are filled with rather low amplitudes, this method will create image artifacts for objects that are not sparse, not very bright, or objects containing spatial frequencies coinciding with the gap. Owing to the contiguity of its support and the strong amplitudes of the OTF, 4Pi confocal microscopy fulfills the preconditions for linear deconvolution. In fact, linear deconvolution along the axial direction based on the separability of the PSF has been applied for the removal of interference artifacts arising in the recording of complex objects. Thus superresolved axial imaging has been shown in the dense filamentous actin[55] and in the microtubular network (Nagorni and Hell, 1998) of a mouse fibroblast cell. We will have a closer look at this important procedure in the subsequent section.

Approximating the 4Pi confocal PSF as in Eq. (10) allows us to perform a computationally inexpensive *one-dimensinal* linear deconvolution that simply eliminates the effect of the lobes in the image. The axial factor of the PSF can basically be decomposed into the convolution of a function $h_p(z)$ describing the shape of a single peak, and a lobe function $l(z)$ containing the position and relative heights of the lobes:

$$h_l(z) \cong h_p(z) \otimes l(z) \tag{11}$$

h_p quantifies the blur and l describes the replication that is responsible for the "ghost images" in unprocessed 4Pi images. The effect of the lobe function can be eliminated using algebraic inversion. We use a discrete notation, where each lobe is represented by a component l_i of the vector $\mathbf{l}$ with l_0 denoting the strength of the central lobe and the index running from $-n$ to n. Negative indices denote lobes to the left of the central peak. If the lobe distance in pixels is denoted by d, the values of the object along the line are given by O_j and those of the image by I_j. Thus, the convolution is given by

$$I_j = \sum_k l_k O_{j-dk} \tag{12}$$

Looking for a filter l^{-1} inverting this convolution we need

$$O_j = \sum_s l_s^{-1} I_{j-ds} = \sum_s l_s^{-1} l_k O_{j-ds-dk} \tag{13}$$

for all possible objects. The inverse filter needs to fulfill the condition

$$\sum_s l_s^{-1} l_{j-s} = \delta_{j0} \tag{14}$$

At this point we can arbitrarily choose the length of the inverse filter, assuming an index running from $-m$ to m. The index j in Eq. (13) can take values from $-m - n$ to $m + n$. Therefore we have a system of $2(m + n) + 1$ equations with $2m + 1$ unknowns, which is usually not solvable. An approximation is found by considering the equations for $j = -m \ldots m$ only. Now, the problem is equivalent to solving a linear Toeplitz problem with the Toeplitz matrix given by the vector $\mathbf{l}$ (Nagorni and Hell, 2001; Press et al., 1993). The approximation is good if the edges of the inverse filters are small, since the remaining equations are nearly satisfied. This holds if the first-order lobes are <45%; in this case, the error is practically not observable. A typical length of the inverse filter is 11. In conjunction with a lobe height of 45%, the edges of the inverse filter feature a modulus below 1% of the filter maximum. For lobes of 35% relative height, this value drops to a value of only about 0.08%. Thus, using this technique, the separability of the 4Pi confocal PSF provides a quick way to obtain a final image that is equivalent to imaging with an "ideal" optical microscope that has a single narrow main maximum at the focal point. The inverse filter is discrete and nonzero only at a few points. This very effective side lobe removal method is therefore referred to as point deconvolution. Exploiting the characteristics of the PSF of the 4Pi microscope, it is applicable only in conjunction with this method.

Axial lobes in the PSF entail suppressed OTF regions along the optic axis. Thus, point deconvolution restores suppressed frequencies by linear deconvolution even though the actual calculation takes place in real space. For PSFs that cannot be written as in Eq. (10) point deconvolution is insufficient, because the transverse directions (x, y) have to be involved as well. In this case, it is mathematically clearly preferable to pass the data through the frequency domain. Established linear deconvolution algorithms rely on inverting Eq. (2) but also accommodate for the vanishing regions of the OTF. An estimate of the frequency spectrum of the object can be obtained by using

$$\hat{E}(k) = \hat{I}(k)o^*(k)/(|o(k)|^2 + \mu) \tag{15}$$

The regularization parameter μ (Bertero et al., 1990) sets a lower threshold on the denominator to avoid amplification of frequencies where the modulus of the OTF is so small that the frequency spectrum of the image is dominated by noise. If the OTF is a convex function that is in the absence of "gaps," the effect of regularization is similar to smoothing. In most cases considered in this chapter, however, the OTF is not convex and the situation is more complicated. Small OTF values are found not only at the OTF boundaries, but also within its region of support, e.g., in the vicinity of the minima (Figure 12–4) or at the frequency gaps (Figure 12–3). If the lobes are too high (typically >50%), the level of the minima of the OTF is comparable or smaller than the noise level. This is definitely the case in the SWM, but also in the presence of slight aberrations in the I^5M, as well as in an aberrated 4Pi microscope. Therefore, the necessity of implementing a certain value of μ is closely connected with the lobe height and with the potential artifacts induced by the lobes.

If the PSF is separable in a peak function and an axial lobe function, the lobe removal can be elegantly targeted:

$$h(\mathbf{r}) = h_p(\mathbf{r}) \otimes l(z) \tag{16}$$

Contrary to the decomposition in Eq. (11), the peak function $h_p(\mathbf{r})$ also contains the dependence of the peak on the lateral coordinates. We no longer require the PSF to separate in a radial and an axial part, yet the lobe function still gives the relative height of the lobes as well as their location:

$$l(z) = \sum_s l_s \delta(z - ds) \tag{17}$$

with d now being the lobe distance in units of length. The frequency spectrum of the image is then given by

$$\hat{I}(\mathbf{k}) = \hat{G}(\mathbf{k}) \cdot o(\mathbf{k}) = \hat{G}(\mathbf{k}) \cdot \hat{h}_p(\mathbf{k}) \cdot \hat{l}(k_z) \tag{18}$$

If the lobe function is symmetric with respect to the focal plane (i.e., for constructive interference), we can write

$$\hat{l}(k_z) = 1 + \sum_{s>0} 2 l_s \cos(sdk_z) \tag{19}$$

This decomposition immediately discloses why the critical lobe height is 50%: If the PSF consists of a main maximum and two primary lobes of 50%, the right-hand side of (19) vanishes for the axial frequencies associated with the distance d. So, if the lobes are >50% the frequency represented by the lobes is not contained in the OTF, and hence not transferred to the image. In reality, the critical lobe height is slightly above 50%. The reason is the influence of the secondary lobes that have been neglected in our reasoning. Nonetheless, the 50% threshold is an excellent rule for the critical lobe height, which applies equally to SWM, I^{5}M, and 4Pi microscopy, for fundamental reasons.

Equation (18) implies that the effect of the lobes can be removed by direct Fourier inversion. This is the case if the Fourier transform of the lobe function remains above the noise level throughout the relevant frequency spectrum. The spectrum of the lobe-free image is then obtained by dividing the image spectrum by the Fourier transform of the lobe function. Figure 12–5 shows a typical data set acquired with a TPE type A 4Pi confocal microscope and illustrates lobe removal both in the spatial and the frequency domain. The avalanche photodiode used as a detector had a typical dark count rate <1 count/pixel, which is negligible (Hell et al., 1997). Hence, the only significant source of noise was the Poisson noise of the photon-counting process, manifesting itself as white noise that is independent of the spatial frequency. Since the primary lobes are well below 50%, the first minima of the OTF are at 19%, which is well above the noise level of typically 0.5–1%. Direct lobe removal is straightforward in this example, underscoring the importance of nonvanishing amplitudes in the OTF. If the OTF exhibited regions close to zero, as is predicted for the I^{5}M, the multiplication with a high number in this region would result in a strong amplification of noise, and compromise the obtained image. In the

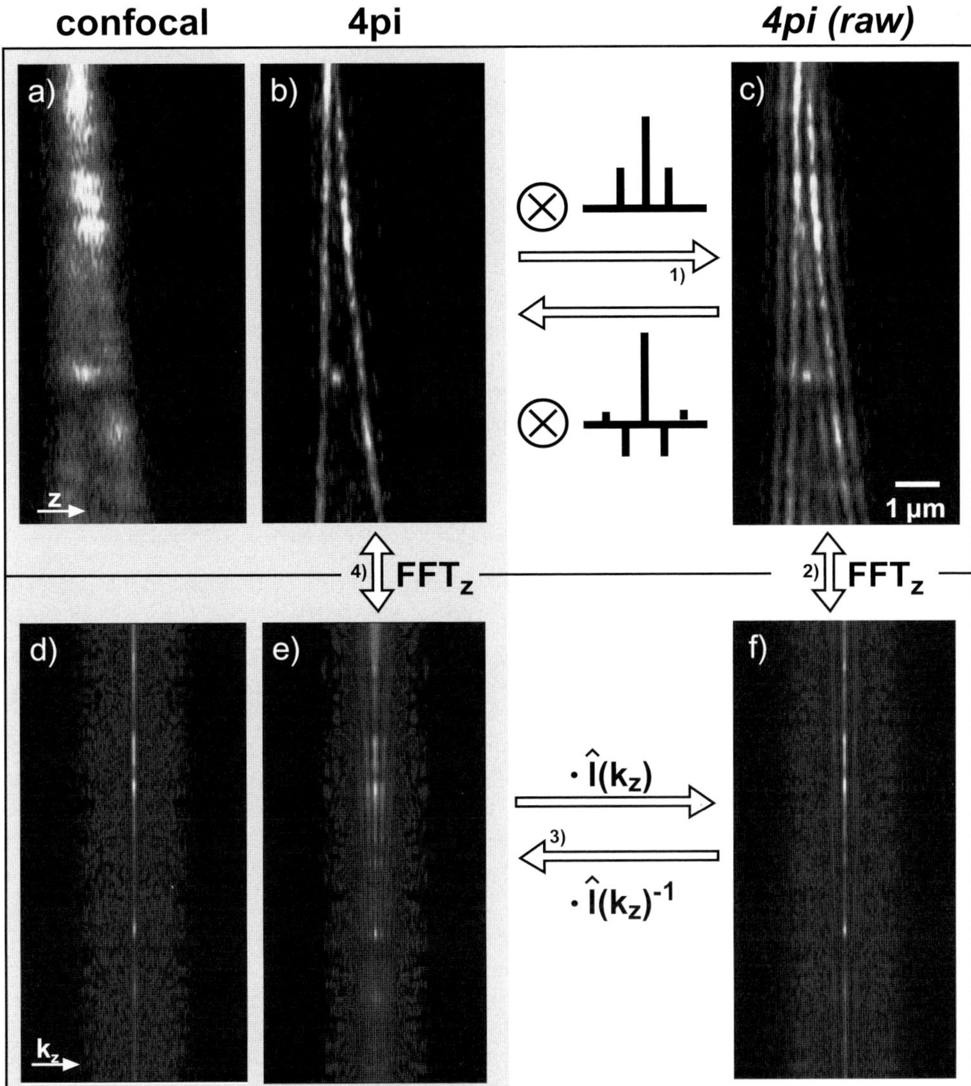

Figure 12–5. Lobe removal and deconvolution in 4Pi microscopy. The figure shows images of the same pair of actin fibers in a fixed mouse fibroblast cell recorded in the TPE confocal (a) and TPE 4Pi type A (b and c) mode. The corresponding Fourier transform along the optic axis is also shown (d, e, and f). The five-fold axial resolution increase (a vs. b) and the correspondingly extended OTF (d vs. e) are immediately visible. The side lobes are well below 50% and the factorization of the PSF's axial and lateral dependence is possible in 4Pi microscopy. Therefore, an inverse discrete filter can be found and its application to the raw data (c) yields a valid and almost artifact-free image (b). Alternatively, lobe removal can be performed in the frequency domain. Equation (16) indicates that the Fourier transforms of the raw data (f) is given by the product of the Fourier transform of the lobe-free image (e) and the lobe function $\hat{l}$ (k_z). Thus, (2) Fourier transforming the raw data, (3) multiplying with the inverse of the lobe function's Fourier transform, $\hat{l}^{-1}(k_z)$, and (4) Fourier backtransforming lead to almost the same lobe-free image. This method can be applied even if the separation of axial and lateral dependence is impossible for the PSF. The Fourier transforms along the axial direction (3 and 4) merely have to be replaced by their 3D counterparts. (See color plate.)

SWM, it is impossible to remove the interference artifacts on a general basis.

While an in-depth analysis of *nonlinear* iterative deconvolution (i.e., image restora-tion) algorithms is far beyond the scope of this chapter, it can be safely stated that these suffer from limitations similar to the linear deconvolution approaches. They cannot really restore information that has been lost in the imaging process. A connected and largely convex OTF that does not imply regions of weak transmission is almost equally important to these algorithms. Nonlinear iterative algorithms take advantage of a priori information about the object that linear deconvolution cannot uncover unambiguously. The simplest a priori information is the positivity of the object and of the image. Carefully applied to 4Pi and I^5M data, nonlinear restoration can yield impressive results, but extreme care must be taken not to compromise the reliability of the outcome by false or even biased assumptions.

3.3 Discussion: Improved Axial Resolution in Practice

While the use of coherent beams from two opposing lenses extends the OTF in the axial direction, a detailed analysis shows that conditions have to be met to exploit this advantage in an effective manner. Reliable and artifact-free restoration is possible only if the region of the OTF is convex, that is, if it does not contain nonnegligible "weak" regions or gaps. It was shown that the OTF of an SWM exhibits marked gaps along the optic axis that make the removal of the interference artifacts almost impossible. It is therefore not surprising that previous experimental studies confirmed that in the SWM it is impossible to unambiguously distinguish two axially separated objects, unless the object is thinner than 50% of the wavelength.[34] Unambiguous resolution of axially extended objects is impossible with SWM for fundamental physical reasons. Nevertheless, both SWM and 4Pi microscopy have been successfully applied to the ultraprecise measurement of axial distances (Albrecht et al., 2002; Schneider et al., 2000; Schmidt et al., 2000) and object sizes (Egner et al., 2000; Failla et al., 2002), which is conceptually and practically less demanding.

The OTF of the I^5M is superior to that of the SWM, because it remains nonzero throughout its support. Still, it is weak over a considerable region when compared to the giant zero-frequency peak, which actually is a singularity. Hence, to benefit from this contiguity, I^5M data must be recorded with a very large signal-to-noise ratio. Besides, the method may be applicable to sparse nonextended objects only, such as points or sparse fine lines. The OTF of the reported 4Pi microscopes are truly contiguous. In the critical regions, the 4Pi confocal OTF (Figure 12–4) exhibits significant values in the 19–23% range. This feature is of key relevance to the removal of the interference artifacts and hence for unambiguous, object-independent 3D microscopy with improved axial resolution.

Another major difference between the effective PSF of the SWM and I^5M on the one hand, and the 4Pi confocal on the other, is the fact that the PSF is spatially much more confined in the latter case. This allows us to reduce the process of lobe removal from a 3D deconvolution to a one-dimensional linear problem with the highly useful side effect that single layers can be acquired and deconvolved while all other methods require acquisition of a full 3D data stack.

The 4Pi concept is the result of rigorously maximizing the aperture angle employed in the imaging process. This results in a superior excitation OTF due to its lateral filling, which is rooted in the fact that the (exciting and detected) light is *focused*. Hence, while scanning with a focused beam inevitably reduces imaging speed, this procedure also results in fundamentally improved imaging properties of the microscope. Flat field standing wave excitation inherent to the I^5M and SWM trades off collected spatial frequencies. The loss of optical frequencies is so significant (Figures 12–3 and 12–4) that the ability to provide unambiguous axial resolution is either put at risk or not viable.

Nevertheless, axial resolution improvement and imaging similar to the 4Pi microscope are reportedly possible when combining I^5M with offline image restoration (Gustafsson et al., 1999). The combination of I^5M with fringe pattern illumination and subsequent image restoration has also been suggested to (1) alleviate the problem of the zero frequency singularity (Gustafsson, 2000; Heintzmann and Cremer, 1998) and (2) to increase the lateral resolution to that of (restored) confocal microscopy. Leaving aside the technical complexity of controlling the interference patterns of typically four pairs of beams, this suggestion confirms that an unambiguous axial resolution requires the employment of a wider angular spectrum. In fact, the spherical beams in a 4Pi microscope can be regarded as a complete spectrum of interfering plane waves coming from all angles available. Hence, from the standpoint of imaging theory, the combination of I^5M with fringe pattern illumination is a modification of I^5M toward a scheme that is more similar to the 4Pi arrangement; the improvements of the OTF are gained by conditions that converge to the "focusing" conditions found in the 4Pi microscope.

In real samples the OTF of the microscopes is compromised by aberrations that were not included in our comparison of the concepts. Residual misalignments of the foci induced by variations of the refractive index in the sample play a role. However, successful 4Pi imaging of the mouse fibroblast cytoskeleton (Nagorni and Hell, 1998) and I^5M imaging of similar structures (Gustafsson, 2000) have revealed that aberration effects are surmountable. Image deconvolution requires prior knowledge of the PSF and hence its explicit determination with a point-like object. This is particularly important since the PSF depends on the relative phase of the two counterpropagating wavefronts. It has been shown that in 4Pi microscopy the type of interference in the focal plane (constructive, destructive, or anything in between) is of lesser importance (Hell and Nagorni, 1998). Depending on its influence on the OTF, the same will also apply to I^5M. So far, the structure of the PSF has been determined by measuring the response of test objects, such as fibers or fluorescent beads. However, it has been shown that the relative phase can be extracted directly from the image data (Blanca et al., 2002). This is of great importance in cases in which the imaged object itself alters the relative phase of the interfering beams.

A filled OTF entails robustness in operation and lower amenability to potential misalignment. Thus 4Pi microscopy allowed superresolved imaging of specimens in aqueous media using water immersion lenses (Bahlmann et al., 2001). This is noteworthy since water lenses feature semiaperture angles of not more than 64° as compared to 68° available for oil or glycerol immersion, therefore increasing side lobes and rendering the problem of missing frequencies in the OTF more severe. Imaging a watery sample is one of the prerequisites of live-cell imaging.

For realistic aperture angles, the inherent lack of contiguity of the OTF in the SWM renders the removal of interference ambiguities impossible. The I^5M becomes more viable through filling the frequency gaps present in the SWM, albeit with values that are weak with respect to the zero-frequency components. Both systems have yet to prove their applicability to live cell imaging. In fact, recent investigations have cast doubt on whether I^5M is able to provide reliable data from live cells with the available water immersion lenses. If water immersion lenses of 70° and higher were available, the robustness and applicability of I^5M would significantly increase. Thus, this method would become viable for a much larger variety of objects, at least for not too dense or not too 3D-convoluted ones. The fact that the enlargement of the semiaperture angle is the key to the performance of the I^5M highlights the exploitation of the entire *spherical* wavefronts as the central physical element. In a sense, the I^5M can be viewed as a 4Pi system that maximizes the degree of parallelization in the focal plane.

Parallelization is readily accomplished in 4Pi and 4Pi confocal microscopy, as well. A multifocal variant of TPE 4Pi confocal microscopy, termed multifocal multiphoton 4Pi microscopy (MMM-4Pi) (Egnor et al., 2002), has indeed translated the typical 140 nm axial resolution of a TPE 4Pi microscope into live cell imaging. Importantly, although present and noticeable, phase changes induced by the cell proved more benign than anticipated. Nevertheless, phase alterations and wavefront aberrations due to refractive index changes within the specimen are likely to confine 4Pi microscopy and related techniques to the imaging of individual cells or thin cell layers. In conjunction with nonlinear image restoration this novel imaging method has displayed ~100-nm 3D resolution under live cell conditions. For example, MMM-4Pi microscopy has provided superior 3D images of the reticular network of GFP-labeled mitochondria in live budding yeast cells.

More recently, 4Pi microscopy has been realized in a compact optical unit that was interlaced with a state-of-the-art single beam scanning confocal fluorescence microscope (Leica TCS-SP2, Mannheim, Germany). Operating in type C mode, i.e., coherent spherical wavefronts both for excitation and for confocal detection, this compact and rugged system displayed a seven-fold improved axial resolution over confocal microscopy in live cells (Gugel et al., 2004). This system is now commercially available as the first far-field optical microscope with axial superresolution.

The fact that *single-photon* excitation 4Pi confocal microscopy has not been realized to date has been due to the superiority of the OTF in the TPE version. Nevertheless, it is clear from imaging theory that a single-photon 4Pi confocal microscope of type C features a more contiguous and better filled OTF than an I^{5}M, which inevitably operates in the single-photon mode. Given further improvements in aberration compensation and lens manufacturing, it is feasible that single-photon excitation 4Pi microscopy (of type C) can also be realized in a reliable manner. An inherent advantage of single-photon excitation over its two-photon counterpart is that the much shorter excitation wavelengths lead to axially narrower focal spots. For example, using a wavelength of ~400 nm, a single-photon 4Pi microscope of type C would deliver an axial resolution of ~73 nm. This number further underscores the potential of 4Pi microscopy to deliver 3D images with axial sections well in the tens of nanometer regime.

In summary, the resolution of a far-field fluorescence light microscope can be improved down to the range of 60–100 nm along the optic axis by coherently adding the focal light fields of two opposing lenses. The addition of the fields leads to improved resolution for both excitation at the focal point and detection of fluorescence at a common point. The axial resolution is improved only if the main maximum of the resulting PSF is at least twice as large as the primary side maxima arising from the coherent addition. The latter condition can be fulfilled only if, at least for one of the processes, the added light field is a *spherical* wavefront covering the aperture angle of the lens. This in turn shows that the key physical element to improving the axial resolution by the coherent use of two opposing lenses is not the production of an interference pattern, but the enlargement of the aperture of the system.

4 Breaking the Diffraction Barrier

Increasing the total aperture of the focusing system with two opposing lenses improves the 3D resolution of a far-field microscope, but does not break the diffraction barrier. On the contrary, 4Pi microscopy exploits the full potential of diffraction-limited imaging. This particularly applies to (multiphoton) type C 4Pi confocal microscopy, which can be regarded as the far-field optical microscope with the largest possible aperture. For a microscope, being subject to the diffraction limit implies that the system features the maximum resolution that cannot be surmounted. On the other hand, breaking the diffraction barrier implies the potential for featuring an infinitely sharp focal spot or an OTF with an infinitely large bandwidth.

The first concept to break the diffraction barrier was STED fluorescence microscopy.[9] The physical concept of STED microscopy was described as early as 1994 and was soon followed by GSD microscopy, a related concept that also entailed diffraction-unlimited resolution (Hell and Kroug, 1995). Whereas STED microscopy utilized stimulated emission to deplete the excited state of the fluorophore, GSD aimed at depleting its ground state by transiently pumping the dye into a long-lived dark state, i.e., its triplet state. Importantly, both concepts share the same principle for breaking Abbe's barrier: a focal intensity distribution featuring a zero point (or at least a strong gradient in space) effects a *saturated depletion* of one of the molecular states that are essential to the fluorescence.[10,11] Following depletion, the state is populated again, that is, the saturated transition is reversible. A saturated transition, such as a depletion of a state, introduces vast nonlinearities that eventually prove essential for breaking the diffraction barrier.

An even closer examination of the underlying concept shows that in fact any saturable transition between two states where the molecule can be returned to its initial state is a potential candidate for breaking the diffraction barrier (Hell, 1997; Hell et al., 2003; Dyba and Hell, 2002). The general concept has therefore been termed RESOLFT. The transition utilized can be selected to match the practical conditions of the imaging problem at hand, such as the required intensities, the available light sources, and the avoidance of photobleaching. An important aspect with respect to biological applications is the compatibility with live cells.

The basic idea of the RESOLFT concept can be understood by considering a molecule with two arbitrary states A and B between which the molecule can be transferred. In fluorophores, typical examples for these states are the ground and first excited electronic states and conformational and isomeric states. The transition A $\rightarrow$ B is induced by light, but no restriction is made about the transition B $\rightarrow$ A. It may be spontaneous, but may also be induced by light, heat, or any other mechanism. The only further assumption is that at least one of the two states is critical to the generation of the signal. In fluorescent microscopy this means that the dye can fluoresce only (or much more intensively) in state A. Such a system may be exploited to generate diffraction-unlimited resolution in fluorescence imaging (or any kind or manipulation, probing, etc. that depends on one of the states), which is illustrated in Figure 12–6.

We begin with all molecules or entities in the sample being in state A. Our goal is to generate a diffraction-*un*limited distribution of molecules in state A. To this end, the sample is illuminated with light that drives the transition from A to B. The intensity distribution $I(r)$ of the illuminat-

Figure 12–6. The RESOLFT principle. Diffraction-unlimited spatial resolution is achieved by saturating a linear but reversible optical transition from state A to state B. The simple explanation: A uniform dye distribution of state A is illuminated by a strongly modulated intensity light distribution that transfers the dye molecules to state B; ideally the modulation is perfect, so that the local intensity minima actually are zeros. Here, we choose the smallest possible modulation $I(r) = \cos^2(2\pi r/\lambda)$. The left-hand panel shows $I(r)$ and the resultant probability $N_A(r)$ of the dye molecules being in state A. The right-hand panel depicts the dependence of $N_A(r)$ on the local intensity $I(r)$, exhibiting the typical saturation behavior. I_{sat}-I_s is defined as the intensity at which half the molecules are transferred to state B. (a) At $I(r) < I_{sat}$, the modulation of the light is replicated in $N_A(r)$. The narrowest distribution of $N_A(r)$ will also be limited by diffraction, because $I(r)$ is diffraction limited and the relationship between I and N_A is basically linear. (b) Increasing the maximum intensity moves the points at which $I(r)$ reaches I_{sat} closer to the local intensity minimum, e.g., the zero. Because these are also the points at which $N_A(r)$ drops to 0.5, the FWHM of $N_A(r)$ is correspondingly reduced. (c) Further increasing the maximum intensity beyond I_{sat} leads to a further reduction of the FWHM. If A is the fluorescent state, $N_A(r)$ is read out as the desired signal stemming from a narrow region. To obtain an image, the local minima are scanned across the sample. If the signal stems from state A, the intensity values $N_A(r)$ are read out subsequently and the image is assembled in a computer. If the signal is generated by the "majority population in state B" (e.g., state B is the fluorescent state) the function $1 - N_A(r)$ is read out and the image must be "inverted" later. This approach is challenged by signal-to-noise issues. (d) There is no theoretical limit to this method and its resolution is ultimately determined by the available laser power and the potentially limiting photodamage.

ing light is of course diffraction limited but features one or several positions with zero intensity. After illumination, the probability, $N_A(r)$, of finding molecules in state A depends on $I(r)$ and displays peaks where $I(r)$ is zero and no transitions to B are induced. If we choose the maximum intensity I_{max} in such a way that it is many times (ζ times, $\zeta = I_{max}/I_s$ is called the saturation factor) higher than the saturation intensity, I_s, of the transition A $\rightarrow$ B (the threshold at which 50% of the molecules are

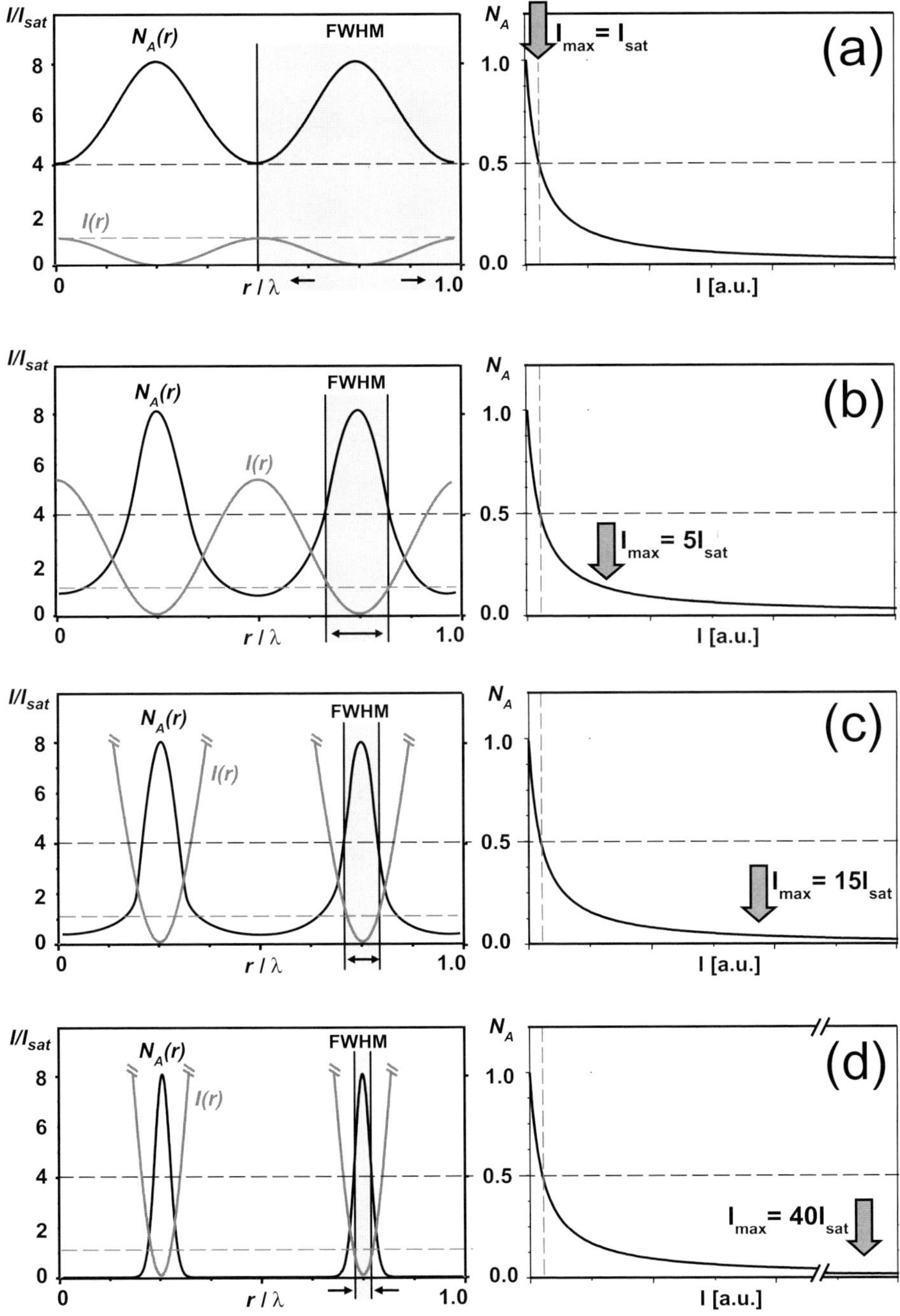

transferred into state B), then molecules will be almost exclusively in state B even at positions where $I(r)$ is only a small fraction (~5/ζ) of $I_{\max}$. This means that while molecules remain in state A in the intensity zeros, they are transferred to state B even at locations in the immediate vicinity of the zero. Thus, the peaks of $N_A(r)$ become very narrow, featuring steep edges at the same time. Figure 12–6 is presented in one dimension for clarity, but this idea is readily extended to all directions in space and hence to 3D imaging.

Their Fourier transform (the effective ex-citation OTF) of an increasingly sharp peak correspondingly becomes wider with an in-creasing $I_{\max}$. A complementary mathematical explanation for this fact is that higher-order contributions that are described by several autoconvolutions of the depletion pattern become more and more important for higher intensities. This is in contrast to very low intensities where the dependence of N_A on the depletion intensity is approximately linear: $N_A(r) \cong 1 - \zeta I(r)$. Here, a single additional convolution is introduced in frequency space and the theoretical cutoff of the support can only be pushed from the $4k$ we derived for conventional imaging to $6k$. Thus, the diffraction barrier is not truly broken for low intensities, but only shifted to a slightly higher value.

The generated probability distribution can be used for high-resolution imaging by applying scanning. Since the fluorescence stems exclusively from the immediate vicinity of the positions where $I(r)$ is zero, scanning the zero(s) through the sample with simultaneous recording of the fluorescence signal allows an image with basically unlimited spatial resolution in the far field to be assembled. For particular conditions, the resolution of this process is determined by the FWHM of the peaks in $N_A(r)$, which in turn is solely determined by the saturation factor. The FWHM can become infinitely narrow, that is down to the size of a single molecule. The reason why Abbe's spatial frequency cutoff does not apply in this case is simply the fact that in the RESOLFT concept the lens merely acts as a condensor collecting the fluorescence. For example, when scanning with a single zero, the detector can be placed right next to the sample to measure $N_A(r)$. Scanning is absolutely mandatory in a far-field optical system with a broken diffraction barrier, because—and Abbe was perfectly correct in this regard—the objective lens cannot transmit the higher spatial frequencies through the lens.

However, the need for scanning does not imply that the scanning system needs to be based on *single*-beam scanning. Parallelization is readily possible and hence detection with a conventional camera is feasible if the nodes are further apart than the classical resolution limit of the microscope. When the sharply localized fluorescence from the nodes is imaged onto the camera, the resulting spot will certainly be blurred as a result of diffraction, and possibly extend over several pixels on the camera. However, if the nodes are further apart than Abbe' diffraction barrier, each of the blurred spots can be assigned to the respective nodes. Sequential read-out of the camera makes it possible to integrate the measured signal of each blurred spot or line separately and to associate it with the corresponding sharply localized region in the sample from which the signal emerged.

Importantly, implementing such a "wide-field" detection in a microscope does not imply that subdiffraction resolution is possible in conventional microscopy, since scanning is still an essential element in the process of image formation. In a sense, "wide-field" is not the most appropriate term in any event, since RESOLFT-based concepts always imply that a part of the sample (at the very least, one single point) is omitted from the saturable transition. Consequently, camera-based RESOLFT approaches are essentially parallelized scanning concepts.

This scanning "wide-field" detection is readily explained in frequency space as well. The PSF describing the generated probability of the molecules of being in state A, $N_A(r)$, consists of periodic sharp peaks, and can be described as a convolution of a single peak with a comb function or with its multidimensional equivalent. In frequency space, this is equivalent to a multiplication of the broad frequency band of a single peak with a comb function. Thus the Fourier transform of $N_A(r)$ which dominates the excitation OTF if given by delta-peaks separated by the inverse distance of the nodes in I(r). The speed with which they drop off towards large frequencies depends on the width of a single peak in $N_A(r)$. The effective OTF is given by the convolution of this series of delta-peaks with the detection OTF. Thus it will be continuous whenever the usable support of the latter is larger than the distance between the delta peaks in the excitation OTF. Not surprisingly, this just means that the intensity zeros of $I(r)$ are far enough apart so that the detection system can separate their fluorescence.

Note that complete depletion of A (or complete darkness of B) is *not* required. It is sufficient that the nonnodal region features a constant, notably lower probability to emit fluorescence, so that it can be distinguished from its sharp counterpart. Even if not A, but B is the brighter state, it is possible to read out B and obtain the same superresolved image after mathematical postprocessing that entails subtraction of signals from each other. While reading out B, in principle, also delivers unlimited resolution, this version is heavily challenged by signal-to-noise issues. This stems from the fact that the "bright" light from the nonnodal regions contributes with a substantial amount of photon shot noise.

It is also important to keep in mind that the nonlinearities introduced in these concepts are not analogous to the nonlinear interactions connected with m-photon excitation, mth harmonics generation, coherent anti-Stokes–Raman scattering (Sheppard and Kompfner, 1978; Shen, 1984), etc. In the latter cases, the nonlinear signal stems from the simultaneous action of *more than one photon* at the sample, which would work only at high focal intensities. In contrast, the nonlinearity brought about by saturation and depletion stems from a change in the *population* of the involved states, which is effected by a *single*-photon process, namely stimulated emission. Therefore, unlike m-photon processes, strong nonlinearities are achieved at comparatively low intensities.

Next, let us derive an estimate for the resolution achievable in such a system at finite depletion intensities (Hell, 2003, 2004). We denote the rates of A $\rightarrow$ B and B $\rightarrow$ A with k_{AB} and k_{BA}, respectively. The time evolution of the normalized populations of the two states n_A and n_B is then given by

$$dn_A/dt = -k_{AB}n_A + k_{BA}n_B = -dn_B/dt \tag{20}$$

Independently from its initial state, after a time

$$t \geq (k_{BA} + k_{AB})^{-1} \tag{21}$$

the equilibrium is approximately reached and the population of state A is given by

$$N_A = k_{BA}/(k_{BA} + k_{AB}) \tag{22}$$

During the process described in Figure 12–6, state A is depleted at a rate $k_{AB} = \sigma I$, where σ denotes the molecular cross section and the intensity I is written as photon flux per unit area. Hence, the equilibrium population is given by

$$N_A(r) = k_{BA}/[\sigma I(r) + k_{BA}] \tag{23}$$

And $N_A = 1/2$ for the saturation intensity

$$I_s = k_{BA}/\sigma \tag{24}$$

From Eq. (23) we see that where $I(r) >> I_s$ all molecules end up in B. Thus, if we choose

$$I(r) = I_{max}f(r) \tag{25}$$

with $I_{max} >> I_s$, molecules in state A are found only in the nodes of the diffraction-limited distribution function $f(r)$. As an example, we choose a sine-square intensity distribution, such as produced by a standing wave

$$f(x) = \sin^2(2\pi nx/\lambda) \tag{26}$$

for illumination, where n denotes the index of refraction of the medium. A simple calculation shows that the FWHM of the peaks of N_A and hence the resolution of the microscope is then given by

$$\Delta x = \frac{\lambda}{\pi n}\arcsin(1/\zeta) \cong \frac{\lambda}{\pi n\sqrt{\zeta}} \tag{27}$$

A saturation factor of $\varsigma = 1000$ yields $\Delta x \approx \lambda/100$, but in principle the spot of "A molecules" can be continuously squeezed by increasing ς. Scanning with such a spot and simultaneous recording of the signal from it deliver diffration-unlimited resolution. Equation (27) quantitatively describes this possibility in a microscope using diffraction-limited beams.

If the intensity distribution $I(r)$ is produced by a lens, the largest frequency will be determined by the numerical aperture of the lens $n\sin\alpha$, due to diffraction. In this case, Eq. (27) changes into

$$\Delta x \cong \frac{\lambda}{\pi n\sin\alpha\sqrt{\zeta}} \tag{27'}$$

If at the same time the molecules are being driven back from B to A by a light intensity distribution following a cosine-square form, we have $k_{BA} = \sigma_{BA}I_{max}^{BA}\cos^2(2\pi n\,\sin\alpha x/\lambda)$, with σ_{BA} denoting the cross section of the transition. By defining the saturation intensity as $I_s = \sigma_{BA}I_{BA}^{max}/\sigma$, we obtain

$$\Delta x \cong \frac{\lambda}{\pi n \sin\alpha \sqrt{1+\zeta}} \qquad (27'')$$

This equation is particularly appealing, since for a vanishing saturation factor $\zeta = 0$ it almost assumes Ernst Abbe's diffraction limited form, whereas for $\zeta \rightarrow \infty$ the spot becomes infinitely small.

There are also adverse effects that need to be taken into account. The first is that state A cannot be completely emptied by even very intense illumination (e.g., because there is an excitation by the same beam) and the second is that state B may also contribute to the signal. Both can be considered by including a constant offset in Eq. (23):

$$N_A(r) = (1 - \delta)k_{BA}/[\sigma I(r) + k_{BA}] + \delta \qquad (28)$$

This would result in the image consisting of a superresolved image plus a (weak) conventional image. The latter does not alter the frequency content of the image. Therefore, given sufficient SNR, the resolution will not deteriorate. In other words, if δ is sufficiently small so as not to swamp the image with noise, the conventional contribution can be subtracted (Hell and Kroug, 1995; Hell, 1997).

A further experimental problem is caused by imperfections of the intensity zeros. Imagine the standing wave is aberrated and approximately described by

$$f(x) = (1 - \gamma)\sin^2(2\pi x/\lambda) + \gamma \qquad (29)$$

Such zeros with insufficient "depth" turn out to have a more serious impact on performance. The maximum signal in the intensity minima drops by a factor $(1 + \zeta\varepsilon\gamma)$, as a result. Following the same calculation as above we obtain

$$\Delta x = \frac{\lambda}{\pi}\sqrt{\gamma + \frac{1}{\zeta}} \qquad (30)$$

and therefore the maximum achievable resolution is given by $\gamma\sqrt{\gamma}\,/\,\pi$. At $\gamma \sim 1\%$ resolutions of $\lambda/20$ at saturation factors of 100 can be achieved without losing more than half the signal in the intensity minima.

While RESOLFT is far more intuitively explained in the way presented above, it is also helpful to take a look at the frequency space to relate these findings to the concepts and results presented in the first part of this chapter. The dependence of the effective excitation PSF, i.e., the distribution of the probability that a molecule actually emits a fluorescence photon, is governed by the saturation level-dependent value of $N_A(r)$ expressed in Eq. (23). If for $I(r) = 0$ the microscope begins, for example, with a conventional PSF $h_c(r)$ used for imaging the distribution $N_A(r)$ onto a camera, the effective PSF of the system is given by

$$h(r) = h_c(r)\frac{1}{[1 + \sigma I(r)/k_{BA}]} \qquad (31)$$

Now let $\sigma I_{max}/k_{BA} = \zeta$ and $I(r) = I_{max}f(r)$ as above, then we can expand (31) in a Taylor series

$$h(r) = \frac{h_c(r)}{1+\varsigma} \sum_v \left[\frac{\varsigma}{\varsigma+1}\right]^v (1-f(r))^v \tag{32}$$

With $g(r) = 1 - f(r)$ and $\xi = \zeta/(\zeta + 1)$ we obtain the OTF after Fourier transformation

$$o(k) = \frac{\hat{h}_c(k)}{1+\zeta} \otimes (\delta(k) + \xi\hat{g} + \xi^2\hat{g} \otimes \hat{g} + \xi^3\hat{g} \otimes \hat{g} \otimes \hat{g} + \ldots) \tag{33}$$

At low intensities, ζ and therefore ξ is so small that only the linear term is relevant and the convolution extends the support to $6k$ as discussed above. The larger the maximum intensity, the more important higher orders of the Taylor series will become. These involve multiple auto-convolutions of the function g extending the support further and further.

While a quantitative treatment in frequency space is more complicated and less intuitive than the one introduced in the previous section, the following analysis gives a feel for the effect of the saturation factor and also illustrates the possible vast expansion of the OTF support. For the sake of simplicity we assume a Gaussian form of the light distribution function

$$f(x) = 1 - \exp(-x^2/2a^2) \tag{34}$$

The properly normalized m-fold auto-convolution of g is then given by

$$\otimes_m \hat{g}(k) = a\sqrt{2\pi/m} \exp(-a^2k^2/(2m)) \tag{35}$$

Now let us assume that the useful support ends at a frequency where the OTF is attenuated to a small fraction ε of its value at small frequencies. For large saturation factors the influence of the convolution with the confocal OTF on the cut-off frequency can be neglected and we have to calculated the sum in brackets in (33). Substituting (35) into equation (33) and approximating the sum by an integral we get for the term in brackets

$$o(k,\varsigma) \cong -i\sqrt{\pi/\ln\xi} \exp\left(-iak\sqrt{2\ln\xi}\right) \tag{36}$$

For large saturation factors we can write $\ln\xi = \ln\zeta - \ln(\zeta + 1) \cong -1/\zeta$ and obtain

$$o(k,\varsigma) \cong \sqrt{\pi\zeta} \exp(-ak\sqrt{2/\varsigma}) \tag{37}$$

This means that the attenuation of the modulus of the OTF at large frequencies is anti-proportional to the square-root of the saturation factor. This is equivalent to saying that the resolution increases with the square root of the saturation factor just as we expected from our previous analysis.

4.1 STED Microscopy

STED microscopy produces subdiffraction resolution and subdiffraction-sized fluorescence volumes in exactly the manner described above by the depletion of the fluorescent state of the dye. Depletion inherently implies saturation of the depleting transition. At present, it is realized in a (partially confocalized) spot-scanning system due to a number of technical advantages, but it has been conceptually clear from the outset that nonconfocalized detection is viable as well (Hell and Wichmann, 1994). The principal idea, a schematic setup and an exemplary measurement of the resolution increase, is shown in Figure 12–7. The fluorophore in the fluorescent state S_1 (state A) is stimulated to the ground

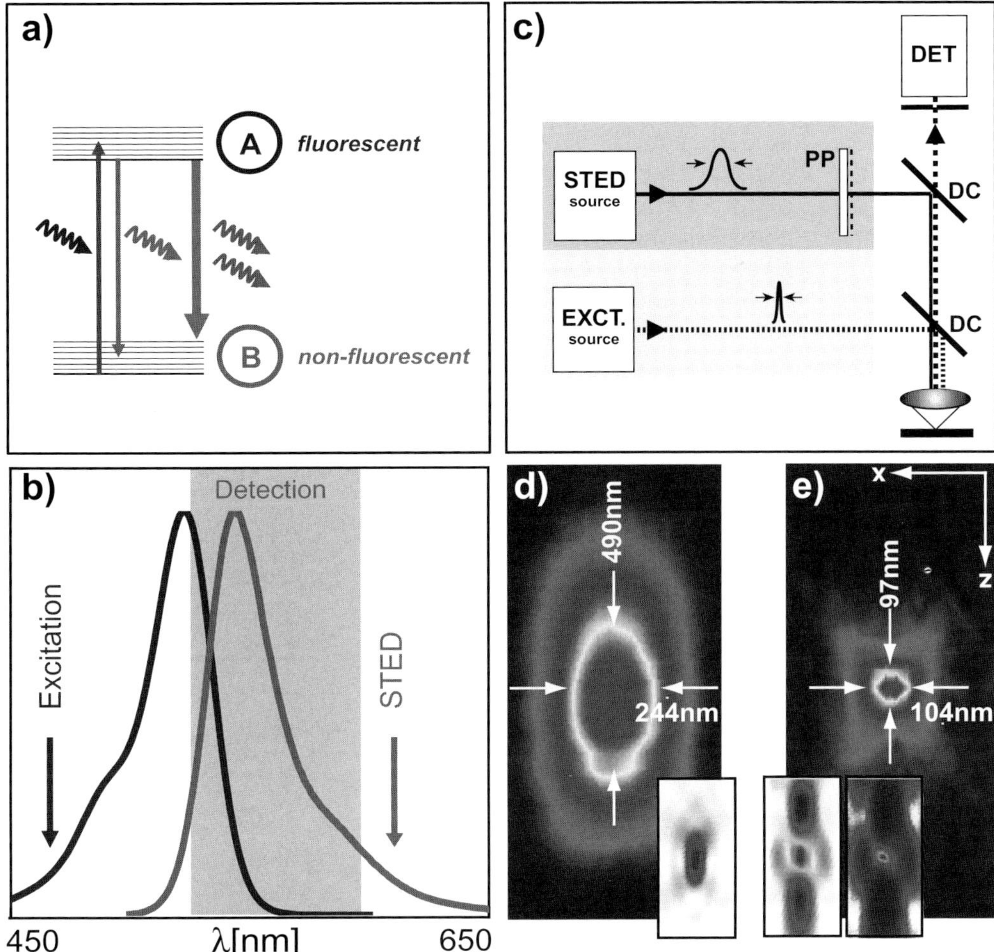

Figure 12–7. Stimulated emission depletion (STED) was the first implementation of the RESOLFT principle. (a) Dye molecules are excited into the S_1 (state A) by an excitation laser pulse. (b) Fluorescence is detected over most of the emission spectrum. However, molecules can be quenched back into the ground state S_0 (state B) using stimulated emission before they fluoresce by irradiating them with a light pulse at the edge of the emission spectrum shortly after the excitation pulse and before they are able to emit a fluorescence photon. Saturation is realized by increasing the intensity of the depletion pulse and consequently inhibiting fluorescence everywhere except at the "zero points" of the focal distribution of the depletion light. (c) Schematic of a point-scanning STED microscope. Excitation and depletion beams are combined using appropriate dichroic mirrors (DC). The excitation beam forms a diffraction-limited excitation spot in the sample (inset in d) while the depletion beam is manipulated using a phase-plate (PP) or any other device to tailor the wavefront in such a way that it forms an intensity distribution with a nodal point in the excitation maximum (left inset in e). The third inlay shows the resulting quenching probability when saturating the depletion process. (d) and (e) show an experimental comparison between the confocal PSF and the effective PSF after switching on the depleting beam. Note the doubled lateral and five-fold improved axial resolution. The reduction in dimensions (x, y, z) yields ultrasmall volumes of subdiffraction size, here 0.67 al (Klar et al., 2000), corresponding to an 18-fold reduction compared to its confocal counterpart. The spot size is not limited on principle grounds but by practical circumstances such as the quality of the zero and the saturation factor of depletion. (See color plate.)

state S_0 (state B) with a doughnut-shaped beam. The saturated depletion of S_1 confines fluorescence to the central intensity zero. With typical
saturation intensities ranging from 1 to $100\,MW/cm^2$, saturation factors
of up to 120 have been reported (Klar et al., 2000, 2001). This should
yield a 10-fold resolution improvement over the diffraction barrier, but
imperfections in the doughnut have limited the improvement to 5 to
7-fold in experiments (Klar et al., 2001).

As already stated, light microscopy resolution can be described either
in real space or in terms of spatial frequencies. In real space, the resolution is assessed by the FWHM of the focal spot. The measurements
depicted in Figure 12–8 were carried out with an excitation wavelength
of $\lambda = 635\,nm$, an oil immersion lens with a numerical aperture of 1.4,
and with the smallest possible probe: a single fluorescent molecule
(Westphal and Hell, 2005; Westphal et al., 2003). Figure 12–8a shows
the measured profile of the PSF in the focal plane (x) for a conventional

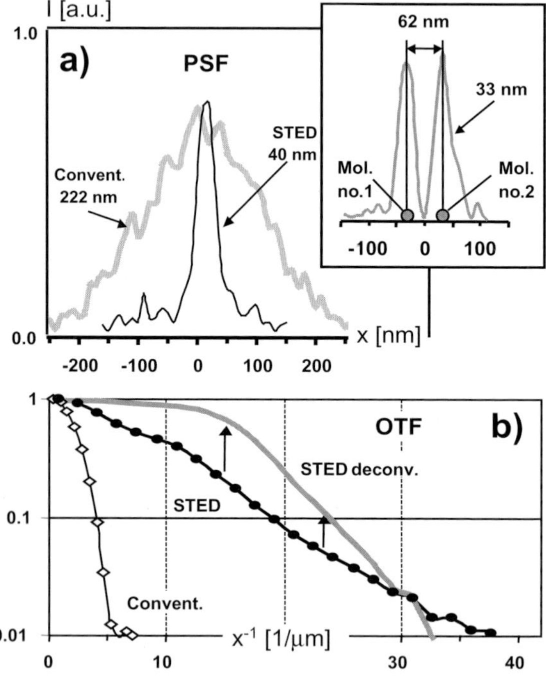

Figure 12–8. (a) Comparison of the effective PSF's lateral intensity profile for
confocal and STED microscopy indicating an ~5.5-fold resolution increase in
the latter. (b) Lateral cuts through the effective OTFs giving the bandwidth of
the lateral spatial frequencies passed to the image. The data plotted in (a) and
(b) are gained by probing the fluorescent spot of a scanning microscope with
a single molecule of the fluorophore JA 26 using a numerical aperture of 1.4
(oil) objective lens and at wavelengths of 635 nm (excitation), 650–720 nm (fluorescence collection), and 790 nm (STED). The inset demonstrates subdiffraction resolution with STED microscopy. Two identical molecules located in the
focal plane that are only 62 nm apart can be entirely separated by their intensity profile in the image. A similar clear separation by conventional microscopy would require the molecules to be at least 300 nm apart. (Date adapted
from Westphal et al.)

fluorescence microscope along with its sharper subdiffraction STED fluorescence counterpart. STED leads to improvement in resolution by a factor of approximately 5.5.

Figure 12–8b shows the OTF of a conventional microscope along with the enlarged OTF of the STED fluorescence microscope. As expected, the effective OTF's support in the confocal case ends at approximately $(2/635\,nm + 2/720\,nm) = 5.91/\mu m$. For STED, we included the OTF after successful linear deconvolution, which restores higher spatial frequencies that are not swamped by noise. The region of usable OTF support is approximately marked by the region where frequencies are enhanced by the deconvolution process without producing artifacts and is ~5.5 times larger than for the confocal case. This marks a fundamental breaking of Abbe's diffraction barrier in the focal plane. The inlay demonstrates the resulting subdiffraction resolution exemplified by the linearly deconvolved STED image of two molecules at a distance of 62 nm. They are distinguished in full by two narrow peaks (Westphal et al., 2003). As a result of deconvolution, the individual peaks are sharper (33 nm FWHM) than the initial peak of 40 nm FWHM.

Very recently, utilizing STED wavelengths of $\lambda = 750$–$800\,nm$, a lateral FWHM of down to 16 nm has been achieved in experiments with single JA 26 molecules spin-coated on a glass slide.[14] Measuring the resolution as a function of the applied STED intensity confirmed the predicted increase of the resolving power with the square root of the saturation factor (see Figure 12–9). Of course the cutoffs presented are based on a somewhat arbitrary definition of what can be considered "usable frequencies" at a certain signal-to-noise ratio. However low this threshold is set, the confocal support cannot extend beyond ~6/μm while the STED-OTF's support is theoretically unlimited.

The one-dimensional (1 D) phase-plate yielding 16 nm FWHM is optimized for maximum resolution improvement in the lateral direction perpendicular to the polarization of the depleting light and leads to an intensity distribution with two strong peaks at either side of the excitation maximum (Keller et al., in preparation). Because the depleting light is polarized, the resolution gain depends on the orientation of the molecules. However, a considerable increase in resolution is still possible for the second phase-plate, which yields a doughnut-shaped intensity distribution and thus an almost isotropic resolution increase in the lateral directions when using circularly polarized light.

To "squeeze" the fluorescence spot in both lateral directions two STED beams aberrated with 1 D phase-plates oriented at 90° to each other can be combined. Together with circularly polarized excitation, almost uniform resolution in the focal plane is achieved as shown in Figure 12–10. A series of xy images acquired with different STED beam powers demonstrate the resolution increase and concomitant widening of the OTF when the applied saturation factor increases (Schönle et al., in preparation). This combination of two incoherent beams causes the resolution to depend on the orientation of the transition dipole and results in spikes along the x and y direction of the OTF when imaging randomly oriented fluorophores (see Figure 12–10). New phase-plates have been proposed to avoid such effects and to improve the effective saturation factors at a given total STED power. Incoherent combination can then be used to improve the resolution in all three spatial dimensions. The resulting PSFs exhibit very weak dependence on dipole orientation (Keller et al., in preparation) and allow application of STED to the imaging of biological specimen and reliable subsequent linear deconvolution (Willing et al., in preparation).

STED microscopy has also been successfully applied to the imaging of biological samples. Subdiffraction images with three-fold enhanced axial and doubled lateral resolution have been obtained with membrane-labeled bacteria and live budding yeast cells (Klar et al., 2000). While there is some evidence for increased nonlinear photobleaching of some dyes when increasing the depletion intensity (Dyba and Hell, 2003), there is no reason to believe that the intensities currently applied would be detrimental to live cells. This is not surprising since the intensities are two to three orders of magnitude lower than those used in multiphoton microscopy (Denk et al., 1990). Moreover, STED has proven to be single molecule sensitive, despite the proximity of the

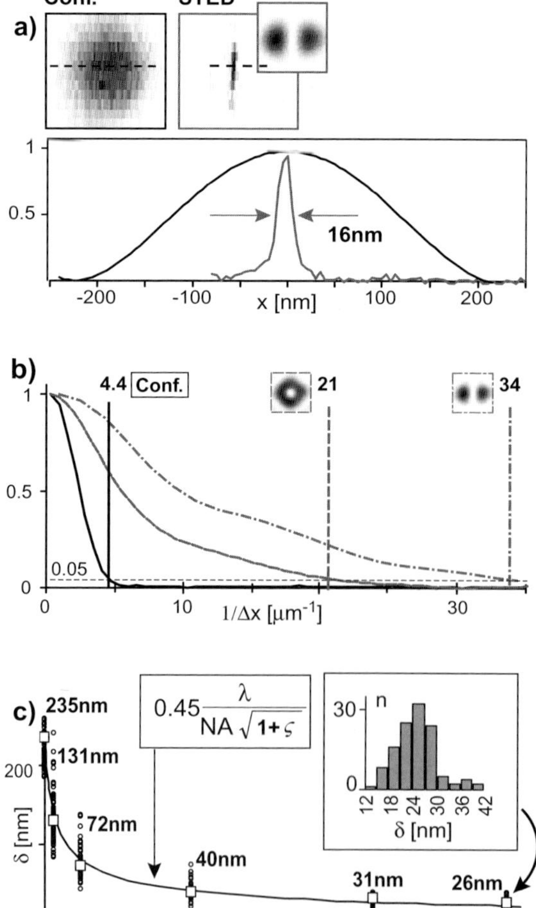

Figure 12–9. STED microscopy reduces the fluorescence focal spot size to a degree far below the diffraction limit: (a) spot of a confocal microscope (left) compared with that in an STED microscope (right) utilizing a y-oriented intensity valley for STED (upper right inset, not to scale) squeezing the spot in the x direction to 16 nm width. (b) As also observed in Figure 12–8, the bandwidth in STED is fundamentally increased over confocal microscopy. The graph shows the normalized magnitude of optical transfer function (OTF) as a function of inverse distance. For the "1D" depletion scheme, the usable support of the OTF is increased by almost a factor of 8. When using a doughnut-shaped depletion beam with a "wider" intensity zero, the OTF support is still extended almost five-fold. (c) The average focal spot size decreases with the STED intensity following a square-root law, in agreement with Eq. (27). Because the resolution depends on molecule orientation, the spot sizes were measured for several tens of single molecules. The curves follow the mean values (squares) and the inset discloses the histogram of the measured spot sizes at 1100 (MW/cm²) with the minimum FWHM at 16 nm and a 26 nm average.

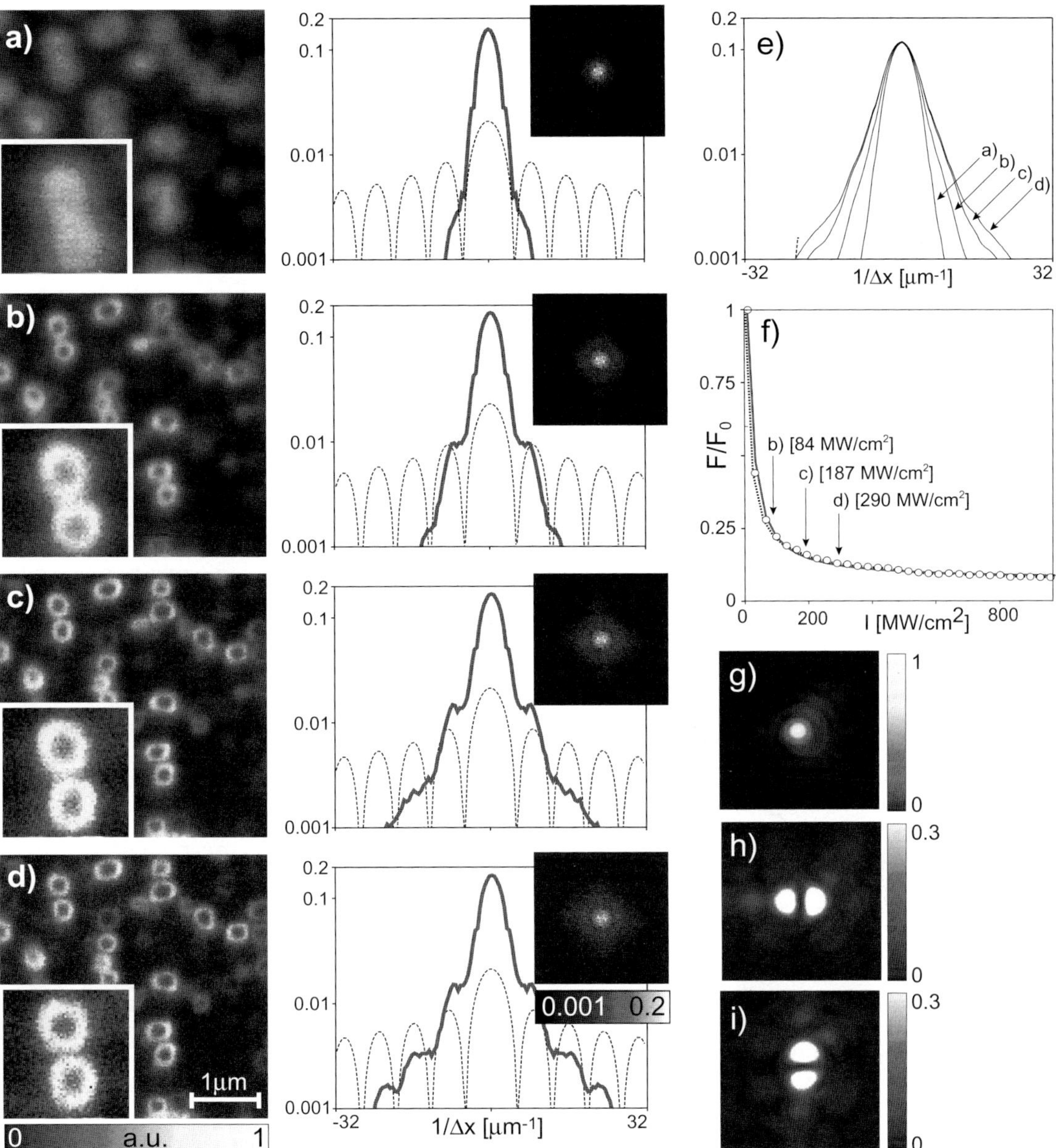

Figure 12–10. Images of a wetted Al_2O_3 matrix featuring z-oriented holes (Whatman plc, Brentford, UK) with a spin cast of a dyed (JA 26) polymethyl methacrylate solution. The rings formed in this way are ~250 nm in diameter and are barely resolved in confocal mode. (a–d) The confocal image (a) and STED images with two depleting beams perpendicularly polarized and aberrated by "1D" phase-plates (b–d). The excitation PSF (g) and the STED PSF for y polarization (h) and x polarization (i) are shown on the right. The STED intensity was chosen at the spots marked in the saturation curve (f). The smaller effective spot size also results in an extended OTF as seen in the second column. Here, the insets show the 2D Fourier transformation of the images in the left and the graphs show a profile along the x direction. Note the logarithmic scales. The Fourier transform of the image is given by the product of OTF and the Fourier transform of the object [Eq. (2)]. For such regular structures, an estimate for the modulus of the OTF can therefore be gained by estimating the latter and solving for the OTF. The dashed line shows the Fourier transform of a ring with a diameter of 275 nm and a width of 50 nm and the estimated OTF is presented in (e). (f) The suppression of fluorescence resulting from stimulated emission. The phase-plates were removed and the ratio of fluorescence without STED light (F_0) and with the STED beams switched on (F) was recorded. The intensities are pulse intensities per beam at the global maximum. (See color plate.)

STED wavelength to the emission peak. In fact, individual molecules have been switched on and off by STED upon command (Kastrup and Hell, 2004; Westphal et al., 2003).

The power of STED and 4Pi microscopy has been synergistically combined to demonstrate for the first time an axial resolution of 30–40 nm in focusing light microscopy (Dyba and Hell, 2002). The intensity distribution of the depleting light is formed by a 4Pi setting with destructive interference at the geometric focus leading to a zero intensity there and two neighboring maxima at a distance of approximately $\lambda/4$. This results in superior xz images, and the technique has initially been successfully applied to membrane-labeled bacteria (Dyba and Hell, 2002). More recently, STED-4Pi microscopy has been extended to immunofluorescence imaging (Figure 12–11). A spatial resolution of ~50 nm has been demonstrated in the imaging of the microtubular meshwork of a mammalian cell.[76] These results indicate that the basic physical obstacles to attaining a 3D resolution of the order of a few tens of nanometers have been overcome. Since the samples were mounted in an aqueous buffer (Dyba and Hell, 2002; Dyba et al., 2003), the results indicate that the optical conditions for obtaining subdiffraction resolution are met under the physical conditions encountered in live cell imaging.

It is to be expected that ultrasmall detection volumes created by STED will also be useful in a number of sensitive bioanalytical techniques. Fluorescence correlation spectroscopy (FCS) (Magde et al., 1972) relies on small focal volumes to detect rare molecular species or interactions in concentrated solutions (Eigen and Rigler, 2001; Elson and Rigler, 2001). While volume reduction can be obtained by nanofabricated structures (Levene et al., 2003), STED may prove instrumental in attaining ultrasmall spherical volumes at the nanoscale inside samples that do not allow for mechanical confinement. The latter fact is particularly important to avoid an alteration of the measured fluctuations by the nanofrabricated surface walls.

In fact, the viability of STED FCS has recently been shown in an experiment (Kastrup et al., 2005). In a particular implementation STED FCS has witnessed a reduction of the focal volume by a factor of five along the optic axis and a concomitant reduction of the axial diffusion time. The initial experiments showed that for particular dyewavelength combinations the evaluation of the STED FCS data might be complicated by a seemingly uncorrelated background at the outer wings of the fluorescence spot where STED may not completely suppress the signal. Further investigations will show whether this challenge is easily overcome in the near future. In any case, published results suggest a further decrease of the volume by another order of magnitude (Westphal et al., 2003; Irie et al., 2002). An inherent disadvantage of STED is the necessity of an additional pulsed light train that is tuned to the red edge of the emission spectrum of the dye. Nevertheless STED is to date the only known method for "squeezing" a fluorescence volume to the zeptoliter scale without making mechanical contact. Thus, the creation of ultrasmall volumes, tens of nanometers in diameter, by STED may be a pathway to improving the sensitivity of fluorescence-based bioanalytical techniques (Weiss, 2000; Laurence and Weiss, 2003).

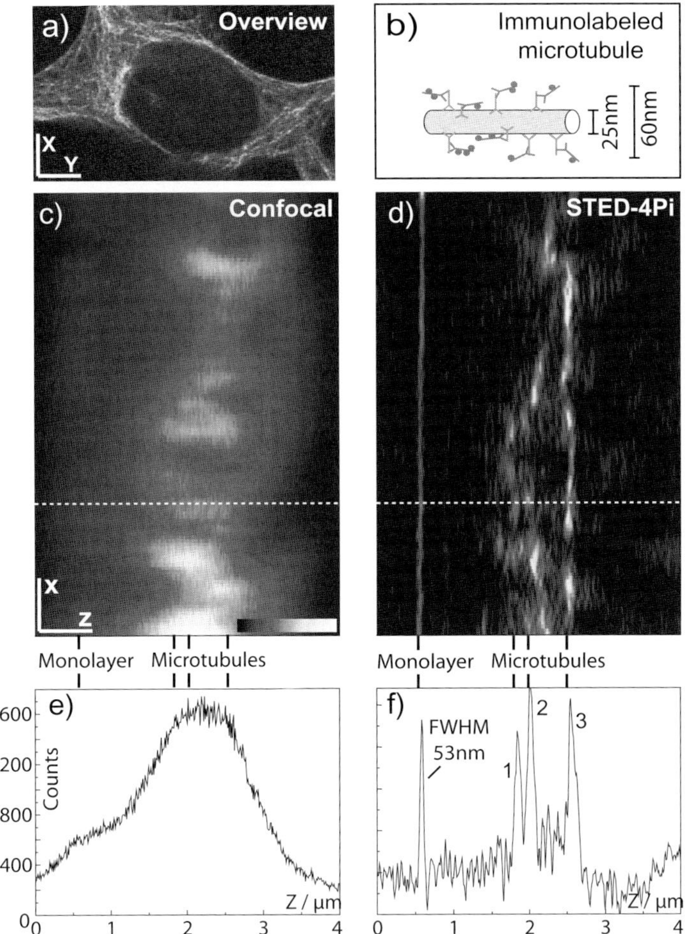

Figure 12–11. Subdiffraction immunofluorescence imaging with STED-4Pi microscopy. (a) Overview image (xy) of the microtubular network of an HEK cell. (b) Sketch of typical dimensions of a labeled microtubule fluorescently decorated via a secondary antibody. (c) and (d) Standard confocal and STED-4Pi xz image recorded at the same site of the cell; the straight line close to the cell stems from a monomolecular fluorescent layer attached to the adjacent coverslip. In both images, the pixel size was 95 × 9.8 nm in the x and z direction, respectively; the dwell time per pixel was 2 ms. Note the fundamentally improved clarity in (d). The STED-4Pi microscope's PSF features two low side lobes caused by the secondary minima STED intensity distribution. These lobes are <25% and were removed in the STED-4Pi image using linear filtering as outlined in the text [see Eq. (16ff)]. (e) and (f) Corresponding profiles of the image data along the dashed lines in (b) and (c) quantify the improved axial resolution of the STED-4Pi microscopy mode (f) over the confocal benchmark. Peaks 1, 2, and 3 due to microtubules are broader than the response to the monolayer. Note the ability of the STED-4Pi microscope to distinguish adjacent features. (See color plate.)

An important step toward far-reaching applicability of STED microscopy was the demonstration of the suitability of laser diodes both for excitation and for depletion (Westphal et al., 2003). However, several issues remain to be addressed. Due to the considerably smaller detection volumes, the signal per pixel is reduced and the amount of pixels to be recorded increases. Therefore, it will be important to incorporate STED into fast, ideally parallelized scanning systems.

While the transition to shorter wavelengths will further increase resolution by a factor of ~1.5, it is most likely that the ultimate resolution limit in STED will be set by the stability of the marker used. The photostability of current markers was considerably improved by stretching the depleting pulse to >300 ps (Dyba and Hell, 2003), but it might not be easily possible to attain saturation factors $\zeta > 200$ in the near future. Nevertheless, according to Eq. (27) $\zeta = 200$ should already yield an improvement by one order of magnitude, provided that the actual intensity value at the "intensity zero" is indeed negligible at this saturation level.

As explained above (Eq. 30), the actual "depth" of the zero codetermines the attainable resolution, because for relatively high saturation factors the saturable transition also becomes effective at the zero point or points. So far, typical depths were in the range of $\gamma = 1$–2.5% of the global maximum of the depleting intensity $I(r)$. The zero could be a single point, as in a single beam scanning system, but in the case of a parallelized system, it may also be a line or an array of points or lines. The actual depth of the zeros will certainly depend on the particular setup and the quality of optical components and proper alignment. Independently of implementation details, active optical elements such as wavefront phase modulators will be a valuable tool to further "deepen" the zeros, which in turn will allow the full potential of the attained saturation level to be exploited for improvement in resolution.

4.2 Variations of RESOLFT Microscopy and Producing Large Saturation Factors at Low Power

At this point, we reiterate that RESOLFT is not restricted to the process of stimulated emission, but can exploit any reversible (linear) transition driven by light; the attainable resolution is determined by the ratio of driving intensity and the competing transition rate k_{BA}. If the applicable intensity is limited by the onset of photodamage to the marker or even to the sample, marker constructs must be found where high saturation levels are attained at lower intensities. This is certainly the case if the rate competing with the transition to be saturated is lower.

One such example is the GSD mentioned earlier. In this version of the RESOLFT concept the ground state (now state A) is depleted by targeting an excited state (B) with a comparatively long lifetime (Hell and Kroug, 1995; Hell, 1997), such as the meta-stable triplet state T_1. In many fluorophores T_1 can be reached through the S_1 with a quantum efficiency of 1–10% (Lakowicz, 1983). Being a forbidden transition, the relaxation of the T_1 is 10^3–10^5 times slower than that of the S_1, thus yielding $I_s = 0.1$–$100 \, kW/cm^2$. The signal to be measured (from the intensity zero) is the fluorescence of the molecules that remained in the singlet system; this measurement can be accomplished through a synchronized further excitation (Hell and Kroug, 1995). For many fluorophores, this approach is not straightforward, because T_1 is involved in the process of photobleaching, but there are potential alternatives such as the meta-stable states of rare earth metal ions that are fed through chelates.

Also proposed has been depleting the ground state S_0 by populating the S_1 (now B) (Heintzmann et al., 2002). This is the technically simplest realization of saturated depletion, since it requires only excitation wavelength matching. However, as the fluorescence emission maps the spatially extended "majority population" in state B, the superresolved images (represented by state A) are negative images hidden under a bright signal from B. Hence photon noise from the large signal might swamp the fluorescence minima that occur when intensity zeros, where no fluorescence is excited, are colocalized with fluorophores. The subsequent computational extraction of the positive image is therefore very dependent on an excellent signal-to-noise ratio. The saturation intensity is of the same order as in STED microscopy, because the saturation of fluorescence also competes with the spontaneous decay of S_1. This probably results in photostability issues similar to the case of STED. In fact, the photobleaching should be exacerbated, since the saturated transition is effected with higher energy photons that are generally more prone to facilitating photochemical reactions. Pumping the dye to a higher state rather than into the ground state also favors photolability. Moreover, the fact that a large number of dye molecules constantly undergo excitation–emission cycles to image a comparatively small spot adds to the problem. Finally, saturation of the S_1 will be possible only if the long-lived triplet state is not allowed to build up during repeated excitation. As most dyes feature a triple relaxation rate of $>1\,\mu s$ (that strongly depends on the environment), effective triplet relaxation requires a pulse repetition rate $<500\,kHz$. Nevertheless, due to the simplicity of raw data acquisition it may remain an attractive method for the imaging of very bright and photostable samples.

One possible solution to the quest for large saturation factors at low intensities should be compounds with two (semi)stable states (Hell et al., 2003; Dyba and Hell, 2002). If the rate k_{BA} (and the spontaneous rate k_{AB}) almost vanish, large saturation factors are attained at very low intensities. The lowest useful intensity is then determined by the slowest acceptable imaging speed, which is ultimately determined by the switching *rate*. A favorable aspect is that in most bistable compounds the speed of the actual switching mechanism, i.e., of the conformational change, is less than a few nanoseconds, which is much faster than the typical pixel dwell time in scanning. In the ideal case, the marker indeed is a bistable fluorescent compound that can be photoswitched at separate wavelengths, from a fluorescent state A to a dark state B, and vice versa, where spontaneous rates will not influence this compromise.

Recently, a photoswitchable coupled molecular system, based on a photochromic diarylethene derivative and a fluorophore, has been reported (Irie et al., 2002). Using the kinetic parameters reported, Eq. (27) predicts that focusing of less than $100\,\mu W$ of deep-blue "switchoff light" to an area of $10^{-8}\,cm^2$ for $50\,\mu s$ should yield better than $5\,nm$ spatial resolution. Future targeted optimization of photochromic or other compounds to fatigue-free switching and visible light operation could therefore open up radically new avenues in microscopy and data storage (Hell et al., 2003).

For live cell imaging, fluorescent proteins have many advantages over synthetic dyes. Many of them feature dark states with light-driven transitions (Hell, 1997; Hell et al., 2003). If the spontaneous lifetimes of these states are longer than 10 ns such proteins may permit much larger saturation factors than those used in STED microscopy today. The most attractive solution, however, involves fluorescent proteins that can be "switched on and off" at different wavelengths (Hell et al., 2003). An example is as FP595 (Lukyanov et al., 2000), insertion of the published data into Eq. (27) predicts saturated depletion of the fluorescence state with intensities of less than a few W/cm^2 and, under favorable switching conditions, spatial resolutions of better than 10 nm.[12] The involved intensities should also enable parallelization of saturation through an array of minima or dark lines.

Initial realization of very low-intensity depletion microscopes may be challenged by switching fatigue (Irie et al., 2002) and overlapping action spectra (Lukyanov et al., 2000). Nevertheless, the prospect of attaining nanoscale resolution with regular lenses and focused light is an incentive to surmount these challenges by strategic fluorophore modification (Hell et al., 2003) and this or similar types of fluorescent proteins are a good starting point for these efforts.

5 Conclusion

The coherent use of opposing lenses enables the axial resolution of a far-field microscope to be improved by a factor of 3–7. The improvement in resolution occurs if the *spherical* wavefronts of illumination are coherently added at the focus, or the emerging spherical wavefronts of fluorescence are coherently added at the detector, because in both cases the total aperture of the system is enlarged. The latter fact is the basic tenet of the concept of 4Pi microscopy. The mere implementation of interference of low aperture (or flat) wavefronts is insufficient, because the "4Pi concept" of aperture increase is the underlying physical element of the improvement in resolution. Consequently, the success of increasing the axial resolution by two opposing lenses fully relies on large aperture angles and on the degree to which the aperture angle is utilized in the particular implementation.

Adding the spherical wavefronts both for illumination and detection, (multiphoton) 4Pi confocal microscopy of type C utilizes the lenses' aperture as much as possible. As a result, this imaging mode features a contiguous and weakly modulated effective OTF that is robust enough for live cell imaging in an aqueous environment. On the other hand, I^5M uses a mutually incoherent set of flat standing waves with varying spatial frequency for illumination. To be viable, I^5M adds the spherical wavefronts of fluorescence emission at the detector; in other words, it uses a 4Pi scheme for detection. The compromise in the illumination path with regard to the "4Pi concept" bestows the OTF of the I^5M with internal regions of very weak frequency transfer. The I^5M is therefore more prone to artifacts and probably not reliably applicable in live cells.

With the 4Pi idea as the key physical element in the process, it is not surprising that both 4Pi microscopy and I^5M would benefit from lenses with higher semiaperture angles than those currently available. An increase of the semiaperture angle even by only a few degrees would make a large difference in I^{5}M. This difference would be even more decisive than in 4Pi microscopy, which already exploits the available aperture to the highest possible degree.

Although 4Pi microscopy has demonstrated a 3D resolution in the 100 nm range following deconvolution, the method is still limited by diffraction. The latter is no longer the case with the emerging approaches exploiting a reversible saturable transition between two states of a marker, which we termed RESOLFT. In a microscope using the RESOLFT principle, the resolution is no longer limited by diffraction but by the attainable level of saturation of a (linear) optical transition in the marker molecule. Therefore the "hard" theoretical resolution barrier is replaced by a "soft" barrier determined solely by practical conditions such as available laser power, cross sections, and the stability of the dye and the sample. The enhanced resolution has already been demonstrated (Hänninen, 2002) in a number of imaging experiments.

It is important to realize that while an effectively nonlinear interaction between light and dye is the basis for this resolution increase, these methods do not require transitions involving more than one photon at a time, such as m-photon excitation, mth harmonics generation, or coherent anti-Stokes–Raman scattering (Sheppard and Kompfner, 1978; Shen, 1984). This means that the required intensities are not determined by the very small cross sections of these processes and the requirement for ultrahigh (peak) intensities. By contrast, the saturation of a linear optical transition depends on the basic kinetics of the population of the involved states. Consequently, pulse-length requirements are less strict and, most importantly, the required intensities can be significantly reduced by choosing appropriate spectroscopic systems.

To date STED microscopy is the most advanced implementation of the RESOLFT principle. It has been well characterized by its application to single fluorescent molecules and has been applied to imaging fixed and live, albeit simple biological specimens. Improvements in resolution by a factor of up to eight have already been demonstrated and the resulting fluorescent volumes are the smallest that have ever been created with focused light. The results hitherto obtained with STED must not be considered as a new limit but as proof of the viability of the concept of RESOLFT and STED microscopy in particular. Because this method is very young, future research on spectroscopy conditions and on practical aspects (Stephens and Allen, 2003) should lead to further improvements.

Nevertheless, due to the rather fast relaxation of the excited state STED requires intensities of the order of at least several tens of MW/cm^2. Hence, there will be a practical limit to the applicable intensity and to the saturation level attainable under practical conditions. Intriguingly, the intensities required for obtaining a high saturation level can be fundamentally lowered in other implementations of the RESOLFT concept. This is particularly true when utilizing optically bistable markers, such as photoswitchable dyes and photochromic fluorescent proteins. Both are very promising candidates for providing high levels of saturation at ultralow intensities of light (Hell et al., 2003). In fact, we expect them to play an important role in translating nanoscale resolution into the noninvasive, all-optical imaging of live cells. Dedicated synthesis or protein engineering might eventually uncover a whole new range of suitable markers.

Light microcopy is still commonly portrayed as fundamentally resolution limited. However, about a decade ago, concepts emerged that broke the diffraction barrier postulated by Abbe in 1873. These developments are poised to radically extend the field of application for far-field light microscopy and eventually lead to far-field "nanoscopes" operating with regular lenses and visible light.

Acknowledgments. The authors thank all members of the Department of NanoBiophotonics for contributions to this work and valuable discussions. Much of the results described in this chapter have been adopted from original work with significant contributions by A. Egner, M. Nagorni (4Pi), M. Dyba, V. Westphal, B. Harke, and J. Keller (STED). We thank J. Jethwa, J. Bewersdorf, and G. Donnert for valuable discussions and critical reading of the manuscript.

References

Abbe, E. (1873). Beiträge zur Theorie des Mikroskops und der mikroskopischen Wahrnehmung. *Arch. Mikr. Anat.* **9**, 413–420.

Albrecht, B., Failla, A.V., Schweizer, A. and Cremer, C. (2002). Spatially modulated illumination microscopy allows axial distance resolution in the nanometer range. *Appl. Opt.* **41**(1), 80–87.

Bahlmann, K. and Hell, S.W. (2000). Polarization effects in 4Pi confocal microscopy studied with water-immersion lenses. *Appl. Opt.* **39**(10), 1653–1658.

Bahlmann, K., Jakobs, S. and Hell, S.W. (2001). 4Pi-confocal microscopy of liver cells. *Ultramicroscopy* **87**, 155–164.

Bailey, B., Farkas, D.L., Taylor, D.L. and Lanni, F. (1993). Enhancement of axial resolution in fluorescence microscopy by standing-wave excitation. *Nature* **366**, 44–48.

Bertero, M., Boccacci, P., Brakenhoff, G.J., Malfanti, F. and Van der Voort, H. T.M. (1990). Three-dimensional image restoration and super-resolution in fluorescence confocal microscopy. *J. Microsc.* **157**, 3–20.

Blanca, C.M., Bewersdorf, J. and Hell, S.W. (2002). Determination of the unknown phase difference in 4Pi-confocal microscopy through the image intensity. *Opt. Commun.* **206**, 281–285.

Bloembergen, N. (1965). *Nonlinear Optics.* (Benjamin, New York).

Born, M. and Wolf, E. (1993). *Principles of Optics*, 6th ed. (Pergamon Press, Oxford).

Carrington, W.A., Lynch, R.M., Moore, E.D.W., Isenberg, G., Fogarty, K.E. and Fay, F.S. (1995). Superresolution in three-dimensional images of fluorescence in cells with minimal light exposure. *Science* **268**, 1483–1487.

Denk, W., Strickler, J.H. and Webb, W.W. (1990). Two-photon laser scanning fluorescence microscopy. *Science* **248**, 73–76.

Dyba, M. and Hell, S.W. (2002). Focal spots of size 1/23 open up far-field fluorescence microscopy at 33 nm axial resolution. *Phys. Rev. Lett.* **88**, 163901.

Dyba, M. and Hell, S.W. (2003). Photostability of a fluorescent marker under pulsed excited-state depletion through stimulated emission. *Appl. Opt.* **42**(25), 5123–5129.

Dyba, M., Jakobs, S. and Hell, S.W. (2003). Immunofluorescence stimulated emission depletion microscopy. *Nature Biotechno.* **21**(11), 1303–1304.

Egner, A., Jakobs, S. and Hell, S.W. (2002). Fast 100-nm resolution 3D-microscope reveals structural plasticity of mitochondria in live yeast. *Proc. Natl. Acad. Sci. USA* **99**, 3370–3375.

Eigen, M. and Rigler, R. (1994). Sorting single molecules: Applications to diagnostics and evolutionary biotechnology. *Proc. Natl. Acad. Sci. USA* **91**, 5740–5747.

Elson, E.L. and Rigler, R. Eds. (2001). *Fluorescence Correlation Spectroscopy. Theory and Applications.* (Springer, Berlin).

Failla, A.V., Spoeri, U., Albrecht, B., Kroll, A. and Cremer, C. (2002). Nanosizing of fluorescent objects by spatially modulated illumination microscopy. *Appl. Opt.* **41**(34), 7275–7283.

Freimann, R., Pentz, S. and Hörler, H. (1997). Development of a standing-wave fluorescence microscope with high nodal plane flatness. *J. Microsc.* **187**(3), 193–200.

Göpper-Mayer, M. (1931). Über Elementarakte mit zwei Quantensprüngen. *Ann. Phys. (Leipzig)* **9**, 273–295.

Goodman, J.W. (1968). *Introduction to Fourier Optics.* (McGraw-Hill, New York).

Gugel, H., Bewersdorf, J., Jakobs, S., Engelhardt, J., Storz, R. and Hell, S.W. (2004). Cooperative 4Pi excitation and detection yields 7-fold sharper optical sections in live cell microscopy. *Biophys. J.* **87**, 4146–4152.

Gustafsson, M.G., Agard, D.A. and Sedat, J.W. (1996). 3D widefield microscopy with two objective lenses: Experimental verification of improved axial resolution. In: *Three-Dimensional Microscopy: Image Acquisition and Processing III.* Proc. SPIE.

Gustafsson, M.G.L. (1999). Extended resolution fluorescence microscopy. *Curr. Opin. Struct. Biol.* **9**, 627–634.

Gustafsson, M.G.L. (2000). Surpassing the lateral resolution limit by a factor of two using structured illumination microscopy. *J. Microsc.* **198**(2), 82–87.

Gustafsson, M.G.L., Agard, D.A. and Sedat, J.W. (1995). Sevenfold improvement of axial resolution in 3D widefield microscopy using two objective lenses. *Proc. SPIE* **2412**, 147–156.

Gustafsson, M.G.L., Agard, D.A. and Sedat, J.W. (1999). I5M: 3 widefield light microscopy with better than 100 nm axial resolution. *J. Microsc.* **195**, 10–16.

Hänninen, P. (2002). Beyond the diffraction limit. *Nature* **419**, 802.

Hänninen, P.E., Lehtelä, L. and Hell, S.W. (1996). Two- and multiphoton excitation of conjugate dyes with continuous wave lasers. *Opt. Commun.* **130**, 29–33.

Hecht, B., Bielefledt, H., Inouyne, Y., Pohl, D.W. and Novotny, L. (1997). Facts and artifacts in near-field optical microscopy. *J. Appl. Phys.* **81**, 1492–2498.

Heintzmann, R. and Cremer, C. (1998). Laterally modulated excitation microscopy: Improvement of resolution by using a diffraction grating. *SPIE Proc.* **3568**, 185–195.

Heintzmann, R., Jovin, T.M. and Cremer, C. (2002). Saturated patterned excitation microscopy—a concept for optical resolution improvement. *J. Opt. Soc. Am. A: Opt. Image Sci. Vision* **19**(8), 1599–1609.

Hell, S. and Stelzer, E.H.K. (1992a). Properties of a 4Pi-confocal fluorescence microscope. *J. Opt. Soc. Am. A* **9**, 2159–2166.

Hell, S.W. (1990). Double-scanning confocal microscope. European Patent.

Hell, S.W. (1997). Increasing the resolution of far-field fluorescence light microscopy by point-spread-function engineering. In *Topics in Fluorescence Spectroscopy* (J.R. Lakowicz, Ed.), 361–422 (Plenum Press, New York).

Hell, S.W. (2003). Toward fluorescence nanoscopy. *Nature Biotechnol.* **21**(11), 1347–1355.

Hell, S.W. (2004). Strategy for far-field optical imaging and writing without diffraction limit. *Phys. Lett. A* **326**(1–2), 140–145.

Hell, S.W. and Kroug, M. (1995). Ground-state depletion fluorescence microscopy, a concept for breaking the diffraction resolution limit. *Appl. Phys. B* **60**, 495–497.

Hell, S.W. and Nagorni, M. (1998). 4Pi-confocal microscopy with alternate interference. *Opt. Lett.* **23**(20), 1567–1569.

Hell, S.W. and Stelzer, E.H.K. (1992b). Fundamental improvement of resolution with a 4Pi-confocal fluorescence microscope using two-photon excitation. *Opt. Commun.* **93**, 277–282.

Hell, S.W. and Wichmann, J. (1994). Breaking the diffraction resolution limit by stimulated emission: Stimulated emission depletion microscopy. *Opt. Lett.* **19**(11), 780–782.

Hell, S.W., Schrader, M., Hänninen, P.E. and Soini, E. (1995). Resolving fluorescence beads at 100–200 distance with a two-photon 4Pi-microscope working in the near infrared. *Opt. Commun.* **117**, 20–24.

Hell, S.W., Jakobs, S. and Kastrup, L. (2003). Imaging and writing at the nanoscale with focused visible light through saturable optical transitions. *Appl. Phys. A* **77**, 859–860.

Hell, S.W., Schrader, M. and van der Voort, H.T.M. (1997). Far-field fluorescence microscopy with three-dimensional resolution in the 100nm range. *J. Microsc.* **185**(1), 1–5.

Holmes, T.J. (1988). Maximum-likelihood image restoration adapted for noncoherent optical imaging. *JOSA A* **5**(5), 666–673.

Holmes, T.J., Bhattacharyya, S., Cooper, J.A., Hanzel, D., Krishnamurthi, V., Lin, W., Roysam, B., Szarowski, D.H. and Turner, J.N. (1995). Light microscopic images reconstruction by maximum likelihood deconvolution. In *Handbook of Biological Confocal Microscopy* (J. Pawley, Ed.) 389–400 (Plenum Press, New York).

Irie, M., Fukaminato, T., Sasaki, T., Tamai, N. and Kawai, T. (2002). A digital fluorescent molecular photoswitch. *Nature* **420**(6917), 759–760.

Kastrup, L. and Hell, S.W. (2004). Absolute optical cross section of individual fluorescent molecules. *Angew. Chem. Int. Ed.* **43**, 6646–6649.

Kastrup, L., Blom, H., Eggeling, C. and Hell, S.W. (2005). Fluorescence fluctuation spectroscopy in subdiffraction focal volumes. *Phys. Rev. Lett.* **94**(17), 178104.

Klar, T.A., Engel, E. and Hell, S.W. (2001). Breaking Abbe's diffraction resolution limit in fluorescence microscopy with stimulated emission depletion beams of various shapes. *Phys. Rev. E* **64**, 1–9.

Klar, T.A., Jakobs, S., Dyba, M., Egner, A. and Hell, S.W. (2000). Fluorescence microscopy with diffraction resolution limit broken by stimulated emission. *Proc. Natl. Acad. Sci. USA* **97**, 8206–8210.

Krishnamurthi, V., Bailey, B. and Lanni, F. (1996). Image processing in 3-D standing wave fluorescence microscopy. *Proc. SPIE* **2655**, 18–25.

Lakowicz, J.R. (1983). *Principles of Fluorescence Spectroscopy.* (Plenum Press, New York).

Lakowicz, J.R., Gryczynski, I., Malak, H. and Gryczynski, Z. (1996). Two-color two-photon excitation of fluorescence. *Photochem. Photobiol.* **64**, 632–635.

Lanni, F. (1986). *Applications of Fluorescence in the Biomedical Sciences*, 1st ed., D.L. Taylor, Ed. 520–521. (Liss, New York).

Laurence, T.A. and Weiss, S. (2003). How to detect weak pairs. *Science* **299**(5607), 667–668.

Levene, M.J., Korlach, J., Turner, S.W., Foquet, M., Craighead, H.G. and Webb, W.W. (2003). Zero-mode waveguides for single-molecule analysis at high concentrations. *Science* **299**, 682–686.

Lukosz, W. (1966). Optical systems with resolving powers exceeding the classical limit. *J. Opt. Soc. Am.* **56**, 1463–1472.

Lukyanov, K.A., Fradkov, A.F., Gurskaya, N.G., Matz, M.V., Labas, Y.A., Savitsky, A.P., Markelov, M.L., Zaraisky, A.G., Zhao, X., Fang, Y., Tan, W. and Lukyanov, S.A. (2000). Natural animal coloration can be determined by a nonfluorescent green fluorescent protein homolog. *J. Biol. Chem.* **275**(34), 25879–25882.

Magde, D., Elson, E.L. and Webb, W.W. (1972). Thermodynamic fluctuations in a reacting system—measurement by fluorescence correlation spectroscopy. *Phys. Rev. Lett.* **29**(11), 705–708.

Nagorni, M. and Hell, S.W. (1998). 4Pi-confocal microscopy provides three-dimensional images of the microtubule network with 100- to 150-nm resolution. *J. Struct. Biol.* **123**, 236–247.

Nagorni, M. and Hell, S.W. (2001a). Coherent use of opposing lenses for axial resolution increase in fluorescence microscopy. I. Comparative study of concepts. *J. Opt. Soc. Am. A* **18**(1), 36–48.

Nagorni, M. and Hell, S.W. (2001a). Coherent use of opposing lenses for axial resolution increase in fluorescence microscopy. II. Power and limitation of nonlinear image restoration. *J. Opt. Soc. Am. A* **18**(1), 48–54.

Pawley, J., Ed. (1995). *Handbook of Biological Confocal Microscopy.* (Plenum Press, New York).

Pohl, D.W. and Courjon, D. (1993). *Near Field Optics.* (Kluwer, Dordrecht).

Press, W.H., Flannery, B.P., Teukolsky, S.A. and Vetterling, W.T. (1993). *Numerical Recipes in C*, 2nd ed. (Cambridge University Press, Cambridge).

Richards, B. and Wolf, E. (1959). Electromagnetic diffraction in optical systems II. Structure of the image field in an aplanatic system. *Proc. R. Soc. Lond. A* **253**, 358–379.

Schmidt, M., Nagorni, M. and Hell, S.W. (2000). Subresolution axial distance measurements in far-field fluorescence microscopy with precision of 1 nanometer. *Rev. Sci. Instrum.* **71**, 2742–2745.

Schneider, B., Albrecht, B., Jaeckle, P., Neofotistos, D., Söding, S., Jäger, T. and Cremer, C. (2000). Nanolocalization measurements in spatially modulated illumination microscopy using two coherent illumination beams. *Proc. SPIE* **3921**, 321–330.

Schönle, A. and Hell, S.W. (1999). Far-field fluorescence microscopy with repetetive excitation. *Eur. Phys. J. D* **6**, 283–290.

Schönle, A. and Hell, S.W. (2002). Calculation of vectorial three-dimensional transfer functions in large-angle focusing systems. *J. Opt. Soc. Am. A* **19**(10), 2121–2126.

Schönle, A., Hänninen, P.E. and Hell, S.W. (1999). Nonlinear fluorescence through intermolecular energy transfer and resolution increase in fluorescence microscopy. *Ann. Phys. (Leipzig)* **8**(2), 115–133.

Schrader, M. and Hell, S.W. (1996). 4Pi-confocal images with axial superresolution. *J. Microsc.* **183**, 189–193.

Schrader, M., Bahlmann, K., Giese, G. and Hell, S.W. (1998). 4Pi-confocal imaging in fixed biological specimens. *Biophys. J.* **75**, 1659–1668.

Shen, Y.R. (1984). *The Principles of Nonlinear Optics*, 1st ed. (John Wiley, New York).

Sheppard, C.J.R. and Kompfner, R. (1978). Resonant scanning optical microscope. *Appl. Optics* **17**, 2879–2882.

Sheppard, C.J.R., Gu, M., Kawata, Y. and Kawata, S. (1993). Three-dimensional transfer functions for high-aperture systems. *J. Opt. Soc. Am. A* **11**(2), 593–596.

Stephens, D.J. and Allen, V.J. (2003). Light microscopy techniques for live cell imaging. *Science* **300**, 82–91.

Toraldo di Francia, G. (1952). Supergain antennas and optical resolving power. *Nuovo Cimento Suppl.* **9**, 426–435.

Weiss, S. (2000). Shattering the diffraction limit of light: A revolution in fluorescence microscopy? *Proc. Natl. Acad. Sci. USA* **97**(16), 8747–8749.

Westphal, V., Blanca, C.M., Dyba, M., Kastrup, L. and Hell, S.W. (2003). Laser-diode-stimulated emission depletion microscopy. *Appl. Phys. Lett.* **82**(18), 3125–3127.

Westphal, V., Kastrup, L. and Hell, S.W. (2003). Lateral resolution of 28 nm ($\lambda/25$) in far-field fluorescence microscopy. *Appl. Phys. B* **77**(4), 377–380.

Westphal, V.H. and Hell, S.W. (2005). Nanoscale resolution in the focal plane of an optical microscope. *Phys. Rev. Lett.* **94**(14), 143903.

Wilson, T. and Sheppard, C.J.R. (1984). *Theory and Practice of Scanning Optical Microscopy.* (Academic Press, New York).

Xu, C., Zipfel, W., Shear, J.B., Williams, R.M. and Webb, W.W. (1996). Multiphoton fluorescence excitation: New spectral windows for biological nonlinear microscopy. *Proc. Natl. Acad. Sci. USA* **93**, 10763–10768.

13

Principles and Applications of Zone Plate X-Ray Microscopes

Malcolm Howells, Chris Jacobsen, and Tony Warwick

1 Introduction

1.1 Background

In the 1949 issue of *Scientific American*, an article by Stanford physicist Paul Kirkpatrick on "The X-ray Microscope" (Kirkpatrick, 1949) was described by the editors as follows:

"It would be a big improvement on microscopes using light or electrons, for X-rays combine short wavelengths, giving fine resolution, and penetration. The main problems standing in the way have now been solved."

With the perspective of a half century, we might change "improvement on" to "complement to" and say that further problems were solved after 1949, but here in essence is the character of X-ray microscopes.

In this chapter, we outline some of the properties of X-ray microscope systems in operation today, and highlight some of their present applications. We will not discuss the history of X-ray microscopes prior to about 1975 but instead refer the reader to a series of conference proceedings known as "X-ray Optics and X-ray Microanalysis," which began in 1956. Originally these had valuable material on X-ray microscopy but this diminished after about 1970. The first five were at Cambridge (1956) (Cosslett et al., 1957), Stockholm (1959) (Engström et al., 1960), Stanford (1962) (Pattee et al., 1963), Orsay (1965) (Castaing et al., 1966) and Tubingen (1968) (Molenstedt et al., 1969). We also recommend the historical perspectives by A. Baez (Baez, 1989, 1997) and the book by Cosslet and Nixon (1960). There is a recognisable thread of continuity between today's status of the field and efforts that began slowly around 1975 (Niemann et al., 1976; Parsons, 1978; Kirz and Sayre, 1980c; Parsons, 1980) and blossomed with the availability of synchrotron light sources and nanofabrication technologies; this thread can be traced in part via the proceedings of another conference series that began in 1984 (Schmahl and Rudolph, 1984a) and has continued until today (Sayre et al., 1988; Michette et al., 1992; Aristov and Erko, 1994; Thieme et al., 1998b; Meyer-Ilse et al., 2000b; Susini et al., 2003). Zone-plate X-ray microscopes now exist at roughly two dozen international synchrotron radiation research centers (see Table 13–3), and commercial lab-based instruments are also available. Three types are in especially widespread use. Transmission X-ray microscopes (TXMs) specialize in the rapid acquisition of 2D images using high flux sources, and in the collection of tilt sequences of projection images for 3D imaging by tomography. Scanning transmission X-ray microscopes (STXMs) specialize in the acquisition of reduced dose images and point spectra with high energy resolution for elemental and chemical state mapping, and require high source brightness. Scanning fluorescence X-ray microprobes (SFXMs) are similar to STXMs except that fluorescence X-rays are collected by energy-resolving detectors for trace element mapping. All three approaches are now working below 100 nm resolution, to the point of reaching 15 nm resolution in some demonstrations (Chao et al., 2005). While many of the new technical developments continue to be pursued by specialists in X-ray optics and microscopy, much of present-day activity comes from scientists in other fields of research who are using X-ray microscopes to address their particular questions. This chapter is mainly aimed at scientists from the latter group as well as those from the other communities represented in the content of this series of books.

1.2 X-Ray Interactions

A microscope requires illumination, magnification, and contrast. The characteristics of X-ray interactions with matter affect all three. In Figure 13–1, we show the cross-section (Hubbell et al., 1980) for photoelectric absorption, coherent (elastic or Thomson) scattering, and inco-

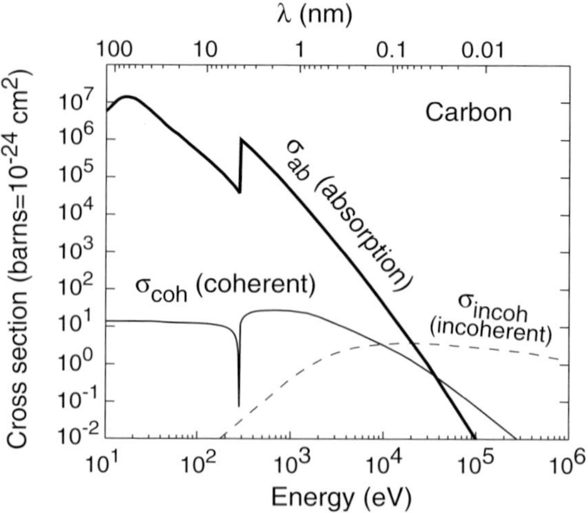

Figure 13–1. X-ray interaction cross sections in carbon. At energies below about 10 keV, absorption dominates so that images are free from the complications of multiple scattering. [Data from Henke et al. (1993) and the NIST XCOM database (Saloman and Hubbell, 1987).]

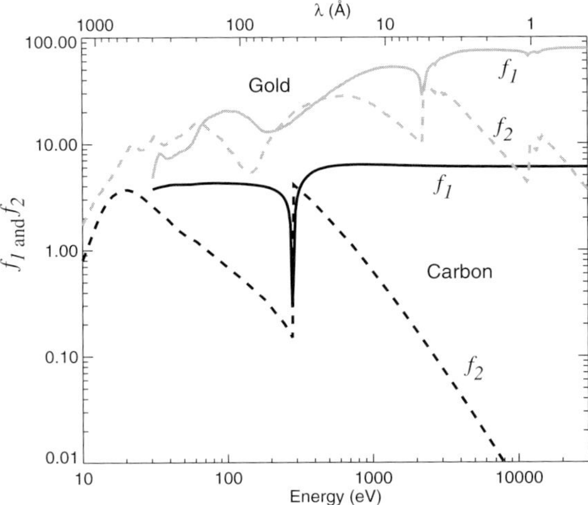

Figure 13–2. The frequency-dependent oscillator strength ($f_1 + if_2$) for carbon and gold. At X-ray absorption edges (such as 290 eV for carbon), f_2 has step increments, while f_1 undergoes anomalous dispersion resonances. (Reprinted from Henke et al., © 1993, with permission from Elsevier.)

herent (inelastic or Compton) scattering for carbon. Below 10 keV, absorption dominates, so multiple X-ray scattering is usually not of concern (an X-ray is much more likely to be absorbed following any scattering event than scattered again) nor is inelastic scattering. However, what is not evident in Figure 13–1 is the fact that the propagation of X-rays in materials can also include refractive effects, and in fact it was Einstein (1918) who first pointed out that the refractive index is slightly less than unity. The X-ray refractive index for a wave forward propagated as $\exp[-i(k\tilde{n}x - \omega t)]$ is often written as $\tilde{n} = 1 - \delta - i\beta$ where δ represents the phase-shifting part of the refractive index and β represents absorption according to a linear coefficient $\mu = 4\pi\beta/\lambda$ in the Lambert-Beer law $I = I_0 \exp[-\mu t]$. In an anomalous dispersion model, the refractive index terms can furthermore be written as $(\delta + i\beta) = \alpha\lambda^2(f_1 + if_2)$ with $\alpha = n_a r_e/2\pi$, where $n_a = \rho N_A/A$ gives the number density of atoms, $r_e = 2.82 \times 10^{-15}$ m is the classical radius of the electron, and ($f_1 + if_2$) represents the frequency-dependent oscillator strength of an atom. This oscillator strength ($f_1 + if_2$) has been tabulated with very good absolute accuracy for all elements by Henke, Gullikson et al. (Henke, 1993) over the energy range 10–30,000 eV (see Figure 13–2). In examining Figure 13–2, two features immediately jump out: f_1 is somewhat constant except near absorption edges, so the thickness $t_\pi = \lambda/2\delta = 1/(2\alpha\lambda f_1)$ needed to provide a phase advance $\exp[ik\delta t_\pi]$ equal to π increases as λ^{-1}, while, because f_2 scales as E^{-2} or λ^2, the thickness $1/\mu = 1/(4\pi\alpha\lambda f_2)$ that produces an attenuation of $1/e$ increases as E^3 or λ^{-3}. As a result, phase contrast becomes the dominant contrast mechanism as one goes to shorter wavelengths (Schmahl and Rudolph, 1987).

Much of modern X-ray microscopy centers on the exploitation of X-ray absorption edges. X-ray absorption edges arise when the X-ray photon reaches the threshold energy needed to completely remove an electron from an inner-shell orbital. The energy at which this occurs is approximately given by the Bohr model as $E_n = (13.6\,\text{eV})(Z\text{-}z_{shield})^2/n^2$, where Z is the atomic number, z_{shield} approximates the partial screening of the nucleus' charge by other inner-shell electrons ($z_{shield} \approx 1$ for K edges), and n is the principal quantum number ($n = 1$ for K edges, 2 for L edges, and so on). This produces the step-like rise in the cross-section for photoelectric absorption that can be seen in the plots of f_2 in Figure 13–2. If one takes one image I_1 at an energy E_1 just below an element's absorption edge where the incident flux is I_{01}, and a similar image I_2 at an energy just above an absorption edge, one can recover the mass per area m_x/A of the element x from (Engström, 1946)

$$\frac{m_x}{A} = \rho\,\frac{(E_1/E_2)^3 \ln(I_1/I_{01}) - \ln(I_2/I_{02})}{\mu_2 - \mu_1 (E_1/E_2)^3} \tag{1}$$

This approach works well for mass concentrations greater than about 1%. Another way in which X-ray absorption edges are exploited is by means of the "water window." At X-ray energies between the carbon and oxygen absorption edges at 290 and 540 eV, respectively, organic materials show strong absorption contrast while water layers up to several μm thick are reasonably transmissive (Wolter, 1952); this is particularly valuable for imaging hydrated biological and environmental science specimens.

For those elements which have absorption edges below the energy of incident X-rays so that inner-shell ionization occurs, the aftermath of absorption involves the emission of either a fluorescent photon or an Auger electron of characteristic energy. The energy of these fluorescent photons, and the fluorescence yield (Krause, 1979) (the fraction of events which result in fluorescence rather than Auger electron emission), are both shown in Figure 13–3. At X-ray energies below 1 keV,

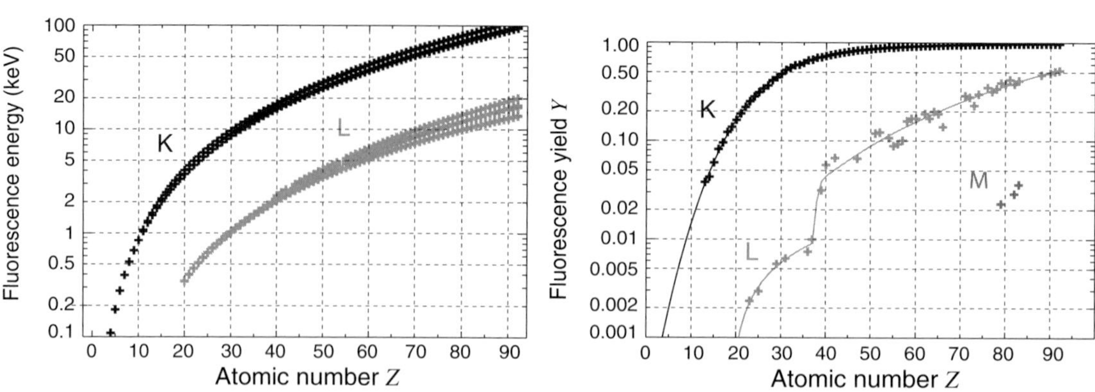

Figure 13–3. Energies (left) and fluorescence yields (right) for K and L edge emission. (Krause, 1979.)

Auger emission dominates, and scanning photoemission microscopes (SPEM) use electron spectrometers to exploit these electrons for surface studies (Ade et al., 1990a; Günther et al., 1997; Warwick et al., 1997; Ko et al., 1998). At higher energies, the fluorescence signal dominates and detection of these characteristic X-rays provides information on the concentration of various elements in the specimen. Most scanning fluorescence X-ray microprobes (SFXM) (Horowitz and Howell, 1972; Sparks, 1980) use energy dispersive detectors where the number of electron-hole pairs created by each fluorescent photon is used to measure its energy, though crystal-based wavelength dispersive spectrometers can also be used. Exact quantitation of the elemental concentration requires accurate knowledge of a number of factors, including the solid angle acceptance of the detector and its quantum efficiency, the degree to which fluorescent photons are reabsorbed in the specimen, and other factors, so that in most cases comparison is made with standards with known elemental concentration and matrix concentration similar to that of the specimen under study. When compared with electron microprobes, X-ray microprobes do not suffer from expansion of the probe beam due to electron scattering, or a large continuum background, so that the sensitivity to trace elements is often in the 100 parts per billion range.

Because X-ray interactions are well understood and do not involve significant complications due to multiple scattering at energies below about 10 keV, reliable predictions of image contrast can be made. If we have a normalized signal I_f from a feature-containing pixel and I_b from a background region, the signal to noise ratio obtained with N illuminating photons is (Glaeser, 1971; Sayre et al., 1977a)

$$\text{SNR} = \frac{\text{Signal}}{\text{Noise}} = \sqrt{N}\,\frac{I_f - I_b}{\sqrt{I_f + I_b}} = \sqrt{N}\Theta \qquad (2)$$

where we have used the Gaussian approximation to Poisson statistics (which is quite good for NI greater than about 10) and the assumption that there are no other noise sources with significant fluctuations. The contrast parameter Θ is different from the usual definition of contrast due to the square root in the denominator. With this definition, the number of photons required to see a feature with a desired signal to noise ratio SNR is given by $N = (\text{SNR})^2/\Theta^2$, and a common choice for the minimum detectable signal to noise ratio is the Rose criterion of SNR = 5 (Rose, 1946). Using this approach, Sayre *et al.* showed that "water window" X-ray microscopes are able to image organic specimens in micrometer-thick water layers with greatly reduced radiation dose compared to electron microscopy (Sayre, 1977b; Sayre, 1977a). This conclusion remains true even when modern energy-filtered electron microscopes are considered (Grimm et al., 1998; Jacobsen et al., 1998) (see Figure 13–4). Other investigators have extended the same approach to include the effects of phase contrast (Rudolph et al., 1990; Gölz, 1992) (see Figure 13–5) and the reduction of modulation transfer at high spatial frequencies (Schneider, 1998), while Kirz *et al.* have used this approach to compare elemental mapping using both differential absorption and X-ray fluorescence (Kirz et al., 1978, 1980a, 1980b).

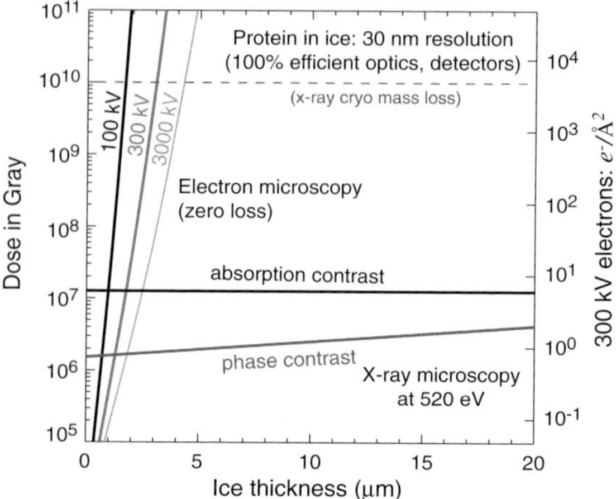

Figure 13–4. Estimated radiation dose required for imaging 30-nm-protein features as a function of ice thickness for 200 keV electron microscopes and 520 eV soft X-ray microscopes. (Reprinted from Jacobsen et al., 1998, with permission from Springer Science+Business Media.)

1.3 Focusing Optics

Microscopes require focusing optics, or some other means to provide a magnified view of the object. X-rays reflect well from single refractive interfaces only at grazing angles of incidence less than a critical angle of $\theta_c \approx \sqrt{2\delta}$ which is typically in the range of 1–5° for soft X-rays. (Once a particular angle has been selected, this same relationship gives a critical energy above which the reflectivity becomes low; this can be used to low-pass-filter the energy spectrum from a radiation source). While a number of labs have explored the use of axially symmetric paraboloid or hyperboloid optics (Wolter, 1952; Aoki, 1994), most present efforts center on the use of two orthogonal cylindrical grazing mirrors in the Kirkpatrick-Baez geometry (Kirkpatrick and Baez, 1948). Advantageous characteristics of these optics include their relatively long focal

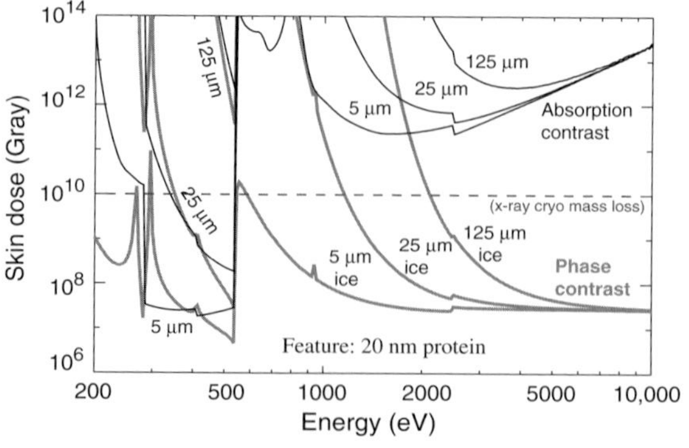

Figure 13–5. Estimated radiation dose required for imaging 20-nm-thick protein features in various ice thicknesses as a function of X-ray energy; 100% efficient optics and detectors are assumed.

length (several centimeters is typical) and their low chromaticity, so that the incident beam energy can be tuned for spectroscopy without any need to adjust the focus on the specimen. Optics of this sort have recently achieved better than 100 nm resolution probe sizes using 12 keV X-rays (Hignette et al., 2003; Yamamura et al., 2003; Mimura et al., 2004), although the profile of the focus always has some degree of "tail" outside of the geometrical image of the source due to scattering from the residual surface roughness of even the best available mirrors. Synthetic multilayer X-ray mirrors (Spiller, 1972; Barbee et al., 1981) can increase the incidence angle well beyond θ_c for narrow-bandwidth radiation, and can achieve good reflection efficiencies for normal incidence reflection at photon energies below about 200 eV. This approach has seen rapid improvements due to the development of EUV projection lithography at 95 eV. However, notwithstanding recent progress in mirror manufacture, it is important to recognize that even a perfectly made Kirkpatrick-Baez mirror system still suffers from aberrations, especially obliquity of field, which severely restrict its field of view and therefore its performance as a microscope. On the other hand, it is still well-able to focus points on or near the optical axis, which has led to a resurgence in its popularity for microprobes and relay mirrors that are imaging small sources such as synchrotrons.

When Röntgen discovered X-rays, he immediately tried to focus them using refractive lenses but without success. The reason for this is now well known: the focal length for a planoconvex lens with radius of curvature R_c is given by $f_R = -R_c/\delta$, so that at 10 keV a glass lens with $R_c = 1$ cm would have a focal length of about 2 km. This does not preclude the usefulness of refractive optics, however; a series of lenses with small R_c can be placed together to produce a significant net focusing effect. One simple way to achieve this result in 1D is to drill a series of holes in a solid block (Snigirev et al., 1996), and more recent work using parabolic optics has demonstrated a resolution of about 100 nm for hard X-ray imaging (Lengeler et al., 2002) with theoretical promise for sub-10 nm resolution imaging (Schroer and Lengeler, 2005). Because the ratio of phase shift to absorption increases with increasing X-ray energy, these optics work primarily at energies above about 5 keV, and at higher energies one will ultimately need to consider the contributions of inelastic scattering to the image due to the overall thickness of the optic. Still, this approach is of interest especially since these optics can be easily water cooled for high power applications.

The third way to focus X-rays is to use diffraction. While bent crystals can provide focused beams of Bragg or Laue diffracted X-rays, most work in X-ray microscopy centers on the use of microfabricated diffractive optics in the form of Fresnel zone plates. Efforts in X-ray microscopy using zone plate optics date back nearly a half century (Baez, 1960, 1961), and X-ray Fresnel zone plates are now benefiting from a high degree of development. Apart from detailed literature that we will cite in Section 2. general reviews can be found in the books by (Michette, 1986; Attwood, 1999). Due to their popularity as high resolution optics for X-ray microscopy, the properties of Fresnel zone plates are described in some detail below.

1.4 Overview and Recent Trends

We now review some of the general trends of X-ray microscopy technology and how they might affect the planning of a new X-ray microscopy program today. In this section, the number of relevant references is essentially unlimited so we cite instead the subsections of this article where many of the references can be found.

Modern X-ray microscopy was based on the twin ideas of the water window (Section 1.1) and microfabricated zone-plate lenses (Section 2). These led initially to life-science experiments consisting of 2D imaging of wet samples in room-temperature air with "naturalness" as the unique capability of the technique. Early versions of both the TXM (Section 3.1.1) and STXM (Section 3.1.3) were capable of such imaging and this style of X-ray microscopy has proved to be good fit to many of the needs of the polymer (Section 4.3) and environmental research communities (Section 4.2) among others. However, the room-temperature approach could not be easily adapted to 3D imaging of radiation-sensitive samples and this was something for which there was, and still is, considerable demand. On the other hand the STXM, even in its earliest realizations, was ready to begin the development of spectromicroscopy (Sections 3.5 and 4). This development has continued over the last two decades or so and spectromicroscopy is now quite a mature field that spans a wide range of application areas.

The step which has brought 3D imaging of biological samples within reach has been the recent move toward X-ray microscopy of hydrated biological samples frozen to cryogenic temperatures. For this type of sample, X-ray microscopy fills an important gap in the range of resolution values and sample thicknesses which can be covered by other methods. The use of cryogenic temperatures is the key step in providing enough radiation-damage protection (see Figure 13–5) to enable tomographic (three-dimensional) imaging (see Section 3.4). However, 3D cryomicroscopy requires significant technical additions to the X-ray microscope including either the use of a vacuum sample chamber or a gas-stream approach to keeping the sample cold plus the mechanical devices required for recording a tilt series.

The interest in 3D imaging is also high in materials science and engineering where there is a similar gap in the coverage by other methods and where generally the samples have higher atomic number and have nonaqueous background materials (microcircuits for example). For these samples the water window has no advantage and X-ray microscopy at higher X-ray energies allows examination of bigger samples and has in fact been going on for some time. Even in biology there are factors pushing in the same direction. Ice is equally transparent at 1.5 keV as in the water window. Moreover, the 1.5–3.0 keV region gives less absorption and more phase contrast so there is only a moderate increase of the required radiation dose compared to the water window (see Figure 13–5). In this energy range zone plates also have longer focal lengths (important for sample tilting) and greater depth of focus (important for 3D reconstruction (Section 3.4.2)). Overall, multi-keV 3D X-ray microscopy has great promise but it poses somewhat different challenges with respect to the resolution/efficiency capabilities of the zone plates and the provision of a suitable condenser.

When possible we have cited publications in widely available journals rather than conference proceedings. However, good snapshots of the field are provided at three-year intervals by the series of conference proceedings dating back to 1984 (Section 1.1). As of the time of writing of this chapter the most recent one (of which the proceedings (Kagoshima et al., 2006) have not yet been published) was in Himeji, Japan, in 2005. We have in some cases cited papers in the proceedings of this conference because they have not appeared yet elsewhere. This is especially true for papers on multi-keV X-ray microscopy which was a growth subject at Himeji. Reports on the high-aspect-ratio zone plates needed for multi-keV X-ray microscopy (Section 2.4.6) were particularly promising. Zone plates of 50 nm outer zone width for 3–10 keV and 30 nm for 1–3 keV were reported with efficiencies of at least 10% (Section 2.4.6). Several of the former are operational in synchrotrons. Equally significantly, images were shown using a 50 nm zone plate in 3rd order to get sub-30-nm resolution (Section 2.4.6).

The resolution values of the best water-window and multi-keV three-dimensional images are both currently around 60 nm. The underlying reasons for this value probably include the depth of focus effect (3.4b) for water-window images and the zone plates for multi-keV ones. If this is true, then the above reports suggest that the tools are now in place for improvements to the 60 nm limit.

On condensers the news was also good. It was announced publicly for the first time that single-reflection monocapillary mirrors have been in use for some time as condensers for multi-keV TXMs and that extensive 3D imaging has already been done using them. The condenser system (Section 2.4.3) has traditionally been the Achilles heel of the TXM and this new generation of condensers is delivering an elegant solution which is described in more detail in Section 2.4.4. Apart from removing various limitations of the condenser-zone-plate, the most important innovation is energy-tunability. This combined with its multiplexed data collection could make the TXM competitive in the spectromicroscopy arena for specimens that are tolerant of its higher dose.

The TXM and STXM have always been seen as complementary devices; roughly equally popular (Table 13–3) and with important advantages on both sides. We do not believe that that general perception is likely to change in the near future. The STXM will always have the advantage in trace element mapping, the ability to instantly switch from high to low magnification, the ability to hold constant magnification even when the X-ray energy is changing, the ability to image at close to the theoretical minimum dose and so forth. On the other hand, the TXM is going through a period of technical enhancement. It has been moving into the multi-keV region and into cryomicroscopy which together with its traditional main asset, multiplexed data collection, is strengthening its capability in tomography which appears to us to be its natural home.

2 Fresnel Zone Plates

2.1 Introduction

A Fresnel zone plate is a circular diffraction grating that can be made to focus light waves in the manner of a lens. It consists of a series of concentric, usually metal, rings alternating with circular slots. Typically the rings are about equal in width to the slots and are fabricated on a thin membrane. The design is based on the idea, that by blocking, say, the even-numbered Fresnel half-period zones (Born and Wolf, 1999), the wavelets from the remaining (odd-numbered) zones will add constructively. To see this quantitatively, consider plane-wave illumination of a zone plate with n zones with radius r_n, outer zone width Δr_n, and a focal length f at wavelength λ (Figure 13–6). To get a first order *diffracted* beam in which the signals from all the open zones reinforce at the focus, we need a path difference λ between neighboring open zones. In other words the optical path $\sqrt{r_n^2 + f^2} - n\lambda/2$ should equal f. Expanding the square root and neglecting terms above fourth order, we obtain

$$\frac{n\lambda}{2} = \frac{r_n^2}{2f} - \frac{r_n^4}{8f^3} + \cdots \tag{3}$$

Evidently the focusing condition is

$$r_n^2 = n\lambda f \tag{4}$$

to second order and

$$r_n^2 = n\lambda f + \frac{n^2\lambda^2}{4} \tag{5}$$

to fourth order. In view of Eq. (7), we can neglect the fourth order (spherical-aberration) term of Eq. (3) if the numerical aperture (NA) $\ll$ 1 which is often the case for X-ray zone plates. If the fourth-order term is significant, then the zone plate can be made according to (5) and will be corrected for spherical aberration but the correction will only apply near the chosen wavelength and conjugate distances (∞ and f). If it can be neglected, we have $r_n = \sqrt{n\lambda f}$, and this defines a zone plate that will

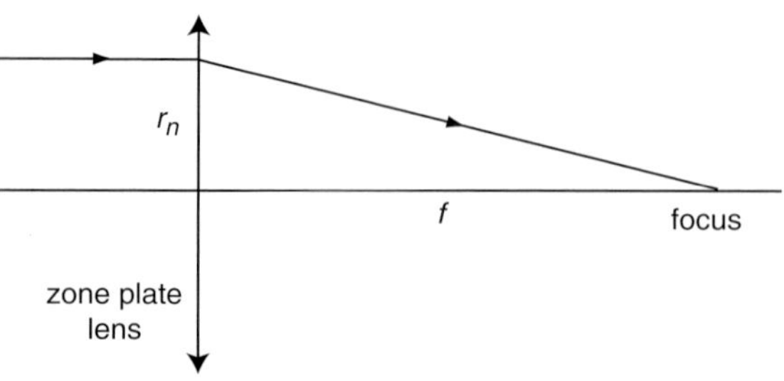

Figure 13–6. Geometry to calculate the radius of the nth half-period zone of a zone plate illuminated by parallel light. The path from the nth zone to the focus must be equal to f plus $n/2$ waves.

Figure 13–7. Diagram showing orders number −5, −3, −1, 1, 3, 5 of a zone plate. Compared to the incident beam, the negative orders diverge, the positive orders converge and the zero order (omitted for clarity) has the same shape. For plane-wave illumination, the virtual foci of the negative order beams and the real foci of the positive order beams are at distances $|f|/m$ from the zone plate where $|f|$ is the focal distance of first order. (From Attwood, © 1999, with permission of Cambridge University Press.)

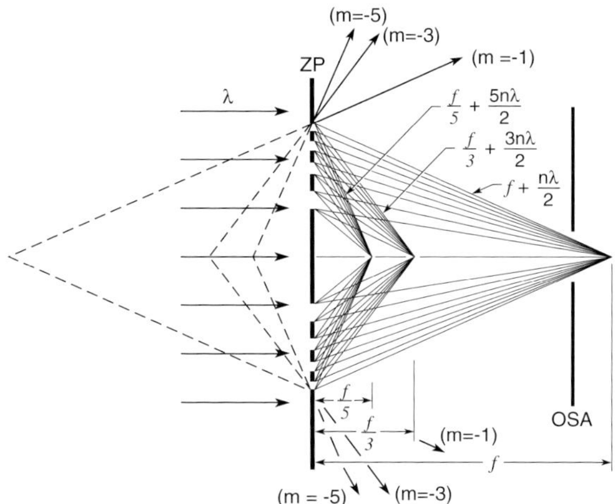

focus well for a range of wavelengths, although the focal length will vary inversely with wavelength. Thus the chromatic aberration of a zone-plate lens is much larger than that of a refractive lens and, to get a good focus, the zone plate needs to be illuminated by monochromatic light. The required degree of monochromaticity for achievement of the diffraction-limited resolution is roughly given by $\Delta\lambda/\lambda \le 1/n$ (Thieme, 1988).

Some useful quantities follow from the fundamental zone-plate Eq. (4). First we can take the difference between the nth and $(n-1)$th equations to get the outer zone width Δr_n

$$\Delta r_n = \frac{\lambda f}{2r_n}. \tag{6}$$

This allows the conclusion that all of the zones have equal area and also gives us the numerical aperture

$$NA \equiv \frac{r_n}{f} = \frac{\lambda}{2\Delta r_n}, \tag{7}$$

and thence the Rayleigh resolution

$$\delta_{Rayleigh} = \frac{0.61\lambda}{NA} = 1.22\Delta r_n. \tag{8}$$

Thus we see that a given zone plate can be specified by its r_n and Δr_n from which the resolution (which is independent of wavelength) and the focal length and numerical aperture at any given wavelength, follow from Eq. (8), Eq. (6) and Eq. (7) respectively. So far we have been discussing the first-order focus but in general, beams of all integral orders may be produced. Thus there is a zero-order (unfocused) beam, a series of positive-order converging beams with focal distance f/m and a series of negative-order diverging beams with focal distance $-f/m$ (Figure 13–7). In mth order the numerical aperture is m times larger and hence the resolution is m times smaller (better) than in first order. As we will see, if the open and opaque zones are of equal width, then the even orders are missing.

2.2 Zone Plate Image Quality

2.2.1 Optical Path Function Analysis

We consider the imaging of a general point A (x,y,z) by a planar zone plate lying in the $y - z$ plane (Figure 13–8) (Kamiya, 1963). We will use the method of the optical path function so we start by calculating the optical path from A to a general point B (x',y',z') that we will later identify as the Gaussian image point. Without loss of generality we set $y = y' = 0$. We calculate the path APB where P $(0,w,l)$ is a general point in the zone plate. The expression for the optical path will be a power series in the aperture coordinates, w and l, and the field angle, z/x and each term in the series will represent a specific aberration. Evidently

$$\mathrm{AP} = \sqrt{x^2 + w^2 + (z - l)^2} \tag{9}$$

so expanding the square root and keeping terms up to fourth order, we have

$$\mathrm{AP} = x\left\{1 + \frac{1}{2}\frac{w^2 + l^2}{x^2} + \frac{1}{2}\frac{z^2}{x^2} - \frac{2zl}{2x^2} - \frac{1}{8}\frac{1}{x^4}\left[(w^2 + l^2)^2 + z^4 + \right.\right.$$

$$\left.\left. 4z^2l^2 - 4z^3l + 2(w^2 + l^2)z^2 - 2(w^2 + l^2)2zl\right] + \cdots\right\} \tag{10}$$

There is an identical series for PB except that, for PB, the x, y and z are replaced by x', y', and z' (Figure 13–8). We are now in a position to write down the optical path function, F. Before doing so we drop terms

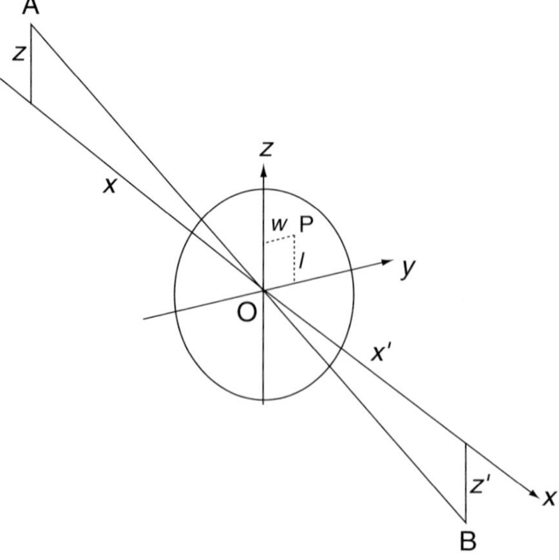

Figure 13–8. Notation for the optical-path-function analysis of a zone plate.

which do not depend on w, l or z/x because they do not represent aberrations, and introduce the term $-nm\lambda/2$ as we did to get Eq. (1). We choose to analyze the case of a parabolic zone plate (that is one built according to Eq. (4)) so initially, we use Eq. (4) to write the optical path function up to third order only, as follows.

$$F = AP + PB - \frac{nm\lambda}{2} = AP + PB - \left| \frac{w^2 + l^2}{2} \frac{1}{f/m} \right|$$

$$= \frac{w^2 + l^2}{2}\left[\frac{1}{x} + \frac{1}{x'} - \frac{1}{f/m} \right] - l\left[\frac{z}{x} + \frac{z'}{x'} \right] + \cdots \tag{11}$$

Specializing to the case when B is the Gaussian image point, the first (defocus) term vanishes and by considering the ray AOB we obtain

$$z/x = -z'/x' \tag{12}$$

so that the second term also vanishes. This leaves only the five fourth-order terms

$$F = -\frac{(w^2 + l^2)^2}{8}\left(\frac{1}{x^3} + \frac{1}{x'^3} \right) - \frac{l^2 z^2}{2x^2} \frac{1}{f/m} + \frac{l}{2}\left(\frac{z^3}{x^3} + \frac{z'^3}{x'^3} \right) -$$

$$\frac{(w^2 + l^2)}{4} \frac{z^2}{x^2} \frac{1}{f/m} + \frac{(w^2 + l^2)}{2} \frac{lz}{x}\left(\frac{1}{x^2} + \frac{1}{x'^2} \right) + \cdots \tag{13}$$

These are the five Seidel aberrations; spherical aberration, astigmatism, distortion, field curvature and coma respectively. Because of Eq. (12), the distortion term vanishes identically which is a useful property of zone-plate lenses and we can therefore turn our attention to the remaining aberrations.

2.2.2 Ray Aberrations

We need to know the ray pattern delivered by the zone plate for a given point object. That is, we want the ray aberrations $\Delta y'$ and $\Delta x'$ relative to the Gaussian image point (coordinates identified by subscript zero). For a normal-incidence optic these are given by the following expression (Born and Wolf, 1999)

$$\Delta y' = x'_0 \frac{\partial F}{\partial w}, \quad \Delta z' = x'_0 \frac{\partial F}{\partial l} \tag{14}$$

We now apply this to the Seidel aberrations individually.

2.2.3 Spherical Aberration

We can rearrange the spherical aberration term using the magnification $M = x'_0/x$

$$F_{\text{sp ab}} = -\frac{(w^2 + l^2)^2}{8f^3} \frac{M^3 + 1}{(M + 1)^3} \tag{15}$$

The last term of this expression, which we will denote by Θ, approaches unity in the cases of interest to us namely M large (a microscope) or M small (a microprobe). If we consider the case that Θ does approach unity, then Eq. (15) reduces to the fourth-order term of Eq. (3). This is expected because we are reverting to the conjugates used to derive (3)

(∞ and f) and the parabolic zone plate we are analyzing is not corrected for spherical aberration. Treating the case of a microscope ($x \approx f$) and using the notation $r = \sqrt{w^2 + l^2}$, we now substitute from (15) into (14) to obtain the ray aberrations

$$\Delta y' = x_0' \frac{1}{8f^3} 2(w^2 + l^2) 2w = \frac{M}{2}\left(\frac{r}{f}\right)^2 w$$
$$\Delta z' = x_0' \frac{1}{8f^3} 2(w^2 + l^2) 2l = \frac{M}{2}\left(\frac{r}{f}\right)^2 l \tag{16}$$

Thus we see that the image-plane figure produced by a point object via the rays passing through the rim of the lens ($r = r_n$), is a circle of diameter D_{SA} where

$$D_{SA} = M(NA)^2 r_n \tag{17}$$

This is produced irrespective of the position of the object point. The presence of uncorrected spherical aberration means there will be an optimum value of the NA, that is the largest NA for which the resolution is still diffraction-limited. This can be estimated (Michette, 1986) by requiring that the path error be less than the Rayleigh quarter-wave limit

$$NA_{opt} = \sqrt[4]{\frac{2\lambda}{\Theta f}} \tag{18}$$

In practice the spherical aberration of a parabolic zone plate is often not negligible in the soft X-ray region (see Figure 13–9 for example) and therefore soft X-ray zone plates are usually made according to Eq. (5), which means that the spherical aberration is corrected for the chosen wavelength. In these cases we may ask what happens when such a zone plate is used with a wavelength other than the correction wavelength. We therefore consider a zone plate corrected for a given wavelength in first positive order with M very small as shown in Figure 13–6 and we assume that the correction is also small ($NA \ll 1$). Now using a subscript zero to represent the properties of the corrected zone plate and subscript one to represent the properties of the same zone plate operating at another wavelength, we can apply Eq. (5) for the nth and $(n-1)$th zone to show that $\Delta r_n \cong (\lambda_0 f_0 + n\lambda_0^2/2)/2r_n$. From that we can use the grating equation to get the ray deviation angle (ε_0) at the zone plate and thus the ray displacement ($\varepsilon_0 f_0$) at the detection plane when the zone plate images a distant axial point (M small):

$$\varepsilon_0 f_0 = r_n\left\{1 - \frac{n\lambda_0}{2f_0}\right\} = r_n - \frac{r_n}{2}(NA_0)^2 , \tag{19}$$

where Eq. (4) and Eq. (7) have been used. For imaging an axial point with M large the ray displacement would be approximately $M\varepsilon_0 f_0$. The first term of these expressions, r_n or Mr_n, is the ray displacement needed for correct imaging of an axial point and the second term is an aberration equal to minus the radius of the spherical aberration disk of a parabolic zone plate (compare Eq. (19) with Eq. (17)). This results in perfect cancellation of the spherical aberration as intended.

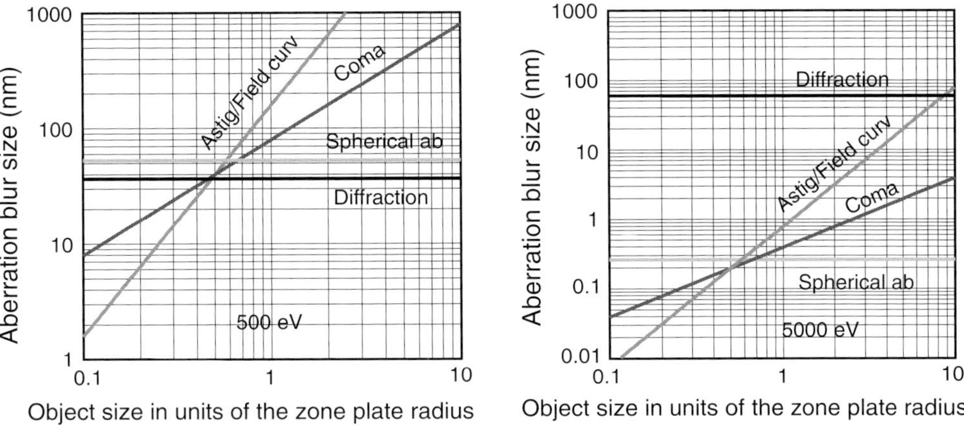

Figure 13–9. Size of the aberrations of a soft X-ray (500 eV; left) and a hard X-ray (5000 eV; right) zone plate as a function of object size. The soft X-ray zone plate has an outer zone width of 30 nm, diameter of 62 μm, and a focal length of 0.75 mm at 500 eV. The hard X-ray one has an outer zone width of 60 nm, diameter 124 μm and focal length 30 mm at 5000 eV. The graphs illustrate the general trend that hard X-ray zone plates have lower aberrations on account of their lower NA. In the example the hard X-ray zone plate has only negligible aberrations up to a sample diameter of about ten radii, which is far beyond the sample size allowed by penetration requirements. On the other hand the resolution of the soft X-ray zone plate is degraded by the field-angle-dependent aberrations for objects of diameter more than about half a radius. This is still good for many experiments but, as the diagram shows, spherical aberration is not negligible for the parabolic soft-X-ray zone plate. Therefore, as explained in Section 2.2.3, zone plates designed for soft-X-ray applications are normally made with built-in spherical-aberration correction according to Eq. (5).

When the same zone plate is used at another wavelength λ_1 with allowance for the change of focal length, Δr_n remains the same so Eq. (19) becomes

$$\varepsilon_1 f_1 = \frac{f_1 \lambda_1}{f_0 \lambda_0} r_n \left\{ 1 - \frac{n \lambda_0}{2 f_0} \right\} = r_n - \frac{r_n}{2}(NA_0)^2 \tag{20}$$

In other words the amount of correction has not changed, whereas for exact correction it should now be $r_n(NA_1)^2/2$. Thus the residual error is an aberration disk of diameter $|r_n[(NA_1)^2 - (NA_0)^2]|$ or a new disk of diameter $|(\lambda_1^2 - \lambda_0^2)/\lambda_0^2|$ times the diameter of the original uncorrected disk. As an example, if the new wavelength differs by 10% from the wavelength of correction, then the aberration disk will be reduced to about 20% of its uncorrected size. This would be enough to make the aberration negligible in the soft X-ray example shown in Figure 13–9. Evidently this general conclusion applies equally to M very small or very large.

2.2.4 Astigmatism and Field Curvature

Taking the astigmatism and field curvature terms together and again calculating the ray aberrations by substituting the path function terms into Eq.(14), we obtain

$$\Delta y' = -\frac{1}{2} x_0' w \frac{z^2}{x^2} \frac{1}{f}$$
$$\Delta z' = -\frac{3}{2} x_0' l \frac{z^2}{x^2} \frac{1}{f}$$

$$(21)$$

Squaring and adding we find that the marginal rays trace out an ellipse of major axis $3k$ and minor axis k

$$\frac{\Delta y'^2}{(k/2)^2} + \frac{\Delta z'^2}{(3k/2)^2} = 1 \quad \text{where} \quad k = M(NA)^2 r_n \bar{z}^2 . \tag{22}$$

The parameter $\bar{z} = z/r_n$ expresses the position of the object point in units of the zone plate radius. Therefore, unlike spherical aberration, the size of this aberration does depend on the position of the object point, and therefore will limit the field of view of the microscope.

2.2.5 Coma
In this case, the above procedure leads to

$$\Delta y' = -2wlQ, \quad \Delta z' = -(w^2 + 3l^2)Q \quad \text{where} \quad Q = \frac{1}{2} \frac{z}{x} \frac{M-1}{f} \tag{23}$$

By expressing w and l in polar coordinates one can show that each circle of radius r in the lens produces an aberration figure

$$(\Delta z' - 2K)^2 + (\Delta y')^2 = K^2 \quad \text{where} \quad K = r^2 Q. \tag{24}$$

This is a series of circles of radius K, each shifted by $2K$ from the origin which is the usual "comet-shaped" figure associated with coma. The outer boundary of the figure has a length $3K_{\max}$ and the largest circle has a diameter $2K_{\max}$ where (assuming M is large so that $M - 1 \approx M$)

$$2K_{\max} = M(NA)^2 r_n \bar{z}. \tag{25}$$

2.2.6 Relative Size of the Aberrations
The relative size of these aberration figures can be obtained from Eq. (17), Eq. (22) and Eq. (25) respectively. The diameter of the spherical aberration circle, the diameter of the largest coma circle and the major axis of the astigmatism ellipse are in the ratio $1 : \bar{z} : 3\bar{z}^2$ (Michette, 1986). For points close to the axis ($\bar{z} \ll 1$) the field-angle-dependent aberrations coma and astigmatism/field curvature will be negligible and the resolution will be determined by diffraction if the numerical aperture is below NA_{opt} or by spherical aberration if it is above. On the other hand for points further from the axis, coma and astigmatism/field curvature become more significant and for points more than about a zone plate radius away from the axis ($\bar{z} \geq 1$) astigmatism/field curvature dominates. The general behavior of these aberrations is shown in Table 13–1, while Figure 13–9 shows examples of parabolic zone plates working at 0.5 keV and 5 keV. However, note that the plots in the figure show the size of the outer boundary of the aberration figures which is a conservative estimate of their contribution to the resolution because the light is somewhat concentrated near the center of the figure. For example, we can deduce from Eq. (17) that half of the light is concentrated in a circle of diameter about one third of D_{SA}.

Table 13–1. Seidel aberrations of a planar zone plate.

Name	Spherical aberration	Astigmatism/field curvature	Distortion	Coma
w, l, z/x term	$(w^2 + l^2)^2$	$(w^2 + 3l^2)(z/x)^2$	$l\{(z/x)^2 + (z'/x')^2\}$	$(w^2 + l^2)l(z/x)$
Aberration figure boundary	circle	ellipse	vanishes identically	two lines through O at $\pm 30°$ to Oz touching a family of circles
y parameter	circle diameter	major axis	na	diameter of the largest circle
Value y parameter	$M(NA)^2 r_n$	$3M(NA)^2 r_n \bar{z}^2$	na	$M(NA)^2 r_n \bar{z}$
z parameter	circle diameter	minor axis	na	distance from the origin to the far side of the largest circle
Value z parameter	same	$M(NA)^2 r_n \bar{z}^2$	na	$\dfrac{3}{2} M(NA)^2 r_n \bar{z}$

2.3 Zone Plate Efficiency

2.3.1 Idealized Structures

One can see from Eq. (4) that an ideal Fresnel zone plate is periodic in r^2 space with period λf, which means that its amplitude transparency function T_{zp} (shown in Figure 13–10) can be written as the Fourier series

$$T_{zp} = \sum_{-\infty}^{+\infty} a_m \exp\left(\frac{im\pi r^2}{\lambda f}\right) \quad \text{where} \quad a_m = q(-1)^m \operatorname{sinc}(mq) \qquad (26)$$

The sinc function is defined by $\operatorname{sinc}(x) = \sin(\pi x)/(\pi x)$ and q is the fraction of each *period* that is opaque. For a classical zone plate, $q = 0.5$, so in that case (26) gives the power in the $\pm m$th harmonic as $|a_m|^2 = 1/(m\pi)^2$ for m odd, zero for m even and 0.25 for $m = 0$. These are therefore the intensity efficiencies of the classical zone plate in those orders. The

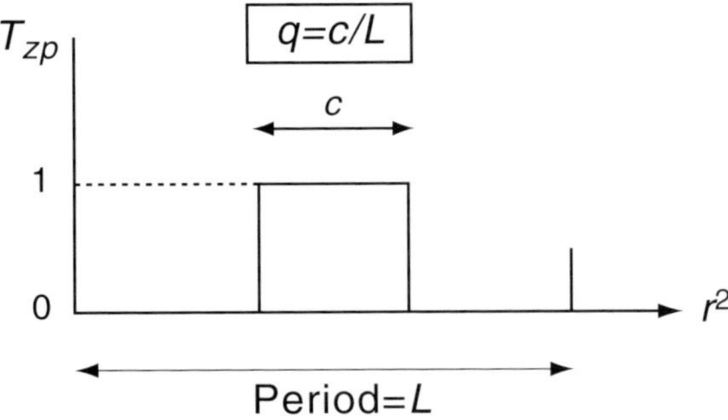

Figure 13–10. The transparency function of one period of a zone plate plotted in r^2 space. The fraction that has transparency unity is $q = c/L$.

main point here is that this type of zone plate has a maximum efficiency in the first order focus of about 10%. By taking the derivative of $|a_m|^2$ we can show that the classical zone plate ($q = 0.5$) is, in fact, the optimum choice of q for first-order efficiency. Similarly, the optimum choice for second order efficiency is $q = 0.25$ or 0.75 which may have some practical significance, as discussed by Simpson and Michette (1984). By applying Parseval's theorem to the series in (26) with $q = 0.5$, one can determine the disposition of the energy for the classical zone plate as shown in Table 13–2.

Another idealized type of zone plate, first proposed by Rayleigh (1888) and implemented by Wood (1898), is the phase plate. Here, the opaque rings are replaced by transparent rings, that impart a phase change of $\phi = \pi$. The phase plate transparency function is $T_{pp} = 1 + (e^{i\phi} - 1)T_{zp}$, so the Fourier series becomes

$$T_{pp} = \sum_{-\infty}^{+\infty} b_m \exp\left(\frac{im\pi r^2}{\lambda f}\right)$$

$$\text{where} \quad b_m = \delta_{m,0} + (e^{i\phi} - 1)q(-1)^m \operatorname{sinc}(mq). \tag{27}$$

This shows that when $\phi = \pi$ and $q = 0.5$, which are the optimum values, the power in the $\pm m$th harmonic is $|b_m|^2 = 4/(m\pi)^2$ for m odd, zero for m even and $|b_0|^2 = |1 + q(e^{i\phi} - 1)|^2 = 0$ for $m = 0$. The efficiency in the first-order focus is now about 40% and the disposition of energy is again shown in Table 13–2.

A third idealized type of zone plate is the Gabor plate (the hologram of a point) in which the rectangular profile of the last two devices is replaced by a sinusoid in r^2 space or a chirp function in r space. An absorption Gabor plate has only three orders, $m = \pm 1$ and 0 of which the +1 order has efficiency 1/16. A phase Gabor plate, with a maximum phase change of 1.84 radians, has all the odd orders and the +1 order receives 34% of the light (Table 13–2).

Table 13–2. Efficiency of various types of zone plate.

Type of zone plate	Fresnel	Rayleigh-Wood	Gabor amplitude	Gabor phase
Type of zones	Opaque, transparent	Phase-change π, transparent	Sine-wave transparency	Sine-wave phase change max = 1.84 rad
Efficiency ±1 order	1/4	$4/\pi^2$	1/16	0.34
Efficiency in general	$1/(m^2\pi^2)$ m odd 0 m even	$4/(m^2\pi^2)$ m odd 0 m even	0	$\neq 0$
Total positive orders ($m \neq 0$)	1/8	1/2	1/16	0.45
Total negative orders ($m \neq 0$)	1/8	1/2	1/16	0.45
Efficiency zero order	1/4	0	1/4	0.10
Absorbed	1/2	0	5/8	0
Overall total of last four rows	1	1	1	1

2.3.2 Real Structures

To make practically useful X-ray lenses, we must extend the treatment given so far to include the optical properties of real materials, suited to manufacturing zone plates. Such an extension was first provided in 1974 in an important paper by Kirz (1974) in which it was demonstrated that

1. Phase zone plates with primary efficiencies of 20–40% can be made from realistic materials.
2. Such zone pates can be designed to reduce or eliminate the zero-order beam and to reduce the absorbed fraction compared to a classical Fresnel zone plate.
3. These improvements can be effected essentially throughout the wavelength range 0.1–80 nm.
4. Realistic fabrication errors lead to only moderate deterioration in the optical performance.

For a phase-reversal zone plate made of a material with complex refractive index (Henke et al., 1993) $\tilde{n} = 1 - \delta - i\beta$, the efficiency $|b_m|^2$ can be found by making the replacement $\phi = \phi_1 + i\phi_2$ in (27) where $\phi_1 = kt\delta$, $\phi_2 = kt\beta$, $k = 2\pi/\lambda$, t is the thickness and we use the shorthand $r_a = e^{-\phi_2}$ for the amplitude attenuation factor. For $m \neq 0$, this gives

$$|b_m|^2 = \frac{1}{(m\pi)^2}(1 + r_a^2 - 2r_a \cos\phi_1) \tag{28}$$

and for $m = 0$

$$|b_0|^2 = \frac{1}{4}(1 + r_a^2 + 2r_a \cos\phi_1) \tag{29}$$

These equations (Kirz's (7) and (10) (Kirz, 1974)) give the efficiency of a planar zone plate of known thickness and refractive index. As discussed by Kirz and later by Michette (Michette, 1986), the optimum phase change is no longer π but is about 10–20% less. However, one can certainly choose a thickness to optimize the efficiency of a planar zone plate based on the use of Eq. (28). Plots of the theoretical efficiency, calculated using Eq. (28), for zone plates made of nickel and gold are shown in Figure 13–11.

2.4 Zone Plates: Fabrication and Examples

2.4.1 Fabrication Technique

The fabrication of X-ray zone plates involves several challenges. For high resolution imaging, one wishes to obtain the smallest possible value for the outermost zone width Δr_n or about 15–80 nm in high resolution modern examples. At the same time, the thickness t of the zone plate should ideally be that required to deliver a phase shift near π so as to maximize efficiency; this generally implies a thickness of greater than 100 nm for soft X-rays of 100–1000 eV energy, and a thickness near 1 μm for zone plates designed to operate at multi keV energies (see Figure 13–11). These two requirements are difficult to meet at the same time, for they lead to a demand for the fabrication of nanostructures

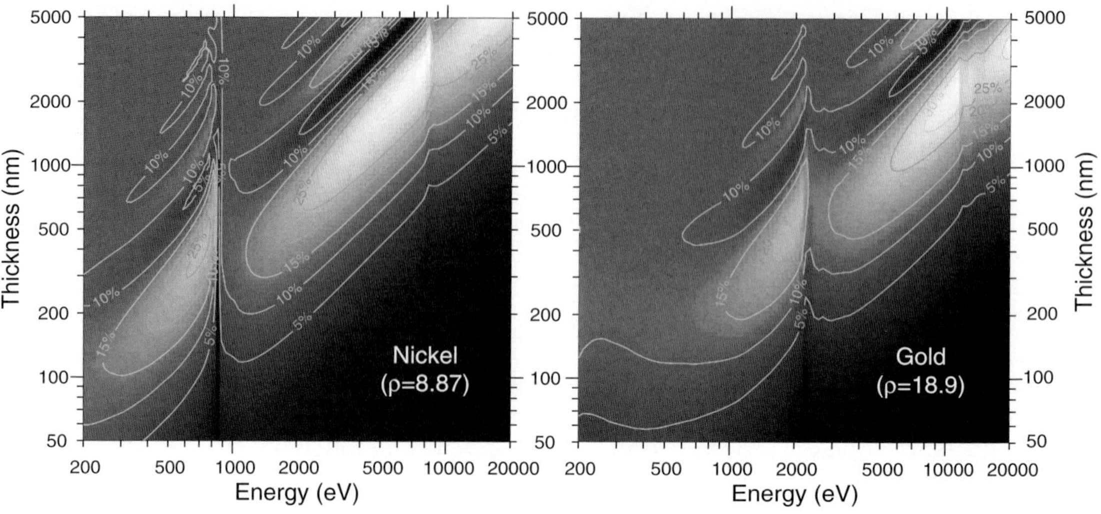

Figure 13–11. Contour plot showing the efficiency of planar zone plates made of Ni (left) and Au (right) according to Eq. (24) as a function of thickness and X-ray energy.

with very high aspect ratios (structure height over width). In addition, to obtain usable focal lengths of order 1 mm or larger, most zone plates used as high resolution objectives have diameters of 50–200 μm yet to minimize loss of efficiency all zones must be placed accurately to roughly one-third of their width (Simpson and Michette, 1983). This implies an absolute accuracy of zone placement of about 0.01% which is quite challenging but which can be achieved in modern 100 keV electron beam lithography systems that incorporate laser interferometer positioning control (Anderson et al., 2000; Tennant et al., 2000). Such zone plates typically have 300–1000 zones, and require corresponding quasimonochromatic illumination with $E/\Delta E > 300$–1000 or better (Thieme, 1988).

Early demonstrations of X-ray zone plate fabrication used optical lithography to create free-standing zone plates with $\Delta r_n = 20$ μm (Baez, 1960). An important early advance was the use of holographic methods to create zone plates with submicron zone widths (Schmahl and Rudolph, 1969; Niemann et al., 1974), eventually leading to outermost zone widths of $\Delta r_n = 56$ nm (Schmahl et al., 1984b). The method used by nearly all laboratories today involves electron beam lithography, which was first suggested by Sayre (Sayre, 1972) and subsequently demonstrated in several laboratories (Shaver et al., 1980; Kern et al., 1984; Buckley et al., 1985). In order to obtain high aspect ratio nanostructures, most laboratories now use some variation of tri-layer resist schemes (Tennant et al., 1981) with electroplating (Schneider et al., 1995) as illustrated in Figure 13–12. This approach allows for the writing of fine, dense features in an electron beam resist which is sufficiently thin that electron side scattering is minimized. A series of reactive ion etches are used to first transfer the e-beam pattern into a hard mask, and to use that hard mask to transfer the pattern into a second polymer used either to define an etch mask for the underlying zone material (Tennant et al., 1991; David et al., 1995, 2000) or, more commonly, as a mold for electroplating of the zone structures (Schneider et al., 1995). Using these approaches, several groups have fabricated zone plates with outermost zone widths of 20 nm or below (Bögli et al., 1988; Schneider et al., 1995; Spector et al., 1997; Anderson et al., 2000; Peuker, 2001; Chao et al., 2003).

A recent promising approach (Chao et al., 2005) has been to write every other zone in one pass, and then to write the other half of the zones in a subsequent processing step (of course extremely high overlay accuracy is required). By increasing the distance from the next zone written in one pass, the proximity effect of electron beam lithography is reduced, leading to higher contrast and thus higher aspect ratio in the developed resist. In addition, the width of the resist used as a plating mold is tripled, thus reducing collapse during processing. This approach has been used to fabricate zone plates with an outermost zone width of 15 nm (Chao et al., 2005) and a theoretical zone efficiency of 6% (see next section). A key challenge lies in maintaining high efficiency as the outermost zone width is decreased. For applications requiring higher efficiency for greater flux in STXM, or reduced radiation dose in TXM, efficiencies of 10–18% have been obtained at 30 nm outermost zone width (Spector et al., 1997; Peuker, 2001). These efficiencies are for zone plates operating at 400–550 eV; zone plates for use at higher X-ray energies are discussed in Section 2.4.5.

2.4.2 Resolution-Determining Zone Plates

While a variety of groups are fabricating high resolution zone plates as described above, we consider here one example which is the effort of the Center for X-ray Optics at Lawrence Berkeley National Laboratory. By using a 100 keV electron beam lithography system with interferometric positioning control and customized circular pattern generation (Anderson et al., 1995), this group has made a series of zone plates with steadily improving resolution (Anderson et al., 2000; Chao et al., 2003) and have developed a sophisticated technique for determining the resolution (Jochum and Meyer-Ilse, 1995; Heck, 1998; Chao

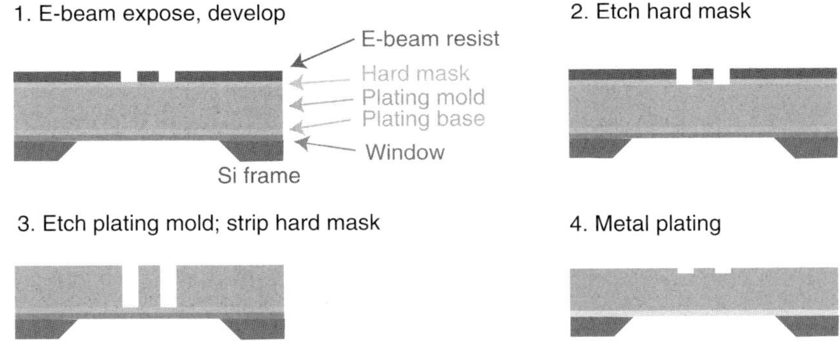

Figure 13–12. Schematic of zone plate fabrication technique as discussed in the text.

et al., 2003). The tests have been done using the XM-1 microscope at beam-line 6.1 at the Advanced Light Source at Berkeley USA (see Microscopes Layouts and Illumination Schemes (Sectin 3.1)). The latest results (Chao et al., 2005), which represent the best zone-plate-microscope resolution that we know of, show a resolution of <15 nm at a photon energy of 815 eV (see Figure 13–13). This measured value depends on the degree of partial coherence of the illumination and we discuss it in that context in a later section. The zone plate had an outer zone width of 15 nm, which, with the XM-1 illumination system, gave a theoretically expected resolution of 12 nm. The zone plate was made by the above-mentioned new process in which the $n = 2, 6, 10, \ldots$ and the $n = 4, 8, 12, \ldots$ opaque zones are made in two separate groups. Chao and colleagues suggest that the new process is not yet at its limit and that 10 nm zone plates should be within reach.

2.4.3 Condenser Zone Plates

Condenser zone plates serve the dual function of imaging the source on to the sample (in critical illumination) and, in combination with a pinhole close to the sample, of acting as a moderate-resolution mono-chromator. Ideally they should deliver a beam which (1) has the same NA as the objective zone plate, (2) exactly fills the sample with light and (3) has a spectral bandwidth equal to the reciprocal of the number of zones of the objective. In practice, due to the fact that bending-magnet synchrotron-radiation sources usually have smaller phase space area than the microscope and due to the fabrication difficulties described below, conditions (1) and (2) cannot be fully met. Moreover, condition (3) implies that the condenser zone plate must be about 5–10 mm in diameter. We discuss the trade-offs involved here in more detail in Section 3.1.1.

The first condenser zone plates were fabricated by the Göttingen group (Schmahl and Rudolph, 1984b; Hettwer, 1998) using roughly the same holographic process then used to make objective zone plates. As a result, the finest line widths (and thus the NAs) of the condenser and

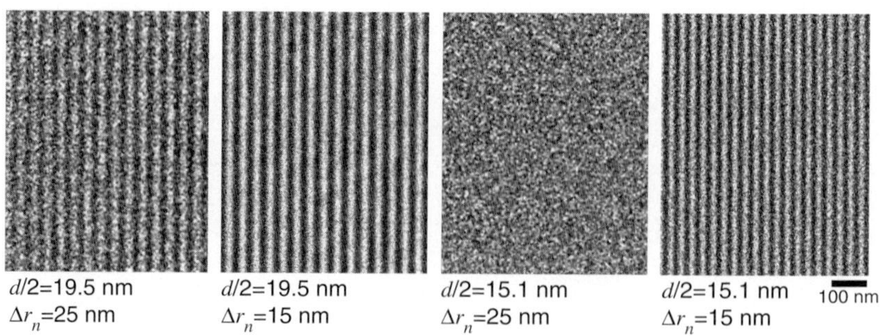

Figure 13–13. The soft X-ray images of square-wave test objects used by Chao and coworkers (2005) to demonstrate a microscope resolution of <15 nm using a 15-nm- and a 25-nm-outer-zone-width zone plate as explained in the text. The half-period of the test object and the outer zone width of the zone plate for the four images are shown. (Reprinted from Chao et al., © 2005, with permission from Nature.)

objective tended to match, which is broadly what is required for best resolution (see for example Fig.10.13 of (Born and Wolf, 1999)). With electron beam lithography, new challenges have arisen. To achieve the same finest zone width and efficiency in the condenser as in the objective, would require the use of the same high-resolution but necessarily slow electron beam resists. Since the area of a 5 mm diameter condenser is 10^4 times larger than a 50 µm objective zone plate, it would take 10^4 times longer to fabricate the condenser with the same process. (Because aberrations on condensers do not degrade image quality, the requirements for zone placement accuracy do not scale up in the same fashion). For these reasons, as the resolution of the objectives has been pushed down below 20 nm, the condenser zone plates have not kept up. Instead, typical condenser zone plates fabricated by electron beam lithography have outer zone widths of 50–60 nm. As one example, the fabrication of a TXM condenser zone plate with 9 mm diameter and 55 nm outer-zone width required a 48-hour writing time (Anderson et al., 2000). Although process improvements have since reduced that time by about a factor of two, such large zone plates are still not widely available. The mismatch of numerical aperture between condenser and objective zone plates limits the modulation transfer at high spatial frequency (Born and Wolf, 1999) and thus limits the ability to detect small structures such as immunogold labels in bright field or dark field modes (Vogt et al., 2001a).

Besides the challenges of matching the NA of the best objectives and their limited availability, condenser zone plates are usually required to operate in an unfriendly environment, relatively near the source in a synchrotron beam line. Even with the protection of an energy-filtering mirror, they still often take significant heat load and the heat removal pathways are poor. This is an undesirable circumstance for optics that take so much effort to build. The problem of power deposition in condenser zone plates can be understood by solving the boundary value problem of a uniformly heated membrane (Howells et al., 2002). Assuming a square membrane of side a, thickness t and conductivity k, the solution for the temperature is

$$T(x, y) = \frac{16Qa^2}{\pi^4 kt} \sum_{\substack{m=1 \\ m,n \ odd}}^{\infty} \sum_{n=1}^{\infty} \frac{1}{mn} \frac{\sin\frac{m\pi x}{a}\sin\frac{n\pi y}{a}}{m^2 + n^2} \tag{30}$$

where Q is the absorbed power density. The double sum is a simple function that has a peak of height 0.448 in the center and is zero at the edges. The center temperature is then given by the useful relation

$$T_{\max} = 7.17 \frac{Qa^2}{\pi^4 kt}. \tag{31}$$

Since the maximum temperature depends on Q/t and, for a simple thin membrane, Q is proportional to t, the temperature is not reduced by making the membrane thicker. However, if the absorption is principally in the zone plate rings then a thicker membrane may help. A smaller or better-conducting membrane always helps.

The list of issues posed by condenser zone plates does not end here. Another serious limitation is that to change the X-ray energy one has to change the condenser zone plate or at least change its object and image distances. This is so difficult that systems fed by a condenser zone plate are effectively not energy tunable. Evidently there are plenty of reasons to seek alternatives to condenser zone plates and to this we now turn.

2.4.4 Alternatives to Condenser Zone Plates

Zone plates deflect the outermost ray by an angle equal to their NA (Eq. (7)). As explained in the previous section, manufacturable NAs of condenser zone plates are too small to match those of the best objectives by a factor of two or more. Now we know that a mirror can deflect an X-ray roughly by $2\sqrt{2\delta}$ or twice its critical angle. Thus we can characterize a mirror by an effective outer zone width $\Delta r_n = \lambda / (4\sqrt{2\delta})$, that a zone plate would need to have to produce a deflection equal to twice the critical angle of the mirror. Since δ is proportional to λ^2 (Section 1.1), one can see that the effective outer zone width will vary rather slowly with both energy and electron density. For example, for a platinum, mirror it varies from 6 nm at 0.5 keV to 4 nm at 5 keV. While for a silicon dioxide mirror it varies from 12 nm to 10 nm over the same energy range. This shows that if a suitable geometry can be arranged a single-reflection mirror system could be a very effective condenser. However it is important to note that a reflective condenser will not provide a monochromatic beam and that either a line source or a monochromator will be required. As we shall see, a separate monochromator coupled with a reflective condenser will have major advantages.

To produce the desired hollow-cone illumination, a grazing incidence ellipsoid of revolution in the form of a hollow tube would be suitable. The input angle should match the angle of the beam from the source or monochromator and the output angle should match or exceed the objective NA. For a synchrotron beam this design will normally lead to under filling of the sample which can be overcome by wobbling. The fabrication accuracy should be such that the point spread function of the mirror is small compared to the size of the object field. Mirrors of this general type have been made for some time (often starting from glass capillaries) and are quite widely used in the hard X-ray research community as reviewed for example by Bilderback (2003).

Single-reflection monocapillary X-ray mirrors have now been in use for microscopes at both laboratory and synchrotron sources (Section 1.4) for the last year or two and have enjoyed considerable success. For example, documentation of the performance of the X-ray microscope at the Taiwan synchrotron, including details of its condenser mirror, is provided by Tang et al. (2006) and Yin et al. (2006). These condenser mirrors have many advantages compared to condenser zone plates. Specifically, capillary-mirror condensers

1. Are more readily available
2. When fed by a separate monochromator allow the TXM to be truly energy-tunable
3. Are able to match the NA of any currently available objective zone plate

4. Are a factor 3–15 times more efficient with no unwanted orders
5. Are more robust, longer lived and easier to clean
6. Are less fragile with respect to thermal or mechanical damage
7. Do not require a band-width-selecting pinhole close to the focus and thus do not limit the size of the sample holder nor its ability to rotate.

As indicated earlier (Section 1.4) we believe that the advent of an energy-tunable TXM is important and may enable the TXM to become competitive for spectromicroscopy.

2.4.5 Zone Plates with Shaped Grooves

Until now we have talked about square-wave zone plates with a gap to period ratio of 0.5 that behaved according to the theory of a thin zone plate, even if the thickness was greater than Δr_n. Just as a blazed reflection grating with a saw-tooth profile has much better efficiency than a square-wave grating (even if the latter is a perfect phase grating with π phase shifts), so one can get higher efficiency from zone plates with shaped groove profiles. Considering that a zone plate is intended to synthesize a smooth spherical wave front from a succession of ring-shaped parts, we might expect that the optimum groove shape will be a parabola that increases the phase shift smoothly across the zone-plate period. In fact the mathematical treatment (Tatchyn et al., 1982; Michette, 1986) shows that the thickness function is

$$t_i(r) = \left\{ \sqrt{f^2 + r^2} - \sqrt{f^2 + r_{i-1}^2} \right\}/\delta \qquad r_{i-1} \leq r < (r_i - d_i)$$
$$= 0 \qquad\qquad\qquad\qquad\qquad\qquad (r_i - d_i) \leq r \leq r_i \qquad (32)$$

where d_i/r_i can be calculated (Tatchyn et al., 1984) and is the outer fraction of the ith period which is to be left open. The first-order efficiency of a nickel zone plate made according to this specification would be about 80% at 7 keV.

It is hard to microfabricate a smooth curve but one can still get much of the advantage of this scheme by approximating the parabolic profile by a stepped structure (Di Fabrizio et al., 1994; Yun et al., 1999). For example Di Fabrizio et al. (1999) have made a nickel zone plate with four equal width steps of optical delay 0, 0.25, 0.5 and 0.75 wavelengths. The measured first order efficiency of this zone plate was 55% at 7 keV, which represents a substantial improvement in efficiency and suppression of unwanted orders compared to traditional soft X-ray performance and shows the benefits of both groove-shaping and phase-plates. Evidently the use of several thickness steps implies that the outermost zone must be several times wider than the finest line width that can be achieved with the particular fabrication process, so that this approach involves a tradeoff between spatial resolution and efficiency.

2.4.6 Hard X-Ray Zone Plates

While much of the effort of the last three decades has gone into developing zone plate microscopy in the 290–540 eV "water window" region for studies of 0.1–10 μm thick specimens, there is increasing activity in hard-X-ray zone-plate imaging at energies of roughly 2–15 keV. Scanning fluorescence X-ray microprobes (SFXM) using zone plate optics are providing new capabilities for trace element mapping, and hard

X-ray transmission X-ray microscopes (TXMs) using absorption or especially phase contrast are able to image much thicker objects than their soft-X-ray counterparts. Zone plates for these energies must be much thicker to achieve good efficiency (see Figure 13–11) which places increasing demands on zone aspect ratio in lithographically patterned zone plates and means that the minimum zone width (and thus first order spatial resolution) were initially in the 50–100 nm range. However, as noted in Section 1.4, several 45 and 50 nm hard X-ray zone plates are now operational at synchrotrons. At the same time, because the ratio of phase shifting to absorption f_1/f_2 improves as the energy is increased, the achievable efficiency becomes much higher and the depth of focus increases considerably (Jacobsen, 1992), which is helpful for applications such as tomography. Quantitatively, the transverse resolution of a zone plate is given by $0.61\lambda/\mathrm{NA} = 1.22\Delta r_n$ (Eq. (8)) and the depth of focus by $2\lambda/(\mathrm{NA})^2 = 8(\Delta r_n)^2/\lambda$. Therefore, a zone plate with $\Delta r_n = 50$ nm has a depth of focus of about 160 μm at 10 keV as opposed to about 8 μm at 500 eV. In addition, the focal length $f = 2r_n(\Delta r_n)/\lambda$ (Eq. (6)) for such a zone plate with 100 μm diameter increases from 2 to 40 mm, which considerably eases some of the challenges of mechanical design for specimen temperature control, insertion of fluorescence detectors, and so on.

In spite of the challenges of fabricating thicker zone plates using lithographic techniques, much success has been achieved. In some cases the initial electron beam lithography write has been transferred into a thicker plating mold using reactive ion etching as described above; this has led to the commercial availability (Xradia, Inc.) of a variety of high-aspect-ratio zone plates including one with outer-zone width 50 nm and thickness 700 nm, or an aspect ratio of 14:1 (an example of an earlier Xradia zone plate is shown in Figure 13–14). Additional parameters of zone plates given at the latest X-ray microscopy conference (Himeji, Japan, June 2005) are given in Section 1.4. Other approaches have involved using an electron-beam-written zone

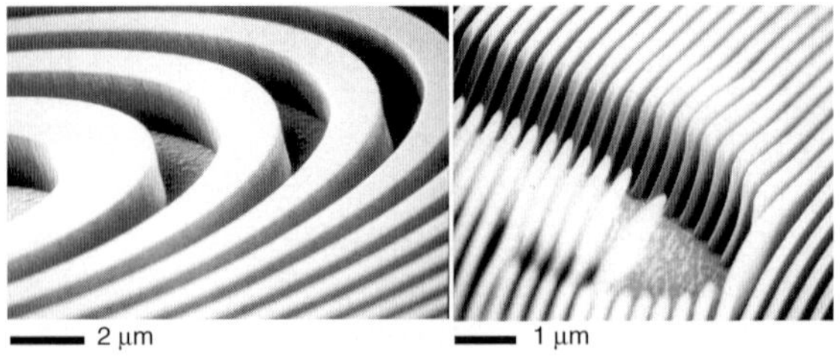

Figure 13–14. A hard X-ray zone plate with 100 nm outermost zone width and 1.6 μm thickness of gold for use at 5.4 keV in a commercial X-ray microscope system (Xradia, Inc.). The simultaneous achievement of narrow zone widths for high spatial resolution, and significant zone thickness so as to achieve a π phase shift, means that the achievement of high aspect ratio nanostructures is important. This zone plate has an aspect ratio of 16:1 and a theoretical focusing efficiency of 31%. (Figure Courtesy W. Yun, Xradia.)

plate as a mask for the subsequent processing of a thicker zone plate using X-ray lithography (Shaver et al., 1980; Lai et al., 1992), including the fabrication of 2.5 μm thick zone plates with a finest zone width of 0.25 μm (Krasnoperova et al., 1993). Another approach is to stack two or more zone plates together (Shastri et al., 2001).

Sputter-sliced or "jelly roll" zone plates (Schmahl et al., 1980) represent a completely different approach in fabrication. The goal of this approach is to start with a rotating wire and then build up alternating layers of weakly and strongly refractive material by sputtering or evaporation. The resulting structure is then sliced to yield zone plates of the appropriate thickness. In this case the achievement of high aspect ratios is not at all challenging; instead, the challenges include avoiding error and roughness accumulation in realizing the proper zone radii, the difficulties of maintaining perfect cylindrical symmetry, and the challenges involved in slicing the structure.

Recent results from the groups involved (Bionta et al., 1994; Tamura et al., 2002; Duvel et al., 2003) show that the technique is making steady progress to the point where a zone plate consisting of 70 Cu/Al layer pairs with outer zone width of 0.16 μm and aspect ratio of more that a thousand has been used to focus a 100 keV beam from Spring 8 to 0.5 μm FWHM (Kamijo et al., 2003). Similarly a sputter-slice soft X-ray zone plate with an aspect ratio of 200, made at Göttingen Germany, had 188 layer pairs of the alloy Ni80-Cr20 and silica. It showed a measured efficiency of 3.8% at 4.1 keV but had a focal spot size considerably greater than the 17-nm outer zone width. These are intriguing results, though for the moment, the sputter-sliced approach has not yet produced optics with an optical performance consistent with the intended geometrical parameters.

2.4.7 Thick Zone Plates

Up until now we have used kinematical diffraction theory to understand the properties of zone plates. In this theory, the incident wave and the diffracted signals from each volume element are all treated as independent. However, in reality, the "incident" wave and the "diffracted" waves in the solid structure are coherently coupled and if the zone plate is thick enough, the effect of this coupling will become evident at the output. Such coupling is known in perfect-crystal diffraction where it leads to anomalous transmission in Laue-geometry experiments (the Borrmann effect), while on a larger size scale, "Bragg-effect" holograms show similar behavior. Zone plates can be designed to exploit coupled-wave effects and these devices offer the possibility of very high efficiency and resolution in high diffraction orders, thus exceeding the resolution limit of the outermost zone width which applies when operating in first order. Theoretical treatments of diffraction by thick periodic structures have been developed in the hard-X-ray community (dynamical diffraction by crystals) (Batterman and Cole, 1964) and the optical-holography community (coupled-wave theory) (Kogelnik, 1969; Solymar and Cooke, 1981). The coupled-wave method has been applied to X-ray gratings and zone plates by Maser (1994) and more recently by Schneider (1997) who has given a solution that includes the case of high orders and gap-to-period ratios other than 0.5. Schneider's solution predicts that, in the soft X-ray region, absolute efficiencies of 30–50% in a single high order are indeed possible with line-to-space ratios of 0.1–0.5 and aspect ratios greater than 30:1. Hambach et al. (2001) have followed up these calculations with a series of experiments, mostly with copolymer gratings with aspect ratio 10:1. The predictions of the theory were broadly confirmed and a maximum efficiency of 15.3% was achieved at 13 nm wavelength. This value was 75% of the prediction of the coupled wave theory and 25 times greater than the prediction of thin-grating (kinematic) theory. The authors suggested that zone plates based on this principle may find application as condensers for table-top microscopes using isotropically emitting sources. A similar verification of dynamical diffraction theory for the case of sectioned multilayers illuminated with 19.5 keV X-rays in Laue geometry has been published recently (Kang et al., 2005). A reflecting efficiency of 70% was observed. Both of these experiments used gratings as being representative of the diffraction-efficiency behavior of a conventional zone plate (Hambach) and a sputter-sliced zone plate (Kang) respectively. However, as pointed out by Maser (1994), thick zone plates, like volume holograms, have a directional selectivity based on Bragg's law. That is, the zones must be oriented so that the incoming wave is locally Bragg-reflected by the zones and there will be a rocking curve outside of which the high efficiency is lost or moved to another order. Eq. (6) can be regarded as showing that Braggs law is obeyed at every zone of a conventional zone plate operating at magnification unity. On the other hand, for high magnification or demagnification applications, the zones must be tilted by an angle that varies with radius. The difficulty of doing this in practice is currently delaying the application of these ideas to practical zone plates.

A promising recent alternative to tilt angles that vary continuously with angle has been to apply the concepts of the "sputter-sliced" zone plate (see previous section) to produce linear half-zone-plates called "multilayer Laue lenses." One starts with an atomically smooth flat and deposits alternating zone materials, starting with the highest-order zones, so that error accumulation mainly affects the coarser, low-order zones. Two multilayer Laue lenses together can achieve two dimensional focusing in the manner of a Kirkparick-Baez mirror pair. Although tilting by a continuously variable angle is not exactly achieved by this approach it can be approximated by tilting the whole lens to a compromise Bragg angle. One dimensional line foci as narrow as 19 nm have already been reported by this method (Maser et al., 2004; Kang et al., 2006) and these authors believe that 5 nm may be possible in future work.

3 X-Ray Microscopes

In the previous sections we have described some of the characteristics of X-ray interactions and focusing optics. We now turn our attention to a discussion of X-ray microscopes currently in operation. They fall into two classes: full-field imaging and scanning, which are both illustrated in Figure 13–15. A large number of microscopes are listed in Table 13–3. We also describe three specific microscopes as examples: a transmission X-ray microscope (TXM) operated at Lawrence Berkeley National Laboratory, a scanning transmission X-ray microscope (STXM) operated at Brookhaven National Laboratory, and a scanning fluorescence X-ray microprobe (SFXM) operated at Argonne National Laboratory.

A key difference between TXM, STXM, and SFXM concerns the illumination phase space that can be accepted. In STXM and in SFXM, the size of the spot delivered by the zone plate objective is a convolution of the geometric image of the source and the point spread function

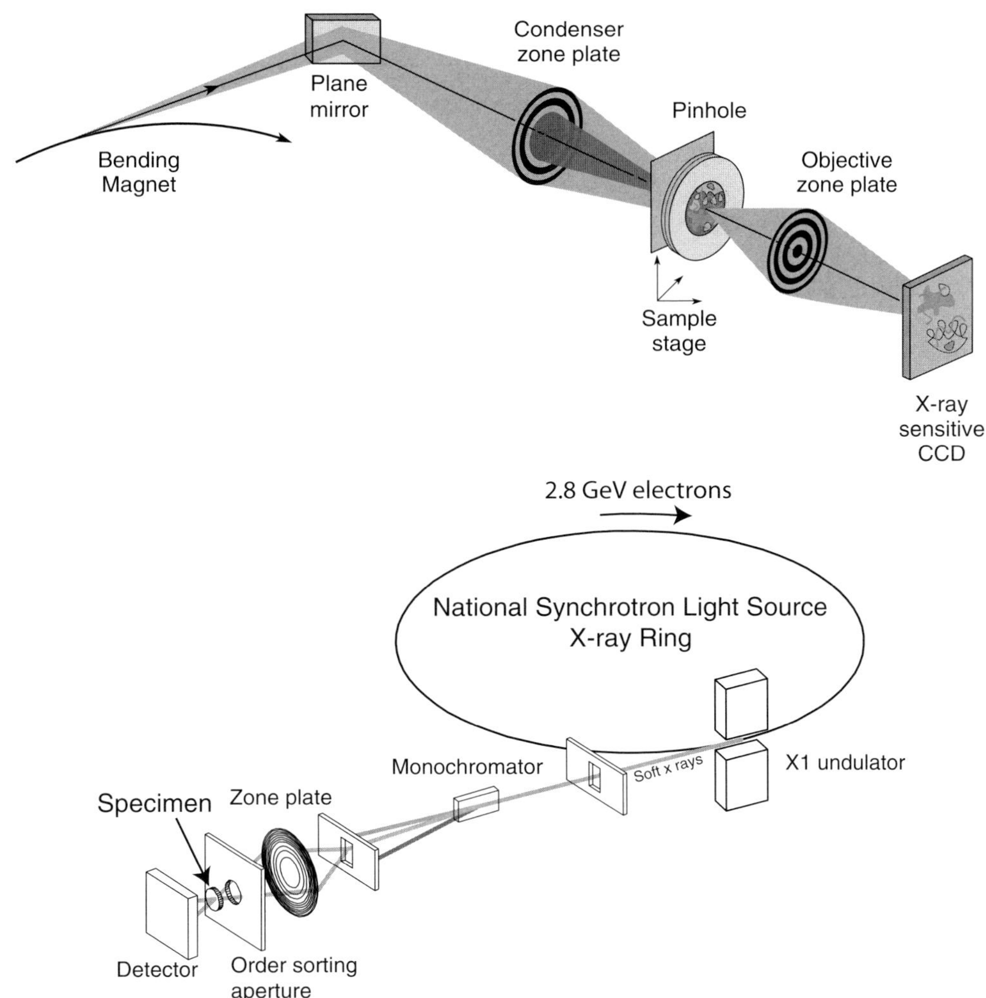

Figure 13–15. Schematic of the main components of a transmission X-ray microscope or TXM (top: courtesy of D. Attwood, Lawrence Berkeley National Laboratory) and a scanning transmission X-ray microscope or STXM (bottom: courtesy of Y. Wang, then of Stony Brook.) (See color plate.)

Table 13–3. Zone plate microscopes.

Microscope/ location	Light source	Illumination/ monochromator	Focusing, imaging	Contrast mechanisms	Techniques, X-ray energy	Citation
MES STXM	ALS undulator	grating	STXM	absorption, magnetization	NEXAFS, MCD 100 to 2000 eV	(Tyliszczak, 2004; Warwick, 2004)
Polymer STXM	ALS bend magnet	grating	STXM	absorption	NEXAFS 250 to 750 eV	(Warwick, 2002; Kilcoyne, 2003)
XM-1 TXM	ALS bend magnet	zone plate condenser/mono	TXM	absorption, magn-etization, phase	Tomog, MCD 200 to 1800 eV	(Meyer-Ilse, 2000a)
XM-2 TXM	ALS bend magnet	zone plate condenser/mono	TXM	absorption, phase	Tomog 200 to 7000 eV	
2-ID-B	APS undulator	multilayer-coated grating	STXM	abs, fluor, phase XANES, tomog	tomography 600 to 4000 eV	(McNulty, 2003a; McNulty, 2003b)
2-ID-D	APS undulator	Crystal/multilayer	STXM	diffraction	strain mapping 6 to 20 keV	(Cai, 2003, McNulty, 2003a)
2-ID-E	APS undulator	Crystal/multilayer	STXM	fluor, XANES diff, microdiff	5 to 35 keV	(McNulty, 2003a)
26-ID	APS undulator	Crystal/multilayer	STXM TXM	abs, fluor, diff XANES	3–30 keV	(McNulty, 2003a)
BL20B2	Spring8 bend magnet	crystal	STXM	absorption	4 to 113 keV opt testing, tomog	(Suzuki, 2003, Takano, 2003)
BL47XU	Spring8 undulator	crystal	TXM	absorption	5–37.7 keV tomography	(Suzuki, 2003, Uesugi, 2003)
BL20XU	Spring8 undulator	crystal—250 m beam line	STXM	absorption	8–37.3, 24–113 keV, μbeams	(Suzuki, 2003)
BL24XU	Spring8 undulator	crystal	TXM	phase contrast	8.77–12.85, 12.4–18.17 keV	(Tsusaka, 2001, Kagoshima, 2003)
BL12	Ritsumeikan bend magnet	zone plate condenser/mono	TXM	absorption	water window,	(Takemoto, 2003)

8A1 U7 SPEM	Pohang undulator	grating	STXM	photoemission	nanoXPS 100 to 1000 eV	(Shin, 2003, Yi, 2005)
1B2 hard xray	Pohang b. magnet	crystal	TXM	absorption	6.95 keV	(Youn, 2005)
U41TXM	BESSYII undulator	zone plate condenser/mono	TXM	absorption, phase	water window, 2D, 3D imaging	(Guttman, 2003, Wiesemann, 2003)
UE46TXM	BESSYII undulator	zone plate condenser	TXM	absorption, magnetization	MCD 0.2 to 2 keV	(Eimüller, 2003)
TWINMIC	ELETTRA Undulator	grating	STXM & TXM	absorption phase contrast	NEXAFS 250 to 2000 eV	(Kaulich, 2003)
BL2.2 ESCA	ELETTRA undulator	grating	STXM	absorption photoemission	nanoXPS 200 to 1400 eV	(Casalis, 1995; Kiskinova, 2003)
KINGS STXM	laser plasma	gas filtered spectrum	STXM	absorption	water window	(Michette, 2000)
X1A STXMs	NSLS undulator	grating	STXM	absorption diffraction phase contrast	NEXAFS, cryomicroscopy 250 to 1000 eV	(Jacobsen, 2000a)
ID21 microscopes	ESRF undulator	grating, crystal	STXM & TXM	absorption fluorescence diffraction phase contrast	NEXAFS 200 to 7000 eV	(Susini, 2000)
ID22 imaging	ESRF undulator	crystal	STXM	absorption fluorescence phase contrast	5 to 70 keV	(Weitkamp, 2000)
Aarhus TXM	ASTRID bend magnet	zone plate condenser/mono	TXM	absorption	typically 517 eV	(Uggerhøj, 2000)
XRADIA	Chromium anode	reflective condenser	TXM	absorption, phase	tomography 5.4 keV	(Scott, 2004)

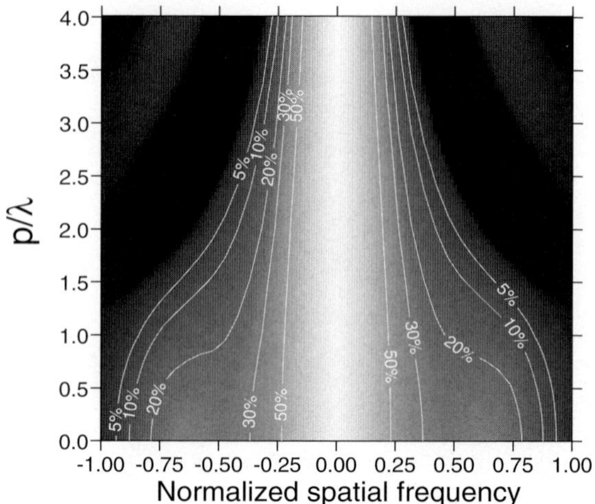

Figure 13–16. The aim in operating a scanning microscope or microprobe is to have a diffraction-limited focus. Therefore the source must be sufficiently demagnified so that it contributes negligibly to the focal width. This contour plot shows how the modulation transfer function (MTF) of an optic with a half-diameter central stop is affected by increasing the phase space parameter p of the source. This parameter $p = w\theta$ (source full width w times the full angle θ accepted by the optic) should be less than about the wavelength λ in both the x and y directions to achieve maximum spatial resolution. The normalized spatial frequency is defined to be unity at the MTF cutoff of $1/\Delta r_n$. (From Winn et al., 2000.)

of the optic. As Figure 13–16 shows, for an objective with diffraction-limited (as opposed to aberration-limited) resolution, the effect of the geometric source size becomes negligible if the product $p = w\theta$ of source width w times the full angle θ accepted by the optic is less than the wavelength λ in each dimension. This is commonly summarized by saying that scanning microscopes require single-mode illumination, although it is understood that a spatially filtered, incoherent source is not the exact equivalent of a single-mode optical cavity. The situation in TXM is much different; for incoherent bright field imaging, each pixel in the object can be imaged independently of its neighbors (within good approximation), so one can illuminate all object pixels simultaneously and with nominally incoherent light. If object resolution elements are imaged in 1:1 correspondence to detector pixels in a TXM, the number of "modes" of phase space p/λ that can be accepted in the x direction is approximately equal to the number of detector pixels in that direction and the same holds for y. As a result, TXMs are often operated with bending magnet synchrotron radiation sources or laboratory sources which deliver high flux (photons per solid angle), while STXMs and SFXMs are often operated with undulator sources which deliver high brightness (photons per solid angle per source area). The issues of microscope illumination and its effects on image formation will be discussed in more detail in Section 3.1.

3.1 Microscope Layouts and Illumination Schemes

3.1.1 Transmission X-Ray Microscope (TXM) Layout

Full-field transmission X-ray microscopes (TXMs) typically use a zone plate to produce a magnified image of the specimen on a 2D detector. This approach was pioneered by the group of G. Schmahl at the Universität Göttingen, who, after initial experiments including reflection-grating monochromators (Niemann et al., 1976) switched to using a condenser zone plate as the sole monochromator (Rudolph et al., 1984). This latter approach is now used by a number of TXMs, including the XM-1 at Lawrence Berkeley Lab (Meyer-Ilse et al., 1994, 2001) for which we provide some example numbers. As shown in Figure 13–15a, the beam from the synchrotron bending magnet source is deflected by a grazing-incidence mirror which filters out the power due to high-energy X-rays, passes through a thin metal filter to remove visible and ultraviolet radiation, and is then imaged by the condenser zone plate onto a pinhole located just upstream of the specimen. As noted in Section 2.4.3 on condenser zone plates, the condenser zone plate (of diameter $D = 9\,\text{mm}$) and the pinhole (of diameter $d \approx 10–20\,\mu\text{m}$), together are equivalent to a monochromator of resolving power equal to $D/(2d)$ (Niemann, 1974). Because the light transmitted by the objective zone plate includes a significant undiffracted (zero order) component which must not reach the detector, the illumination of the sample needs to be hollow-cone and this is achieved by means of a stop built into the condenser, blocking a central circle of radius about one third to one half of the condenser radius. The objective zone plate used by XM-1 in the resolution test described above had the following characteristics: outer zone width $\Delta r_n = 15\,\text{nm}$, diameter $d = 30\,\mu\text{m}$, 500 zones of 80 nm thick gold (giving a maximum aspect ratio of 5:1), and focal length $f = 0.3\,\text{mm}$ at 815 eV. This is the highest resolution zone plate used to date and slightly larger outer zone widths (25–30) are used for routine user operations. The vertical phase space area of the synchrotron source is generally smaller than its horizontal phase-space area and smaller than that of the microscope (which equals object full-width d times twice the objective NA). Since the condenser zone plate cannot expand the phase space, both the object width and the numerical aperture of the objective of a TXM will generally be underfilled. To counter the under filling of the object field, the condenser is usually "wobbled" up and down during the course of an exposure. This type of microscope layout, in which the source is imaged on to the sample, is known as "critical illumination" (Born and Wolf, 1999) and is widely used for amplitude contrast.

Traditionally, the specimen has been placed in an atmospheric pressure environment and to accomplish this, thin vacuum windows (100 nm Si_3N_4 or Si are common) can be used between the condenser and the specimen, and also between the specimen and the objective zone plate. Because the focal length of the objective zone plate is quite small (for example, in the case of a 25-nm-outermost-zone-width, 60-μm-diameter zone plate operating at 530 eV it would be 1.3 mm.), the specimen region lying between these two windows is quite constrained.

The beam then re-enters a vacuum environment where the objective zone plate and the image detector are located. At energies below a few keV, the most common detector is a backside-thinned CCD which is directly illuminated by the X-ray beam; at higher energies, phosphor screens imaged by a visible light lens onto a CCD detector are commonly used. Because of the desire to deliver 10–50 nm resolution using detectors with 1–20 µm pixel size, the distance from the zone plate objective to the detector is often in the range 1–2 m to give acceptably high optical magnification.

The approach described above is commonly used with bending magnet and laboratory sources. Particular challenges arise when the source phase space area is dramatically smaller than desired, which is the case for undulator sources on low emittance storage rings. As an example, Niemann has studied a variety of solutions for the condenser of the current Berlin TXM which is illuminated by a BESSY II undulator (Niemann, 1998). The adopted solution (Niemann et al., 2000) involves a zone plate segment, and three flat mirrors. The magnification of the zone plate is chosen to fully illuminate the object field. The first mirror is fixed but must be tilted for a change of wavelength. The other two mirrors are mounted in a structure that rotates and delivers an incoherent hollow-cone beam of which the inner-to-outer angular difference ($\Delta\vartheta$ say) is determined by phase-space matching and the outer angle (the NA) is determined by the last-mirror reflection angle. Thus the illuminated area and the NA can both be chosen, no wobbling is required and Liouville's theorem (which states that the phase space area of an optical beam is a conserved quantity) is respected by allowing $\Delta\vartheta$ to float. This system represents an elegant optical solution, though it is mechanically quite complex. Other strategies to expand the phase space of an XUV beam have been explored by the microfabrication community who are concerned about "fringing" in XUV lithography (Murphy et al., 1993; White et al., 1995). One such approach is to design a pseudorandom diffractive optic specifically to "spoil" the phase space of a beam and match the object size and the NA (David et al., 2003); such optics must meet the challenge of evenly filling both the object plane and the back focal plane with light.

3.1.2 TXM Phase Contrast Layout

As noted above, phase contrast plays an important role in X-ray microscopy, particularly at higher photon energies. In order for phase variations at the specimen plane to produce intensity variations at the detector, some method of mixing the wave diffracted by the specimen with an undiffracted phase-reference wave must be employed. The most common approach in X-ray microscopes is that of Zernike. In light microscopes, Köhler illumination is provided by using a relay lens to image the source on to the front focal plane of the condenser. Points at this front focal plane deliver parallel beams to the object plane which (if undeviated by the object) are focused on to a phase ring at the back focal plane of the objective where they are phase shifted usually by $\pm\pi/2$. Thus the ring aperture at the front focal plane of the condenser provides a narrow, hollow cone of illumination of the specimen, and is conjugate to the phase ring. At the same time, light originating from a point scatterer in the object is focused to the detector, where it interferes with the phase-shifted unscattered light (see Figure 13–17).

In X-ray microscopes, it is more difficult to use a relay optic to work in the Köhler illumination condition; instead, the front focal plane ring aperture is illuminated by nearly parallel light from the source (i.e., critical illumination) so that a much smaller area of the condenser is illuminated. Because the light from the front aperture is much more collimated than would have been the case with Köhler illumination, the longitudinal location of the aperture and its corresponding phase ring is much less critical than it is with visible light microscopes so the

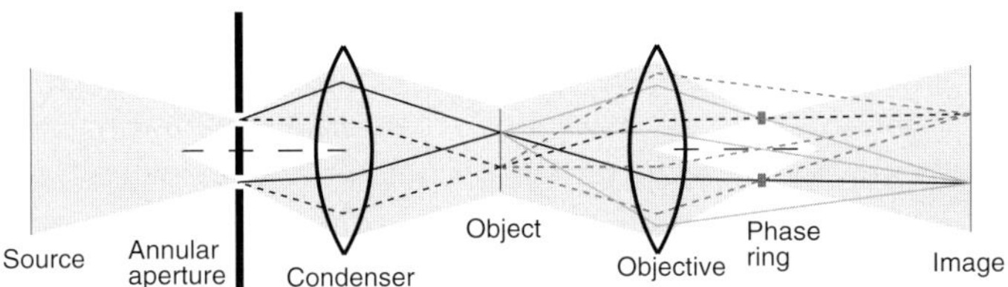

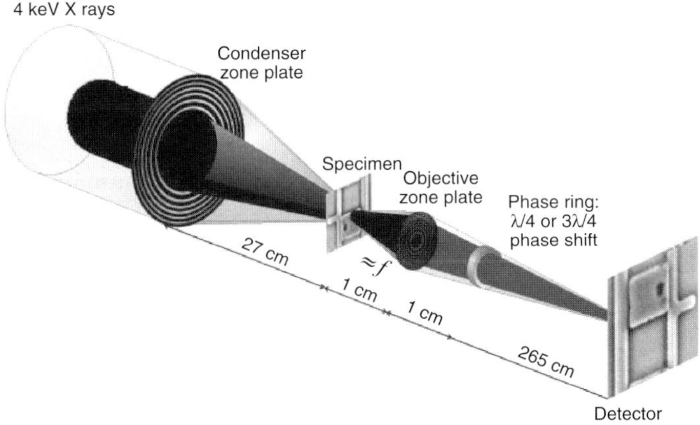

Figure 13–17. Two illustrations of a Zernike phase contrast optical system. The upper one shows the classical scheme used in light microscopes based on Köhler's illumination (Born and wolf, 1999). Light from each point of the annular aperture, placed in the front focal plane of the condenser, is delivered to the object as a parallel beam. Two object points are shown receiving example rays from the source. The rays may be undeviated by the object, in which case they are seen to pass through the phase ring on their way to the detector. On the other hand the rays may be deviated (diffracted) by the object in which case they reach the detector without passing through the phase ring. Interferences between these two types of optical signal result in a mapping of the object phase variations into an intensity pattern on the detector. The lower diagram shows a practical synchrotron radiation implementation of a Zernike-phase-contrast TXM at the 4.1 keV beam line on ID21 at the European Synchrotron Radiation Facility in Grenoble. An undulator X-ray source is followed by a crystal monochromator illuminating a condenser zone plate (which can be small since it does not have to act as a linear monochromator). The condenser provides critical illuminated to the sample rather than Köhler and, due to the good collimation of the incoming beam, the illuminated area of the condenser is projected on to a phase ring in the back-focal plane of the objective zone plate so as to provide phase contrast. The tradeoffs involved in choosing the illuminated area of the condense pupil are discussed in Section 3.3.6 (Courtesy J. Susini, ESRF.)

primary alignment requirement is transverse to the X-ray beam direction. For computations it is convenient to consider the two rings as built into the lens pupil functions. Evidently when the source is well-collimated, as in the case of a synchrotron, there is no need for the two rings to be exactly conjugate. The use of the ring aperture obviously makes the illumination more coherent. The question of the best choice of width and radius for the two rings or, equivalently, how much coherence to have, we defer until later (Section 3.3.6). The phase ring itself is constructed out of a material with a large phase shift per absorption length $f_1/(2f_2)$, such as any material used at energies just below an absorption edge.

Zernike phase contrast X-ray microscopy was pioneered by Schmahl and Rudolph (1987) and Ruldoph et al. (1990), and the Göttingen group has shown impressive results in phase contrast for water-window imaging of biological specimens. The ring aperture and phase ring used in these experiments could be rapidly inserted or retracted (Schmahl et al., 1994, 1995; Schneider, 1998). Phase contrast is arguably even more important at higher X-ray energies where it is the dominant contrast mechanism. For example a hard X-ray (4 keV) phase-contrast microscope, illuminated by a system using a crystal monochromator followed by a condenser zone plate, is operating at the ID21 beam line at the European Synchrotron Radiation Facility in Grenoble, France (see Figure 13–17). Imaging of functioning integrated circuits at 60 nm resolution has been demonstrated (Neuhäusler, 2003); see Section 4 for more information on this. In another example of this configuration, the National Synchrotron Radiation Research Center in Hsinchu, Taiwan has installed a TXM built by Xradia. This instrument (alluded to in Sections 1.4 and 2.4.4) has been used to demonstrate the long-existing idea of using zone plate higher focal orders for imaging. In water-window instruments which typically have focal lengths on the order of 1–2 mm in first order (and therefore 1/3–2/3 mm in third order), the idea has not been readily adopted due to practical considerations of working distance. However, in the Taiwan experiment (Tang et al., 2006) a phase contrast image of a fabricated test object was made at 8 keV using a 50 nm outer-zone-width zone plate in third order. Lines of minimum width 30 nm were imaged clearly and the authors estimate a resolution below 25 nm. This is evidently a most important development (see Section 5).

3.1.3 Scanning Ttransmission X-Ray Microscope (STXM) Layout

Scanning transmission X-ray microscopes (STXMs) typically use a zone plate to demagnify a pinhole source to a small focus spot through which the specimen is scanned. While initial demonstrations using synchrotron radiation used pinhole optics (Horowitz and Howell, 1972; Rarback et al., 1980), the use of zone plate optics in scanning microscopes was pioneered by Rarback et al. (1984) and later by (Niemann, 1987; Niemann et al., 1988). (Normal incidence optics with synthetic multilayer reflective coatings have also been used in the 50–120 eV range (Haelbich, 1980a; Haelbich et al., 1980b; Ng et al., 1990)). Since scanning microscopes require coherent illumination to reach their maximum resolution, they have often used undulators as high brightness sources (Rarback et al., 1988; Kenney et al., 1989; Morrison et al., 1989b) though excellent performance has also been obtained using bending magnet sources on low emittance storage rings (Kilcoyne et al., 2003). While a large number of STXMs are now in operation, we describe here the characteristics of the most recent in a series (Rarback et al., 1988; Jacobsen et al., 1991; Feser et al., 1998, 2000) of undulator-based scanning microscopes built at Stony Brook University for operation at the National Synchrotron Light Source at Brookhaven National Laboratory in New York. A soft X-ray undulator plus spherical grating monochromator with an energy resolution that can be as good as 0.06 eV at 290 eV (Winn et al., 2000) is used to deliver soft X-rays to a 2D exit slit which can limit the beam size in the range 25–120 µm in both x and y. This slit then serves as a secondary radiation source for zone plates of either 80 or 160 µm diameter and zone widths of 30–45 nm (Spector et al., 1997; Tennant et al., 2000), producing a focal spot of 36–54 nm Rayleigh resolution. The beam emerges from the ultra high vacuum synchrotron beam line into an atmospheric pressure environment by passing through a 100 nm thick Si_3N_4 window. The zone plate includes a central stop of about half the zone plate diameter; this stop must be made quite thick (0.3 µm gold is common for soft X-ray applications) so that the undiffracted light transmitted through the large central stop is kept to a very small level compared to the flux in the focused X-ray beam. The zone plate is then followed by an order sorting or selecting aperture (OSA) so that a pure first-order focal spot is obtained.

While steering mirrors are used to scan the beam in visible light scanning microscopes, it is easier to maintain signal uniformity by keeping the beam and zone plate fixed and scanning the specimen through the focal spot. This is accomplished using an X-Y-Z stack of stepping motor stages for large motion with 1 µm precision, and a piezo scanning stage for 50–100 µm range and nanometer precision. Because piezos have nonlinearities and hysteresis in their response to scan voltages, some form of closed-loop feedback is generally used, based on position signals such those provided by linear voltage differential transformers (Kenney et al., 1985), capacitance micrometers (Jacobsen et al., 1991), or laser interferometers (Shu et al., 1988; Kilcoyne et al., 2003); the latest Stony Brook STXM allows the user to choose between capacitive or laser interferometer feedback. The specimen is then followed by a high efficiency X-ray detector; common choices include the use of gas-based proportional counters which offer extremely high efficiency of detection for those X-rays that make it through a thin entrance window (Rarback et al., 1980; Kenney et al., 1985; Feser et al., 2000) but which suffer from a count-rate limit of about 1 MHz. Alternatives are phosphor-coated screens followed by photomultipliers to detect the resulting visible light (Maser et al., 2000), and solid state detectors which are capable of significantly higher signal rates (Barrett et al., 1998; Wiesemann et al., 2000; Feser et al., 2001, 2003; Guttmann et al., 2001). In the Stony Brook STXM, the user can choose between proportional counter and segmented silicon detectors, and a visible light microscope is also placed on the detector stage with X-Y-Z motorized motion so as to pre-locate desired regions of the specimen. Another approach, which works for either a TXM or a STXM, is co-indexing of off-line light microscopes with the X-ray microscope (Meyer-Ilse et al., 1994, 2001; Kilcoyne et al., 2003).

Scanning microscopes offer different characteristics than full-field imaging systems do. These include the ability to quickly change from scanning very large areas at low resolution to taking high resolution, small field scans, and reduced radiation dose because the 5–20% efficient zone plate is located upstream of the specimen rather than downstream. Because of the need to mechanically scan the specimen in most present microscopes, and the need for coherent illumination, imaging times are generally longer (in the range of one or a few minutes, rather than seconds in the case of many TXMs). At the same time, the requirement for coherent illumination means that the etendue or phase space that the monochromator must accept is greatly reduced, so that aberrations are reduced and it is relatively easy to obtain very high spectral resolution. These characteristics make scanning transmission X-ray microscopes especially well suited to low-dose spectromicroscopy applications, as will be described below.

Phase contrast has historically seen less use in scanning transmission X-ray microscopes. However, refractive and diffractive effects by the specimen lead to a redistribution of signal on the detector which can be interpreted to give phase contrast images (as will be discussed below). The ultimate approach is to use a 2D detector (such as a CCD camera) to detect the entire intensity distribution at each pixel of a scanned image; Chapman has used this in an impressive demonstration of Wigner deconvolution microscopy to recover the phase and magnitude distribution of the specimen as well as the zone plate objective (Chapman, 1996a), while Morrison et al. have used this to obtain first moment images which reveal the dominant phase gradient at each pixel location (Eaton et al., 2000; Morrison et al., 2002). Coupled with the potential power of these approaches are significant challenges: the readout time of large pixel detectors is often not in the few or sub millisecond pixel timescale required for fast scanning, and the resulting 4D data files are quite large. More fundamental is the question of statistical significance in each pixel of a large array detector when radiation dose to the specimen must be considered; in some cases it may be preferable to divide a weaker signal into fewer detector segments. This approach has been used by Feser et al. (Figure 13–18) who have used a detector with only 8 segments to obtain quantitative phase contrast images while operating at per-pixel acquisition times of a few milliseconds and producing data files of manageable size (Feser et al., 2003) (see Figure 13–18). Additional approaches to obtaining phase contrast in scanning microscopes include the use of zone plate doublets (Kaulich et al., 2002) or phase modifiers (Polack et al., 2000) to produce differential interference contrast. Undoubtedly different experiments will involve different choices in the tradeoff of the fineness of segmentation of scanning microscope detectors, but in any case

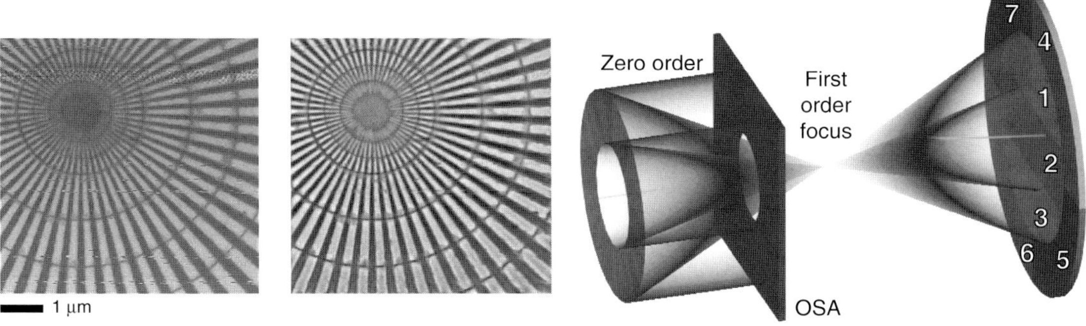

Figure 13–18. Amplitude (left) and phase (right) contrast images of a germanium test pattern imaged using a scanning transmission X-ray microscope with a segmented detector. Also shown is a schematic view of the detector. The undeflected beam from the first order focus is directed into bright field segments 1, 2 and 3 (these segments also allow differential interference contrast). The deflected beam is detected in the angular segments 4, 5, 6, and 7 for dark field imaging and differential phase contrast. (Reprinted from Feser et al., © 2003, with permission of EDP Sciences.)

it is clear that phase contrast plays an interesting role in scanning transmission X-ray microscopy as well as in microprobes as will be noted below.

3.1.4 Scanning Fluorescence X-Ray Microprobe (SFXM) Layout

Scanning fluorescence X-ray microprobes (SFXM) use a focused X-ray beam to stimulate the emission of characteristic fluorescence X-rays from specific elements in the specimen. When linearly polarized radiation (such as is usually obtained from synchrotron sources) is used, a fluorescence detector placed 90° to the beam in the polarization plane will detect a minimum of coherent scattering signal; this detector must then have some means of discriminating between different X-ray emission energies. Energy-dispersive detectors accomplish this by measuring the number of electron-hole pairs created by each X-ray in a semiconductor material, while wavelength-dispersive detectors use a crystal optic or a grating to separate the X-ray energies. Energy dispersive detectors generally have large solid angle collection, and multi-element detectors can be used to overcome the ~50 kHz count rate limit determined by charge readout time, while wavelength dispersive detectors offer better separation between nearby spectral lines and larger dynamic range for detecting low concentration elements amongst other fluorescing elements of higher concentration. There is a long and rich history of synchrotron-based microprobes (Horowitz and Howell, 1972; Sparks, 1980; Rivers et al., 1988; Thompson et al., 1988; Hayakawa et al., 1989), and a variety of optical approaches including the use of compound refractive lenses and Kirkpatrick-Baez mirror optics are now achieving submicron resolution. We outline here some of the characteristics of microprobes using zone plate optics (Barrett et al., 1998; Yun et al., 1998a; Suzuki et al., 2001; Kamijo et al., 2003) by considering the example of the 2-ID-E microprobe at the Advanced Photon Source at Argonne National Laboratory near Chicago.

This microprobe operates using a side-deflecting crystal monochromator to transfer an off-axis part of the central cone produced by a hard X-ray undulator. In the vertical direction, the objective zone plate images the source directly onto the specimen, while in the horizontal direction the variable width monochromator exit slit is imaged. Astigmatism effects are avoided in the resulting focused beam by the fact that the depth of focus is much larger than the difference between the positions of the horizontal and vertical foci of the zone plate (in addition, the zone plate can be tilted to compensate for more severe source astigmatism). Zone plates of diameter 160–320 µm and outermost zone width of 100 nm are typically used, giving focal lengths of 12–25 cm at 10 keV. While the probe size can be as small as 150 nm, a larger horizontal source size is often chosen to give more flux at the cost of resolution. The specimen is mounted at 15° to the incident beam to provide access to both the incident X-ray beam and the fluorescence detector, and it is scanned by motor-driven stages with 0.1 µm step size. A multi-element germanium fluorescence detector is used to collect the fluorescent signal; one can either record the signal in a limited number of pre-defined energy windows for rapid analysis with modest data file size, or record the full fluorescence spectrum per pixel for improved quantitation of elements with closely spaced fluorescence energies. The region consisting of the specimen and detector is located inside a glovebox which can be purged with helium to eliminate fluorescence from argon in air which would otherwise obscure a number of low-Z elements, and to reduce the absorption of low-Z fluorescence signals by air. Per-pixel dwell times are on the order of one second, so that the experimenter must be judicious in the choice of scan area (the use of common position indexing between a visible light microscope and the microprobe aids in rapid specimen location).

Zone plates of diameter 50–100 µm and outermost zone width of 100–300 nm are typically used, giving focal lengths of 5–30 mm. While the probe size can be as small as the Rayleigh resolution of 1.22 times the outermost zone width, a larger virtual source size is often chosen to give more flux at the cost of resolution. The specimen is mounted at 45° to the incident beam to provide access to both the incident X-ray beam and the fluorescence detector, and it is scanned by motor-driven stages with 0.1 µm step size. A multi-element germanium fluorescence detector is used to collect the fluorescent signal; one can either record the signal in a limited number of pre-defined energy windows for rapid analysis with modest data file size, or record the full fluorescence spectrum per pixel for improved quantitation of elements with closely spaced fluorescence energies. The region consisting of the zone plate, specimen, and detector is all located inside a glovebox which can be purged with helium to eliminate fluorescence from argon in air which would otherwise obscure a number of low-Z elements. Per-pixel dwell times are on the order of one second, so that the experimenter must be judicious in the choice of scan area (the use of common position indexing between a visible light microscope and the microprobe aids in rapid specimen location).

Trace element mapping by fluorescence detection with sensitivities down to about 100 parts per billion, or about 10^{-17} grams of iron within a $(200\,\text{nm})^2$ spot, represents the majority of microprobe applications. However, X-ray microprobes can be used in a number of other ways as well, including measurements of crystal strain in small regions (Rebonato et al., 1989; Cai et al., 1999; Soh et al., 2002) and differential-aperture measurements of microstructure and strain (Larson et al., 2002). The phase contrast methods described above for STXM are equally applicable in SFXM, and offer a much-needed way to image the overall mass and ultrastructure of specimens while simultaneously forming trace element or strain maps.

3.2 Fundamentals of Contrast in the TXM

It is useful to have an analytical treatment that provides insight into the way a microscope produces contrast and at the same time allows simple calculations to assess experimental plans. This was provided by Rudolph and coworkers (1990) in a form that allows amplitude-contrast, Zernike-phase-contrast and dark-field imaging, to be included in a unified description, that is largely independent of the microscope design. Assuming only that we have an *imaging* microscope, we consider first the Zernike phase-contrast TXM.

We are interested in the contrast C or the contrast parameter Θ (see Section 1) between an interesting feature F and a background feature B generated via the phase shifter S. F and B are defined to have the same thickness but in reality the background material (water for example) may be thicker than the feature so we allow for that by adding a layer L. If we define the complex transmission factors of F, B, S and L as $a_F \cdot p_F \equiv [\exp\{-2\pi\beta_F t_F/\lambda\}] \cdot [\exp\{2\pi i\delta_F t_F/\lambda\}]$ etc where $1-\delta-i\beta$ is the refractive index and t is the thickness, then following Rudolph et al. (1990) we can obtain the image intensities (I_F and I_B) and thence C and Θ

$$C = \frac{(I_F - I_B)}{(I_F + I_B)}, \qquad \Theta = \frac{(I_F - I_B)}{\sqrt{I_F + I_B}}$$

$$I_F = \left\{ a_B^2 a_S^2 + 2a_F a_B a_S \,\text{Re}\left[p_F p_B^* p_S^* \right] - 2a_B^2 a_S \,\text{Re}\left[p_S^* \right] \right.$$
$$\left. + a_F^2 - 2a_F a_B \,\text{Re}\left[p_F p_B^* \right] + a_B^2 \right\} a_L^2.$$

$$I_B = a_B^2 a_S^2 a_L^2$$

The dose D (the energy deposited per unit mass of sample) needed to detect a feature of area d^2, thickness $t_F = d$ and density ρ with signal-to-noise ratio S/N can now be calculated (Rudolph et al., 1990) as follows

$$D = \left(\frac{S}{N}\right)^2 \frac{hc}{\lambda\rho d^3} \frac{1 - a_F^2 a_L^2}{\Theta^2}$$

where $hc = 1240\,\text{eV-nm}$ represents the product of Planck's constant and the velocity of light. The above relations are convenient because, in

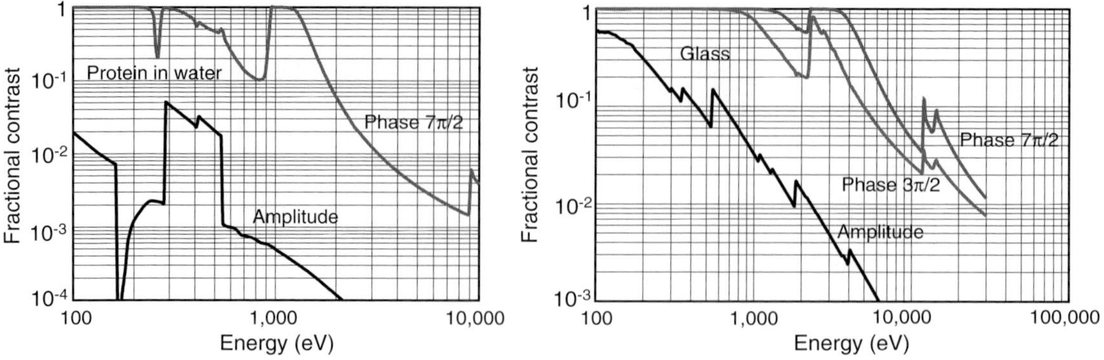

Figure 13–19. Intrinsic amplitude and Zernike phase contrast for two types of sample of thickness 30 nm relative to a background material of the same thickness: protein in water (left) and vacuum in glass, representing a crack (right). The protein is modeled assuming a density of 1.35 g/cm^3, composition of $H_{50}C_{30}N_9O_{10}S$ and a phase ring made of copper. Glass is modeled assuming a density of 2.5 g/cm^3, composition $Si_{16}Na_{12}K_1Ca_7Mg_6P_1O_{57}$ and a phase ring made of gold. It is noteworthy that the phase contrast can be much greater than the amplitude contrast even in the 290–540 eV water window.

addition to phase contrast, they also describe amplitude-contrast ($t_s =$ 0) and dark-field ($t_s =$ large) experiments. The formula for D yields dose plots like Figure 13–5 and also tells us that the number of X-rays (of energy E) per unit area required to make the measurement with the given resolution and signal-to-noise ratio is $D\rho/(\mu E)$, where μ is the X-ray absorption coefficient.

In the multi-keV X-ray energy range, the phase contrast is substantially larger than the absorption contrast for suitable choices of the thickness of the phase shifter. The best result is typically achieved by attenuating the direct beam by the phase plate so that its amplitude is comparable to that of the scattered signal, resulting in an interference of two beams of similar amplitude. The available choices of phase plate thickness to optimize this are positive phase contrast (phase shift = $\pi/2$, $5\pi/2$, . . .) or negative phase contrast (phase shift = $3\pi/2$, $7\pi/2$, . . .). Figure 13–19 shows contrast plots of some of these possibilities. Although these plots are useful for providing comparative information, they represent a considerable idealization; the phase shift is assumed to be applied to 100% of the undiffracted light and 0% of the diffracted light, the thickness of the phase shifter is chosen, at each energy, to give the stated phase shift and the optical system is assumed 100% efficient. Under these assumptions, the dark-field contrast is identically equal to one. This suggests that dark-field has a dose advantage that will be dependent on the practical value of the nominally zero signal due to the undiffracted light and on the strength of the dark-field signal (Chapman et al., 1996c; Vogt et al., 2001a).

3.3 Partial Coherence

3.3.1 History

The resolution of microscopes, including X-ray microscopes, depends on the angular widths of the light beams delivered to, and collected from, the sample . The analysis of this effect was pioneered in the 1950s by Hopkins, Wolf and others and was part of a movement to apply the linear-systems ideas, widely used by the engineering community, in the optical arena. This work has been reviewed by Hopkins (1957), Thompson (1969), and in various texts (Wilson and Sheppard, 1984; Goodman, 1985; Born and Wolf, 1999). The main point is that the finest features (highest spatial frequencies) in the sample diffract the illuminating beam by the largest angles θ. The best geometry to include such large deflection angles is therefore one that has a wide-angle beam both inward to, and outward from, the sample. This implies broadly that in a TXM, a large-area source providing *spatially incoherent* illumination gives better resolution than a point source giving *coherent* illumination (Figure 13–20), although such a comparison is not as simple as it sounds (Goodman, 1968). It would be wrong to conclude from this that STXMs which use coherent illumination have intrinsically worse resolution than TXMs. In fact, scanning microscopes with large area detectors and transmission microscopes with large illumination angles both deliver incoherent bright-field images with the same resolution and transfer function provided only that they use objective lenses of equal resolution.

The first application of linear-systems concepts in X-ray microscopy was in the analysis of STXM images (Jacobsen et al., 1991; Zhang et al.,

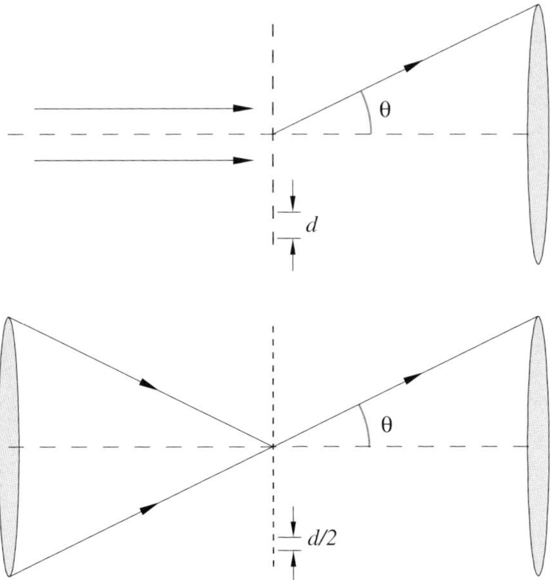

Figure 13–20. Schematic showing why the transfer function for incoherent imaging extends to twice the spatial frequency of coherent imaging for a given optic numerical aperture. For coherent imaging (top) the marginal ray is deviated by an angle theta due to diffraction by the sample periodicity d. For incoherent imaging (bottom) some light is deviated by 2θ due to the sample periodicity $d/2$. Both TXM and STXM with large area detector deliver incoherent bright-field images. (After Jacobsen et al., 1992b.)

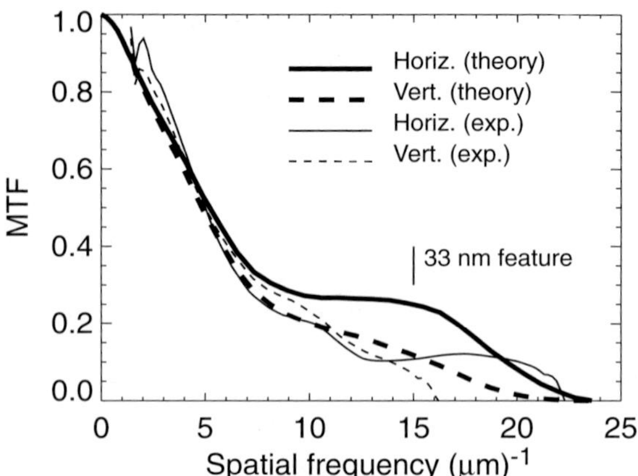

Figure 13–21. Measured and calculated modulation-transfer function for STXM imaging with a nickel zone plate of outer zone width 45 nm and diameter 100 nm. The calculated curve was derived from the known zone plate aperture function and the size and distance of the source pinhole. (Reprinted from Jacobsen, © 1991, with permission from Elsevier.)

1992) in which the intensity point spread function and its Fourier transform, the optical transfer function (OTF) were calculated. In fact, the magnitude of the OTF, known as the modulation transfer function (MTF), was both calculated and measured for the Stony Brook STXM and good agreement was obtained (Figure 13–21). Similar analysis has been provided for TXMs (Jochum and Meyer-Ilse, 1995; Niemann et al., 2000). Jochum and Meyer-Ilse provided a fairly general treatment of the application of coherence theory to X-ray microscopy including imaging of two-point and step objects by a realistic TXM in bright-field amplitude contrast. Other discussions of coherence issues have been provided by (Heck et al., 1998; Chao et al., 2003).

3.3.2 Fourier Optics Treatment

Partially coherent imaging by a microscope can be described generally by the methods of Fourier optics (Wilson et al., 1984; Goodman 1985; Born and Wolf, 1999). This method uses a real-space and a frequency-space description of waves in which frequencies (u) are closely related to directions (θ) according to the general (1D) relation $u = \sin\theta/\lambda$. Following Chapman et al. (1996c), we consider first a STXM with amplitude point spread function $h(\mathbf{x})$ imaging a sample of amplitude transparency $t(\mathbf{x})$. Using capital letters to represent Fourier transforms, the pupil function of the lens is $H(\mathbf{u})$ where $\mathbf{u}$ is the general frequency coordinate, that is conjugate to the object-plane spatial coordinate $\mathbf{x}$ and has a maximum value of NA/λ where NA refers to the beam-limiting lens. Any point in the lens pupil or the detection plane can be represented by a $\mathbf{u}$ value. Since the detector in a STXM is placed in the far field of the X-ray focal spot, the diffraction pattern formed in the detection plane in the absence of a sample will be $H^*(-\mathbf{u})$. When the sample is present and the spot is at $\mathbf{x}_s$, the wave field immediately behind the sample is $h(\mathbf{x})t(\mathbf{x} - \mathbf{x}_s)$ and the field in the far-field detection plane is given by the Fourier transform of that. The detected intensity is therefore

$$F(\mathbf{u},\mathbf{x}_s) = |H(\mathbf{u})\otimes_{\mathbf{u}}T(\mathbf{u})e^{2\pi i x_s \cdot \mathbf{u}}|^2 \qquad (33)$$

where $\otimes$ represents convolution and the convolution and shift theorems have been used. The same quantity $F(\mathbf{u},\mathbf{x})$ can also be represented in another useful way. By inserting the representations of $H(\mathbf{u})$ and $T(\mathbf{u})$ as Fourier integrals into the convolution integral Eq. (33) and using the Fourier-integral definition of the delta function (Born and Wolf, 1999) and then its sifting property, we obtain (Chapman et al., 1996c)

$$F(\mathbf{u},\mathbf{x}) = |h(\mathbf{x})\otimes_{\mathbf{x}}t(\mathbf{x})e^{-2\pi i x \cdot \mathbf{u}}|^2, \qquad (34)$$

The first of the above two equations represents the diffraction pattern formed in the detection plane by a STXM at each scan position as $F(\mathbf{u},\mathbf{x}_s)$, regarded as a function of $\mathbf{u}$ for a given $\mathbf{x}_s$. The second equation represents a coherent image in a TXM, for illumination direction $\mathbf{u}$, as $F(\mathbf{u},\mathbf{x})$, regarded as a function of $\mathbf{x}$ for a given $\mathbf{u}$. In the first case the exponential represents the scan shift $\mathbf{x}_s$ and in the second case it represents the incoming plane wave at direction $\mathbf{u}$. This optical equivalence of the STXM and TXM is known as "reciprocity" and is discussed further in Section 3.3.4. To get the delivered intensity image $I(\mathbf{x})$ in either case one has to integrate the signal in the detection plane over the particular distribution of $\mathbf{u}$ values that are used. That is in STXM we integrate over the intensity response function of the detector $|D(\mathbf{u})|^2$ while in TXM we similarly integrate over the intensity distribution in $\mathbf{u}$ delivered by the source $|S(\mathbf{u})|^2$.

$$I_{\mathrm{STXM}}(\mathbf{x}) = \int_{\mathrm{DET}} F(\mathbf{u},\mathbf{x})|D(\mathbf{u})|^2\,d\mathbf{u}$$

and $\qquad (35)$

$$I_{\mathrm{TXM}}(\mathbf{x}) = \int_{\mathrm{SOURCE}} F(\mathbf{u},\mathbf{x})|S(\mathbf{u})|^2\,d\mathbf{u}.$$

If the condenser is approximately incoherently illuminated (as specified in §10.5.1 Eq. 13 of (Born and Wolf, 1999) for example), which is often the case for TXMs, (Schneider, 1998; Vogt et al., 2001a), then the effective source (Hopkins, 1957), $|S(\mathbf{u})|^2$, will be the condenser lens aperture function. The expression for the fully incoherent bright field image ($|D(\mathbf{u})|^2 = 1$ or $|S(\mathbf{u})|^2 = 1$) is the same in both TXM and STXM and is obtained by inserting Eq. (33) into Eq. (35) and applying Parseval's theorem (Chapman et al., 1996c)

$$I_{\mathrm{BF}}(\mathbf{x}) = |h(\mathbf{x})|^2\otimes_{\mathbf{x}}|t(\mathbf{x})|^2 \qquad (36)$$

The coherent bright field image is available from a STXM by using an axial point detector and from a TXM by using an axial point source. Both are given by $F(0,\mathbf{x})$ although neither is widely used in X-ray microscopy. The process of integrating over S or D, which is carried out automatically by the hardware of the TXM or STXM, is generally convenient but it destroys potentially useful information about the sample. A procedure for capturing this information, by storing the full detection-plane pattern at every pixel position of the STXM image, has been described and implemented to obtain phase- and amplitude-contrast images by Chapman etal. (1996a). The speed of the procedure was limited by the speed of early 1990s computers but, given the improvement of computers since then, it may well be time to re-examine this approach (see Section 3.1.3). Eq. (34) and Eq. (36) show respectively that coherent imaging is linear in the amplitude and incoherent imaging is linear in the intensity. On the other hand, as we see below, *partially coherent* imaging is not linear in either.

3.3.3 Contrast Transfer

In the case that we do not have $|D(\mathbf{u})|^2 = 1$ or $|S(\mathbf{u})|^2 = 1$, the above procedure used to obtain Eq. (36) does not lead to such a simple result but rather to the following expression representing partially coherent imaging in a STXM (Kintner et al., 1978; Wilson and Sheppard, 1984; Born and Wolf, 1999).

$$I(\mathbf{x}) = \int\limits_{-\infty}^{+\infty} \int \iint C(\mathbf{m};\mathbf{p})\, T(\mathbf{m})\, T^*(\mathbf{p})\, e^{-2\pi i[(\mathbf{m}-\mathbf{p})\cdot \mathbf{x}]} d\mathbf{m}d\mathbf{p} \qquad (37)$$

$$C(\mathbf{m};\mathbf{p}) = \int\limits_{-\infty}^{+\infty} \int |D(\mathbf{u})|^2 H(\mathbf{u}-\mathbf{m})\, H^*(\mathbf{u}-\mathbf{p})\, d\mathbf{u} \qquad (38)$$

For a TXM S replaces D in the last equation. The integration variables $\mathbf{m}$ and $\mathbf{p}$ in Eq. (37) are frequencies similar to $\mathbf{u}$ but $\mathbf{m}$ represents a ray incident on the sample while $\mathbf{p}$ represents a ray emerging from it. The ranges of frequencies included in these beams by the form of S or D determine the range of periodicities $(\mathbf{m}-\mathbf{p})$ in the sample that contribute to the image and thus determine the extent of the MTF in frequency space. The function $C(\mathbf{m}; \mathbf{p})$ is known in optics as the transmission cross coefficient (Born and Wolf, 1999) or the partially coherent transfer function (Wilson and Sheppard, 1984) and provides a *sample-independent* description of the effect of both the illumination and the optical system on the transfer of information from object to image. It is not a true transfer function, since the transfer is not linear, but is a member of a wider class of "bilinear transfer functions." Such functions are described, for example, by (Saleh, 1979) and have been applied to partially coherent X-ray imaging by Vogt et al. (2001a).

$C(\mathbf{m}; \mathbf{p})$ is widely used in the optical and electron microscopy communities and its properties have been worked out in detail; see for example (Sheppard and Wilson, 1980; Wilson and Sheppard, 1984). It is normally a 4D function but in the case of a 1D object it becomes the 2D function $C(m;p)$. The value of $C(m;p)$ is then equal to the overlap integral of the three appropriately shifted aperture functions in the integrand of (38) (Kintner et al., 1978; Wilson and Sheppard, 1984; Born and Wolf, 1999). For many cases of interest in both TXM and STXM, all three are circular disks or annuli (Figures 13–22 and 13–23). For the ideally incoherent bright-field image the value of D or S is taken to be unity for all frequencies and the overlap then depends only on the

Figure 13–22. Imaging of a one-dimensional cosine amplitude grating object using a STXM. The top row shows the signal in the detection plane. The center row shows gray-level plots of $C(m, p)$ in m–p space overlaid with the support boundary of $C(m, p)$ (solid lines) and the spectra $T(m)T^*(p)$ as in Eq. (37) (spots). The middle row shows bright field and the bottom row dark field. The three columns correspond to grating frequencies that are (a) >2, (b) between 1 and 2 and (c) <1 expressed in units of the maximum values of m and p which are both equal to NA/λ. (Reprinted from Chapman et al., © 1996c, with permission from Elsevier.)

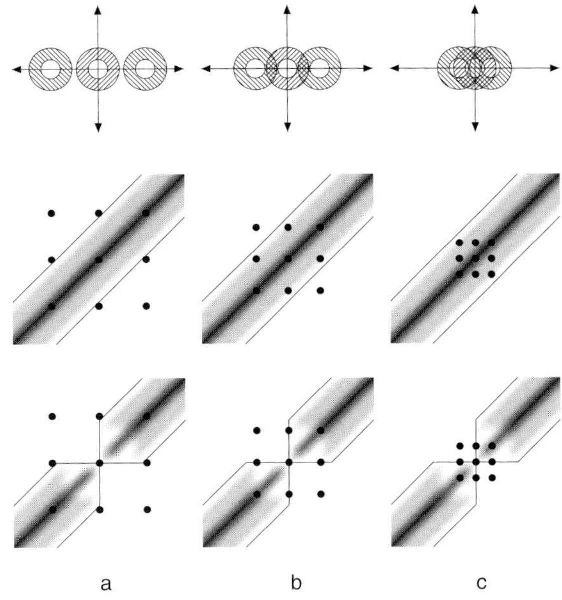

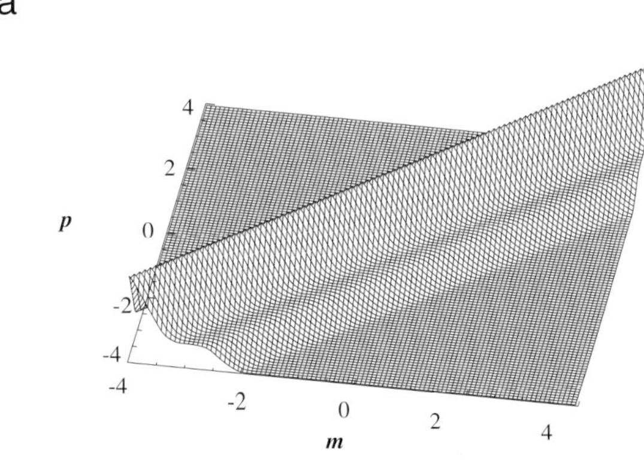

Figure 13–23. Bright-field (a) and dark-field (b) partially coherent transfer functions for a 1D object and an annular lens with inner radius equal to 0.44 times the outer (after Chapman et al., 1996c). The function $C(m, p)$ is plotted against m and p, expressed as multiples of their maximum value NA/λ. (Reprinted from Chapman et al., © 1996c, with permission from Elsevier.)

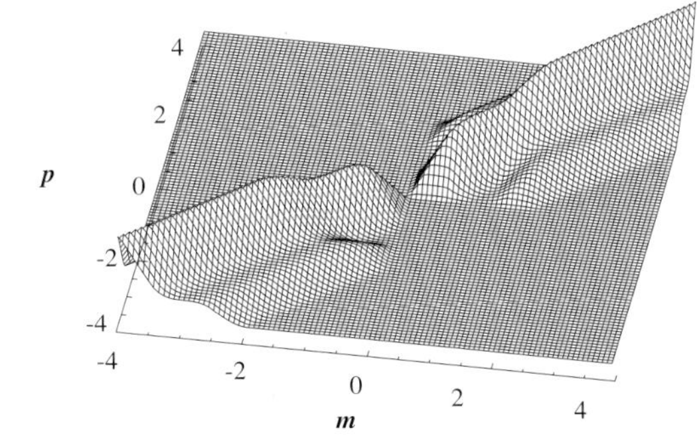

difference $m–p$ of the shifts of H and H^*. That is, there is only *one* response to the sample frequency $s = m − p$ irrespective of m, which indicates a linear system with MTF equal to $C(m–p;0)/C(0;0)$. For forms of D or S corresponding to partial coherence, the system is not linear. On the other hand a dark-field configuration must have detector (or source) and lens aperture functions which have zero overlap at $m = p = 0$. For example D or S could be the Babinet inverse of H. For circular functions of the latter type $C(m;p) = 0$ if sign$(m) \neq$ sign(p). Examples of both bright- and dark-field transfer functions for aperture geometries that are representative of a STXM and that show the above characteristics are given by Chapman et al. (1996c) (Figure 13–23). The response of the same systems to a grating-like object are also given (Figure 13–22). Dark-field STXM is particularly well suited to imaging samples with small features such as gold labels that scatter by large angles (Chapman et al., 1996b). The procedures outlined above allow the calculation of the MTF and the resolution behavior of both types of X-ray microscope based on a knowledge of the resolution-determining lens and the geometry of the source or detector. It is noteworthy that, as in other types of microscope, the resolution does not depend on aberrations of the condenser if there is one. As noted in Section 3.1.2 and illustrated in Figure 13–17, the placement of a ring aperture and phase ring to get Zernike phase contrast in a TXM may be modeled as modifications of the source and lens aperture functions. By this means the above method of analysis may be applied to this case as well (Mondal and Slansky, 1970; Sheppard and Wilson, 1980; Morrison, 1989a).

3.3.4 Reciprocity

The general conclusion of the above analysis is that the optical systems of the TXM and STXM are the same with the position of the lens, before or after the sample, interchanged and the role of the source and detector interchanged. This is the "reciprocity" relationship (Zeitler and Thomson, 1970) that has long been recognized in the visible-light and electron imaging communities and has been explained in the context of X-ray imaging by Morrison (1989a; Morrison et al., 2002). Thus we might expect that, given identical resolution-determining lenses, a TXM and a STXM (both operating in incoherent bright-field mode) could equally well utilize wide-angle beams and get good resolution. For TXM the requirement would be that the condenser should deliver a wide angle to the sample and for STXM that the detector should collect a wide angle from the sample. However, in the past, the practical realization of a wide-angle condenser for a TXM has been much harder than a wide-angle detector for a STXM as we discussed in the condenser zone plate section above.

In practice the TXM/STXM relationship is not quite as symmetrical as the above account suggests because of the general use of objective zone plates with a central stop for STXM (Section 3.1.3) but not for TXM. The stop produces a point-spread function, which has a narrower central peak but larger side lobes. As a consequence the frequency response (the MTF, see Figure 13–21) is increased in the high-frequency and decreased in the low-frequency region.

3.3.5 The Influence of Coherence on Resolution

Calculations of the transfer function as an overlap area of three aperture functions in the integrand of Eq. (38) were discussed in Section 3.3.3. For standard TXM, these functions are circular and two of them are the same. One can therefore follow (Hopkins, 1957) and characterize the illumination by a coherence parameter σ defined by the ratio of the condenser and objective numerical apertures or $\sigma = NA_c/NA_o$. Full coherence is represented by $\sigma = 0$ and full incoherence by $\sigma = \infty$, although $\sigma = 1$ is usually sufficient to get close to fully incoherent behavior. We start by considering the modulation (the percent dip in the valley between the peaks) for a two-point object with separation $0.61\lambda/NA_o$. The modulation is 26.5% for incoherent illumination (Born and Wolf, 1999) and according to the Rayleigh criterion, the two points are just resolved. It is common practice (Born and Wolf, 1999) to extend the Rayleigh criterion to other pairs of objects and define them as resolved if the modulation is at least 26.5%. An example is the two-point object with in-phase coherent illumination for which the just-resolvable separation is $0.82\lambda/NA_o$. Further detail of this can be seen from the plots in Figure 13–24 (Jochum and Meyer-Ilse, 1995). An important example for X-ray imaging is the modulation due to a periodic object, in particular a square-wave transmission object. Such an object can either be prepared by standard lithography methods (Jacobsen et al., 1991) or, for finer line widths, by preparing thin cross-sections of synthetic multilayers (Chao et al., 2003), to yield resolution test patterns for an X-ray microscope. If the resolution is defined as the half period of the finest square wave that can be imaged with 26.5% modulation and is expressed as $k_1\lambda/NA_o$, then according to (Chao et al., 2005), the diffraction-limited value of k_1 is 0.5 for a coherent system ($\sigma = 0$) and 0.4 for $\sigma = 0.38$ (the actual value for XM-1 illuminating the 15 nm zone plate). Thus the diffraction-limited resolution of their experiment was $0.8\Delta r_n$ while the

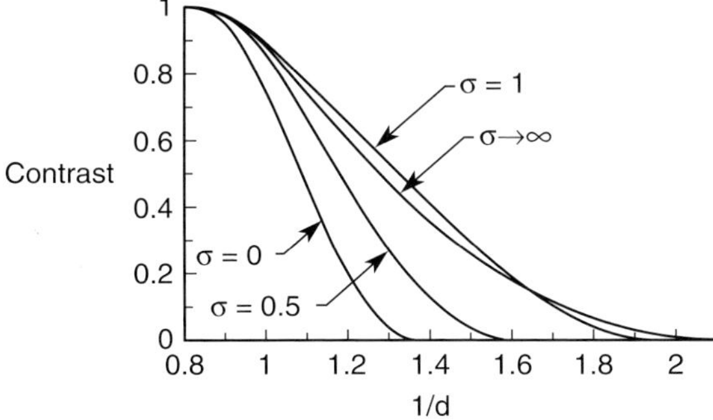

Figure 13–24. Image contrast as a function of $1/d$ where d is the point separation of a two-point object imaged by a lens with a circular pupil and coherence parameter σ equal to 0 (coherent case), 0.5, 1 and infinity (incoherent case). d is expressed in units of λ/NA. Note that the Rayleigh resolution corresponds to 15.3% intensity contrast (defined as $(I_{max} - I_{min})/(I_{max} + I_{min})$), which is the same as 26.5% modulation. It occurs at $d = 0.61\lambda/NA$ for both $\sigma = 1$ and $\sigma \to$ infinity. (Reprinted from Jochum and Meyer-Ilse, © 1995, with permission from Optical Society of America.)

achieved value was about $1.0\Delta r_n$ (<15 nm). The data demonstrating the achieved resolution are shown in Figure 13–13. By reciprocity, the same arguments about the value of the diffraction-limited resolution would apply for a STXM in incoherent bright field using the same zone plate. With a large enough detector one would achieve $\sigma = 1$ and the same diffraction-limited resolution as a TXM with $\sigma = 1$.

3.3.6 Coherence in Zernike Phase Contrast

We return now to the question of the choice of width and radius for the two rings in a Zernike phase-contrast configuration of the TXM, or, equivalently, how much coherence is desirable in this case. This choice has been discussed by (Mondal and Slansky, 1970) and is essentially a trade off between light collection and the distorting effects of the fact that the phase ring must have finite area, which applies an unintended phase change to a certain portion of the diffracted light causing the so-called halo effect (Wilson and Sheppard, 1981). It is generally thought that one needs very little coherence, that is, the ring aperture can leave a large fraction of the condenser area open. This is true if the requirement is merely to make otherwise invisible phase features, especially phase jumps, become visible. However, with low coherence, a phase step is rendered as a double-peaked zero-crossing function and a phase rect function is rendered as two such double peaks. How much coherence do we need to get anything resembling a faithful rendition of the object? We have not found much attention to this point in the literature but the treatment by Martin (1966) provides an answer, which is confirmed by our own computer modeling. To get a rendition of a rect function that looks like the original function one needs to have the coherence width $w_c = \lambda f_{cond}/\Delta r_{ring}$ of light arriving at the sample at least equal to the width of the rect function.

3.3.7 Propagation-Based Phase Contrast

Another way to achieve phase contrast is to exploit the $\exp[i\pi(x^2 + y^2)/\lambda z]$ phase shifts that occur in the (x, y) plane as a result of the propagation of a coherent wavefield through a distance z. This is exploited in X-ray holographic microscopy which has had many successes (Aoki and Kikuta, 1974; Joyeux, et al. 1988; Jacobsen et al., 1990; McNulty et al., 1992; Snigirev et al., 1995; Lindaas et al., 1996; Eisebitt et al., 2004) but which is so far not used for routine X-ray imaging. The exception is in the use of holography for phase contrast tomography at higher X-ray energies, where Cloetens and coworkers have achieved considerable success in routine micrometer-resolution tomography using a phosphor/lens/CCD detector system (Cloetens et al., 1999). While it lies beyond the scope of the present article's emphasis on zone plate X-ray microscopy, this unique approach is providing impressive 3D reconstructions of difficult specimens including foams.

3.4 Tomography in X-Ray Microscopes

3.4.1 Principle of Operation

As was noted in Section 2.4.5 on hard X-ray zone plates, the transverse resolution of a zone plate operated in first diffraction order is given by

$0.61\lambda/\mathrm{NA} = 1.22\Delta r_n$, and the depth of focus is $2\lambda/(\mathrm{NA})^2 = 8(\Delta r_n)^2/\lambda$. Since present zone plates have outermost zone widths Δr_n that are much larger than the wavelength λ, this means that the depth of focus is necessarily large compared to the resolution as can be seen by the illustration of the 3D modulation transfer function in Figure 13–25. This provides an opportunity for 3D imaging if the object is smaller than the depth of focus because a 2D image can then be interpreted as providing a simple projection through the specimen, which is precisely what conventional tomography requires at each viewing angle. Tomography with electron microscopes is long established, and following earlier demonstrations by Haddad et al. (1994) using the Stony Brook STXM and Lehr (1997) using the Göttingen TXM, a number of groups are now using X-ray microscopes for tomographic studies of frozen hydrated biological specimens and integrated circuits, among other applications that will be described below in Section 4. However, although the technique used in these studies is improving, they have still not reached the resolution achieved by the same microscopes in 2D.

3.4.2 The Depth-of-Focus Limit

Given the successes achieved and the amount of current interest, what are the issues to be faced in improving the resolution of tomography

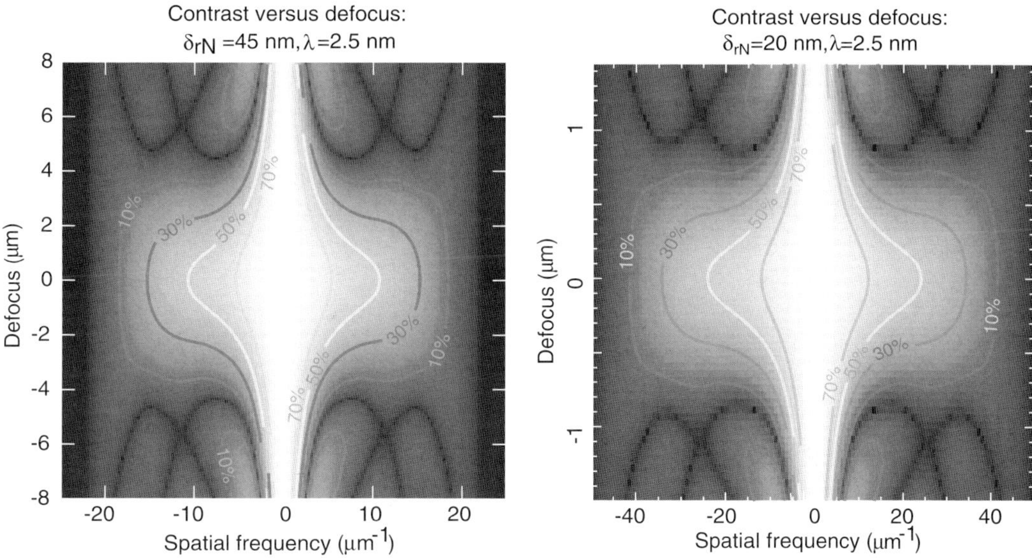

Figure 13–25. Properties of soft X-ray tomography using zone plate optics. At the left is shown the 3D modulation transfer function for monochromatic, spatially incoherent bright-field imaging with a 45-nm-outer-zone-width zone plate with a half-diameter central stop, as a function of depth. One can see that the good in-focus frequency response is preserved over a total depth of about 8 μm. This is useful for many tomography experiments that rely on the fact that the delivered image is a *projection* of the object. At the right is the same information for a 20-nm-zone plate; as can be seen, the figure scales quite well from the figure at the left according to the square of the ratio of the finest zone widths. With a 20-nm-zone plate and monochromatic illumination, good frequency response is only preserved over a depth of about 0.5–1 μm which is much more restrictive, illustration the challenges of improved resolution in TXM zone-plate tomography. (Reprinted from Wang et al., © 2000, with permission from Blackwell Publishing.)

in X-ray microscopes? One of them concerns the same depth of focus that makes such tomography straightforward. To our knowledge all demonstrations of soft X-ray tomography reconstructions have used zone plates with outermost zone width Δr_n no finer than 35 nm, so that the depth of focus in the water window region is at least 4 μm which has been comparable to the specimen size. As higher resolution zone plates are employed, the depth of focus will decrease as the square of improvements in transverse resolution, so that a 15 nm outermost-zone-width zone plate would have a depth of focus of about 0.8 μm. This approaches the ~0.5 μm thickness accessible to cryo electron tomography of frozen hydrated cell regions at 6–8 nm resolution (Grimm et al., 1998; Medalia et al., 2002) and thus would seem to leave a much-reduced "window of opportunity" for soft X-ray tomographers and apparently deny them the important goal of high-resolution, 3D imaging of intact eucaryotic cells.

The depth-of-focus limit evidently poses a considerable challenge to those who would exploit zone plate X-ray microscopes for tomography. Accordingly we discuss in the sections that follow several different approaches to dealing with the challenge.

3.4.3   Avoiding the Depth-of-Focus Limit by Reconstructing a Partial Volume

Zone-plate-microscope X-ray tomography as described above is based on geometrical optics and is mathematically identical to the technique; Computed Axial Tomography (CAT). Therefore, following an established CAT procedure (see for example Ritman et al., 1997), it is possible to reconstruct partial volumes that are embedded in a larger overall object. This has been done in the X-ray microscope context by Xradia Inc. In this case the depth-of-focus limit applies to the partial volume, rather than the whole volume, and the signals generated by the out-of focus parts of the sample reduce, after combining all members of the tilt series, to a smooth background. The size of the region that is always in focus will be roughly a sphere of diameter equal to the depth of focus.

3.4.4   Avoiding the Depth-of-Focus Limit by Wide-Band Illumination

In fact, the depth of focus calculation given above is for monochromatic illumination, which is applicable to demonstrations of tomography of frozen hydrated cells in a STXM with a high-energy-resolution mono-chromator (Wang et al., 2000). In some existing soft X-ray TXMs using the zone plate condenser as a linear monochromator, an energy resolving power of $E/\Delta E \approx 200$ has been used for objective zone plates with $N = 375$ zones (the recommended bandwidth for such an objective would usually have been $E/\Delta E > N$, (Thieme, 1988). However, Weiss et al.(2002) have shown that use of $E/\Delta E \approx 200$ leads to significant changes of the modulation transfer function (MTF) as a function of defocus (similar results for different cases have been given by Schneider et al. (2002a)). In their calculations for a zone plate with 40 nm outermost zone width, the MTF at a spatial frequency of 15 μm^{-1} (corresponding to a spatial half-period of 33 nm) declines from about 0.28 in focus to about 0.015 at a defocus of 4 μm in the monochromatic case. When the bandwidth is increased to $E/\Delta E \approx 200$, the MTF at 15 μm^{-1} is reduced in focus to only about 0.11 but at the same 4 μm defocus it degrades much less to 0.08. Therefore these calculations show that, although the transverse resolution and efficiency for the collection of structural information, are both made worse by the use of high-bandwidth radiation, the depth of field is improved. (To our knowledge no calculations have yet addressed the possibly interesting question of how this tradeoff with nonmonochromatic illumination compares with the tradeoff resulting from using a lower or higher resolution zone plate). In summary, it will require further investigation to determine whether manipulation of the bandwidth can enable significant improvements to the tomographic resolution without reduction of the reconstructible sample volume.

3.4.5 Avoiding the Depth-of-Focus Limit by Through-Focus Deconvolution

One possible approach to beat the depth of focus limit is to use through-focus deconvolution as is done in light and electron microscopy. In electron microscopy, the recording of defocus image sequences is routine; each defocus provides positive and negative phase contrast at various bands of spatial frequencies along with zeroes in the transfer function, and the combination of several images can provide a complete image of the specimen (Reimer, 1984). In fluorescence light microscopy, through-focus image sequences can yield a high quality 3D image through the use of deconvolution of the 3D point spread function (Agard and Sedat, 1983; Carrington et al., 1995). However, there are important differences between these examples and the situation present in X-ray microscopy. In electron microscopy, this approach is usually applied to thin samples for which phase contrast dominates (indeed the specimen focus can be quickly estimated by looking for a minimum in image contrast). In light microscopy, the use of fluorescence means that the object is a sparse, pure-real function (incoherent emission from independent fluorescence emitters with no sensitivity to the relative phase of the illumination) so that the deconvolution can be done based on the intensity point spread function. While some form of generalized through-focus deconvolution may provide the needed breakthrough, quantitatively reliable results such as are needed for assembly into a tomographic reconstruction will have to account for the fact that biological specimens imaged at water-window energies produce both absorption and phase contrast so one will require exact knowledge of the complex bilinear transfer function of the zone plate optic and illumination system. In other words, the problem of 3D deconvolution of a strongly absorbing, optically thick, complex object with partial coherence is much more difficult than the cases in which optical sectioning is typically used at present.

3.4.6 Avoiding the Depth-of-Focus Limit by Use of Higher Energy X-Rays

The use of shorter wavelengths (higher X-ray energies) to increase the monochromatic depth of field $8(\Delta r_n)^2/\lambda$ is a guaranteed way to extend the depth of focus. The associated questions of what will be the cost in resolution, contrast and efficiency are also now beginning to be answered favorably. This approach has been used to obtain sub-100 nm resolution tomographic reconstructions of metallic layers within thinned integrated circuits using a laboratory X-ray microscope operating at 5.4 keV (Wang, 2002) as will be described in Section 4.3. For lower density specimens, the use of hard X-rays naturally leads to the use of phase contrast which is far more dose-efficient than absortion contrast in this energy region. As shown in Figure 13–5, this enables multi-keV imaging at similar dose levels to the water window. As noted in Section 3.3.7 there have already been quite successful demonstrations of phase contrast tomography using hard X-rays (Cloetens et al., 1999) in the past though not (to our knowledge) at the sub-100 nm resolution level accessible to zone plate microscopes. Now the group at NSRRC, Taiwan have recently used their 8 KeV phase-contrast TXM with a 50 nm zone plate to produce 3D images of a microcircuit with defects at 60 nm resolution (Yin et al., 2006) (see Sections 1.4 and 2.4.4).

3.4.7 *Avoiding the Depth-of-Focus Limit by Lens-Free Imaging*

The challenges of achieving the highest possible resolution in 3D imaging have led to the consideration of lens-free imaging. Of course crystallography is able to obtain exquisite 3D maps of the electron density of a unit cell in a crystal by interpretation of a tilt series of Bragg diffraction patterns. In the case of a noncrystalline specimen, one obtains a continuous rather than Bragg-sampled diffraction pattern but there has been considerable recent progress in obtaining X-ray images through the application of iterative phasing algorithms to diffraction data from objects known to be limited in size (Miao et al., 2002; Marchesini et al., 2003; Williams et al., 2003; Shapiro et al., 005; Chapman et al., 2006). At the moment the data collection time in 3D experiments of this type is 10–20 hours and 10 nm resolution images of materials-science samples in 3D and 30-nm-resolution images of biological specimens in 2D have been obtained. Moreover, a modern beam line designed specifically for this type of experiment would easily bring image acquisition times down to a convenient level. It is noteworthy that the phasing algorithms depend on use of the Born approximation which sets an upper limit to the sample size which may become significant at low X-ray energies such as those in the water window.

3.5 X-Ray Spectromicroscopy

As noted in Section 1, X-ray absorption edges arise when the X-ray photon reaches the threshold energy needed to completely remove an electron from an inner-shell orbital. At photon energies within about 10 eV of the edge, electrons can also be promoted to unoccupied or partially occupied molecular orbitals (see Figure 13–26); photons over a narrow energy range are sometimes able to excite inner-shell electrons into such orbitals, giving rise to absorption resonances. This so-called X-ray absorption near-edge structure (XANES) or near-edge X-ray absorption fine structure (NEXAFS) is highly sensitive to the

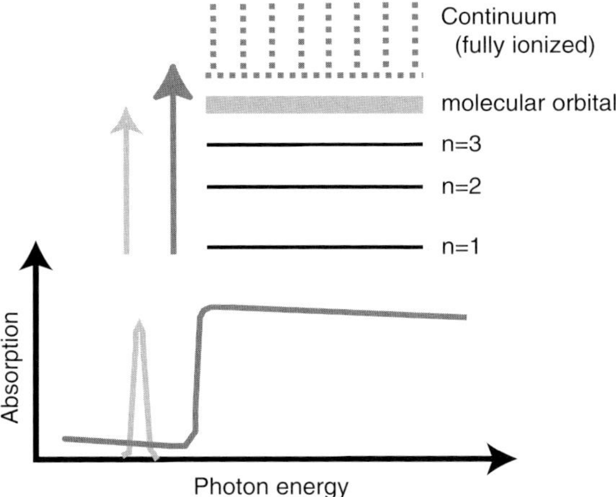

Figure 13–26. Schematic of an X-ray absorption edge, which involves the removal of an inner-shell electron, and a near-edge absorption resonance in which the electron is promoted to a partially occupied or vacant molecular orbital. These resonances are referred to as X-ray absorption near-edge structure (XANES) or near-edge X-ray absorption fine structure (NEXAFS).

local chemical bonding state of the atom in question (Stöhr, 1992) (see Figure 13–26).

One can exploit these resonances as an additional contrast mechanism in soft X-ray imaging. In electron energy loss spectroscopy (EELS), the equivalent contrast mechanism is known as ELNES for energy-loss near-edge structure and its use in energy-loss spectrum imaging (Jeanguillaume and Colliex, 1989; Hunt and Williams, 1991) is described elsewhere in this volume. Early efforts in X-ray imaging included the use of XANES resonances to enhance the sensitivity of differential absorption measurements of calcium in bone (Kenney et al., 1985), spectral imaging (King et al., 1989) and microspectroscopy in photoelectron microscopes (Harp et al., 1990), and photoelectron and transmission imaging at selected photon energies (Ade et al., 1990b, 1992). It is now common to take image sequences across X-ray absorption edges (Jacobsen et al., 2000b) yielding data sets with a full near-edge spectrum per pixel. When comparing spectrum imaging in electron versus X-ray microscopes, a few comments are in order:

- ELNES is typically done using a fixed electron energy in the range 80–200 keV. The ideal specimen thickness is under 100 nm in most cases.
- In ELNES, one gets spectroscopic information over a wide range of energies, including plasmon energies of ~10 eV, in a single measurement. However, plural inelastic scattering dominates the signal at higher energies (for example, electrons can lose 300 eV once, or 50 eV six times, etc.) resulting in poorer signal-to-background.
- In X-ray absorption spectroscopy, one must tune the incident X-ray energy across each absorption edge of interest. The optimum

specimen thickness of about $1/\mu(E)$ changes accordingly, so that in the ideal case one would require samples of several different thicknesses to study chemical speciation of several elements. However, X-rays suffer almost no plural inelastic scattering, which leads to improved signal-to-background.

- It is common to find scanning X-ray microscopes operating with monochromators with an energy resolution of 0.1 eV or better. Most electron microscopes have an energy resolution of 0.5–0.7 eV which leads to "blurring" of near-edge spectral features, although a limited number of higher energy resolution systems are starting to become available.

- Using XANES, one can exploit the favorable characteristics of X-ray microscopes including the ability to study hydrated specimens and/or specimens in an ambient atmosphere environment.

In X-ray microscopes, we obtain images (maps of transmitted flux I) according to the Lambert-Beer law for absorption: $I = I_0\exp(-\mu t)$ where I_0 is the incident X-ray flux, μ is an absorption coefficient for a specific material, as discussed in Section 1.1, and t is thickness of that material. The value of $\mu(E)$ for near-edge absorption resonances can be calculated based on the electronic structure of specific molecules, and this has been employed in detailed studies via microscopy of the absorption spectra of polymers (Urquhart and Ade, 2002; Dhez et al., 2003) and amino acids (Kaznacheyev et al., 2002) (see Figure 13–27), to name two recent examples.

For a thickness t of a single material, a measurement of the transmitted flux $I(E)$ relative to the incident flux $I_0(E)$ provides a means to calculate the energy-dependent optical density $D(E) = -\ln(I(E)/I_0(E)) = \mu(E)t$. If, however, we measure the optical density not over a continuous energy range E but at some set of $n = 1 \ldots N$ discrete energies E_n, we then measure

$$D_n = \mu_n t$$

for each of the $n = 1 \ldots N$ photon energies. Let us next consider a mixture of $s = 1 \ldots S$ different materials with partial thicknesses t_s; our

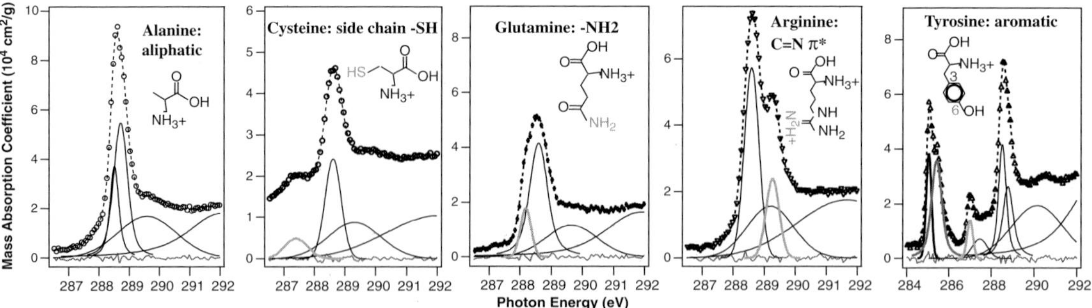

Figure 13–27. Near-carbon-edge absorption spectra of several amino acids, showing the effects of various molecular bonds in the absorption spectrum. These resonances can be used for chemical contrast in X-ray microscopy. (Reprinted from Kaznacheyev et al., © 2002, with permission from American Chemical Society.) (See color plate.)

total measurement of optical density D_n at one photon energy is given by the combined absorption of all the materials, or

$$D_n = \mu_{n1}t_1 + \mu_{n2}t_2 + \cdots + \mu_{NS}t_S.$$

Finally, if we carry out this measurement not from a single homogeneous uniform film, but from heterogeneous pixels $p = 1 \ldots P$ indexed by $p = i_{\text{column}} + (i_{\text{row}} - 1) \cdot (\# \text{ columns})$ in an image, the optical density measured at one pixel p is given by

$$D_{np} = \mu_{n1}t_{1p} + \mu_{n2}t_{2p} + \cdots + \mu_{nS}t_{Sp}.$$

When all N photon energies are considered, we see that we have a data matrix D_{NP} of

$$
\begin{bmatrix} D_{11} & \cdots & D_{1P} \\ \vdots & \ddots & \vdots \\ D_{N1} & \cdots & D_{NP} \end{bmatrix} = \begin{bmatrix} \mu_{11} & \cdots & \mu_{1S} \\ \vdots & \ddots & \vdots \\ \mu_{N1} & \cdots & \mu_{NS} \end{bmatrix} \cdot \begin{bmatrix} t_{11} & \cdots & t_{1P} \\ \vdots & \ddots & \vdots \\ t_{S1} & \cdots & t_{SP} \end{bmatrix}
$$

or $D_{N \times P} = \mu_{N \times S} \cdot t_{S \times P}$. In other words the data represent a series of spectral signatures $\mu_{N \times S}$ and thickness maps $t_{S \times P}$.

When we acquire a series of images at different photon energies N, we are in fact measuring the data matrix $D_{N \times P}$. If we know the exact absorption spectrum μ_{Ns} for each of the $s = 1 \ldots S$ components in the sample, then we can find the spatially resolved thicknesses $t_{S \times P}$ of the components by matrix inversion:

$$t_{S \times P} = \mu_{S \times N}^{-1} \cdot D_{N \times P}$$

The inversion of the matrix of spectra from all known components $\mu_{N \times S}$ can be accomplished in a robust fashion using singular value decomposition (Zhang et al., 1996; Koprinarov et al., 2002). This approach, as well as approaches which involve pixel-by-pixel least squares fits of all reference spectra, work well with specimens that involve mixtures of components that can all be measured separately. Examples using this approach are shown in the chemical imaging section of this chapter.

In many areas of research, such as biology or environmental science, the complexity of the specimen and the possibility of reactions between components means that one cannot know in advance the set of all absorption spectra $\mu_{N \times S}$ present in the specimen. In this case, one approach that has yielded recent success is to first use principal component analysis (King et al., 1989; Osanna and Jacobsen, 2000) to orthogonalize and noise-reduce the data matrix $D_{N \times P}$, and then use cluster analysis (a method of unsupervised pattern recognition) to group pixels together based on similarity of spectral signatures (Lerotic et al., 2004, 2005) (see Figure 13–34 below). This method yields a set of absorption spectra $\mu_{N \times S}$ where S now indexes the set of characteristic spectra found from the data. The power of this approach lies in its ability to improve the signal-to-noise of spectra of heterogeneous specimens by averaging noncontiguous pixels, to find even quite small regions with distinct spectroscopic signatures, and to deliver continuous "thickness" maps based on the distribution of the discovered signature spectra.

For studies at the carbon edge, one can characterize the observed set of near-edge resonances in terms of a limited number of functional group types (see e.g., Scheinost et al., 2001). While there are a number of open questions regarding this approach (for example, how many resonances should be used, with what range of allowed center photon energies, and what range of energy widths?), confidence in it can be enhanced by correlation with other spectroscopies such as solid-state nuclear magnetic resonance (Scheinost et al., 2001; Schumacher et al., 2005) and Fourier transform infrared (Solomon et al., 2005).

4 Applications

Two decades ago, nearly all research using X-ray microscopes was done by the groups that had developed the instruments. Today, most X-ray microscopes are operated as user facilities at synchrotron radiation research centers, and are used both by their developers but also by a wider community of scientists. As a result, while it was originally possible to see the major applications of X-ray microscopes in conference proceedings (Schmahl and Rudolph, 1984a; Sayre et al., 1988; Michette et al., 1992), papers in which X-ray microscopes were used to address the problem of interest now appear across a very wide array of scientific journals. In what follows, we do not presume to be exhaustive in coverage of all research using X-ray microscopes; instead, we will briefly highlight a few examples from some of the areas of present activity.

4.1 Biology

X-ray microscopes using zone plates and synchrotron radiation have been used for studies of biological specimens from the start (Niemann et al., 1976; Rarback et al., 1980), and a number of reviews have concentrated on biological applications of X-ray microscopes (see for example Kirz et al., 1995) for background information and older results, or Abraham-Peskir, (2000). One emphasis has been on high resolution imaging of whole cells at "water window" wavelengths (see Figure 13–28), including studies of human sperm (Chantler and Abraham-Peskir, 2004), malaria in red blood cells (Magowan et al., 1997), Kupffer cells (Scharf and Schneider, 1999) and COS cells (Yamamoto et al., 1998) from liver, protists (Abraham-Peskir, 1998), and chromosomes (Guttmann et al., 1992; Williams et al., 1993; Kinjo et al., 1994) among other examples. As soft X-ray microscopes push to higher spatial resolution, views through whole cells will involve a great deal of overlap of structure, but several developments offer information beyond two-dimensional images with natural contrast. One of these is to use molecular labeling methods to tag specific proteins (such as is done with great success in visible light microscopy). Several groups have demonstrated the use of gold labeling in X-ray microscopes, including detection by dark field (Chapman et al., 1996c) (see Figure 13–29) and bright field (Meyer-Ilse, 2001 et al.,; Vogt et al., 2001b) approaches. One of the challenges faced thus far is that the label must

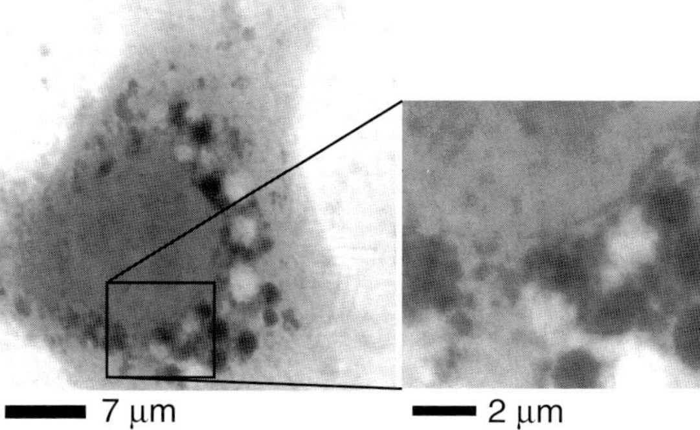

Figure 13–28. Whole fibroblast imaged in the frozen-hydrated state. The cell was cultured on a formvar-coated gold electron microscope grid, and rapidly frozen by plunging into liquid ethane. It was then imaged using a cryo STXM operated at 516 eV. In addition to this 2D image, 3D reconstructions were also obtained using tomography (Wang et al., 2000). (Reprinted from Maser et al., © 2000, with permission from Blackwell Publishing.)

be comparable in size to the resolution of the microscope for efficient detection (Chapman et al., 1996c; Vogt et al., 2001a), which means that in all studies carried out thus far the cell membrane has been permeabilized by agents such as methanol to allow relatively large labels to reach the cell's interior and this step must be preceded by chemical fixation. As a result, future improvements in X-ray microscope resolution will not only lead to improved visualization of unlabeled

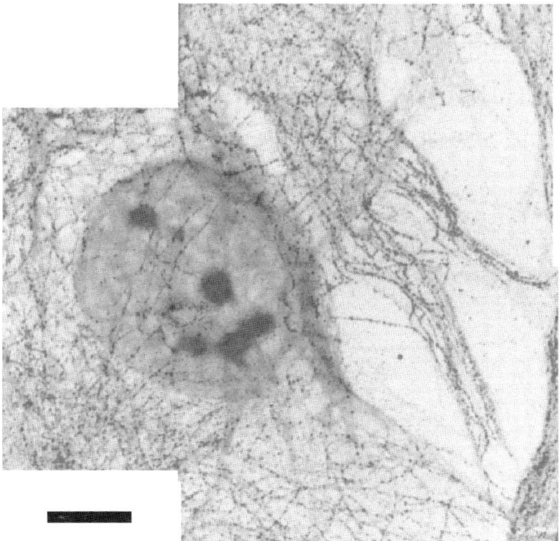

Figure 13–29. Human fibroblast with immunogold labeling for tubulin. This is a composite of two images: a bright field image (gray tones) to image overall mass, and a dark field image (red tones) to selectively imaging the silver-enhanced gold labels. This whole-mount cell was fixed and then permeabolized to allow for introduction of the immunogold labels, after which it was air dried. (From Chapman et al., © 1996b,c, courtesy of the Microscopy Society of America.) (See color plate.)

ultrastructure but will also make it possible to use smaller immunolabels with more "natural" preparation protocols.

Another approach to exploit the characteristics of X-ray microscopes is to go beyond two-dimensional imaging. One approach is to use XANES spectromicroscopy for mapping chemical speciation in bacteria and cells (Ito et al., 1996; Zhang et al., 1996; Lerotic et al., 2005) and biomaterials (Hitchcock et al., 2002) using the approaches outlined in Section 3.6 above. Another involves the use of tomography as has been discussed in Section 3.5. This was first used to study algae in a thin capillary by Weiss et al. (2000) (Figure 13–30) and to study whole-mount eukaryotic cells by Wang et al. (2000), followed by studies of yeast in capillaries (Larabell et al., 2004) (see Figure 13–31). In all of these cases, cells were studied in the frozen hydrated state for reasons that will be discussed in the following paragraph. A third approach beyond two-dimensional imaging is to use X-ray microscopes (Kenney et al., 1985; De Stasio et al., 1996; Buckley etal., 1997) or X-ray microprobes (Kawai et al., 2001; Ortega et al., 2003; Paunesku et al., 2003; Kemner et al., 2004; Behets et al., 2005; Wagner et al., 2005) to study elemental content and distribution in bacteria and cells, particularly in the case of calcium in bone and metals that regulate various biological activities.

When studying biological specimens, attention must be paid to the limitations set by radiation damage. Basic considerations of signal-to-noise and absorption indicate that the radiation dose that is necessarily imparted for X-ray imaging at 50 nm or better resolution is in excess of 10^6 Gray (Sayre et al., 1977a; Schneider, 1998). This is well in excess of

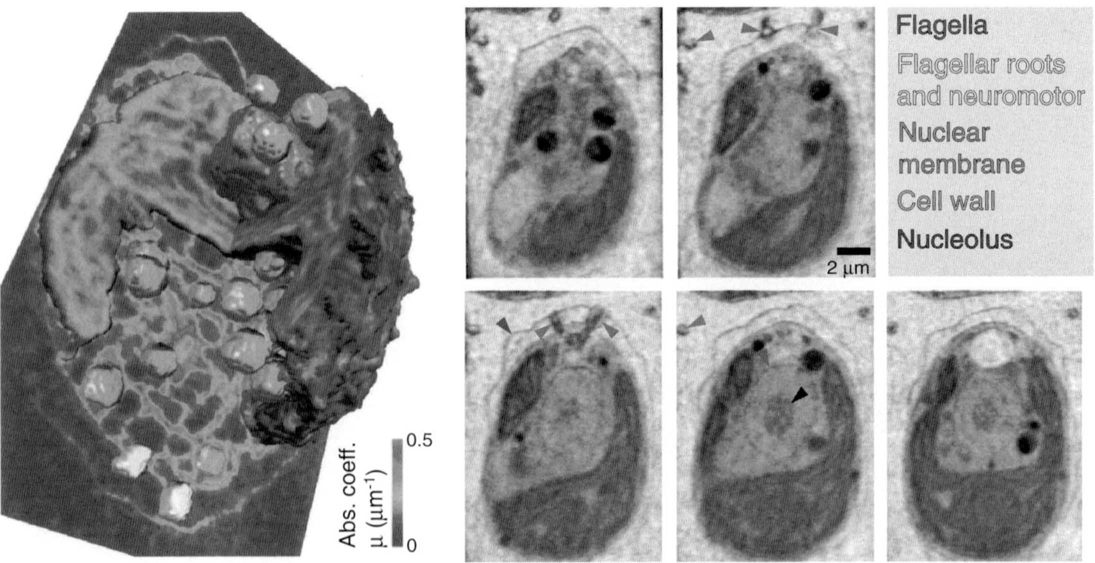

Figure 13–30. 3D rendering (left) and reconstruction slices (right) of the algae *Chlamydomonas reinhardtii* viewed by soft X-ray tomography at the BESSY I synchrotron. This alga was plunge-frozen in liquid ethane, and imaged over 180° rotation sequence. The reconstruction is given in terms of the quantitative linear absorption coefficient for 517 eV X-rays. (Reprinted from Wei et al., © 2000, with permission from Elsevier.) (See color plate.)

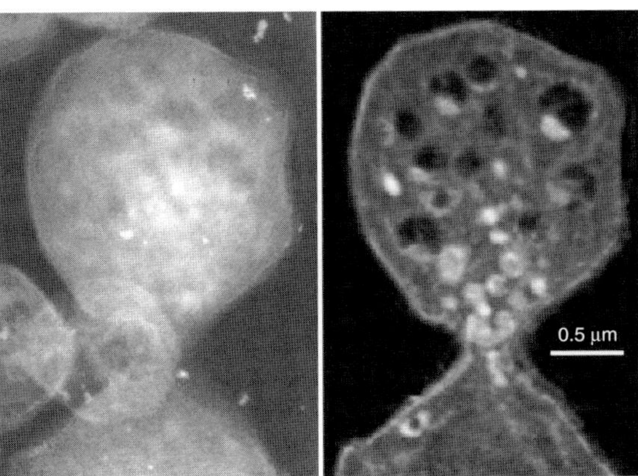

Figure 13–31. Single projection image (left) and slice from a tomographic reconstruction (right) of a frozen hydrated yeast *Saccharomyces cerevisiae*. A number of cells were loaded into a thin-walled, 10 µm diameter glass capillary and rapidly frozen using a jet of helium gas cooled by liquid nitrogen. A series of 45 images through a 180° tilt range was then acquired using the XM-1 TXM at the Advanced Light Source. This illustrates the ability of soft X-ray tomography to image the interior detail of cells rapidly frozen from a living state. (From Larabell et al., 2004. Reprinted from Molecular Biology of the Cell, with permission of the American Society for Cell Biology.)

the <10 Gray (1 Gray = 100 rad) dose that is lethal to humans when received over a short time interval. Studies of intially living cells have shown that doses of 10^6 Gray are at the approximate threshold for producing immediate changes in bacteria and are well above the dose needed to affect more complex cells in X-ray microscopy investigations (Gilbert et al., 1992; Pine and Gilbert, 1992; Bennett et al., 1993; Kirz et al., 1995). One of the main damage mechanisms is the creation of radiolytical free radicals in water. Some but not all chemically fixed, hydrated biological specimens will show effects such as mass loss, shrinkage, and the loss of ultrastructural information at these radiation doses as well (Ford et al., 1992; Williams et al., 1993) (of course, chemical fixation produces its own changes on many specimens (Coetzee and van der Merwe, 1984, 1989; Stead et al., 1992; O'Toole et al., 1993; Jearanaikoon and Abraham-Peskir, 2005). Fortunately, a ready solution was developed some years ago by electron microscopists: the use of rapidly frozen specimens in cryomicroscopy (Taylor and Glaeser, 1976; Steinbrecht and Zierold, 1987; Echlin, 1992). In X-ray microscopes, frozen hydrated biological specimens have been shown (Schneider, 1998; Maser et al., 2000) to be well preserved and free of easily visible structural changes and mass loss at radiation doses up to about 10^{10} Gray thus providing the required conditions for a variety of biological studies. The situation for spectroscopy is not yet so clear; cryo methods have been shown to be less effective in preserving XANES resonances in dry polymers (Beetz and Jacobsen, 2003) but they may be more advantageous in studies of frozen hydrated organic specimens due to the inactivation of the diffusion of free radicals ("cage" effect) (Schneider, 1998).

4.2 Environmental Science

Environmental science using synchrotron radiation is a broad topic, as discussed in a recent review (Brown, 2002); we point out here just a few examples using X-ray microscopes.

By placing microliter drops between two silicon nitride windows which are then drawn together by surface tension and some sort of seal, it is straightforward to make a specimen chamber with micrometer-thick water layers and study samples wet and at room temperature (Neuhäusler et al., 2000) (see Figure 13–32). Using this approach, one can use soft X-ray spectromicroscopy to study the role of bacteria and their biofilms in changing the reduction/oxidation state or sequestration of various elements in the environment (Lawrence et al., 2003; Yoon et al., 2004; Hitchcock et al., 2005) (see Figure 13–33), the growth of crystaline materials (Chan etal., 2004), and other geochemical reactions (Myneni et al., 1997; Tonner et al., 1999; Pecher et al., 2003). Spectromicroscopy at the carbon edge can be used to study a variety of organic processes, ranging from the diagenetic breakdown of organic material over geological timescales and its presence and preservation in fossilized plants and wood (Cody, 2000; Boyce et al., 2002, 2004) and coals (Botto et al., 1994; Cody et al., 1995), and the role of natural organic matter in the properties of soils (Thieme et al., 1994; Thieme and Niemeyer, 1998a; Scheinost et al., 2001; Schäfer et al., 2003; Solomon et al., 2005) including its role in the groundwater transport of radionuclides (Schäfer et al., 2005) (see Figure 13–34). Tomography has also been used to study bacterial "microhabitats" (Thieme et al., 2003). Other studies have considered the functional groups present in the soot produced by combustion in diesel engines (Braun et al., 2004).

The trace element mapping capabilities of X-ray microprobes are also very useful for studies in environmental science. Low concentration of iron sets a biotic limit to carbon uptake in the southern Pacific; Twining et al. (2003) have used microprobe studies to investigate this on a cell-by-cell basis (see Figure 13–35) since bulk chemistry measurements do

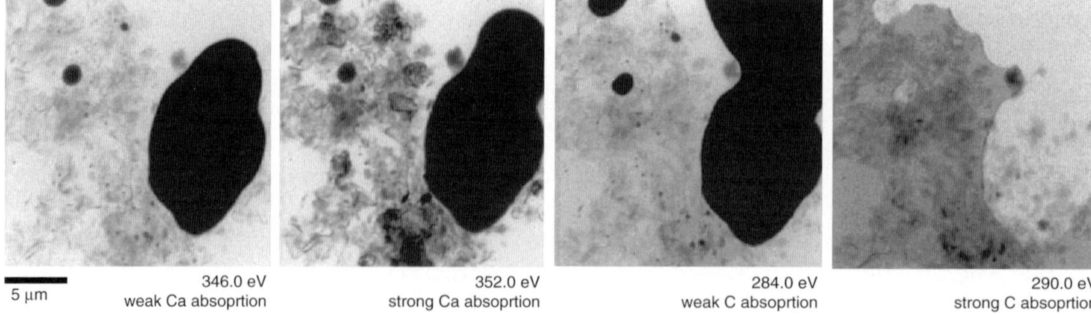

Figure 13–32. Images of a colloidal chemistry sample consisting of oil in water with clays and calcium-rich layered double hydroxides used to "cage" the oil droplet where present (left and bottom edges of the droplet). This illustrates the ability to highlight various elemental components in a room temperature wet specimen. (Courtesy of Neuhäusler, 1999.)

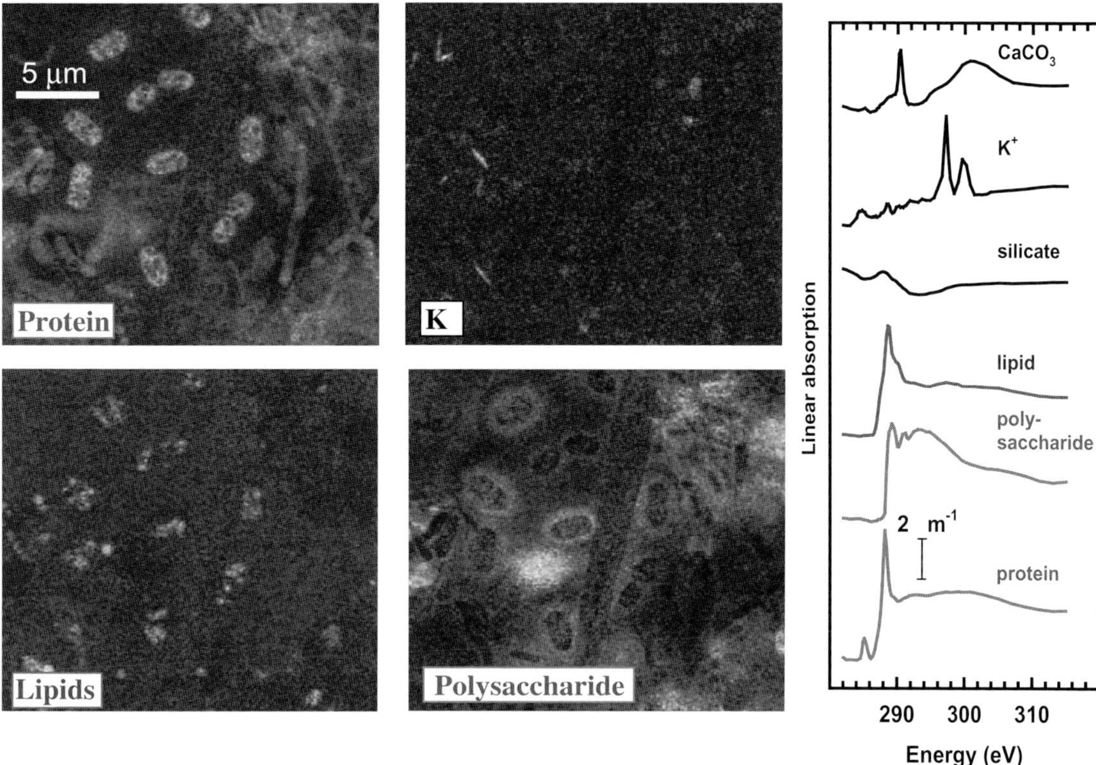

Figure 13–33. Quantitative chemical maps of protein, K^+, lipids, and polysaccharides from a wet microbial colony from the South Saskatchewan river, derived from STXM images (880 × 880 pixels) and image sequences (52 energies, 230 × 230 pixels). The spatial distributions of the various chemical species are determined by fitting the spectra from each pixel with a linear combination of the absorption spectra of the constituents. X-ray absorption spectra in the C 1s region are shown for $CaCO_3$, K^+, silicate, lipid, polysaccharides, and protein. The spectrum of $CaCO_3$ is from pure material. Those of the other five species are derived from the C 1s image sequence recorded from this biofilm using pixel identification and (for lipid, polysaccharides) spectral subtractions based on fits of the image sequence to the spectra of pure reference materials. (Courtesy A.P. Hitchcock, McMaster University.)

not allow one to differentiate between protist types and particulate matter at the same size scale. Other studies using zone plate microprobes have concentrated on the speciation of metals near the roots of healthy and diseased plants (Yun et al., 1998b), the presence of metals in soil bacteria (Kemner et al., 2004), sulfur speciation in bacteria (Labrenz et al., 2000), in natural silicate glasses (Bonnin-Mosbah et al., 2002), and in microbial filaments (Foriel et al., 2004), and elemental concentrations in atmospheric particles (Ma et al., 2004). These represent only early examples, as the number of projects being carried out using zone plate microprobes is increasing rapidly.

4.3 Materials Science

Applications of X-ray microscopes to material science include four broad categories of study: chemical state mapping in polymer systems

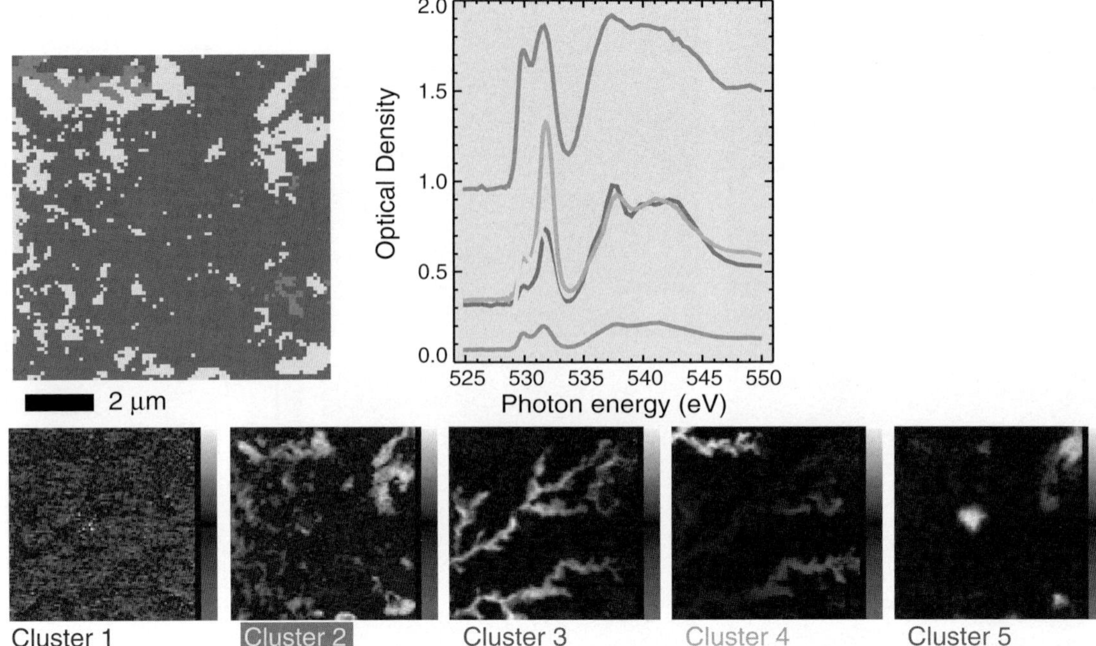

Figure 13–34. Cluster analysis in a spectromicroscopy study of lutetium in hematite. Lutetium is serving as a homologue to americium in an investigation of the uptake and transport of nuclear waste products in groundwater colloids. By using a pattern recognition algorithm to search for pixels with spectroscopic similarities, a set of signature spectra is automatically recovered from the data (shown here in a color-coded classification map) and thickness maps can be formed based on these signature spectra. Analysis at the oxygen edge reveals two different phases of reactivity for lutetium with hematite. Analysis by Lerotic (2004), from a study by T. Schäfer, INE Karlsruhe. (See color plate.)

using spectromicroscopy, imaging of the structure and electromigration failure of integrated circuits, measurements of strain in crystalline materials using microdiffraction, and studies of surface properties using photoelectrons. Studies of polymer systems represent one of the

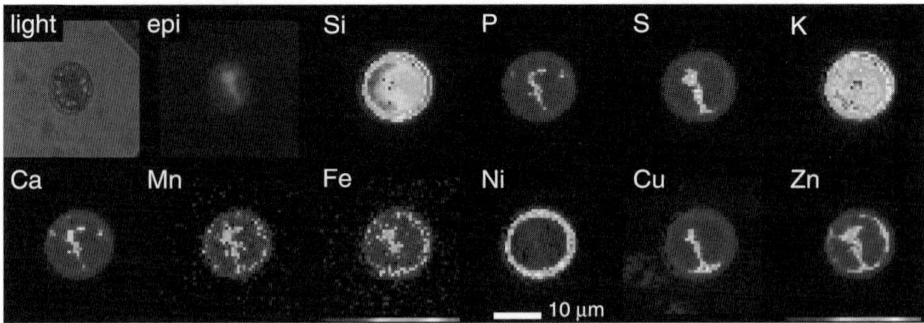

Figure 13–35. Visible light and epifluorescence micrographs, and false color X-ray fluorescence element maps of a centric diatom collected from the southern Pacific. In this region of the ocean, iron availability is a biolimiter with an impact on oceanic uptake of carbon dioxide from the atmosphere. X-ray microprobes allow one to study iron content on a protist-specific basis. (Reprinted from Twining et al., © 2003, with permission from American Chemical Society.) (See color plate.)

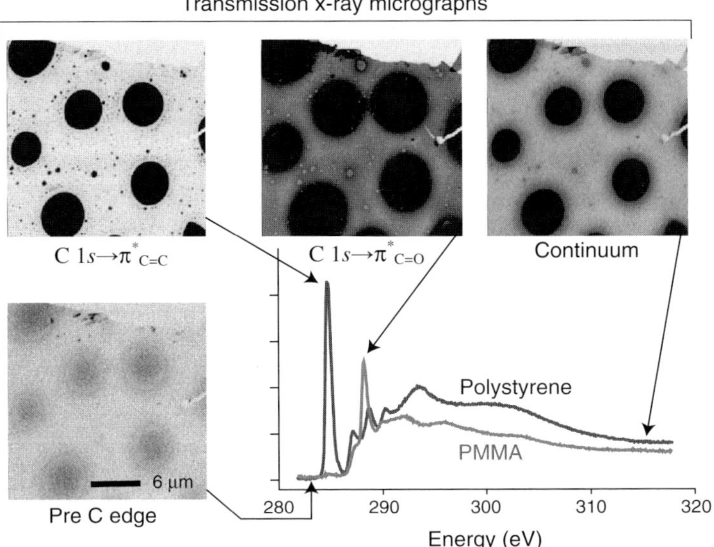

Figure 13–36. One of the first applications of X-ray transmission spectromicroscopy was to the study of polymers, where the chemical selectivity of near-edge absorption resonances allows one to make maps based on XANES spectral signatures. In this example, polymethylmethacrylate (PMMA) was spun cast with polystyrene (PS) before annealing, giving rise to phase segregation. Images acquired at specific absorption resonances show very different contrast and can be used to form compositional maps of the polymers. (Courtesy of D.A. Winest, NCSU.)

first uses of zone plate spectromicroscopy (Ade, 1992) (see Figure 13–36), and subsequent work has ranged from exploring fundamental questions such as confinement-induced miscibility (Zhu et al., 1999) to studies of specific industrially useful materials using both absorption contrast (Smith et al., 2001; Rightor et al., 2002; Hitchcock et al., 2003; Croll et al., 2005) and linear dichroism (Ade an dHsiao, 1993). Other studies have measured the degree to which polymers can seep into wood at the cellular level in particleboard (Buckley et al., 2002). These represent only a few examples; a much wider survey is given in recent reviews (Ade, 1998; Urquhart et al., 1999).

Modern integrated circuits are incredibly intricate, with oxidation layers sometimes only a few molecular layers thick, and metallization planes and vias which connect them, having dimensions in the 100 nm range. The ability of X-ray microscopes to image thick specimens (especially using phase contrast at higher energies; see Figure 13–37) is well suited to studies of the properties and failure modes of such circuits. As one example, Schneider et al. (2002b, 2002c, 2003) have studied electromigration failures as they take place, leading to observations of the propagation of voids from the point of their original formation (see Figures 13–38 and 13–39), while Levine et al. have done tomographic imaging of electromigration voids using a STXM (Levine et al., 2000). For industrial applications of chip inspection, a very significant development has been the recent commercial availability (Xradia, Inc.) of

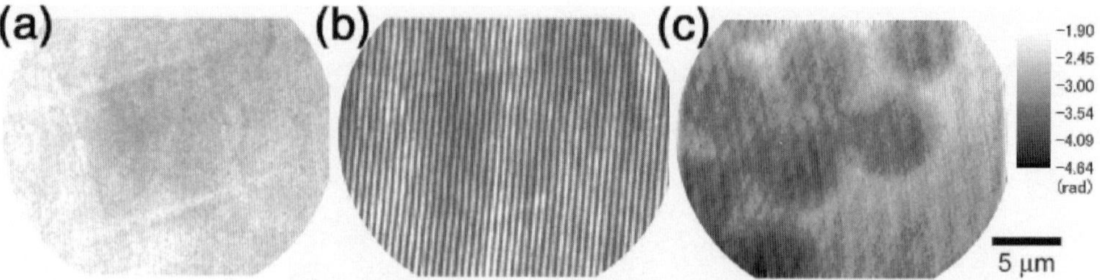

Figure 13–37. Interferometric TXM imaging of polystyrenes at 9 KeV. In these experiments, a hard X-ray micro-interferometer has been constructed by using two overlying objective zone plates with a slight transverse offset to produce an interferometric fringe pattern as shown in (b). Compared to the single-objective image (a), interference fringes with a visibility of as high as 60% can clearly be seen. Analysis of the interferometric image (b) is then used to obtain the quantitative contrast image of the polystyrene spheres shown in (c). This example shows how low absorption contrast objects can be imaged in hard X-ray microscopes. (Courtesy of T. Koyama, Himeji Institute of Technology.)

laboratory-based tomography systems using zone plate optics and operating at a sufficiently high energy (5.4 keV) so as to allow tomographic data sets to be acquired and reconstructed; this allows one to study various metallization layers in intact, working chips (Wang et al., 2002) (see Figure 13–40).

Another way in which zone plate X-ray microscopes are used to study material properties is through microdiffraction, where one

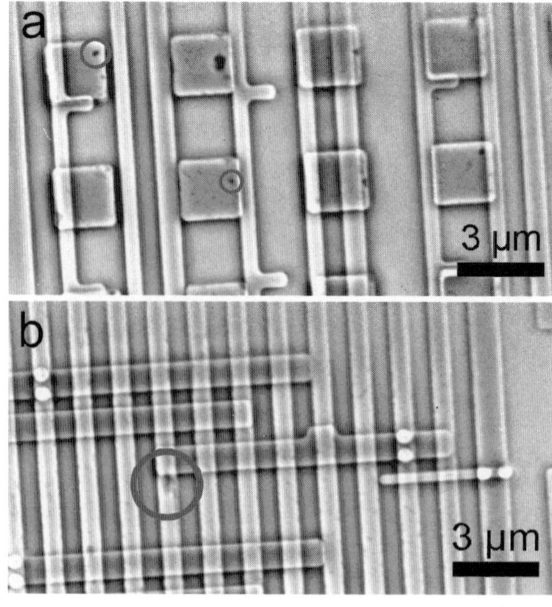

Figure 13–38. Zernike phase contrast provides one means to image the metallic layers of integrated circuits in regions where the underlying silicon wafer has been thinned. A common failure mode in integrated circuits is electromigration in which voids in a conducting layer or via are formed. These images obtained using a TXM operating at 4 keV at the European Synchrotron Radiation Research Facility or ESRF, show what appear to be such voids (circles) within test structures for advanced microprocessors. (Reprinted from Schneider et al., © 2003, with permission from Elsevier.)

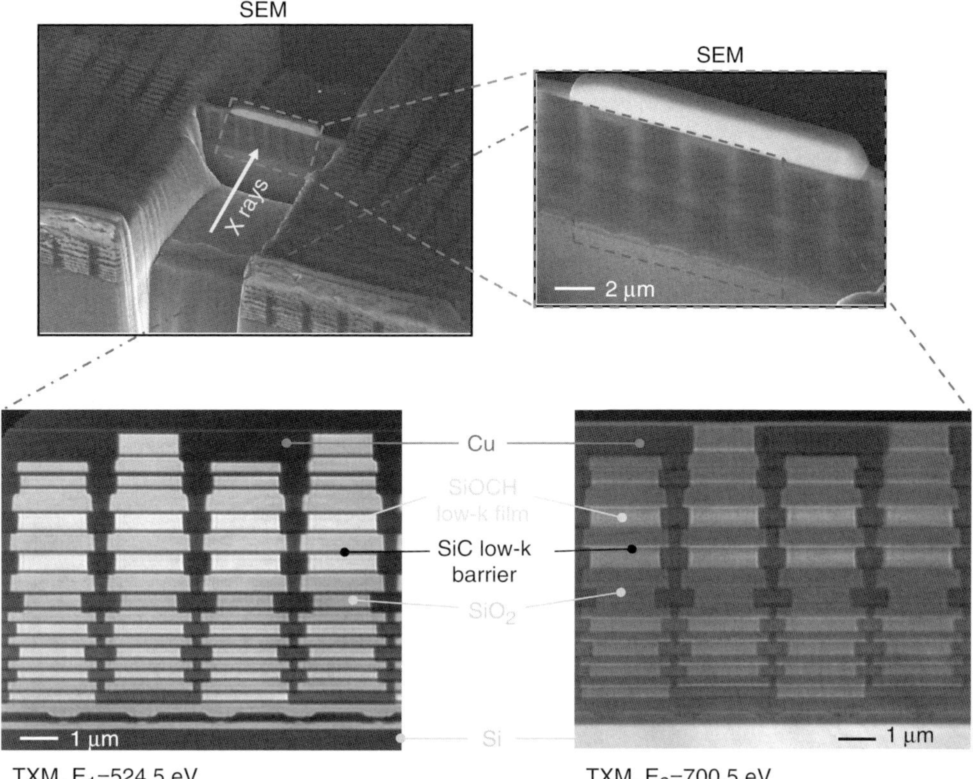

Figure 13–39. In studies of integrated circuits, it is often important to study changes in metallization from layer-to-layer by imaging in cross section. In this example, a focused ion beam (FIB) system was used to prepare a thinned, fully passivated cross section of copper interconnect structures within an electrically functional test structure as shown in the two scanning electron micrographs at top. A soft X-ray TXM at the BESSY II synchrotron facility in Berlin was then used to image this cross section at 525.5 eV (left) and 700.5 eV (right) to highlight different metal and dielectric layers in the chip, with features as fine as 20 nm visible. (Courtesy of G. Schneider, BESSY.)

Figure 13–40. Tomographic imaging of an integrated circuit done with a commercial laboratory X-ray microscope (Xradia). An integrated circuit had the silicon wafer underneath a region of interest thinned to about 15 µm, after which a tilt series of TXM images was acquired over 8 hours using a rotating anode source operation at 5.4 keV. The figure shows slices extracted at depths corresponding to the center of three Cu interconnect layers in the tomographic reconstruction with an estimated resolution of 60 nm in the transverse dimension and 90 nm in depth. This system can be used for chip inspection at a chip fab plant, among other applications. (Courtesy W. Yun, Xradia.)

examines not the undeviated transmission image through the specimen, but the signal that is Bragg diffracted (usually in the Laue geometry) by specific crystalline regions within the specimen (see for example (Engström et al., 1995)). Measurement of the position of the Bragg peaks can give values of the local lattice constants so that repetition of the measurement over a grid of points provides a strain map of the sample. This has been applied to optoelectronic devices (Cai et al., 1999), magnetic domain evolution (Evans et al., 2002) as well as for examination of the strain at the midpoint and edges of mesoscopic structures (Murray et al., 2005) (see Figure 13–41).

Since photoelectrons emerge only from within the top 100 nm or so of a bulk specimen, methods that use photoelectron detection are ideal for studies of surface phenomena. Photoelectron emission microscopes using X-ray illumination of a broad area and sub-30 nm resolution electron optics are beyond the scope of our concentration on zone plate microscopes, though we note that they are used with great success and at very high spatial resolution (see Figure 13–42). Another type of photoelectron microscope is a Scanning PhotoEmission Microscope or SPEM using a zone plate to produce a fine focus and an electron spectrometer for signal detection (Ade et al., 1990b; Ko et al., 1998; Yi et al., 2005) (see Fig. 43); activities in this area were recently reviewed by Günther et al. (2002).

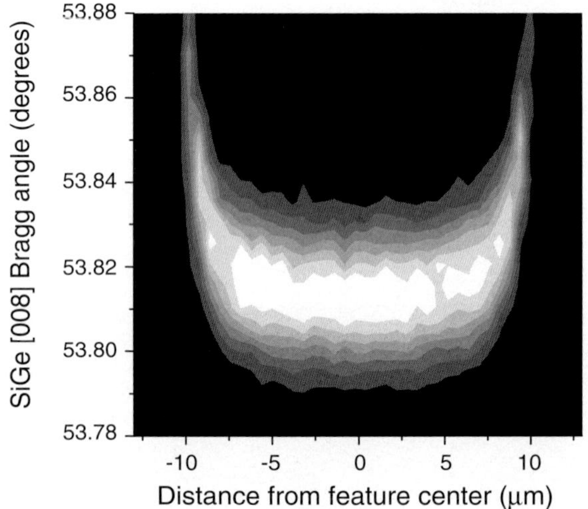

Figure 13–41. In X-ray microdiffraction, a detector is set to collect diffraction from small, crystalline features of the specimen that can be selectively illuminated by the microfocus beam. Local variations from perfect crystal order are seen as changes in the width or angle of the diffraction peaks. In this example, a 20 μm wide, 0.24 μm thick $Si_{0.86}Ge_{0.14}$ pseudomorphically strained film is located on a Si [001] surface. A determination of the angle of the SiGe [008] diffraction peak as a function of position on the sample reveals elastic relaxation at the free edges of the SiGe feature, and demonstrates the ability of a zone-plate STXM to study the strain distribution of patterned microstructures. (Reprinted from Murray et al., © 2005, with permission from American Institute of Physics.) (See color plate.)

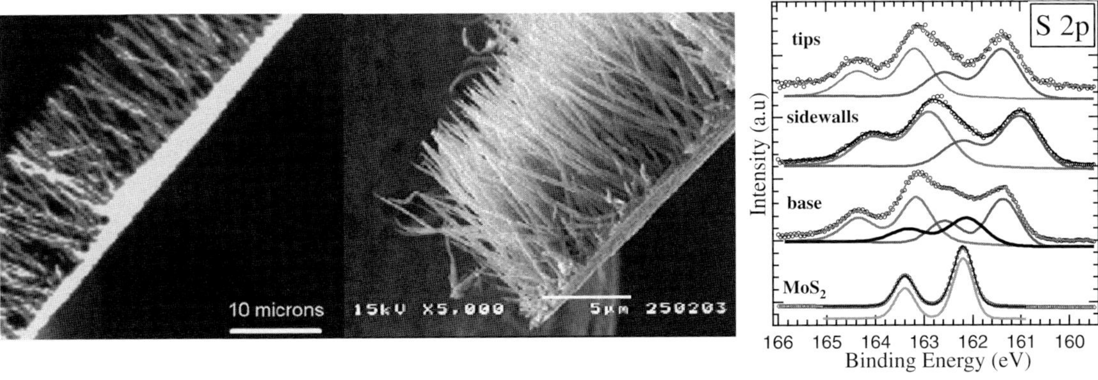

Figure 13–42. In Electron Spectroscopy for Chemical Analysis or ESCA microscopy, a monochromatic beam is used to illuminate a region several micrometers across; electron optics are then used to image a tunable electron ejection energy to reveal surface chemistry. Though this does not involve zone plate imaging, we include it here due to its widespread use with tunable X rays. In this case a 90-nm resolution ESCA microscope was used to locate aligned MoS_2 nanotube bundles and select certain areas along the axes of the tubes for detailed examination. The image at left was acquired using Mo $3d$ electrons, while S $2p$, Mo $3d$, and valence band spectra taken at the tips and sidewalls and the growth base from the Si wafer appear strongly affected by the low dimensionality of the nanotubes and differ significantly from the corresponding spectra taken on a reference MoS_2 crystal. (Reprinted from Kiskinova et al., © 2003, with permission of EDP Sciences.)

4.4 Magnetic Materials

X-ray magnetic circular dichroism (XMCD) exploits changes in absorption due to the relative orientation of magnetic domains and incident circularly polarized radiation. It draws upon the fact that in magnetic

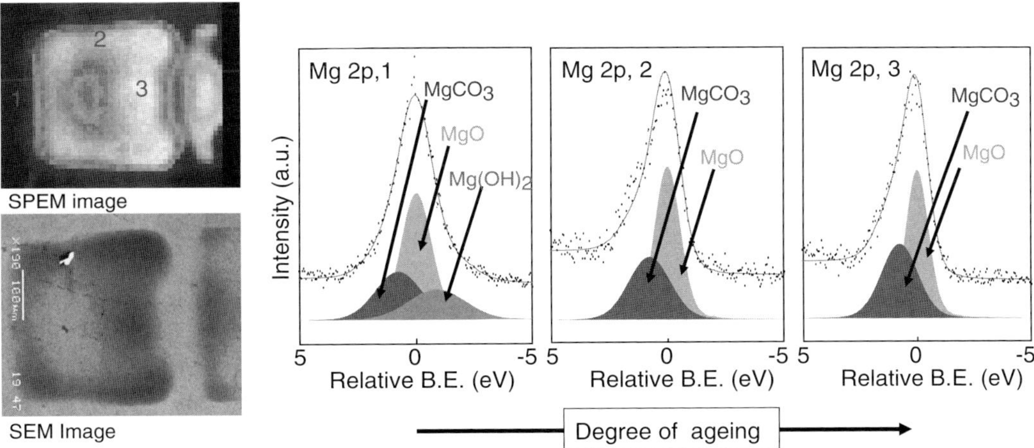

Figure 13–43. Scanning photoemission microscope (SPEM) study of a plasma display cell. In this microscope the specimen is scanned through the zone plate focus while photoelectrons are collected by an electron spectrometer. This figure shows a SPEM image, a scanning electron micrograph, and photoelectron spectra from several regions of the sample. In a plasma display cell, light of the appropriate color emerges through a front glass window which is protected from plasma damage by a composite insulating layer including MgO. The photoelectron spectra show aging in the $Mg(OH)_2$ component of the layer over the life of the display cell. (Reprinted from Yi et al., © 2005, with permission from the Institute of Pure and Applied Physics.) (See color plate.)

materials the density of certain electronic states is different for electrons of spin parallel to the magnetization, compared to electrons of spin anti-parallel. The absorption of circularly polarized photons selects between electron spins, and depends on the component of spin parallel to the helicity of the photon (the direction of the photon beam). Images taken with a particular polarization of the illumination beam, at saturated magnetization states, or at L_2 versus L_3 absorption edges can by themselves show magnetic contrast effects, while difference images between two polarization states at an absorption edge can be used to obtain element-specific images of magnetic contrast only. While much work has been done using photoemission microscopes (Stöhr et al., 1993) and there are recent exciting results using X-ray holography (Eisebitt et al., 2004), zone plate microscopy provides two primary approaches. One method is to use a TXM with a large-angle-collection condenser zone plate and exploit the fact that the radiation from synchrotron bending magnet sources is circularly polarized above and below the synchrotron plane; this was the first method demonstrated (Fischer et al., 1996) and it has led to considerable success for the study of out-of-specimen-plane magnetism (Fischer et al., 2001a) (see Figure 13–44) and has recently been extended to the study of in-specimen-

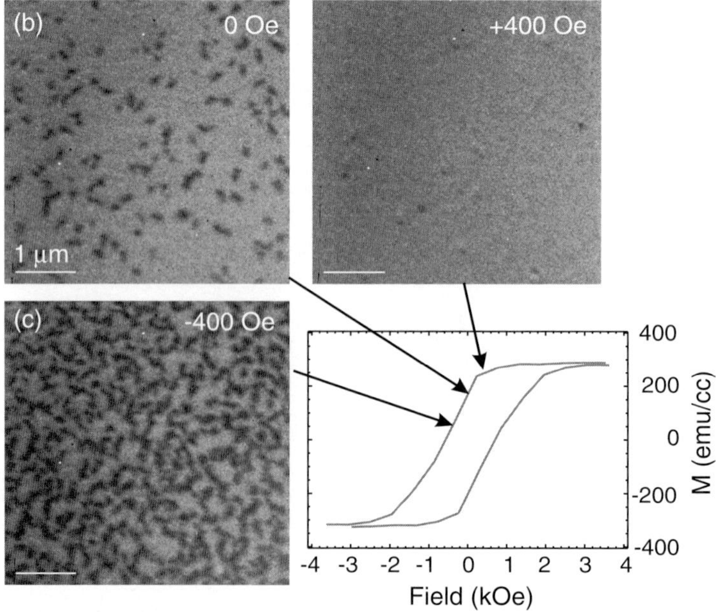

Figure 13–44. X-ray magnetic circular dichroism (XMCD) images of the magnetic domain structure of a 50-nm thick $(Co_{83}Cr_{17})_{87}Pt_{13}$ alloy film recorded at the Co L_3 absorption edge (777 eV) and in an external field of (a) +400 Oe, (b) 0 Oe, and (c) −400 Oe. (d) M vs. H hysteresis loop obtained via VSM measurement. The arrows indicate the point in the reversal cycle at which each image is recorded. Domain structure is apparent as the magnetization of the film is driven around the hysteresis loop and the net magnetization reversal can be seen to be the average of the reversal of individual domains, with the number of reversed domains increasing as the strength of the applied field is increased. (Reprinted from Im et al., © 2003, with permission from American Institute of Physics.)

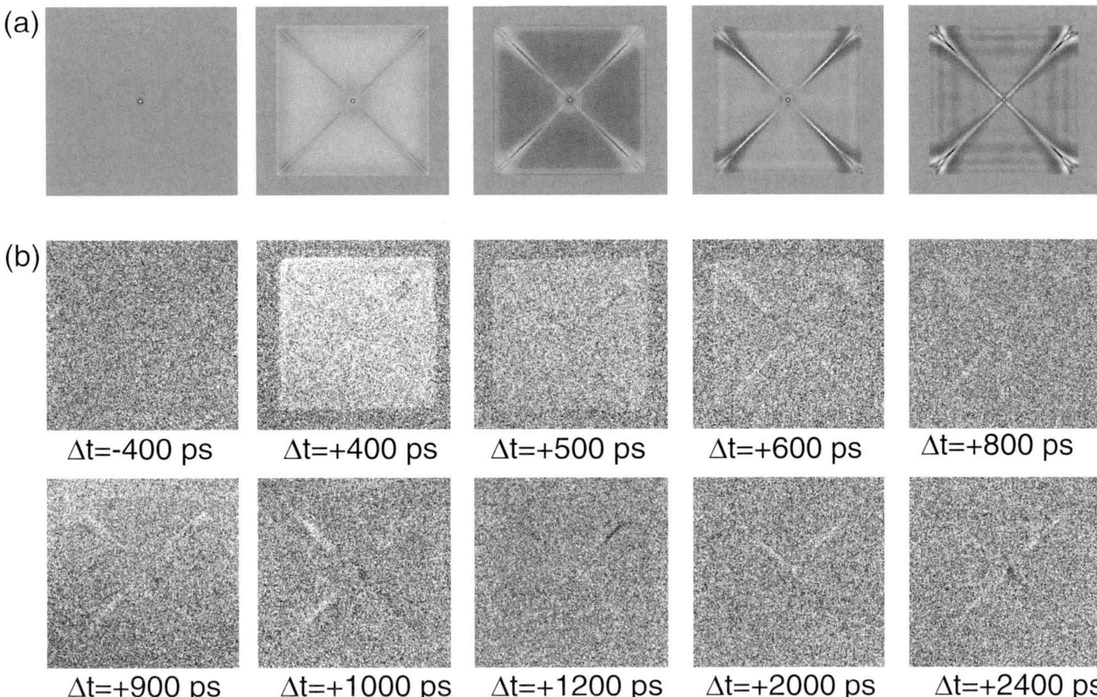

Figure 13–45. Time-resolved XMCD imaging of a magnetized Ni-Fe film patch as the magnetization is reversed in an applied magnetic field. a) The z-component of the dynamic magnetization at selected time delays obtained from micro-magnetic simulations (OOMMF). b) XMCD images from the XM-1 TXM taken with various time delays between the application of the pulsed magnetic field and the arrival of radiation from electron bunches in the storage ring. By integrating over many bunches with a particular time delay, one can study the temporal evolution of the z-component of the magnetization at delay times varying from probe pulse 400 ps before the pump, up to 2400 ps after the pump. (Reprinted from Stoll et al., © 2004, with permission from American Institute of Physics.)

plane magnetic structure as well (Fischer et al., 2001b). Another more recent approach is to use a STXM with a variable polarization undulator source. In either case, the pulsed nature of synchrotron radiation from electron bunches means that one can cycle an applied magnetic field in synchrony with the arrival of short (~100 psec) pulses of X-rays, and thereby accumulate images corresponding to controlled time delays before and after application of the pulsed field (Stoll et al., 2004) (see Figure 13–45). A more extended discussion of magnetic contrast X-ray microscopy is provided in a recent review by Fischer (Fischer, 2003).

5 Conclusion

In this chapter, we have outlined some of the principles and characteristics of X-ray microscopes using zone plate optics, and have attempted to convey an incomplete but representative survey of their applications in scientific studies. We have seen that the resolution and efficiency of zone plates has improved considerably over the lifetime of the field, although, in spite of constant efforts and the application of the best technology, the rate of improvement has been slow. For some time the "Moore's Law" graph for zone plate resolution has had a slope of about a factor of two per decade. However, as we have seen, this area of development has been especially active in recent times. There is now some optimism that the 10nm barrier may be broken and the present art is nowhere close to hitting fundamental limits. Resolution is not the whole story, however; many applications are combining imaging with tilt of the specimen for tomography, with energy tunability for spectromicroscopy, and with fluorescence detection for elemental identification. These represent the application of zone plate optics to extend the boundaries of previously existing techniques with active communities, so these areas are likely to expand. Another general trend of the last few years has been the growth in hard X-ray applications of zone plate imaging. This has been especially beneficial for tomography and microanalysis and, as recent experiments have shown, the use of hard X-ray zone plates in high order may soon approach the best resolution of soft X-ray zone plates in first order. At the time of this writing it seems that technical developments in X-ray microscopy and its marriage with promising application areas is happening at an ever-increasing pace and we can now forecast that these activities have a bright future with more confidence than ever before.

Acknowledgments. Naturally an enterprise like writing this review depends greatly on the willingness of our colleagues around the X-ray microscopy community to provide us with advice information and images and we thank the many people who have done that. We especially thank Janos Kirz and Henry Chapman for reading the manuscript and our immediate colleagues at Stony Brook, Brookhaven and Berkeley for many helpful discussions. Work by MH and TW was supported by the Director, Office of Energy Research, Office of Basics Energy Sciences, Materials Sciences Division of the U. S. Department of Energy, under Contract No. DE-AC03-76SF00098. Work by CJ was supported by the National Institutes of Health under grant R01 EB00479-01A1, and the National Science Foundation under grants DBI-9986819, ECS-0099893, and CHE-0221934.

References

Abraham-Peskir, J. (1998). Structural changes in fully hydrated *Chilomonas paramecium* exposed to copper. *Eur. J. Protisto.* **34**, 51–57.

Abraham-Peskir, J. (2000). X-ray microscopy with synchrotron radiation: Applications to cellular biology. *Cel. Molec. Biol.* **46**(6), 1045–1052.

Ade, H. (1998). X-ray spectromicroscopy. *Experimental Methods in the Physical Sciences.* **32**, 225–262 (R. Celotta and T. Lucatorto, Eds.). (Academic Press New York).

Ade, H. and Hsiao, B. (1993). X-ray linear dichroism microscopy. *Science* **262**, 1427–1429.

Ade, H., Kirz, J., Hulbert, S., Johnson, E., Anderson, E. and Kern, D. (1990a). Scanning photoelectron microscope with a zone plate generated microprobe. *Nucl. Inst. Methods Phys. Res. A* **291**, 126–131.

Ade, H., Kirz, J., Hulbert, S.L., Johnson, E., Anderson, E. and Kern, D. (1990b). X-ray spectromicroscopy with a zone plate generated microprobe. *Appl. Phys. Lett.* **56**, 1841–1843.

Ade, H., Zhang, X., Cameron, S., Costello, C., Kirz, J. and Williams, S. (1992). Chemical contrast in X-ray microscopy and spatially resolved XANES spectroscopy of organic specimens. *Science* **258**, 972–975.

Agard, D. and Sedat, J. (1983). Three-dimensional architecture of a polytene chromosome. *Nature* **302**, 676–681.

Anderson, E.H., Bögli, V. and Muray, L.P. (1995). Electron beam lithography digital pattern generator and electronics for generalized curvilinear structures. *J. Vacuum Sci. Tech. B* **13**, 2529–2534.

Anderson, E.H., Olynick, D.L., Harteneck, B., Veklerov, E., Denbeaux, G., Chao, W., Lucero, A., Johnson, L. and Attwood, D. (2000). Nanofabrication and diffractive optics for high-resolution X-ray applications. *J. Vacuum Sci. Tech. B* **18**(6), 2970–2975.

Aoki, S. (1994). Grazing incidence X-ray microscope with a Wolter type mirror. In *Radiation in the Biosciences* (B. Chance, D. Deisenhofer, S. Ebashi, D.T. Goodhead, J.R. Helliwell, H.E. Huxley, T. Iizuka, J. Kirz, T. Mitsui, E. Rubenstein, N. Sakabe, T. Sasaki, G. Schmahl, H. Sturhmann, K. Wüthrich and G. Zaccai, Eds.) (Oxford University Press, New York).

Aoki, S. and Kikuta, S. (1974). X-ray holographic microscopy. *Japan. J. Appl. Phys.* **13**, 1385–1392.

Aristov, V.V. and Erko, A.I., Eds. (1994). *X-ray Microscopy IV.* (Bogorodskii Pechatnik, Chernogolovka, Russia).

Atlwood, D. (1999). *Soft X-rays and Extreme Ultraviolet Radiation.* (Cambridge Univ. Press, Cambridge).

Baez, A.V. (1960). A self-supporting metal Fresnel zone-plate to focus extreme ultra-violet and soft X-rays. *Nature* **186**, 958.

Baez, A.V. (1961). Fresnel zone plate for optical image formation using extreme ultraviolet and soft X radiation. *J. Opt. Soc. Amer.* **51**, 405–412.

Baez, A.V. (1989). The early days of X-ray optics: A personal memoir. *J. X-ray Sci. Tech.* **1**, 3–6.

Baez, A.V. (1997). Anecdotes about the early days of X-ray optics. *J. X-ray Sci. Tech.* **7**(2), 90–97.

Barbee, T.W., Jr. (1981). Sputtered layered synthetic microstructure (LSM) dispersion elements. In *Proceedings of the Conference on Low Energy X-ray Diagnostics* (American Institute of Physics, Monterey, CA).

Barrett, R., Kaulich, B., Oestreich, S. and Susini, J. (1998). Scanning microscopy end station at the ESRF X-ray microscopy beamline. In *X-ray Microfocusing: Applications and Techniques* (I. McNulty, Ed.). *Proceedings of the SPIE*, Vol. 3449, 80–90 (SPIE, Bellingham, WA).

Batterman, B. and Cole, H. (1964). Dynamical diffraction of X-rays by perfect crystals. *Rev. Mod. Phys.* **36**, 681–717.

Beetz, T. and Jacobsen, C. (2003). Soft X-ray radiation-damage studies in PMMA using a cryo-STXM. *J. Synchrotron Rad.* **10**(3), 280–283.

Behets, G.J., Verberckmoes, S.C., Oste, L., Bervoets, A.R., Salome, M., Cox, A.G., Denton, J., De Broe, M.E. and D'Haese, P.C. (2005). Localization of lanthanum in bone of chronic renal failure rats after oral dosing with lanthanum carbonate. *Kidney Int.* **67**(5), 1830–1836.

Bennett, P.M., Foster, G.F., Buckley, C.J. and Burge, R.E. (1993). The effect of soft X-radiation on myofibrils. *J. Microsc.* **172**, 109–119.

Bilderback, D.H. (2003). Review of capillary X-ray optics from the second international capillary optics meeting. *X-ray Spectrom.* **32**(3), 195–207.

Bionta, R.M., Skulina, K.M. and Weinberg, J. (1994). Hard X-ray sputtered-sliced phase zone plates. *Appl. Phys. Lett.* **64**(8), 945–947.

Bögli, V., Unger, P., Beneking, H., Greinke, B., Guttman, P., Neimann, B., Rudolf, D. and Schmahl G. (1988). Microzoneplate fabrication by 100 keV electron beam lithography. In *X-ray Microscopy II* (Springer-Verlag, New York).

Bonnin-Mosbah, M., Metrich, N., Susini, J., Salome, M., Massare, D. and Menez, B. (2002). Micro X-ray absorption near edge structure at the sulfur and iron K-edges in natural silicate glasses. *Spectrochimica Acta Part B* **57**(4), 711–725.

Born, M. and Wolf, E. (1999). *Principles of Optics.* (Cambridge University Press, Cambridge).

Botto, R.E., Cody, G.D., Kirz, J., Ade, H., Behal, S. and Disko, M. (1994). Selective Chemical Mapping of Coal Microheterogeneity by Scanning Transmission X-ray Microscopy. *Energy and Fuels* **8**, 151–154.

Boyce, C.K., Cody, G.D., Feser, M., Jacobsen, C., Knoll, A.H. and Wirick, S. (2002). Organic chemical differentiation within fossil plant cell walls detected with X-ray spectromicroscopy. *Geology* **30**(11), 1039–1042.

Boyce, C.K., Zwieniecki, M.A., Cody, G.D., Jacobsen, C., Wirick, S., Knoll, A.H. and Holbrook, N.M. (2004). Evolution of xylem lignification and hydrogel transport regulation. *Proc. Nat. Acad. Sci.* **101**(50), 17555–17558.

Braun, A., Shah, N., Huggins, F.E., Huffman, G.P., Wirick, S., Jacobsen, C., Kelly, K. and Sarofim, A.F. (2004). A study of diesel PM with X-ray microspectroscopy. *Fuel* **83**, 997–1000.

Brown, G.E. and Sturchio, N.C. (2002). An overview of synchrotron radiation applications to low temperature geochemistry and environmental science. In *Applications of Synchrotron Radiation in Low-Temperature Geochemistry and Environmental Sciences. Rev. Mineral Geochem.*, **49**, 1–115.

Buckley, C.J., Browne, M.T. and Charalambous, P.S. (1985). Contamination lithography for the fabrication of zone plate X-ray lenses. In *Electron Beam, X-ray and Ion-Beam Techniques for Submicrometer Lithographies* (P.D. Blair, Ed.). *Proceedings of the SPIE*, Vol. 537, 213–217 (SPIE, Bellingham, WA).

Buckley, C.J., Khaleque, N., Bellamy, S.J., Robins, M. and Zhang, X. (1997). Mapping the organic and inorganic components of tissue using NEXAFS. *J. Phys. IV* **7**, 83–90.

Buckley, C.J., Phanopoulos, C., Khaleque, N., Engelen, A., Holwill, M.E.J. and Michette, A.G. (2002). Examination of the penetration of polymeric methylene di-phenyl-di-isocyanate (pMDI) into wood structure using chemical-state x-ray microscopy. *Holzforschung* **56**(2), 215–222.

Cai, Z., Lai, B., Xiao, Y. and Xu, S. (2003). An X-ray diffraction microscope at the Advanced Photon Source. *J. Phys. IV* **104**, 17–20.

Cai, Z.H., Rodrigues, W., Ilinski, P., Legnini, D., Lai, B., Yun, W., Isaacs, E.D., Lutterodt, K.E., Grenko, J., Glew, R., Sputz, S., Vandenberg, J., People, R., Alam, M.E., Hybertsen, M. and Ketelsen, L.J.P. (1999). Synchrotron x-ray microdiffraction diagnostics of multilayer optoelectronic devices. *App. Phys. Lett.* **75**(1), 100–102.

Carrington, W.A., Lynch, R.M., Moore, E.D.M., Isenberg, G., Fogarty, K.E. and Fay, F.S. (1995). Superresolution three-dimensional images of fluorescence in cells with minimal light exposure. *Science* **268**, 1483–1487.

Casalis, L., Jark, W., Kiskinova, M., Lonza, D., Melpignano, P., Morris, D., Rosei, R., Savoia, A., Abrami, A., Fava, C., Furlan, P., Pugliese, R., Vivoda, D., Sandrin, G., Wei, F.Q., Contarini, S., DeAngelis, L., Gariazzo, C., Nataletti,

P. and Morrison, G. (1995). ESCA Microscopy beamline at ELETTRA. *Rev. Sci. Inst.* **66**, 4870–4875.

Castaing, R., Deshamps, P. and Philibert, J., Eds. (1966). *X-ray Optics and X-ray Microanalysis* (Herman, Paris).

Chan, C.S., De Stasio, G., Welch, S.A., Girasole, M., Frazer, B.H., Nesterova, M.V., Fakra, S. and Banfield, J.F. (2004). Microbial polysaccharides template assembly of nanocrystal fibers. *Science* **303**, 1656–1658.

Chantler, E. and Abraham-Peskir, J.V. (2004). Significance of midpiece vesicles and functional integrity of the membranes of human spermatozoa after osmotic stress. *Andrologia* **36**(2), 87–93.

Chao, W., Anderson, E., Denbeaux, G.P., Harteneck, B., Liddle, J.A., Olynick, D.L., Pearson, A.L., Salmassi, F., Song, C.Y. and Attwood, D.T. (2003) 20-nm-resolution soft x-ray microscopy demonstrated by use of multilayer test structures, *Opt. Lett.* **28**(21), 2019–2021.

Chao, W., Harteneck, B., Liddle, A., Anderson, E. and Attwood, D. (2005). Soft X-ray microscopy at a spatial resolution better than 15 nm. *Nature* **345**, 1210–1213.

Chapman, H.N. (1996a). Phase retrieval X-ray microscopy by Wigner-distribution deconvolution. *Ultramicroscopy* **66**, 153–172.

Chapman, H.N., Fu, J., Jacobsen, C. and Williams, S. (1996b). Dark-field X-ray microscopy of immunogold-labeled cells. *J. Microsc. Soc. Amer.* **2**(2), 53–62.

Chapman, H.N., Jacobsen, C. and Williams, S. (1996c). A characterisation of dark-field imaging of colloidal gold labels in a scanning transmission x-ray microscope. *Ultramicroscopy* **62**(3), 191–213.

Chapman, H.N., Barty, A., Beetz, T., Cui, C., He, H., Howells, M.R., Marchesini, S., Noy, A., Spence, J.C.H. and Weierstall, U. (2006). High-resolution three-dimensional X-ray diffraction microscopy. *J. Opt. Soc. Amer.* **23**, 1179–1201.

Cloetens, P., Ludwig, W., Baruchel, J., Van Dyck, D., Van Landuyt, J., Guigay, J.P. and Schlenker, M. (1999). Holotomography: Quantitative phase tomography with micrometer resolution using hard synchrotron radiation x rays. *App. Phys. Lett.* **75**(19), 2912–2914.

Cody, G., Botto, R.E., Ade, H. and Wirick, S. (1995). Soft x-ray microscopy and microanalysis: Applications in organic geochemistry. In *X-ray Microbeam Technology and Applications* (W. Yun, Ed.). *Proceedings of the SPIE*, Vol. 2516, 185–196 (SPIE, Bellingham, WA).

Cody, G.D. (2000). Probing chemistry within the membrane structure of wood with soft X-ray spectral microscopy. In *X-ray Microscopy: Proceedings of the Sixth International Conference* (W. Meyer-Ilse, T. Warwick and D. Attwood, Eds.), (American Institute of Physics, Melville, NY).

Coetzee, J. and van der Merwe, C.F. (1984). Extraction of substances during glutaraldehyde fixation of plant cells. *J. Microsc.* **135**, 147–158.

Coetzee, J. and Merwe, C.F.V.D. (1989). Extraction of carbon 14 labeled compounds from tissue during processing for electron microscopy. *J. Electron Microsc. Tech.* **11**, 155–160.

Cosslett, V.E. and Nixon, W.C. (1960). *X-ray Microscopy* (Cambridge University Press, London).

Cosslett, V.E., Engström, A. and Pattee, H.H. Eds. (1957). *X-ray Optics and X-ray Microanalysis* (Academic Press, New York).

Croll, L.M., Stover, H.D.H. and Hitchcock, A.P. (2005). Composite tectocapsules containing porous polymer microspheres as release gates. *Macromolecules* **38**(7), 2903–2910.

David, C., Kaulich, B., Medenwaldt, R., Hettwer, M., Fay, N., Diehl, M., Thieme, J. and Schmahl, G. (1995). Low-distortion electron-beam lithography for fabrication of high-resolution germanium and tantalum phase zone plates. *J. Vacuum Sci. Tech. B* **13**(6), 2762–2766.

David, C., Kaulich, B., Barrett, R., Salome, M. and Susini, J. (2000). High-resolution lenses for sub-100 nm x-ray fluorescence microscopy. *App. Phys. Lett.* **77**(23), 3851–3853.

David, C., Nöhammer, B., Solak, H.H., Glaus, F., Haas, B., Grubelnik, A., Dolocan, A., Ziegler, E., Hignette, O., Burghammer, M., Kaulich, B., Susini, J., Bongaerts, J. and Veen, J.V.D. (2003). Diffractive soft and hard x-ray optics. *J. Phys. IV* **104**, 171–176.

De Stasio, G., Mercanti, D., Ciotti, M.T., Droubay, T.C., Perfetti, P., Margaritonda, G. and Tonner, B.P. (1996). Synchrotron spectromicroscopy of cobalt accumulation in granule cells, glial cells and GABAergic neurons. *J. Phys. D: App. Phys.* **29**, 259–262.

Dhez, O., Ade, H. and Urquhart, S.G. (2003). Calibrated NEXAFS spectra of some common polymers. *J. Electron Spectrosc. Rel. Phenomena* **128**(1), 85–96.

Di Fabrizio, E., Gentili, M., Grella, L., Baciocchi, M., Krasnoperova, A., Cerrina, F., Yun, W., Lai, B. and Gluskin, E. (1994). High-performance multilevel blazed x-ray microscopy Fresnel zone plates: Fabricated using x-ray lithography. *J. Vacuum Sci. Tech. B* **12**(6), 3979–3985.

Di Fabrizio, E., Romanato, F., Gentili, M., Cabrini, S., Kaulich, B., Susini, J. and Barrett, R. (1999). High-efficiency multilevel zone plate for keV x-rays. *Nature* **401**(6756), 895–898.

Duvel, A., Rudolph, D. and Schmahl, G. (2003). Fabrication of thick zone plates for multi-kilovolt x-rays. *J. Phys. IV* **104**, 607–614.

Eaton, W.J., Morrison, G.R. and Waltham, N.R. (2000). Configured detector for STXM imaging. In *X-ray Microscopy: Proceedings of the Sixth International Conference* (W. Meyer-Ilse, T. Warwick and D. Attwood, Eds.). (American Institute of Physics, Melville, NY).

Echlin, P. (1992). *Low Temperature Microscopy and Analysis.* (Plenum Publishing, New York).

Eimüller, T., Niemann, B., Guttman, P., Fischer, P., Englisch, U., Vatter, R., Wolter, C., Seiffert, S., Schmahl, G. and Schütz, G. (2003). The magnetic transmission X-ray microscope project at BESSY II. *J. Phys. IV* **104**, 91–94.

Einstein, A. (1918). Lassen sich Brechungsexponenten der Korper für Röntgenstrahlen experimentell ermitteln? *Verhandlungen Deutsche Physikalische Gesellschaft* **20**, 86–87.

Eisebitt, S., Lüning, J., Schlotter, W.F., Lörgen, M., Hellwig, O., Eberhardt, W. and Stöhr, J. (2004). Lensless imaging of magnetic nanostructures by X-ray spectro-holography. *Nature* **432**, 885–888.

Engström, A. (1946). Quantitative micro- and histochemical elementary analysis by Roentgen absorption spectrography. *Acta Radiologica (Supplementum LXIII)* **63**, 1–106.

Engström, A., Cosslett, V.E. and Pattee, J.H.H., Eds. (1960). *X-ray Microscopy and X-ray Microanalysis.* (Elsevier, Amsterdam).

Engström, P., Fiedler, S. and Riekel, C. (1995). Microdiffraction instrumentation and experiments on the microfocus beamline at the ESRF. *Rev. Sci. Inst.* **66**(2), 1348–1350.

Evans, P.G., Isaacs, E.D., Aeppli, G., Cai, Z. and Lai, B. (2002). X-ray microdiffraction images of antiferromagnetic domain evolution in chromium. *Science* **295**, 1042–1045.

Feser, M., Carlucci-Dayton, M., Jacobsen, C.J., Kirz, J., Neuhäusler, U., Smith, G. and Yu, B. (1998). *Applications and Instrumentation Advances with the Stony Brook Scanning Transmission X-ray Microscope.* In *X-ray Microfocusing: Appli-*

cations and Techniques (I. McNulty, Ed.). *Proceedings of the SPIE*, Vol. 3449, 19–29. (SPIE, Bellingham, WA).

Feser, M., Beetz, T., Carlucci-Dayton, M. and Jacobsen, C. (2000). Instrumentation advances and detector development with the Stony Brook scanning transmission X-ray microscope. In *X-ray Microscopy: Proceedings of the Sixth International Conference* (W. Meyer-Ilse, T. Warwick and D. Attwood, Eds.), (American Institute of Physics, Melville, NY).

Feser, M., Jacobsen, C., Rehak, P., DeGeronimo, G., Holl, P. and Strüder, L. (2001). Novel integrating solid state detector with segmentation for scanning transmission soft x-ray microscopy. *X-ray Micro- and Nano-focusing: Applications and Techniques II* (I. McNulty, Ed.). *Proceedings of the SPIE*, Vol. 4499, 117–125. (SPIE, Bellingham, WA).

Feser, M., Jacobsen, C., Rehak, P. and De Geronimo, G. (2003). Scanning transmission x-ray microscopy with a segmented detector. *J. Phys. IV* **104**, 529–534.

Fischer, P., Schutz, G., Schmahl, G., Guttmann, P. and Raasch, D. (1996). Imaging of magnetic domains with the x-ray microscope at BESSY using x-ray magnetic circular dichroism. *Z. Phys. B* **101**, 313–316.

Fischer, P., Eimüller, T., Schütz, G., Denbeaux, G., Pearson, A., Johnson, L., Attwood, D., Tsunashima, S., Kumazawa, M., Takagi, N., Köhler, M., Bayreuther, G. (2001a). Element-specific imaging of magnetic domains at 25 nm spatial resolution using soft x-ray microscopy. *Rev. Sci. Instrum.,* **72**(5), 2322–2324.

Fischer, P., Eimüller, T., Schütz, G., Köhler, M., Bayreuther, G., Denbeaux, G. and Attwood, D. (2001b). Study of in-plane magnetic domains with magnetic transmission x-ray microscopy, *J. App. Phys.* **89**, 7159–7161.

Fischer, P. (2003). Magnetic soft X-ray transmission microscopy, *Curr. Opin. Solid State Mater. Sci.* **7**, 173–179.

Ford, T.W., Page, A.M., Foster, G.F. and Stead, A.D. (1992). Effects of soft x-ray irradiation on cell ultrastructure. In *Soft X-ray Microscopy* (C. Jacobsen and J. Trebes, Eds.). *Proceedings of the SPIE*, Vol. 1741, 325–332. (SPIE, Bellingham, WA).

Foriel, J., Philippot, P., Susini, J., Dumas, P., Somogyi, A., Salome, M., Khodja, H., Menez, M., Fouquet, Y., Moreira, D. and Lopez-Garcia, P. (2004). High-resolution imaging of sulfur oxidation states, trace elements, and organic molecules distribution in individual microfossils and contemporary microbial filaments. *Geochimica et Cosmochimica Acta* **68**(7), 1561–1569.

Gilbert, J.R., Pine, J., Kirz, J., Jacobsen, C., Williams, S., Buckley, C.J. and Rarback, H. (1992). Soft x-ray absorption imaging of whole wet tissue culture cells. In *X-ray Microscopy III* (A.G. Michette, G.R. Morrison and C.J. Buckley, Eds.). (Springer-Verlag, Berlin).

Glaeser, R.M. (1971). Limitations to significant information in biological electron microscopy as a result of radiation damage. *J. Ultrastruct. Res.* **36**, 466–482.

Gölz, P. (1992). Calculations on radiation dosages of biological materials in phase contrast and amplitude contrast x-ray microscopy. In *X-ray Microscopy III.* (A.G. Michette, G.R. Morrison and C.J. Buckley, Eds.). (Springer-Verlag, Berlin).

Goodman, J.W. (1968). *Introduction to Fourier Optics.* (McGraw-Hill, San Francisco).

Goodman, J.W. (1985). *Statistical Optics.* (John Wiley & Son, New York).

Grimm, R., Singh, H., Rachel, R., Typke, D., Zillig, W. and Baumeister, W. (1998). Electron tomography of ice-embedded prokaryotic cells. *Biophys. J.* **74**, 1031–1042.

Günther, S., Kolmakov, A., Kovac, J., Casalis, L., Gregoratti, L., Marsi, M. and Kiskinova, M. (1997). Scanning photoelectron microscopy of a bimetal/Si interface: Au coadsorbed on Ag/Si(111). *Surf. Sci.* **377**, 145–149.

Günther, S., Kaulich, B., Gregoratti, L. and Kiskinova, M. (2002). Photoelectron microscopy and applications in surface and materials science. *Prog. Surf. Sci.* **70**(4–8), 187–260.

Guttmann, P., Schneider, G., Robert-Nicoud, M., Niemann, B., Rudolph, D., Thieme, J., Jovin, T.M. and Schmahl, G. (1992). X-ray microscopic investigations on giant chromosomes isolated from salivary glands of *Chironomus thummi* larvae. In *X-ray Microscopy III.* (A.G. Michette, G.R. Morrison and C.J. Buckley, Eds.). (Springer-Verlag, Berlin).

Guttmann, P., Niemann, B., Thieme, J., Hambach, D., Schneider, G., Wiesemann, U., Rudolph, D. and Schmahl, G. (2001). Instrumentation advances with the new X-ray microscopes at BESSY II. *Nucl. Inst. Methods Phys. Res. Sect. a* **467**, 849–852.

Guttmann, P., Niemann, B., Rehbein, S., Knoechel, C., Rudolph, D. and Schmahl, G. (2003). The transmission X-ray microscope at BESSY II. *J. Phys. IV* **104**, 85–90.

Haddad, W.S., McNulty, I., Trebes, J.E., Anderson, E.H., Levesque, R.A. and Yang, L. (1994). Ultra high resolution x-ray tomography. *Science* **266**, 1213–1215.

Haelbich, R.-P. (1980). A scanning ultrasoft X-ray microscope with multilayer coated reflection optics: First test with synchrotron radiation around 60 eV photon energy. *Scanned Image Microscopy.* (E.A. Ash, Ed.). (Academic Press, London).

Haelbich, R.-P., Staehr, W. and Kunz, C. (1980). A scanning ultrasoft x-ray microscope with large aperture reflection optics for use with synchrotron radiation. *Ann. NY Acad. Sci.* **342**, 148–157.

Hambach, D., Schneider, G. and Gullikson, E. (2001). Efficient high-order diffraction of extreme-ultraviolet light and soft x-rays by nanostructured volume gratings. *Opt. Lett.* **26**(15), 1200–1202.

Harp, G.R., Han, Z.-L. and Tonner, B.P. (1990). Spatially-resolved x-ray absorption near-edge spectroscopy of silicon in thin silicon-oxide films. *Phys. Scripta* **T31**, 23–27.

Hayakawa, S., Iida, A., Aoki, S. and Gohshi, Y. (1989). Development of a scanning x-ray microprobe with synchrotron radiation. *Rev. Sci. Inst.* **60**, 2452–2455.

Heck, J., Attwood, D., Ilse, W.M. and Anderson, E. (1998). Resolution determination in x-ray microscopy: An analysis of the effects of partial coherence and illumination spectrum. *J. X-ray Sci. Tech.* **8**, 95–104.

Henke, B.L., Gullikson, E.M. and Davis, J.C. (1993). X-ray interactions: Photoabsorption, scattering, transmission and reflection at E = 50–30,000 eV, Z = 1–92. *Atom. Dat. Nucl. Dat. Tables* **54**, 181–342.

Hettwer, M. and Rudolph, D. (1998). Fabrication of the x-ray condenser zone plate KZP7. In *X-ray Microscopy and Spectromicroscopy* (J. Thieme, G. Schmahl, D. Rudolph, E. Umbach, Eds.). (Springer, Berlin.)

Hignette, O., Cloetens, P., Lee, W.-K., Ludwig, W. and Rostaing, G. (2003). Hard x-ray microscopy with reflecting mirrors status and perspectives of the ESRF technology. *J. Phys. IV* **104**, 231–234.

Hitchcock, A.P., Morin, C., Heng, Y.M., Cornelius, R.M. and Brash, J.L. (2002). Towards practical soft X-ray spectromicroscopy of biomaterials. *J. Biomat. Sci.-Poly.* **13**(8), 919–937.

Hitchcock, A.P., Araki, T., Ikeura-Sekiguchi, H., Iwata, N. and Tani, K. (2003). 3d chemical mapping of toners by serial section scanning transmission X-ray microscopy. *J. Phys. IV* **104**, 509–512.

Hopkins, H.H. (1957). Applications of coherence theory in microscopy and interferometry. *J. Opt. Soc. Am.* **47**(6), 508–526.

Horowitz, P. and Howell, J.A. (1972). A scanning x-ray microscope using synchrotron radiation. *Science* **178**, 608–611.

Howells, M.R., Charalambous, P., He, H., Marchesini, S. and Spence, J.C.H. (2002). An off-axis zone-plate monochromator for high-power undulator radiation. In *Design and Microfabrication of Novel X-ray Optics* (D. Mancini, Ed.), *Proceedings of the SPIE*, Vol. 4783, 65–73 (SPIE, Bellingham, WA).

Hubbell, J.H., Gimm, H.A. and Øverbø, I. (1980). Pair, triplet, and total atomic cross sections (and mass attenuation coefficients) for 1 MeV–100 GeV photons in elements $Z = 1$–100. *J. Phys. Chem. Ref. Data* **9**, 1023–1147.

Hunt, J.A. and Williams, D.B. (1991). Electron energy-loss spectrum-imaging. *Ultramicroscopy* **38**(1), 47–73.

Im, M., Fischer, P., Eimuller, T., Denbeaux, G. and Shin, S. (2003). Magnetization reversal study of CoCrP alloy thin films on a nanogranular length scale using magnetic transmission soft X-ray microscopy. *Appl. Phys. Lett.*, **83**, 4589–4591.

Ito, A., Shinohara, K., Nakano, H., Matsumura, T. and Kinoshita, K. (1996). Measurement of soft x-ray absorption spectra and elemental analysis in local regions of mammalian cells using an electronic zooming tube. *J. Microsc.* **181**, 54–60.

Jacobsen, C. (1992). Making soft x-ray microscopy harder: Considerations for sub-0.1 micron resolution imaging at ~4 Å wavelengths. In *X-ray Microscopy III* (A.G. Michette, G.R. Morrison and C.J. Buckley, Eds.). (Springer-Verlag, Berlin).

Jacobsen, C., Howells, M., Kirz, J. and Rothman, S. (1990). X-ray holographic microscopy using photoresists. *J. Opt. Soc. Amer. A* **7**, 1847–1861.

Jacobsen, C., Williams, S., Anderson, E., Browne, M.T., Buckley, C.J., Kern, D., Kirz, J., Rivers, M. and Zhang, X. (1991). Diffraction-limited imaging in a scanning transmission x-ray microscope. *Opt. Comm.* **86**, 351–364.

Jacobsen, C., Kirz, J. and Williams, S. (1992). Resolution in soft x-ray microscopes. *Ultramicroscopy* **47**, 55–79.

Jacobsen, C., Medenwaldt, R. and Williams, S. (1998). A perspective on biological x-ray and electron microscopy. *X-ray Microscopy and Spectromicroscopy* (J. Thieme, G. Schmahl, E. Umbach and D. Rudolph, Eds.). (Springer-Verlag, Berlin).

Jacobsen, C., Abend, S., Beetz, T., Carlucci-Dayton, M., Feser, M., Kaznacheyev, K., Kirz, J., Maser, J., Neuhäusler, U., Osanna, A., Stein, A., Vaa, C., Wang, Y., Winn, B. and Wirick, S. (2000a). Recent developments in scanning microscopy at Stony Brook. In *X-ray Microscopy: Proceedings of the Sixth International Conference* (American Institute of Physics, Berkeley, CA).

Jacobsen, C., Flynn, G., Wirick, S. and Zimba, C. (2000b). Soft x-ray spectroscopy from image sequences with sub-100 nm spatial resolution. *J. Microsc.* **197**, 173–184.

Jeanguillaume, C. and Colliex, C. (1989). Spectrum-image: The next step in EELS digital acquisition and processing. *Ultramicroscopy* **28**, 252–257.

Jearanaikoon, S. and Abraham-Peskir, J. (2005). An x-ray microscopy perspective on the effect of glutaraldehyde fixation on cells. *J. Micros.* **218**(2), 185–192.

Jochum, L. and Meyer-Ilse, W. (1995). Partially coherent image formation with x-ray microscopes. *App. Opt.* **34**(22), 4944–4950.

Joyeux, D., Lowenthal, S., Polack, F. and Bernstein, A. (1988). X-ray microscopy by holography at LURE. In *X-ray Microscopy II.* (D. Sayre, M.R. Howells, J. Kirz and H. Rarback, Eds.). (Springer-Verlag, Berlin).

Kagoshima, Y., Yokoyama, Y., Nimi, T., Koyama, T., Tsusaka, Y., Matsui, J. and Takai, K. (2003). Hard X-ray phase contrast microscope for observing transparent specimens. *J. Phys. IV* **104**, 49–52.

Kagoshima, Y., Ed. (2006). Proceedings of the 8th International Conference on X-ray Microscopy, Vol. 7, IPAP Conference Series, Tokyo.

Kamijo, N., Suzuki, Y., Takano, H., Tamura, S., Yasumoto, M., Takeuchi, A. and Awaji, M. (2003). Microbeam of 100 keV x ray with a sputtered-sliced Fresnel zone plate. *Rev. Sci. Inst.* **74**(12), 5101–5104.

Kamiya, K. (1963). Theory of Fresnel zone plate. *Sci. Light* **12**(2), 35–49.

Kang, H., Stephenson, G., Liu, C., Conley, R., Macrander, A., Maser, J., Bajt, S. and Chapman, H. (2005). High-efficiency diffractive x-ray optics from sectioned multilayers. *App. Phys. Lett.* **86**, 151109.

Kang, H.C., Maser, J., Stephenson, G.B., Liu, C., Conley, R., Macrander, A.T. and Vogt, S. (2006). Nanometer linear focusing of hard x-rays by a multilayer Laue lens. *Phys. Rev. Lett.* **96**, 127401.

Kaulich, B., Wilhein, T., Di Fabrizio, E., Romanato, F., Altissimo, M., Cabrini, S., Fayard, B. and Susini, J. (2002). Differential interference contrast x-ray microscopy with twin zone plates. *J. Opt. Soc. Amer. A* **19**(4), 797–806.

Kaulich, B., Bacescu, D., Cocco, D., Susini, J., Salomé, M., Dhex, O., David, C., Weitkamp, T., Fabrizio, E.D., Cabrini, S., Morrison, G., Charalambous, P., Thieme, J., Wilhein, T., Kovac, J., Podnar, M. and Kiscinova, M. (2003). Twinmic: A European twin microscope station combining full-field imaging and scanning microscopy. *J. Phys. IV* **104**, 103–107.

Kawai, J., Takagawa, K., Fujisawa, S., Ektessabi, A. and Hayakawa, S. (2001). Microbeam XANES and X-ray fluorescence analysis of cadmium in kidney. *J. Trace Microprobe Tech.* **19**(4), 541–546.

Kaznacheyev, K., Osanna, A., Jacobsen, C., Plashkevych, O., Vahtras, O., Ågren, H., Carravetta, V. and Hitchcock, A.P. (2002). Innershell absorption spectroscopy of amino acids. *J. Phys. Chem. A* **106**(13), 3153–3168.

Kemner, K.M., Kelly, S.D., Lai, B., Maser, J., O'Loughlin, E.J., Sholto-Douglas, D., Cai, Z.H., Schneegurt, M.A., Kulpa, C.F. and Nealson, K.H. (2004). Elemental and redox analysis of single bacterial cells by X-ray microbeam analysis. *Science* **306**, 686–687.

Kenney, J.M., Jacobsen, C., Kirz, J., Rarback, H., Cinotti, F., Thomlinson, W., Rosser, R. and Schidlovsky, G. (1985). Absorption microanalysis with a scanning soft x-ray microscope: Mapping the distribution of calcium in bone. *J. Microsc.* **138**, 321–328.

Kenney, J.M., Morrison, G.R., Browne, M.T., Buckley, C.J., Burge, R.E., Cave, R.C., Charalambous, P.S., Duke, P.D., Hare, A.R., Hills, C.P.B., Michette, A.G., Ogawa, K. and Rogoyski, A.M. (1989). Evaluation of a scanning transmission x-ray microscope using undulator radiation at the SERC Daresbury Laboratory. *J. Phys. E* **22**, 234–238.

Kern, D., Coane, P., Acosta, R., Chang, T.H.P., Feder, R., Houzego, P., Molzen, W., Powers, J., Speth, A., Viswanathan, R., Kirz, J., Rarback, H. and Kenney, J. (1984). Electron beam fabrication and characterization of Fresnel zone plates for soft x-ray microscopy. In *Science with Soft X-Rays* (F.J. Himpsel and R.W. Klaffky, Eds.). *Proceedings of the SPIE*, Vol. 447, 204–213 (SPIE, Bellingham, WA).

Kilcoyne, A.L.D., Tyliszczak, T., Steele, W.F., Fakra, S., Hitchcock, P., Franck, K., Anderson, E., Harteneck, B., Rightor, E.G., Mitchell, G.E., Hitchcock, A.P., Yang, L., Warwick, T. and Ade, H. (2003). Interferometer-controlled scanning transmission X-ray microscopes at the Advanced Light Source. *J. Synchrotron Radiat.* **10**(2), 125–136.

King, P.L., Browning, R., Pianetta, P., Lindau, I., Keenlyside, M. and Knapp, G. (1989). Image-processing of multispectral x-ray photoelectron-spectroscopy images. *J. Vac. Sci. Tech. A* **7**(6), 3301–3304.

Kinjo, Y., Shinohara, K., Ito, A., Nakano, H., Watanabe, M., Horiike, Y., Kikuchi, Y., Richardson, M.C. and Tanaka, K.A. (1994). Direct imaging in a water layer of human chromosome fibres composed of nucleosomes and their higher order structures by laser plasma X-ray contact microscopy. *J. Microsc.* **176**, 63–74.

Kintner, E.C. (1978). Method for the calculation of partially coherent imagery. *App. Opt.* **17**(17), 2747.

Kirkpatrick, P. (1949). The x-ray microscope. *Sci. Amer.* **180**, 44–47.

Kirkpatrick, P. and Baez, A.V. (1948). Formation of optical images by x-rays. *J. Opt. Soc. Amer.* **38**, 766–774.

Kirz, J. (1974). Phase zone plates for X rays and the extreme UV. *J. Opt. Soc. Amer.* **64**, 301–309.

Kirz, J. (1980a). Mapping the distribution of particular atomic species. *Ann. NY Acad. Sci.* **342**, 273–287.

Kirz, J. (1980b). Specimen damage considerations in biological microprobe analysis. *Scanning Electron Microscopy 1980*, **2**: 239–249 (Scanning Electron Microscopy International: Chicago).

Kirz, J. and Sayre, D. (1980). Soft x-ray microscopy of biological specimens. In *Synchrotron Radiation Research* (H. Winick and S. Doniach, Eds.). (Plenum, New York).

Kirz, J., Sayre, D. and Dilger, J. (1978). Comparative analysis of x-ray emission microscopies for biological specimens. *Ann. NY Acad. Sci.* **306**, 291–305.

Kirz, J., Jacobsen, C. and Howells, M. (1995). Soft x-ray microscopes and their biological applications. *Q. Rev. Biophys.* **28**(1), 33–130.

Kiskinova, M. (2003). Chemical, electronic and magnetic properties of surfaces and interfaces probed with X-ray microcopes at ELETTRA. *J. Phys. IV* **104**, 453–458.

Ko, C.H., Klauser, R., Wei, D.-H., Chan, H.-H. and Chuang, T.J. (1998). The soft x-ray scanning photoemission microscopy project at SRRC. *J. Synchrotron Radiat.* **5**, 299–304.

Kogelnik, H. (1969). Coupled wave theory for thick hologram gratings. *Bell Syst. Tech. J.* **48**, 2909–2930.

Koprinarov, I.N., Hitchcock, A.P., McCrory, C.T. and Childs, R.F. (2002). Quantitative mapping of structured polymeric systems using singular value decomposition analysis of soft x-ray images. *J. Phys. Chem. B* **106**, 5358–5364.

Krasnoperova, A.A., Xiao, J., Cerrina, F., Di Fabrizio, E., Luciani, L., Figliomeni, M., Gentili, M., Yun, W., Lai, B. and Gluskin, E. (1993). Fabrication of hard x-ray phase zone plate by x-ray lithography. *J. Vac. Sci. Tech. B* **11**, 2588–2591.

Krause, M.O. (1979). Atomic radiative and radiationless yields for K and L shells. *J. Phys. Chem. Ref. Data* **8**, 307–327.

Labrenz, M., Druschel, G.K., Thomsen-Ebert, T., Gilbert, B., Welch, S.A., Kemner, K.M., Logan, G.A., Summons, R.E., De Stasio, G., Bond, P.L., Lai, B., Kelly, S.D. and Banfield, J.F. (2000). Formation of sphalerite (ZnS) deposits in natural biofilms of sulfate-reducing bacteria. *Science* **290**, 1744–1747.

Lai, B., Yun, W.-B., Legnini, D., Xiao, Y., Chrzas, J., Viccaro, P.J., White, V., Bajikar, S., Denton, D., Cerrino, F., Di Fabrizio, E., Gentili, M., Grella, L. and Baciocchi, M. (1992). Hard x-ray phase zone plate fabricated by lithographic techniques. *App. Phys. Lett.* **61**, 1877–1879.

Larabell, C.A. and Le Gros, M.A. (2004). X-ray tomography generates 3-D reconstructions of the yeast, *Saccharomyces cerevisiae*, at 60-nm resolution. *Molec. Biol. Cell* **115**, 957–962.

Larson, B.C., Yang, W., Ice, G.E., Budai, J.D. and Tischler, J.Z. (2002). Three-dimensional X-ray structural microscopy with submicrometre resolution. *Nature* **415**, 887–890.

Lawrence, J.R., Swerhone, G.D.W., Leppard, G.G., Araki, T., Zhang, X., West, M.M. and Hitchcock, A.P. (2003). Scanning transmission X-ray, laser scanning, and transmission electron microscopy mapping of the exopolymeric matrix of microbial biofilms. *App. Env. Microbiol.* **69**(9), 5543–5554.

Lehr, J. (1997). 3D x-ray microscopy: Tomographic imaging of mineral sheaths of bacteria *Leptothrix ochracea* with the Göttingen x-ray microscope at BESSY. *Optik* **104**(4), 166–170.

Lengeler, B., Schroer, C.G., Benner, B., Gerhardus, A., Günzler, T.F., Kuhlmann, M., Meyer, J. and Zimprich, C. (2002). Parabolic refractive X-ray lenses. *J. Synchrotron Radiat.* **9**, 119–124.

Lerotic, M., Jacobsen, C., Schäfer, T. and Vogt, S. (2004). Cluster analysis of soft x-ray spectromicroscopy data. *Ultramicroscopy* **100**, 35–57.

Lerotic, M., Jacobsen, C., Gillow, J.B., Francis, A.J., Wirick, S., Vogt, S. and Maser, J. (2005). Cluster analysis in soft X-ray spectromicroscopy: Finding the patterns in complex specimens. *J. Electron Spectrosc. Rel. Phenomena* **144–147**C, 1137–1143.

Levine, Z.H., Kalukin, A.R., Kuhn, M., Frigo, S.P., McNulty, I., Retsch, C.C., Wang, Y., Arp, U., Lucatorto, T.B., Ravel, B.D. and Tarrio, C. (2000). Tomography of an integrated circuit interconnect with an electromigration void. *J. App. Phys.* **87**(9), 4483–4488.

Lindaas, S., Howells, M., Jacobsen, C. and Kalinovsky, A. (1996). X-ray holographic microscopy by means of photoresist recording and atomic-force microscope readout. *J. Opt. Soc. Amer. A* **13**(9), 1788–1800.

Ma, C.J., Tohno, S., Kasahara, M. and Hayakawa, S. (2004). Determination of the chemical properties of residues retained in individual cloud droplets by XRF microprobe at SPring-8. *Nucl. Inst. Methods Phys. Res. Section B* **217**(4), 657–665.

Magowan, C., Brown, J.T., Liang, J., Heck, J., Coppel, R.L., Mohandas, N. and Meyer-Ilse, W. (1997). Intracellular structures of normal and aberrant *Plasmodium falciparum* malaria parasites imaged by soft x-ray microscopy. *Proc. Nat. Acad. Sci.* USA **94**, 6222–6227.

Marchesini, S., He, H., Chapman, H.N., Hau-Riege, S.P., Noy, A., Howells, M.R., Weierstall, U. and Spence, J.C.H. (2003). X-ray image reconstruction from a diffraction pattern alone. *Phys. Rev. B* **68**, 140101.

Martin, L.C. (1966). *The Theory of the Microscope.* (American Elsevier, New York).

Maser, J. (1994). Theoretical description of the diffraction properties of zone plates with small outermost zone width. *X-ray Microscopy IV*. (V. Aristov and A. Erko, Eds.). (Bogorodski Pechatnik, Chernogolovka, Russia).

Maser, J., Osanna, A., Wang, Y., Jacobsen, C., Kirz, J., Spector, S., Winn, B. and Tennant, D. (2000). Soft x-ray microscopy with a cryo STXM: I. Instrumentation, imaging, and spectroscopy. *J. Microsc.* **197**, 68–79.

Maser, J., Stephenson, G.B., Vogt, S., Yun, W., Macrander, A., Kang, H.C., Liu, C. and Conley, R. (2004). Multilayer Laue lenses as high-resolution x-ray optics. In *Design and Microfabrication of Novel X-ray Optics II* (A.A. Snigirev, D.C. Mancini, Eds). *Proceedings of the SPIE*, Vol. 5539, 185–194 (SPIE, Bellingham, WA).

McNulty, I., Kirz, J., Jacobsen, C., Anderson, E., Kern, D. and Howells, M. (1992). High-resolution imaging by Fourier transform x-ray holography. *Science* **256**, 1009–1012.

McNulty, I., Lai, B., Maser, J., Paterson, D., Evans, P., Heald, S., Ice, G., Isaacs, E., Rivers, M. and Sutton, S. (2003a). X-ray microscopy at the advanced photon source. *Synchrotron Radiat. News* **16**, 34–39.

McNulty, I., Paterson, D., Arko, J., Erdmann, M., Frigo, S.P., Goetze, K., Ilinski, P., Krapf, N., Mooney, T., Retsch, C.C., Stampfl, A.P.J., Vogt, S., Wang, Y. and Xu, S. (2003b). The 2-ID-B intermediate-energy scanning X-ray microscope at the APS. *J. Phys. IV* **104**, 11–15.

Medalia, O., Weber, I., Frangakis, A., Nicastro, D., Gerisch, G. and Baumeister, W. (2002). Macromolecular architecture in eukariotic cells visualized by cryoelectron tomography. *Science* **298**, 1209–1213.

Meyer-Ilse, W., Attwood, D. and Koike, M. (1994). The X-ray Microscopy Resource Center at the Advanced Light Source. *Synchrotron Radiation in the Biociences*. (B. Chance, D. Deisenhober, S. Ebashi, D.T. Goodhead, J.R. Helliwell, H.E. Huxley, T. Iizuka, J. Kirz, T. Mitsui, E. Rubenstein, N. Sakabe, T. Sasaki, G. Schmahl, H. Sturhmann, K. Wüthrich and G. Zaccai, Eds.). (Oxford University Press, New York).

Meyer-Ilse, W., Denbeaux, G., Johnson, L.E., Bates, W., Lucero, A. and Anderson, E.H. (2000a). The high resolution x-ray microscope XM-1. *X-ray Microscopy: Proceedings of the Sixth International Conference*. (W. Meyer-Ilse, T. Warwick and D. Attwood, Eds.). (American Institute of Physics, Melville, NY).

Meyer-Ilse, W., Warwick, T. and Attwood, D., Eds. (2000). *X-ray Microscopy: Proceedings of the Sixth International Conference*. (American Institute of Physics, Melville, NY).

Meyer-Ilse, W., Hamamoto, D., Nair, A., Lelievre, S.A., Denbeaux, G., Johnson, L., Pearson, A.L., Yager, D., Legros, M.A. and Larabell, C.A. (2001). High resolution protein localization using soft X-ray microscopy. *J. Microsc.* **201**, 395–403.

Miao, J.W., Ishikawa, T., Johnson, B., Anderson, E.H., Lai, B. and Hodgson, K.O. (2002). High resolution 3D x-ray diffraction microscopy. *Phys. Rev. Lett.* **89**(8), 088303.

Michette, A.G. (1986). *Optical Systems for Soft X Rays*. (Plenum, New York).

Michette, A.G., Morrison, G.R. and Buckley, C.J., Eds. (1992). *X-ray Microscopy III*. Springer Series in Optical Sciences. (Springer-Verlag, Berlin).

Michette, A., Buckley, C., Pfauntsch, S., Arnot, N., Wilkinson, J., Wang, Z., Khaleque, N. and Dermody, G. (2000). The Kings College Laboratory Scanning X-ray Microscope, In *Proceedings of the Sixth International Conference on Synchrotron Radiation Instrumentation* (American Institute of Physics Melville, NY).

Mimura, H., Yamauchi, K., Yamamura, K., Kubota, A., Matsuyama, S., Sano, Y., Ueno, K., Endo, K., Nishino, Y., Tamasaku, K., Yabashi, M., Ishikawa, T. and Mori, Y. (2004). Image quality improvement in a hard X-ray projection microscope using total reflection mirror optics. *J. Synchrotron Radiat.* **11**, 343–346.

Molenstedt, G. and Gaukler, K.H., Eds. (1969). *X-ray Optics and X-ray Microanalysis*. (Springer-Verlag, Berlin).

Mondal, P.K. and Slansky, S. (1970). Influence de la position de l'anneau de phase sur le contraste de l'image dans un objectif à contraste de phase. *App. Opt.* **9**(8), 1879–1882.

Morrison, G.R. (1989). Some aspects of quantitative x-ray microscopy. In *X-ray Instrumentation in Medicine and Biology, Plasma Physics, Astrophysics, and Synchrotron Radiation*. (R. Benattar, Ed.), *Proceedings of the SPIE*, Vol. 1140, 41–49 (SPIE, Bellingham, WA).

Morrison, G.R., Bridgwater, S., Browne, M.T., Burge, R.E., Cave, R.C., Charalambous, P.S., Foster, G.F., Hare, A.R., Michette, A.G., Morris, D.,

Taguchi, T. and Duke, P. (1989). Development of x-ray imaging at the Daresbury SRS. *Rev. Sci. Inst.* **60**, 2464–2467.

Morrison, G.R., Eaton, W.J. and Charalambous, P. (2002). STXM imaging with a configured detector. *J. Phys. IV* **104**, 547–550.

Murphy, J., White, D., MacDowell, A. and Wood, O. (1993). Synchrotron radiation sources and condensers for projection lithography. *App. Opt.* **32**(34), 6920–6928.

Murray, C.E., Yan, H.-F., Noyan, I.C., Cai, Z. and Lai, B. (2005). High-resolution strain mapping in heteroepitaxial thin film features. *J. App. Phys.* **98**, 013504.

Myneni, S.C.B., Tokunaga, T.K. and Brown, G.E. Jr. (1997). Abiotic selenium redox transformations in the presence of Fe(II,III) oxides. *Science* **278**, 1106–1109.

Neuhäusler, U. Soft X-ray spectromicroscopy on hydrated colloidal and environmental science specimens, PhD thesis, Georg-August-Universität zu Göttingen, Germany, 1999.

Neuhäusler, U., Jacobsen, C., Schulze, D., Stott, D. and Abend, S. (2000). A specimen chamber for soft x-ray spectromicroscopy on aqueous and liquid samples. *J. Synchrotron Radiat.* **7**, 110–112.

Neuhäusler, U., Schneider, G., Ludwig, W., Meyer, M., Zschech, E. and Hambach, D. (2003). X-ray microscopy in Zernike phase contrast mode at 4 keV photon energy with 60 nm resolution. *J. Phys. D: App. Phys.* **36**, A79–A82.

Ng, W., Ray-Chaudhuri, A.K., Crossley, S., Crossley, D., Gong, C., Guo, J., Hansen, R.W.C., Margaritondo, G., Cerrina, F., Underwood, J.H., Perera, R.C.C. and Kortright, J. (1990). The photoemission spectromicroscope multiple-application x-ray imaging undulator microscope (MAXIMUM). *J. Vacuum Sci. Tech.* A**8**, 2563–2565.

Niemann, B. (1987). The Göttingen scanning x-ray microscope. In *Soft x-ray optics and technology.* (E.-E. Koch and G. Schmahl, Eds.), *Proceedings of the SPIE*, Vol. 733, 422–427 (SPIE, Bellingham, WA).

Niemann, B. (1998). High numerical-aperture x-ray condensers for transmission x-ray microscopes. In *X-ray Microscopy and Spectromicroscopy.* (J. Thieme, G. Schmahl, E. Umbach and D. Rudolph, Eds.). (Springer-Verlag, Berlin).

Niemann, B., Rudolph, D. and Schmahl, G. (1974). Soft x-ray imaging zone plates with large zone numbers for microscopic and spectroscopic applications. *Opt. Comm.* **12**, 160–163.

Niemann, B., Rudolph, D. and Schmahl, G. (1976). X-ray microscopy with synchrotron radiation. *App. Opt.* **15**, 1883–1884.

Niemann, B., Guttmann, P., Hilkenbach, R., Thieme, J. and Meyer-Ilse, W. (1988). The Göttingen scanning x-ray microscope. In *X-ray Microscopy II.* (D. Sayre, M.R. Howells, J. Kirz and H. Rarback, Eds.). (Springer-Verlag, Berlin).

Niemann, B., Guttman, P., Hambach, D., Schneider, G., Weiss, D. and Schmahl, G. (1999). The condenser-monochromator with dynamical aperture synthesis for the TXM at an undulator beam line at BESSY II. In *X-ray Microscopy: Sixth International Conference.* (W. Meyer-Ilse, T. Warwick and D. Attwood, Eds.). (American Institute of Physics, Melville, NY).

Ortega, R., Deves, G., Fayard, B., Salome, M. and Susini, J. (2003). Combination of synchrotron radiation X-ray microprobe and nuclear microprobe for chromium and chromium oxidation states quantitative mapping in single cells. *Nucl. Inst. Methods Phys. Res. B* **210**, 325–329.

Osanna, A. and Jacobsen, C. (2000). Principal component analysis for soft x-ray spectromicroscopy. In *X-ray Microscopy: Proceedings of the Sixth International*

Conference. (W. Meyer-Ilse, T. Warwick and D. Attwood, Eds.). (American Institute of Physics, Melville, NY).

O'Toole, E., Wray, G., Kremer, J. and McIntosh, J.R. (1993). High voltage cryo-microscopy of human blood platelets. *J. Struct. Biol.* **110**, 55–66.

Parsons, D., Ed. (1978). *Short Wavelength Microscopy*. (Academy of Sciences, New York, NY).

Parsons, D., Ed. (1980). *Ultrasoft X-ray Microscopy: Its Application to Biological and Physical Sciences*. (Academy of Sciences, New York).

Pattee, H.H., Cosslett, V.E. and Engström, A., Eds. (1963). *X-ray Optics and X-ray Microanalysis*. (Academic Press, New York).

Paunesku, T., Rajh, T., Wiederrecht, G., Maser, J., Vogt, S., Stojićević, N., Protić, M., Lai, B., Oryhon, J., Thurnauer, M. and Woloschak, G. (2003). Biology of TiO_2 oligonucleotide nanocomposites. *Nature Mater* **2**, 343–346.

Pecher, K., McCubbery, D., Kneedler, E., Rothe, J., Bargar, J., Meigs, G., Cox, L., Nealson, K. and Tonner, B. (2003). Quantitative charge state analysis of manganese biominerals in aqueous suspension using scanning transmission X-ray microscopy (STXM). *Geochimica et Cosmochimica Acta* **67**(6), 1089–1098.

Peuker, M. (2001). High-efficiency nickel phase zone plates with 20 nm minimum outermost zone width. *App. Phys. Lett.* **78**(15), 2208–2210.

Pine, J. and Gilbert, J. (1992). Live cell specimens for x-ray microscopy. In *X-ray Microscopy III*. (A.G. Michette, G.R. Morrison and C.J. Buckley, Eds.). (Springer-Verlag, Berlin).

Polack, F., Joyeux, D., Feser, M., Phalippou, D., Carlucci-Dayton, M., Kaznacheyev, K. and Jacobsen, C. (2000). Demonstration of phase contrast in scanning transmission x-ray microscopy: Comparison of images obtained at NSLS X-1A with numerical simulations. In *X-ray Microscopy: Proceedings of the Sixth International Conference*. (W. Meyer-Ilse, T. Warwick and D. Attwood, Eds.). (American Institute of Physics, Melville, NY).

Rarback, H., Kenney, J., Kirz, J. and Xie, X.S. (1980). Scanning X-ray Microscopy—First Tests with Synchrotron Radiation. In *Scanned Image Microscopy* (E.A. Ash, Ed.). (Academic Press: London).

Rarback, H., Kenney, J.M., Kirz, J., Howells, M.R., Chang, P., Coane, P.J. Feder, R., Houzego, P.J., Kern, D.P. and Sayre, D. (1984). Recent results from the Stony Brook scanning microscope. In *X-ray Microscopy* (G. Schmahl and D. Rudolph, Eds.). (Springer-Verlag, Berlin).

Rarback, H., Shu, D., Feng, S.C., Ade, H., Kirz, J., McNulty, I., Kern, D.P., Chang, T.H.P., Vladimirsky, Y., Iskander, N., Attwood, D., McQuaid, K. and Rothman, S. (1988). Scanning x-ray microscope with 75-nm resolution. *Rev. Sci. Inst.* **59**, 52–59.

Rayleigh, L. (1888). Wave theory of light. In *Scientific Papers by Lord Rayleigh*, Vol. III, 47–187 (Dover, New York).

Rebonato, R., Ice, G.E., Habenschuss, A. and Bilello, J.C. (1989). High-resolution microdiffraction study of notch-tip deformation in Mo single crystals using X-ray synchrotron radiation. *Philos. Mag. A* **60**(5), 571–583.

Reimer, L. (1984). *Transmission Electron Microscopy: Physics of Image Formation and Microanalysis*. (Springer-Verlag, Berlin).

Rightor, E.G., Urquhart, S.G., Hitchcock, A.P., Ade, H., Smith, A.P., Mitchell, G.E., Priester, R.D., Aneja, A., Appel, G., Wilkes, G. and Lidy, W.E. (2002). Identification and quantitation of urea precipitates in flexible polyurethane foam formulations by X-ray spectromicroscopy. *Macromolecules* **35**(15), 5873–5882.

Ritman, E.L., Jorgensen, S.M., Lundo, P.E., Thomas, P.J., Dunsmuir, J.H., Romero, J.C., Tumerc, R.T. and Bolander, M.E. (1997). Synchrotron-based micro-CT of in situ biological basic functional units and their integration. In *Developments in X-ray Tomography* (U. Bonse, Ed.). *Proceedings of the SPIE*, Vol. 3149, 13–24 (SPIE, Bellingham, WA).

Rivers, M.L., Sutton, S.R. and Smith, J.V. (1988). A synchrotron X-ray-fluorescence microprobe. *Chem. Geol.* **70**(1–2), 179.

Rose, A. (1946). Unified approach to performance of photographic film, television pickup tubes, and human eye. *J. Soc. Motion Pict. Eng.* **47**, 273–294.

Rudolph, D., Niemann, B., Schmahl, G. and Christ, O. (1984). The Göttingen x-ray microscope and x-ray microscopy experiments at the BESSY storage ring. In *X-ray Microscopy*. (G. Schmahl and D. Rudolph, Eds.). (Springer-Verlag, Berlin).

Rudolph, D., Schmahl, G. and Niemann, B. (1990). Amplitude and phase contrast in x-ray microscopy. In *Modern Microscopies*. (P.J. Duke and A.G. Michette, Eds.). (Plenum, New York).

Saleh, B.E.A. (1979). Optical bilinear transformations: General properties. *Opt Acta* **26**(6), 777–799.

Saloman, E.B. and Hubbell, J.H. (1987). Critical analysis of soft X-ray cross section data. *Nucl. Inst. Methods Phys. Res. A* **255**, 38–42.

Sayre, D. (1972). Proposal for the utilization of electron beam technology in the fabrication of an image forming device for the soft x-ray region. Report RC 3974 (#17965) (IBM Research, Yorktown Heights, NY).

Sayre, D., Kirz, J., Feder, R., Kim, D.M. and Spiller, E. (1977a). Transmission microscopy of unmodified biological materials: Comparative radiation dosages with electrons and ultrasoft x-ray photons. *Ultramicroscopy* **2**, 337–341.

Sayre, D., Kirz, J., Feder, R., Klim, D.M., Spiller, E. (1977b) Potential operating region for ultrasoft x-ray microscopy of biological specimens. *Science,* **196**, 1339–1340.

Sayre, D., Howells, M.R., Kirz, J. and Rarback, H. (1988). *X-ray Microscopy II.* (Springer-Verlag, Berlin).

Schäfer, T., Claret, F., Bauer, A., Griffault, L., Ferrage, E. and Lanson, B. (2003). Natural organic matter (NOM)-clay association and impact on Callovo-Oxfordian clay stability in high alkaline solution: Spectromicroscopic evidence. *J. Phys. IV* **104**, 413–416.

Schäfer, T., Buckau, G., Artinger, R., Kim, J.I., Geyer, S., Wolf, M., Bleam, W.F., Wirick, S. and Jacobsen, C. (2005). Origin and mobility of fulvic acids in the Gorleben acquifer system: Implications from isotopic data and carbon/sulfur XANES. *Org. Geochem.* **36**, 567–582.

Scharf, J.-G. and Schneider, G. (1999). Ultrastructural characterization of isolated rat Kupffer cells by transmission x-ray microscopy. *J. Microsc.* **193**(3), 250–256.

Scheinost, A.C., Kretzschmar, R., Christl, I. and Jacobsen C. (2001). Carbon Group Chemistry of Humic and Fulvic Acid: A Comparison of C-1s NEXAFS and ^{13}C-NMR Spectroscopies. In *Humic Substances: Structures, Models, and Functions.* (E.A. Ghabbor and G. Davies, Eds.) (Royal Society of Chemistry, London).

Schmahl, G. and Rudolph, D. (1969). Lichtstarke Zonenplatten als abbildende Systeme für weiche Röntgenstrahlung. *Optik* **29**, 577–585.

Schmahl, G. and Rudolph, D., Eds. (1984). *X-ray Microscopy.* Springer Series in Optical Sciences. (Springer-Verlag, Berlin).

Schmahl, G. and Rudolph, D. (1987). Proposal for a phase contrast x-ray microscope. *X-ray Microscopy: Instrumentation and Biological Applications* (P.C. Cheng and Jan, G.J. Eds.). (Springer-Verlag, Berlin).

Schmahl, G., Rudolph, D. and Niemann, B. (1980). Imaging and scanning soft x-ray microscopy with zone plates. In *Scanned Image Microscopy* (E.A. Ash, Ed.). (Academic Press: London).

Schmahl, G., Rudolph, D., Guttmann, P. and Christ, O. (1984). Zone plate lenses for x-ray microscopy. In *X-ray Microscopy*. (G. Schmahl and D. Rudolph, Eds.). (Springer-Verlag, Berlin).

Schmahl, G., Rudolph, D., Schneider, G., Guttmann, P. and Niemann, B. (1994). Phase contrast x-ray microscopy studies. *Optik* **97**, 181–182.

Schmahl, G., Rudolph, D., Guttmann, P., Schneider, G., Thieme, J. and Niemann, B. (1995). Phase contrast studies of biological studies with the x-ray microscope at BESSY. *Rev. Sci. Inst.* **66**(2), 1282–1286.

Schneider, G. (1997). Zone plates with high efficiency in high orders of diffraction described by dynamical theory. *App. Phys. Lett.* **71**(16), 2242–2244.

Schneider, G. (1998). Cryo x-ray microscopy with high spatial resolution in amplitude and phase contrast. *Ultramicroscopy* **75**, 85–104.

Schneider, G., Schliebe, T. and Aschoff, H. (1995). Cross-linked polymers for nanofabrication of high-resolution zone plates in nickel and germanium. *J. Vac. Sci. Tech. B* **13**(6), 2809–2812.

Schneider, G., Anderson, E., Vogt, S., Knöchel, C., Wliss, D., Lebros, M. and Larabell, C. (2002a). Computed tomograghy of cryogenic cells. *Surface Rev Lett.* **9**, 177–183.

Schneider, G., Denbeaux, G., Anderson, E.H., Bates, B., Pearson, A., Meyer, M.A., Zschech, E., Hambach, D. and Stach, E.A. (2002b). Dynamical x-ray microscopy investigation of electromigration in passivated inlaid Cu interconnect structures. *App. Phys. Lett.* **81**(14), 2535–2537.

Schneider, G., Meyer, M.A., Denbeaux, G., Anderson, E., Bates, B., Pearson, A., Knochel, C., Hambach, D., Stach, E.A. and Zschech, E. (2002c). Electromigration in passivated Cu interconnects studied by transmission x-ray microscopy. *J. Vac. Sci. Tech. B* **20**(6), 3089–3094.

Schneider, G., Denbeaux, G., Anderson, E., Bates, W., Salmassi, F., Nachimuthu, P., Pearson, A., Richardson, D., Hambach, D., Hoffmann, N., Hasse, W. and Hoffmann, K. (2003). Electromigration in integrated circuit interconnects studied by X-ray microscopy. *Nucl. Inst. Methods Phys. Res. B* **199**, 469–474.

Schroer, C.G. and Lengeler, B. (2005). Focusing hard x rays to nanometer dimensions by adiabatically focusing lenses. *Phys. Rev. Lett.* **94**, 054802.

Schumacher, M., Christl, I., Scheinost, A.C., Jacobsen, C. and Kretzschmar, R. (2005). C-1s XANES spectroscopy of humic substances: Comparison with ^{13}C-NMR results. *Eur. J. Soil Sci.* in press.

Scott, D., Duewer, F., Kamath, S., Lyon, A., Trapp, D., Wang, S. and Yun, W. (2004). *A Novel X-Ray Microtomography System with High Resolution and Throughput for Non-Destructive 3D Imaging of Advanced Packages*. Thirtieth International Symposium on Testing and Failure Analysis. (ASM Interational).

Shapiro, D., Thibault, P., Beetz, T., Elser, V., Howells, M., Jacobsen, C., Kirz, J., Lima, E., Miao, H., Neiman, A. and Sayre, D. (2004). Biological Imaging by Soft X-ray Diffraction Microscopy. *Proc. Nat. Acad. Sci. USA* **102**, 15343–15346.

Shastri, S.P., Maser, J.M., Lai, B. and Tys, J. (2001). Microfocusing of 50~kev undulator radiation with two stacked zone plates. *Optics comm*, **197**, 9–14.

Shaver, D.C., Flanders, D.C., Ceglio, N.M. and Smith, H.I. (1980). X-ray zone plates fabricated using electron-beam and x-ray lithography. *J. Vac. Sci. Tech.* **16**, 1626–1630.

Sheppard, C. and Wilson, T. (1980). Fourier imaging of phase information in scanning and conventional optical microscopes. *Philos. Trans. R. Soc.* **295**(1415), 513–536.

Shin, H.J., Lee, M.K., Kim, G., Hong, C., Lee, J., Kim, J., Park, S., Roh, Y. and Leong, K. (2003). A scanning photoelectron microscope for materials science spectromicroscopy at the Pohang light source. *J. Phys. IV* **104**, 67–70.

Shu, D., Siddons, D.P., Rarback, H. and Kirz, J. (1988). Two-dimensional laser interferometric encoder for the soft x-ray scanning microscope at the NSLS. *Nucl. Inst. Methods Phys. Res. A* **266**, 313–317.

Simpson, M.J. and Michette, A.G. (1983). The effects of manufacturing inaccuracies on the imaging properties of Fresnel zone plates. *Opt. Acta* **30**, 1455–1462.

Smith, A.P., Ade, H., Koch, C.C. and Spontak, R.J. (2001). Cryogenic mechanical alloying as an alternative strategy for the recycling of tires. *Polymer* **42**(9), 4453–4457.

Snigirev, A., Snigireva, I., Kohn, V., Kuznetsov, S. and Schelokov, I. (1995). On the possibility of x-ray phase contrast microimaging by coherent high-energy synchrotron radiation. *Rev. Sci. Inst.* **66**(12), 5486–5492.

Snigirev, A., Kohn, V., Snigireva, I. and Lengeler, B. (1996). A compound refractive lens for focusing high energy x-rays. *Nature* **384**, 49–51.

Soh, Y.A., Evans, P.G., Cai, Z., Lai, B., Kim, C.Y., Aeppli, G., Mathur, N.D., Blamire, M.G. and Isaacs, E.D. (2002). Local mapping of strain at grain boundaries in colossal magnetoresistive films using x-ray microdiffraction. *J. App. Phys.* **91**(10), 7742–7744.

Solomon, D., Lehmann, J., Kinyangi, J., Liang, B. and Schäfer, T. (2005). Carbon K-edge NEXAFS and FTIR-ATR spectroscopic investigation of organic carbon speciation in soils. *Soil Sci. Soc. Amer. J.* **67**, 1721–1731.

Solymar, L. and Cooke, D.J. (1981). *Volume Holography and Volume Gratings.* (Academic Press, London).

Sparks, J., C.J. (1980). X-ray fluorescence microprobe for chemical analysis. *Synchrotron Radiation Research.* (H. Winick and S. Doniach, Eds.). (Plenum Press, New York).

Spector, S., Jacobsen, C. and Tennant, D. (1997). Process optimization for production of sub-20 nm soft x-ray zone plates. *J. Vac. Sci. Tech. B* **15**(6), 2872–2876.

Spiller, E. (1972). Low-loss reflection coatings using absorbing materials. *App. Phys. Lett.* **20**, 365–367.

Stead, A.D., Cotton, R.A., Page, A.M., Dooley, M.D. and Ford, T.W. (1992). Visualization of the effects of electron microscopy fixatives on the structure of hydrated epidermal hairs of tomato (Lycopersicum peruvianum) as revealed by soft x-ray microscopy. In *Soft X-ray Microscopy* (C. Jacobsen and J. Trebes, Eds.). *Proceedings of the SPIE*, Vol. 174, 351–362 (SPIE, Bellingham, WA).

Steinbrecht, R.A. and Zierold, K. (1987). *Cryotechniques in Biological Electron Microscopy.* (Springer-Verlag, Berlin).

Stöhr, J. (1992). *NEXAFS Spectroscopy.* (Springer-Verlag, Berlin).

Stöhr, J., Wu, Y., Hermsmeier, B.D., Samant, M.G., Harp, G.R., Koranda, S., Dunham, D., and Tonner, B.P. (1993). Element-specific magnetic microscopy with circularly polarized x-rays. *Science* **259**, 658–661.

Stoll, H., Puzic, A., vonWayenberge, B., Fischer, P., Raabe, J., Buess, M., Huag, T., Hollinger, R., Back, C.H., Weiss, D. and Denbeaux, G. (2004). High resolution imaging of fast magnetization dynamics in magnetic nanostructures. *App. Phy. Lett.* **84**, 3328–3330.

Susini, J., Barrett, R., Kaulich, B., Oestreich, S. and Salome, M. (2000). The X-ray microscopy facility at the ESRF: A status report. In *X-ray Microscopy: Proceed-*

ings of the Sixth International Conference (W. Meyer-Ilse, T. Warwick, D. Attwood, Eds.). (American Institute of Physics, Melville, NY).

Susini, J., Joyeux, D. and Polak, F., Eds. (2003). X-ray Microscopy 2002. *J. Phys. IV* (EDP Sciences, Les Ulis).

Suzuki, Y., Takeuchi, A., Takano, H., Ohigashi, T. and Takenaka. (2001). Diffraction-limited microbeam with Fresnel zone plate optics in hard x-ray regions. *J. App. Phys.* **40**, 1508–1510.

Suzuki, Y., Awaji, M., Takeuchi, A., Takano, H., Uesugi, K., Kohmura, Y., Kamijo, N., Yasumoto, M. and Tamura, S. (2003). Hard X-ray microscopy activities at SPring-8. *J. Phys. IV* **104**, 35–40.

Takano, H., Uesigi, K., Takeuchi, A., Takai, K. and Suzuki, Y. (2003). High-Sensitive Imaging with Scanning Transmission Hard X-ray Microscope. *J. Phys. IV* **104**, 41–44.

Takemoto, K., Mizuno, T., Yoshikawa, T., Mishibata, H., Ueki, T., Uyama, T., Miyoshi, T., Sawa, D., Matsumoto, T., Wada, N., Onoda, H., Kojima, K., Niemann, B., Hettwer, M., Rudolph, D., Anderson, E., Attwood, D., Kern, D.P., Iwasaki, H. and Kihara, H. (2003). Xray Microscopy in Ritsumeikan Synchrotron Radiation Center. *J. Phys. IV* **104**, 57–61.

Tamura, S., Yasumoto, M., Kamijo, N., Suzuki, Y., Awaji, M., Takeuchi, A., Takano, H. and Handa, K. (2002). Development of a multilayer Fresnel zone plate for high-energy synchrotron radiation X-rays by DC sputtering deposition. *J. Synchrotron Radiat.* **9**, 154–159.

Tang, M.-T., Yin, G.-C., Song, Y.-F., Chen, J.-H., Tsang, K.-L., Liang, K.S., Chen, F.-R., Duewer, F. and Yun, W. (2006). Hard X-ray microscopy with sub-30 nm spatial resolution at NSRRC. In *Proceedings of the 8th International Conference on X-ray Microscopy* (Y. Kagoshima, Ed.). (IPAP, Tokyo).

Tatchyn, R., Csonka, P.L. and Lindau, I. (1982). Outline of a variational formulation of zone-plate theory. *J. Opt. Soci. Amer.* **72**, 1630–1638.

Tatchyn, R., Csonka, P.L. and Lindau, I. (1984). The constant-thickness zone plate as a variational problem. *Opt. Acta* **31**, 729–733.

Taylor, K.A. and Glaeser, R.M. (1976). Electron microscopy of frozen hydrated biological specimens. *J. Ultrastruct. Res.* **55**, 448–456.

Tennant, D., Jackel, L.D., Howard, R.E., Hu, E.L., Grabbe, P., Capik, R.J. and Schneider, B.S. (1981). 25 nm features patterned with trilevel e-beam resist. *J. Vac. Sci. Tech.* **19**, 1304–1307.

Tennant, D.M., Gregus, J.E., Jacobsen, C. and Raab, E.L. (1991). Construction and test of phase zone plates for x-ray microscopy. *Opt. Lett.* **16**, 621–623.

Tennant, D., Spector, S., Stein, A. and Jacobsen, C. (2000). Electron beam lithography of Fresnel zone plates using a rectilinear machine and trilayer resists. In *X-ray Microscopy: Proceedings of the Sixth International Conference* (W. Meyer-Ilse, T. Warwick and D. Attwood, Eds.). (American Institute of Physics, Melville, NY).

Thieme, J. (1988). Theoretical investigations of imaging properties of zone plates and zone plate systems using diffraction theory. In *X-ray Microscopy II* (D. Sayre, M.R. Howells, J. Kirz and H. Rarback, Eds.), Springer Series in Optical Sciences. (Springer-Verlag, Berlin).

Thieme, J. and Niemeyer, J. (1998). Interaction of colloidal soil particles, humic substances and cationic detergents studied by x-ray microscopy. *Prog. Colloid Polymer Sci.* **111**, 193–201.

Thieme, J., Wilhein, T., Guttmann, P., Niemeyer, J., Jacob, K.-H. and Dietrich, S. (1994). Direct visualization of iron and manganese accumulating microorganisms by x-ray microscopy. In *X-ray Microscopy IV.* (V. Aristov and A. Erko, Eds.). (Bogorodskii Pechatnik, Chernogolovka, Russia).

Thieme, J., Schmahl, G., Umbach, E. and Rudolph, D. Eds. (1998). *X-ray Microscopy and Spectromicroscopy.* (Springer-Verlag, Berlin).

Thieme, J., Schneider, G. and Knochel, C. (2003). X-ray tomography of a microhabitat of bacteria and other soil colloids with sub-100 nm resolution. *Micron* **34**(6–7), 339–344.

Thompson, A.C., Underwood, J.H., Wu, Y., Giauque, R.D., Jones, K.W. and Rivers, M.L. (1988). Elemental measurements with an X-ray microprobe of biological and geological samples with femtogram sensitivity. *Nucl. Inst. Methods Phys. Res. A* **266**, 318–323.

Thompson, B.J. (1969). Image formation with partially coherent light. In *Progress in Optics* (E. Wolf, Ed.). (North Holland, Amsterdam).

Tonner, B.P., Droubay, T., Denlinger, J., Meyer-Ilse, W., Warwick, T., Rothe, J., Kneedler, E., Pecher, K., Nealson, K. and Grundl, T. (1999). Soft x-ray spectroscopy and imaging of interfacial chemistry in environmental specimens. *Surface Interface Anal.* **27**, 247–258.

Tsusaka, Y., Yokoyama, K., Takeda, S., Takai, K., Kagoshima, Y. and Matsui, J. (2001). Hyogo beamline at SPring-8: multiple station beamline with the TROIKA concept. *Nucl. Instruments Meth.* **A**, 467–468, 670–673.

Twining, B.S., Baines, S.B., Fisher, N.S., Maser, J., Vogt, S., Jacobsen, C., Tovar-Sanchez, A. and Sañudo-Wilhelmy, S.A. (2003). Quantifying trace elements in individual aquatic protist cells with a synchrotron x-ray fluorescence microprobe. *Anal. Chem.* **75**, 3806–3816.

Tyliszczak, T., Warwick, T., Kilcoyne, A., Fakra, S., Yoon, D.S., Brown, G., Andrews, S., Chembrolu, V., Strachan, J. and Acremann, Y. (2004). Soft X-ray scanning transmission microscope working in an extended energy range at the Advanced light source. In *Proceedings of the Eighth International Conference on Synchrotron Radiation Instrumentation.* (American Institute of Physics, Melville, NY).

Uesugi, K., Tsuchiyama, A., Yasuda, H., Nakamura, M., Nakano, T., Suzuki, Y. and Yagi, N. (2003). Micro-tomographic imaging for material sciences at BL47XU in Spring-8. *J. Phys. IV* **104**, 45–48.

Uggerhoj, E. and Abraham-Peskir, J. (2000). X-ray microscopy in Aarhus. In *X-ray Microscopy: Proceedings of the Sixth International Conference* (W. Meyer-Ilse, T. Warwick and D. Attwood, Eds.). (American Institute of Physics, Melville, NY).

Urquhart, S.G. and Ade, H. (2002). Trends in the carbonyl core (C 1s, O 1s) → pi*c = o transition in the near-edge X-ray absorption fine structure spectra of organic molecules. *J. Phys. Chem. B* **106**(34), 8531–8538.

Urquhart, S.G., Hitchcock, A.P., Smith, A.P., Ade, H.W., Lidy, W., Rightor, E.G. and Mitchell, G.E. (1999). NEXAFS spectromicroscopy of polymers: Overview and quantitative analysis of polyurethane polymers. *J. Electron Spectrosc. Rel. Phenomena* **100**, 119–135.

Vogt, S., Chapman, H.N., Jacobsen, C. and Medenwaldt, R. (2001a). Dark field X-ray microscopy: The effects of condenser/detector aperture. *Ultramicroscopy* **87**(1–2), 25–44.

Vogt, S., Jager, M., Schneider, G., Schulze, E., Saumweber, H., Rudolph, D. and Schmahl, G. (2001b). Visualizing specific nuclear proteins in eukaryotic cells using soft X-ray microscopy. *Nucl. Inst. Methods Phys. Res. A* **467**, 1312–1314.

Wagner, D., Maser, J., Lai, B., Cai, Z.H., Barry, C.E., Bentrup, K.H.Z., Russell, D.G. and Bermudez, L.E. (2005). Elemental analysis of Mycobacterium avium-, Mycobacterium tuberculosis-, and Mycobacterium smegmatis-containing phagosomes indicates pathogen-induced microenvironments within the host cell's endosomal system. *J. Immunol.* **174**(3), 1491–1500.

Wang, S., Duewer, F., Kamath, S., Kelly, C., Lyon, A., Nill, K., Pombo, P., Scott, D., Trapp, D., Yun, W., Neogi, S., Kuhn, M., Bennet, C., Coon, P. and Yan, S. (2002). A transmission x-ray microscope (TXM) for non-destructive 3D imaging of ICs at sub-100 nm resolution. In *Proceedings of the 28th International Symposium for Testing and Failure Analysis*. (AMS International).

Wang, Y., Jacobsen, C., Maser, J. and Osanna, A. (2000). Soft x-ray microscopy with a cryo STXM: II. Tomography. *J. Microsc.* **197**, 80–93.

Warwick, T., Ade, H., Hitchcock, A.P., Padmore, H., Rightor, E.G. and Tonner, B.P. (1997). Soft x-ray spectromicroscopy development for materials science at the Advanced Light Source. *J. Electron Spectrosc. Rel. Phenomena* **84**, 85–98.

Warwick, T., Ade, H., Kilcoyne, D., Kritscher, M., Tylisczcak, T., Fakra, S., Hitchcock, A., Hitchcock, P. and Padmore, H. (2002). A new bend-magnet beamline for scanning transmission X-ray microscopy at the Advanced Light Source. *J. Synchrotron Radiat.* **9**, 254–257.

Warwick, T., Andresen, N., Comins, J., Kaznacheyev, K., Kortright, J., McKean, J., Padmore, H., Shuh, D., Stevens, T. and Tyliszczak, T. (2004). New implementation of an SX700 undulator beamline at the advanced light source. In *Proceedings of the Eighth International Conference on Synchrotron Radiation Instrumentation*. (American Institute of Physics, Melville, NY).

Weisemann, U., Thieme, J., Guttmann, P., Frueke, R., Rehbein, S., Neimann, B., Rudolph, D. and Schmahl, G. (2003). First results of the new scanning transmission X-ray microscope at BESSY II. *J. Phys. IV* **104**, 95–98.

Weiss, D., Schneider, G., Niemann, B., Guttmann, P., Rudolph, D. and Schmahl, G. (2000). Computed tomography of cryogenic biological specimens based on x-ray microscopic images. *Ultramicroscopy* **84**, 185–197.

Weitkamp, T., Drakopoulos, M., Leitenberger, W., Raven, C., Schroer, C., Simionivici, A., Snigireva, I. and Snigirev, A. (2000). High resolution X-ray imaging and tomography at the ESRF beamline ID22. In *X-ray Microscopy: Proceeding of the Sixth International Conference on X-ray Microscopy* (W. Meyer-Ilse, T. Warwick and D. Attwood, Eds.). (American Institute of Physics, Melville, NY).

White, D.L., O.R.W. II, Bjorkholm, J.E., Spector, S., MacDowell, A.A. and LaFontaine, B. (1995). Modification of the coherence of undulator radiation. *Rev. Sci. Inst.* **66**(2), 1930–1933.

Wiesemann, U., Thieme, J., Guttmann, P., Niemann, B., Rudolph, D. and Schmahl, G. (2000). The new scanning transmission x-ray microscope at BESSY II. In *X-ray Microscopy: Proceedings of the Sixth International Conference* (W. Meyer-Ilse, T. Warwick and D. Attwood, Eds.). (American Institute of Physics, Melville, NY).

Williams, G.J., Pfeifer, M.A., Vartanyants, I.A. and Robinson, I.K. (2003). Three-dimensional imaging of microstructure in Au nanocrystals. *Phys. Rev. Lett.* **90**, 175501.

Williams, S., Zhang, X., Jacobsen, C., Kirz, J., Lindaas, S., Hof, J.V.T. and Lamm, S.S. (1993). Measurements of wet metaphase chromosomes in the scanning transmission x-ray microscope. *J. Microsc.* **170**, 155–165.

Wilson, T. and Sheppard, C. (1981). The halo effect of image processing by spatial frequency filtering. *Optik* **59**, 119–123.

Wilson, T. and Sheppard, C. (1984). *Theory and Practice of Scanning Optical Microscopy*. (Academic Press, London).

Winn, B., Ade, H., Buckley, C., Feser, M., Howells, M., Hulbert, S., Jacobsen, C., Kaznacheyev, K., Kirz, J., Osanna, A., Maser, J., McNulty, I., Miao, J., Oversluizen, T., Spector, S., Sullivan, B., Wang, Y., Wirick, S. and Zhang, H. (2000). Illumination for coherent soft X-ray applications: The new X1A beamline at the NSLS. *J. Synchrotron Radiat.* **7**, 395–404.

Wolter, H. (1952). Spiegelsysteme streifenden Einfalls als abbildende Optiken für Röntgenstrahlen. *Ann. Phys.* **10**, 94–114, 286.

Wood, R.W. (1898). Phase-reversal zone-plates, and diffraction telescopes. *Philos. Mag.* **45**, 511–522.

Yamamoto, A., Masaki, R., Guttmann, P., Schmahl, G. and Kihara, H. (1998). Studies on intracellular structures of COS cells by x-ray microscopy. *J. Synchrotron Radiat.* **5**, 1105–1107.

Yamamura, K., Yamauchi, K., Mimura, H., Sano, Y., Saito, A., Endo, K., Souvorov, A., Yabashi, M., Tamasaku, K., Ishikawa, T. and Mori, Y. (2003). Fabrication of elliptical mirror at nanometer-level accuracy for hard x-ray focusing by numerically controlled plasma chemical vaporization machining. *Rev. Sci. Inst.* **74**(10), 4549–4553.

Yi, Y., Cho, S., Noh, M., Wang, C., Jeong, K. and Shin, H. (2005). Characterization of surface chemical states of a thick insulator: Chemical state imaging on MgO surface. *J. J. App. Phys.* **44**, 861–864.

Yin, G.-C., Tang, M.T., Song, Y.-F., Chen, F.-R., Liang, K.S., Duewer, F.W., Yun, W., Ko, C.-H., Shieh, H.-P.D. (2006). An energy-tunable transmission x-ray microscope for differential contrast imaging with near 60-nm resolution tomography. *Appl. Phys. Lett.* (accepted).

Yoon, T.H., Johnson, S.B., Benzerara, K., Doyle, C.S., Tyliszczak, T., Shuh, D.K. and Brown, G.E. (2004). *In situ* characterization of aluminum-containing mineral-microorganism aqueous suspensions using scanning transmission X-ray microscopy. *Langmuir* **20**(24), 10361–10366.

Youn, H., Baik, S. and Chang, C. (2005). Hard X-ray microscopy with a 130 nm spatial resolution. *Rev. Sci. Inst.* **76**, 23702.

Yun, W. (2005). Private communication.

Yun, W., Pratt, S.T., Miller, R.M., Cai, Z., Hunter, D.B., Jarstfer, A.G., Kemner, K.M., Lai, B., Lee, H.R., Legnini, D.G., Rodrigues, W. and Smith, C.I. (1998). X-ray imaging and microspectroscopy of plants and fungi. *J. Synchrotron Radiat.* **5**, 1390–1395.

Yun, W., Lai, B., Krasnoperova, A.A., Di Fabrizio, E., Cai, Z., Cerrina, F., Chen, Z., Gentili, M. and Gluskin, E. (1999). Development of zone plates with a blazed zone profile for hard x-ray applications. *Rev. Sci. Inst.* **70**, 3537–3541.

Zeitler, E. and Thomson, M.G.R. (1970). Scanning transmission electron microscopy. *Optik* **31**, 258–280, 359–366.

Zhang, X., Jacobsen, C. and Williams, S. (1992). Image enhancement through deconvolution. In *Soft X-ray Microscopy* (C. Jacobsen and J. Trebes, Eds.). *Proceedings of the SPIE*, Vol. 1741, 251–259 (SPIE, Bellingham, WA).

Zhang, X., Balhorn, R., Mazrimas, J. and Kirz, J. (1996). Measuring DNA to protein ratios in mammalian sperm head by XANES imaging. *J. Struc. Biol.* **116**, 335–344.

Zhu, S., Liu, Y., Rafailovich, M.H., Sokolov, J., Gersappe, D., Winesett, A. and Ade, H. (1999). Confinement-induced miscibility in polymer blends. *Nature* **400**, 49–51.

Part III

NEAR-FIELD SCANNING PROBES

14

Scanning Probe Microscopy in Materials Science

Maxim P. Nikiforov and Dawn A. Bonnell

1 Introduction

The quest toward understanding the behavior of condensed matter has relied on measuring structure, bonding, and properties at increasingly local levels. This has driven advances in techniques that probe both soft and hard materials directly as well as indirectly. Many of these advances are described in other chapters of this volume. While structure and bonding-based probes have accessed molecular and atomic scales for decades, local determination of properties has been elusive. The emergence of scanning probes filled this gap to some extent.

There are three major classes of scanning probe techniques that access electronic, magnetic, optical, and mechanical properties. *Scanning tunneling microscopy* (STM) was the first and is based on electrons tunneling between a metal tip and a sample. The distance sensitivity of tunneling imparts intrinsically high spatial resolution and the voltage dependence yields local density of states. It is, however, applicable only to conducting materials. Another class of techniques is based on local optical responses induced and/or detected with a very fine optical fiber. These techniques, e.g., *near field optical microscopy*, find extensive application in organic and biological systems.

The focus of this chapter will be those probes that exploit the interactions of a small tip with a surface that are detected via the properties of a cantilever to which the tip is attached. The original cantilever probe, *atomic force microscopy* (AFM), is based on van der Waals interactions at the tip/surface junction. As a cantilever is mechanically oscillated near its resonant frequency (usually 10–500 kHz) and, at relatively small sample-tip separations (usually 0.5–100 nm), the van der Waals interaction causes a force that alters the oscillation. Cantilever motion is detected with laser reflection into a photodiode. If this measurement is made at every point as the tip is scanned across a surface, and a signal is used to maintain a profile at constant force, the topographic structure of the surface is mapped. It was almost immediately understood that other interactions can be detected with this scheme and, furthermore, that these interactions might be distinguished by their distance dependence. For example, at a 10-nm sample/tip separation the van der Waals interactions can be used to determine the topographic structure, while at 200 nm an electrostatic force (EFM) or magnetic force (MFM) would dominate the measurement.

The utility of local probes is illustrated by the fact that even though the field is relatively young, upward of 2700 papers per year are published that cite "AFM or STM" as a key word.[1] Several monographs have summarized the state of this field and various books are available that provide an introduction to the field and general overviews.[2–5] Most applications utilize SPM as a straightforward qualitative mapping tool. Some researchers interested in complex behavior of solids have examined fundamental tip–surface interactions and extended SPM to probe local electronic transport, dielectric, ferroelectric, and magnetic properties. Rather than address conventional STM, AFM, MFM, or EFM, this chapter will describe recent advances in nanometer probes of complex properties, highlighting potential insight, as well as remaining challenges. First, factors leading to atomic resolution imaging based on forces will be described along with some example applications. This is followed by summaries of approaches to probe spatially localized electrical and dielectric properties. Finally we will present of view of where this segment of the field is headed.

2 Imaging at Atomic Resolution with Force Interactions

There are a number of so-called "imaging modes" in AFM based on various combinations of force detection and feedback mechanism. Confusion can arise since there is no universal naming convention among microscope suppliers. Generally the cantilever is oscillating. From a fundamental perspective the force regimes categorize three imaging modes: **contact AFM** [the tip applies a small force (1–10 nN) normal to the surface of the sample], **intermittent contact AFM** (the oscillating cantilever tip is brought close to the sample so that it barely hits, or "taps" the sample at the bottom of the excursion), and **noncontact AFM** (the tip never contacts the surface). Superimposed upon this scheme are the tip responses that can be monitored to detect the tip–surface interaction: oscillation amplitude, frequency, or phase. The instrument configurations are illustrated in Figure 14–1. Atomic resolution imaging can be achieved in conventional AFM in contact mode for some materials and in noncontact mode under certain conditions. The latter is a relatively recent advance and offers the potential to investigate materials properties at the atomic scale, which is of importance not only from a fundamental, but also from a technological, point of view.

2.1 Operational Principles of Noncontact Atomic Force Microscopy

Noncontact atomic force microscopy (NC-AFM) is the general name for the group of techniques in which the tip of the cantilever oscillates in close proximity to the surface but never makes contact. Several review articles present details of imaging mechanisms, which we summarize

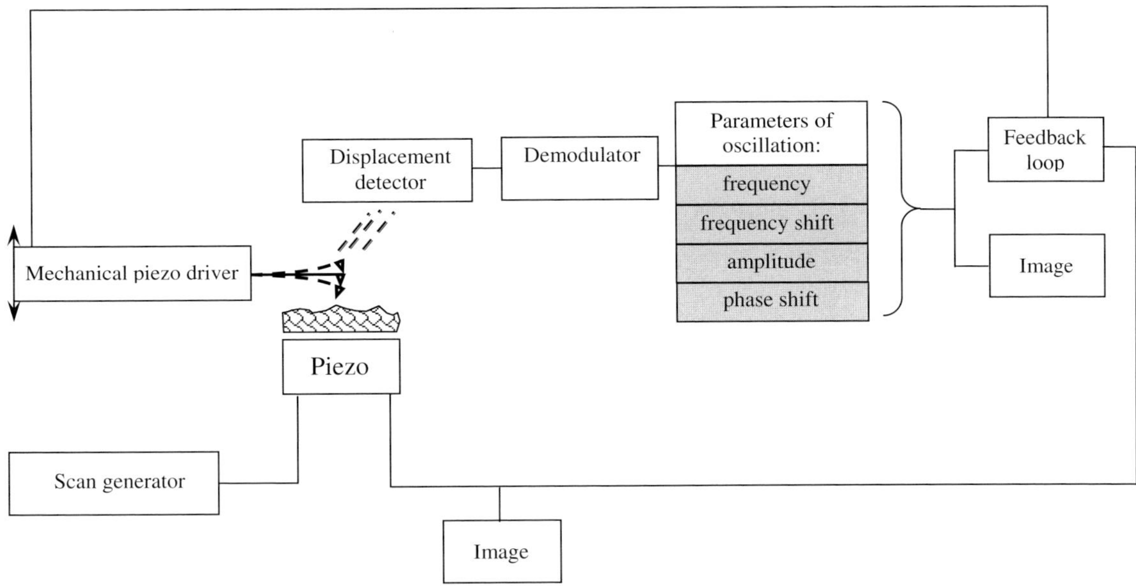

Figure 14–1. Basic block diagram for surface imaging with the NC-AFM. Varying one of the oscillation parameters and maintaining the others constant leads to the different NC-AFM techniques described in the text.

here.[6–8] It should be noted that only specific types of NC-AFM (frequency modulated NC-AFM) achieve atomic resolution for the simple reason that the interactions between the sample and tip, i.e., long-range van der Waals and electrostatic forces, are power law functions of order 2. This is not sufficiently sensitive to track distances of fractions of angstroms. If, however, the geometry is configured such that the tip experiences short-range van der Waals and/or bonding interactions during a substantial part of the cantilever oscillation, sensitivity is enhanced. In this situation the local force between the sample and tip is the sum of the electrostatic, F_{elec}, van der Waals, F_{vdW}, and bonding $F_{bonding}$:

$$F_{tot} = F_{elec} + F_{vdW} + F_{bonding}$$

Figure 14–2 illustrates the range over which these forces operate based on the following estimations/assumptions.

$$F_{tot} = -\frac{q_{surf}q_{tip}}{4\pi\varepsilon_0 r^2} + \frac{1}{2}\frac{\partial C}{\partial r}V^2 + -\frac{AR}{6r^2} + F_{shortrange}$$

where q_{surf} is the electrostatic charge accumulated on the surface, q_{tip} is the electrostatic charge accumulated on the tip apex, ε_0 is the dielectric constant of the vacuum, r is the distance from the tip apex to the surface, C is the capacitance between the tip and surface, V is the electrostatic potential difference between the tip and surface, A is the Hamaker constant, and R is the tip radius.

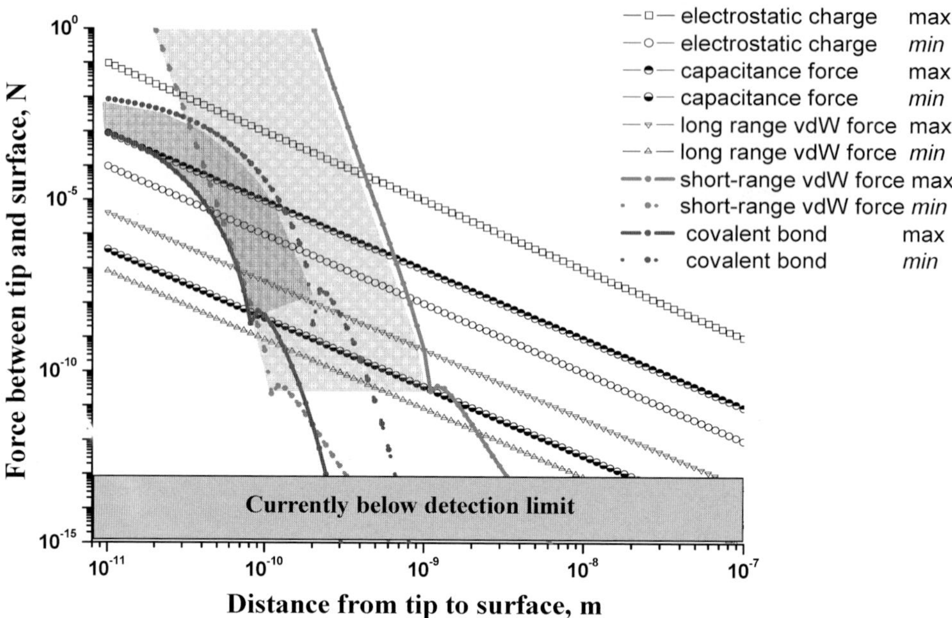

Figure 14–2. Distance dependencies of short- and long-range forces. Forces acting between the tip and surface during AFM operation in the noncontact mode are estimated.

Short-range forces play a significant role for separations less then 10 nm. For an illustration of the magnitude of the various forces we consider two types of short-range interactions: van der Waals and covalent bonding. The nature of short-range van der Waals forces is dipole–dipole interactions, which can be approximated by a Lenard–Jones potential.

$$F_{\text{srvdW}}(r) = 24\frac{E_{\text{bond}}}{r}\left[2\left(\frac{r_{eq}}{r}\right)^{12} - \left(\frac{r_{eq}}{r}\right)^{6}\right]$$

where E_{bond} is the equilibrium bond energy (usual values 0.1–0.005 eV) and r_{eq} is the equilibrium distance between atoms (usual values 1–3 Å). A fair approximation for covalent bonding is the Morse potential. Lim showed that the Morse potential could be written as a function of two parameters $(E_{\text{bond}}, r_{\text{eq}})$:[9]

$$F_{\text{Morse}}(r) = \frac{12D}{r_{eq}}\left[\exp\left(-12\frac{r - r_{eq}}{r_{eq}}\right) - \exp\left(-6\frac{r - r_{eq}}{r_{eq}}\right)\right]$$

The typical parameters used in these calculations are $r_{\text{eq}} = 0.8$ Å and $E_{\text{bond}} = 1$ eV for the Morse *min* and $r_{\text{eq}} = 2$ Å and $E_{\text{bond}} = 10$ eV for the Morse *max*.

Figure 14–2 shows that $F_{electrost}$, $F_{capacitance}$, and F_{vdW} (large-range forces) are attractive for any tip–surface distances. At the same time short-range van der Waals forces and covalent forces (Morse potential) are attractive at large distances and repulsive at small distances. The repulsive part of the interactions is shaded with light gray for van der Waals forces and with dark gray for Morse forces in Figure 14–2. Although these considerations are based on rather simple approximations, Figure 14–2 illustrates that for short-range interactions to be the same order of magnitude as long-range interactions, a requirement for high resolution, oscillations should be within tens of nanometers of the surface.

Despite the fact that relatively simple theories qualitatively explain experimental results, a complete self-consistent theory that relates contrast to atomic structure has not been developed. Hofer et al.[8] concluded that although consistent methods for simulating the basic aspects of scanning probe microscopy (SPM) exist, no model can reproduce all features of the experimental tip–surface interaction. Each model depends crucially on a set of assumptions, which is for scanning probe microscopy in general:

1. The ground-state properties of a system are equal to the properties at finite temperature (e.g., in an ambient environment).
2. There is a hierarchy of interactions that allows the separation of different effects in the theoretical models (no inclusion of, for example, cumulative or time-dependent interactions).

For *scanning tunneling microscopy* specifically:

1. Macroscopic interactions do not affect the tunneling current.
2. The resistance in the STM circuit is due only to the tunnel barrier.
3. The current cross section is centered at the apex atom of the tip.
4. Current flow does not change the properties of a system.

The last point has recently been analyzed by Todorov et al.,[10] who found that the current flow slightly changes the position of the surface atoms.

For *scanning force microscopy* (SFM) specifically:

1. Charge transfer between the tip and surface is not a significant component of the interactions for an insulating surface.
2. The effects of the system electronics, such as apparent dissipation,[6,11,12] do not affect the physics of the tip–surface interaction.

There has been one study, the adsorption of formate ions on TiO_2 (110),[13] in which a chemical force model was used to establish chemical identification fairly conclusively. Theoretical interpretation of STM images helped to interpret noncontact AFM images of the same surface. Recent STM and noncontact (NC)-SFM theoretical modeling[14] has confirmed the original experimental interpretation. Thus, the combined use of STM and AFM can resolve issues of contrast interpretation; however, by the nature of STM it is limited to surfaces that can be made to conduct. Further development of the theory of scanning force microscopy is needed.

2.2 NC-AFM Techniques

Mechanical oscillations of the cantilever lie at the heart of NC-AFM. Forces acting between the tip and surface change the oscillation parameters and these parameters are sensitive to atomic scale resolution. During operation three sets of parameters play the most important roles: cantilever-related (spring constant of the cantilever, the eigenfrequency of the cantilever, the quality factor of the cantilever), oscillation-related (amplitude, frequency shift), and image-related (either constant separation or constant height operation mode) parameters. Cantilever-related parameters are predetermined by the tip and are not variable during scanning. Changes in the other parameters result in different AFM techniques, as shown in Table 14–1. It should be noted that NC-AFM experiments are usually carried out in high vacuum or in liquid to avoid water condensation between the tip and surface.

2.2.1 Amplitude-Modulated AFM (AM-AFM)

As the name implies, in **AM-AFM** the amplitude of the tip oscillation is monitored while frequency shift and separation are kept constant by feedback. Usually AM-AFM is done at the resonant frequency of the cantilever; since this is the maximum oscillation amplitude. This mode is analogous to constant current STM, despite the difference in the physics of image formation. To date, atomic resolution has not been reported using this technique even on atomically smooth surfaces. A possible explanation is that the sensitivity is low because the amplitude depends on the tip–surface force averaged over the oscillation cycle [Eq. (1)].[15]

$$A \cong A_0 \left[1 - 4\left(\frac{\langle F_{ts}\rangle}{F_0}\right)^2 \right]^{1/2} \tag{1}$$

where $\langle F_{ts}\rangle$ is the average force over the oscillation cycle and F_0 is the driving force. According to Eq. (1) AM-AFM senses the change of average force over the oscillation cycle at constant separation. The typical driving force might be estimated as $F_0 \sim kA_0 \sim 10\,\text{N/m}\ 50\,\text{nm} = 5 \times 10^{-7}\,\text{N}$. The change in average force that produces atomic resolution is usually on the order of 10^{-8}–$10^{-10}\,\text{N}$. Thus, the smaller the oscillation amplitude, the better the tip "senses" the surface because of the decrease in F_0. To achieve atomic resolution (vertical ~0.01 nm and lateral ~0.1 nm) the oscillation amplitude must be kept on the order of 1–10 nm. Small oscillation amplitude can be achieved either by decreasing the driving force or increasing the spring constant. A decrease of driving force

Table 14–1. Operational modes in noncontact AFM.

	Amplitude of tip oscillations	Frequency shift of tip oscillations	Tip–surface separation
AM-AFM	*Monitored (varied)*	Constant	Constant
FM-AFM (1)	Constant	*Monitored (varied)*	Constant
FM-AFM (2)	Constant	Constant	*Monitored (varied)*

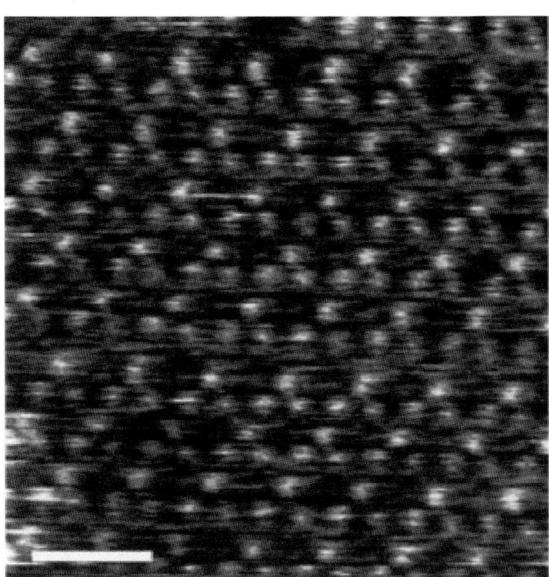

Figure 14–3. AM-AFM topographic image of the extracellular purple membrane surface in buffer solution. The scale mark is 10 nm. [Courtesy of Moller et al. Reprinted with permission from Biophysical Journal, 77(2), 1150–1158, 1999.]

leads to a decrease in the oscillation stability. A substantial increase of the spring constant is not easy due to manufacturing issues. At this time silicon cantilevers with spring constants from 0.1 to 100 N/m are on the market.

With spatial resolution on the order of ~2 nm, AM-AFM is ideal for many biological applications. In addition, the relatively small force minimizes destructive imaging of soft samples.[16] Figure 14–3 illustrates AM-AFM on a purple membrane of *Halobacterium salinarum* stain ET1001, isolated as described in Oesterhelt and Stoekenius[17] and imaged in 300 mM KCl, pH 7.8, 10 mM Tris-HCl. The authors observed the topography of the extracellular purple membrane surface, which exhibits a trimeric structure protruding 0.4 ± 0.1 nm above a lipid bilayer. The trimers are arranged in a trigonal lattice of 6.2 ± 0.2 nm length. In this experiment high lateral resolution (1.1–1.2 nm) was achieved. More examples of biological application AM-AFM are found in a recent review,[6] which also describes the detailed theory of AM-AFM.

Similar resolution has been demonstrated on inorganic surfaces. Resolution close to atomic level was demonstrated on calcite with a modified AM-AFM technique. F.M. Ohnesorge[18,19] operated the microscope in the separation region close to that for snap-into-contact instability. In this region feedback with an inverted sign provides reasonably stable images of high resolution.

2.2.2 Frequency-Modulated AFM (FM-AFM)

There are two FM-AFM techniques (**FM-AFM(1)** and **FM-AFM(2)** in Table 14–1). For these two techniques the oscillation amplitude is constant and either the frequency shift of oscillation (in constant height mode) or the change in piezo movement (in constant force gradient mode) is varied. Atomic resolution was obtained almost simultaneously by Kitamura and Iwatsuki and Giessibl in 1995 using FM-AFM. Both groups operated the microscope in constant force gradient mode. The difference in the approaches was in the cantilever excitation: Kitamura and Iwatsuki used constant excitation mode and Giessibl used an automatic gain control of the excitation signal to maintain constant oscillation amplitude. Figure 14–4 shows one of the first images of the (7×7) reconstruction on silicon (111) obtained by Kitamura and Iwatsuki. The authors noted that in this early attempt the image stability is poor, but subsequent improvement in the frequency stability of the demodulator resulted in routine high-quality imaging of this surface. At this point the (7×7) reconstruction on silicon (111) has become a standard for NC-AFM calibration. From a practical perspective the quality of an NC-AFM image depends not only on the surface quality, but also on the quality of tip. Tips sputtered with an ion gun often provide the best results.

Note that an FM-AFM image is not only an aesthetically pleasing image, it is a quantitative description of tip–surface interactions, hence, mathematical modeling of these interactions is often required to interpret the image contrast. The basic equation for image formation is: $\Delta f(z_c) = (f_0/2k)k_{ts}(z_c)$, where Δf is the frequency shift, z_c is the lift height, f_0 is the base frequency, k is the cantilever spring constant, and k_{ts} is the force gradient.[20] Therefore the image in **FM-AFM(1)** mode is a map of force gradient over the surface at constant separation, and the image in **FM-AFM(2)** mode is a map of the heights of constant tip–surface force gradient. For stable imaging several requirements must be fulfilled:

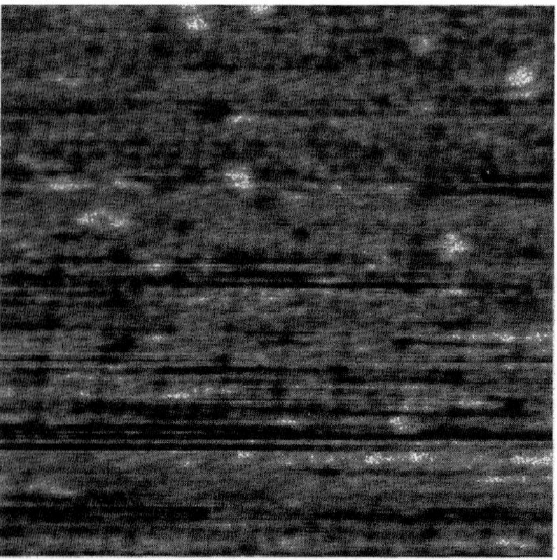

Figure 14–4. First NC-AFM image of the Si (111)7 × 7 reconstruction. [Courtesy of Kitamura and Iwatsuki. Reprinted with permission from Japanese Journal of Applied Physics Part 2—Letters, 34(1B), L145–L148, 1995.]

$$\max \left| \frac{d^2 V_{ts}}{d^2 z^2} \right| = k_{ts}^{max} < k \text{ and}$$

$$\max \left| -\frac{dV_{ts}}{dz} \right| = \left| F_{ts}^{max} \right| < kA_0$$

where V_{ts} is the tip–surface potential, k_{ts} is the tip–surface force gradient, k is the cantilever spring constant, F_{ts}^{max} is the tip–surface force, and A_0 is the oscillation amplitude. For atomic resolution stiff, high-frequency cantilevers are required to provide stable oscillation with a good quality factor. The basis for atomic resolution in FM-AFM is that the first derivative of the tip–surface force is monitored, in contrast to AM-AFM, in which the force itself is monitored. Experimental data obtained in FM-AFM(1) can be converted into FM-AFM(2) data only when the distance dependence of the tip–surface potential is known. Experimental parameters such as cantilever resonant frequency, spring constant, and oscillation amplitude vary from one experiment to another. To compare experimental results obtained under different conditions the concept of *normalized frequency shift* was developed by Giessibl.[20] Normalized frequency shift, $\gamma = (kA^{3/2}/f_0)\Delta f$, depends on the tip–surface potential and not on oscillation amplitude or oscillation frequency. Analytical solutions for basic tip–surface interactions (power law dependence and exponential decay)[21] have also been developed. This allows the magnitude of the tip–surface force to be related to the image formation mechanism. The concept of normalized frequency shift will be used in subsequent discussions of various applications.

2.3 Role of the Tip in Image Interpretation

2.3.1 General Approach for Tip Modeling

Classical physics fails to quantitatively describe interactions between the tip and surface, while analytical expressions for a quantum mechanical treatment have not yet been developed. In other words at this point numerical calculations are required to calculate the quantum mechanical part of the problem. The currently used approach is to develop "nanotip" models based on a cluster at the end of the tip, which interacts with the surface obeying quantum mechanical laws. Using this model, "chemical" interactions between the "nanotip" and surface can be calculated based on the pair potentials between tip and surface atoms. Since numerical calculations are time consuming the "nanotip" is kept as small as possible. Obviously, the "nanotip" does not represent the entire tip–surface interaction. Among the models developed to incorporate long-range forces the most popular is the representation of the tip as a cone with a hemispherical end. A complete description of the tip then includes a cone with a hemispherical cap (analytical description) with a "nanotip" at the end of the hemisphere (numerical modeling). Some information regarding the tip shape, conductivity, and charging is inherent in the experimental dependence of the cantilever frequency change on tip–surface separation measured before and just after imaging.[22–25]

2.3.2 *"Nanotip" Models*

In SPM simulations, the most common conception for Si and other semiconductor surfaces has been that the main component of the tip–surface interaction is the interaction of a dangling Si bond at the end of the tip with the surface atoms. This dangling bond can be well described using relatively small 4- or 10-atom Si clusters saturated by H atoms.[26] Another approach is to assume from the outset that the tip is ionic (MgO).[27,28] Calculations showed that if the bottom of the tip was flat, i.e., no nanotip (only a macroscopic tip), then the interaction with the surface was averaged over several tip ions, and no contrast was produced. When a nanotip was included, it had to extend significantly beyond the main part of the tip to achieve atomic resolution.

Silicon cluster tip models have been successful in developing a qualitative understanding of the origins of image contrast in SPM on metals, semiconductors, and insulators; however, they fail in most cases to qualitatively reproduce image contrast. The solution is to compare images to simulations with a number of different tip models. Silicon tips under experimental conditions are likely to be contaminated by residual oxide, adsorbed hydrogen, and water,[29] or even materials transferred from the surface. To compare the properties of clean and contaminated silicon tips, the electronic structures of Si10 clusters with adsorbed contaminant species were calculated using the density functional theory by Sushko et al.[30] The results clearly showed that adsorbed hydrogen has no effect on the potential gradient from the uncontaminated silicon cluster; however, adsorbed oxygen and hydroxyl groups cause a significant change in the potential gradient. Both potentials decayed over a much longer distance than did that of the uncontaminated cluster, and the stronger gradients suggested a much stronger interaction with the surface. Interestingly, the potential gradient from an MgO cube corner with an O atom at the end was very similar to that of the oxygen-contaminated silicon cluster, a strong negative potential. An MgO cluster is a good model of a hard oxide tip, and has the important advantage that reliable interatomic potentials exists for MgO, alkali halides, and other oxides.

In an interesting experiment Bennewitz and co-workers atomically resolved a copper (111) substrate, as well as a unit cell thick NaCl grown on that surface.[31] An MgO "nanotip" of only a few atoms would not atomically resolve the features observed in the experimental images. Different "nanotip" models for SFM were compared to determine which most closely matched experimental results.[31] It was found that a 64-atom MgO "nanotip" with an oxygen atom at the very end of the tip imbedded in a macroscopic tip gives quantitative agreement with image contrast. The MgO cube can also be oriented with the Mg ion down, providing a strong positive potential. For many SFM experiments on insulators, the ionic MgO tip model provides excellent quantitative agreement with the image. Recent achievements in instrumental control reduce the possibility of tip contamination by sample material. It has also become important to use conducting tip models when studying insulating surfaces.

A good illustration of the importance of the "nanotip" structure is given by Hembacher and co-workers.[32] The molecular orbitals of a W atom at the end of a tip were "imaged" using an sp^2 orbital of flat graphite as a probe. This experiment demonstrates that current instruments are capable of resolving not only atoms but atomic orbitals as well. Although orbitals are not immediately obvious in the image contrast, Hembacher et al.[32] filtered the first harmonic of the signal and then summed all other harmonics with weighted coefficients to eliminate topographic contributions to the image, and the contrast due to the atomic orbitals became evident.

2.4 Applications of NC-AFM

In spite of the relatively short time since the demonstration of the atomic imaging with NC-AFM, its potential for characterizing nonconducting materials has motivated numerous studies. Illustrative examples of several classes of materials are summarized below.

2.4.1 Oxides: SrTiO$_3$ and TiO$_2$

Atomic resolution imaging of many oxide surfaces including Al_2O_3,[33] TiO_2,[34] $SrTiO_3$,[35] NiO,[36] CeO_2,[37] mica,[38] and MoO_3[39] has been demonstrated using frequency modulation (FM)-AFM. Two representative examples, $SrTiO_3$ and TiO_2, are illustrated here. The (100) surface of $SrTiO_3$ attracts much attention not least because it is the best substrate for the deposition of epitaxial films of superconductive oxides. Another attractive feature of $SrTiO_3$ is the ability to dope the surface with oxygen to very high concentrations ($\sim 10^{18}\,cm^{-3}$)[40] suggesting the potential of applications as channel layers of field-effect transistor.

Unit cell resolution was obtained by Kubo and Nozoye[35] on an $SrTiO_3$ (100) single crystal, on which they observed the $\sqrt{5} \times \sqrt{5}$ reconstruction, using the FM-AFM(2) mode. The details of frequency shift were not provided so an estimation of the nature of the forces is not possible. W_2C-coated conductive tips were used to eliminate electrostatic charge on the tip and chemical interaction between the tip and surface, and to nullify the capacitance forces. As a result, unit cell resolution was achieved. The authors describe the $\sqrt{5} \times \sqrt{5}$ surface re-construction as an ordered array of Sr adatoms. Figure 14–5 compares STM and NC-AFM images of the surface. The STM resolves only the Sr adatoms, while the NC-AFM resolves both the adatom and the underlying lattice. STM contrast on oxides contains both a geometric and electronic structure contribution, the relative magnitude of which cannot be determined a priori. In this case the geometric height of the Sr adatom appears to preclude resolution of the Ti lattice in unfilled state images. The AFM contrast, in principle, contains integrated charge density of all atoms so the oxygen sublattice can be imaged simultaneously. Other orientations of $SrTiO_3$ single crystals were examined by NC-AFM; atomic rows were observed on $SrTiO_3$ (110)[41] and atomic steps on $SrTiO_3$ (111).[42] It should be noted that unit cell resolution is routinely achieved with STM as well, when the sample is either doped or significantly reduced to provide sufficient conductivity ($SrTiO_3$ is an insulator).

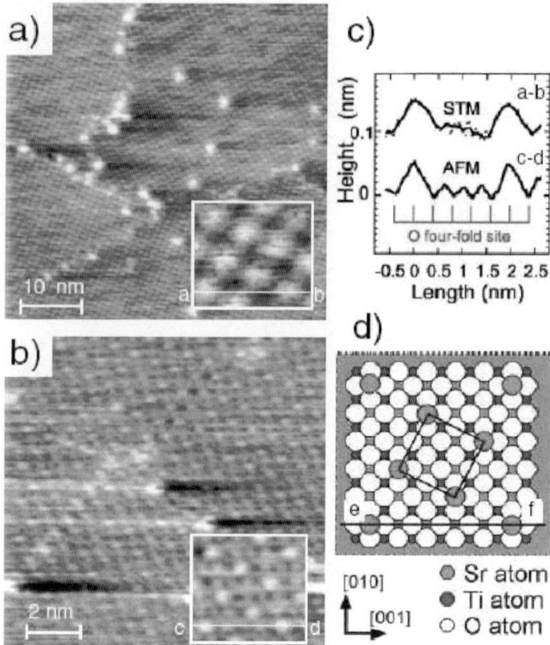

Figure 14–5. SPM on (100) SrTiO$_3$. (a) STM image ($V = +0.65$ V, $I = 0.11$ nA); (b) NC-AFM image ($V = -1.49$ V); (c) line profile comparison for STM and NC-AFM; note that STM shows the Sr adatoms on the surface while NC-AFM resolves unit cells underneath; (d) the model for $\sqrt{5} \times \sqrt{5}$ Sr adatoms on SrTiO$_3$. [Courtesy of Kubo and Nozoye. Reprinted with permission from Physical Review Letters, 86(9), 1801–1804, 2001.]

As discussed earlier, the details of formate adsorption on TiO$_2$ (110) were demonstrated with atomic resolution STM and FM-AFM(2) with a conductive silicon tip. An untreated (and probably oxidized) Si tip was used to image the (110) surface.[13,43] The (100) surface of TiO$_2$ is rather more complex than is the (110) surface, exhibiting a (1 × 3) micro-faceted structure at high temperatures and a combination (1 × 1) and (1 × 3) structure at intermediate temperatures. Raza and co-workers compared the intermediate structure of the TiO$_2$ (100) surface with STM and AFM.[44] Figure 14–6 illustrates the level of detail evident in the AFM image contrast. The normalized frequency shift, γ, in this case is 0.147 fN/m$^{1/2}$, implying that short-range van der Waals or covalent (Morse potential) forces were accessed. It was noted that the lateral dimensions of the features in the AFM and STM images agree but the apparent corrugation amplitude is much larger in AFM (0.5 nm compared to 0.1 nm). This illustrates the importance of height calibration. In this case the (1 × 3) microfacet was used as an internal standard.

2.4.2 FM-AFM of Semiconductors: Si and GaAs

The Si (111) (7 × 7) reconstruction has long been a standard in surface science. The structure based on minimization of the number of dangling bonds took 20 years to solve. The "Dimer–Adatom–Staking Fault" model proposed by Takayanagi et al.[45] is the currently accepted structure for the silicon (111) 7×7 surface. Immediately after the invention of the STM, Binnig and co-workers imaged the 7×7 reconstruction on Si (111) in real space by STM.[46] It took more then 10 years before the first NC-AFM images of the 7×7 reconstruction were obtained simultaneously by Kitamura and Iwatsuki[47] and Giessibl.[48] Both groups used FM-AFM(2) to obtain atomic resolution as described earlier (Figure 14–4).

A number of semiconductor surfaces have been imaged since then, including Si (111), (100), (110); GaAs (100); and InAs (110).[49] Several representative examples are presented here. The most technologically relevant Si (100) undergoes a 2×1 reconstruction. Figure 14–7 shows an atomic resolution FM-AFM(1) image by Yokoyama and co-workers.[50] The normalized frequency shift calculated for these images is ~$4.35 \times 10^{-14}\,N/m^{1/2}$, a magnitude that suggests that the forces are either short-range van der Waals or Morse forces, hence atomic resolution. Although the surface is known to adopt the (2×1) reconstruction, the distance between the "dimer-like" features in Figure 14–7 is larger (0.35 nm) that

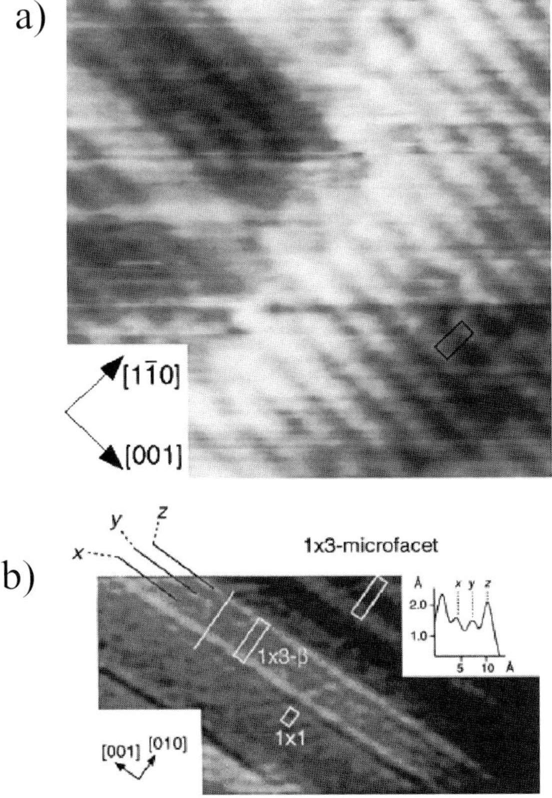

Figure 14–6. NC-AFM image of 1×1 reconstruction on TiO_2 (100) (a); facetted structure with 1×3 and 1×1 reconstruction on TiO_2 (100) (b). [Courtesy of Ashino et al. Reprinted with permission from Physical Review Letters, 86(19), 4334–4337, 2001.]

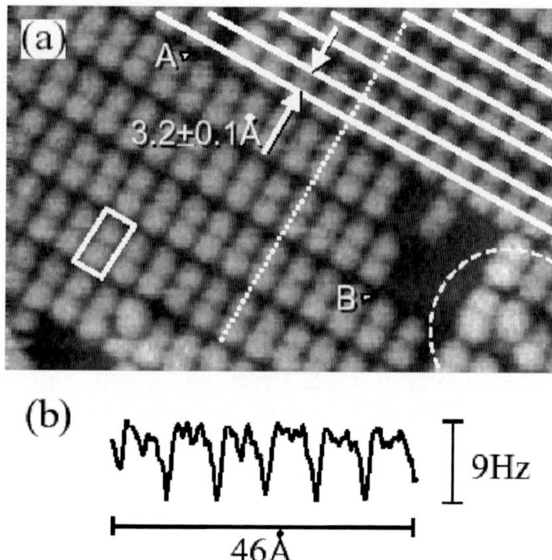

Figure 14–7. (a) NC-AFM image of Si (100) 2 × 1 reconstructed surface. One 2 × 1 unit cell is outlined by a box. (b) Line profile of the NC-AFM image across the dotted line. [Courtesy of Yokoyama et al. Reproduced with permission from Japanese Journal of Applied Physics Part 2—Letters, 39(2A), L113–L115, 2000.]

of the expected structure (0.23–0.20 nm). This result was compared with an NC-AFM image of the hydrogen passivated surface on which the lateral dimensions agree with the expected positions of hydrogen atoms. This dilemma is resolved if the image mechanism is that of a Si dangling bond of the tip interacting with the surface. In the case of the clean (2 × 1) surface the contrast then relates to surface dangling bonds rather than dimer bonds and in the case of the passivated surface the contrast relates to the dangling bond–hydrogen interaction. These results further emphasize the role of chemical/bonding forces in the image mechanism.

The surface of GaAs affords an opportunity to examine differences in tip–surface interactions on the same surface. Uehara et al.[51] compare the corrugations of As atoms and Ga atoms on a GaAs (110) surface at different tip–surface distances. Again FM-AFM(2) was used to obtain atomic resolution. Unfortunately the oscillation amplitude is not specified so a quantitative analysis of the forces is not possible. In this case, though, the apparent distance between As atoms did not change with tip–sample distance, while the distance between Ga atoms did. This is explained by considering that the force on the surface atom increases with decreasing tip–sample distance. Therefore, the As feature is associated with valence charge distribution and is not sensitive to this range of forces, while the Ga feature is associated with a dangling bond, which can be displaced by the interaction with the tip. This is another case of using atomic bonding interactions for chemical specificity in AFM image contrast.

2.4.3 Layered Materials: Graphite

Numerous layered compounds have been studied by NC-AFM, including mica,[52] MoO_3,[39] etc. Highly oriented pyrolytic graphite is perhaps the most general in that it is used as a calibration standard for SPM and as an atomically flat substrate in numerous studies. As with most layered compounds the graphite is easily prepared as a clean atomically flat surface by cleavage. Despite the proverbial ease of imaging graphite by STM with atomic resolution, every second atom in the hexagonal surface unit cell is usually unresolved. Giessibl and co-workers[53] demonstrated that a low-temperature FM-AFM(1) with pico-Newton force sensitivity reveals the hidden surface atom (Figure 14–8). This is another clear example that despite similarity in implementation and that combined AFM/STM microscopes are manufactured, differences between the image formation mechanisms of STM and NC-AFM are very important. In STM the tip images the density of states at the tip bias, while in NC-AFM the tip images an equiforce surface. With an oscillation amplitude of 300 pm, a cantilever spring constant of 1800 N/m, and an eigenfrequency of 18,076.5 Hz, the normalized frequency shift, γ, is $-2.7\,fN/m^{0.5}$. The magnitude of the normalized frequency shift indicates that the contrast in FM-AFM(1) is due either to short-range van der Waals or to Morse forces for the image in Figure 14–8.

2.4.4 Dielectrics: KBr and CaF₂

The ability to image insulating surfaces of high resistance has been noted as an advantage of NC-AFM. Of course the oxides presented in Section 2.4.1 are insulating compounds when fully oxidized and undoped. Here we illustrate compounds that are not made conducting. Benewitz and co-workers present an illustrative example[54] by imaging the potassium chloride–bromide surface. True atomic resolution was

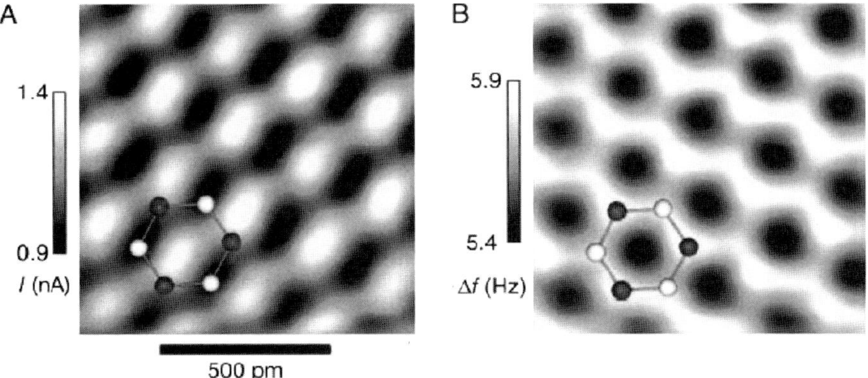

Figure 14–8. Low temperature STM (A) and NC-AFM (B) images on graphite. One hexagon of the graphite structure is superimposed on STM and NC-AFM images. One atom per unit cell for STM and two for NC-AFM. [Courtesy of Hembacher et al. Reproduced with permission from Proceedings of the National Academy of Sciences USA, 100(22), 12539–12542, 2003.]

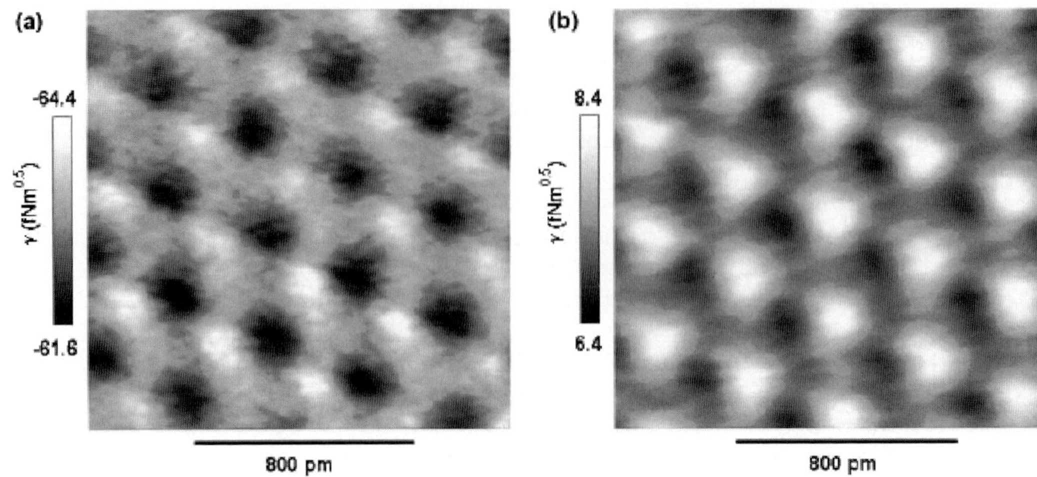

Figure 14–9. High-resolution constant height images of CaF2 (111) in (a) the attractive and (b) the repulsive mode of FM-AFM(1) using qPlus sensor as canteliver. (Courtesy of Giessibl and Reichling. Reproduced with permission from Nanotechnology, 16, (2005) S118–S124, 2005.

not achieved in their analysis of K(Cl,Br) single crystals, but atomic corrugations with spacing equal to the size of unit cell were resolved. This difference in corrugation amplitude was attributed to anion positions on the surface, providing an additional example of chemical specificity in NC-AFM.

Giessibl and Reichling[55] imaged the CaF_2 (111) surface with atomic resolution in both attractive and repulsive modes (Figure 14–9). The level of spatial resolution was facilitated by the use of a quartz tuning fork to oscillate the tip. The normalized frequency shift was $-62\,fN/m^{0.5}$ for attractive and $7.4\,fN/m^{0.5}$ for repulsive interactions, suggesting that van der Waals forces are responsible for image formation. The quartz tuning fork has a high spring constant (~1000 N/m compared to ~10 N/m for silicon cantilever), resulting in high stability of the small-amplitude oscillations. The quality of these images emphasizes the importance of small oscillation amplitudes and the proximity of these oscillations to the surface. Also, imaging in repulsive mode is challenging but yields high spatial resolution.

3 Imaging Properties: Advanced SPM Techniques

In addition to mapping topographic and atomic structure, variants of SPM probe local properties. The most common of these are EFM and magnetic force microscopy (MFM). These are straightforward measurements in which the tip is lifted above the surface to a distance at which only long-range forces are detected. By expanding the mechanisms of force detection and utilizing all possible image modes, the list of properties that can be probed is extensive (Table 14–2). The cantilever can be driven mechanically or electrically; this can be done in contact or noncontact regimes; and not only amplitude, but also first, second, and third harmonics of amplitude (or phase) can be detected. The relation between these operational regimes is shown in Figure 14–10. For example, conventional AFM is a noncontact, mechanically driven oscillation with amplitude or phase-based detection. The newer piezoelectric force microscopy (PFM) is in contact mode, electrically driven, and with phase-based detection. Static or periodic electric or magnetic fields can be applied to the sample, independent of the tip signal. An underlying theme of the newest developments is the use of multiple signal modulations or high order harmonics of modulated signals. Within this framework several techniques that address transport and dielectric properties will be reviewed.

3.1 Advanced SPM Techniques for Transport Properties

Perhaps the most common of the so-called "advanced" SPM techniques is scanning surface potential microscopy (SSPM), sometimes referred to as Kelvin probe force microscopy (KPFM), which maps the work function of surfaces.[56,57] In SSPM, the cantilever oscillation is driven electrically. This is a noncontact (generally 50–200 nm separation) probe, with feedback on the first harmonic (Figure 14–11). An ac voltage

Table 14–2. Properties accessible with scanning probe microscopy.

Technique	Mode	Property	References
AFM	nc/ic, mech, phase/amp	vdW interaction, topography	[2–5]
EFM	nc, mech, phase/amp	Electrostatic force	[2–5]
MFM	nc, mech, phase/amp	Magnetic force, current flow	[2–5]
SSPM (KPM)	nc, elec, 1st harmonic	Potential, work function, adsorbate enthalpy/entropy	[2–5, 56–75]
SCM	c, F, cap sensor	Capacitance, relative dopant density	[2–5, 82, 84–87, 89–92]
SCFM	c, elec, 3rd harmonic	dC/dV, dopant profile	
SSRM	c, F, dc current	Resistivity, relative dopant density	[2–5, 81–83, 87]
SGM	nc, elec, amp	Current flow, local band energy, contact potential variation	[2–5, 97]
SIM	nc, elec, phase/amp	Interface potential, capacitance, time constant, local band energy, potential, current flow (in combination with SSPM)	[93, 97, 99]
NIM	c, F, freq spectrum	Interface potential, capacitance, time constant, dopant profiling	[102–104]
PFM	c, elec, phase/amp	d33	[63, 105–111, 125, 126, 128, 129]
NPFM	c, elec, 2nd harmonic	Switching dynamics, relaxation time and domain nucleation	[130]
SNDM	c, F, 1st or 3rd harmonic	dC/dV, dielectric constant	[133, 134]
NFMM	c, F, phase	Microwave losses, d33	[135–138]

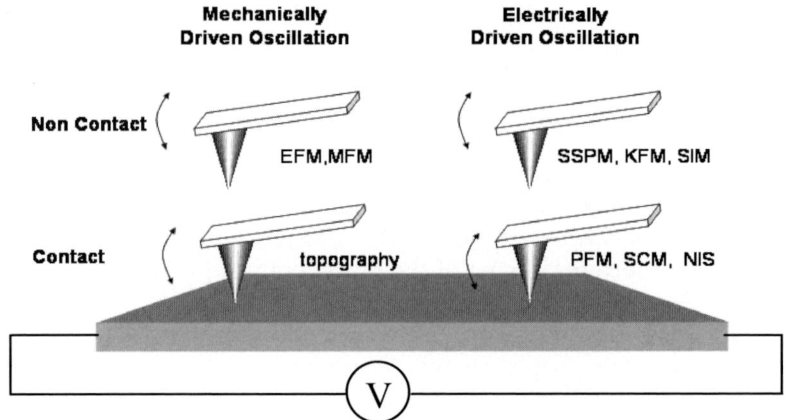

Figure 14–10. Variations of SPM are separated into four operational regimes distinguished by cantilever driving force and condition of contact.

is applied to the tip. V_{dc} is varied until it matches the surface potential, V_{surf}, at which point the first harmonic of the force F (with frequency ω) is nullified. Thus, the applied V_{dc} is equal to the surface potential. The tip potential is expressed as $V_{tip} = V_{dc} + V_{ac}\sin\omega t$, and the force acting between the tip and surface $F(z) = \frac{1}{2}(\partial C/\partial z)[V_{tip} - V_{surf}]^2$, so the first harmonic of the force is $F_{1w}(z) = (\partial C/\partial z)[V_{dc} - V_{surf}]V_{ac}$.

The apparent ease with which potential variations can be mapped belies the challenges in interpreting images on electrically[58,59] and topographically inhomogeneous surfaces. In the cases of semiconductor and dielectric surfaces the electrostatic properties of a surface are not characterized solely by intrinsic potential and topography. SSPM images must be interpreted in terms of *effective* surface potential that includes capacitive interactions, surface and volume bound charges, double layers, and remnant polarization.[60–63] For semiconductor surfaces tip-induced band bending[64] can offset local surface potential.[65] Despite these challenges the obvious need to examine variations in

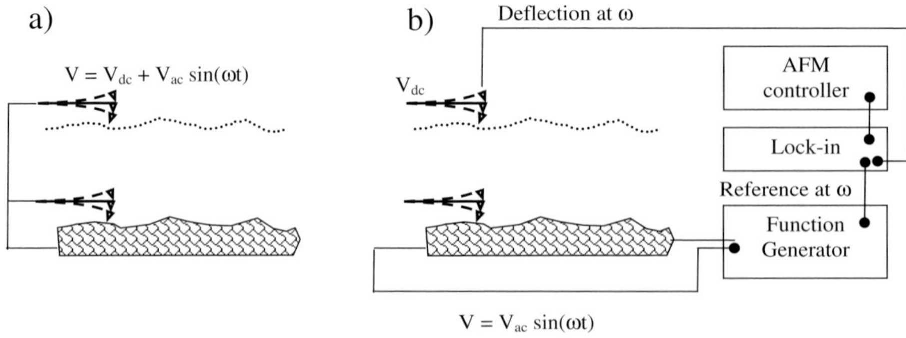

Figure 14–11. Comparison of working principles for SSPM (KFM) and SIM. In SSPM (a) the oscillating potential is applied between the sample and tip; in SIM (b) the oscillating potential is applied across the sample.

local potential in electronic nanodevices spurred efforts to overcome some of the obstacles with careful analytical treatments that determined limits in quantification. In the late 1990s SSPM was applied to semiconductor,[66,67] organic,[68] and ferroelectric[69,70] surfaces, as well as to defects,[71,72] and photoinduced[73,74] and thermal phenomena.[75] It is fair to say that at this point that absolute values of potential cannot be quantified, but variations in potential can be determined with energy resolution of 2–4 meV and spatial resolution of the order of 50–100 nm.

One interesting application is the measurements of surface potential of self-assembled monolayers (SAM). Figure 14–12 shows a typical potential map of two SAMs patterned by microcontact printing. In this case one SAM is an alkanethiol, while the other is a conjugated conductive molecule. The difference in surface potential is obvious but the basis of the difference is not entirely clear. There are, in principle, four contributions to the potential of SAM: the substrate work function, the dipole in the surface bond (usually S—M or SiO_x—Si), the dipole in the molecule, and the terminating end group. Eng and co-workers showed that the potential of alkanethiols on Au increases with the increase of chain length.[76] Sugimura et al. calculated dipole moments in a sequence of siloxane-coupled molecules and the trends agreed with the experiment.[77] These results provide evidence for the contribution of the molecular dipole to the measured surface potential. Alvarez and Bonnell[78] showed that for the SAM of alkanes on metal, the S—M bond dipole also affects the measured surface potential. While work remains to quantify the relative contribution of SAMs to surface potential, SSPM easily quantifies variations and is routinely used to characterize complex films.

On samples with morphological variations, such as grains, SSPM in the presence of a lateral field can elucidate the behavior of individual microstructural features. This is illustrated in Figure 14–13, which shows the behavior of a polycrystalline ZnO surface under different applied lateral biases. The topography contains features due to contaminants and inter- and intragranular pores. On application of 5 V

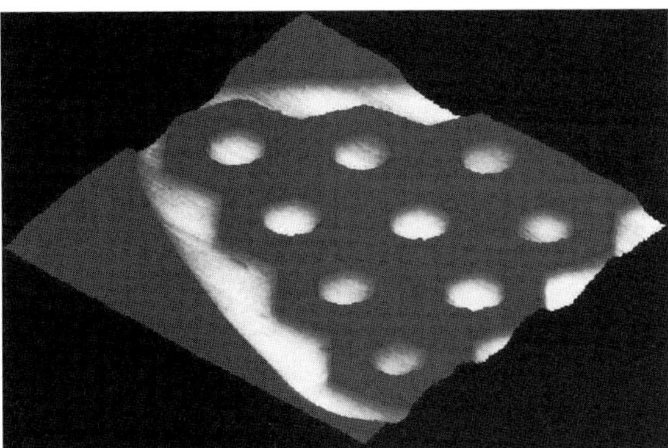

Figure 14–12. SSPM image of SAM of alkanethiol and conjugated molecules self-assembled on Au, patterned with a microcontact printing technique. (Courtesy of Rodolfo Alvarez.)

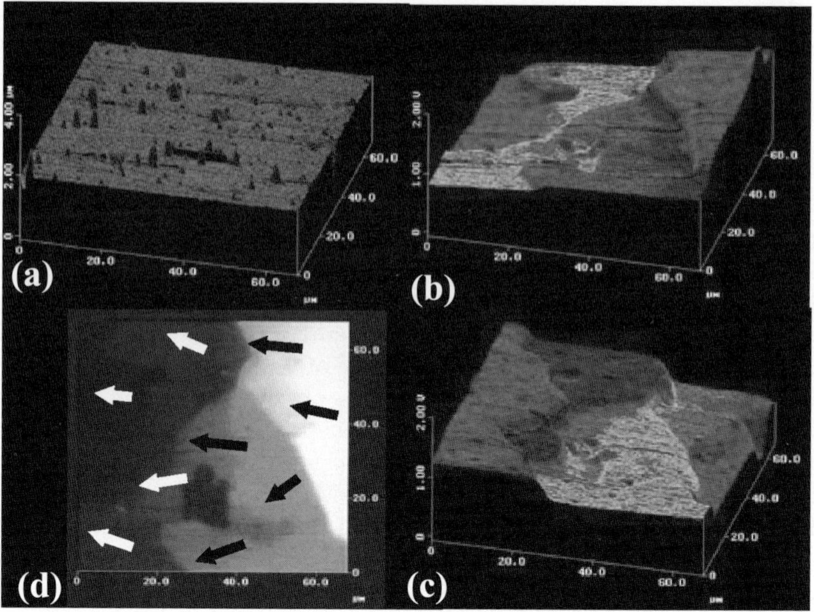

Figure 14–13. Transport imaging in polycrystalline ZnO. (a) Surface topography. (b and c) SSPM images under lateral bias exhibit potential drops at grain boundaries, indicative of grain boundary resistive behavior. Note that the direction of potential drops inverts with bias. (d) Current maps for positive and negative bias polarity. (Partially reproduced from Kalinin and Bonnell, *Zeitschrift fur Metallkunde*, 90, 983–989, 1999.) (See color plate.)

lateral bias the potential steps at grain boundaries become evident (Figure 14–13b), and the contrast inverts when the direction of the applied bias is reversed (Figure 14–13c). Clearly the behavior of individual grains and grain boundaries is distinguished and the voltage dependence of all microstructural constituents can be examined separately. Furthermore, taking directional derivatives allows the current flow to be determined at all points on the surface, as shown in Figure 14–13d.

Although not explicitly stated, much SSPM is done in ambient conditions. Indeed the ease of measurement is a strong advantage for its use as a qualitative characterization tool. As with all ambient measurements, the possible interaction of the environment with the surface must be considered. There are, in fact, situations in which these effects dominate. Some of the early studies involved imaging ferroelectric domains in air[79] and it was eventually understood that the strong field due to domain polarization at a surface is almost completely screened, presumably by adsorption. Domains are easily imaged but quantifying the magnitude of the surface potential is impossible. In fact the sign of the measured potential is usually opposite that of the domain potential. Ferroelectric domains represent an extreme case of local electric fields, but even for grain boundaries intersecting a surface the effect is observed. Figure 14–14 compares the potential of an atomically abrupt boundary in $SrTiO_3$ measured in air and in ultrahigh vacuum (UHV).[80]

The boundary contains an intrinsic charge, which results in a potential variation at the intersection with the surface. In ambient conditions the charge attracts compensating adsorbates and the measured difference in surface potential is under 10 mV. In UHV, the uncompensated difference in potential is on the order of 90 mV and opposite in sign.

Two contact techniques, scanning spreading resistance microscopy (SSRM)[81–83] and scanning capacitance microscopy (SCM),[84–86] have been developed to characterize semiconductors. In SSRM, a conducting tip is biased with respect to the sample and the dc current through the tip/surface contact is detected under force feedback control. The amount of current is determined by the local spreading resistance of the surface, which is related to the local conductivity. Because the native oxide layer on Si hinders tip/surface contact, carrier profiling in Si-based devices usually requires a tip coated with a hard material and a high spring constant cantilever to provide strong indentation forces ($\sim$20 μN).[87] In SCM, a high-frequency capacitance sensor detects the tip/sample capacitance as the tip is scanned across the sample, which allows a smaller indentation force than is used in SSRM. Diamond-coated tips are used in spreading resistance measurements, but lower indentation force makes metal-coated probes suitable for SCM.[88] An ac voltage applied to the tip induces the depletion and accumulation of carriers, resulting in a change in capacitance, ΔC. In a semiconductor, the depletion/accumulation width is inversely related to the carrier concentration, so mapping $\Delta C/\Delta V$ yields a carrier concentration profile. Difficulties in quantification arise if the dopant concentration is non-uniform, and when spatial resolution decreases due to low dopant concentrations. A recent modification incorporates an additional feedback that adjusts ΔV to maintain a constant ΔC during the scan, maintaining a constant depletion width.[88] Analysis of SCM results are mathematically

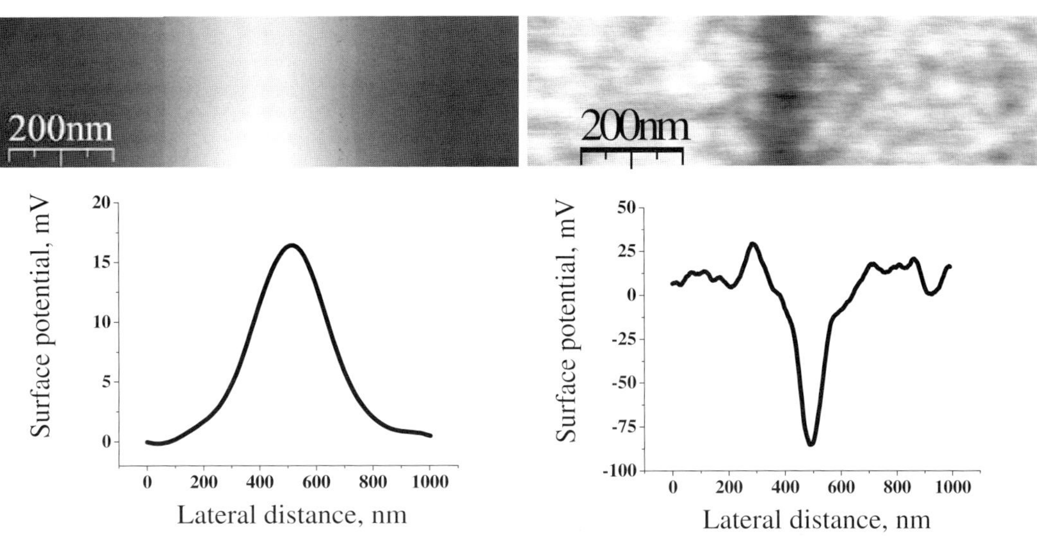

Figure 14–14. KFM of the Σ5 grain boundary in SrTiO$_3$ (100). Left, in air; right, in UHV. (Courtesy of Rui Shao.)

challenging, usually requiring 2D or 3D numerical approaches.[89–92] Note that SCM distinguishes the difference between n- and p-dopants in the sample, whereas SSRM does not. On the other hand metallic and insulating samples give no contrast in an SCM image, while SSRM images defects in insulations and in metallic films.[88]

Another technique for electronic property characterization on the nanoscale is scanning gate microscopy (SGM). SGM is a two-scan technique. During the first scan topography of the sample is monitored in intermittent contact mode. During the second scan an ac voltage is applied across the sample; a biased tip is brought within proximity of the sample, and the magnitude of the current *across the sample* is measured as a function of tip position. The first implementation of SGM was demonstrated on a single wall carbon nanotube in 2000.[93] The tip acted as a gate electrode in this nanotube circuit. It was found that the Fermi surface of the nanotube was not uniform along its length. Re-searchers from Stanford University,[94] University of California at Berkeley[95] and University of Pennsylvania[96,97] used SGM to good effect in nanotube-based electronic circuits illustrating the ability to quantify local properties in individual nanotubes and to demonstrate threeterminal device behavior. Recently researchers from the University of California at Irvine[98] measured the gating effect of a biased AFM tip on ZnO nanowires and found it to be 200 times smaller than that of single-wall carbon nanotubes.[94] The change of resistance in these nanowires differs because of the difference in the widths of the single-wall carbon nanotube (1–2 nm) and ZnO nanowires (~400 nm).

Research in the group of R.M. Westervelt extended the concept of using an SPM tip as a perturbing probe to visualize the flow of electron waves in a two-dimensional electron gas. They produced a planar electron gas 57 nm below the surface in a GaAs/AlGaAs heterostructure with a quantum point contact formed with a pair of gates on the surface. A biased tip near the surface will capacitively couple to the underlying electron gas. As the tip is scanned across the surface, it decreases the conductance through the contacts when it is over a region of high electron flow and has no effect over a region with low electron flow. Similar to SGM, the image is the current or conductance across the gap as a function of tip position on the surface. Figure 14–15 illustrates the geometry of the quantum point contact and the pattern of electron flow that results from a system with a density of $4.5 \times 10^{11}\,\mathrm{cm}^{-2}$, a mobility of $1 \times 10^{6}\,\mathrm{cm}^{2}/\mathrm{V\,s}$, a mean free path of 11 μm, and a Fermi wavelength $\lambda_F = 37$ nm. Images were recorded with the sample and the SPM at $T = 1.7$ K.

Many phenomena underlying materials behavior involve time-dependent processes that could be accessed if signals applied to the sample and/or tip spanned a wide frequency space. The first introduction of frequency dependence in scanning probes is referred to as scanning impedance microscopy (SIM).[99] This is a noncontact, first harmonic detection in which the oscillating electrical signal is applied to the sample instead of the tip. The implementation of SIM is compared to SSPM in Figure 14–11. The tip can act as a nonperturbing probe or as a local gate in a configuration that allows both the amplitude, which is related to potential, and the phase, which is related to

Figure 14–15. Experimental images (outside) and theoretical simulations (inside) of the flow of electron waves through a quantum point contact. Fringes spaced by half the Fermi wavelength demonstrate coherence in the flow. [Courtesy of Westervelt et al. Reproduced with permission from Physica E, 24, 63–69, 2004.] (See color plate.)

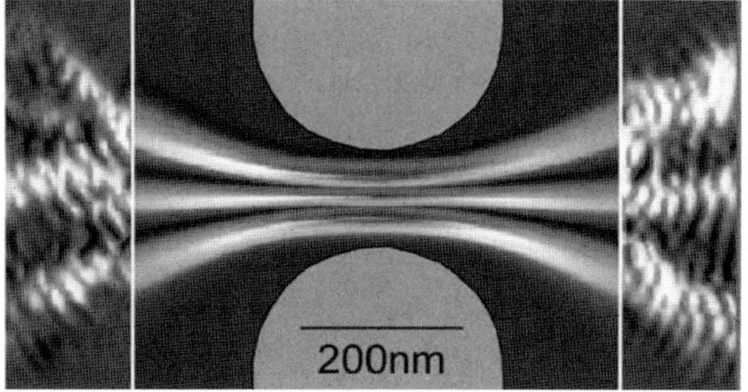

loss, to be quantified. The frequency dependence can be used to isolate relaxation associated with electron traps at interfaces and defects.

Figure 14–16a and b illustrates SIM of an interface and grain boundary. For a prototypical ideal sample composed of a single electroactive

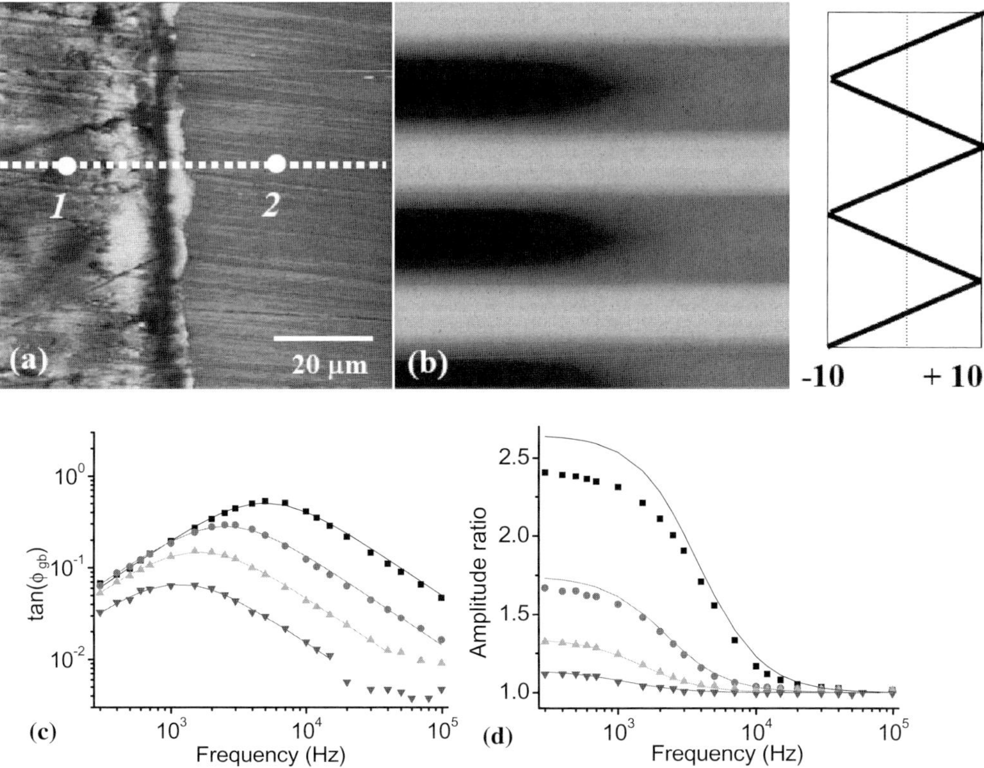

Figure 14–16. Surface topography (a) of the cross-sectioned diode. The potential profiles were acquired along the dotted line. The changes of potential, phase, and amplitude were determined from positions 1 and 2. Surface potential (b) during a 0.002-Hz triangular voltage ramp to the sample for $R = 500\,\Omega$. The scale is 300 nm (a) and 10 V (b). (c) Frequency dependence of SIM phase shift and (d) amplitude ratio of grain boundary in an Nb-doped $\Sigma5$ $SrTiO_3$ bicrystal. Solid lines on (c) are fits for frequency independent grain boundary resistance and capacitance and on (d) the calculations are from using interface resistance and capacitance from phase data. Data are shown for circuit terminations $148\,\Omega$ (■), $520\,\Omega$ (●), $1.48\,k\Omega$ (▲), and $4.8\,k\Omega$ (▼). (Courtesy of Kalinin and Bonnell. Reprinted with permission from Physical Review B 70, 235304, 2004).

interface with resistance, R_d, and capacitance, C_d, in series with two current-limiting resistors, R, the phase shift across the interface as a function of frequency is $\tan(\varphi_d) = (\omega C_d R_d^2)/(R + R_d) + R\omega^2 C_d^2 R_d^2$, where $\omega = 2\pi f$ and f is the oscillation frequency. See Figure 14–11 for the exact configuration. For high frequencies the interface phase shift has the simple form $\tan(\varphi_d) = 1/\omega C_d R$ and interface capacitance can be calculated directly from the frequency shift. At low frequencies the amplitude ratio across the interface, $A_1/A_2 = (R + R_{gb})/R$. In addition, a dc potential can be applied across the surface establishing a potential step at the interface, which can then be measured by SSPM. As the dc bias across the interface of a metal/silicon diode is switched from positive to negative, the forward/reverse bias behavior is evident in the potential image (Figure 14–16a). The phase image of an atomically abrupt $SrTiO_3$ grain boundary (Figure 14–16b) shows the large effect of a 2D defect. The frequency dependences of both the amplitude (which is related to potential) and phase shift (which is related to interface processes) of the $SrTiO_3$ boundary are shown in Figure 14–16c and d. The combination of SIM and SSPM allows local transport spectroscopy of individual interfaces, e.g., direct measurement of the spatially resolved interface C–V characteristics. The quantitative nature of this approach has been confirmed on studies of Si/metal diodes.[100]

SIM also contains an intrinsic improvement in spatial resolution relative to scanning surface potential or SGM. In combination with other SPM techniques, SIM has been used to map the rectifying behavior of a Schottky junction, and characterize defect-mediated transport in a single nanotube,[97] as well as provide transport measurements of idealized grain boundaries. Recently SIM was used to image current transport in carbon nanotube networks imbedded in a polymer.[101] Similar to SSPM, the spatial resolution in SIM is ultimately limited by capacitive tip–surface interactions.

A second approach to accessing frequency-dependent transport expands the frequency range to eight orders of magnitude and provides higher spatial resolution. Nanoimpedance microscopy and spectroscopy (NIM)[102–104] is a contact probe with force feedback, in which the oscillating bias signal is applied to the tip; current phase and amplitude are detected at the sample. Figure 14–17a illustrates the local impedance between a tip and lateral electrode, across a ZnO grain boundary. By accessing this range of frequencies, the data can be presented in a conventional Cole–Cole plot that describes both the real and imaginary components of impedance. In these plots each of the semicircles corresponds to the properties of a microstructural element. The interpretation is somewhat model dependent, but for sample configurations such as thin films with bottom electrodes or single boundaries between electrodes, quantitative characterization of the local properties is possible. The local boundary potential, capacitance, charge, and depletion lengths can be extracted with spatial resolution on the order of tens of nanometers. In a configuration with an electrode under the sample, the impedance of individual grains can be imaged, as shown in Figure 14–17b.

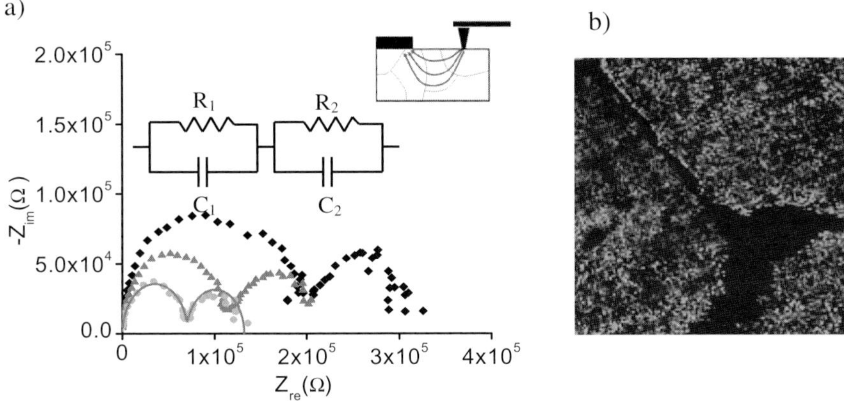

Figure 14–17. (a) Cole–Cole plot of local impedance spectra between the SPM tip and the top electrode (nanoimpedance spectroscopy) at tip/sample biases of +5 V (●), +3 V (△), and +2 V (◆) for fixed tip location. Solid line is the fitting of impedance data to the equivalent circuit of two RC elements in series (——). Note that interface resistance decreases with dc bias, indicative of varistor behavior. (b) Representative NIM phase image; the dark region corresponds to nonconductive inclusion in the sample (phase $\theta = -90°$).

3.2 Advanced SPM Techniques for Dielectric Properties

In principle, utilizing higher order harmonic signals and clever detector design allows dielectric constant, electrostriction, and piezoelectric properties to be detected (Table 14–2). In practice a number of probes have been developed to quantify linear and nonlinear dielectric properties locally. These have focused on electromechanical coupling coefficients, hysteretic ferroelectric domain switching, etc.

PFM is a scanning probe that is used increasingly to determine electromechanical coupling coefficients at local scales. It is a contact, electrically driven probe technique with feedback based on phase lag. Like NIM, an oscillatory electric field is applied to the tip, which is in contact with the sample (Figure 14–18). If the material is piezoelectric,

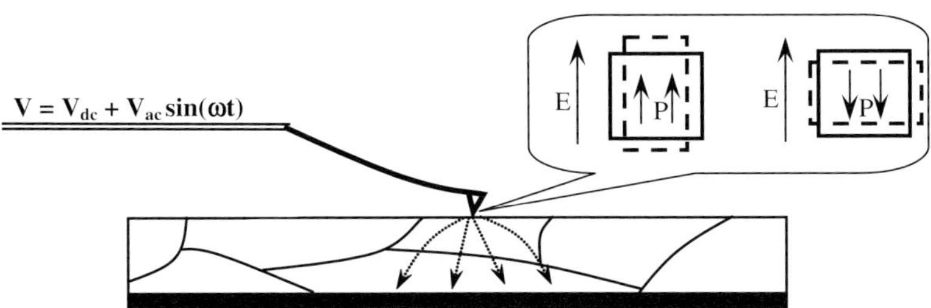

Figure 14–18. Working principle of PFM. An ac bias is applied to the tip, which his in contact with the sample, and an electric field is generated. If the material is ferroelectric the size of the region around the tip will change. A schematic of the responses of polarized domains is presented in the inset.

the field results in a local deformation of the surface that oscillates the tip, i.e., piezoresponse (PR).[105,106] In a ferroelectric material domains with downward polarization vector contract with a positive voltage, producing a phase shift of $\delta = 180°$. For upward oriented domains, the situation is reversed, and $\delta = 0°$, because the deformation is in phase with the field. The phase, therefore, indicates the orientation of polarization. The piezoresponse amplitude, $A = A_{1\omega}/V_{ac}$, defines the magnitude of the interaction. For the ideal case of a (100) surface of a tetragonal compound, $A = \alpha d_{33}$, where α is proportionality coefficient close to unity,[107] the piezoelectric constant, d_{33}, is related to the polarization, P, as $d_{33} = \varepsilon\varepsilon_0 Q_{33}P$, where Q is the second-order electromechanical coefficient. For the general case there are in-plane components of polarization that can be accessed by measuring the lateral response of the tip to a field variation.[108,109] Furthermore, the piezoelectric response is a tensorial function, the complexity of which depends on the symmetry of the compound and the orientation of the grain or crystal. Harnagea et al.[110] have shown that even for $BaTiO_3$ with relatively high symmetry either the grain orientation or the in-plane component must also be known to determine domain orientation.

This is illustrated nicely by Gruverman and co-workers[111] who undertook the three-dimensional high-resolution reconstruction of the polarization vectors in a (111)-oriented $Pb(Zr,Ti)O_3$ ferroelectric capacitor by detecting the in-plane and out-of-plane polarization components using PFM. Figure 14–19a and b shows the vertical and lateral contributions to the phase images of a region that is nominally poled in the vertical direction. In spite of exhibiting uniform vertical contrast, the lateral component exhibits significant variation. Knowing that this material is oriented in the (111) direction, the individual domain orientation can be determined, as shown on Figure 14–19c. In this case the vertical components of domains polarization have the same magnitude, while the lateral component varies between domains.

In PFM voltage spectroscopy,[112] piezoresponse, $A_{1\omega}$, and phase, f, are measured as a function of dc potential offset, V_{dc}, on the tip.[109,113] PFM spectroscopy yields local electromechanical hysteresis loops, quantifying remnant response and coercive bias, on the 20–50 nm level. Piezoresponse force spectroscopy of $PbTiO_3$ is illustrated in Figure 14–20. These hysteresis curves are acquired on a $PbTiO_3$ film with 100 nm grain size. The phase signal is related to polarization and therefore has the shape of a conventional P–E hysteresis curve. The amplitude signal shows that strain, in addition to switching, occurs with voltage and again traces the conventional butterfly shape of a curve. These curves illustrate the critical tip bias required to achieve polarization switching. A number of attempts to relate local hysteresis to crystallographic orientation and piezoresponse amplitude have been reported.[114]

One critical issue with regard to quantification and spatial resolution limit is the question of what volume is accessed by PFM. The answer is found by determining the decay of the electric field below the probe tip, which in turn depends on the dielectric constant and conductivity of the material. For oxide ferroelectric compounds the volume in on the order of 20–200 nm.[115,116] Consequently, for films thinner than this,

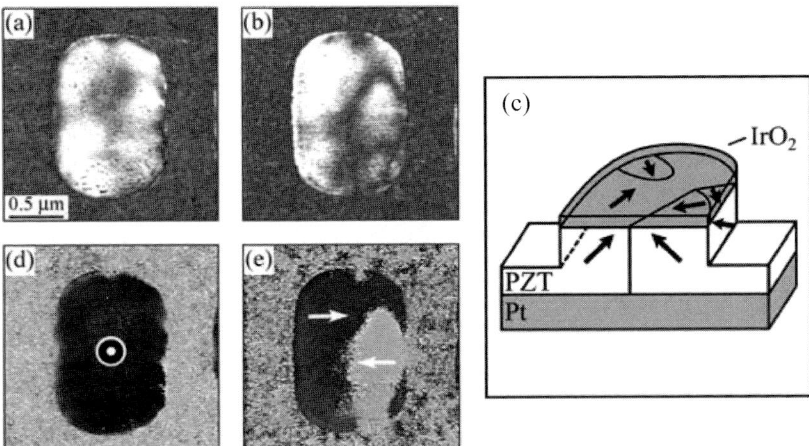

Figure 14–19. Investigation of a PZT capacitor with vertical and lateral force PFM. Images of amplitude (a) and phase (d) in vertical PFM and amplitude (b) and phase (e) in lateral PFM are presented. (c) Schematics of the domain orientation in the sample followed from PFM measurements. Note that only a combination of three PFM modes gives complete information about the direction of the polarization vector in the sample (only two are presented on the picture). [Courtesy of Rodriques et al. Reproduced with permission from Journal of Applied Physics, 95(4), 1958–1962, 2004.]

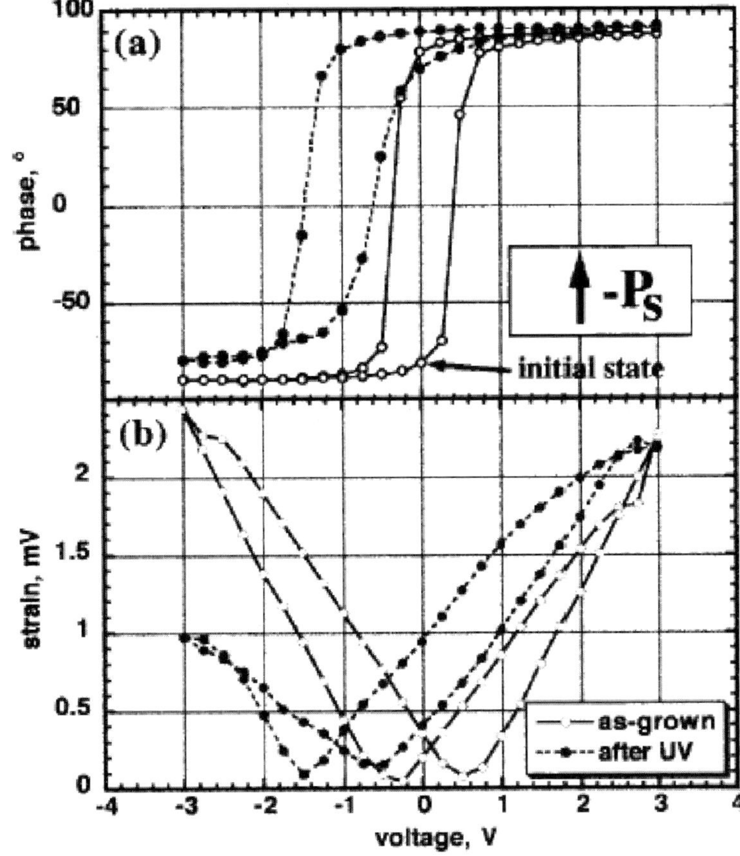

Figure 14–20. (a) Vertical phase and (b) amplitude hysteresis loops of PbTiO₃ thin films. (Courtesy of Gruveman and co-workers. Reprinted with permission from Journal of Applied Physics, 92, 2734, 2002).

or inclined domain structures, quantification requires accounting for this volume.

The question arises as to whether the PFM influences the properties it is measuring. Recently, it was shown that the mechanical strain produced by the tip can suppress local polarization[117] or induce local ferroelectroelastic polarization switching.[118,119] A quantitative analysis of the tip-induced potential and stress distribution is required to characterize local ferroelectric properties by SPM.[120–124] A rigorous treatment of the image contrast includes simultaneous electrostatic and electromechanical interactions. One complication in PFM is that both long-range electrostatic forces and the electroelastic response of the surface contribute to the PFM signal.[63,125] Even under optimal conditions, the basis of the electroelastic contribution, A_{piezo}, is not straightforward because of the complex geometry of the tip–surface junction. Some progress in the quantitative understanding of PFM has been achieved.[126–129] Depending on the tip radius of curvature and the indentation force, PFM may correspond to the electroelastic response of the surface induced by the contact area (strong indentation limit) or be dominated by the electroelastic response of the surface due to the field produced by the spherical part of the tip (weak indentation limit). In these cases, the magnitude of surface and tip displacements is determined by the electromechanical coupling in the material. Alternatively, the signal can be dominated by the electrostatic tip–surface interactions (electrostatic limit) and have little relation to the properties of the material. Taking an approach familiar to materials scientists, the analytical solutions of these interactions can be presented as contrast mechanism maps that relate experimental conditions to properties of the material and delineate the conditions under which quantitative measurements can be obtained (Figure 14–21).[126]

The relationship between higher order harmonics of the PR function and time dependence of domain switching has been developed into a probe of switching dynamics called second harmonic piezo force microscopy (SH-PFM).[130] When the field and the measured electrostrictive strain are in the z direction, electrostriction is expressed in terms of the field-induced polarization P as $x = Q_{33}P^2$. For a ferroelectric with spontaneous polarization P^S and field-induced polarization P^E, the strain becomes $x = Q_{33}(P^S)^2 + 2Q_{33}P^SP^E + Q_{33}(P^E)^2$, where the second and third terms are the piezoelectric response (because $d_{33} = 2Q_{33}P^S/\varepsilon$) and the electrostrictive response, respectively. Under external field, $E_3(\omega) = E_3\cos\omega t$, induced

$$x = Q_{33}\left[(P^S)^2 + \frac{(P^E)^2}{2}\right] + 2Q_{33}P^SP^E\cos\omega t + \frac{1}{2}Q_{33}(P^E)^2\cos2\omega t$$

In macroscopic measurements electrostriction is quantified with interferometry.[131,132] In SH-PFM the cantilever oscillation in contact mode determines the local electrostrictive properties.

A complementary strategy to accessing linear and nonlinear dielectric properties is referred to as scanning nonlinear dielectric microscopy[133,134] and near-field microwave micro-scopy (NFMM).[135,136] The

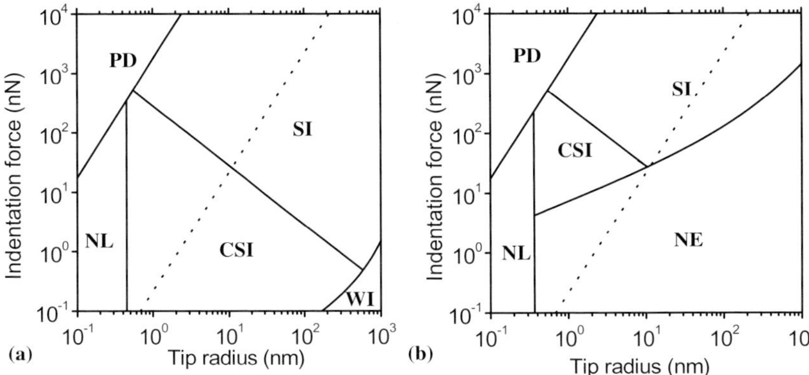

Figure 14–21. Contrast mechanism maps of piezoresponse force microscopy. SI, strong indentation regime; CSI, contact-limited strong indentation; WI, weak indentation regime; NE, nonlinear electrostatic regime; NL, nonlocal interactions; PD, plastic deformation. The dotted line delineates the region where stress-induced switching is possible. (a) $w = 0.1$ nm, $\Delta V = V_{\text{tip}} - V_s = 0$ V; (b) $w = 0.1$ nm, $\Delta V = 5$ V. (Courtesy of Kalinin and Bonnell. Reprinted with permission from Physical Review B, 65, 125408, 2002.)

approach utilizes a coaxial probe in which a sharp, center conductor "tip" protrudes. The probe is actually the end of a transmission line resonator, which is coupled to a microwave source. The concentration of the microwave fields at the tip changes the boundary condition of the resonator, and, hence, the resonant frequency and quality factor. The magnitude of the perturbation depends on the dielectric properties of the sample. Specifically, the change in resonant frequency, $\Delta f_0/f_0 = g\Delta\varepsilon'$, where g is a constant and $\Delta\varepsilon'$ is the real part of the complex dielectric constant; $\Delta(1/Q) = q\Delta\varepsilon''$, where q is a constant and $\Delta\varepsilon''$ is the imaginary part of the complex dielectric constant. The spatial resolution of the microscope in this mode of operation is about $1\,\mu$m. NFMM with coaxial resonators has been used successfully to quantitatively image sheet resistance, dielectric constant, dielectric polarization, topography, magnetic permeability, and Hall effect. Lu et al.[137] have used NFMM on ferroelectrics (001) $LiNbO_3$ single crystal to distinguish variations in dielectric constant and ferroelectric domains (Figure 14–22). The growth process of this crystal results in a periodic composition change that should alter the local dielectric constant. Different ferroelectric domains in this crystal have the same dielectric constant. The two periodic variations in composition and ferroelectric domain orientation are not coincident. Since $\Delta f_0/f_0$ relates to dielectric constant and ΔQ relates to loss associated with ferroelectric domain boundaries, the images should demonstrate different periodicities. This is illustrated in Figure 14–22a and b. Lu et al.[137] claimed that their NFMM has submicrometer lateral resolution, so the prognosis for further improvement of special resolution is good. Lee and Anlage[138] demonstrated that high-order harmonic powers acquired by NFMM can be used to spatially resolve the local nonlinearity. In their work, the grain boundary

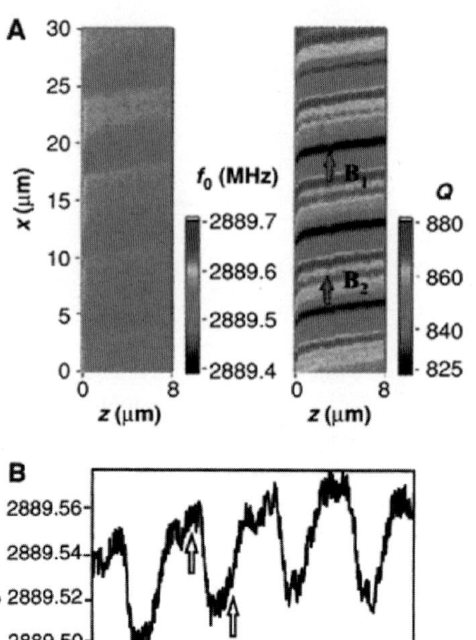

Figure 14–22. (A) Images of the profiles of periodic dielectric constant (on the left) and ferroelectric domain boundaries (on the right). (B) Line profiles of frequency (top) and quality factor (bottom) images. (See color plate.)

area of superconducting YBCO thin film deposited on an $SrTiO_3$ bicrystal was spatially resolved from the ratio of the second and the third power.

4 Future Trends

The rapid pace of advances in SPM shows no sign of slowing. Atomic resolution imaging is becoming ever more common and the range of local properties that can be quantified is expanding. In the next few years, several areas of research will likely push the limits of SPM even further.

Combinations of multiple modulations and high order harmonics will continue to be used in new ways to access complex properties. Table 14–2 illustrates the potential of such combinations but is not a comprehensive list of all possibilities. Neither the advanced magnetic probes nor the near-field optical probes are included. Frequency mixing has not yet been exploited.

There is recent evidence that some of the property probes may achieve atomic resolution. Eguchi et al. obtained atomic scale imaging in an SPPM (KFM).[139] Figure 14–23 compares NC-AFM and SSPM of a Ge/Si (105) surface with a model of the atomic structure. An in another community, Rugar et al. demonstrated the measurement of a single spin of an electron.[140] While further work is needed to confirm the absence of artifacts in these measurements, these observations are exciting in that they portend a future of atomic resolution property imaging. Advances in high precision cantilevers and tips could unite NC-AFM imaging of structure with similar resolution imaging of potential, work function, dielectric function, etc. Concomitant advances in the physics of tip–surface interactions will be necessary when we need to interpret, for example, atomic resolution "dielectric constant."

Finally, in the ideal world it would be useful to combine multiple probes with a variety of options for sample stimulation for comprehensive analysis. This vision is schematically illustrated in Figure 14–24. Scanning probes with multiple transport-related tips are commercially

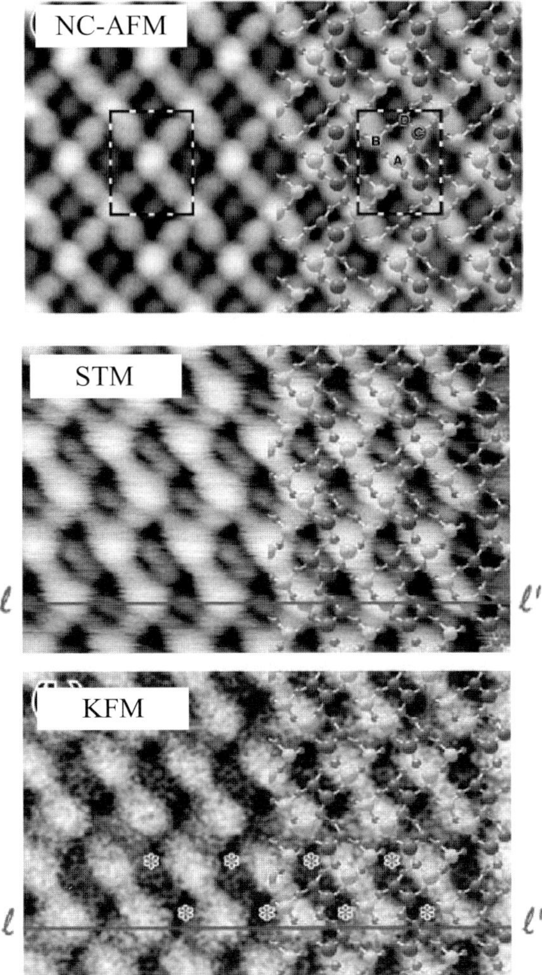

Figure 14–23. Atomic resolution NC-AFM, STM, and KFM on Ge/Si (105). [Courtesy of Eguchi et al. Reproduced with permission from Physical Review Letters, 93(26), 2004.] (See color plate.)

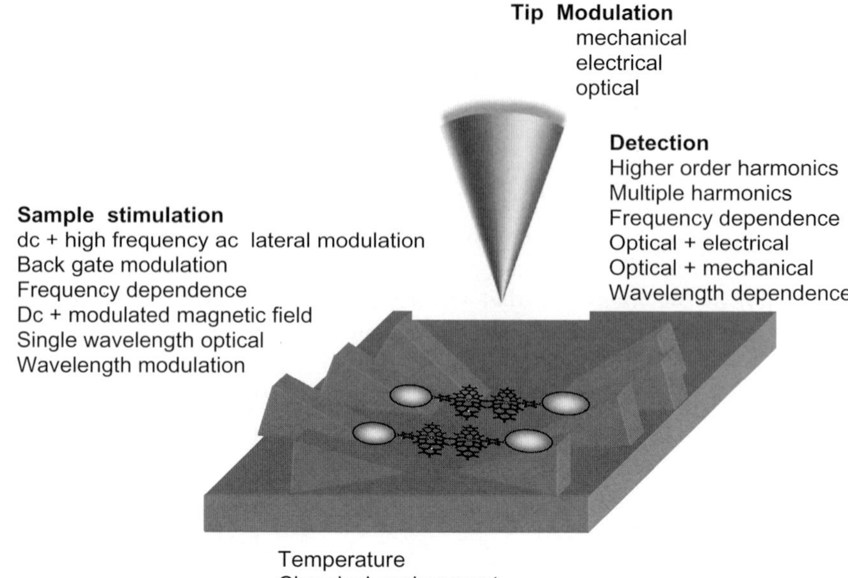

Figure 14–24. Generalized approach to SPM. An ideal probe system. Tip modulation, detection, sample stimulation, and environment might be changed to characterize the system for each specific case.

available, but there is a long way to go before combined electrical, optical, and magnetic probes become available. The gap between invention/discovery and the availability of these tools through commercial vendors is on the order of 8–10 years. Minimizing this gap would advance numerous fields that rely on highresolution structure/ property characterization.

Acknowledgments. The authors would like to acknowledge financial support from NSF (DMR05-20020, DMR-0425780), DoE (DE-FG02-00 ER45813-A000). Dr. Sergei Kalinin is gratefully acknowledged for extremely helpful and informative discussion.

References

Abplanalp, M., Eng, L.M. and Gunter, P. (1998). Mapping the domain distribution at ferroelectric surfaces by scanning force microscopy. *App. Phys. a— Mat. Sci. Proc.* **66**, S231–S234.

Abplanalp, M., Fousek, J. and Gunter, P. (2001). Higher order ferroic switching induced by scanning force microscopy. *Phys. Rev. Lett.* **86**, 5799–5802.

Alexe, M., Gruverman, A., Harnagea, C., Zakharov, N.D., Pignolet, A., Hesse, D. and Scott, J.F. (1999). Switching properties of self-assembled ferroelectric memory cells. *App. Phys. Lett.* **75**, 1158–1160.

Alexe, M., Harnagea, C., Hesse, D. and Gosele, U. (1999). Patterning and switching of nanosize ferroelectric memory cells. *App. Phys. Lett.* **75**, 1793–1795.

Alvarez, R.A. and Bonnell, D.A. (2005). Unpublished (presented on Fall MRS 2004).

Ashino, M., Sugawara, Y., Morita, S. and Ishikawa, M. (2001). Atomic resolution noncontact atomic force and scanning tunneling microscopy of $TiO_2(110)$-(1×1) and -(1×2): Simultaneous imaging of surface structures and electronic states. *Phys. Rev. Lett.* **86**, 4334–4337.

Bachtold, A., Fuhrer, M.S., Plyasunov, S., Forero, M., Anderson, E.H., Zettl, A. and McEuen, P.L. (2000). Scanned probe microscopy of electronic transport in carbon nanotubes. *Phys. Rev. Lett.* **84**, 6082–6085.

Barrett, R.C. and Quate, C.F. (1991). Charge storage in a nitride-oxide-silicon medium by scanning capacitance microscopy. *J. App. Phys.* **70**, 2725–2733.

Belaidi, S., Girard, P. and Leveque, G. (1997). Electrostatic forces acting on the tip in atomic force microscopy: Modelization and comparison with analytic expressions. *J. App. Phys.* **81**, 1023–1030.

Bennewitz, R., Foster, A.S., Kantorovich, L.N., Bammerlin, M., Loppacher, C., Schar, S., Guggisberg, M., Meyer, E. and Shluger, A.L. (2000). Atomically resolved edges and kinks of NaCl islands on Cu(111): Experiment and theory. *Phys. Rev. B* **62**, 2074–2084.

Bennewitz, R., Pfeiffer, O., Schar, S., Barwich, V., Meyer, E. and Kantorovich, L.N. (2002). Atomic corrugation in nc-AFM of alkali halides. *App. Surface Sci.* **188**, 232–237.

Binnig, G., Rohrer, H., Gerber, C. and Weibel, E. (1983). 7×7 reconstruction on Si(111) resolved in real space. *Phys. Rev. Lett.* **50**, 120–123.

Bonnell, D.A. (2000). *Scanning Probe Microscopy and Spectroscopy: Theory, Techniques and Applications*, 2nd ed. (Willey VCH, New York).

Bottomley, L.A. (1998). Scanning probe microscopy. *Anal. Chem.* **70**, 425R–475R.

Bridger, P.M., Bandic, Z.Z., Piquette, E.C. and McGill, T.C. (1999). Measurement of induced surface charges, contact potentials, and surface states in GaN by electric force microscopy. *App. Phys. Lett.* **74**, 3522–3524.

Chavezpirson, A., Vatel, O., Tanimoto, M., Ando, H., Iwamura, H. and Kanbe, H. (1995). Nanometer-scale imaging of potential profiles in optically-excited N-I-P-I heterostructure using Kelvin probe force microscopy. *App. Phys. Lett.* **67**, 3069–3071.

Chen, X.Q., Yamada, H., Horiuchi, T., Matsushige, K., Watanabe, S., Kawai, M. and Weiss, P.S. (1999). Surface potential of ferroelectric thin films investigated by scanning probe microscopy. *J. Vac. Sci. Tech. B* **17**, 1930–1934.

Cho, Y.S., Kirihara, A. and Saeki, T. (1996). Scanning nonlinear dielectric microscope. *Rev. Sci. Inst.* **67**, 2297–2303.

Coleman, R.V., Xue, Q., Gong, Y. and Price, P.B. (1993). Atomic-force microscope study of etched tracks of low-energy heavy-ions in mica. *Surface Sci.* **297**, 359–370.

Compendex, as determined from COMPENDEX for 2004, 2004.

Cunningham, S., Larkin, I.A. and Davis, J.H. (1998). Noncontact scanning probe microscope potentiometry of surface charge patches: Origin and interpretation of time-dependent signals. *App. Phys. Lett.* **73**, 123–125.

De Wolf, P., Brazel, E. and Erickson, A. (2001). Electrical characterization of semiconductor materials and devices using scanning probe microscopy. *Mat. Sci. Semicond. Proc.* **4**, 71–76.

De Wolf, P., Clarysse, T., Vandervorst, W. and Hellemans, L. (1998). Low weight spreading resistance profiling of ultrashallow dopant profiles. *J. Vac. Sci. Tech. B* **16**, 401–405.

De Wolf, P., Snauwaert, J., Hellemans, L., Clarysse, T., Vandervorst, W., D'Olieslaeger, M. and Quaeyhaegens, D. (1995). Lateral and vertical dopant profiling in semiconductors by atomic force microscopy using conducting tips. *J. Vac. Sci. Tech. a—Vac. Surfaces Films* **13**, 1699–1704.

De Wolf, P., Stephenson, R., Trenkler, T., Clarysse, T., Hantschel, T. and Vandervorst, W. (2000). Status and review of two-dimensional carrier and dopant profiling using scanning probe microscopy. *J. Vac. Sci. Tech. B* **18**, 361–368.

Donolato, C. (1996). Electrostatic tip-sample interaction in immersion force microscopy of semiconductors. *Phys. Rev. B* **54**, 1478–1481.

Durkan, C. and Welland, M.E. (2000). Investigations into local ferroelectric properties by atomic force microscopy. *Ultramicroscopy* **82**, 141–148.

Eguchi, T., Fujikawa, Y., Akiyama, K., An, T., Ono, M., Hashimoto, T., Morikawa, Y., Terakura, K., Sakurai, T., Lagally, M.G. and Hasegawa, Y. (2004). Imaging of all dangling bonds and their potential on the Ge/Si(105) surface by noncontact atomic force microscopy. *Phys. Rev. Lett.* **93**, Art. no. 266102.

Eng, L.M., Bammerlin, M., Loppacher, C., Guggisberg, M., Bennewitz, R., Luthi, R., Meyer, E., Huser, T., Heinzelmann, H. and Guntherodt, H.J. (1999). Ferroelectric domain characterisation and manipulation: A challenge for scanning probe microscopy. *Ferroelectrics* **222**, 411–420.

Eng, L.M., Guntherodt, H.J., Schneider, G.A., Kopke, U. and Saldana, J.M. (1999). Nanoscale reconstruction of surface crystallography from three-dimensional polarization distribution in ferroelectric barium-titanate ceramics. *App. Phys. Lett.* **74**, 233–235.

Fan, Z. and Lu, J.G. (2005). Electrical properties of ZnO nanowire field effect transistors characterized with scanning probes. *App. Phys. Lett.* **86**, 032111.

Franke, K., Huelz, H. and Weihnacht, M. (1998). How to extract spontaneous polarization information from experimental data in electric force microscopy. *Surface Sci.* **415**, 178–182.

Freitag, M., Johnson, A.T., Kalinin, S.V. and Bonnell, D.A. (2002). Role of single defects in electronic transport through carbon nanotube field-effect transistors. *Phys. Rev. Lett.* **89**, Art. no. 216801.

Friedbacher, G. and Fuchs, H. (1999). Classification of scanning probe microscopies (Technical Report). *Pure App. Chem.* **71**, 1337–1357.

Fujihira, M. (1999). Kelvin probe force microscopy of molecular surfaces. *Ann. Rev. Mat. Sci.* **29**, 353–380.

Fukui, K., Namai, Y. and Iwasawa, Y. (2002). Imaging of surface oxygen atoms and their defect structures on $CeO_2(111)$ by noncontact atomic force microscopy. *App. Surface Sci.* **188**, 252–256.

Fukui, K., Onishi, H. and Iwasawa, Y. (1997). Atom-resolved image of the $TiO_2(110)$ surface by noncontact atomic force microscopy. *Phys. Rev. Lett.* **79**, 4202–4205.

Fukui, K., Onishi, H. and Iwasawa, Y. (1997). Imaging of individual formate ions adsorbed on $TiO_2(110)$ surface by non-contact atomic force microscopy. *Chem. Phys. Lett.* **280**, 296–301.

Ganpule, C.S. (2001). Nanoscale phenomena in ferroelectric thin films. Ph.D. thesis. (University of Maryland, College Park).

Ganpule, C.S., Stanishevsky, A., Aggarwal, S., Melngailis, J., Williams, E., Ramesh, R., Joshi, V. and de Araujo, C.P. (1999). Scaling of ferroelectric and piezoelectric properties in $Pt/SrBi_2Ta_2O_9/Pt$ thin films. *App. Phys. Lett.* **75**, 3874–3876.

Gao, C. and Xiang, X.D. (1998). Quantitative microwave near-field microscopy of dielectric properties. *Rev. Sci. Inst.* **69**, 3846–3851.

Garcia, R. and Perez, R. (2002). Dynamic atomic force microscopy methods. *Surface Sci. Rep.* **47**, 197–301.

Gauthier, M. and Tsukada, M. (1999). Theory of noncontact dissipation force microscopy. *Phys. Rev. B* **60**, 11716–11722.

Giessibl, F.J. (1995). Atomic-Resolution of the Silicon (111)-(7 × 7) Surface by atomic-force microscopy. *Science* **267**, 68–71.

Giessibl, F.J. (1997). Forces and frequency shifts in atomic-resolution dynamic-force microscopy. *Phys. Rev. B* **56**, 16010–16015.

Giessibl, F.J. (2003). Advances in atomic force microscopy. *Rev. Mod. Phys.* **75**, 949–983.

Giessibl, F.J. and Bielefeldt, H. (2000). Physical interpretation of frequency-modulation atomic force microscopy. *Phys. Rev. B* **61**, 9968–9971.

Giessibl, F.J. and Reichling, M. (2005). Investigating atomic details of the CaF_2 (111) surface with a qPlus sensor. *Nanotechnology* **16**, S118–S124.

Gruverman, A., Kolosov, O., Hatano, J., Takahashi, K. and Tokumoto, H. (1995). Domain-Structure and Polarization Reversal in Ferroelectrics Studied by Atomic-Force Microscopy. *J. Vac. Sci. Tech. B* **13**, 1095–1099.

Gruverman, A., Rodriguez, B.J., Nemanich, R.J. and Kingon, A.I. (2002). Nanoscale observation of photoinduced domain pinning and investigation of imprint behavior in ferroelectric thin films. *J. App. Phys.* **92**, 2734–2739.

Guggisberg, M., Bammerlin, M., Loppacher, C., Pfeiffer, O., Abdurixit, A., Barwich, V., Bennewitz, R., Baratoff, A., Meyer, E. and Guntherodt, H.J. (2000). Separation of interactions by noncontact force microscopy. *Phys. Rev. B* **61**, 11151–11155.

Gunhold, A., Beuermann, L., Gomann, K., Borchardt, G., Kempter, V., Maus-Friedrichs, W., Piskunov, S., Kotomin, E.A. and Dorfman, S. (2003). Study of the electronic and atomic structure of thermally treated $SrTiO_3$(110) surfaces. *Surface Interface Anal.* **35**, 998–1003.

Guy, I.L. and Zheng, Z. (2001). Piezoelectricity and electrostriction in ferroelectric polymers. *Ferroelectrics* **264**, 1691–1696.

Hantschel, T., Niedermann, P., Trenkler, T. and Vandervorst, W. (2000). Highly conductive diamond probes for scanning spreading resistance microscopy. *App. Phys. Lett.* **76**, 1603–1605.

Harnagea, C. (2001). Local piesoelectric response and domain structures in ferroelectric thin films investigated by voltage modulated force microscopy. Dr. Rer. Nat. thesis. Halle: Martin-Luther-Universitaet Halle Wittenberg.

Harnagea, C., Pignolet, A., Alexe, M. and Hesse, D. (2001). Piezoresponse scanning force microscopy: What quantitative information can we really get out of piezoresponse measurements on ferroelectric thin films. *Integ. Ferroelectrics* **38**, 667–673.

Hembacher, S., Giessibl, F.J. and Mannhart, J. (2004). Force microscopy with light-atom probes. *Science* **305**, 380–383.

Hembacher, S., Giessibl, F.J., Mannhart, J. and Quate, C.F. (2003). From the cover: Revealing the hidden atom in graphite by low-temperature atomic force microscopy. *Proc. Nat. Acad. Sci. USA* **100**, 12539–12542.

Henning, A.K. and Hochwitz, T. (1996). Scanning probe microscopy for 2-D semiconductor dopant profiling and device failure analysis. *Mat. Sci. Eng. B—Solid State Mat. Adv. Tech.* **42**, 88–98.

Hochwitz, T., Henning, A.K., Levey, C., Daghlian, C., Slinkman, J., Never, J., Kaszuba, P., Gluck, R., Wells, R., Pekarik, J. and Finch, R. (1996). Imaging integrated circuit dopant profiles with the force-based scanning Kelvin probe microscope. *J. Vac. Sci. Tech. B* **14**, 440–446.

Hofer, W.A., Foster, A.S. and Shluger, A.L. (2003). Theories of scanning probe microscopes at the atomic scale. *Rev. Mod. Phys.* **75**, 1287–1331.

Hong, S., Woo, J., Shin, H., Jeon, J.U., Pak, Y.E., Colla, E.L., Setter, N., Kim, E. and No, K. (2001). Principle of ferroelectric domain imaging using atomic force microscope. *J. App. Phys.* **89**, 1377–1386.

Hosoi, H., Sueoka, K., Hayakawa, K. and Mukasa, K. (2000). Atomic resolved imaging of cleaved NiO(100) surfaces by NC-AFM. *App. Surface Sci.* **157**, 218–221.

Huang, Y., Williams, C.C. and Wendman, M.A. (1996). Quantitative two-dimensional dopant profiling of abrupt dopant profiles by cross-sectional scanning capacitance microscopy. *J. Vac. Sci. Tech. a—Vac. Surfaces Films* **14**, 1168–1171.

Jacobs, H.O., Leuchtmann, P., Homan, O.J. and Stemmer, A. (1998). Resolution and contrast in Kelvin probe force microscopy. *J. App. Phys.* **84**, 1168–1173.

Kalinin, S.V. (2002). Ph.D. thesis. in *Materials Science and Engineering.* (University of Pennsylvania, Philadelphia).

Kalinin, S.V. and Bonnell, D.A. (1999). Dynamic behavior of domain-related topography and surface potential on the BaTiO$_3$ (100) surface by variable temperature scanning surface potential microscopy. *Z. Metallkunde* **90**, 983–989.

Kalinin, S.V. and Bonnell, D.A. (2001). Local potential and polarization screening on ferroelectric surfaces. *Phys. Rev. B* **63**, 125411.

Kalinin, S.V. and Bonnell, D.A. (2001). Scanning impedance microscopy of electroactive interfaces. *App. Phys. Lett.* **78**, 1306–1308.

Kalinin, S.V. and Bonnell, D.A. (2002). Imaging mechanism of piezoresponse force microscopy of ferroelectric surfaces. *Phys. Rev. B* **65**, 125408.

Kalinin, S.V. and Bonnell, D.A. (2002). Scanning impedance microscopy of an active Schottky barrier diode. *J. App. Phys.* **91**, 832–839.

Kalinin, S.V., Bonnell, D.A., Freitag, M. and Johnson, A.T. (2002). Tip-gating effect in scanning impedance microscopy of nanoelectronic devices. *App. Phys. Lett.* **81**, 5219–5221.

Kalinin, S.V., Gruverman, A. and Bonnell, D.A. (2004). Quantitative analysis of nanoscale switching in SrBi$_2$Ta$_2$O$_9$ thin films by piezoresponse force microscopy. *App. Phys. Lett.* **85**, 795–797.

Kalinin, S.V., Jesse, S., Shin, J., Baddorf, A.P., Guillorn, M.A. and Geohegan, D.B. (2004). Scanning probe microscopy imaging of frequency dependent electrical transport through carbon nanotube networks in polymers. *Nanotechnology* **15**, 907–912.

Kalinin, S.V., Johnson, C.Y. and Bonnell, D.A. (2002). Domain polarity and temperature induced potential inversion on the BaTiO$_3$(100) surface. *J. App. Phys.* **91**, 3816–3823.

Kalinin, S.V., Karapetian, E. and Kachanov, M. (2004). Nanoelectromechanics of piezoresponse force microscopy. *Phys. Rev. B* **70**, Art. no. 266102.

Kantorovich, L.N., Foster, A.S., Shluger, A.L. and Stoneham, A.M. (2000). Role of image forces in non-contact scanning force microscope images of ionic surfaces. *Surface Sci.* **445**, 283–299.

Ke, S.H., Uda, T. and Terakura, K. (2002). First principles studies of tip-sample interaction and STM-AFM image formation on TiO$_2$(110)-1 × 1 and TiO$_2$(110)-1 × 2 surfaces. *Phys. Rev. B* **65**, 125417.

Kholkin, A.L., Shvartsman, V.V., Emelyanov, A.Y., Poyato, R., Calzada, M.L. and Pardo, L. (2003). Stress-induced suppression of piezoelectric properties in PbTiO$_3$: La thin films via scanning force microscopy. *App. Phys. Lett.* **82**, 2127–2129.

Kitamura, S. and Iwatsuki, M. (1995). Observation of 7 × 7 Reconstructed structure on the silicon (111) surface using ultrahigh-vacuum noncontact atomic-force microscopy. *JPN J. App. Phys. Part 2—Lett.* **34**, L145–L148.

Kubo, T. and Nozoye, H. (2001). Surface structure of SrTiO$_3$(100)-(root 5 × root 5)-R26.6 degrees. *Phys. Rev. Lett.* **86**, 1801–1804.

Labardi, M., Polop, C., Likodimos, V., Pardi, L., Allegrini, M., Vasco, E. and Zaldo, C. (2003). Surface deformation and ferroelectric domain switching induced by a force microscope tip on a Lamodified $PbTiO_3$ thin film. *App. Phys. Lett.* **83**, 2028–2030.

Lanyi, S., Torok, J. and Rehurek, P. (1996). Imaging conducting surfaces and dielectric films by a scanning capacitance microscope. *J. Vac. Sci. Tech. B* **14**, 892–896.

Lee, S.C. and Anlage, S.M. (2003). Spatially-resolved nonlinearity measurements of $YBa_2Cu_3O_7$-delta bicrystal grain boundaries. *App. Phys. Lett.* **82**, 1893–1895.

Leng, Y., Williams, C.C., Su, L.C. and Stringfellow, G.B. (1995). Atomic ordering of Gainp studied by Kelvin probe force microscopy. *App. Phys. Lett.* **66**, 1264–1266.

Lim, T.C. (2003). The relationship between Lennard-Jones (12-6) and Morse potential functions. *Z. Naturforsc. Sect. a-a J. Phys. Sci.* **58**, 615–617.

Loppacher, C., Bammerlin, M., Guggisberg, M., Schar, S., Bennewitz, R., Baratoff, A., Meyer, E. and Guntherodt, H.J. (2000). Dynamic force microscopy of copper surfaces: Atomic resolution and distance dependence of tip-sample interaction and tunneling current. *Phys. Rev. B* **62**, 16944–16949.

Lu, J., Delamarche, E., Eng, L., Bennewitz, R., Meyer, E. and Guntherodt, H.J. (1999). Kelvin probe force microscopy on surfaces: Investigation of the surface potential of self-assembled monolayers on gold. *Langmuir* **15**, 8184–8188.

Lu, X.M., Schlaphof, F., Grafstrom, S., Loppacher, C., Eng, L.M., Suchaneck, G. and Gerlach, G. (2002). Scanning force microscopy investigation of the Pb(Zr0.25Ti0.75)O-3/Pt interface. *App. Phys. Lett.* **81**, 3215–3217.

Lu, Y.L., Wei, T., Duewer, F., Lu, Y.Q., Ming, N.B., Schultz, P.G. and Xiang, X.D. (1997). Nondestructive imaging of dielectric-constant profiles and ferroelectric domains with a scanning-tip microwave near-field microscope. *Science* **276**, 2004–2006.

Marchiando, J.F. and Kopanski, J.J. (2002). Regression procedure for determining the dopant profile in semiconductors from scanning capacitance microscopy data. *J. App. Phys.* **92**, 5798–5809.

Matey, J.R. and Blanc, J. (1985). Scanning capacitance microscopy. *J. App. Phys.* **57**, 1437–1444.

Meoded, T., Shikler, R., Fried, N. and Rosenwaks, Y. (1999). Direct measurement of minority carriers diffusion length using Kelvin probe force microscopy. *App. Phys. Lett.* **75**, 2435–2437.

Moller, C., Allen, M., Elings, V., Engel, A. and Muller, D.J. (1999). Tapping-mode atomic force microscopy produces faithful high-resolution images of protein surfaces. *Biophys. J.* **77**, 1150–1158.

Morita, S., Wiesendanger, R. and Meyer, E. (2002). *Non-Contact Atomic Force Microscopy.* (Springer, Berlin).

Nonnenmacher, M., Oboyle, M.P. and Wickramasinghe, H.K. (1991). Kelvin probe force microscopy. *App. Phys. Lett.* **58**, 2921–2923.

O'Hayre, R., Feng, G., Nix, W.D. and Prinz, F.B. (2004). Quantitative impedance measurement using atomic force microscopy. *J. App. Phys.* **96**, 3540–3549.

O'Hayre, R., Lee, M. and Prinz, F.B. (2004). Ionic and electronic impedance imaging using atomic force microscopy. *J. App. Phys.* **95**, 8382–8392.

Oesterhelt, D. and Stoeckenius, W. (1974). Isolation of the cell membrane of *Halobacterium halobium* and its fraction into red and purple membrane. *Methods Enzymol.* **31**, 667–678.

Ohnesorge, F.M. (1999). Towards atomic resolution non-contact dynamic force microscopy in a liquid. *Surface Interface Anal.* **27**, 797–797.

Ohnesorge, F.M. (1999). Towards atomic resolution non-contact dynamic force microscopy in a liquid. *Surface Interface Anal.* **27**, 379–385.

Pang, C.L., Raza, H., Haycock, S.A. and Thornton, G. (2000). Imaging reconstructed TiO_2 surfaces with non-contact atomic force microscopy. *App. Surface Sci.* **157**, 233–238.

Perez, R., Stich, I., Payne, M.C. and Terakura, K. (1998). Surface-tip interactions in noncontact atomic-force microscopy on reactive surfaces: Si(111). *Phys. Rev. B* **58**, 10835–10849.

Rodriguez, B.J., Gruverman, A., Kingon, A.I., Nemanich, R.J. and Cross, J.S. (2004). Three-dimensional high-resolution reconstruction of polarization in ferroelectric capacitors by piezoresponse force microscopy. *J. App. Phys.* **95**, 1958–1962.

Roelofs, A., Bottger, U., Waser, R., Schlaphof, F., Trogisch, S. and Eng, L.M. (2000). Differentiating 180 degrees and 90 degrees switching of ferroelectric domains with three-dimensional piezoresponse force microscopy. *App. Phys. Lett.* **77**, 3444–3446.

Roytburd, A.L., Alpay, S.P., Nagarajan, V., Ganpule, C.S., Aggarwal, S., Williams, E.D. and Ramesh, R. (2000). Measurement of internal stresses via the polarization in epitaxial ferroelectric films. *Phys. Rev. Lett.* **85**, 190–193.

Rugar, D., Budakian, R., Mamin, H.J. and Chui, B.W. (2004). Single spin detection by magnetic resonance force microscopy. *Nature* **430**, 329–332.

San Paulo, A. and Garcia, R. (2001). Tip-surface forces, amplitude, and energy dissipation in amplitude-modulation (tapping mode) force microscopy. *Phys. Rev. B* **64**, 193411.

Sasahara, A., Uetsuka, H. and Onishi, K. (2000). Noncontact-mode atomic force microscopy observation of alpha-Al_2O_3(0001) surface. *JPN J. App. Phys. Part 1* **39**, 3773–3776.

Schwarz, A., Allers, W., Schwarz, U.D. and Wiesendanger, R. (1999). Simultaneous imaging of the In and As sublattice on InAs(110)-(1×1) with dynamic scanning force microscopy. *App. Surface Sci.* **140**, 293–297.

Sekiguchi, S., Fujimoto, M., Kang, M.G., Koizumi, S., Cho, S.B. and Tanaka, J. (1998). Structure analysis of $SrTiO_3$ (111) polar surfaces. *JPN J. App. Phys. Part 1* **37**, 4140–4143.

Seo, Y., Choe, H. and Jhe, W. (2003). Atomic-resolution noncontact atomic force microscopy in air. *App. Phys. Lett.* **83**, 1860–1862.

Shao, R. and Bonnell, D.A. (2004). Scanning probes of nonlinear properties in complex materials. *JPN J. App. Phys. Part 1* **43**, 4471–4476.

Shao, R. and Bonnell, D.A. (2005). Unpublished.

Shao, R., Kalinin, S.V. and Bonnell, D.A. (2003). Local impedance imaging and spectroscopy of polycrystalline ZnO using contact atomic force microscopy. *App. Phys. Lett.* **82**, 1869–1871.

Shluger, A.L. and Rohl, A.L. (1996). A model of the interaction of ionic tips with ionic surfaces for interpretation of scanning force microscope images. *Top. Catalysis* **3**, 221–247.

Shluger, A.L., Livshits, A.I., Foster, A.S. and Catlow, C.R.A. (1999). Models of image contrast in scanning force microscopy on insulators. *J. Phys.-Cond. Matt.* **11**, R295–R322.

Sounilhac, S., Barthel, E. and Creuzet, F. (1999). The electrostatic contribution to the long-range interactions between tungsten and oxide surfaces under ultrahigh vacuum. *App. Surface Sci.* **140**, 411–414.

Steinhauer, D.E., Vlahacos, C.P., Dutta, S.K., Wellstood, F.C. and Anlage, S.M. (1997). Surface resistance imaging with a scanning near-field microwave microscope. *App. Phys. Lett.* **71**, 1736–1738.

Steinhauer, D.E., Vlahacos, C.P., Wellstood, F.C., Anlage, S.M., Canedy, C., Ramesh, R., Stanishevsky, A. and Melngailis, J. (1999). Imaging of microwave permittivity, tunability, and damage recovery in (Ba, Sr)TiO$_3$ thin films. *App. Phys. Lett.* **75**, 3180–3182.

Sugimura, H., Hayashi, K., Saito, N., Nakagiri, N. and Takai, O. (2002). Surface potential microscopy for organized molecular systems. *App. Surface Sci.* **188**, 403–410.

Sushko, P.V., Foster, A.S., Kantorovich, L.N. and Shluger, A.L. (1999). Investigating the effects of silicon tip contamination in noncontact scanning force microscopy (SFM). *App. Surface Sci.* **145**, 608–612.

Suzuki, S., Ohminami, Y., Tsutsumi, T., Shoaib, M.M., Ichikawa, M. and Asakura, K. (2003). The first observation of an atomic scale noncontact AFM image of MoO$_3$(010). *Chem. Lett.* **32**, 1098–1099.

Takayanagi, K., Tanishiro, Y., Takahashi, M. and Takahashi, S. (1985). Structural-analysis of Si(111)7 × 7 by Uhv-transmission electron-diffraction and microscopy. *J. Vac. Sci. Tech. a—Vac. Surfaces Films* **3**, 1502–1506.

Tanimoto, M. and Vatel, O. (1996). Kelvin probe force microscopy for characterization of semiconductor devices and processes. *J. Vac. Sci. Tech. B* **14**, 1547–1551.

Tans, S.J. and Dekker, C. (2000). Molecular transistors—Potential modulations along carbon nanotubes. *Nature* **404**, 834–835.

Todorov, T.N., Hoekstra, J. and Sutton, A.P. (2000). Current-induced forces in atomic-scale conductors. *Philos. Magn. B—Phys. Cond. Matt. Stat. Mech. Electr. Opt. Magn. Properties* **80**, 421–455.

Tombler, T.W., Zhou, C.W., Kong, J. and Dai, H.J. (2000). Gating individual nanotubes and crosses with scanning probes. *App. Phys. Lett.* **76**, 2412–2414.

Tybell, T., Ahn, C.H. and Triscone, J.M. (1999). Ferroelectricity in thin perovskite films. *App. Phys. Lett.* **75**, 856–858.

Uehara, N., Hosoi, H., Sueoka, K. and Mukasa, K. (2004). Non-contact atomic force microscopy observation on GaAs(110) surface with tip-induced relaxation. *JPN J. App. Phys. Part 1* **43**, 4676–4678.

Ueno, K., Inoue, I.H., Akoh, H., Kawasaki, M., Tokura, Y. and Takagi, H. (2003). Field-effect transistor on SrTiO$_3$ with sputtered Al$_2$O$_3$ gate insulator. *App. Phys. Lett.* **83**, 1755–1757.

Urban, J.J., Spanier, J.E., Lian, O.Y., Yun, W.S. and Park, H. (2003). Single-crystalline barium titanate nanowires. *Adv. Mat.* **15**, 423–426.

Weaver, J.M.R. and Abraham, D.W. (1991). High-resolution atomic force microscopy potentiometry. *J. Vac. Sci. Tech. B* **9**, 1559–1561.

Wiesendanger, R. (1994). *Scanning Probe Microscopy and Spectroscopy—Methods and Applications*. (Cambridge University Press, Cambridge, UK).

Xu, Q. and Hsu, J.W.P. (1999). Electrostatic force microscopy studies of surface defects on GaAs/Ge films. *J. App. Phys.* **85**, 2465–2472.

Yang, J. and Kong, F.C.J. (2002). Simulation of interface states effect on the scanning capacitance microscopy measurement of p-n junctions. *App. Phys. Lett.* **81**, 4973–4975.

Yokoyama, K., Ochi, T., Yoshimoto, A., Sugawara, Y. and Morita, S. (2000). Atomic resolution imaging on Si(100)2 × 1 and Si(100)2× 1: H surfaces with noncontact atomic force microscopy. *JPN J. App. Phys. Part 2—Lett.* **39**, L113–L115.

Yun, W.S., Urban, J.J., Gu, Q. and Park, H. (2002). Ferroelectric properties of individual barium titanate nanowires investigated by scanned probe microscopy. *Nano Lett.* **2**, 447–450.

Zaibi, M.A., Lacharme, J.P. and Sebenne, C.A. (1997). Water vapour adsorption on the Si(111)-(7 × 7) surface. *Surface Sci.* **377**, 639–643.

Zhang, Q.M., Pan, W.Y. and Cross, L.E. (1988). Laser interferometer for the study of piezoelectric and electrostrictive strains. *J. App. Phys.* **63**, 2492–2496.

15

Scanning Tunneling Microscopy in Surface Science

Peter Sutter

1 Introduction

After its invention in 1982 by Gerd Binnig and Heinrich Rohrer (Binnig et al., 1982, 1983), who were awarded the 1986 Nobel Price for this discovery, scanning tunneling microscopy (STM) has rapidly become a standard technique for high-resolution imaging of conducting surfaces. As such, it has revolutionized the way surface science is conducted. Having relied almost entirely on diffraction methods for determining surface structures, and electron spectroscopy for measuring surface chemistry and electronic structure, surface scientists immediately embraced the powerful imaging and spectroscopy capabilities of STM. Earlier techniques for high-resolution surface microscopy include field ion microscopy (Muller and Tsong, 1969), low-energy electron microscopy (Bauer, 1998), ultrahigh-vacuum scanning electron microscopy (Venables, 2000), and reflection electron microscopy (Yagi, 1982). Field ion microscopy, the first surface imaging technique to routinely provide atomic resolution on metals over small sample areas, has recently seen a revival in the form of modern atom probes that provide tomographic images of three-dimensional sample volumes (Cerezo et al., 2001), nicely complementing the surface imaging capabilities of STM. The other techniques, none of which achieves atomic resolution, have been developed further for specific applications. As an example, low-energy electron microscopy has become a powerful technique for real-time microscopy of fast surface processes, such as epitaxial growth or surface reactions (Bauer, 1998). STM itself has recently been at the center of yet another scientific revolution, enabling systematic studies on individual structures composed of a small number of atoms or molecules, and on the order of a few nanometers in size.

This chapter attempts to provide an introduction of the basic concepts of STM, together with illustrations of applications of the technique. Over two decades after its invention, the applications of STM for imaging, spectroscopy, and manipulation at the atomic level have clearly become too numerous to allow for a comprehensive review. In view of the wide range of applications and overall maturity of the technique, a historic overview may also be of limited value. Hence, I have tried to identify representative examples in the current literature, which are discussed briefly to illustrate the different uses of STM. The connection to historic developments can in most cases be made easily via literature references. For a more in-depth discussion of basic concepts and application examples, several dedicated monographs on STM and other scanning probe techniques (Chen, 1993; Wiesendanger, 1994; Güntherodt and Wiesendanger, 1992; Bonnell, 2001) represent an invaluable resource.

This chapter is organized as follows. In Section 2 basic principles of STM imaging are introduced, and the imaging methodology as well as practical and instrumentation requirements are discussed. The approach taken in surface imaging by STM is illustrated by the example of silicon surfaces. Section 3 highlights an application of STM that has gained ever-increasing importance: atomic scale spectroscopy. It provides a survey of the spectroscopy capabilities of STM, and introduces a variety of techniques for local spectroscopy and spectroscopic imaging. In addition, pathways toward obtaining chemical and element specificity at the atomic scale—traditionally a weakness of STM—are discussed. The extension of the operating conditions of STM to high and low temperatures has opened up new avenues of investigation. Variable temperature STM of dynamic surface processes as well as atom and molecule manipulation at cryogenic temperatures are the topics of Section 4. Section 5 discusses STM imaging and spectroscopy on subsurface structures, using ballistic electrons to probe buried interfaces or cross-sectional STM on cleavage faces of III–V semiconductors to image embedded nanostructures. The chapter concludes with a brief discussion of STM image simulation techniques in Section 6.

2 Basic Principles of STM Imaging

2.1 Elastic Vacuum Tunneling and Scanning Tunneling Microscopy

STM is based on the vacuum tunneling of electrons between two solids, one of which is a sharp tip and the other a sample. To remove an electron from a solid and bring it into vacuum with zero kinetic energy requires energy equal to the work function $\phi = E_{Vac} - E_F$, i.e., the difference between the vacuum level, E_{vac}, and the Fermi energy of the material, E_F. To move an electron from one solid to another, it has to cross the same vacuum barrier. If the two solids are separated by a microscopic distance of the order of the decay length of electronic states into the vacuum (typically about 1–5 nm), electrons can cross the barrier by quantum mechanical tunneling (Figure 15–1).

If, say, two metals are brought to within tunneling distance, a rapid charge transfer takes place between them until an equilibrium state is established in which their Fermi levels are aligned. Once in equilibrium, no net charge is transferred on average. Consider now a slightly modified situation, in which metal A is biased relative to B by applying

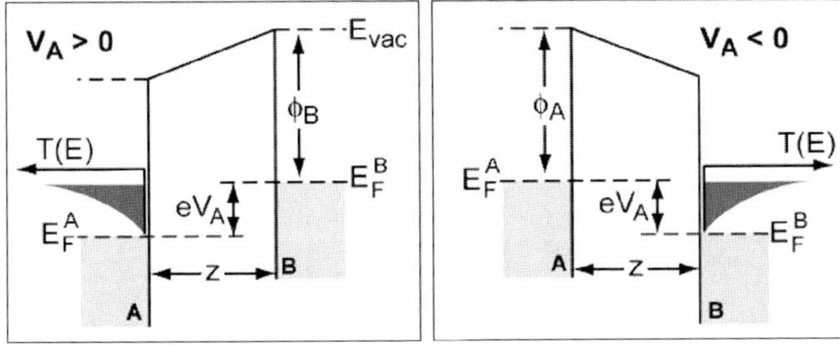

Figure 15–1. Band diagrams (energy vs. distance along the tunneling direction) for vacuum tunneling between two metal electrodes, A and B.

a voltage, V_A. The bias offsets the Fermi energies on either side of the vacuum barrier by eV_A. Under steady-state conditions, this potential difference establishes a net tunneling current, I, between A and B.

In a simple planar tunneling model using the Wentzel–Kramers–Brillouin (WKB) approximation, the tunneling current is given by an integral over the energy range between the Fermi energies on either side of the vacuum gap,

$$I = \int_0^{eV_A} \rho_A(E)\rho_B(E - eV_A)T(E,eV_A)\,dE \tag{1}$$

where ρ_A and ρ_B denote the local density of states at energy E at the surface of A and B, respectively, and $T(E, eV_A)$ is the tunneling transmission probability for electrons with energy E and applied bias V_A:

$$T(E,eV_A) = \exp\left(-\frac{2z\sqrt{2m}}{\hbar}\right)\sqrt{\frac{\phi_A + \phi_B}{2} + \frac{eV_A}{2} - E} \tag{2}$$

Here, $\phi_{A,B}$ are the work functions of the two solids, z is their separation, and m is the mass of the electron. Evaluation of T for positive ($V_A > 0$) and negative bias ($V_A < 0$) shows that the transmission probability is highest for electrons at the Fermi energy of the material that is negatively biased, and falls off exponentially for lower energies down to a lower cutoff at the Fermi energy of the positively biased electrode. This general observation has important consequences in tunneling spectroscopy and spectroscopic imaging, as discussed in Section 3.

The tunneling current depends strongly on the separation, z, between the two solids. To illustrate this fact, we simplify Eq. (1) by assuming constant densities of states, independent of energy. In the limit of low bias voltage, $V_A/\phi_{A,B} \ll 1$, the tunneling current is then given by

$$I = \rho_A\rho_B V_A \exp\left(-2\sqrt{m(\phi_A + \phi_B)}/\hbar^2 z\right) \tag{3}$$

i.e., it depends exponentially on the separation. It is from this strong z dependence of the tunneling current that STM derives its exquisite height resolution, typically of the order of 0.1 pm, or below 1/100 monolayer for most solids.

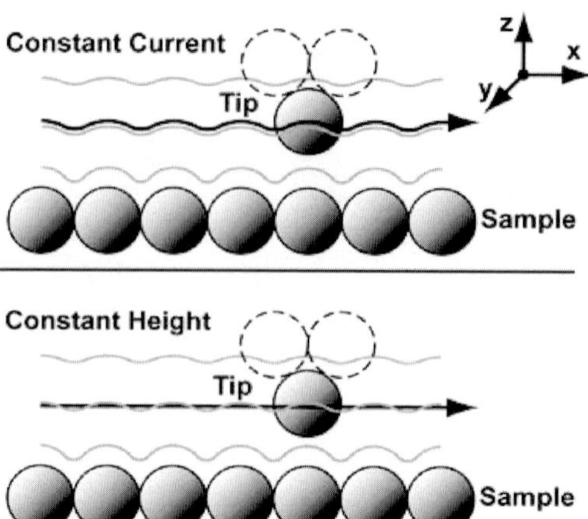

Figure 15–2. Schematic tip trajectory for STM imaging in constant current and constant height mode.

In actual STM imaging, an atomically sharp tip is used to obtain laterally confined tunneling at a well-defined position of a sample. A piezoelectric positioning element controls the in-plane (x, y) position and height (z) of the tip with picometer resolution. While a constant bias voltage is applied between sample and tip, the tip is scanned in a line-by-line fashion to continuously change the tunneling position and build up an image of a chosen area of interest on the sample. The strong dependence of the tunneling current on the tip–sample separation can be used for two basic modes of STM imaging: constant-current and constant-height imaging (Figure 15–2).

In constant-current imaging, the tunneling current is measured at each pixel of the scan and is compared with a chosen current set point. Deviations between the measured tunneling current and the set point are corrected by applying an appropriate voltage to the z-piezo, thus adjusting the tip–sample separation. This feedback mechanism maintains a constant tunneling current during the scan, while the trajectory $z(x, y)$ followed by the tip is used to generate a map of the sample surface or, more accurately, a map of a particular charge density contour above the surface.

In constant-height mode, the tip height is not modified during the scan. The tunneling current is again measured at each image pixel, but is now used directly as a representation of the sample surface via a current map, $I(x, y)$. The current maps represent a cut through charge density contours in the plane in which the tip is scanned above the sample. Obviously, since the tip–sample separation is uncontrolled during the scan, constant-height imaging is limited to samples with low corrugation and/or small scan sizes. A more practical implementation of constant-height STM uses a slow feedback system to adjust the tip–sample separation on time scales that are long compared to the residence time at each pixel, thus compensating for sample surface topography or sample tilt. As the current comparison and tip z correction are eliminated and only the tunneling current is measured, data acquisition can be faster than in constant-current imaging. The constant-height mode is thus preferred for fast STM image acquisition.

2.2 Inelastic Vacuum Tunneling

The above considerations assumed elastic tunneling between two materials, e.g., a sharp tip and the sample. In this case, the energy of a tunneling electron is conserved in the transit through the vacuum barrier. Only inside the opposite electrode the electron thermalizes to the Fermi energy by phonon emission. Tunneling electrons can, however, undergo inelastic scattering during the tunneling process. Vibrational modes of molecules adsorbed on the sample surface, for example, can be excited by inelastic scattering of a small fraction of the tunneling electrons (Figure 15–3).

The onset of inelastic tunneling at characteristic energies can then be detected and used as a fingerprint to identify individual molecular bonds. Such STM-based single molecule vibrational spectroscopy will be discussed in Section 3. In addition, controlled amounts of energy can be deposited locally into individual adsorbates, important for the manipulation of adsorbed atoms or molecules, as discussed in Section 4.

2.3 Practical Requirements: Tips, Samples, and Operating Environment

Obtaining a highly localized tunneling contact requires very sharp probe tips, ideally terminated by a single atom or a small cluster. D-band metals (e.g., W) or alloys (PtIr) are generally thought to be superior for obtaining high spatial resolution due to the strongly directional nature of the d-wave function. However, atomic resolution has also been obtained with Au tips, i.e., tip materials with primarily s-electrons at the Fermi level.

Electrochemically etched tungsten wires are often used for high-resolution STM in vacuum, while less reactive PtIr or Au tips are preferred for imaging under ambient conditions. Sharp W tips are produced by electrochemical etching in ~2 M KOH or NaOH solution, using a stainless steel or Pt foil as a counter electrode. dc or ac etching at typical bias voltages of 5–10 V can be used. Preferential etching at the meniscus of the solution leads to a tapering of the immersed wire, causing one end to eventually break off. To prevent further etching that would blunt the tip apex, an electronic circuit detects the break-off

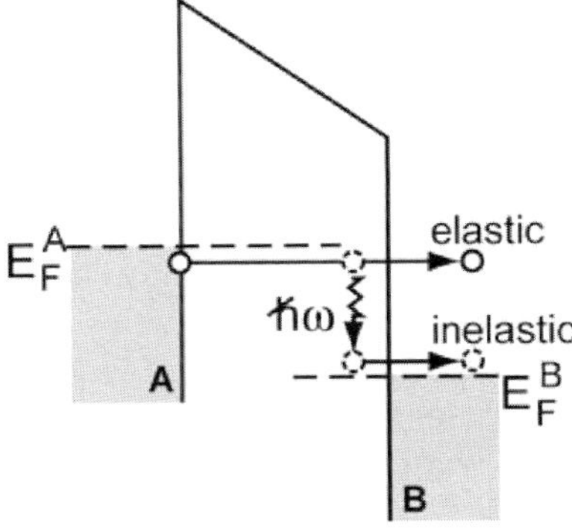

Figure 15–3. Band diagrams for elastic and inelastic electron tunneling.

current and switches the bias to zero within a few microseconds. Etched tips are rinsed in deionized water to remove traces of the etch solution, and loaded into ultrahigh vacuum (UHV). Contaminants and surface oxide are desorbed from the tip by a heat treatment in UHV prior to mounting the tip into the STM. Electron beam bombardment of the tip apex provides an efficient local cleaning of the thin part of the tip, and the applied electric field in this procedure may lead to an additional tip sharpening (Kuk and Silverman, 1986). Alternative tip cleaning and modification techniques while in tunneling contact involve voltage pulses of the order of 4–10 V while scanning the sample, or tip forming in the field emission regime (i.e., at tip–sample large separation) with applied bias voltages between 10 and 100 V and currents of several nA. Ideally, the atomic configuration of the tip would be characterized by field ion microscopy (FIM) prior to its use in the STM. Attempts have been made to combine FIM with STM. However, in practice such efforts are of limited benefit since tip changes occur frequently during scanning, rendering a complete characterization of the tip impractical.

To allow the controlled biasing of tip and sample, and to prevent charging, conventional direct current STM is limited to metal or semiconductor samples. Alternating current STM (Kochanski, 1989) has been demonstrated on insulators and organic layers, but is not widely used. Clean and ordered crystalline surfaces with low surface roughness are preferred for most studies, in particular for atomic-resolution imaging. Reactive semiconductor or metal surfaces are best prepared and imaged in UHV, but metal surfaces are also prepared and imaged routinely in electrolytes (Gewirth and Niece, 1997). In vacuum, sample preparation often involves a thermal cleaning step to remove contaminants. Flash cleaning to 1200°C for a brief period of time (few seconds) while keeping the vacuum in the low 10^{-9} torr range removes the native oxide and provides atomically clean Si samples with a low density of SiC contaminants. Metal surfaces (e.g., Cu, Al, Pt, Au), and some semiconductors (e.g., Ge) are prepared in a two-step procedure involving repeated cycles of Ar^+ ion sputtering and annealing. In refractory metals such as Ru, C contamination is removed by many (several hundred) cycles of oxygen adsorption near room temperature and flashing to 1500°C. *In situ* cleavage can be a very effective method for preparing sample surfaces that cannot be sputtered or heated to high temperatures. Samples best prepared by cleavage include III–V compound semiconductors (GaAs, InAs), as well as high-T_C superconductors ($Yba_2Cu_3O_{7-x}$, $BiSr_3Cu_2O_{8+x}$). *In situ* cleavage is also the preparation method of choice for cross-sectional STM (see Section 5), commonly performed on (110) cleavage planes of III–V semiconductors. Bulk insulating metal oxides, most prominent among them TiO_2, can be imaged successfully by STM following Ar^+ ion sputtering and annealing, which generates oxygen vacancies and renders the surface region sufficiently conductive for stable STM imaging. Au and highly oriented pyrolitic graphite (HOPG) are among the few materials on which atomic resolution is obtained under ambient conditions. Well-ordered, (111) oriented surfaces of Au single crystals and thin films on mica can be prepared by flame annealing in air (Robinson et al., 1992). Preparation of layered materials, notably HOPG, is particularly simple. It involves the removal of bundles of graphene layers by adhesive tape from the crystal. Since the exposed graphene sheet has no dangling bonds and is very inert, such samples can be imaged with atomic resolution in air for extended time periods.

For many materials, in particular metals and alloys, electrochemical STM is a powerful alternative to imaging in UHV. In this environment, almost the entire portion of the tip that is immersed in the solution must be coated by an insulating layer (e.g., wax or glass) to minimize the faradaic (i.e., ionic) current, which is typically much larger than the tunneling current and would otherwise dominate the measured signal. Only a small fraction near the apex of the tip remains uncoated to allow for tunneling between the tip and sample.

2.4 STM Instrumentation

In tunneling contact the probe tip is only a distance of the order of 1 nm away from the sample. As the tunneling current is exponentially sensitive on the tip–sample separation, any mechanical vibrations of the tip relative to the sample have to be minimized. To this end, a two-fold strategy is commonly pursued. The microscope head itself is built as compact and stiff as possible by closely integrating sample and tip on a rigid platform. This results in a high resonance frequency for oscillations of the tip relative to the sample. In addition, the entire microscope is isolated mechanically from outside vibration sources, e.g., by suspending it on soft extension springs or long bungy cords. The overall goal is to achieve a maximal mismatch between the mechanical modes of a soft suspension system and the high resonance frequency of the stiff microscope head. In addition, efficient vibration damping is required, which can be achieved by magnetically induced eddy currents. The springs used in many suspension systems have their own mechanical modes that can be damped by polymer strips woven into their coils. Direct coupling of acoustic noise into the microscope, finally, is a problem that is best solved by operating the STM in vacuum (Figure 15–4).

A few additional components are part of any practical tunneling microscope. A coarse approach mechanism has to be in place, which moves the tip from an initial distance of several millimeters into tunneling range without making physical contact with the sample. Although this has been achieved by mechanical approach mechanisms (Smith and Binnig, 1986), piezoelectric driven stick-slip motors are now preferred since they allow the entire coarse approach to be performed automatically under computer control. For the positioning and scanning of the tip during STM imaging, a segmented piezoelectric tube scanner is used (Binnig and Smith, 1986). By applying high voltages to the individual segments, the scanner is independently actuated in the (x, y) plane parallel to the sample surface and along the perpendicular z axis with little crosstalk. The tunneling current in the pA to nA range is converted into a voltage by a sensitive, low-noise amplifier

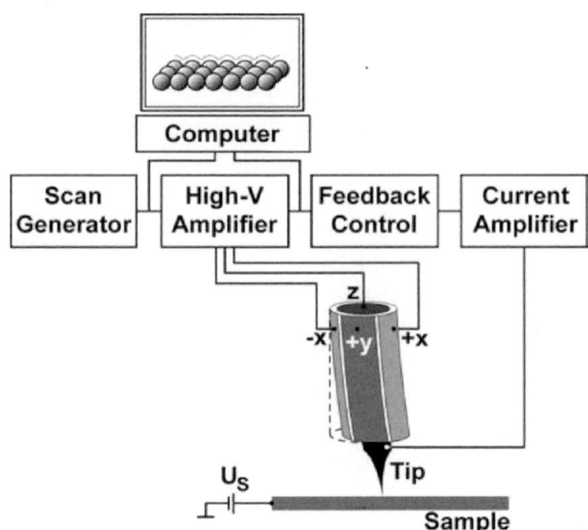

Figure 15–4. Schematic illustration of the various components of an STM system, including piezoelectric tube scanner for tip positioning, and control electronics (current amplifier, feedback, scan generator, high-voltage amplifier).

with adjustable transconductance of 10^7–10^{10} V/A, which provides the input signal for the tunneling current feedback electronics. Modern STM controllers typically use digital feedback and scan modules implemented by specific software programs running on fast digital signal processor (DSP) chips. The measured analog tunneling current signal is digitized using an analog-to-digital (A/D) converter, and is compared with the current set point by the feedback program on the DSP. Necessary corrections in the tip height z are computed, and are converted into analog voltages by digital-to-analog (D/A) converters. These voltages may be amplified further by a high-voltage amplifier stage to levels suitable to drive the tube scanner. The scan signal, i.e., independent voltage ramps driving the piezo scanner along the in-plane x and y directions, is also generated digitally on a DSP, followed by conversion and amplification steps analogous to those of the z signal. A program running on a host computer and communicating with the DSP, finally, acquires, displays, and stores the measured tip trajectory $z(x, y)$ (in constant-current imaging) or position-dependent tunneling current $I(x, y)$ (in constant-height mode).

2.5 STM Imaging: Application Examples

High-resolution imaging of surface structure is probably the most widespread application of STM. As a direct imaging technique, STM has added significantly to the diffraction techniques used traditionally to determine the structure of clean and adsorbate covered surfaces. Examples of typical applications include the determination of the structure of reconstructed surfaces of clean semiconductors, for instance vicinal Si surfaces (Erwin et al., 1996) and oxides, such as TiO_2 (Diebold, 2003), and of adsorbate-induced reconstructions on metals (e.g., O/Ru(0001); Meinel et al., 1997); the mapping of the evolution of surface morphology during epitaxial growth (e.g., metal epitaxy; Chambliss et al, 1995), etching (Boland and Weaver, 1998), and energetic particle bombardment (e.g., electrons on Si; Nakayama and Weaver, 1999); and the imaging and structural identification of non-periodic structures such as surface defects, steps, as well as surface-supported solid (e.g., silicide nanowires on Si; Chen et al., 2000) and molecular nanostructures [e.g., molecular rotors on Cu(100); Gimzewski et al., 1998]. Although electrically conducting samples are a prerequisite for stable tunneling, high-resolution imaging is feasible on ultrathin insulating films (for a recent review, see Schintke and Schneider, 2004) and self-assembled organic monolayers [e.g., alkane-thiol monolayers on Au(111); Cygan et al., 1998) supported by metal substrates.

In view of the large number of materials systems to which STM has been applied, a comprehensive survey would be beyond the scope of this chapter. Instead, we discuss in some detail recent contributions of STM imaging to one specific material: silicon.

2.5.1 STM Imaging—Silicon Surfaces

Due to the important role of Si in electronics, extensive STM studies on the structure of clean Si surfaces with different orientation have contributed to making Si one of the best-characterized materials system. Important early milestones in the characterization of clean Si surfaces include the determination of surface bonding and reconstructions on Si(111) (Binnig et al., 1983) and on Si(001) (Tromp et al., 1985) used as a substrate in microelectronics. STM imaging served to establish the thermodynamics of terraces and steps (Swartzentruber et al., 1990; Men et al., 1988) on these surfaces, to study electromigration and step-bunching (Yang et al., 1996), and to survey the stable facets involved in the equilibrium shape of Si crystals (Gai et al., 1998).

Figure 15–5 shows an example of constant current STM images on Si(111), a complex reconstructed surface whose geometric (Binnig and

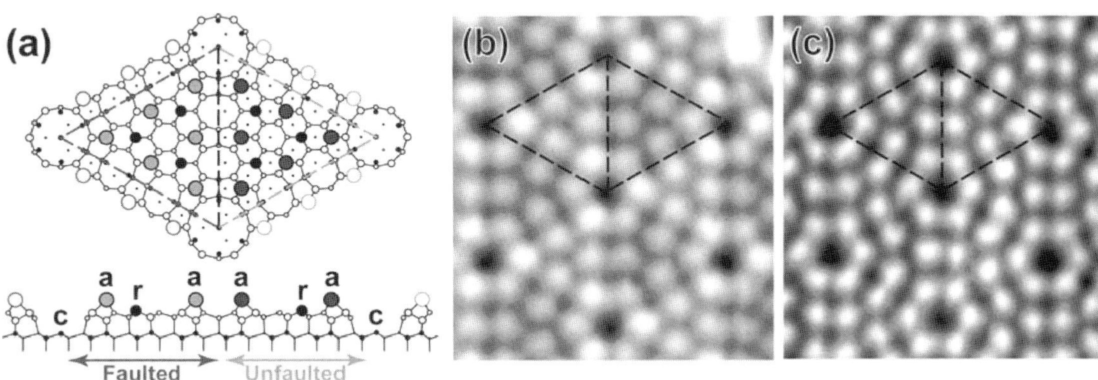

Figure 15–5. (a) Ball-and-stick model of the Si(111)–(7 × 7) surface. Top: top view; bottom: side view. a, adatom; r, rest atom; c, corner hole. (b) Filled-state constant-current STM on Si(111)–(7 × 7); $V = -1.0\,V$; $I = 0.4\,nA$. (c) Empty-state constant-current STM; $V = +1.2\,V$; $I = 0.4\,nA$.

Rohrer, 1983) and electronic (Hamers et al., 1986) structure was solved in pioneering STM experiments, and which continues to play an important role in STM technique development. Figure 15–5a shows a structure model of the (7×7) reconstruction (Takayanagi et al., 1985). Figure 15–5b, obtained at negative sample bias ($V = -1.0\,\text{V}$), i.e., with electrons tunneling from occupied sample states to unoccupied tip states, shows the corrugation associated with filled states of the sample. Conversely, Figure 15–5c, obtained at positive sample bias ($V = +1.2\,\text{V}$), maps the empty states of the sample. Both images show atomic resolution, clearly resolving the twelve adatoms ("a") per unit cell. In addition characteristic deep "corner holes" ("c") bounding the diagonals of the 4.6 nm × 2.9 nm rhombohedral unit cell are imaged.

While all adatoms are mapped uniformly in the empty state image, the filled state scan shows one-half of the unit cell somewhat higher than the other, an effect on the charge density due to the different stacking sequence of atomic layers in the two halves of the unit cell. A comparison of the images obtained at opposite bias polarity suggests that in addition to surface topography, the electronic structure of the sample surface adds substantially to the contrast observed in STM imaging. This is implicit also in Eq. (1) via the dependence of the tunneling current on the local densities of states of both tip and sample at the tunneling contact. Note also that the rest atoms ("r"), a second near-surface species with dangling bonds protruding into the vacuum, are imaged neither at positive nor at negative sample bias. A detailed discussion of bias-dependent imaging and other tunneling spectroscopy methods used to assess the local electronic structure of a sample surface is given in Section 3.

Apart from the structure of clean Si surfaces, Si-based surface chemistry has been studied widely by STM. Initial studies provided a knowledge base for technological processes, such as reactive ion etching, doping, and chemical vapor deposition (CVD). STM was used to probe interactions of Si with halogens (Boland, 1993) and with small molecules involved in surface passivation (H_2, H; Laracuente and Whitman, 2001), doping (PH_3; Wang et al., 1994a), and CVD growth (Si_2H_6; Wang et al., 1994b). More recent research directions include the integration of molecular electronic elements with Si. Under this perspective, the covalent bonding of a wide variety of organic molecules on Si has been studied (for a review, see Wolkow, 1999), including small molecules such as ethylene (Mayne et al., 1993) and acetylene (Li et al., 1997), simple alkenes (e.g., propylene; Lopinski et al., 1998) and polyenes (e.g., 1,3-cyclohexadiene; Hovis and Hamers, 1997), pentacene (Kasaya et al, 1998), and benzene (Borovski et al., 1998). On clean Si(001), adsorption is from the gas phase in UHV. H-passivated Si can also be modified by *ex situ* wet-chemical techniques, and reintroduced into UHV for STM imaging.

Traditionally, an important aspect of the surface science of semiconductors has been epitaxial growth. STM observations at initial growth stages, i.e., at coverages of fractions of a monolayer (ML) up to several ML, can provide direct insight into fundamental processes such as adatom diffusion, incorporation into steps, and nucleation of monolayer islands (for a review, see Zhang and Lagally, 1998). Observations of growth and equilibrium structures allow identifying kinetic and thermodynamic factors affecting the growth process, as well as the role of defects, of elastic strain, etc.

STM studies of Si epitaxy are commonly performed in either of two modes. Conceptually preferred is the dynamic observation, via time-lapse STM "movies," of the growth surface with the sample held in the microscope at high temperature and in the presence of a deposition flux. This difficult technique will be discussed in more detail in Section 4. As an alternative, STM imaging can be performed at room temperature on samples that have been quenched at key stages in the growth process. If measures are taken to verify that the quenched-in morphology is indeed representative of that at high temperature during growth, this "quench-and-look" technique offers the advantage of higher spatial resolution and larger scan sizes over high-temperature STM.

Early STM experiments focused on the initial stages of Si homoepitaxy and Ge/Si heteroepitaxy, mostly on the technologically important (111) or (001) surfaces. Key results include the identification of regimes of step-flow growth and nucleation, determination of the activation energy and atomistic pathway of surface diffusion (Mo et al., 1991), explanation of growth and equilibrium shapes of two-dimensional (2D) islands (Mo et al., 1989), and exploration of the modification of the growth process by surfactants (Horn-von Hoegen, 1994). Figure 15–6 illustrates the identification of the initial stage of island formation in

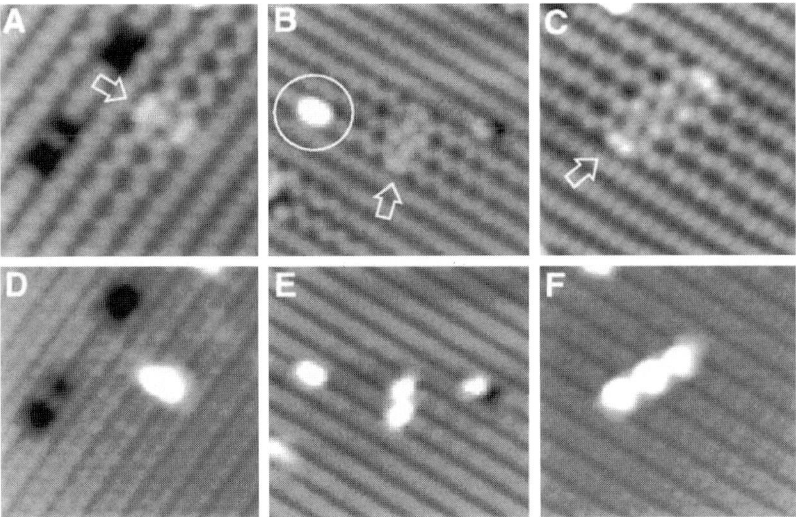

Figure 15–6. STM images of 0.01 monolayers Ge on Si(001). The long diagonal bands are substrate dimer rows. Filled-state images (A–C; $V = -2\,V; I = 0.2\,nA$) and corresponding empty-state images (D–F; $V = +2\,V; I = 0.2\,nA$), showing rows of symmetric (bean-shaped) and asymmetric (buckled) substrate dimers, and Ge adatom-induced chain-like structures (marked by arrows). The chain-like paired adatom structures are seen as metastable precursors to the formation of monolayer islands in Ge/Si homoepitaxy (Reprinted with permission from Qin and Lagally, © 1997 AAAS).

Ge/Si(001) epitaxy (Qin and Lagally, 1997). Classic theories of nucleation in thin film growth are based on the notion of a "critical nucleus," defined as the structural entity for which the addition of one more atom will for the first time reduce the free energy. Employing very low Ge coverages and high-resolution dual bias STM imaging to uncover the link between monomer adsorption and initial 2D growth islands, this study shows that island formation may not involve a "critical nucleus," but instead may proceed via a family of metastable structures of varying size, with significant consequences on many growth models that are based on nucleation and require the size of a critical nucleus as input.

More recently, interest has concentrated on strained layer heteroepitaxial growth, e.g., of $Si_{1-x}Ge_x$ alloys on Si(001), as an elegant way of producing large arrays of nanostructures. Si and Ge are miscible over the entire composition range, and the lattice mismatch in the $Si_{1-x}Ge_x$/Si(001) system can be tuned between 0 and 4% by alloying. Over a wide range of compositions, $Si_{1-x}Ge_x$ alloys initially wet the Si substrate, and at higher coverage develop coherent (i.e., dislocation free) faceted three-dimensional (3D) islands that lower the free energy by relaxing part of the lattice mismatch strain. Potentially useful in electronics or optoelectronics, these faceted nanostructures have shown a strikingly complex array of growth phenomena, which make them interesting for fundamental growth studies.

Constant current STM images with atomic resolution typically encompass fields of view of few tens of nanometers. Combining a drift-stable microscope with state-of-the-art control electronics, substantially larger atom-resolved images have become feasible. The ability to obtain image sizes in excess of $1\,\mu m^2$ with high resolution is potentially powerful for imaging growth processes, since it provides access to all relevant length scales as well as image statistics far superior to that of conventional small scans. Figure 15–7 illustrates this capability with an STM image of 1.5 monolayers Ge on Si(111) with a field of view of $0.75\,\mu m$ and $0.05\,nm$ pixel size. The large scan provides an excellent overview of the step and island structures resulting from monolayer Ge deposition. At progressive zoom into the image, individual terraces, and, ultimately, single surface defects and reconstruction domains (here coexisting 7×7 and 5×5 domains) can be examined.

Figure 15–8 gives an example in which the superior statistics resulting from large atom-resolved images was used successfully to pinpoint the mechanism of periodic surface roughening in Ge/Si(001) growth (Sutter et al., 2003a). Via a sequence of "quench-and-look" experiments at different Ge coverages, short-range interactions between surface steps and vacancy lines, periodic arrays of linear chains of dimer vacancies, are shown to cause a highly correlated surface roughness in the form of anisotropic 2D islands with progressively higher aspect ratio (a–c). Finally, images obtained at a critical Ge coverage of four atomic layers (d) show the transition from 2D to 3D growth by formation of the first faceted islands on the rough wetting layer, and identify the atomic-scale pathway of the 2D–3D transition.

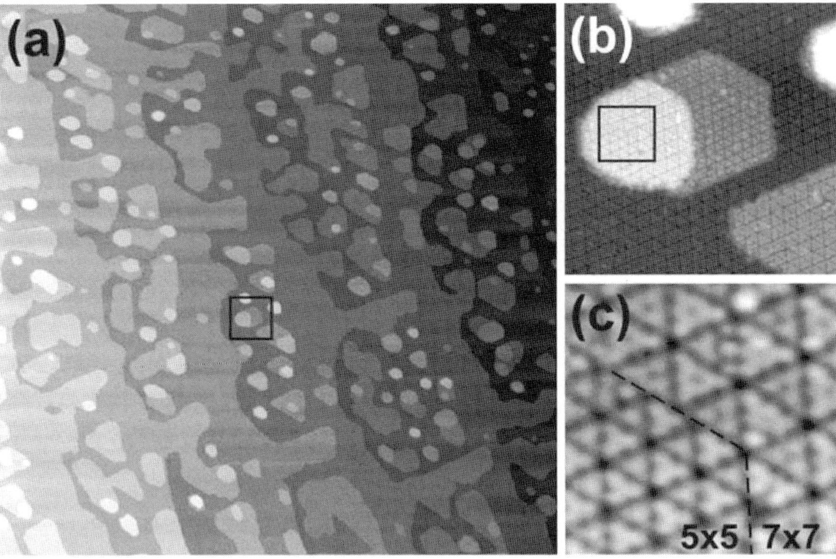

Figure 15–7. Large-area atom-resolved STM. (a) STM image with 0.75-μm field of view and 0.05-nm pixel size, showing the surface morphology of Si(111) after deposition of 1.5 ML Ge at 550°C. An array of parallel steps, running diagonally from the upper left to lower right, coexists with islands nucleated during Ge deposition. (b) Zoomed-in view (50 nm) of the area marked by a square in (a), showing one of the islands. (c) Further zoom-in (10 nm), showing the coexistence of (7×7) and (5×5) reconstructed domains in the area marked in (b).

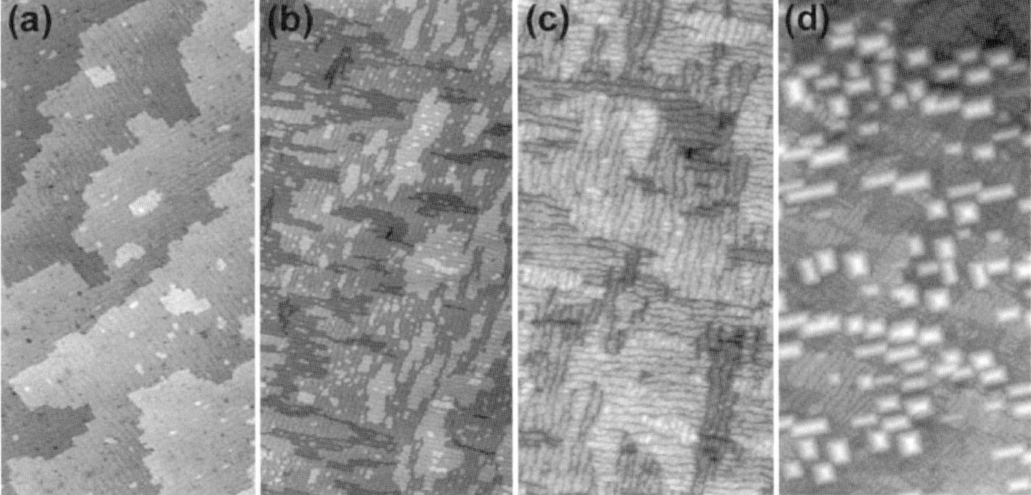

Figure 15–8. Transition from 2D to 3D morphology in the growth of Ge on Si(001). Large-area atom-resolved STM images (250 nm × 125 nm sections are shown here) illustrate the surface morphology at (a) 1.5 ML, (b) 2.3 ML, (c) 3.5 ML, and (d) 4.0 ML Ge coverage. The transition to 3D growth is preceded by the formation of highly correlated, anisotropic surface roughness. With the superior statistics of large-area high-resolution images, the repulsive interaction between surface steps and defects (dimer vacancy lines) was identified as the origin of the ordering of this surface roughness. (Reprinted with permission from Sutter et al., © 2003a by the American Physical Society.)

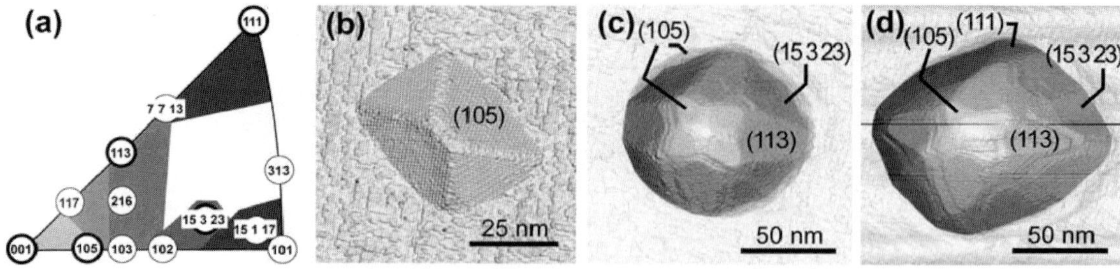

Figure 15–9. Shapes of coherent (i.e., dislocation-free) 3D islands induced by lattice mismatch strain during Si$_{1-x}$Ge$_x$/Si(001) heteroepitaxy. (a) Unit stereographic triangle for Si, showing the major and minor stable facets (Gai et al., 1998). (b) STM top view, showing a shallow (105) faceted "hut." (c) Multifaceted "dome" shape, terminated by (105), (113), and (15 3 23) facets. (d) "Barn"-shaped 3D island, observed for low-misfit Si$_{1-x}$Ge$_x$ alloys on Si(001), transformed from a "dome" by introduction of additional (111) facets (Reprinted with permission from Sutter et al., © 2004 American Institute of Physics.).

Initial 3D islands with shallow facets, such as those shown in Figure 15–8d, evolve into more complex shapes involving steeper facets, which give rise to more efficient strain relaxation (Figure 15–9). For Si$_{1-x}$Ge$_x$ epitaxy on Si(001), the facets bounding the islands identified by STM are to a large extent major stable facets of Si and Ge (Figure 15–9a). At early stages of 3D growth, shallow (105) faceted "huts" are invariably observed (Figure 15–9b; Mo et al., 1990). These islands then undergo a shape transformation to become multifaceted "domes" (Figure 15–9c; Medeiros-Ribeiro et al., 1998). At later stages in the shape evolution, elastic strain relaxation increasingly competes with plastic relaxation via dislocations. For Ge/Si(001) dislocations are introduced before the "dome" shape can transform further by introducing additional steeper facets. By reducing the Ge concentration of Si$_{1-x}$Ge$_x$ alloys, and thus lowering the lattice mismatch strain, the coherent shape evolution has been extended beyond the "dome" shape to the end-point in the stereographic triangle, where steep (111) facets are introduced (Sutter et al., 2004).

3 Tunneling Spectroscopy

Equation (1) gives an expression for the tunneling current I in the WKB approximation for planar tunneling. In this approximation the tunneling current is an integral, over the energy range of width $|eV|$ in which tunneling can occur, of a product of the densities of states (DOS) of tip [$\rho_T(E)$] and sample [$\rho_S(E)$] multiplied by the transmission probability, $T(E, eV)$. While this simple approximation cannot address the high spatial resolution of STM—a more elaborate theory capable of addressing this aspect will be discussed in Section 6—it demonstrates that the tunneling current is affected by the electronic structure of both the tip and the sample. In a large number of applications it would be desirable to merely map the atomic scale "topography" of a sample. The sensitivity of the tunneling current to the DOS can then be a disadvantage, since it often prevents a simple interpretation of a tunneling image as a "topographic" map. For many metals with small variations in the DOS near the Fermi energy such a simple interpretation is often possible, i.e., height maxima in a constant current STM image are correlated with atomic positions. However, for semiconductors with strongly varying DOS near E_F, the simple picture frequently breaks down.

An example is the (001) surface of Si. Geometrically, this surface consists of Si dimers, in which each atom is bonded to two atoms in the second layer and to its dimer partner. Constant-current STM images obtained on Si(001) at positive sample bias indeed show two maxima in each dimer (Figure 15–10B), which could be interpreted as the "atomic positions." At negative sample bias, however, only a single broader bean-shaped structure is observed (Figure 15–10A), which extends across each dimer. In fact, the tunneling current in empty-state STM (positive sample bias) reflects the charge distribution in a π^* antibonding state just above the Fermi energy, whose wave function has a node at the center of the dimer. Even in this case in which the STM image shows two maxima per dimer, these maxima do not reflect the atomic positions in the dimer. Worse yet, in filled-state STM (negative sample bias) a π bonding state is imaged whose charge density peaks at the center of the dimer bond, thus giving rise to a single "topographic" maximum at that position.

The sensitivity to the electronic structure of tip and sample, however, is not merely a complication in STM imaging. It is also one of the major strengths of the technique, since it allows measuring and mapping surface electronic structure with very high spatial resolution well below 1 nm. While most other techniques, such as ultraviolet or x-ray photoelectron spectroscopy or Auger electron spectroscopy, average over macroscopic sample areas, recently developed spectromicroscopy techniques, such as synchrotron photoelectron emission microscopy, provide spectroscopic imaging with 10 nm spatial resolution (see Chapter 8, this volume). However, measurements of surface electronic

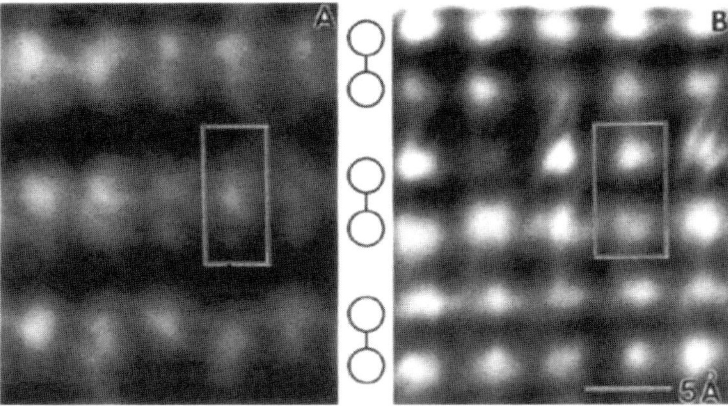

Figure 15–10. Constant-current STM images of Si(001) at negative [$V = -1.6$ V (A)] and positive [$V = +1.6$ V, (B)]. The (2 × 1) unit cell is outlined, and the locations of the dimers are shown in the center. (Reprinted with permission from the Annual Review of Physical Chemistry, Volume 40 © 1989 by Annual Reviews www.annualreviews.org.)

structure "atom-by-atom," summarized under the term scanning tunneling spectroscopy, are feasible only by STM.

3.1 Bias-Dependent Imaging

The simplest scenario for gathering electronic structure information involves the successive measurement of constant-current STM images at different sample bias. As discussed in Section 1, the tunneling transmission probability is generally highest at the Fermi energy of the negatively biased electrode, and falls off exponentially at lower energies. For positive sample bias, i.e., empty-state STM, elastic tunneling occurs mainly from states near the tip Fermi energy into unoccupied sample states. In this polarity, varying the bias voltage probes different surface states of the sample, particularly if the electronic structure of the tip varies slowly in the energy range defined by the bias voltages used. A series of constant-current STM images at different bias can thus be used to map the spatial distribution of unoccupied surface states of the sample.

For negative voltages applied to the sample, i.e., filled-state STM, the situation is not as fortunate. The tunneling current is now dominated by states near the Fermi level of the sample. As a result, the occupied sample state with highest energy is invariably imaged preferentially (Becker et al., 1989). Even if lower-lying surface states exist, as indicated schematically in Figure 15–11, constant-current STM imaging is rather insensitive to those states.

As a possible solution to this problem in bias-dependent filled-state imaging it has been suggested that metal probe tips with smoothly varying density of states near the Fermi energy could be replaced by a semiconductor tip with a strongly modulated DOS. Band structure effects have long been recognized in the interaction of low-energy electrons with metal surfaces (Bauer, 1994). As an example, the high normal incidence reflectivity on W(110) at low electron energies is due to a (110) projected band gap extending over 5 eV, i.e., a lack of extended bulk states in the energy range of this gap. For W(100), such a gap does

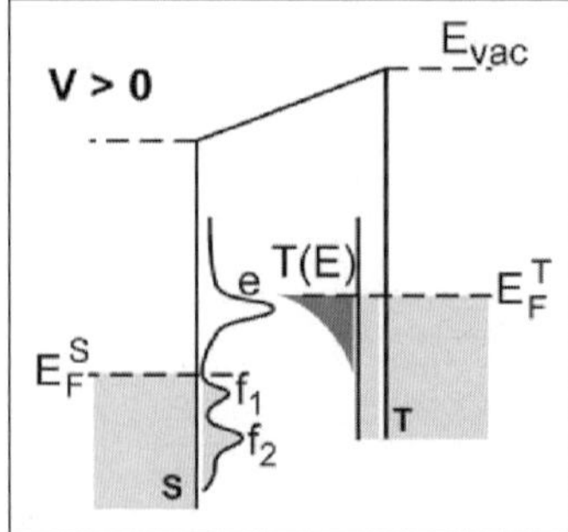

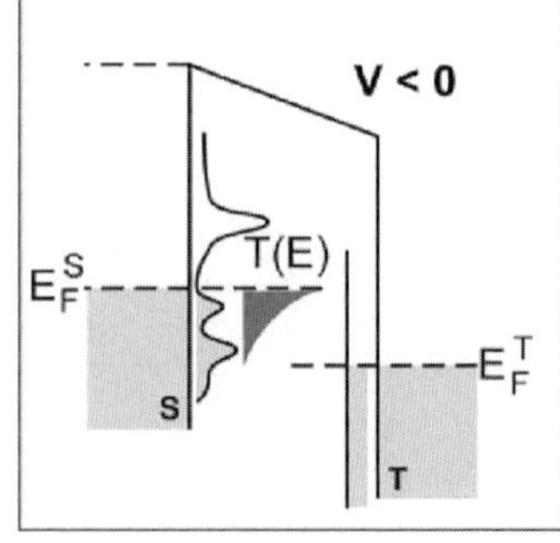

Figure 15–11. Band diagrams illustrating the bias-dependent tunneling between tip (T) and sample (S). Sample bias **V > 0**: empty-state STM. **V < 0**: filled-state STM. *T(E)* shows schematically the transmission probability as a function of energy for states participating in the tunneling process.

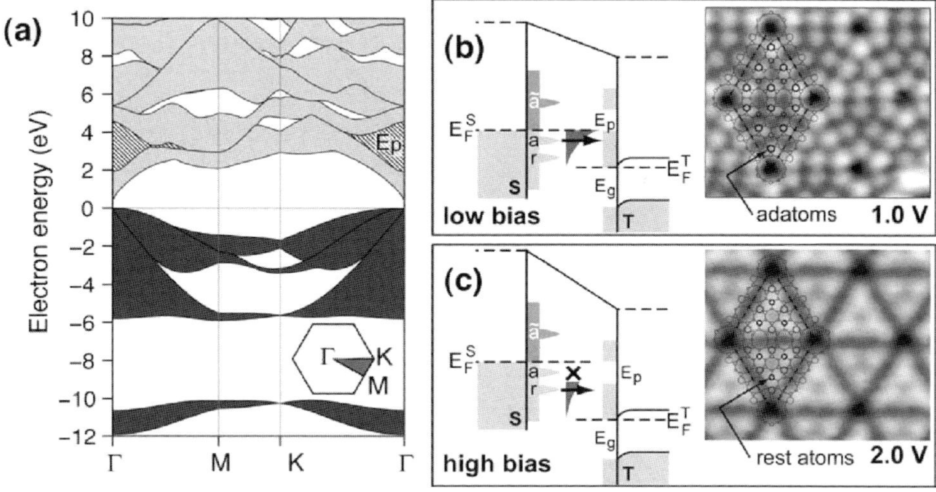

Figure 15–12. Energy-filtered STM. (a) Bulk band structure of monocrystalline InAs probe tips, projected along the (111) zone axis parallel to the tunneling direction. E_p denotes a projected gap in the bulk conduction band. (b) Band diagram and STM image for low-bias filled state STM with InAs tips, demonstrating preferential imaging of the adatom state on Si(111)-(7 × 7). (c) High-bias filled state InAs-tip STM on Si(111)-(7 × 7): alignment of the adatom state with the projected tip gap reduces its contribution to the tunneling current, giving rise to preferential imaging of the energetically lower rest-atom state (Reprinted with permission from Sutter et al., © 2003b by the American Physical Society).

not exist. Low-energy electrons can propagate into extended states, and the reflectivity in the same energy range is low. Similarly, it can be expected that a lack of final states due to projected band gaps in the STM tip material can significantly affect the tunneling probability in the energy range spanned by such gaps. This modulation of the transmission probability could be used for efficient energy filtering of the tunneling current in constant-current STM, specifically in filled-state imaging.

Calculations by density functional theory (DFT) show a number of gaps in the [111]-projected band structure of III–V compound semiconductors such as InAs. III–V compound semiconductors cleave easily at (110) planes, and cleavage corners bounded by {110} and (100) planes are sufficiently sharp for atomic-resolution STM (Nunes and Amer, 1993). In addition, InAs(110) has the advantage of an unpinned surface with minimal band bending and slight electron accumulation. The feasibility of energy-filtered STM has been demonstrated by using a cleaved InAs tip to selectively image different occupied surface states on Si(111)-(7 × 7) (Sutter et al., 2003b). In these experiments, summarized in Figure 15–12, a [111]-projected gap in the InAs conduction band, centered at Γ and about 2.6 eV wide, is used to modulate the tunneling current in filled-state constant current STM. The "adatom" state ("a") on Si(111)-(7 × 7), which lies only 0.35 eV below the Fermi energy and is imaged by STM using a metal tip, dominates the tunneling current at low tip–sample bias. At higher bias, this state aligns with a projected gap in the InAs conduction bands and its tunneling probability is reduced sharply. Under these conditions the "rest atom"

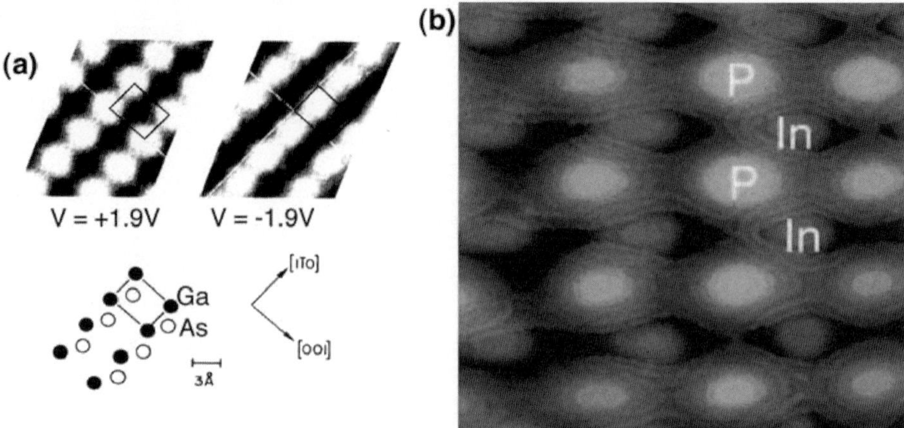

Figure 15–13. (a) Bias-dependent STM on GaAs(110): selective imaging of Ga and As sublattices at positive and negative sample bias, respectively (Reprinted with Permission from Feenstra et al., ©1987 by the Armeucion Physics Socuity). (b) Compound STM image of the InP(110) surface, assembled from separate positive and negative bias scans (Reprinted from Ebert et al., ©1992 with permission from Elsevier). (See color plate.)

state ("r"), a surface state 0.8 eV below E_F that is not accessible in conventional STM imaging, contributes a majority of the tunneling current and can be imaged selectively. Importantly, these results suggest that robust bulk band structure effects, which are independent of the atomic structure of the tip, can be used to modulate the transmission probability in vacuum tunneling. Once developed into a routine imaging technique, energy-filtered STM has the potential to provide rich spectroscopic contrast within the relatively simple framework of bias-dependent constant current STM.

A classic example of a scenario in which conventional bias-dependent imaging provides directly interpretable spectroscopic information is imaging of cation and anion sites on (110) cleavage planes of compound semiconductors such as GaAs (Feenstra et al., 1987) or InP (Ebert et al., 1992) (Figure 15–13). For (110) surfaces of compound semiconductors, such as GaAs, the occupied state density (imaged at negative sample voltage) is localized on the anions (As, P), while the unoccupied state density (positive sample voltage) is localized on the cations (Ga, In). As a result, bias-dependent STM imaging can be used to selectively image the cation and anion sublattices on these surfaces.

3.2 Local I-V Measurements and Current Imaging Tunneling Spectroscopy

In many instances, a complete electronic structure map is not required, but one would like to determine, at one or several positions on a sample, the surface density of states in an interval of a few eV around the Fermi energy. Such local measurements can, for example, serve to determine if a specific surface reconstruction or a small cluster is semiconducting or metallic, to measure the band gap at the surface of semiconductor samples, and to determine for a given location the energies of surface states.

Local current–voltage [$I(V)$] spectra serve this purpose. The basic information content of a tunneling $I(V)$ spectrum can be demonstrated by differentiating Eq. (1) [$V > 0$]:

$$\frac{dI}{dV} = e\rho_s(eV)\rho_T(0)T(eV,eV) + \int_0^{eV} [e\rho_s(E)\rho_T(E-eV)\partial T(E,eV)/$$
$$d(eV) - e\rho_s(E)\rho'_T(E-eV)\tilde{T}(E,eV)]dE \qquad (4)$$

For a given energy eV, the tunneling conductance dI/dV reflects the density of states of the sample at that energy, $\rho_S(eV)$.

Feenstra et al. (1987) noted difficulties with conductance spectroscopy, resulting from the fact that the expression for dI/dV diverges exponentially both in voltage (V) and separation (z). These divergences can be eliminated by normalizing dI/dV by the conductance of the tunneling junction, $[dI/dV]/[I/V] \approx d(\ln I)/d(\ln V)$. In this form, spectra obtained at different tip–sample separation z can be compared directly.

Local spectroscopy is quite simple to perform. The STM tip is positioned and stabilized over a chosen point on the sample. The feedback loop is switched off while the sample voltage is varied over the desired values, and a local measurement of tunneling current as a function of bias voltage, $I(V)$, is performed. As an alternative, the tunneling conductance, $dI(V)/dV$, can be measured by ramping the bias voltage with a small ac voltage added, and measuring the ac component of the tunneling current at the modulation frequency of the ac part by lock-in techniques. After the $I(V)$ or $dI(V)/dV$ spectrum is recorded, the tip feedback is reactivated, and STM imaging can resume.

Figure 15–14a shows tunneling conductance spectra obtained on p- and n-type GaAs(110) (Feenstra et al., 1987), which provide a measure

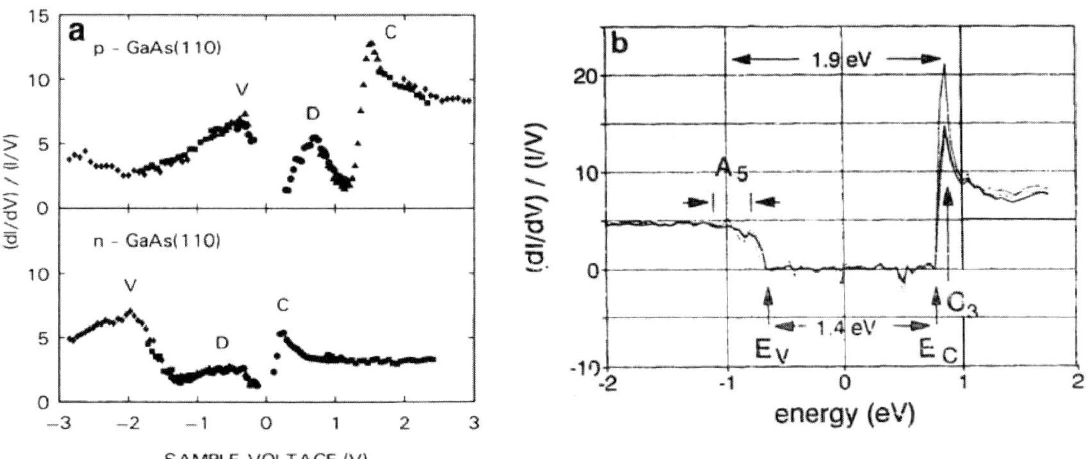

Figure 15–14. (a) Local normalized conductance spectra obtained on n- and p-doped GaAs(110), showing three peaks (V, D, C) originating from surface states (Feenstra et al., 1987). (b) Normalized conductance spectra obtained on InP(110), illustrating the valence and conduction band edges delimiting the bulk bandgap (1.4 eV), and the A_5 and C_3 surface states defining the surface gap (1.9 eV) (Reprinted from Ebert et al., © 1992 with permission from Elsevier).

of the surface state density. The spectra show three distinct peaks associated with surface states. Peak C is related to a state in the GaAs conduction band, localized on the Ga atoms, while peak V stems from a valence state localized on the As anions. Peak D is associated with tunneling of dopant-induced carriers within the bulk band gap. The separation between the leading edges of the C and V peaks is close to the bulk band gap of GaAs ($E_g = 1.43\,eV$), i.e., in this system tunneling spectroscopy can be used to determine the band gap. Figure 15–14b) shows tunneling conductance spectra on the (110) cleavage surface of another III–V compound, InP. Also for InP, the band gap (~1.4 eV) is evident from sharp onsets of significant state density at the valence and conduction band edges. However, for this system it was argued that a wider surface band gap (~1.9 eV) is indirect, delimited by a surface state C_3 at the edge of the surface Brillouin zone and a broadened state A5 at the zone center.

A natural, albeit experimentally much more complex extension of local $I(V)$ spectroscopy is the measurement of tunneling spectra, as discussed above, at each image pixel of a constant-current STM scan. This measuring scheme is called current-imaging tunneling spectroscopy (CITS). Since complete $I(V)$ spectra are obtained at each image point, the corresponding data sets can provide a full range of spectroscopic information. As an example, current maps $I(x, y)$ at fixed tip–sample bias can be produced. Instead, the voltage dependent tunneling conductance $dI(V)/dV$ can be calculated numerically and mapped as a function of sample position. If the chosen voltage corresponds to the energy of a surface state of the sample, conductance maps will provide a direct image of the spatial distribution of that state. Due to the complexity of the data acquisition, the experimental requirements for CITS are quite stringent. Since complete $I(V)$ curves are measured at each image pixel, requiring the stopping of a scan and deactivation of the feedback loop for a fraction of a second per spectrum, a very high stability of the tunneling gap and low lateral drift of the tip relative to the sample are of key importance. These conditions are more easily fulfilled at cryogenic temperatures, where a z-stability of the order of 1 pm and lateral drift velocities of the order of few Å/hour are possible. Low temperatures also reduce the thermal broadening of the tunneling transmission coefficient $T(E)$, and thus narrow the linewidth of features in the tunneling spectra.

A classic example of the application of CITS, the measurement of the electronic structure of Si(111)-(7 × 7), is shown in Figure 15–15 (Hamers et al., 1986). Shown are a constant current image of the surface at +2 V sample bias (a), as well as current images acquired during the same scan at bias voltages of +1.45 V and –1.45 V, showing spatial maps of unoccupied (+)/occupied (–) sample states at energies up to 1.45 eV above/below the Fermi energy. CITS difference images, a precursor to modern dI/dV maps, discussed below, calculated by numerically subtracting current images at energies bracketing those of specific surface states, allowed the mapping of the adatom state, dangling bond, and backbond state on this complex reconstructed surface.

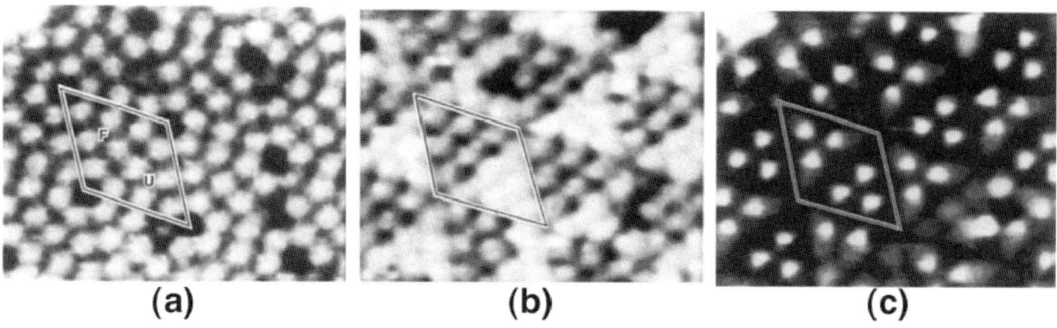

Figure 15–15. Current-imaging tunneling spectroscopy on Si(111)-(7 × 7) (Hames et al., 1986). (a) Constant-current STM image (V = +2 V), with one (7 × 7) unit cell outlined (F, faulted; U, unfaulted half). (b and c) Current images obtained at +1.45 V and –1.45 V, respectively. Note the strong similarity between the rest-atom contrast in the current image (c), and the contrast obtained in rest-atom imaging by energy-filtered constant-current STM (Figure 15–12).

More recently, this complex data acquisition mode has been used, for example, to map modifications in the electronic structure of individual carbon nanotubes arising with the supramolecular assembly of C_{60} molecules inside the hollow tube to form nanotube "peapods" (Hornbaker et al., 2002). dI/dV spectra obtained along the symmetry axis of C nanotubes show distinct peaks in the density of unoccupied states with a spatial periodicity corresponding to the average spacing of embedded C_{60} molecules observed by transmission electron microscopy, suggesting that the insertion of C_{60} causes a significant modulation of the electronic structure. A control measurement was constructed by using the STM tip to push the C_{60} away, and remeasuring dI/dV spectra along the same nanotube section without embedded C_{60}. The empty tube obtained in this way shows small and smooth variations in the density of states along the nanotube axis. Qualitatively, the electronic structure of the peapods was explained by the single wall C-nanotube acting as a conduit that enhances the coupling between C_{60} molecules nested inside it, and the Bragg scattering of nanotube states due to the periodic arrangement of the C_{60}. Calculations of the peapod electronic structure confirm the formation of a hybrid electronic band, which derives its character from both the nanotube states and the C_{60} molecular orbitals (Figure 15–16).

3.3 Differential Conductance (dI/dV) Mapping

For a given energy eV, the tunneling conductance dI/dV reflects the density of states of the sample at that energy, $\rho_S(eV)$. As a result, it can be expected that high-resolution maps of tunneling conductance can provide detailed information on spatial variations in sample electronic structure. For example, by plotting dI/dV at an energy corresponding to a localized state at the sample surface, the spatial distribution of that sample state can be mapped.

In a typical experiment a constant current image, at a specific bias voltage and tunneling current used to stabilize the tunneling gap, is

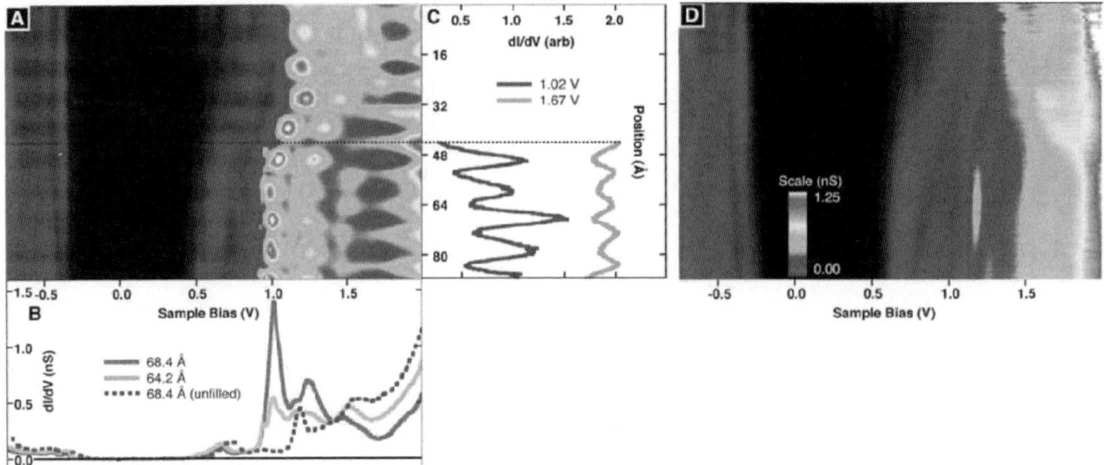

Figure 15–16. $I(V)$ tunneling spectroscopy on C_{60}/C-nanotube "peapods" (Reprinted with permission from Hornbaker et al., © 2002 AAAS). (A) Map of an array of full dI/dV spectra along the axis of a C-nanotube "peapod." Sample bias voltage is plotted on the horizontal axis and displacement along the tube on the vertical axis. (B) Representative dI/dV spectra at selected positions along the tube. Large conductance peaks are found at positions of embedded C_{60} molecules. (C) Variation of tunneling conductance along the tube axis. (D) Reference spectroscopic map on an empty C-nanotube section without embedded C_{60}, in which no strong modulation of the tunneling conductance is observed. (See color plate.)

measured simultaneously with dI/dV maps at one or several voltages/energies. To obtain the spectroscopic data, the feedback loop is turned off at each point of the STM scan, and the selected bias voltages are applied. The tunneling conductance is measured directly using a modulation technique. At the end of the measurement, the feedback is reactivated, and the tip is moved to the next image pixel.

If a small modulation voltage $dV(t) = a\cos\omega t$ with amplitude a and frequency ω is added to the dc sample bias V_0, the resulting tunneling current

$$I(t) = I_0 + dI(t) = I_0 + \frac{dI}{dV}\Big|_{V_0} dV(t) + \cdots$$
$$= I_0 + \frac{dI}{dV}\Big|_{V_0} a\cos\omega t + \cdots \tag{5}$$

has a small ac component. Its amplitude at the modulation frequency ω, which can be measured using a lock-in amplifier, is proportional to $dI/dV\big|_{V_0}$. Via measurements at different dc bias V_0, a set of tunneling conductance values at different energies can be determined at each pixel during an STM scan, and can later be displayed as maps of the sample density of states at those energies. Although quite simple conceptually, the long dwell times at each image pixel in dI/dV mapping pose stringent demands on microscope stability, which are best met at cryogenic temperatures. Under these conditions, thermal broadening of features in the density of states is also minimized.

dI/dV mapping has been applied in a wide variety of measurements of electronic states at surfaces and in nanostructures, such as atomic chains (Crain and Pierce, 2005) or single molecules (Repp et al., 2005). A particularly widespread application is the imaging of quasiparticle interference. In an ideal metal, the Landau-quasiparticle eigenstates are Bloch wavefunctions with specific wavevectors $\mathbf{k}$ and energies ε. Their dispersion cannot be measured directly by STM. However, in the presence of disorder, e.g., impurities or crystal defects, elastic scattering mixes eigenstates with different $\mathbf{k}$ that are located on the same quasiparticle contour of constant energy in k-space. When states $\mathbf{k}_1$ and $\mathbf{k}_2$ are mixed by scattering, an interference pattern with wavevector $\mathbf{q} = \mathbf{k}_2 - \mathbf{k}_1$ appears in the norm of the quasiparticle wavefunction. The interference leads to modulations in the local density of states with wavelength $\lambda = 2\pi/|\mathbf{q}|$, which can be observed as modulations in the differential tunneling conductance. Via differential conductance mapping, interference of electronic eigenstates has been probed in metals (e.g., Cu(111), Crommie et al., 1993; Au(111), Hasegawa and Avouris, 1993) and semiconductors (e.g., InAs(110), Wittneven et al., 1998). When density of states modulations are detected by STM, certain contours of constant energy in $\mathbf{k}$-space can be reconstructed by analyzing the Fourier transform of the real-space density of states map. Such Fourier transform spectroscopy simultaneously yields real-space and momentum-space information on wavefunctions, scattering processes, and quasiparticle dispersion.

Elastic scattering of Bogoliubov quasiparticles in superconductors can also give rise to conductance modulations (Hoffman et al., 2002a), suggesting that STM could be used as a local probe to study electron correlation in superconductors. Quasiparticle inteference has indeed been demonstrated for cuprate high-temperature superconductors, such as $Bi_2Sr_2CaCu_2O_{8+\delta}$ (Hoffman et al., 2002a; Vershinin et al., 2004) and $Ca_{2-x}Na_xCuO_2Cl_2$ (Hanaguri et al., 2004). On $Bi_2Sr_2CaCu_2O_{8+\delta}$ crystals cleaved at the BiO plane at cryogenic temperature in UHV, a constant-current image (Figure 15–17A) and differential conductance maps (Figure 15–17B) are obtained simultaneously in a low-temperature

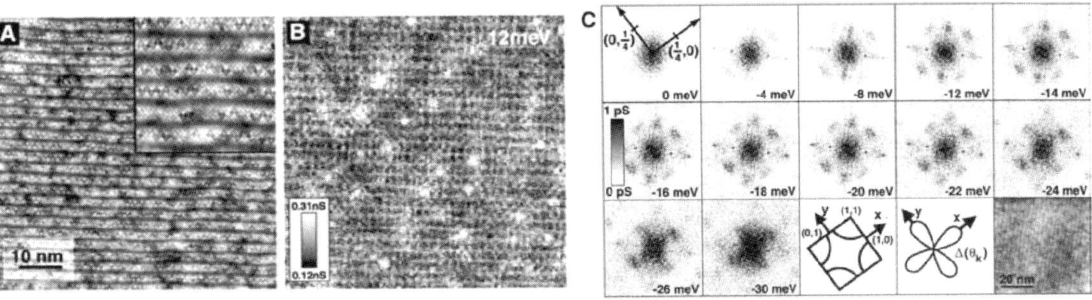

Figure 15–17. Fourier transform spectroscopy on a cleaved $Bi_2Sr_2CaCu_2O_{8+\delta}$ high-temperature superconductor. (A) Constant-current STM image (65 nm, 0.13 nm resolution). (B) Simultaneously acquired differential conductance map (~ local DOS) at an energy of 12 meV below E_F, showing a pronounced modulation of the electronic structure. dI/dV is measured with 2 mV modulation. (C) Fourier transforms of DOS maps similar to (B), for energies up to 30 meV, representing maps of quasiparticle scattering wave vectors in this energy range. (Reprinted with permission from Hoffman et al., © 2002a AAAS.)

STM. The conductance map shows a checkerboard-like modulation of the local density of states. Fourier transforms of dI/dV maps obtained over a range of energies show dominant **q**-vectors associated with quasiparticle scattering. Different spatial patterns and wavelengths are observed at different energies. Applying a magnetic field perpendicular to the sample, the quasiparticle scattering can be modified by the controlled introduction of vortices (Hoffman et al., 2002b). Although the interpretation of these complex data sets is still somewhat controversial (Vershinin et al., 2004), these experiments clearly demonstrate the power of an STM-based approach for studying electronic structure and ordering phenomena in high-temperature superconductors.

A second example in which dI/dV mapping plays a key role is spin-polarized STM. A magnetic probe tip, e.g., a conventional etched W tip coated with a thin Fe film and polarized in a magnetic film perpendicular to the tip axis, is used for tunneling on a magnetic sample. Figure 15–18 illustrates the principle of spin-polarized STM for the example of a ferromagnetic Gd(0001) sample and a hard magnetic Fe tip (Bode et al., 1998). In a weak external magnetic field applied parallel to the sample surface, Gd(0001) has an exchange-split surface state with majority and minority parts at binding energies of –220 meV and +500 meV, respectively (Figure 15–18a). In the experiment, the magnetization direction of the Fe tip remains fixed independent of the direction of the applied magnetic field, while the magnetization of the Gd sample can be switched by reversing the direction of the applied field. Due to a spin valve effect, the tunneling current of the surface state spin component parallel to the tip magnetization is enhanced. If the majority spin is aligned with the tip magnetization (blue curve in Figure 15–18b) the tunneling conductance of the occupied majority

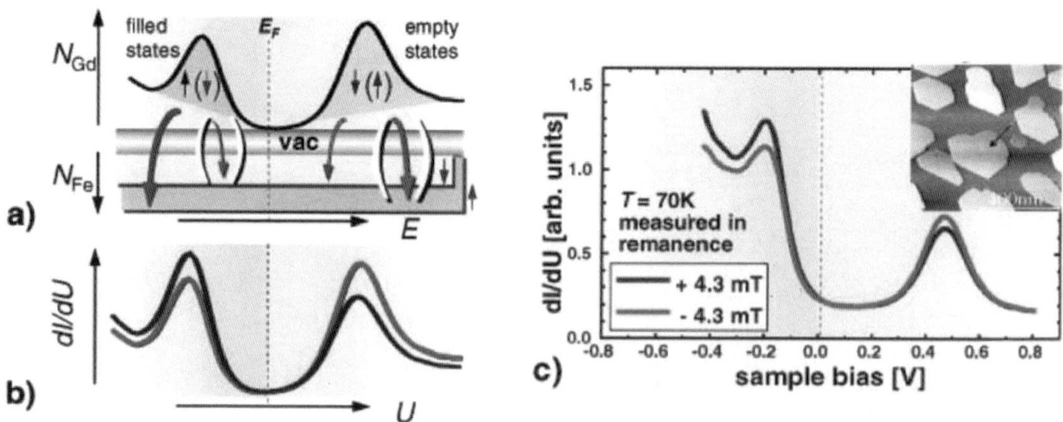

Figure 15–18. Spin-polarized STM on a Gd(0001) sample with an exchange-split surface state and a magnetic Fe tip with constant spin polarization close to E_F. (a) Due to the spin-valve effect the tunneling current of the surface state spin component parallel to the tip magnetization is enhanced. (b) Illustration of the reversal in the dI/dV signal at the surface state peak position upon switching the sample magnetically. (c) Experimental observation of this reversal in tunneling into an isolated Gd island (Reprinted with permission from Bode et al., ©1998 by the American Physical Society). (See color plate.)

state will be high, while that of the minority state is low. After reversal of the applied magnetic field, the minority spin will align with the tip magnetization. The conductance of the minority state thus increases, while that of the majority state is reduced. Difference spectra obtained by subtracting the traces, measured at two polarities of the external field, provide a direct measure of the sample magnetization, and can be used to build maps of magnetic order. Spin polarized STM has been shown to be a very powerful technique for mapping magnetic and antiferrromagnetic order with atomic resolution, as demonstrated by studies on 2D antiferromagnetic ordering in Mn/W(110) (Heinze et al., 2000) and on magnetic hysteresis in Fe/W(110) (Pietzsch et al., 2001).

3.4 Inelastic Tunneling: Vibrational Spectroscopy

A major drawback of conventional STM imaging is its lack of chemical and element specificity, which stems from the fact that only states within few eV of the Fermi energy can be probed. As a result, characteristic core levels, which could provide element specificity, are not accessible. For single molecules, an alternative pathway to chemically specific imaging employs vibrational signatures, which are characteristic of specific molecular bonds.

The use of inelastic electron tunneling spectroscopy (IETS) for vibrational fingerprinting on molecules embedded in planar tunnel junctions dates back to the 1960s (Jaklevic and Lambe, 1966). The technique is based on the detection of a slight increase in tunneling conductance at an electron energy $eV = \hbar\omega$, corresponding to a vibrational mode with frequency ω of an embedded molecule. When the applied bias reaches this threshold, an inelastic channel for tunneling opens up in addition to the elastic tunneling at lower bias. Inelastically scattered electrons make up only a small fraction of the total tunneling current (typically a few percent). Nevertheless, the second derivative of the tunneling current, (d^2I/dV^2), usually shows clear signatures of inelastic tunneling in the form of peaks at the characteristic bias voltages corresponding to different vibrational modes. As in dI/dV maps, d^2I/dV^2 can be measured by lock-in techniques, by adding a small modulation voltage to the tunneling bias V_0. The signal amplitude measured at twice the modulation frequency is proportional to $d^2I/dV^2|_{V_0}$.

Conventional IETS in planar oxide tunnel junctions samples large ensembles of embedded molecules. Soon after the invention of STM, it was suggested that IETS should be feasible in local tunneling between a tip and a sample in STM (Binnig et al., 1985). A first convincing experimental demonstration of inelastic tunneling through few molecules was obtained in a setup of two crossed Au wires with finely adjustable normal force, allowing the trapping and IET spectroscopy at cryogenic temperatures on a small number of hydrocarbons (Gregory, 1990).

IETS on single molecules in a low-temperature STM was finally demonstrated by Ho and co-workers in 1998 (Stipe et al., 1998a). A beautiful experiment on single acetylene molecules adsorbed on Cu(100) not only provides clear evidence for inelastic tunneling at

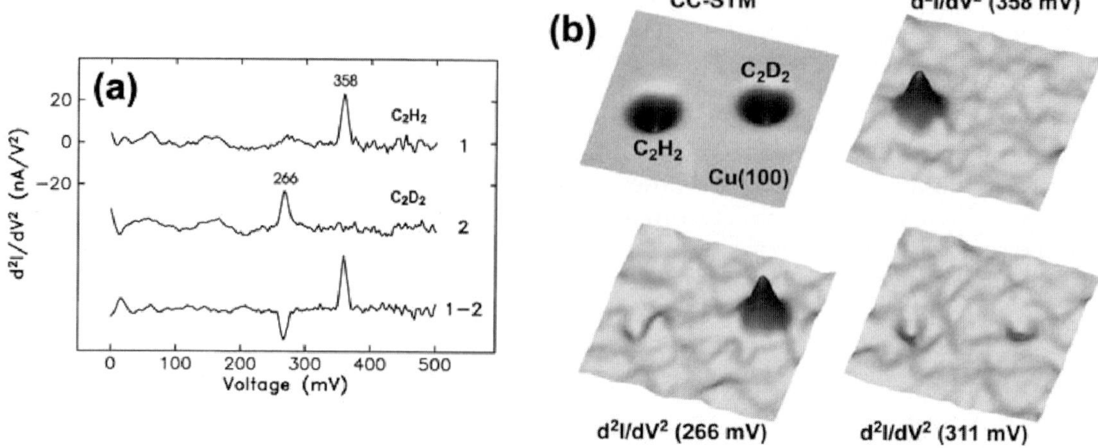

Figure 15–19. Single molecule vibrational spectroscopy and microscopy. (a) d^2I/dV^2 recorded over individual C_2H_2 and C_2D_2 molecules on Cu(100) at 8 K. (1) Signature of inelastic electron tunneling by exciting the C—H stretch mode with an energy of 358 meV. (2) Isotope shifted (266 meV) inelastic tunneling peak of the C—D stretch mode. (b) Constant current (CC, 4.8 nm) STM of adjacent C_2H_2 and C_2D_2 molecules on Cu(100), along with simultaneously acquired d^2I/dV^2 maps obtained at 358 mV and 266 mV, imaging the C—H and C—D bonds selectively. The symmetric round appearance is attributed to rotations of the molecules during the experiment. A d^2I/dV^2 map obtained at intermediate energy (311 mV) shows no vibrational contrast. (Reprinted with permission from Stipe et al., © 1998 AAAS.)

energies close to vibrational frequencies determined in the gas phase, but it also shows the expected isotope shift between measurements on normal and deuterated acetylene. Robust increases in differential conductance of the order of 3–12% are observed for this system.

In addition to inelastic tunneling spectra on individual molecular bonds, as shown in Figure 15–19a, inelastic electron tunneling microscopy with molecular resolution was demonstrated on small molecules, such as acetylene (Figure 15–19b). In constant-current STM imaging, both C_2H_2 and C_2D_2 molecules are imaged as identical depressions, i.e., are indistinguishable. On the other hand, C_2H_2 molecules are selectively imaged in (d^2I/dV^2) maps obtained at the characteristic energy of the C—H stretch mode (358 meV), while C_2D_2 molecules are imaged at the isotope shifted energy (266 meV). In a control experiment at an energy between these two characteristic values, none of the two isotopes gives rise to contrast.

The development of single-molecule vibrational spectroscopy and microscopy provides a powerful tool for performing and viewing chemistry at the ultimate limit of concentration (individual molecules) and spatial resolution (atomic). Local energy input by inelastic tunneling allows the precise manipulation of individual molecules (rotation, translation, see Section 4) and the cleavage and formation of individual molecular bonds, as well as the characterization of the individual and final configurations on the surface.

3.5 Element-Specific Imaging

In contrast to chemical specificity at the level of individual small molecules, which IETS can provide independent of the particular sample structure, no universal approach has been identified to obtain element specific contrast in STM. Various techniques have been suggested to achieve generic element specificity, among them a scanning tunneling atom probe operated by controlled atom transfer to the STM tip and desorption into a time-of-flight chamber (Weierstall and Spence, 1998), and a scanning probe energy loss spectrometer using an STM at high tip–sample bias in combination with an electron spectrometer for local probe electron energy loss spectrometry (Festy and Palmer, 2004). None of the techniques demonstrated to date provided a universal and practical pathway toward element specific imaging at the atomic scale. However, a number of sample–tip combinations have been identified, in which element contrast is indeed achieved on a case-by-case basis. Prime examples are bimetallic metal surfaces, on which different atomic species can be discriminated in constant current STM images.

In a few simple cases, a mere difference in atomic size can give rise to contrast, e.g., for submonolayer coverages of Pb/Cu(111) with a difference in metallic radii of nearly 50 pm (Nagl et al., 1994). In a broader class of metal alloys, electronic effects, i.e., differences in local density of states above different atoms, give rise to element contrast. Examples of systems with substantial density of states differences near the Fermi energy include alloys between transition metals with a partially filled d-shell and noble metals (Au, Ag, Cu). STM images of the (111) surface of AgPd alloys (Wouda et al., 1998) indeed show the Ag atoms with lower apparent height than the transition metal atoms, as expected based on a simple density of states argument. For alloys of transition metals (e.g., PtRh; Wouda et al., 1996), the prediction of the STM contrast is less straightforward, and generally requires *ab initio* calculations of the local density of states. For the PtRh(100) surface shown in Figure 15–20a, a local density of states difference at low tunneling resistance produces a height difference of about 20 pm between Pt and Rh, with Rh imaged at larger apparent height. In some systems, where sufficient differences in local density of states to provide STM contrast do not exist, such differences can be induced by suitable modification of the sample surface. In an important semiconductor system, $Si_{1-x}Ge_x$ alloys, Si and Ge atoms are indistinguishable in conventional constant current STM, which severely hampers the understanding of epitaxial growth since crucial local composition information is lacking. Covering a heterogeneous (111) surface composed of Si- and Ge-rich areas with 1 atomic layer of Bi induces robust differences in apparent height of nearly 0.1 nm between the chemically different regions, thus providing a means for imaging and characterizing lateral heterostructures of monolayer-thick alternating epitaxial Si and Ge stripes (Kawamura et al., 2003).

A third mechanism that can give rise to atomic-scale contrast between different elements is thought to be based on a direct interaction between tip and sample, in what could be described as a "precursor" chemical

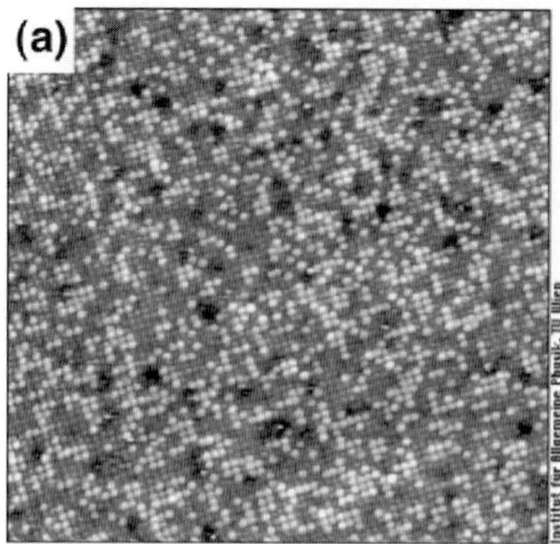
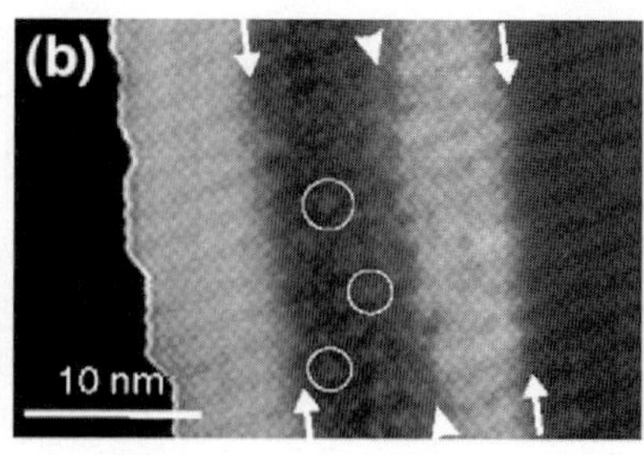

Figure 15–20. Element-specific imaging with atomic resolution by STM. (a) PtRh(100): for a bulk composition of 50% Pt and Rh, there are approximately 31% Rh (bright) and 69% Pt atoms (darker) at the surface. Black spots are caused by residual carbon. Pt and Rh atoms tend to cluster in small groups of the same species (Reprinted with permission from Wouda et al., © 1996 with permission from Elsevier). (b) Atomically and chemically resolved image of Ge (bright)/Si (dark) nanowires formed by sequential deposition of submonolayer amounts of Ge and Si in step flow mode on Bi-terminated Si(111). The initial Si/Ge boundary (right arrows) is nearly atomically sharp, while interdiffusion is observed at the Ge/Si boundary (arrowheads) (Reprinted with permission from Kawamura et al., © 2003 by the American Physical Society).

bond. With the tip–sample distance too large to allow true chemical bonding, either atomic relaxation or the local charge density can change as the tip is moved across different atomic species, which in turn affects the tunneling current and gives rise to apparent height differences in constant current STM. Qualitatively, surface atoms with higher chemical affinity to the tip atom should have higher tunneling conductance and thus will be imaged at larger apparent height. Several metal alloy systems, such as PtNi(111) and PtRh(111), are believed to show element contrast that is based on tip–sample interaction. Clearly, the nature of the tip apex would be of key importance for this type of contrast. This may explain why for some systems element contrast is obtained only after deliberate chemical modification of the STM tip.

4 STM at High and Low Temperatures

STM combines unique capabilities in both high-resolution imaging and spectroscopy at surfaces. Some form of controlled environment—such as UHV, an inert gas or fluid—may be required, mainly to ensure reproducible sample surface conditions. But many basic microscopy tasks, including a broad range of imaging and spectroscopy experiments, are performed conveniently at room temperature. Soon after the invention of STM, however, it was recognized that control over the sample temperature enables a wide variety of new experiments that are impossible to perform at room temperature. As an example, the structure or chemical reactivity of a given surface, e.g., of a catalyst, may depend on temperature. To characterize its properties in a particular temperature regime, the sample has to be heated or cooled during

STM imaging. Further, though imaging of thin film growth in a "quench and look" mode is possible, as discussed previously, studying growth processes directly at relevant sample temperatures would often be preferred. And finally, investigation of surface diffusion, chemical reactions, or molecular dissociation at surfaces often requires cooling of the sample to cryogenic temperatures to slow the rates of these thermally activated processes sufficiently to map them by relatively slow STM imaging. Probably the best known example of the benefits of cryogenic STM is atom and molecule immobilization for atomic or molecular manipulation, which allows artificial structures to be built from individual atoms and can provide highly idealized structures for probing chemistry and condensed matter physics at the ultimate spatial limit. Below, the technical challenges involved in temperature-dependent STM are discussed briefly, and selected examples of the main classes of STM experiments at high and low temperature—the imaging of dynamic surface processes via STM movies and the assembly of nanostructures "one atom at a time" by atomic/molecular manipulation with the STM tip—are presented.

4.1 STM at High and Low Temperatures: Motivation and Issues

Compared to room temperature operation, STM experiments at high, low, or even variable temperatures involve a number of additional issues. A primary experimental difficulty arises from thermal drift due to incomplete thermalization and the resulting variations in thermal expansion across the microscope, which cause relative motion between the STM tip and sample. Thermal expansion coefficients are temperature dependent, and generally approach zero as $T \to 0$. As a result, thermal drift rates tend to be low and the overall instrument stability is improved in low-temperature STM. Stable low-temperature microscopes reach lateral drift rates of 0.1 nm/h (Meyer and Rieder, 1997), and values of vertical stability of the tunneling gap approaching 1 pm have been reported (Stipe et al., 1998a). Beyond instrument stability, there are strong additional incentives for operation at cryogenic temperatures. The rates of thermally activated processes, such as surface diffusion of adsorbed atoms or molecules, can be slowed considerably at low temperatures. The freeze-out of thermal motion paves the way for time-lapse movies documenting diffusion processes, and for controlled atom or molecule manipulation using the STM tip. On the other hand, the thermal broadening of the tunneling distribution and of the electronic structure can be minimized at cryogenic temperatures, which results in higher energy resolution in spectroscopy.

STM imaging at high temperatures is generally affected by high drift rates, which may require active drift compensation between consecutive scans to allow imaging a given region on the sample for extended periods of time. Intrinsic drift rates of 3 nm/min without and 0.2 nm/min with drift correction have been reported for dedicated high-temperature microscopes (Voigtländer, 2001). Additional practical difficulties involve sample heating, temperature measurement, as well as the need for high scan speeds to capture dynamic phenomena at high temperatures.

4.2 Capturing Dynamic Surface Processes: STM Movies and Atom Tracking

The most popular approach for imaging dynamic processes involves repeated scanning of the same sample area to generate a time-lapse STM "movie." This capability has been used to study a wide variety of phenomena, including step dynamics (Kuipers et al., 1993), epitaxial growth (Voigtländer and Zinner, 1993), self-diffusion on semiconductor (Borovski et al., 1997) and metal surfaces (Horch et al., 1999), as well as diffusion of adsorbates (Wintterlin et al., 1997) and large molecules (Schunack et al., 2002). If rates of dynamic processes, such as surface diffusion, are measured at several temperatures, activation energies of these processes can be determined. Complex reaction pathways, including adsorption, diffusion, dissociation, etc., can be elucidated at the atomic scale, and active sites can be identified.

As a near-field microscopy technique STM is based on raster scanning a probe tip. The scan speed is limited by the resonance frequency of the scanning element, typically below about 10 kHz, and by the electronic bandwidths of the tunneling current amplifier and feedback circuit. Frame rates in STM movies can be expected to be inherently lower than those in parallel imaging techniques, such as transmission electron microscopy or low-energy electron microscopy (Bauer, 1998; Tromp, 2000). To achieve adequate time resolution, the scan sizes in STM movies tend to be small, of the order of 10^4 pixels. Additional measures are typically necessary to deal with slow imaging rates and capture the detailed time evolution of a given process. Whenever a phenomenon under consideration permits, cooling of the substrate can be employed to slow the rates of thermally activated processes. In thin film growth, slow evaporation rates are employed. In studying surface reactions involving adsorption from the gas phase, e.g., on catalysts, low partial pressures of the reactive species in the gas phase are maintained. As an alternative to these often rather restrictive provisions, the development of high-speed microscopes and control electronics has been a focus of active research recently. As an example, scanners with mechanical resonances in the range between 50 and 100 kHz have been developed. Combined with a fast current amplifier (600 kHz bandwidth) and feedback loop (1 MHz), atomically resolved images have been obtained at rates approaching 200 frames/s (Rost et al., 2005). Such exciting instrument developments may in the future allow the imaging of new classes of dynamic surface phenomena at relaxed ambient conditions (higher temperatures, pressure, and growth rates) by STM.

To illustrate the use of STM movies at cryogenic temperatures to identify and quantify dynamic surface processes, we discuss the example of rutile $TiO_2(110)$, which has emerged as the prototypical system for fundamental surface science studies of transition metal oxides (Figure 15–21). TiO_2 has numerous applications in areas as diverse as heterogeneous catalysis, solar cells, photocatalysis, and organic waste remediation, and STM plays a key role in elucidating its fundamental surface processes. The $TiO_2(110)$-(1×1) surface consists

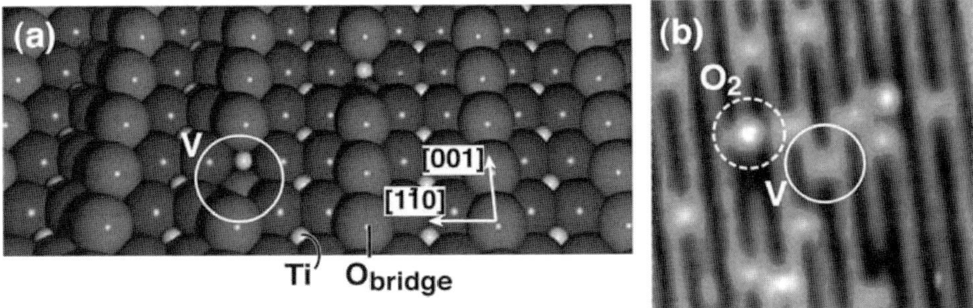

Figure 15–21. STM on Rutile TiO$_2$(110). (a) Structure of the TiO$_2$(110) surface. V, a single oxygen vacancy. (b) Constant-current STM on TiO$_2$(110). Bright rows are assigned to five-fold coordinated Ti atoms and dark rows to bridging oxygen atoms. Oxygen vacancies are imaged as protrusions between the Ti rows. Adsorbed O$_2$ molecules are associated with bright protrusions on the Ti rows. (Reprinted with permission from Schaub et al., © 2003 AAAS.)

of alternating rows of Ti and O atoms aligned along the [001] direction. Due to electronic effects, STM images Ti as protruding rows and bridging oxygen rows, which are geometrically highest on the surface by about 0.12 nm, as troughs.

The reactivity of TiO$_2$(110) is affected to a great extent by the presence of oxygen vacancies, which are generated in the process of reducing the surface by annealing in vacuum. Using STM movies, the reaction mechanisms at these defect sites were studied for model reactions such as water dissociation (Brookes et al., 2001; Schaub et al., 2001). The diffusion of oxygen vacancies follows an intriguing mechanism mediated by O$_2$ molecules (Figure 15–22) (Schaub et al., 2003). Oxygen vacancies

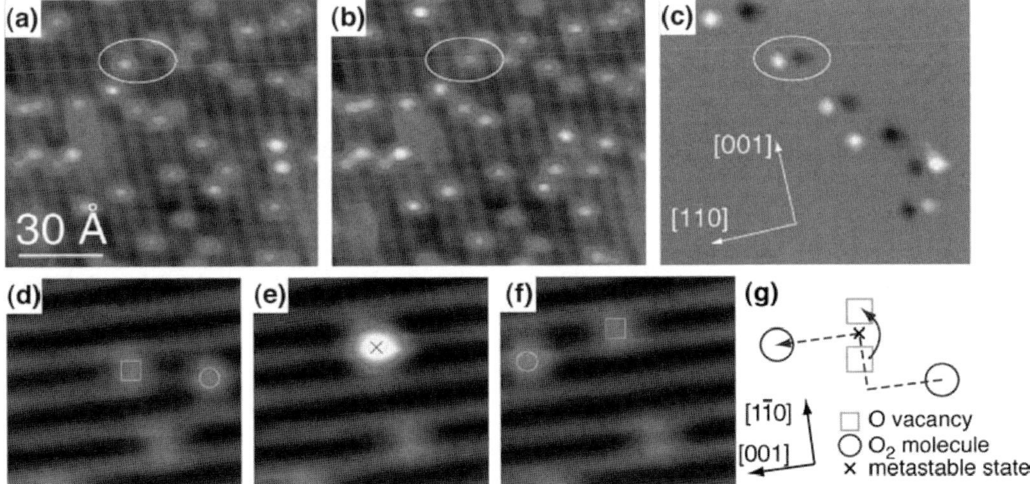

Figure 15–22. O$_2$-mediated diffusion of oxygen vacancies on TiO$_2$(110). (a and b) Consecutive frames of an STM movie on the motion of oxygen vacancies on TiO$_2$(110) (T = 300 K, 8.5 s/frame). (c) Difference image of (a) and (b), highlighting changes due to the diffusion of single oxygen vacancies. Vacancies diffuse perpendicular to the bridging oxygen rows. (d–f) Time-lapse STM of the O$_2$-assisted diffusion of an individual oxygen vacancy (T = 230 K, 1.1 s/frame). (g) Schematic illustration of the O$_2$-mediated vacancy diffusion mechanism. (Reprinted with permission from Schaub et al., © 2003 AAAS.)

are immobile on the adsorbate-free surface. Vacancy diffusion is greatly enhanced by adsorbed O_2. At temperatures sufficiently low that the diffusion of adsorbed O_2 molecules can be captured by STM, the subtraction of consecutive frames in time-lapse STM shows that single oxygen vacancies diffuse along $[1\bar{1}0]$ from one bridging O row to the next, always in the presence of neighboring O_2 molecules.

The role of O_2 molecules in the vacancy diffusion process is established from detailed investigation of single vacancy hops, again based on time-lapse STM movies at low temperature. As an O_2 molecule diffusing along a Ti row approaches an oxygen vacancy, it dissociates and contributes one oxygen atom toward healing the vacancy, thus creating a metastable intermediate consisting of a single O atom. The O adatom is highly reactive, as corroborated in separate experiments involving dosing of atomic oxygen. It rapidly recombines with a bridging O atom and emerges as an O_2 molecule. If in this process the bridging O atom is removed from one of the adjacent rows, the net result is a diffusion jump of an oxygen vacancy by one bridging oxygen row. Given this O_2-mediated mechanism of oxygen vacancy diffusion, the rate of diffusion events is expected to scale linearly with O_2 coverage. STM movies obtained at different O_2 exposure show that this is indeed the case.

While early imaging of dynamic surface processes was performed almost invariably in UHV, several applications require STM imaging in what is seen as more "realistic" environments for those applications. A prominent example is heterogeneous catalysis. It has been recognized that actual reactions under technologically relevant conditions, often involving elevated temperatures and pressures at or above atmospheric pressure, can involve surface structures and compositions, and entire reaction mechanisms that differ substantially, even qualitatively, from those of "simulated" reactions running in UHV, a situation commonly termed the "pressure gap" problem of heterogeneous catalysis. To address the need for imaging with high spatial and temporal resolution at elevated pressure, a family of dedicated STM instruments was developed (Rasmussen et al., 1998; Jensen et al., 1999; Lægsgaard et al., 2001; Rößler et al., 2005). These instruments allow sample preparation and surface analysis in UHV, followed by exposure to reactants at high pressure and simultaneous STM imaging. A particularly elegant implementation of this concept is the "reactor STM," allowing dynamic STM imaging of surfaces exposed to reactants in a compact catalytic flow reactor in combination with the simultaneous analysis of the reaction products by mass spectrometry.

Figure 15–23 shows an example of a complex dataset obtained during high-pressure CO oxidation on Pt(110) (Hendriksen and Frenken, 2002). The upper panel traces mass spectrometer signals for O_2, CO, and CO_2, showing initial exposure to CO, followed by the introduction of molecular oxygen into the reactor. The panel below shows representative STM images obtained at specific stages of the reaction, during which the sample is kept at a constant temperature of 425 K. Images A, B, E, F, and H show flat terraces separated by steps, representing the metallic, CO-covered Pt(110) surface. Image C shows the change in

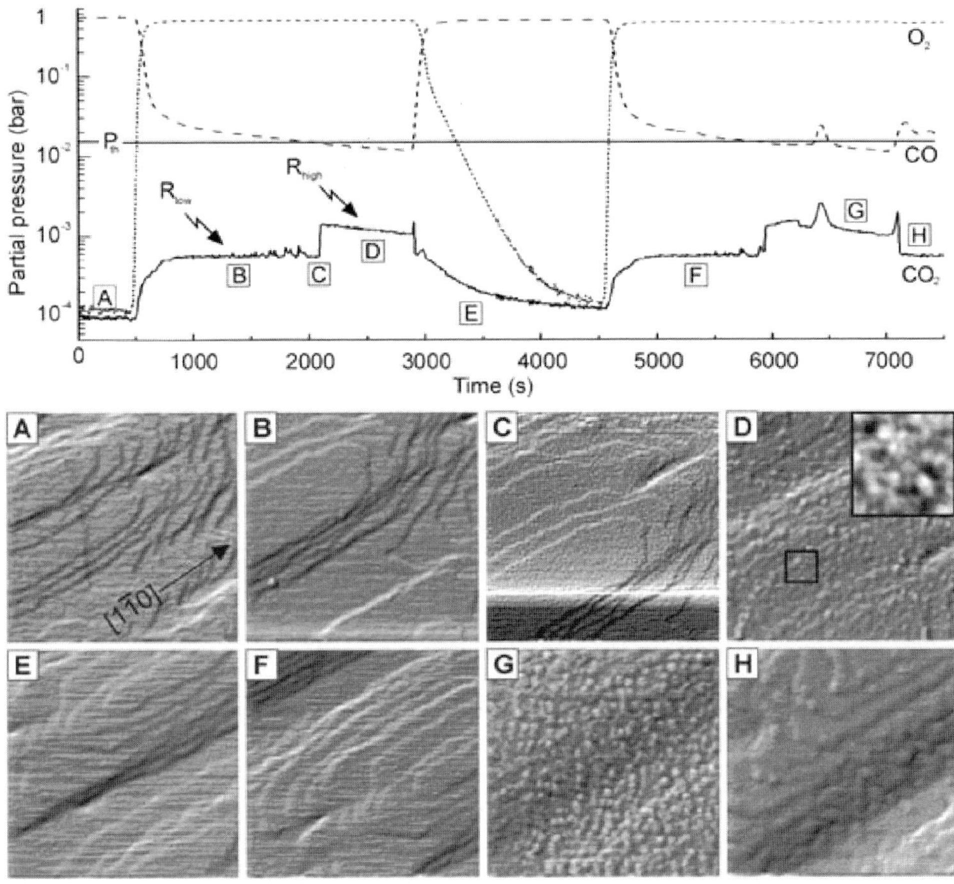

Figure 15–23. Simultaneous mass spectrometry and STM imaging in a catalytic flow reactor: CO oxidation on Pt(110). (Top) Mass spectrometer signals of O_2, CO, and CO_2 measured at the output of the flow reactor cell. (Bottom) STM images on the Pt(110) catalyst surface acquired at selected stages of the process: CO adsorption (A), O_2 flow (B–D), and repetition of the sequence (E–H). Note the strong surface roughening in (D), associated with increased CO_2 evolution. (Reprinted with permission from Hendriksen and Frenken, © 2002 by the American Physical Society.)

surface morphology, a pronounced surface roughening, during a step in activity giving rise to a sudden increase in CO_2 evolution, demonstrating a direct link between surface roughness and activity for this surface involving a mechanism that is not observed at low pressure.

Although capable of mapping dynamic surface phenomena, STM movies have obvious limitations in imaging fast dynamic processes, such as surface diffusion at room temperature or above. A possible solution is the development of novel approaches and instruments for very high-speed STM imaging. As an alternative, frame-by-frame imaging can be abandoned altogether if only a map of the trajectory of the diffusing species is desired. The recognition of this fact led to the development of atom tracking STM (Swartzentruber, 1996). In atom tracking, the STM tip is locked onto a diffusing surface species using a 2D lateral feedback mechanism. Once locked, the feedback maintains the tip over that species and tracks its coordinates as it diffuses over the substrate. In the atom-tracking mode, the STM spends all of its time measuring the diffusion trajectory, which results in substantially improved time-resolution compared to time-lapse STM movies.

Atom tracking STM has been used to measure the diffusion kinetics of Si (Swartzentruber, 1996) and SiGe (Qin et al., 2000) dimers on Si(001) above room temperatures, of water molecules on Pd(111) (Mitsui et al., 2002), and of Pd atoms in a Pd/Cu(001) surface alloy (Grant et al., 2001). The latter example is illustrated in Figure 15–24. Pd atoms in the surface alloy are imaged as protrusions in STM. Locking the STM tip onto individual Pd atoms, their diffusion pathway can be tracked. Analyzing the residence time (the time between hops) and jump length leads to the conclusion that there is no time correlation between individual Pd diffusion events, and that the diffusion is medi- ated by surface vacancies rather than Cu adatoms, i.e., involves rapidly diffusing vacancies visiting Pd atoms in the surface layer. From the temperature dependence of the Pd hop rate—determined from temperature-dependent tracking experiments—the activation energy of the overall process, in this case equal to the sum of the vacancy for- mation energy and the energy barrier for lateral Pd-vacancy exchange, can be measured.

In contrast to time-lapse STM, in which the tip is scanned rapidly across a larger field of view, the tip maintains close contact with the diffusing entity in atom tracking. Hence, the observed diffusion process could be affected by tip–sample interactions. While this question has to be studied on a case-by-case basis, at least one system [SiGe dimer

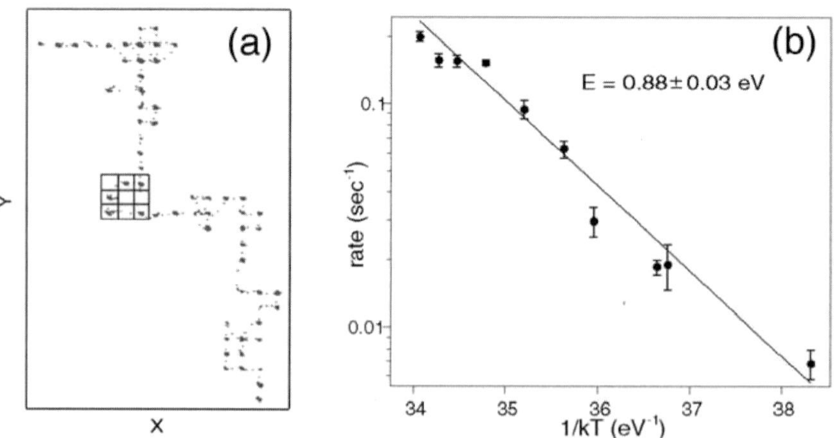

Figure 15–24. Diffusion kinetics of Pd atoms in the Pd/Cu(001) surface alloy. (a) Site visitation map of an individual Pd atom, obtained by atom-tracking STM at a temperature of 62°C. The square array marks the position of the Cu(001) unit mesh. In this dataset the atom hopped 853 times in a time interval of 5557 s. (b) Temperature dependence between 31 and 69°C of the average hop rate of an incorporated Pd atom. The average residence time of Pd atoms decreases from 145.3 to 5.0 s in this temperature range. The data follow an Arrhenius form with an activation energy of 0.88 eV and measured prefactor of $10^{12.4\pm0.4}$ Hz. (Reprinted with Permission from Grant et al., © 2001 by the American Physical Society.)

diffusion on Si(001)] has been identified in which the diffusion mechanism depends on the sign of the electric field between tip and sample (Sanders et al., 2003), indicating that the presence of the probe tip can indeed affect the measurement.

4.3 Atom and Molecule Manipulation

Atom and molecule manipulation (Hla and Rieder, 2003) experiments utilize the sharp STM probe tip and the ability to control it laterally and vertically with picometer resolution to build and modify nanometer-scale structures, typically in combination with their analysis by STM imaging and tunneling spectroscopy. While conceptually related to atom tracking, manipulation experiments are performed in a different regime: the sample is cooled to temperatures low enough that all thermal diffusion is frozen out, and the tip is typically brought into close contact to induce strong tip–adatom/molecule interactions.

Since the first demonstration of atomic manipulation of Xe/Ni(110) by Eigler and Schweizer (1990), manipulation experiments have branched out considerably, and now encompass scenarios as diverse as controlled chemical reactions of individual molecules (Lee and Ho, 1999), contacting single molecules with atomic metal wires (Nazin et al., 2003), construction of mechanical logic gates from adsorbed molecules (Heinrich et al., 2002), and control of excess charge on individual atoms at insulator surfaces (Repp et al., 2004). While generally based on some form of tip–sample interaction, atom or molecule manipulation can involve a number of different physical mechanisms: close proximity and short-range chemical interaction between a tip atom and adatom to affect the potential landscape seen by the adatom on the surface, vibrational excitation within adsorbed molecules or between substrate and adatom, electric field, or direct charging by tunneling electrons. Different patterns of adsorbate motion can be induced, including lateral hopping between adsorption sites on the substrate, vertical transfer between sample and tip, rotation, as well as the controlled making and breaking of individual bonds. Substrate surfaces with relatively low symmetry, e.g., (110) surface orientations or stepped surfaces, are often used to establish one-dimensional diffusion pathways between substrate adsorption sites that help guide the manipulation process.

Lateral manipulation is based on establishing close proximity between an adatom and the STM tip to increase the interaction between them. The approach process is controlled via the tunneling resistance, which is lowered from typically several hundred $M\Omega$ during imaging to values of the order of $100\,k\Omega$ for manipulation. Depending on the particular system, the lateral force needed to move an adsorbate between adjacent adsorption sites can be either repulsive or attractive, giving rise to manipulation by pushing and pulling, respectively. The manipulation process itself is monitored by measuring the tip height (or z-piezo voltage) during the lateral motion of the tip, as illustrated in Figure 15–25 for Pb/Cu(211) (pulling) and CO/Cu(211) (pushing) (Meyer and Rieder, 1998).

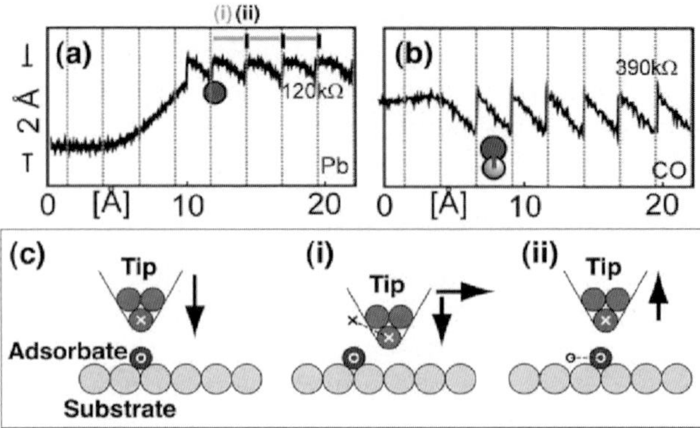

Figure 15–25. Mechanisms of lateral atom manipulation. (a) "Pulling" manipulation via attractive interaction between a Pb atom on Cu(211) and the W tip. (b) "Pushing" manipulation via repulsive interaction between a CO adsorbate and the W tip. In both cases a sawtooth-like tip height profile with the periodicity of the substrate adsorption sites is observed (Reprinted with permission from Meyer and Rieder, © 1998 The Materials Research Society). (c) Illustration of the tip and adsorbate motion during manipulation via attractive interaction. (i) Lateral motion of the tip, accompanied by an approach toward the substrate. (ii) Hopping of the adsorbate, followed by a sharp retraction of the tip.

The adsorbate motion in both pushing and pulling modes is strongly influenced by the preferred adsorption sites defined by the substrate. A sawtooth-like vertical motion of the tip accompanies each jump of the adsorbate between neighboring adsorption sites (Figure 15–25a and b). The origin of this tip motion is illustrated in Figure 15–25c for the example of an attractive tip–adsorbate interaction. Following the approach to the adsorbate, the tip is moved laterally. Initially, the adsorbate, held in the potential well of a substrate adsorption site, does not follow. The tunneling gap thus increases, and the tip moves forward to maintain a constant tunneling current (i). Ultimately, the interaction with the tip induces a lateral jump of the adsorbate into a neighboring potential minimum. The tunneling gap closes abruptly and the feedback loop causes the tip to retract.

The precision and complexity of structures achievable by lateral manipulation is illustrated in Figure 15–26 (Heinrich et al., 2002). CO molecules adsorbed in monomer, dimer, or trimer configurations on Cu(111) give rise to distinct contrast in constant current STM. While monomers and dimers are stable at low temperature (5 K in this example), there are three distinct configurations for trimers: a stable three-fold symmetric "close-packed" arrangement, and metastable "straight-line" and "bent-line" ("chevron") configurations, which relax with time constants of the order of seconds into the stable state. Complex structures consisting of up to 545 CO molecules were built with atomic precision such that any neighboring molecules were initially in stable dimer configurations, but could be changed to an unstable "chevron" configuration with a place exchange of a single molecule in the vicinity. In this way, molecular cascades, similar to toppling rows of standing dominoes, could be constructed. Mechanical computation was demonstrated by molecular cascades set up as logic AND gates, two-input, and three-input sorters.

Apart from lateral manipulation, vertical manipulation, i.e., transfer of adsorbates between sample and tip, is possible. Vertical manipulation of CO molecules, for example, has been used for controlled modification of the STM tip and to induce single molecule reactions (Lee and Ho, 1999). Vertical transfer is also important for moving adatoms or molecules across obstacles such as step edges, if they cannot be surmounted by controlled lateral hops. Manipulation mechanisms other than direct chemical interaction between the STM tip and adsorbates have received increased interest recently. Notably the different roles of vibrational excitations are actively investigated (Komeda et al., 2002; Stroscio and Celotta, 2004). Inelastic tunneling can, for instance, drive controlled rotations of adsorbed molecules (Stipe et al., 1998b). An intriguing example of atomic manipulation based on inelastic tunneling, the controlled manipulation of excess charge on single Au atoms, is illustrated in Figure 15–27 (Repp et al., 2004). Individual Au

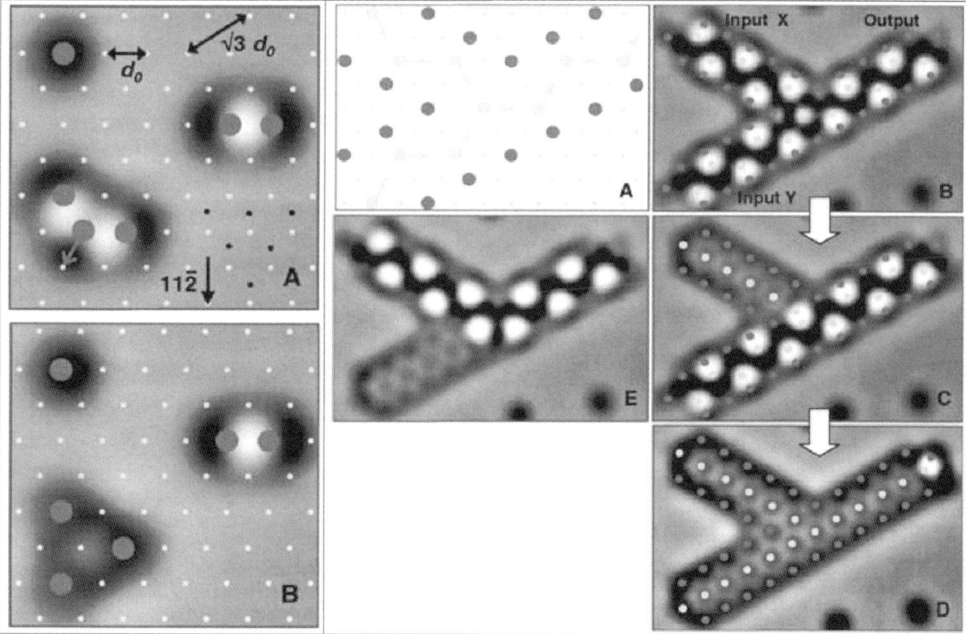

Figure 15–26. Molecule cascades. (Left) (A) Configurations of neighboring CO molecules on Cu(111): isolated CO molecule, dimer, and trimer in a "chevron" configuration. Large circles mark the positions of the molecules and small dots indicate the surface layer Cu atoms (1.9 nm scans; $T = 5$ K). (B) The same area after one CO molecule in the trimer has hopped to generate a stable close-packed trimer. (Right) Demonstration of a molecular mechanical AND gate, implemented via cascades of "chevron"-type CO trimers. (A) Model of the AND gate. (B–D) Sequence of STM images (5.1 nm by 3.4 nm) showing the operation of the AND gate: (B) Initialization. (C) Result after input X was triggered with the STM tip. (D) When input Y was triggered, the cascade propagated all the way to the output. (E) Result of triggering only input Y. (Reprinted with permission from Heinrich et al., © 2002 AAAS.)

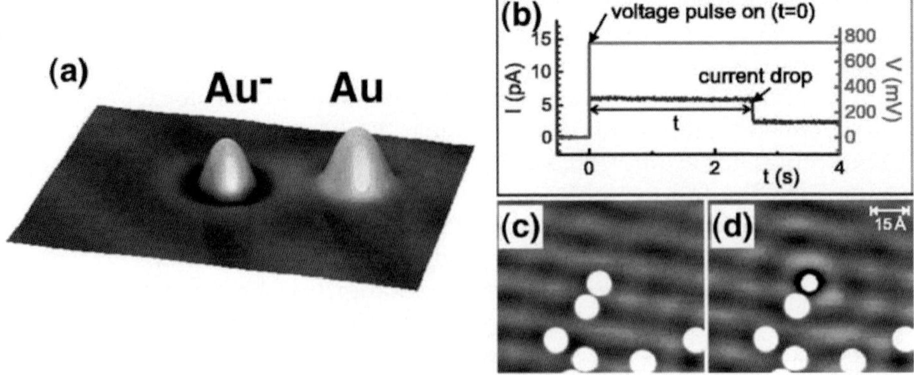

Figure 15–27. Charging of individual Au atoms on ultrathin NaCl/Cu(111). (a) Constant-current STM on a pair of Au atoms, one of which was charged negatively with the other remaining neutral. (b) Signature of the charging event. Successful charging, achieved by a voltage pulse while maintaining the tip above a single Au atom, is indicated by a sudden drop in tunneling current. (c and d) Quasi-particle scattering: neutral Au adatoms do not scatter NaCl/Cu(111) interface-state electrons (c), whereas the negatively charged adatom acts as a scatterer (d). (Reprinted with permission from Repp et al., © 2004 AAAS.)

atoms are deposited on ultrathin, insulating NaCl supported by Cu(111). Electrons can tunnel between the metal substrate and the STM tip through the ultrathin insulator, i.e., STM imaging, spectroscopy, and manipulation are possible for this system. However, the coupling of the Au adsorbate to the substrate is affected profoundly by the insulating support. Voltage pulses can be used as a means to reversibly deposit an excess electron on individual Au atoms. The charging occurs via an inelastic electron tunneling mechanism enabled by a weak coupling of the adatom and Cu substrate electronic states. The data suggest that the coupling is so weak that the lifetime of a negative ion resonance state of the adatom is in the range of ionic vibrational periods for the NaCl interlayer, allowing the relaxation of the NaCl lattice, shift of the negative ion resonance state below the Fermi energy, and capture of the electron. The excess charge is maintained due to the substantial relaxation of the underlying NaCl lattice. This stabilization prevents discharging by tunneling into the metal, and the electron resides on the Au atoms until removed by a voltage pulse of opposite sign. The charged Au atom has a distinct signature in constant-current STM imaging, and the long-range electric field due to the charged Au atom strongly scatters electrons in interface states at the Cu/NaCl interface. Charging individual adsorbed atoms is a potentially powerful approach to tuning their physical properties. For Au/NaCl, significant differences in surface diffusivity were identified, with the onset temperature of significant surface diffusion lowered from 60 K (Au) to 40 K (Au⁻). Other possible modifications due to controlled charging at the atomic level include changes in catalytic activity and the controlled manipulation of the net spin magnetic moment of individual atoms.

5 Heterostructures and Buried Interfaces: BEEM, Quantum Size Effects, and Cross-Sectional STM

The exponential dependence of the tunneling current on the tip–sample separation makes STM a versatile tool for atomic-resolution microscopy and spectroscopy at surfaces. As a result of its unique resolution and surface sensitivity STM has brought important advances in fields such as catalyis or epitaxial growth, in which surface processes play a key role. For many technological applications of semiconductor materials, e.g., in device structures such as transistors, detectors, or lasers, the active regions encompass complex heterostructures, and device performance is affected much less by the free surface than by buried interfaces. Therefore, a need arises for a technique capable of probing subsurface structures, electronic properties, and carrier transport. The mapping of interfacial and transport properties should occur with nanometer spatial resolution to provide data that are relevant for semiconductor devices with progressively reduced dimensions. Since the early 1990s, several STM-based techniques have been developed to provide high-resolution imaging and spectroscopy of buried interfaces and heterostructures. Ballistic electron emission microscopy (BEEM) operates in a three-terminal configuration, in which the STM tip, whose height is controlled by a constant-current feedback loop, injects hot carriers into a thin film or heterostructure while an additional collector contact measures the current of carriers that are transmitted through buried interfaces. Electron interference in a thin metal film, again with carriers injected from an STM tip and traveling ballistically in the metal, provides detailed thickness maps of metallic overlayers, and can—under favorable circumstances—even map the atomic structure of a buried metal–semiconductor interface. Cross-sectional STM, finally, is used to image complex heterostructures such as superlattices, embedded self-assembled quantum dots, or substitutional magnetic impurities, on nonpolar (110) cleavage planes of III–V compound semiconductors.

5.1 Probing Buried Interfaces (I)—BEEM

BEEM was invented by Kaiser and Bell (1988) to probe Schottky contacts, i.e., metal–semiconductor interfaces with high spatial resolution (for a recent review, see Narayanamurti and Kozhevnikov, 2001). Beyond Schottky barriers, the technique can determine the height of other subsurface potential barriers and it has also been applied to measure band offsets between different semiconductors, the energies of transmission resonances in semiconductor quantum wells and superlattices, and bound states in buried quantum wires and dots. In addition, electron scattering at subsurface linear and point defects has been characterized.

We illustrate the operating principle of BEEM using the example of a metal–semiconductor junction. The technique operates in a three-terminal configuration, i.e., adds an additional collector electrode to the STM tip and sample contact, as shown schematically in Figure 15–28.

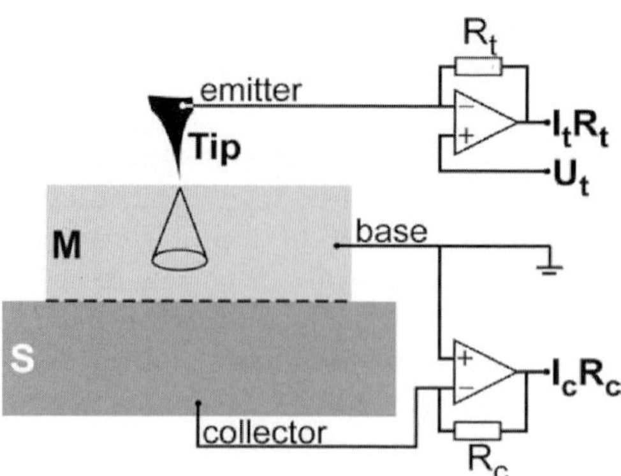

Figure 15–28. Schematic setup of a typical BEEM experiment on a metal (M)–semiconductor (S) heterostructure.

The STM tip is scanned at constant current I_t, measured between the tip and base electrode, over the surface of the base layer. The tip serves as a point source of hot carriers, electrons or holes, injected into the metal base. If the metal thickness is small compared to the mean free path for electron–phonon, electron–electron, or impurity scattering, most carriers reach the metal–semiconductor interface *ballistically*, i.e., with their original energy and momentum distribution. In metals, the mean free path for electron–phonon scattering of electrons with energies of a few eV is typically of the order of 10 nm. In significantly thicker films, electron–phonon scattering would broaden the momentum distribution of the injected carriers, i.e., mostly affect the spatial resolution of BEEM, while having little effect on the energy distribution. Hence, the injected ballistic carriers generally have a well-defined energy distribution and can be used to perform spectroscopic measurements on buried potential barriers.

At the metal–semiconductor interface, ballistic carriers with energies below a threshold equal to the Schottky barrier height $e\phi_B$ are reflected back into the metal base, while carriers with energy above the threshold are transmitted into the semiconductor collector (Figure 15–29). If

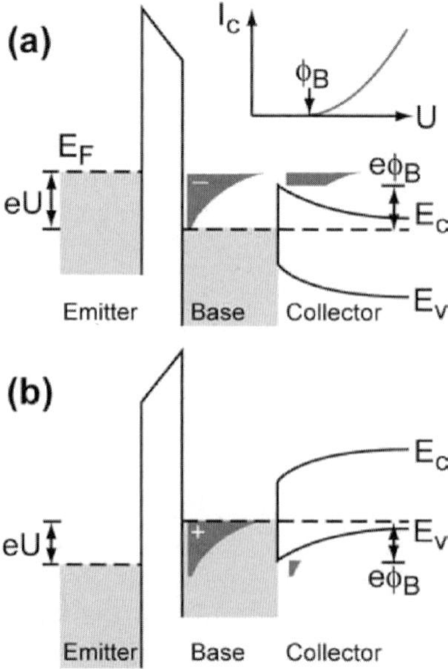

Figure 15–29. Band diagrams for (a) ballistic electron and (b) ballistic hole injection into the metal base on an n- and p-type semiconductor, respectively. The dark shaded areas indicate the energy distribution of the injected carriers, as well as of those transmitted into the collector. The inset in (a) illustrates the onset of the collector current at a threshold bias corresponding to the Schottky barrier height, $e\phi_B$. (After Bell et al., 1991.)

other barriers exist in the collector layer, the transport through that material can involve additional interfacial reflections. Carriers that are transmitted through the entire collector layer contribute to the collector current (or BEEM current), which is measured between base and collector contacts.

The energy of the injected carriers is varied by changing the voltage applied between the STM tip and metal base. For low bias, none of the injected carriers is transferred across the metal–semiconductor interface. At voltages above the threshold, some carriers have energy above the Schottky barrier and can cross the interface, causing an increase in collector current with increasing tip–sample bias above threshold. From the onset of the BEEM current, the barrier height $e\phi_B$ can be determined, e.g., by fitting the measured $I_c(V)$ to a theoretical expression. In a simple one-dimensional theory,

$$I_c(V) = RI_t \int dE \left[f(E) - f(E - eV) \right] \times \theta \left[E - (E_F - eV + e\phi_B) \right] \tag{6}$$

where R is a bias independent parameter and $f(E)$ is the Fermi function.

Figure 15–30 shows experimental BEEM spectra on an Au/n-Si(001) junction, measured at room temperature in N_2 atmosphere, fitted using Eq. (6) to determine a Schottky barrier $e\phi_B = 0.92\,\text{eV}$. Due to its low reactivity and oxidation resistance, Au was used in many of the early BEEM experiments as a nonepitaxial metal base on a variety of other semiconductors. Schottky barrier heights were determined, for example, for Au/n-GaAs(001) ($e\phi_B = 0.70\,\text{eV}$ at 77 K) (Bell et al., 1990) and Au/n-GaP(110) ($e\phi_B = 1.41\,\text{eV}$) (Prietsch and Ludeke, 1991). For Au/n-CdTe(001) ($e\phi_B = 0.69$–$1.07\,\text{eV}$) (Fowell et al., 1990), strong lateral variations in the

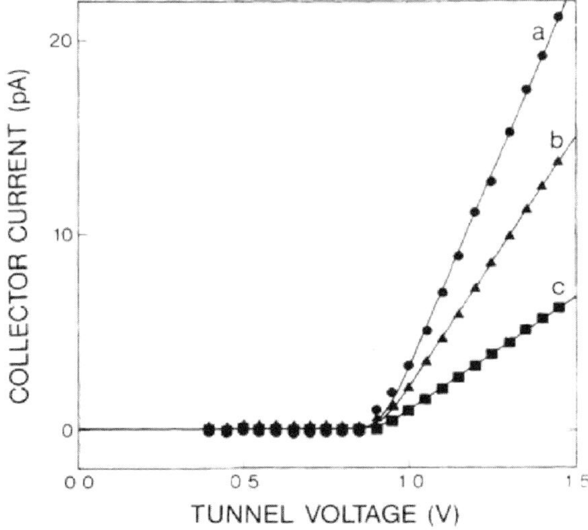

Figure 15–30. BEEM spectra obtained on a polycrystalline Au/n-Si(001) junction. Spectra a–c (symbols) are measured at different tunneling currents. The calculated spectra (solid lines) correspond to a common Schottky barrier height value $\phi_B = 0.92\,\text{eV}$ and R value of $0.045\,\text{eV}^{-1}$. (Reprinted with permission from Kaiser and Bell, © 1988 by the American Physical Society.)

measured Schottky barrier height were detected, emphasizing the need for high-resolution maps of interfacial transport.

Carrier transport in BEEM is affected by all stages of the injection pathway: (1) tunneling from the tip into the metal base layer, (2) hot carrier transport through the base, (3) transmission across the metal–semiconductor interface, and (4) transport in the semiconductor, including transmission through additional heterostructure interfaces. In addition to local Schottky barrier heights, the technique is therefore sensitive to a number of factors, e.g., surface topography, elastic or inelastic scattering in the different layers or at interfaces, and the band structures of the junction partners, which affect not only interface transmission probabilities but also the spatial resolution obtainable in BEEM via metal band structure-induced focusing or defocusing.

BEEM current maps can be acquired simultaneously with constant-current STM images by measuring I_c spatially resolved during a constant-current scan. Figure 15–31 shows STM and BEEM current images obtained on 2.5 nm epitaxial $CoSi_2/Si(111)$ (Sirringhaus et al., 1994). The lattice mismatch of 1.2% between the metallic silicide and the Si substrate is accommodated by an interfacial dislocation network. In STM topography, the location of these line defects is detected via their elastic strain field at the $CoSi_2$ surface. Strikingly, the defects are mapped in BEEM as sharply localized regions with increased collector current with a width of only 0.8 nm, as expected for ballistic transport of carriers with a strongly forward focused tunneling momentum distribution, i.e., $k_\parallel \sim 0$. While the Schottky barrier was found to be uniform, as expected for an epitaxial interface, the sharp local increase in transmissivity of the $CoSi_2/Si$ interface was explained by a significant increase in in-plane momentum due to scattering by the dislocation core, facilitating the transmission into the Si conduction band minima.

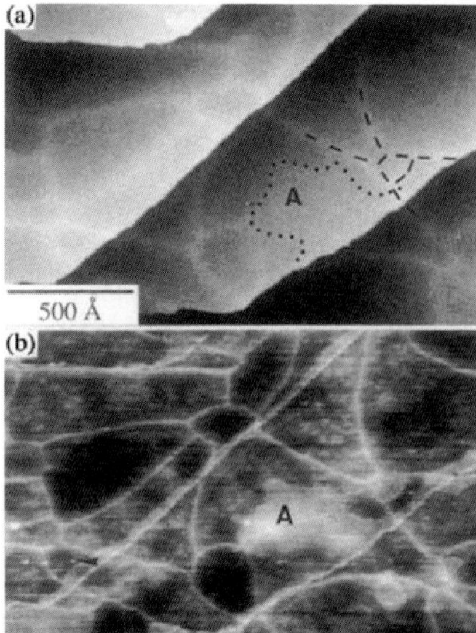

Figure 15–31. (a) Constant-current STM image and (b) corresponding BEEM image obtained on epitaxial 2.5 nm $CoSi_2/Si(111)$. A dislocation network at the silicide/silicon interface is indicate by dashed lines. In the BEEM image brighter areas indicate regions of higher collector current. Hot electron scattering at the interfacial dislocations causes a sharply localized increase in collector current.

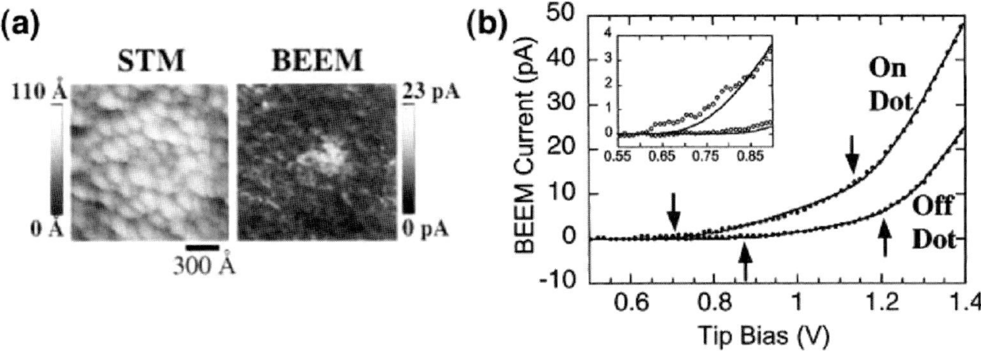

Figure 15–32. BEEM on self-assembled quantum dots. (a) STM and BEEM images of an individual InAs/GaAs quantum dot under a polycrystalline Au base. (b) Comparison of BEEM spectra obtained by ballistic electron injection into an InAs dot and into the wetting layer between dots. (Reprinted with permission from Rubin et al., © 1996 by the American Physical Society.)

In experiments such as those on $CoSi_2$/Si(111), growth and probing of the entire heterostructure in the same UHV environment is critical to achieve ordered interfaces and reliable measurements on reactive metal surfaces. For the microscopy itself, low temperatures have several practical advantages, such as improved energy resolution due to reduced thermal broadening of the tunneling distribution, and reduced thermal drift of the STM. An additional benefit is a lower thermal noise due to a reduction of thermionic emission across interfacial barriers, which allows shallower potential barriers to be probed. As a result, semiconductor band offsets, resonant transport in semiconductor heterostructures, and interfacial barriers between an organic layer and a metal base (Troadec et al., 2005) can be measured successfully by low-temperature BEEM. Examples are embedded self-assembled quantum dots (Rubin et al., 1996), lithographically patterned quantum wires (Eder et al., 1996), double-barrier resonant tunneling structures (Sajoto et al., 1995), and superlattices (Heer et al., 1998). Figure 15–32 shows the band profile for carrier injection into an individual InAs quantum dots in GaAs (Rubin et al., 1996). Also shown are STM and BEEM images, as well as BEEM spectra obtained with the tip positioned on top and next to the dot. The semiconductor heterostructure was coated with a polycrystalline Au base layer, whose morphology largely dominates the contrast in STM. A strong enhancement in BEEM current in a circular region with 30 nm diameter is associated with carrier transport through a single InAs quantum dot. A comparison of BEEM spectra obtained on and between dots shows signatures of transport through two zero-dimensional states of the dot.

Ballistic carrier transport and scattering can, in principle, be used to probe a wide variety of other systems, beyond Schottky barriers and semiconductor heterostructures. Examples are metal–insulator–semiconductor structures (Cuberes et al., 1994) or magnetic multilayers (Rippard and Buhrmann, 2000). In the latter, an Au/Si(111) Schottky barrier is used as an analyzer for spin-dependent scattering of carriers in a Co/Cu/Co trilayer structure, used in spin-valve devices, integrated on top of the Au base layer. Unpolarized carriers injected from an STM tip into this magnetic multilayer base will undergo spin-dependent scattering in the ferromagnetic Co layers. In areas in which the Co layers couple ferromagnetically, only one spin component will be scattered heavily and a large fraction of unscattered carriers is transmitted across the Au/Si interface. If the magnetization direction of the two Co layers is misaligned, both spin components are strongly attenuated by scattering, causing a sharp drop in collector current.

5.2 Probing Buried Interfaces (II)—Quantum Size Effects

In BEEM hot carriers, locally injected into a metal/semiconductor heterostructure, are used to measure the transmitted current across a buried interface. Instead, carriers that impinge with energies below threshold and are reflected back into the metal film can also be considered. Due to the long mean free path, these carriers can set up standing waves in the metal film. An STM tip, probing the evanescent tails into the vacuum, can then be used to image fringes due to interference of these electrons. The concept of electron interference in metal films dates back to Jaklevic et al. (1971), and interference effects have been observed by angle-resolved photoemission, low-energy electron diffraction and low-energy electron microscopy.

An implementation of an STM experiment to probe interference effects in the system Pb/Si(111) is shown in Figure 15–33. Pb evaporated onto a stepped Si(111) substrate forms (111)-oriented islands whose surface is atomically flat. Due to the substrate vicinality, the Pb

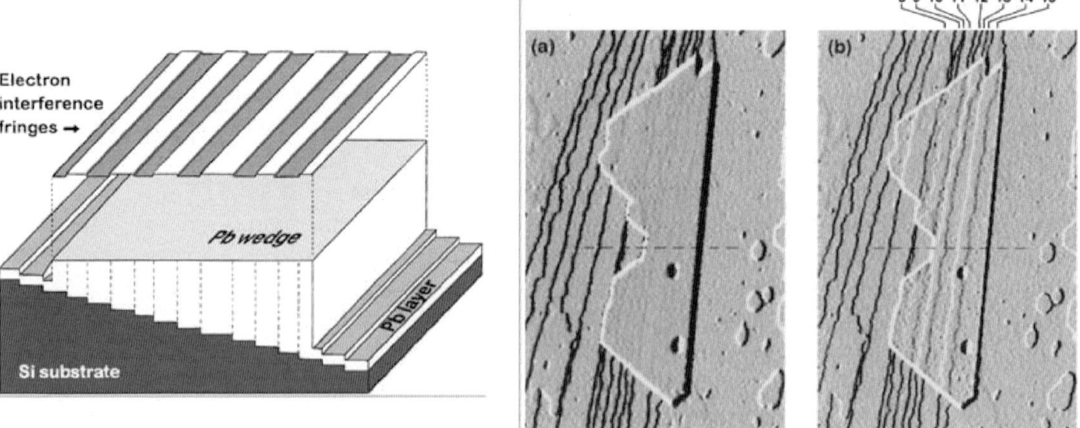

Figure 15–33. Electron interference in a Pb "quantum wedge" on Si(111). (Left) Geometry of the Pb wedge on a stepped Si(111) substrate. The Pb thickness varies in increments of roughly one Si(111) step height, giving rise to electron interference fringes in the direction of the substrate steps. (Right) Constant-current STM images (730 nm × 1100 nm) at −5 V (a) and +5 V (b) sample bias. The image obtained at positive bias shows apparent height changes at the surface of the wedge due to electron interference in the Pb film. (Reprinted with permission from Altfeder et al., © 1997 by the American Physical Society.)

island is wedge shaped, with a thickness that changes in integer multiples of the Si(111) step height from one substrate terrace to the next. While STM at positive sample bias, i.e., injection of electrons from the tip, images the atomically smooth Pb surface, STM images at opposite bias show bands of apparent terraces and steps, aligned with steps of the Si substrate, at the surface of the Pb wedge. These images, and associated tunneling spectra, are interpreted as signatures of quantum well states in the Pb wedge (Altfeder et al., 1997), and can be used to determine the position of subsurface steps as well as the position-dependent absolute thickness of the Pb film. Bands with constructive and destructive interference alternate with a thickness change d_0 of one Pb(111) monolayer if $d_0 \approx \lambda_F/4$, where λ_F denotes the Fermi wavelength. Similar interference effects were also observed by STM for epitaxial silicide layers, such as $CoSi_2/Si(111)$ (Lee et al., 1994) and $NiSi_2/Si(111)$ (Kubby and Greene, 1992).

Strikingly, electron interference can even be used to image interfacial atomic structures buried under as much as 10 nm of metal. In the Pb/Si(111) system, the (7×7) reconstruction of the Si(111) surface remains essentially intact upon low temperature evaporation of Pb, except for some intermixing by replacing Si adatoms by Pb (Altfeder et al., 1998). Due to the topology of the Fermi surface of Pb, in particular a large mismatch between the electron effective mass in in-plane and normal directions, the quantized electron states in the Pb film can be used to map interfacial structure with a resolution of 0.6 nm for overlayer thicknesses exceeding 10 nm, or roughly 10 times the Fermi wavelength in the metal.

5.3 Imaging Buried Heterostructures—Cross-Sectional STM

The strong interest in low-dimensional semiconductor structures—quantum wells, wires, and dots—has stimulated widespread activity in nanoscale imaging of electronic materials with reduced dimensionality. Recent efforts have focused on self-assembled quantum dots, generated by lattice mismatched heteroepitaxial growth. Semiconductor quantum dots with lateral size in the 10–100 nm range are readily imaged by STM if they are exposed as islands on a free surface. Consequently, a large number of studies have been devoted to studying epitaxial growth and quantum dot self-assembly by conventional STM imaging. However, almost any technological applications of self-assembled quantum dots require embedding in a matrix, often consisting of the substrate material. The embedding process causes significant modifications to the dots that include segregation and intermixing, shape changes, dopant redistribution, and adjustments to the local strain field in the dot and in the surrounding material. All these factors make it desirable to image embedded rather than exposed nanostructures.

Cross-sectional STM (X-STM), originally demonstrated by Feenstra et al. (1987) for imaging and spectroscopy on (110) surfaces of III–V compound semiconductors, offers an elegant solution to this challenge. Semiconductors that cleave easily, as most III–V compounds do, are used as a substrate for the growth of embedded self-assembled quantum dots. A sample is then cleaved in UHV, and STM imaging is performed on the cleavage face, typically a nonpolar (110) plane, which provides large step-free areas for atomic resolution imaging. In this geometry, X-STM gives access to subsurface structures over the entire thickness of the epitaxial layer and the underlying substrate.

X-STM has been used to image a variety of buried semiconductor heterostructures, such as superlattice structures in GaAs/AlGaAs (Salemink and Albrektsen, 1991), InAs/GaSb (Feenstra et al., 1994), and GaN/GaAs (Goldman et al., 1996), showing atomic-scale composition fluctuations, interface roughness, lateral stain variations, and phase separation in these systems. The technique has seen a strongly revived interest with the advent of self-assembled quantum dots and quantum dot superlattices (Legrand et al., 1998). Specifically for the imaging of quantum dots, X-STM relies on the fact that the dots are small and form rather dense populations. Hence, a random cleavage will cut through a large number of these nanostructures and provide a cross-sectional view of their atomic structure on the cleavage plane (Figure 15–34).

Much of the power of X-STM imaging derives from the fact that bias-dependent imaging provides chemical contrast on (110) cleavage faces of III–V compounds (Feenstra et al., 1987). Empty-state imaging gives atomically resolved maps of the cation sublattice and allows a direct identification of atomic species, e.g., indium atoms in a GaAs matrix (Pfister et al., 1995). This electronic structure effect not only provides strong contrast to determine the shape of individual buried InAs quantum dots and their stacking in multilayer structures, but can also be used to quantify interface roughness, intermixing, and surface segregation, near the dots and the wetting layer, with atomic precision.

To access this information in high-resolution images, a background subtraction has to be performed to remove topographic contrast of the

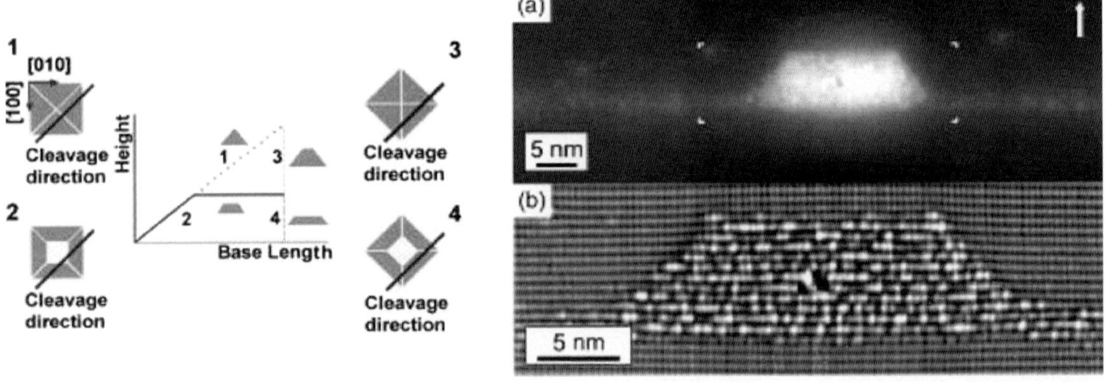

Figure 15–34. Cross-sectional STM on self-assembled quantum dots. (Left) Illustration of the possible apparent geometries observed due to cleavage at random positions through quantum dots with pyramidal and truncated pyramidal shape. (Right) Filled-state constant-current STM of an InAs quantum dot embedded in GaAs. Part of the image in (a) is treated by a local mean equalization filter to accentuate the atomic corrugations in the dot and the surrounding GaAs matrix, as shown in (b). (Reprinted with permission from Bruls et al., © 2002 American Institute of Physics; reprinted with permission from Gong et al., © 2004 Amercian Institute of Physics.)

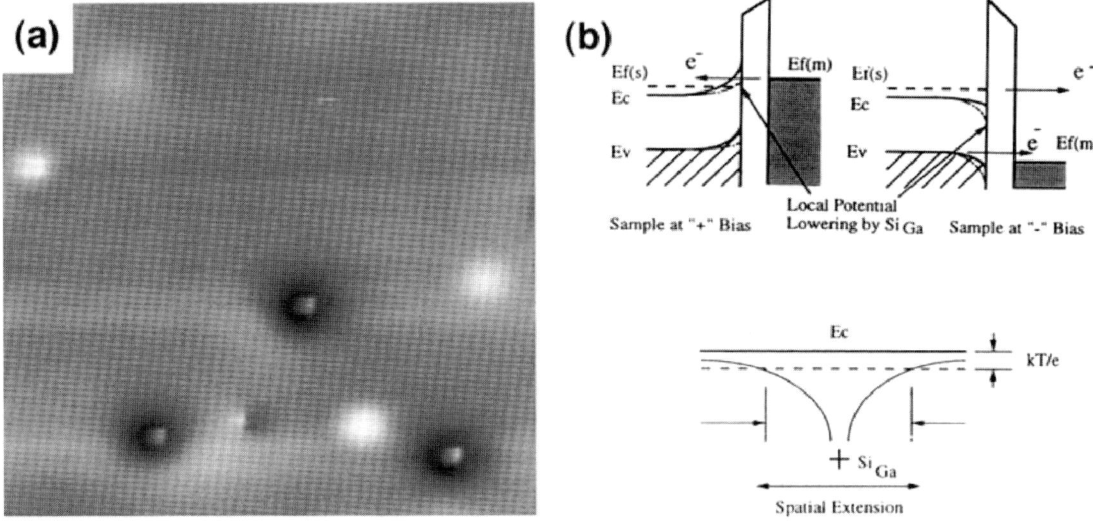

Figure 15–35. Imaging of individual dopants by X-STM. (a) Filled-state constant-current STM image ($V = -1.6\,$V) of cleaved GaAs(110). Individual Si_{Ga} substitutional donors appear as protrusions of different apparent height. (b) Band diagram illustrating the electron tunneling process between a metal tip and the GaAs(110) surface in the presence of tip-induced band bending. The Coulomb potential of a donor ion locally alters the band bending, causing an increase in tunneling current above the donor. (Reprinted with permission from Zheng et al., © 1994 by the American Physical Society.)

pronounced local deformation of the cleavage plane due to the relaxation of the strained dots and matrix (Davies et al., 2002). Conversely, when combined with modeling, the elastic relaxation at the free cleavage surface itself can be used to quantitatively determine the strain field in and around individual dots.

Apart from the imaging of buried semiconductor quantum structures, an application of X-STM that has been receiving increasing attention is the mapping of electronic dopants for electronics (Figure 15–35) and, more recently, of magnetic dopants for spintronics applications. Buried substitutional Zn and Be acceptors in p-GaAs have been imaged as protrusions in filled-state constant-current STM via a contrast mechanism assigned to an increased local state density directly above a dopant (Johnson et al., 1993). Si donors at Ga sites (Si_{Ga}) in GaAs have been imaged on GaAs(110) as protrusions, a few nanometers in size and with discrete values of apparent height, superimposed on the background lattice. The observed contrast in filled-state constant current STM was attributed to a local perturbation of the near-surface band bending by the Coulomb potential of the Si_{Ga}. The discrete apparent heights were interpreted as a consequence of a distribution of the donor atoms over five subsurface layers beneath the cleavage surface (Zheng et al., 1994).

Deep acceptor states due to magnetic impurities in III–V semiconductors, such as substitutional Mn_{Ga} in GaAs, are expected to play an important role in the hole-mediated coupling between magnetic impurities, and thus determine the magnetic properties of hole-mediated ferromagnetic semiconductors such as $Ga_{1-x}Mn_xAs$, important for emerging spintronics applications. X-STM at room temperature was used to map the wave function of the hole bound to an individual Mn acceptor in GaAs (Yakunin et al., 2004). Via bias-induced changes to the local band bending, the acceptor could be imaged in both the neutral ($U_S = +0.6\,V$) and ionized state ($U_S = -0.7\,V$). The acceptor ground state has a highly anisotropic structure due to a significant contribution of d-wave envelope functions, which is well-reproduced by simulated images based on a tight-binding model of the Mn acceptor structure (Tang and Flatté, 2004).

6 STM Image Simulation

Although STM is a very powerful experimental technique in its own right, geometric and spectroscopic contrast tend to be difficult to separate, as discussed previously. In many cases, the unequivocal identification of surface structures or of adsorption geometries requires a comparison of experimental STM images with contrast simulations. The conventional approach to STM image simulations follows a two-step process. Relaxed atomic positions of candidate surface structures are calculated by *ab initio* theoretical methods, such as density functional theory. The actual STM contrast is then computed using the Tersoff–Hamann theory of STM (Tersoff and Hamann, 1983, 1985). By combining bias-dependent atomic-resolution STM images with simulated images for various candidate structures, even completely unknown surface structures can in principle be "solved" on the basis of STM imaging alone. The difficulty lies less with the microscopy than with the identification of plausible candidate surface structures. In the past, this key step has typically been approached intuitively, i.e., by guessing structures based on minimizing the density of dangling bonds, starting from a truncated bulk structure. Recently developed systematic techniques, based on stochastic optimization, for generating comprehensive sets of candidate structures promise to become a powerful tool for solving a wide range of surface structures by STM imaging and image simulation (Ciobanu and Predescu, 2004).

The Tersoff–Hamann theory, the basis of most STM image simulations to date, provides a transfer matrix formalism to calculate the tunneling matrix element and tunneling current for realistic configurations of tip and sample. For values of tip–sample separation typical in STM imaging (~1 nm), the coupling between tip and sample wave functions is weak, and the tunneling process can be treated by first-order perturbation theory. Within this framework the tunneling current is given by

$$I = \frac{2\pi e}{\hbar} \sum_{\mu,\nu} [f(E_\mu) - f(E_\nu)] |M_{\mu\nu}| \delta(E_\nu + V - E_\mu)$$

$$\approx \frac{2\pi}{\hbar} e^2 V \sum_{\mu,\nu} |M_{\mu\nu}|^2 \delta(E_\mu - E_F) \delta(E_\nu - E_F) \tag{7}$$

where $f(E)$ is the Fermi function, V denotes the tunneling bias, $M_{\mu v}$ the tunneling matrix element between a tip state ψ_μ and a sample state ψ_v, and $E_{\mu(v)}$ the energy of state $\psi_{\mu(v)}$. In the second line, the Fermi function was replaced by a step function (zero-T approximation), and the limit of low tunneling bias has been assumed. If, after Bardeen (1961), the tunneling matrix element is written as a surface integral, and the sample and tip wavefunctions are expanded in plane waves, one obtains

$$M_{\mu v} = -\frac{4\pi^2\hbar^2}{m}\int dq\, a_q b_q^* e^{-\kappa_q z_t} e^{i\mathbf{q}\cdot\mathbf{x}_t} \tag{8}$$

Here $\mathbf{q}$ is the Fourier wavevector and a_q, b_q are expansion coefficients in the plane wave expansion of the sample and tip wavefunctions, x_t and z_t denote the lateral and vertical tip position, and κ_q is the decay constant. Given the wavefunctions of sample and tip, this expression provides the tunneling matrix element, from which the tunneling current can be calculated via Eq. (7).

In practice, a difficulty arises from the fact that the atomic-scale geometry and chemical composition of the tip are unknown. Thus, assumptions have to be made as to the tip wavefunctions (i.e., b_q). For an s-wavefunction of the tip (as for an ideal "point"–tip), the tunneling current is

$$I \propto \sum |\psi_v(\mathbf{r}_t)|^2 \delta(E_v - E_F) \equiv \rho(\mathbf{r}_t, E_F) \tag{9}$$

i.e., is proportional to the local density of states of the sample at the Fermi energy, a property of the sample surface alone! For finite tunneling bias, V, the tunneling current can be written as an integral over the energy range between the Fermi energy E_F and $E_F + V$:

$$I = \int_0^{eV} \rho_t(E)\rho_s(E-eV)T(E,eV)\,dEt \tag{10}$$

and the tunneling conductance at bias V is proportional to the position-dependent density of states of the sample at energy eV from the Fermi energy

$$\left(\frac{dI}{dV}\right)\bigg|_V \approx \rho_S(eV)\rho_T(0)T(eV,V) \tag{11}$$

Contrast calculations thus involve the computation of the position- and energy-dependent sample DOS to obtain simulated maps of tunneling conductance $dI/dV(x,y)$ or tunneling current $I(x,y)$.

Although originally derived for small voltages, as used for STM imaging on metals, with this extension the Tersoff–Hamann formalism can be applied to simulate images at higher bias as well, and has proven a powerful simulation tools for a wide range of imaging scenarios. Tromp et al. (1986) were the first to show that it could be applied to semiconductor surfaces, as demonstrated by their successful contrast simulation for the Si(111)-(7×7) reconstruction. The systems simulated since then include surface structures of semiconductors (Fujikawa et al., 2002; Klijn et al., 2003), ultrathin insulators (Olsson et al., 2005), as well as adsorbed atoms (Repp et al., 2004) and molecules (Olsson et al., 2003; Kühnle et al., 2002).

Figures 15–36 and 15–37 illustrate two of the numerous examples of this approach in the literature.

Shallow Ge quantum dots self-assembled during heteroepitaxy on Si(001) invariably have surface termination by (105) facets, with surface normal just 11° away from [001]. These (105) facets appear extremely stable, likely due to a low density of dangling bonds and additional strain stabilization by surface strain compensating for some of the 4% lattice mismatch strain between the Ge overlayer and the Si substrate. To calculate the surface energy, and thus explain the stability of the facet, a detailed knowledge of the surface structure is necessary. Dual bias constant-current STM combined with STM contrast calculations was used to identify a best match with one of two proposed structures of the (105) surface: "paired dimer" (Structure A) and "rebonded step" (Structure B) (Fujikawa et al., 2002). The STM contrast is clearly identified as that of a "rebonded step" structure, the structure that not only has the lowest dangling bond density but also causes tensile surface strain, key to the strain stabilization of the (105) facet.

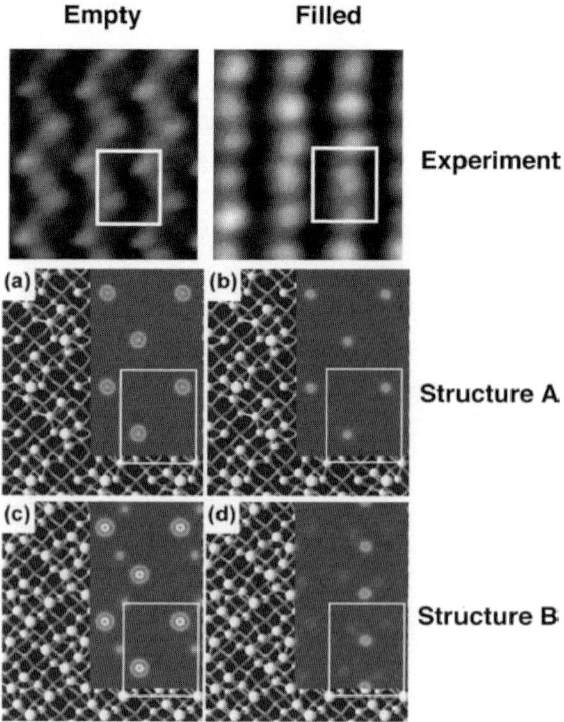

Figure 15–36. Identification of the surface structure of Ge(105). Empty- and filled-state constant-current STM images are compared with two candidate structures, the paired dimer Structure A and rebonded step Structure B (c and d). A Tersoff–Hamann calculation of the STM contrast for these two structures clearly shows that the experimental STM images arise from the rebonded step structure. (Reprinted with permission from Fujikawa et al., © 2002 by the American Physical Society.)

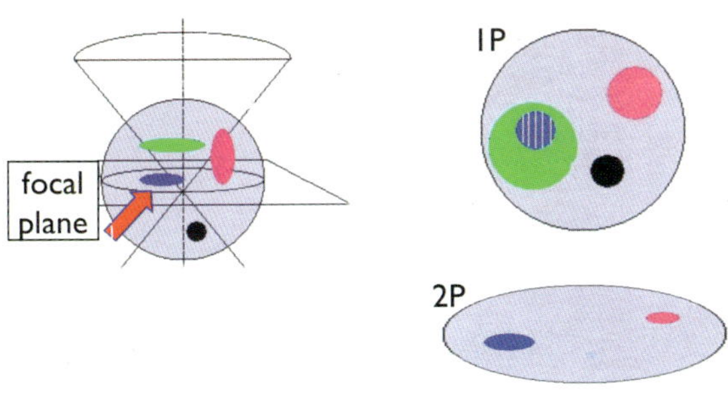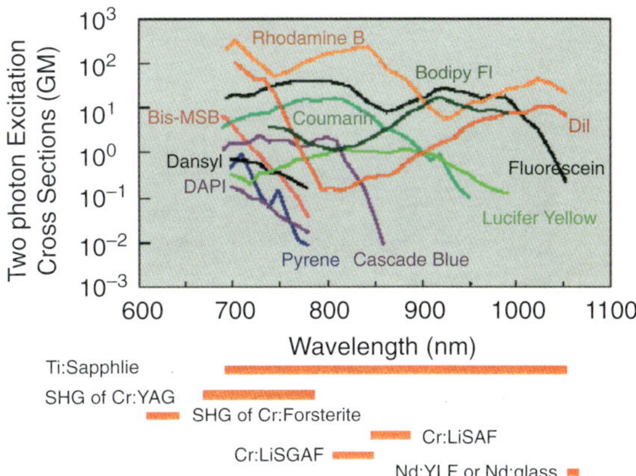

Figure 11–2. Comparison between conventional (1P) and two-photon (2P) excitation with respect to image formation. When focusing on the actual focal plane under 1P, a contribution from adjacent planes that are physically excluded in the 2P process is obtained, as happens in a confocal setup. (From Giuseppe Vicidomini, LAMBS, MicroScoBio, University of Genoa.)

Figure 11–6. Two-photon cross-sections for popular fluorescent molecules as a function of the excitation wavelength. Red bars indicate the emission range of some common laser sources utilized in TPE microscopy and spectroscopy.

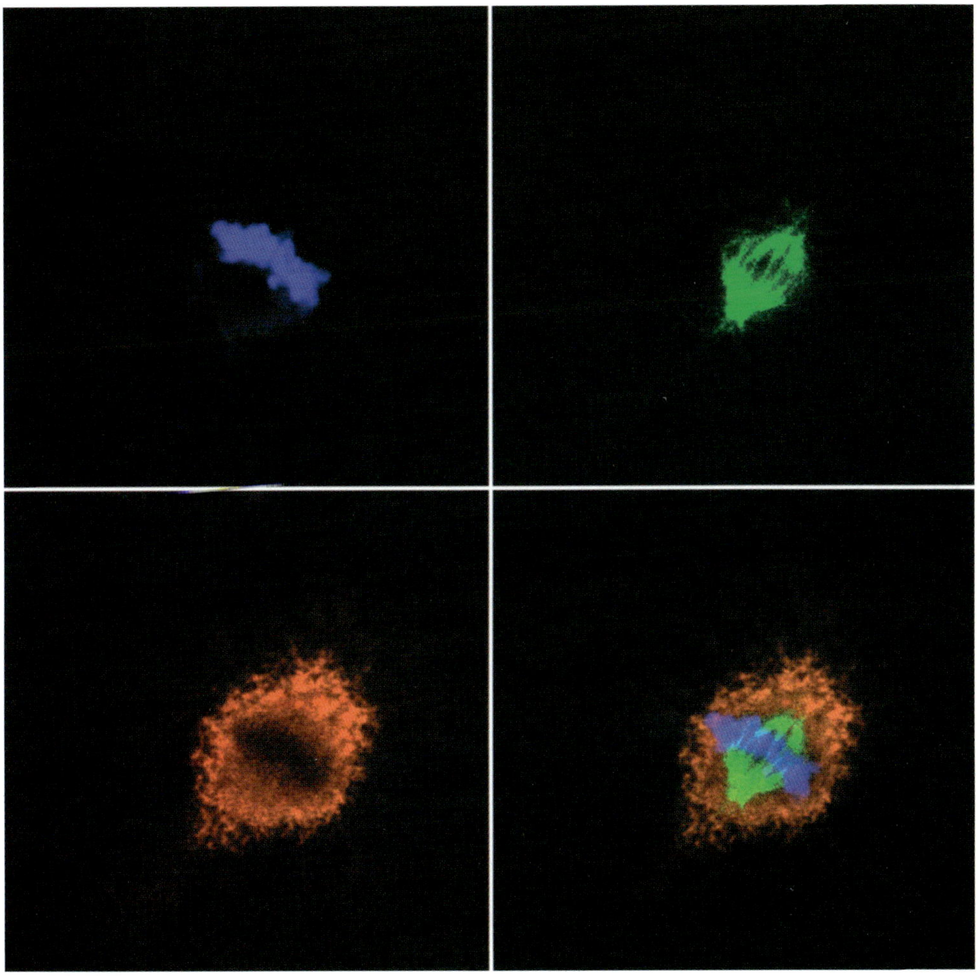

Figure 11–12. Multiple excitation of three fluorescent dyes using 740 nm under a TPE regime. The conventional excitation would have required the utilization of 360 or 405 nm, 488 nm, and 543 nm laser lines. The final image (lower right quadrant) is realized by merging the three subsets. (This image has been acquired by students of the Biotechnology School during the course of Advanced Microscopy Techniques activated at the University of Genoa, academic year 2005. Advisors: Grazia Tagliafierro and Alberto Diaspro.)

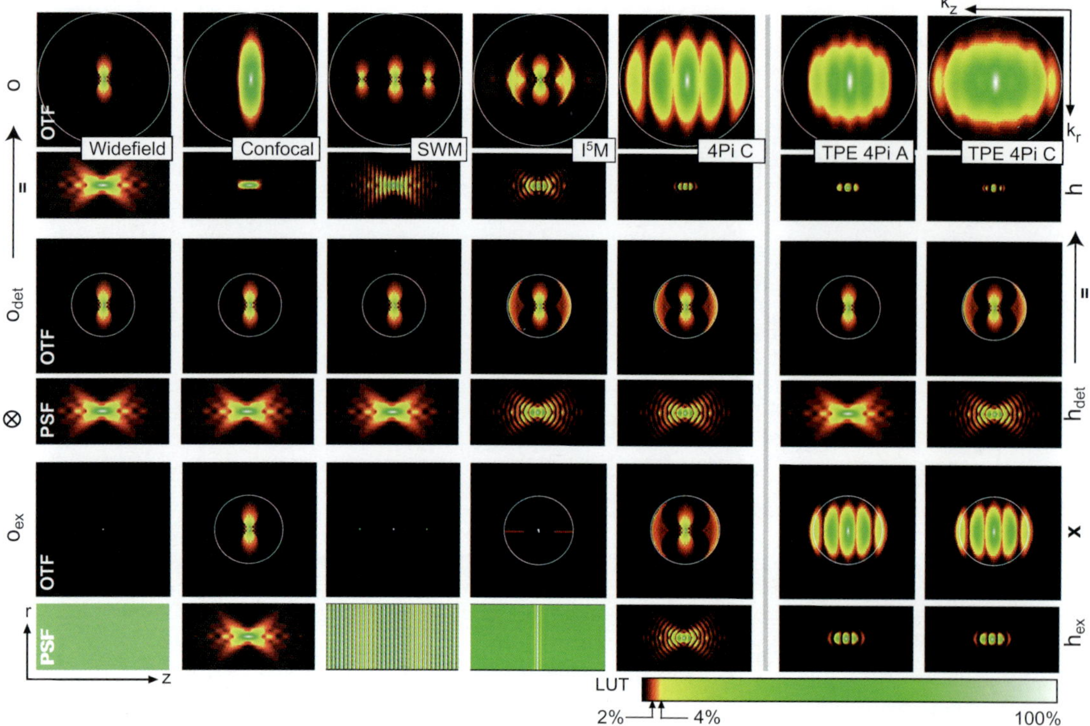

Figure 12–4. Overview of the excitation (bottom), detection (center), and effective (top) OTFs' modulus and the corresponding PSFs of the wide-field, confocal, standing wave (SWM), I^{5}M, 4Pi confocal type C, TPE 4Pi confocal type A, and TPE 4Pi confocal type C microscopes. The color look-up table (LUT) has been designed to emphasize the important weak OTF regions. The OTFs are shown in the squares above the corresponding PSFs; the zero frequency point is in the center and the largest frequency displayed is $2\pi/80\,\mathrm{nm}^{-1}$. The circles represent the maximum possible carrier as explained in Figure 12–3. For TPE, the excitation OTFs slightly extend over these circles because the excitation wavelength of 800 nm is less than double the one-photon excitation wavelength of 488 nm. While all these methods extend the OTF along the axial direction, they fundamentally differ in contiguity and absolute strength within the support region. For example, there are pronounced frequency gaps for the SWM and depressions for the I^{5}M. The rectangular images of the PSFs represent a region of $5 \times 2.5\,\mu\mathrm{m}$ with the geometric focus in the center.

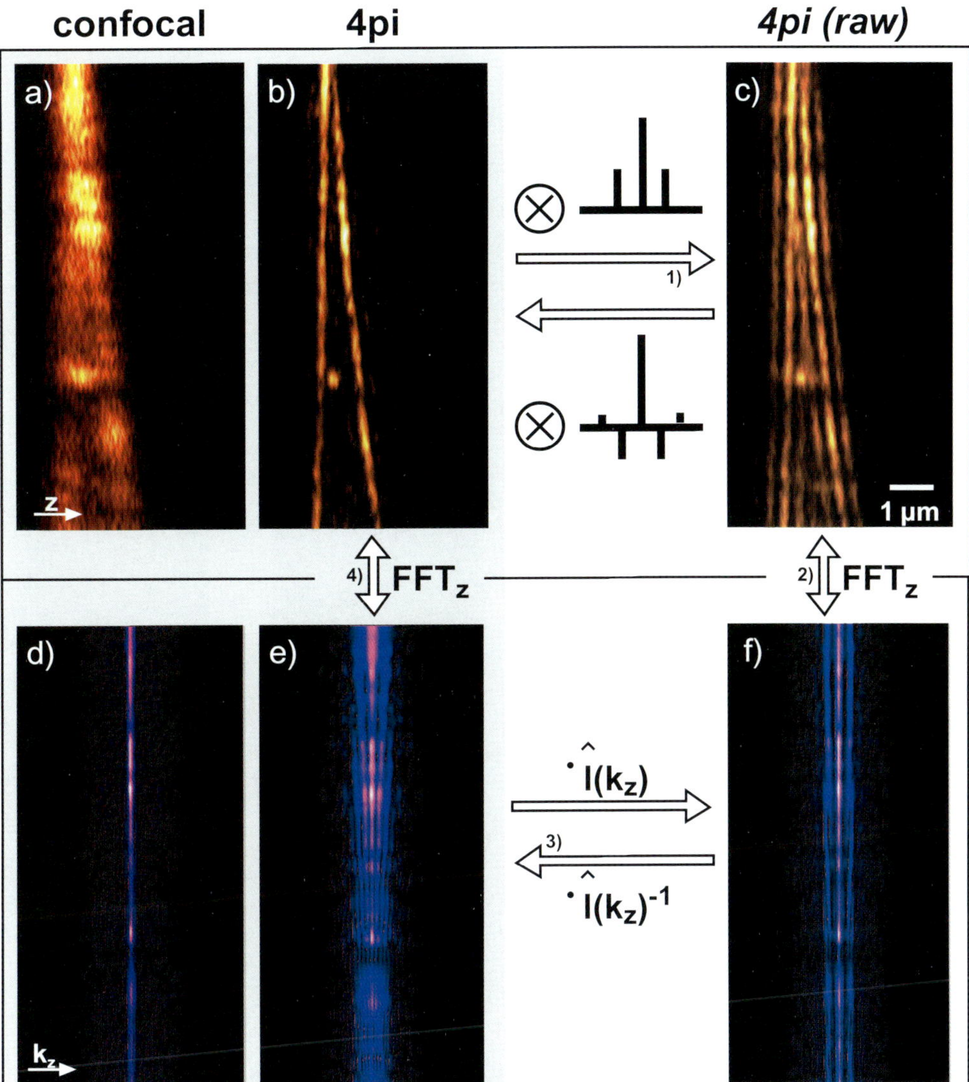

Figure 12–5. Lobe removal and deconvolution in 4Pi microscopy. The figure shows images of the same pair of actin fibers in a fixed mouse fibroblast cell recorded in the TPE confocal (a) and TPE 4Pi type A (b and c) mode. The corresponding Fourier transform along the optic axis is also shown (d, e, and f). The five-fold axial resolution increase (a vs. b) and the correspondingly extended OTF (d vs. e) are immediately visible. The side lobes are well below 50% and the factorization of the PSF's axial and lateral dependence is possible in 4Pi microscopy. Therefore, an inverse discrete filter can be found and its application to the raw data (c) yields a valid and almost artifact-free image (b). Alternatively, lobe removal can be performed in the frequency domain. Equation (16) indicates that the Fourier transforms of the raw data (f) is given by the product of the Fourier transform of the lobe-free image (e) and the lobe function $\hat{l}(k_z)$. Thus, (2) Fourier transforming the raw data, (3) multiplying with the inverse of the lobe function's Fourier transform, $\hat{l}^{-1}(k_z)$, and (4) Fourier backtransforming lead to almost the same lobe-free image. This method can be applied even if the separation of axial and lateral dependence is impossible for the PSF. The Fourier transforms along the axial direction (3 and 4) merely have to be replaced by their 3D counterparts.

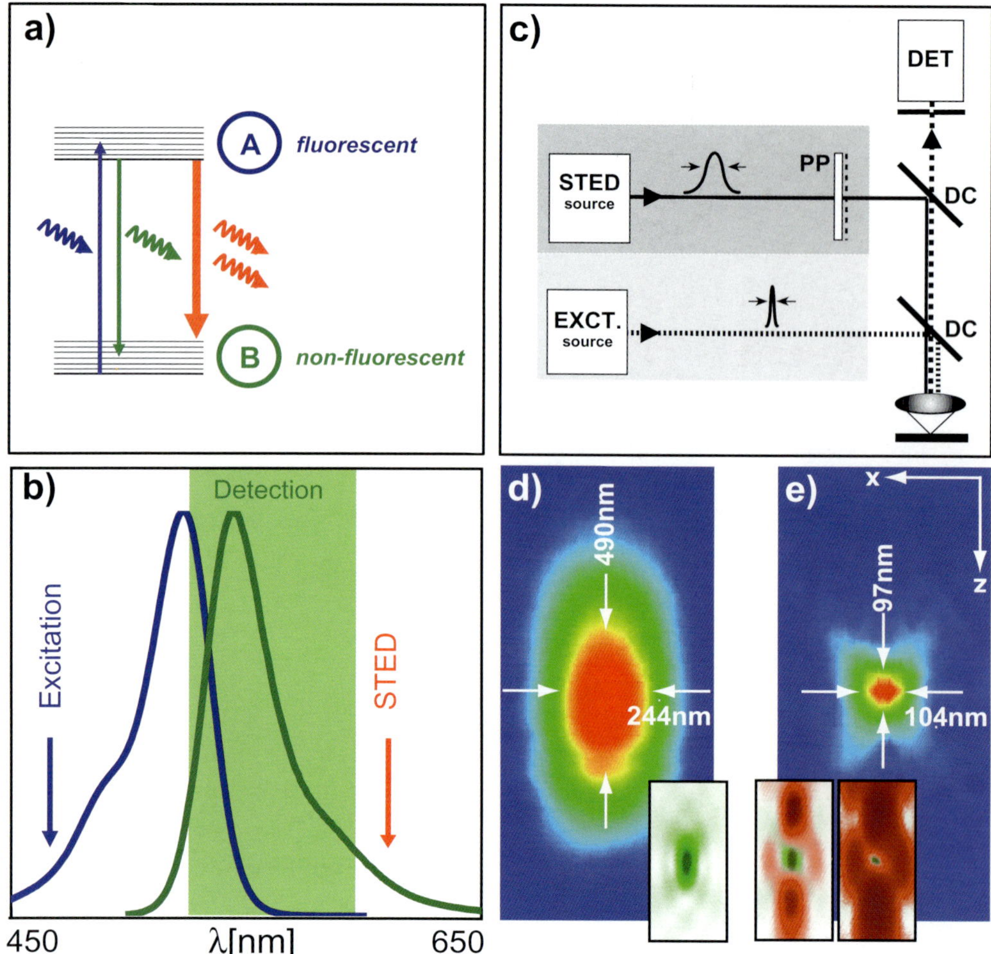

Figure 12–7. Stimulated emission depletion (STED) was the first implementation of the RESOLFT principle. (a) Dye molecules are excited into the S_1 (state A) by an excitation laser pulse. (b) Fluorescence is detected over most of the emission spectrum. However, molecules can be quenched back into the ground state S_0 (state B) using stimulated emission before they fluoresce by irradiating them with a light pulse at the edge of the emission spectrum shortly after the excitation pulse and before they are able to emit a fluorescence photon. Saturation is realized by increasing the intensity of the depletion pulse and consequently inhibiting fluorescence everywhere except at the "zero points" of the focal distribution of the depletion light. (c) Schematic of a point-scanning STED microscope. Excitation and depletion beams are combined using appropriate dichroic mirrors (DC). The excitation beam forms a diffraction-limited excitation spot in the sample (inset in d) while the depletion beam is manipulated using a phase-plate (PP) or any other device to tailor the wavefront in such a way that it forms an intensity distribution with a nodal point in the excitation maximum (left inset in e). The third inlay shows the resulting quenching probability when saturating the depletion process. (d) and (e) show an experimental comparison between the confocal PSF and the effective PSF after switching on the depleting beam. Note the doubled lateral and five-fold improved axial resolution. The reduction in dimensions (x, y, z) yields ultrasmall volumes of subdiffraction size, here 0.67 al (Klar et al., 2000), corresponding to an 18-fold reduction compared to its confocal counterpart. The spot size is not limited on principle grounds but by practical circumstances such as the quality of the zero and the saturation factor of depletion.

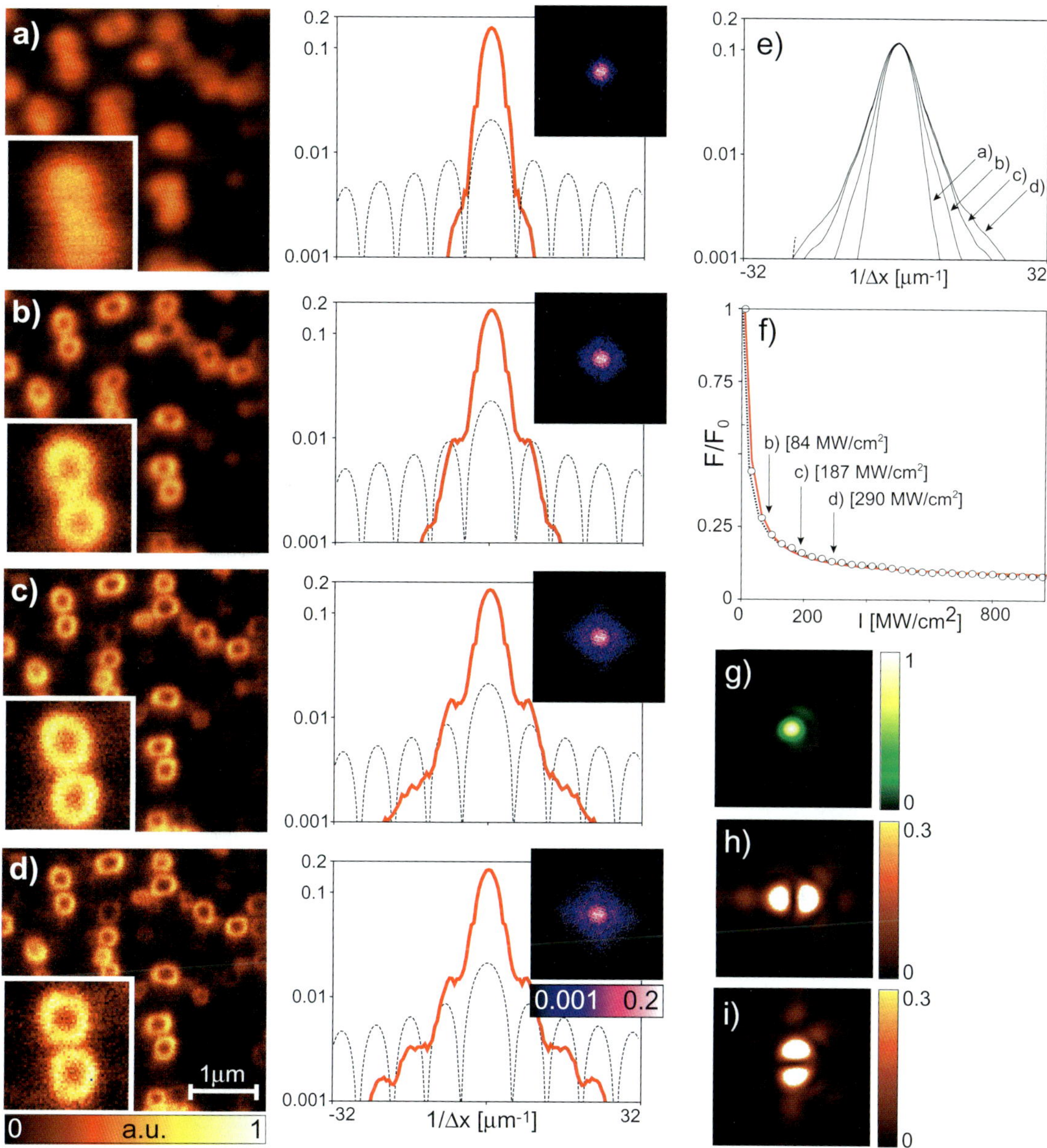

Figure 12–10. Images of a wetted Al_2O_3 matrix featuring z-oriented holes (Whatman plc, Brentford, UK) with a spin cast of a dyed (JA 26) polymethyl methacrylate solution. The rings formed in this way are ~250 nm in diameter and are barely resolved in confocal mode. (a–d) The confocal image (a) and STED images with two depleting beams perpendicularly polarized and aberrated by "1D" phase-plates (b–d). The excitation PSF (g) and the STED PSF for y polarization (h) and x polarization (i) are shown on the right. The STED intensity was chosen at the spots marked in the saturation curve (f). The smaller effective spot size also results in an extended OTF as seen in the second column. Here, the insets show the 2D Fourier transformation of the images in the left and the graphs show a profile along the x direction. Note the logarithmic scales. The Fourier transform of the image is given by the product of OTF and the Fourier transform of the object [Eq. (2)]. For such regular structures, an estimate for the modulus of the OTF can therefore be gained by estimating the latter and solving for the OTF. The dashed line shows the Fourier transform of a ring with a diameter of 275 nm and a width of 50 nm and the estimated OTF is presented in (e). (f) The suppression of fluorescence resulting from stimulated emission. The phase-plates were removed and the ratio of fluorescence without STED light (F_0) and with the STED beams switched on (F) was recorded. The intensities are pulse intensities per beam at the global maximum.

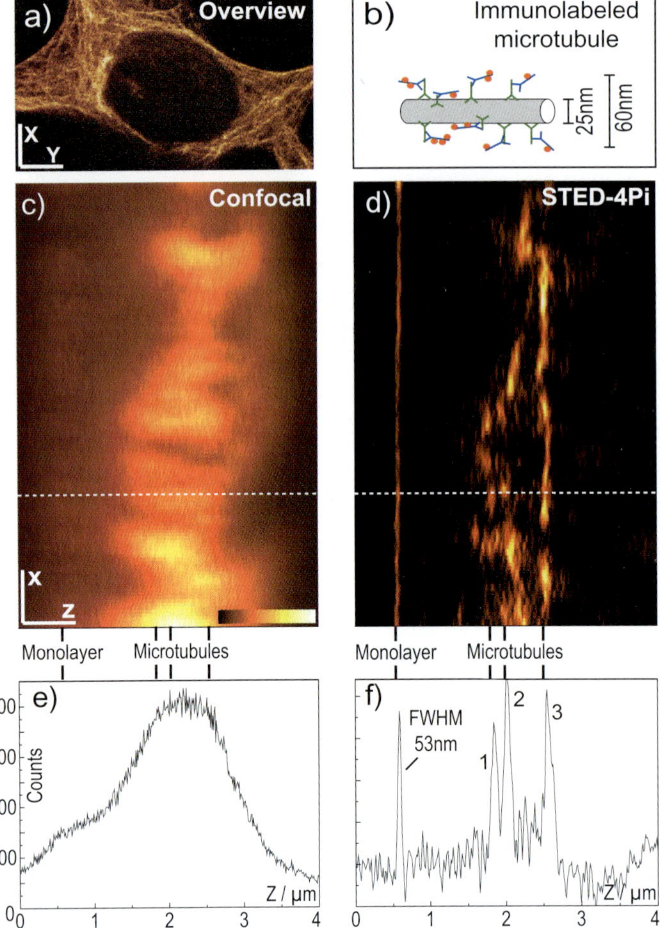

Figure 12–11. Subdiffraction immunofluorescence imaging with STED-4Pi microscopy. (a) Overview image (*xy*) of the microtubular network of an HEK cell. (b) Sketch of typical dimensions of a labeled microtubule fluorescently decorated via a secondary antibody. (c) and (d) Standard confocal and STED-4Pi *xz* image recorded at the same site of the cell; the straight line close to the cell stems from a monomolecular fluorescent layer attached to the adjacent coverslip. In both images, the pixel size was 95×9.8 nm in the x and z direction, respectively; the dwell time per pixel was 2 ms. Note the fundamentally improved clarity in (d). The STED-4Pi microscope's PSF features two low side lobes caused by the secondary minima STED intensity distribution. These lobes are <25% and were removed in the STED-4Pi image using linear filtering as outlined in the text [see Eq. (16ff)]. (e) and (f) Corresponding profiles of the image data along the dashed lines in (b) and (c) quantify the improved axial resolution of the STED-4Pi microscopy mode (f) over the confocal benchmark. Peaks 1, 2, and 3 due to microtubules are broader than the response to the monolayer. Note the ability of the STED-4Pi microscope to distinguish adjacent features.

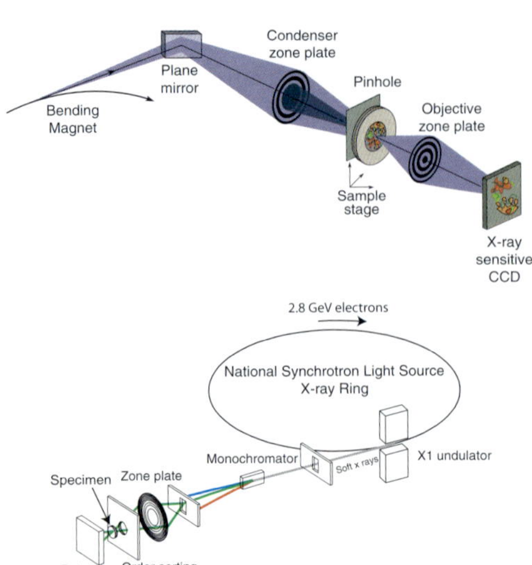

Figure 13–15. Schematic of the main components of a transmission X-ray microscope or TXM (top: courtesy of D. Attwood, Lawrence Berkeley National Laboratory) and a scanning transmission X-ray microscope or STXM (bottom: courtesy of Y. Wang, then of Stony Brook.)

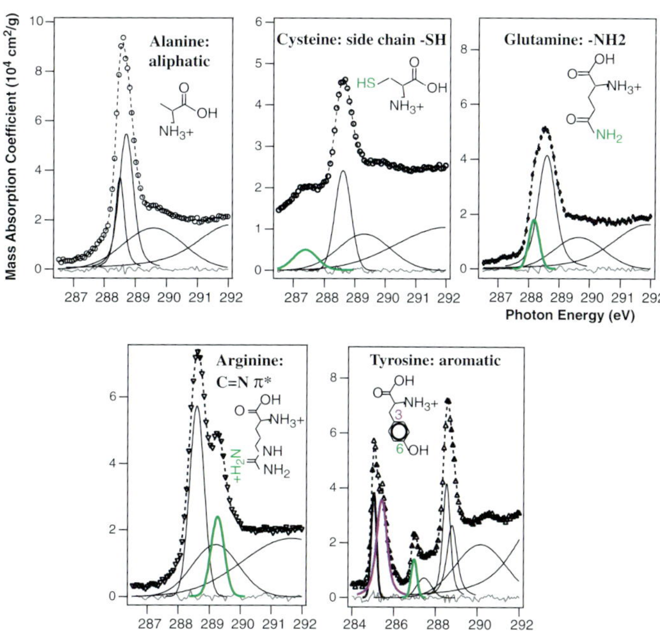

Figure 13–27. Near-carbon-edge absorption spectra of several amino acids, showing the effects of various molecular bonds in the absorption spectrum. These resonances can be used for chemical contrast in X-ray microscopy. (Reprinted from Kaznacheyev et al., © 2002, with permission from American Chemical Society.)

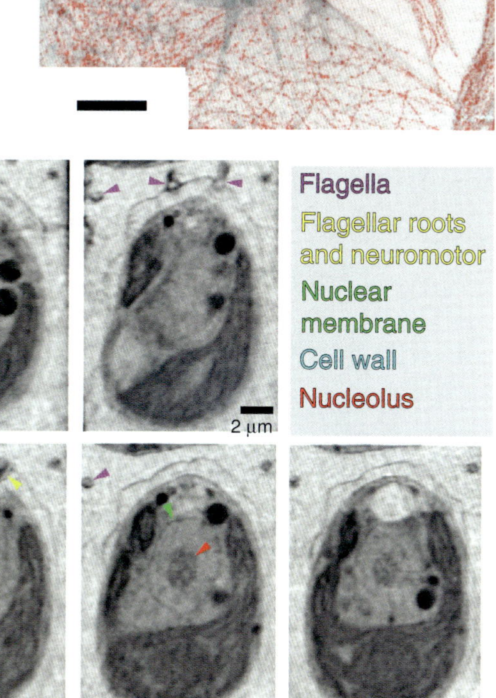

Figure 13–29. Human fibroblast with immunogold labeling for tubulin. This is a composite of two images: a bright field image (gray tones) to image overall mass, and a dark field image (red tones) to selectively imaging the silver-enhanced gold labels. This whole-mount cell was fixed and then permeabolized to allow for introduction of the immunogold labels, after which it was air dried. (From Chapman et al., © 1996b,c, courtesy of the Microscopy Society of America.)

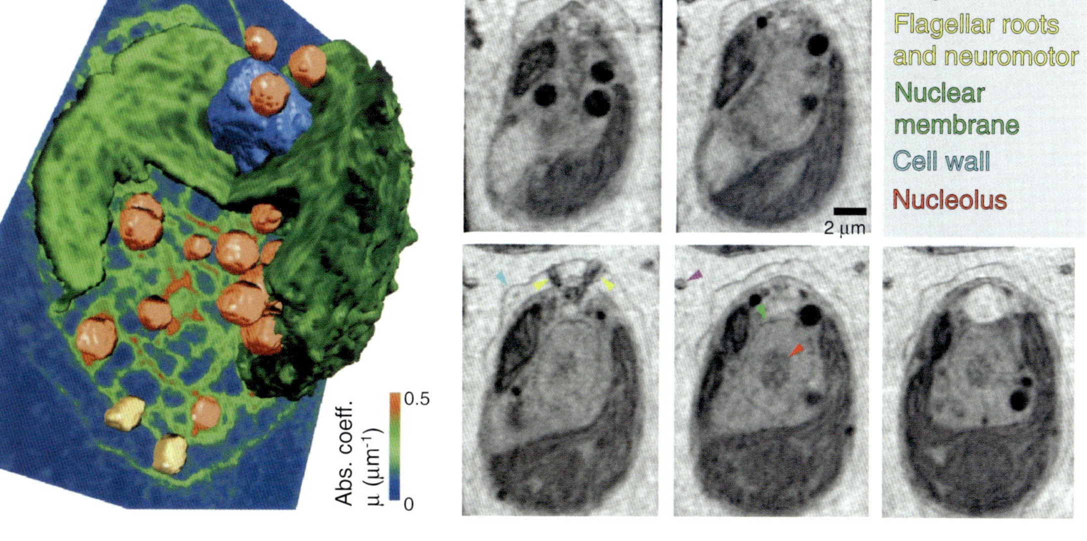

Figure 13–30. 3D rendering (left) and reconstruction slices (right) of the algae *Chlamydomonas reinhardtii* viewed by soft X-ray tomography at the BESSY I synchrotron. This alga was plunge-frozen in liquid ethane, and imaged over 180° rotation sequence. The reconstruction is given in terms of the quantitative linear absorption coefficient for 517 eV X-rays. (Reprinted from Wei et al., © 2000, with permission from Elsevier.)

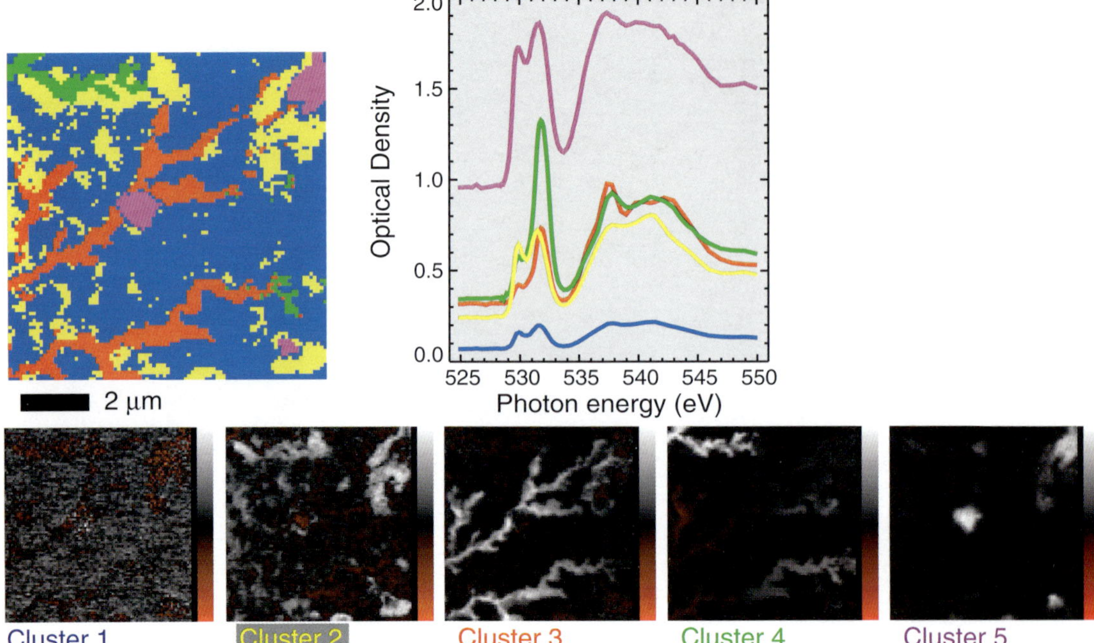

Figure 13–34. Cluster analysis in a spectromicroscopy study of lutetium in hematite. Lutetium is serving as a homologue to americium in an investigation of the uptake and transport of nuclear waste products in groundwater colloids. By using a pattern recognition algorithm to search for pixels with spectroscopic similarities, a set of signature spectra is automatically recovered from the data (shown here in a color-coded classification map) and thickness maps can be formed based on these signature spectra. Analysis at the oxygen edge reveals two different phases of reactivity for lutetium with hematite. Analysis by Lerotic (2004), from a study by T. Schäfer, INE Karlsruhe.

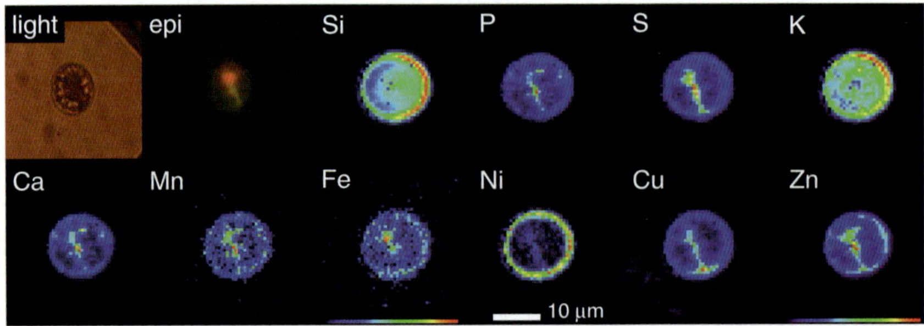

Figure 13–35. Visible light and epifluorescence micrographs, and false color X-ray fluorescence element maps of a centric diatom collected from the southern Pacific. In this region of the ocean, iron availability is a biolimiter with an impact on oceanic uptake of carbon dioxide from the atmosphere. X-ray microprobes allow one to study iron content on a protist-specific basis. (Reprinted from Twining et al., © 2003, with permission from American Chemical Society.)

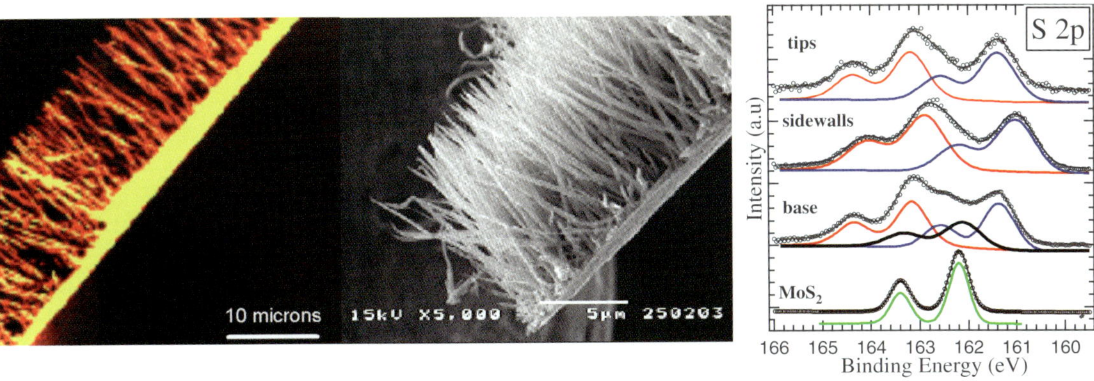

Figure 13–42. In Electron Spectroscopy for Chemical Analysis or ESCA microscopy, a monochromatic beam is used to illuminate a region several micrometers across; electron optics are then used to image a tunable electron ejection energy to reveal surface chemistry. Though this does not involve zone plate imaging, we include it here due to its widespread use with tunable X rays. In this case a 90-nm resolution ESCA microscope was used to locate aligned MoS_2 nanotube bundles and select certain areas along the axes of the tubes for detailed examination. The image at left was acquired using Mo $3d$ electrons, while S $2p$, Mo $3d$, and valence band spectra taken at the tips and sidewalls and the growth base from the Si wafer appear strongly affected by the low dimensionality of the nanotubes and differ significantly from the corresponding spectra taken on a reference MoS_2 crystal. (Reprinted from Kiskinova et al., © 2003, with permission of EDP Sciences.)

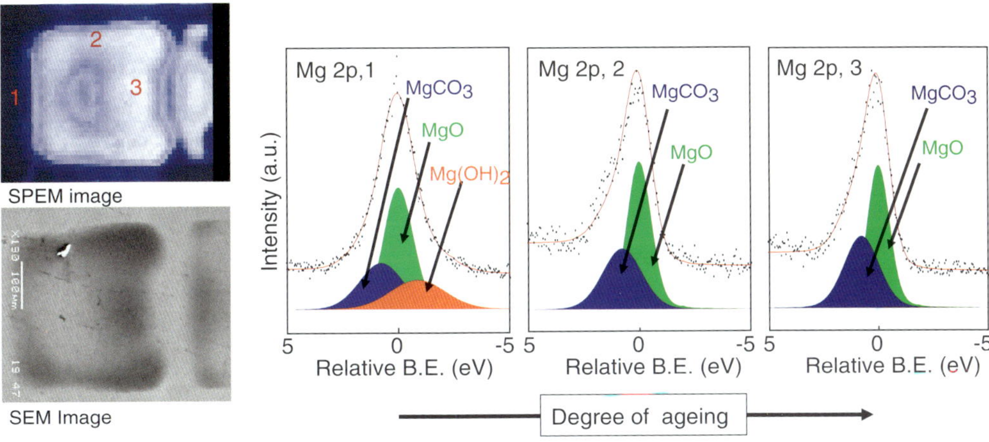

Figure 13–43. Scanning photoemission microscope (SPEM) study of a plasma display cell. In this microscope the specimen is scanned through the zone plate focus while photoelectrons are collected by an electron spectrometer. This figure shows a SPEM image, a scanning electron micrograph, and photoelectron spectra from several regions of the sample. In a plasma display cell, light of the appropriate color emerges through a front glass window which is protected from plasma damage by a composite insulating layer including MgO. The photoelectron spectra show aging in the $Mg(OH)_2$ component of the layer over the life of the display cell. (Reprinted from Yi et al., © 2005, with permission from the Institute of Pure and Applied Physics.)

Figure 14–13. Transport imaging in polycrystalline ZnO. (a) Surface topography. (b and c) SSPM images under lateral bias exhibit potential drops at grain boundaries, indicative of grain boundary resistive behavior. Note that the direction of potential drops inverts with bias. (d) Current maps for positive and negative bias polarity. (Partially reproduced from Kalinin and Bonnell, *Zeitschrift fur Metallkunde*, 90, 983–989, 1999.)

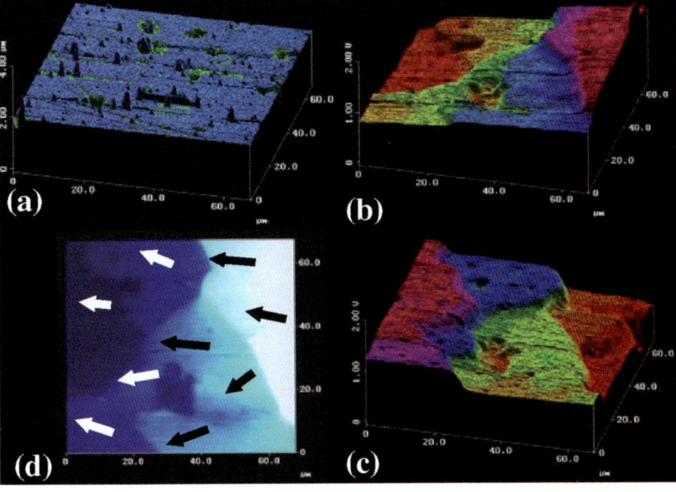

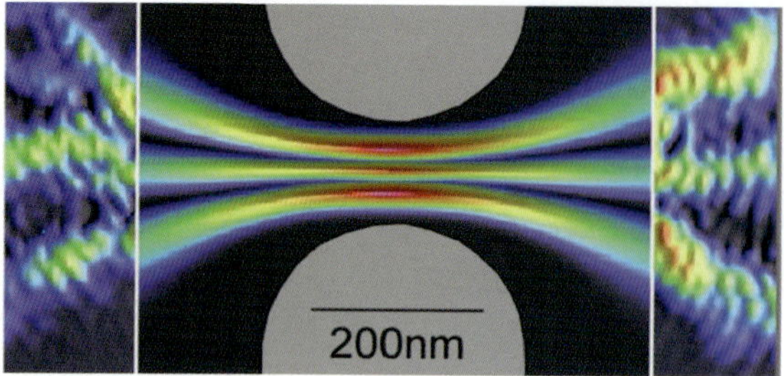

Figure 14–15. Experimental images (outside) and theoretical simulations (inside) of the flow of electron waves through a quantum point contact. Fringes spaced by half the Fermi wavelength demonstrate coherence in the flow. [Courtesy of Westervelt et al. Reproduced with permission from Physica E, 24, 63–69, 2004.]

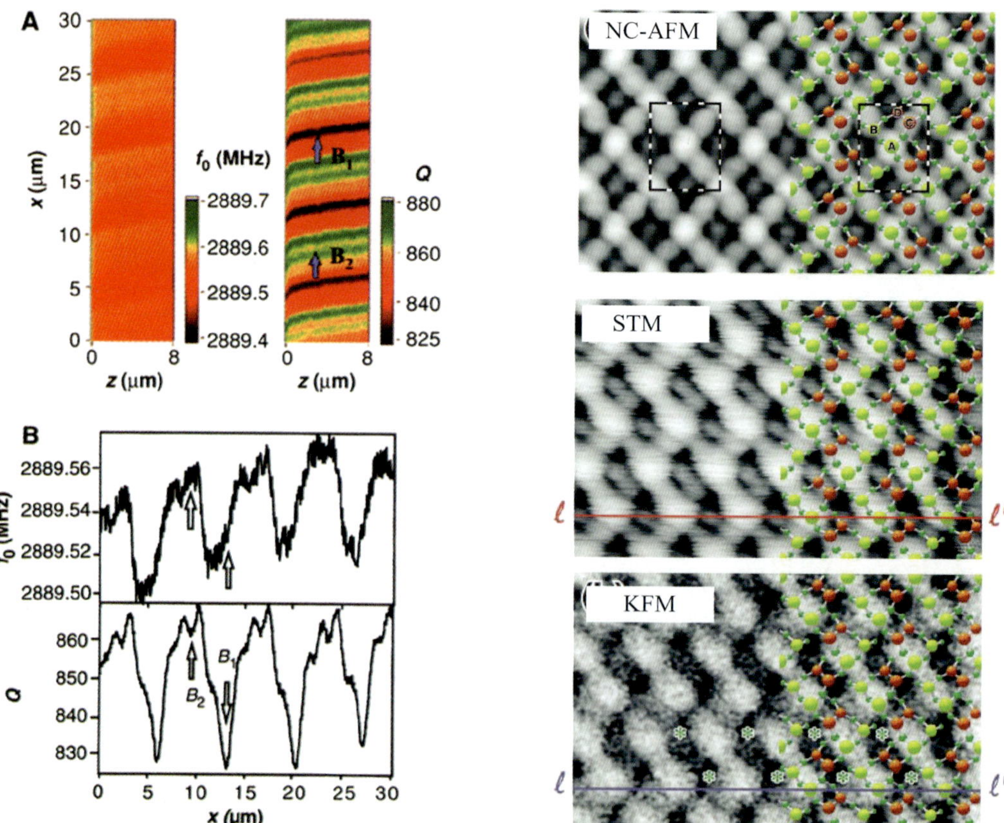

Figure 14–22. (A) Images of the profiles of periodic dielectric constant (on the left) and ferroelectric domain boundaries (on the right). (B) Line profiles of frequency (top) and quality factor (bottom) images.

Figure 14–23. Atomic resolution NC-AFM, STM, and KFM on Ge/Si (105). [Courtesy of Eguchi et al. Reproduced with permission from Physical Review Letters, 93(26), 2004.]

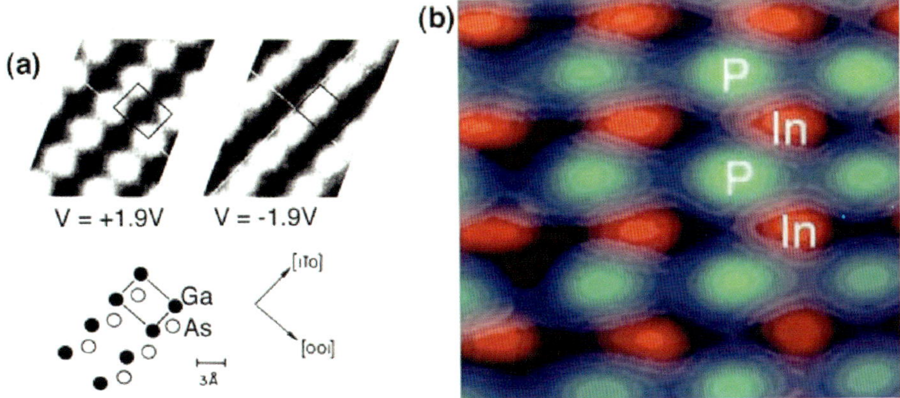

Figure 15–13. (a) Bias-dependent STM on GaAs(110): selective imaging of Ga and As sublattices at positive and negative sample bias, respectively (Reprinted with Permission from Feenstra et al., ©1987 by the Armeucion Physics Socuity). (b) Compound STM image of the InP(110) surface, assembled from separate positive and negative bias scans (Reprinted from Ebert et al., ©1992 with permission from Elsevier).

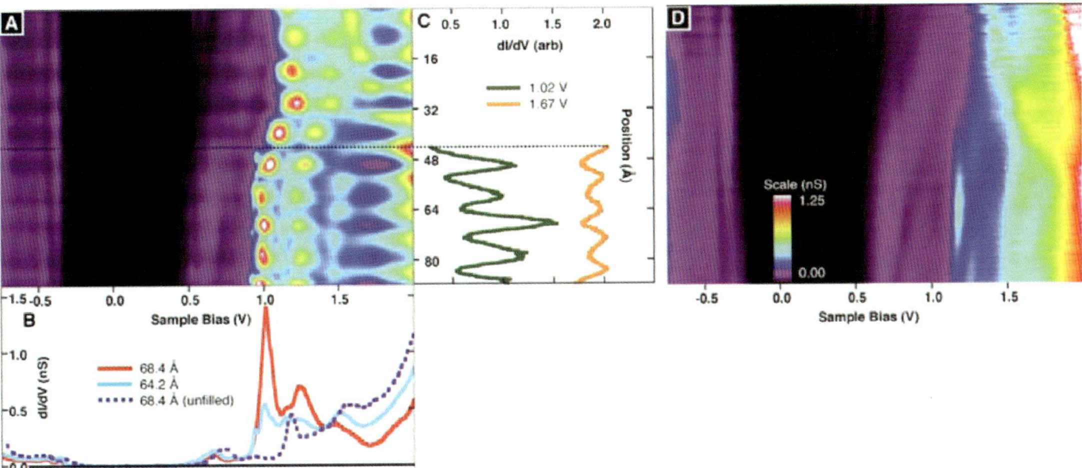

Figure 15–16. $I(V)$ tunneling spectroscopy on C_{60}/C-nanotube "peapods" (Reprinted with Permission from Hornbaker et al., ©2002 AAAS). (A) Map of an array of full dI/dV spectra along the axis of a C-nanotube "peapod." Sample bias voltage is plotted on the horizontal axis and displacement along the tube on the vertical axis. (B) Representative dI/dV spectra at selected positions along the tube. Large conductance peaks are found at positions of embedded C_{60} molecules. (C) Variation of tunneling conductance along the tube axis. (D) Reference spectroscopic map on an empty C-nanotube section without embedded C_{60}, in which no strong modulation of the tunneling conductance is observed.

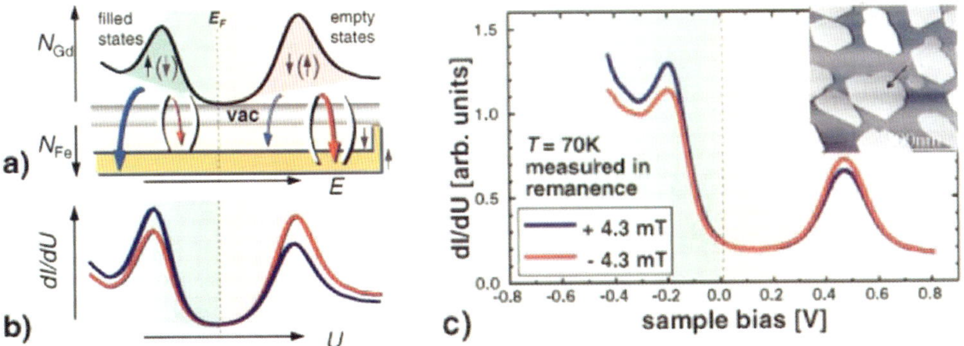

Figure 15–18. Spin-polarized STM on a Gd(0001) sample with an exchange-split surface state and a magnetic Fe tip with constant spin polarization close to E_F. (a) Due to the spin-valve effect the tunneling current of the surface state spin component parallel to the tip magnetization is enhanced. (b) Illustration of the reversal in the dI/dV signal at the surface state peak position upon switching the sample magnetically. (c) Experimental observation of this reversal in tunneling into an isolated Gd island (Reprinted with permission from Bode et al., ©1998 by the American Physical Society).

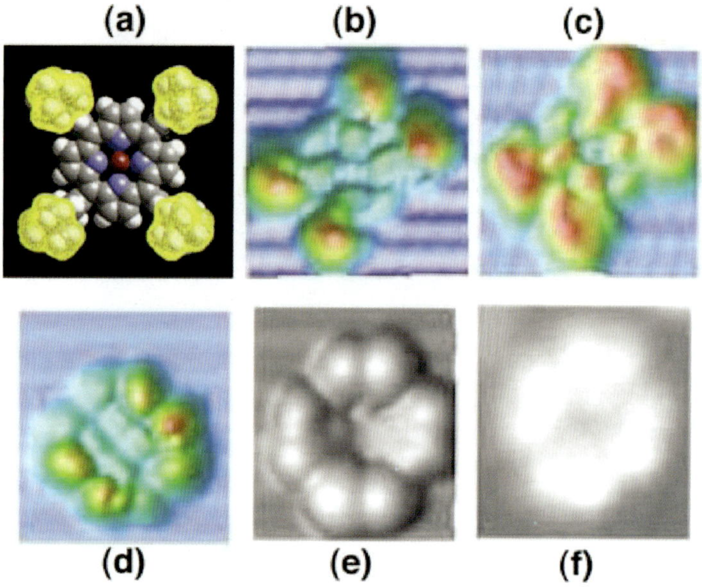

Figure 15–37. Identification of the molecular conformation of Cu-DTBPP on Cu(211). (a) Molecular model of Cu-DTBPP, with the four-lobed pattern observed by STM marked in yellow. (b–e) STM contrast calculation for different angles between the four legs and the substrate: (b) 60°, (c) 45°, (d) 30°, and (e) 10°. (f) Experimental STM image of the molecule. (Reprinted from Moresco et al., ©2002 with permission from Elsevier.)

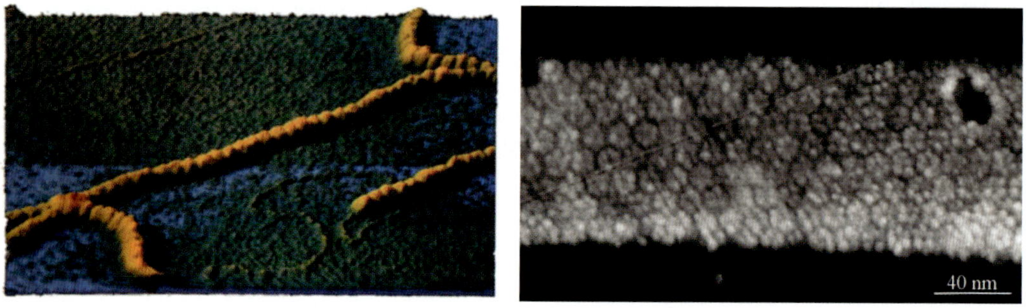

Figure 16–1. STM topographies of Rec-DNA complex (left) and an aberrant bacteriophage capside (right). In the upper right corner, a hole has been created applying a voltage pulse between the tip and the sample. Both images immediately reveal the molecular arrangement of the respective complex structure. (Left, from Amrein et al., 1988, reproduced with permission. Right, from Amrein et al., 1998b, reproduced with permission.)

Figure 16–3. Neuronal cell, cultured on an electronic chip. The chip is designed such as to pick up an action potential of the cell. The image demonstrates proper tracing of the cell surface. In a future application, an appropriately designed stylus might be used as an additional electrode to excite or record an action potential at any location of the cell body or a process of the neuronal cell.

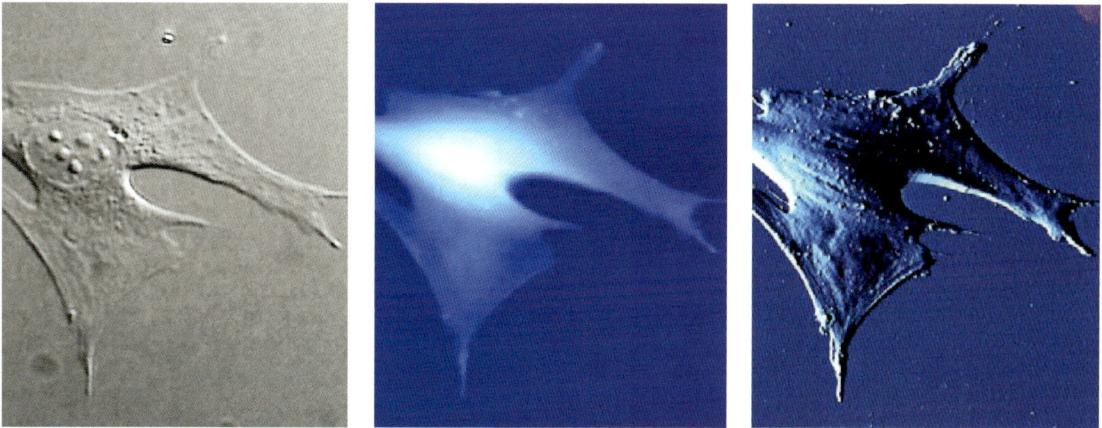

Figure 16–18. Images of a fixed 3T3 cell in buffer, on a coverslip. Left to right: Optical (differential interference contrast, DIC) and simultaneous AFM (JPK Instruments, contact mode) topography and feedback signal.

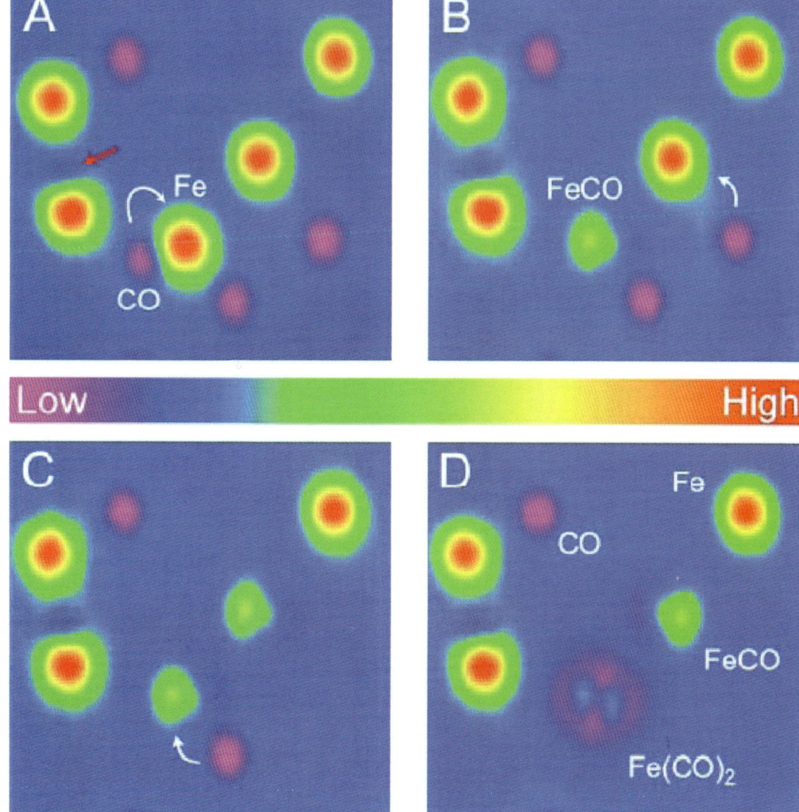

Figure 17–36. Bond formation induced with STM tip. A sequence of STM constant-current images at 13 K showing the formation of Fe–CO bonds by vertical manipulation. Fe atoms are imaged as protrusions and CO molecules as depressions. The white arrows indicate the pair of adsorbates involved in each bond formation step. In (B) and (C) a CO molecule has been picked up and bonded to an Fe atom to form Fe(CO). In (D) a second CO molecule has been bonded to Fe(CO) to form Fe(CO)$_2$. (From Lee and Ho, 1999.)

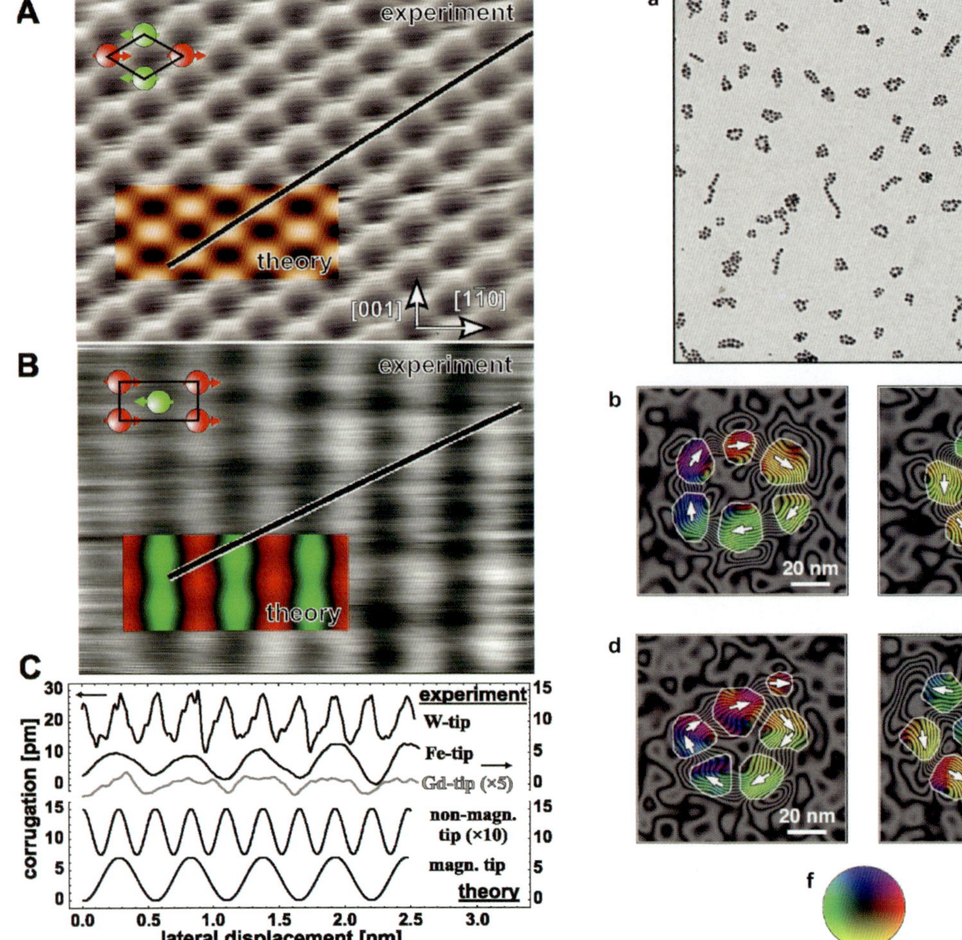

Figure 17–43. Atomic resolution magnetic imaging with SP-STM. (A) Constant-current image of one monolayer of Mn on W(110) recorded with a non-magnetic W tip at 16 K. (B) Image recorded with a magnetic Fe tip showing an antiferromagnetic configuration as predicted by theory. The colored insets show calculated STM images. (C) Experimental and theoretical line sections from (A) and (B). The image size is 2.7 nm by 2.2 nm. (From Heinze et al., 2000.)

Figure 18–9. (a) Low magnification brightfield image of self-assembled Co nanoparticle rings and chains deposited onto an amorphous carbon support film. Each Co particle has a diameter of between 20 and 30 nm. (b–e) Magnetic phase contours (128× amplification; 0.049 radian spacing), formed from the magnetic contribution to the measured phase shift, in four different nanoparticle rings. The outlines of the nanoparticles are marked in white, while the direction of the measured magnetic induction is indicated both using arrows and according to the color wheel shown in (f) (red = right, yellow = down, green = left, blue = up). (Reprinted from Dunin-Borkowski et al., 2004b.)

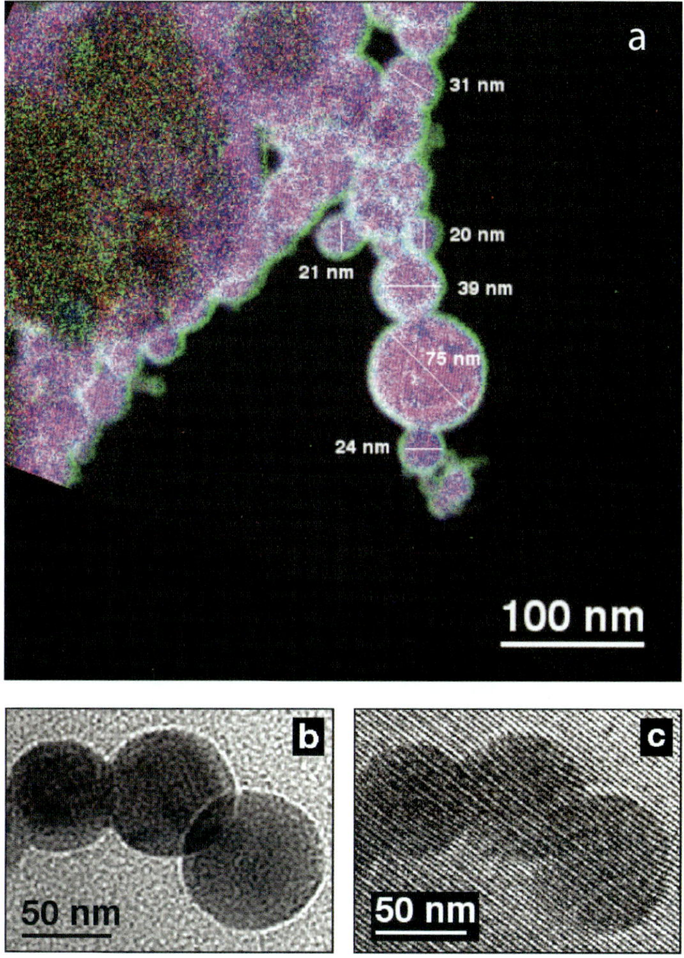

Figure 18–10. (a) Chemical map of $Fe_{0.56}Ni_{0.44}$ nanoparticles, obtained using three-window background-subtracted elemental mapping with a Gatan imaging filter, showing Fe (red), Ni (blue), and O (green). (b) Bright-field image and (c) electron hologram of the end of a chain of $Fe_{0.56}Ni_{0.44}$ particles. The hologram was recorded using an interference fringe spacing of 2.6 nm. (Reprinted from Dunin-Borkowski et al., 2004b.)

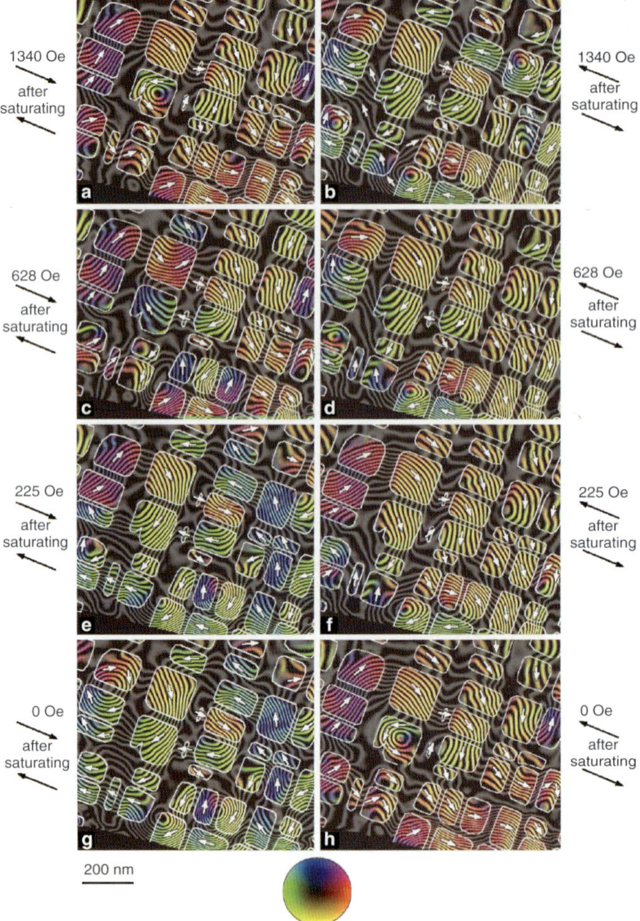

Figure 18–13. Magnetic phase contours from the region shown in Figure 8–12, measured using electron holography. Each image was acquired with the specimen in magnetic field-free conditions. The outlines of the magnetite-rich regions are marked in white, while the direction of the measured magnetic induction is indicated both using arrows and according to the color wheel shown at the bottom of the figure (red = right, yellow = down, green = left, blue = up). Images (a), (c), (e), and (g) were obtained after applying a large (>10,000 Oe) field toward the top left, then the indicated field toward the bottom right, after which the external magnetic field was removed for hologram acquisition. Images (b), (d), (f), and (h) were obtained after applying identical fields in the opposite directions. (Reprinted from Harrison et al., 2002.)

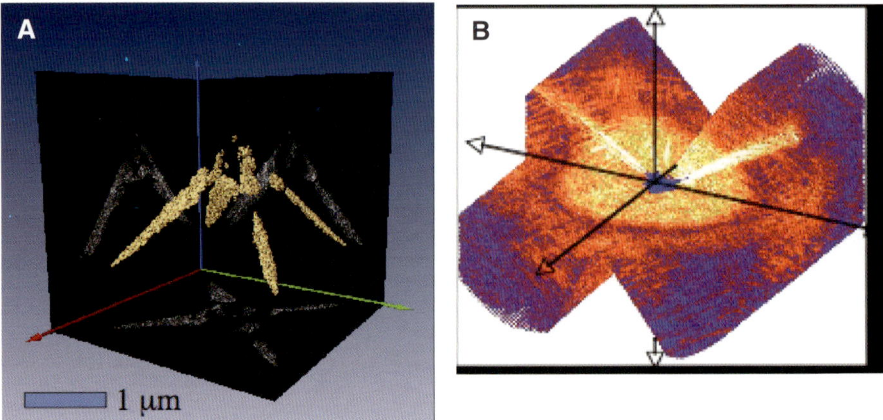

Figure 19–5. (A) Tomographic reconstruction from a soft X-ray diffraction pattern shown in (B). The object consists of gold balls (50 nm diameter) lying along the edges of a pyramidal-shaped silicon nitride structure. This is one image from a rotation series. From the complete series, three-dimensional surfaces of constant density can be constructed. (B) The volume of soft X-ray diffraction data collected to obtain the three-dimensional reconstruction in (A).

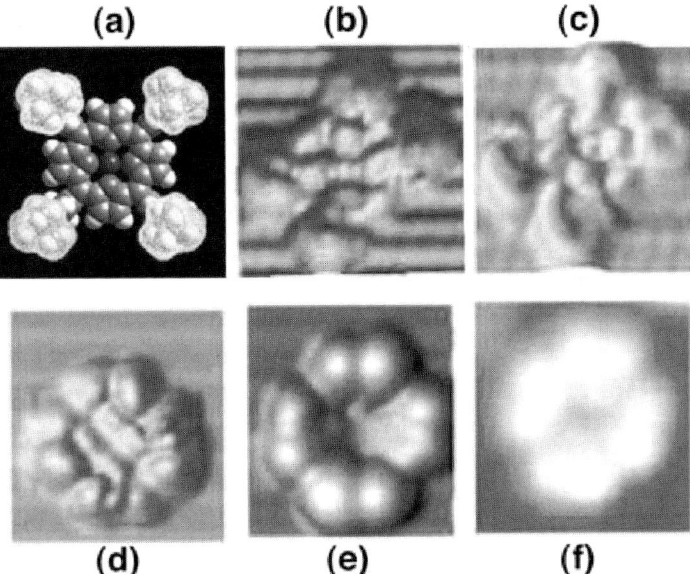

Figure 15–37. Identification of the molecular conformation of Cu-DTBPP on Cu(211). (a) Molecular model of Cu-DTBPP, with the four-lobed pattern observed by STM marked in yellow. (b–e) STM contrast calculation for different angles between the four legs and the substrate: (b) 60°, (c) 45°, (d) 30°, and (e) 10°. (f) Experimental STM image of the molecule. (Reprinted from Moresco et al., ©2002 with permission from Elsevier.) (See color plate.)

Simulations of STM contrast of adsorbed molecules pose a somewhat different problem than simulations of surface structures. For solid surfaces, the combination of STM imaging with image simulation can discriminate between several candidate structures, as discussed in the previous example. For molecules, the structure is typically well known. However, the adsorption geometry (site, orientation) as well as the conformation for larger, more flexible molecules enter as unknowns into the simulation. A relatively simple example is the stiff molecular cage structure of fullerene C_{60} (Hou et al., 1999; Pascual et al., 2000; Lu et al., 2003). In this case, the structure and conformation are well known, and the only parameter of the contrast calculation is the orientation. A more complex system, Cu-tetra(3,5-di-*tert* butyl phenyl) porphyrin (Cu-DTBPP) on Cu(211), is illustrated in Figure 15–37.

Cu-DTBPP consists of a central porphyrin ring with four symmetrically attached di-*tert*-butyl phenyl (DTBP) groups (Figure 15–37a). These four bulky groups determine the shape of the molecule, which is imaged in STM as a four-leaf clover structure, and define the interaction with the metal substrate (Jung et al., 1996). The orientation of the side groups is the main parameter that needs to be optimized to determine the conformation of the adsorbed molecule, and to simulate the experimental STM images. Figure 15–37 shows several candidate conformations, with leg angles of 60° (b), 45° (c), 30° (d), and the optimized angle of 10° relative to the substrate (e), which was found to best reproduce the experimental STM image of the molecule.

Acknowledgments. The compilation of this chapter was supported by the U.S. Department of Energy under Contract No. DE-AC02-98CH10886.

References

Altfeder, I.B., Matveev, K.A. and Chen, D.M. (1997). *Phys. Rev. Lett.* **78**, 2815.

Altfeder, I.B., Chen, D.M. and Matveev, K.A. (1998). *Phys. Rev. Lett.* **80**, 4895.

Bardeen, J. (1961). *Phys. Rev. Lett.* **6**, 57.

Bauer, E. (1994). *Rep. Prog. Phys.* **57**, 895.

Bauer, E. (1998). *Surf. Rev. Lett.* **5**, 1275.

Becker, R.S., Swartzentruber, B.S., Vickers, J.S. and Klitsner, T. (1989). *Phys. Rev. B* **39**, 1633.

Bell, L.D., Hecht, M.H., Kaiser, W.J. and Davis, L.C. (1990). *Phys. Rev. Lett.* **64**, 2679.

Bell, L.D., Kaiser, W.J., Hecht, M.H. and Davis, L.C. (1991). *J. Vac. Sci. Technol. B* **9**, 594.

Binnig, G. and Rohrer, H. (1983). *Surf. Sci.* **126**, 236.

Binnig, G., Rohrer, H., Gerber, Ch. and Weibel, E. (1982). *Appl. Phys. Lett.* **40**, 178.

Binnig, G., Rohrer, H., Gerber, Ch. and Weibel, E. (1983). *Phys. Rev. Lett.* **50**, 120.

Binnig, G., Garcia, N. and Rohrer, H. (1985). *Phys. Rev. B* **32**, 1336.

Binnig, G. and Smith, D.P.E. (1986). *Rev. Sci. Instrum.* **57**, 1688.

Bode, M., Getzlaff, M. and Wiesendanger, R. (1998). *Phys. Rev. Lett.* **81**, 4256.

Boland, J.J. (1993). *Science* **262**, 1703.

Boland, J.J. and Weaver, J.H. (1998). *Phys. Today* **51**, 34.

Bonnell, D.A., Ed. (2001). *Scanning Probe Microscopy and Spectroscopy—Theory, Techniques, and Applications*, 2nd ed., Wiley-VCH, New York.

Borovski, B., Krueger, M. and Ganz, E. (1997). *Phys. Rev. Lett.* **78**, 4229.

Borovski, B., Krueger, M. and Ganz, E. (1998). *Phys. Rev. B* **57**, 4269.

Brookes, I.M., Muryn, C.A. and Thornton, G. (2001). *Phys. Rev. Lett.* **87**, 266103.

Bruls, D.M., Vugs, J.W.A.M., Roenraad, P.M., Salemink, H.W.M., Wolter, J.H., Hopkinson, M., Skolnick, M.S., Long, F. and Gill, S.P.A. (2002). *Appl. Phys.* **81**, 1708.

Cerezo, A., Larson, D.J. and Smith, G.D.W. (2001). *MRS Bull.* **26**, 102.

Chambliss, D.D., Wilson, R.J. and Chiang, S. (1995). *IBM J. Res. Dev.* **39**, 639.

Chen, C.J. (1993). *Introduction to Scanning Tunneling Microscopy* (Oxford University Press, Oxford).

Chen, Y., Ohlberg, D.A.A., Medeiros-Ribeiro, G., Chang, Y.A. and Williams, R.S. (2000). *Appl. Phys. Lett.* **76**, 4004.

Ciobanu, C.V. and Predescu, C. (2004). *Phys. Rev. B* **70**, 085321.

Crain, J.N. and Pierce, D.T. (2005). *Science* **307**, 703.

Crommie, M.F., Lutz, C.P. and Eigler, D.M. (1993). *Nature* **363**, 524.

Cuberes, M.T., Bauer, A., Wen, H.J., Prietsch, M. and Kaindl, G. (1994). *J. Vac. Sci. Technol. B* **12**, 2646.

Cygan, M.T., Dunbar, T.D., Arnold, J.J., Bumm, L.A., Shedlock, N.F., Burgin, T.P., Jones, L., Allara, D.L., Tour, J.M. and Weiss, P.S. (1998). *J. Am. Chem. Soc.* **120**, 2721.

Davies, J.H., Bruls, D.M., Vugs, J.W.A.M. and Koenraad, P.M. (2002). *J. Appl. Phys.* **91**, 4171.

Diebold, U. (2003). *Surf. Sci. Rep.* **48**, 53.

Ebert, Ph., Cox, G., Poppe, U. and Urban, K. (1992). *Surf. Sci.* **271**, 587.

Eder, C., Smoliner, J. and Strasser, G. (1996). *Appl. Phys. Lett.* **68**, 2876.

Eigler, D.M. and Schweizer, E.K. (1990). *Nature* **344**, 524.

Erwin, S.C., Baski, A.A. and Whitman, L.J. (1996). *Phys. Rev. Lett.* **77**, 687.

Feenstra, R.M., Stroscio, J.A., Tersoff, J. and Fein, A.P. (1987). *Phys. Rev. Lett.* **58**, 1192.

Feenstra, R.M., Collins, D.A., Ting, D.Z.-Y., Wang, M.W. and McGill, T.C. (1994). *Phys. Rev. Lett.* **72**, 2749.

Festy, F. and Palmer, R.E. (2004). *Appl. Phys. Lett.* **85**, 5034.

Fowell, A.E., Williams, R.H., Richardson, B.E. and Shen, T.-H. (1990). *Semicond. Sci. Technol.* **5**, 348.

Fujikawa, Y., Akiyama, K., Nagao, T., Sakurai, T., Lagally, M.G., Hashimoto, T., Morikawa, Y. and Terakura, K. (2002). *Phys. Rev. Lett.* **88**, 176101.

Gai, Z., Li, X., Zhao, R.G. and Yang, W.S. (1998). *Phys. Rev. B* **57**, 15060.

Gewirth, A.A. and Niece, B.K. (1997). *Chem. Rev.* **97**, 1129.

Gimzewski, J.K., Joachim, C., Schlittler, R.R., Langlais, V., Tang, H. and Johannsen, I. (1998). *Science* **281**, 531.

Goldman, R.S., Feenstra, R.M., Briner, B.G., O'Steen, M.L. and Hauenstein, R. J. (1996). *Appl. Phys. Lett.* **69**, 3698.

Gong, Q., Offermans, P., Nötzel, R., Koenraod, P.M. and Loolter, J.H. (2004). *Appl. Phys.* **85**, 5697.

Grant, M.L., Swartzentruber, B.S., Bartelt, N.C. and Hannon, J.B. (2001). *Phys. Rev. Lett.* **86**, 4588.

Gregory, S. (1990). *Phys. Rev. Lett.* **64**, 689.

Güntherodt, H.-J. and Wiesendanger, R., Eds. (1992). *Scanning Tunneling Microscopy Vol. I & II* (Springer, Berlin).

Hamers, R.J. (1989). *Annu. Rev. Phys. Chem.* 531.

Hamers, R.J., Tromp, R.M. and Demuth, J.E. (1986). *Phys. Rev. Lett.* **56**, 1972.

Hanaguri, T., Lupien, C., Kohsaka, Y., Lee, D.-H., Azuma, M., Takano, M., Takagi, H. and Davis, J.C. (2004). *Nature* **430**, 1001.

Hasegawa, Y. and Avouris, Ph. (1993). *Phys. Rev. Lett.* **71**, 1071.

Heer, R., Smoliner, J., Strasser, G. and Gornik, E. (1998). *Appl. Phys. Lett.* **73**, 3138.

Heinrich, A.J., Lutz, C.P., Gupta, J.A. and Eigler, D.M. (2002). *Science* **298**, 1381.

Heinze, S., Bode, M., Kubetzka, A., Pietzsch, O., Nie, X., Blügel, S. and Wiesendanger, R. (2000). *Science* **288**, 1805.

Hendriksen, B.L.M. and Frenken, J.W.M. (2002). *Phys. Rev. Lett.* **89**, 046101.

Hla, S.-W. and Rieder, K.-H. (2003). *Annu. Rev. Phys. Chem.* **54**, 307.

Hoffman, J.E., McElroy, K., Lee, D.-H., Lang, K.M., Eisaki, H., Uchida, S. and Davis, J.C. (2002a). *Science* **297**, 1148.

Hoffman, J.E., Hudson, E.W., Lang, K.M., Madhavan, V., Eisaki, H., Uchida, S. and Davis, J.C. (2002b). *Science* **295**, 466.

Horch, S., Lorensen, H.T., Helveg, S., Lægsgaard, E., Stensgaard, I., Jacobsen, K.W., Nørskov, J.K. and Besenbacher, F. (1999). *Nature* **398**, 134.

Hornbaker, D.J., Kahng, S.-J., Misra, S., Smith, B.W., Johnson, A.T., Mele, E.J., Luzzi, D.E. and Yazdani, A. (2002). *Science* **295**, 828.

Horn-von Hoegen, M. (1994). *Appl. Phys. A* **59**, 503.

Hou, J.G., Jinlong, Y., Haiqian, W., Qunxiang, L., Changgan, Z., Hai, L., Wang, B., Chen, D.M. and Qinshi, Z. (1999). *Phys. Rev. Lett.* **83**, 3001.

Hovis, J.S. and Hamers, R.J. (1997). *Surf. Sci.* **402**, 1.

Jaklevic, R.C. and Lambe, J. (1966). *Phys. Rev. Lett.* **17**, 1139.

Jaklevic, R.C., Lambe, J., Mikkor, M. and Vassell, W.C. (1971). *Phys. Rev. Lett.* **26**, 88.

Jensen, J.A., Rider, K.B., Chen, Y., Salmeron, M. and Somorjai, G.A. (1999). *J. Vac. Sci. Technol. B* **17**, 1080.

Johnson, M.B., Albrektsen, O., Feenstra, R.M. and Salemink, H.W.M. (1993). *Appl. Phys. Lett.* **63**, 2923.

Jung, T.A., Schlittler, R.R., Gimzewski, J.K., Tang, H. and Joachim, C. (1996). *Science* **271**, 181.

Kaiser, W.J. and Bell, L.D. (1988). *Phys. Rev. Lett.* **60**, 1406.

Kasaya, M., Tabata, H. and Kawai, T. (1998). *Surf. Sci.* **400**, 367.

Kawamura, M., Paul, N., Cherepanov, V. and Voigtländer, B. (2003). *Phys. Rev. Lett.* **91**, 096102.

Klijn, J., Sacharov, L., Meyer, C., Blügel, S., Morgenstern, M. and Wiesendanger, R. (2003). *Phys. Rev. B* **68**, 205327.

Kochanski, G.P. (1989). *Phys. Rev. Lett.* **62**, 2285.

Komeda, T., Kim, Y., Kawai, M., Persson, B.N.J. and Ueba, H. (2002). *Science* **295**, 2055.

Kubby, J.A. and Greene, W.J. (1992). *Phys. Rev. Lett.* **68**, 329.

Kühnle, A., Linderoth, T.R., Hammer, B. and Besenbacher, F. (2002). *Nature* **415**, 891.

Kuipers, L., Hoogeman, M.S. and Frenken, J.W.M. (1993). *Phys. Rev. Lett.* **71**, 3517.

Kuk, Y. and Silverman, P.J. (1986). *Appl. Phys. Lett.* **48**, 1597.

Lægsgaard, E., Österlund, L., Rasmussen, P.B., Stensgaard, I. and Besenbacher, F. (2001). *Rev. Sci. Instrum.* **72**, 3537.

Laracuente, A. and Whitman, L.J. (2001). *Surf. Sci.* **476**, L247.

Lee, E.Y., Sirringhaus, H. and von Känel, H. (1994). *Phys. Rev. B* **50**, 5807.

Lee, H.J. and Ho, W. (1999). *Science* **286**, 1719.

Legrand, B., Grandidier, B., Nys, J.P., Stiévenard, D., Gérard, J.M. and Thierry-Mieg, V. (1998). *Appl. Phys. Lett.* **73**, 96.

Li, L., Tindall, C., Takaoka, O., Hasegawa, Y. and Sakurai, T. (1997). *Phys. Rev. B* **56**, 4648.

Lopinski, G.P., Moffatt, D.J., Wayner, D.D.M. and Wolkow, R.A. (1998). *Nature* **392**, 909.

Lu, X., Grobis, M., Khoo, K.H., Louie, S.G. and Crommie, M.F. (2003). *Phys. Rev. Lett.* **90**, 096802.

Mayne, A.J., Avery, A.R., Knall, J., Jones, T.S., Briggs, G.A.D. and Weinberg, W.H. (1993). *Surf. Sci.* **284**, 247.

Medeiros-Ribeiro, G., Bratkovski, A.M., Kamins, T.I., Ohlberg, D.A.A. and Williams, R.S. (1998). *Science* **279**, 353.

Meinel, K., Wolter, H., Ammer, Ch., Beckmann, A. and Neddermeyer, H. (1997). *J. Phys. Condens. Matter* **9**, 4611.

Men, F.K., Packard, W.E. and Webb, M.B. (1988). *Phys. Rev. Lett.* **61**, 2469.

Meyer, G. and Rieder, K.-H. (1997). *Surf. Sci.* **377–379**, 1087.

Meyer, G. and Rieder, K.-H. (1998). *MRS Bull.* **23**, 28.

Mitsui, T., Rose, M.K., Fomin, E., Ogletree, D.F. and Salmeron, M. (2002). *Science* **297**, 1850.

Mo, Y.-W., Swartzentruber, B.S., Kariotis, R., Webb, M.B. and Lagally, M.G. (1989). *Phys. Rev. Lett.* **63**, 2393.

Mo, Y.-W., Savage, D.E., Swartzentruber, B.S. and Lagally, M.G. (1990). *Phys. Rev. Lett.* **65**, 1020.

Mo, Y.-W., Kleiner, J., Webb, M.B. and Lagally, M.G. (1991). *Phys. Rev. Lett.* **66**, 1998.

Moresco, F., Meyer, G., Rieder, K.-H., Ping, J., Tang, H. and Joachim, C. (2002). *Surf. Sci.* **499**, 94.

Muller, E.W. and Tsong, T.T. (1969). *Field Ion Microscopy* (Elsevier, New York).

Nagl, C., Haller, O., Platzgummer, E., Schmid, M. and Varga, P. (1994). *Surf. Sci.* **321**, 237.

Nakayama, K. and Weaver, J.H. (1999). *Phys. Rev. Lett.* **82**, 980.

Narayanamurti, V. and Kozhevnikov, M. (2001). *Phys. Rep.* **349**, 447.

Nazin, G.V., Qiu, X.H. and Ho, W. (2003). *Science* **302**, 77.

Nunes, G., Jr. and Amer, N.M. (1993). *Appl. Phys. Lett.* **63**, 1851.

Olsson, F.E., Lorente, N. and Persson, M. (2003). *Surf. Sci.* **522**, L27.

Olsson, F.E., Persson, M., Repp, J. and Meyer, G. (2005). *Phys. Rev. B* **71**, 075419.

Pascual, J.I., Gomez-Herrero, J., Rogero, C., Baro, A.M., Sanchez-Portal, D., Artacho, E., Ordejon, P. and Soler, J.M. (2000). *Chem. Phys. Lett.* **321**, 78.

Pfister, M., Johnson, M.B., Alvarado, S.F., Salemink, H.W.M., Marti, U., Martin, D., Morier-Genoud, F. and Reinhard, F.K. (1995). *Appl. Phys. Lett.* **67**, 1459.

Pietzsch, O., Kubetzka, A., Bode, M. and Wiesendanger, R. (2001). *Science* **292**, 2053.

Prietsch, M. and Ludeke, R. (1991). *Phys. Rev. Lett.* **66**, 2511.

Qin, X.R. and Lagally, M.G. (1997). *Science* **278**, 1444.

Qin, X.R., Swartzentruber, B.S. and Lagally, M.G. (2000). *Phys. Rev. Lett.* **85**, 3660.

Rasmussen, P.B., Hendriksen, B.L.M., Zeijlemaker, H., Ficke, H.G. and Frenken, J.W.M. (1998). *Rev. Sci. Instrum.* **69**, 3879.

Repp, J., Meyer, G., Olsson, F.E. and Persson, M. (2004). *Science* **305**, 493.

Repp, J., Meyer, G., Stojkovic, S.M., Gourdon, A. and Joachim, C. (2005). *Phys. Rev. Lett.* **94**, 026803.

Rippard, W.H. and Buhrmann, R.A. (2000). *Phys. Rev. Lett.* **84**, 971.

Robinson, K.M., Robinson, I.K. and O'Grady, W.E. (1992). *Surf. Sci.* **262**, 387.

Rößler, M., Geng, P. and Wintterlin, J. (2005). *Rev. Sci. Instrum.* **76**, 023705.

Rost, M.J., Crama, L., Schakel, P., van Tol, E., van Velzen-Williams, G.B.E.M., Overgauw, C.F., ter Horst, H., Dekker, H., Okhuijsen, B., Seynen, M., Vijftigschild, A., Han, P., Katan, A.J., Schoots, K., Schumm, R., van Loo, W., Oosterkamp, T.H. and Frenken, J.W.M. (2005). *Rev. Sci. Instrum.* **76**, 053710.

Rubin, M.E., Medeiros-Ribeiro, G., O'Shea, J.J., Chin, M.A., Lee, E.Y., Petroff, P.M. and Narayanamurti, V. (1996). *Phys. Rev. Lett.* **77**, 5268.

Sajoto, T., O'Shea, J.J., Bhargava, S., Leonard, D., Chin, M.A. and Narayanamurti, V. (1995). *Phys. Rev. Lett.* **74**, 3427.

Salemink, H. and Albrektsen, O. (1991). *J. Vac. Sci. Technol. B* **9**, 779.

Sanders, L.M., Stumpf, R., Mattson, T.R. and Swartzentruber, B.S. (2003). *Phys. Rev. Lett.* **91**, 206104.

Schaub, R., Thostrup, P., Lopez, N., Stensgaard, I., Nørskov, J.K. and Besenbacher, F. (2001). *Phys. Rev. Lett.* **87**, 266104.

Schaub, R., Wahlström, E., Rønnau, A., Lægsgaard, E., Stensgaard, I. and Besenbacher, F. (2003). *Science* **299**, 377.

Schintke, S. and Schneider, W.-D. (2004). *J. Phys: Cond. Matter* **16**, R49.

Schunack, M., Linderoth, T.R., Rosei, F., Lægsgaard, E., Stensgaard, I. and Besenbacher, F. (2002). *Phys. Rev. Lett.* **88**, 156102.

Sirringhaus, H., Lee, E.Y. and von Känel, H. (1994). *Phys. Rev. Lett.* **73**, 577.

Smith, D.P.E. and Binnig, G. (1986). *Rev. Sci. Instrum.* **57**, 2630.

Stipe, B.C., Rezaei, M.A. and Ho, W. (1998a). *Science* **280**, 1732.

Stipe, B.C., Rezaei, M.A. and Ho, W. (1998b). *Science* **279**, 1907.

Stroscio, J.A. and Celotta, R.J. (2004). *Science* **306**, 242.

Sutter, P., Schick, I., Ernst, W. and Sutter, E. (2003a). *Phys. Rev. Lett.* **91**, 176102.

Sutter, P., Zahl, P., Sutter, E. and Bernard, J.E. (2003b). *Phys. Rev. Lett.* **90**, 166101.

Sutter, E., Sutter, P. and Bernard, J.E. (2004). *Appl. Phys. Lett.* **84**, 2262.

Swartzentruber, B.S. (1996). *Phys. Rev. Lett.* **76**, 459.

Swartzentruber, B.S., Mo, Y.-W., Kariotis, R., Lagally, M.G. and Webb, M.B. (1990). *Phys. Rev. Lett.* **65**, 1913.

Takayanagi, K., Tanishiro, Y., Takahashi, S. and Takahashi, M. (1985). *Surf. Sci.* **164**, 367.

Tang, J.-M. and Flatté, M.E. (2004). *Phys. Rev. Lett.* **92**, 047201.

Tersoff, J. and Hamann, D.R. (1983). *Phys. Rev. Lett.* **50**, 1998.

Tersoff, J. and Hamann, D.R. (1985). *Phys. Rev. B* **31**, 805.

Troadec, C., Kunardi, L. and Chandrasekhar, N. (2005). *Appl. Phys. Lett.* **86**, 72101.

Tromp, R.M. (2000). *IBM J. Res. Dev.* **44**, 503.

Tromp, R.M., Hamers, R.J. and Demuth, J.E. (1985). *Phys. Rev. Lett.* **55**, 1303.

Tromp, R.M., Hamers, R.J. and Demuth, J.E. (1986). *Phys. Rev. B* **34**, 1388.

Venables, J.A. (2000). *Introduction to Surface and Thin Film Processes* (Cambridge University Press, Cambridge).

Vershinin, M., Misra, S., Ono, S., Abe, Y., Ando, Y. and Yazdani, A. (2004). *Science* **303**, 1995.

Voigtländer, B. (2001). *Surf. Sci. Rep.* **43**, 127.

Voigtländer, B. and Zinner, A. (1993). *Appl. Phys. Lett.* **63**, 3055.

Wang, Y., Chen, X. and Hamers, R.J. (1994a). *Phys. Rev. B* **50**, 4534.

Wang, Y., Bronikowski, M.J. and Hamers, R.J. (1994b). *Surf. Sci.* **311**, 64.

Weierstall, U. and Spence, J. (1998). *Surf. Sci.* **398**, 267.

Wiesendanger, R. (1994). *Scanning Probe Microscopy and Spectroscopy—Methods and Applications* (Cambridge University Press, Cambridge).

Wintterlin, J., Trost, J., Renisch, S., Schuster, R., Zambelli, T. and Ertl, G. (1997). *Surf. Sci.* **394**, 159.

Wittneven, Chr., Dombrowski, R., Morgenstern, M. and Wiesendanger, R. (1998). *Phys. Rev. Lett.* **81**, 5616.

Wolkow, R.A. (1999). *Annu. Rev. Phys. Chem.* **50**, 413.

Wouda, P.T., Niewenhuys, B.E., Schmid, M. and Varga, P. (1996). *Surf. Sci.* **359**, 17.

Wouda, P.T., Schmid, M., Nieuwenhuys, B.E. and Varga, P. (1998). *Surf. Sci.* **417**, 292.

Yagi, K. (1982). *Scan. Electron Microsc.* **4**, 1421.

Yakunin, A.M., Silov, A.Yu., Koenraad, P.M., Wolter, J.H., Van Roy, W., De Boeck, J., Tang, J.-M. and Flatté, M.E. (2004). *Phys. Rev. Lett.* **92**, 216806.

Yang, Y.-N., Fu, E.S. and Williams, E.D. (1996). *Surf. Sci.* **356**, 101.

Zhang, Z.Y. and Lagally, M.G. (1998). *Science* **276**, 377.

Zheng, J.F., Liu, X., Newman, N., Weber, E.R., Ogletree, D.F. and Salmeron, M. (1994). *Phys. Rev. Lett.* **72**, 1490.

16

Atomic Force Microscopy in the Life Sciences

Matthias Amrein

He sees the face and the moving hands, even hears it ticking. If he is ingenious he may form some picture of a mechanism, which could be responsible for all the things he observes, but he may never be quite sure his picture is the only one that could explain his observations. He will never be able to compare his picture with the real mechanism and he cannot even imagine the possibility of the meaning of such a comparison.
Albert Einstein, describing a man trying to understand a closed watch.

1 Introduction

Cells are highly organized in compartments and functional units down to the macromolecular level. The frameworks of these functional and structural units are either protein complexes or complexes of proteins with nucleic acids or with lipids. The recently completed map of the human genome and the systematic mapping of the proteins expressed in tissues and cells (*proteomics*) are part of a concerted effort to rapidly advance the understanding of the functions of macromolecular units and the cell. But proteomics reveals only the basic inventory of a cell and the inventory is insufficient to explain the function of each element and the orchestration of the components. As with Einstein's image of the closed watch, understanding life is inconceivable without observing the structures behind function. Microscopy plays an important role by placing the molecular elements into a structural context. Because the pace of discovery of these elements has been increased substantially by proteomics, the need for more sophisticated microscopy has also substantially increased in recent years. Atomic force microscopy (AFM) plays a specific role in life science microscopy by allowing imaging to be combined with locally probing functions of macromolecular elements.

Most microscopes depend on radiation, emitted and recorded at a distance from the sample, using lenses. The resolution power of these microscopes in the life sciences is limited by diffraction and/or damage to the sample by the illuminating beam. In contrast, the scanning tunneling microscope (STM) has been the first microscope to rely on an effect only present in the immediate vicinity of a physical probe and the sample. In an STM, an electrically conductive needle approaches the sample until current flows. The current is based on a quantum-mechanical tunneling effect and flows before the tip and sample physically touch. During scanning, the tip–sample distance remains constant by keeping the tunneling current constant. The movement of the tip relative to the sample reflects the sample topography and is recorded. Not only has the atomic surface lattice structure been revealed for many crystalline samples, but also single atom defects imaged directly. The STM immediately created great interest among biologists because of its outstanding resolution power and the absence of radiation damage (Amrein et al., 1989, 1993; (Baró et al., 1985; Voelker et al., 1988; Guckenberger et al., 1989; Stemmer et al., 1989; Welland et al., 1989; Miles et al., 1990; Arscott and Bloomfield, 1993; Lindsay and Tao, 1993; Miles and McMaster, 1993). But the application of STM in the life sciences suffered from the poor electrical conductivity of most biological matter, and reproducible images were obtained only after metal coating the specimens (Travaglini et al., 1987; Zasadzinski et al., 1988; Blackford and Jericho, 1991; Amrein et al., 1988, 1989b, 1993). Nevertheless, early applications demonstrated the resolution power and high signal-to-noise ratio at the macromolecular level offered by the new microscopy. Although unintentional at first, it also became clear that the probe could be used to manipulate macromolecular structures individually (Figure 16–1).

The basic principles of the STM were soon extended to form a suite of new devices, the scanning probe microscopes (SPM). For all SPM, a physical probe is scanned over the sample in nanometer distance at most. Sample properties are mapped from a very small interaction volume in the near field of the sample and the probe. The meaning of the term of *near field* is different for the various types of SPM. In an AFM, short-range forces between a tip and the sample take the role of the tunneling current in the STM, thus making the new paradigm applicable to electrically nonconductive samples. This is how AFM gained ground in the life sciences.

The direct measurement of the position of the surface means that the achieveable resolution is not limited by the environment—the AFM lends itself well to imaging molecular and cellular specimens under physiological conditions, in buffer. In many cases, the sample prepara-

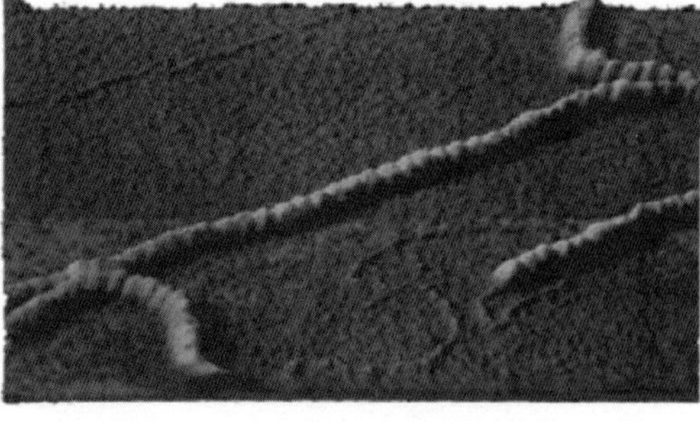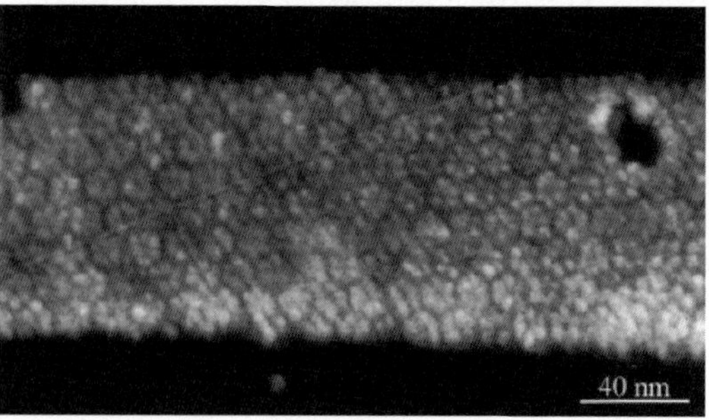

Figure 16–1. STM topographies of Rec-DNA complex (left) and an aberrant bacteriophage capside (right). In the upper right corner, a hole has been created applying a voltage pulse between the tip and the sample. Both images immediately reveal the molecular arrangement of the respective complex structure. (Left, from Amrein et al., 1988, reproduced with permission. Right, from Amrein et al., 1998b, reproduced with permission.) (See color plate.)

tion is very straightforward, and at a basic level requires only the immobilization of the sample to be imaged. In the case of adherent cells, for instance, the sample can be imaged directly without any special preparation. The local measurement at the sample surface also allows direct quantitative measurement of dimensions and forces. In addition to the topographic image that is built up by scanning the tip over the surface, AFM can simultaneously probe the mechanical properties of the surface, by dynamically oscillating the cantilever that supports the tip. Tip–sample adhesion can also be probed, and maps of surface (visco)elasticity or adhesion built up. One application is to coat the tip with appropriate molecules, such as antibodies, other proteins, or sugars to generate maps of specific ligand–receptor binding over the surface. The AFM is also well suited to combination with other techniques—a combination with optical microscopy is often particularly worthwhile for life science research. Other techniques, such as electrochemistry or electrophysiology, can also be used simultaneously, and *in situ* probing of cell response to electrical, chemical, or mechanical stimulation is possible.

The foundation of AFM in the life sciences has now been laid by a profound understanding of the interactions between the tip and the sample and the possibility to subtly control the interactions by choosing an appropriate sensor and by adjusting the imaging environment (Weisenhorn et al., 1992; Butt et al., 1995; Heuberger et al., 1996).

To appreciate the contribution of AFM to the life sciences, it is necessary to consider it in relation to other microscopy techniques currently in use. In the realm of single molecules, the motivation for using AFM comes mainly from the high resolution achievable and the direct (real-space) measurement of samples without coating. Most of the knowledge so far gained by AFM in the life sciences comes from studying single macromolecules or macromolecular complex structures. Biological membranes, for example, can be imaged in their native state at a lateral resolution of 0.5–1 nm and a vertical resolution of 0.1–0.2 nm. Conformational changes that are related to functions can be resolved to a similar resolution. In the study of macromolecular structure, AFM competes with transmission electron microscopy. In many cases, the outcome is comparable and either microscope can be used with equal success. For example, measuring the contour length of a plasmid DNA or determining the binding site of a protein to a DNA may be performed by either microscope. However, solving the three-dimensional structure of macromolecules or macromolecular complexes that cannot be solved by X-ray crystallography or by nuclear magnetic resonance (NMR) is better accomplished by (cryo)electron microscopy, because AFM reveals only a topographical view of the structures. On the other hand, when a molecular assembly is not highly defined structurally and individual units differ from each other substantially, AFM may allow a specific question to be answered in a straightforward manner, because of its uniquely high signal-to-noise ratio at molecular dimensions.

The strength of AFM does not lie in imaging alone, but in the possibility of combining microscopy with an experiment at the molecular level. In a number of cases, molecular images have been obtained with

Figure 16–2. The ATP synthase is a proton-driven molecular motor with a stator (seen here) and a rotor. Simply counting the number of individual proteins inside the ring structure has not been possible from electron micrographs. (From Seelert et al., 2000, reprinted with permission.)

sufficient resolution to individually recognize single macromolecules and then address the molecules individually with the stylus of the AFM (Figure 16–2).

Most prominently, single molecule force spectroscopy combined with single molecule imaging has provided unprecedented possibilities for analyzing intramolecular and intermolecular forces including the study of the folding pathway of proteins. Probing the self-assembly of macromolecular complexes and measuring the mechanical properties of macromolecular "springs" are other examples in which AFM has made substantial contributions in the life sciences.

Imaging living cells with AFM is also performed in conjunction with local probing of the sample. The images by themselves usually are used only to obtain proper experimental control. The extreme flexibilty of the cell membrane means that the images often show a combination of the cell topography with the mechanical stiffness of cytoskeletal fibers and vesicles below the cell surface. An important example for a meaningful application of AFM in living cells is the measurement of the constrained diffusion of elements of the plasma membrane. These measurements have substantially contributed to establishing the nature of lipid rafts. In other applications, the cellular response on defined mechanical stress has been studied in conjunction with hearing or with the myogenic effect (Figure 16–3).

Future applications and future instrument developments in life sciences AFM are certain to capitalize even more strongly on performing local spectroscopy or manipulation of samples at the nanoscopic scale in addition to imaging. Such experiments will certainly extend the possibilities of probing function directly in many ways. But future AFM technology might also contribute to revealing the spatial relationships of the macromolecules of cells and tissues. Today, immunohistochemistry and immunocytochemistry are mostly in use to establish these structural relationships in cells and tissues. This implies that macromolecular identity is revealed only after labeling. However, the number of different proteins and other macromolecules at work at any time in a cell makes it inconceivable to extend this approach to determine even a substantial subset of all of its macromolecular relation-

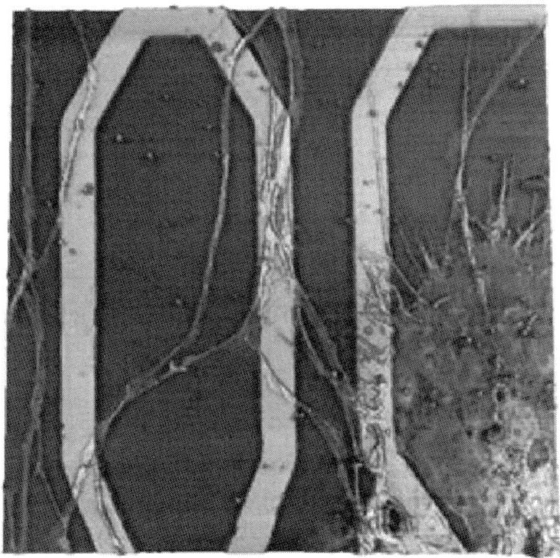

Figure 16–3. Neuronal cell, cultured on an electronic chip. The chip is designed such as to pick up an action potential of the cell. The image demonstrates proper tracing of the cell surface. In a future application, an appropriately designed stylus might be used as an additional electrode to excite or record an action potential at any location of the cell body or a process of the neuronal cell. (See color plate.)

ships by current techniques. AFM might offer a means to overcome the need for labeling. It has recently been demonstrated that thin sections of fixed and embedded tissue and cell samples make the cell's interior accessible to AFM. High-resolution cell images give a clear picture of the cell's ultrastructure, sometimes down to the molecular level. Although AFM will not likely allow identifying macromolecules by their shape, it could be combined with powerful local spectroscopy. It is tempting to speculate that spectroscopy might then enable microscopy to identify macromolecular specimens under the tip directly (Figure 16–4).

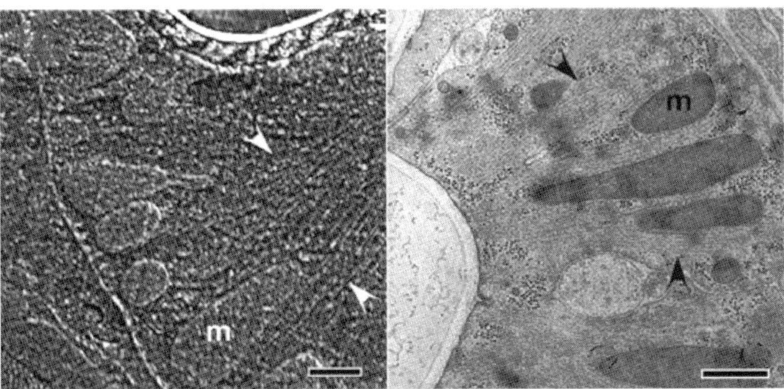

Figure 16–4. AFM image of the block face surface (left) and TEM image of an ultrathin section (right) from the nematode *Caenorhabditis elegans* embedded in epoxy resin. Scale bars equal 500 nm. Arrows point to actin filaments, m, mitochondria; G, gut. (From Matsko and Mueller, 2004, reprinted with permission.)

In the following section, AFM and its elements are described. The basics for preparing macromolecular and cellular samples are then described. Finally, a few select examples highlight AFM experiments, in which a combination of imaging with sample manipulation has been used to understand macromolecular or cellular function. A comprehensive review all AFM applications in the life sciences is beyond the scope of this chapter.

2 Instrumentation and Imaging

2.1 Introduction

In AFM, the topology of the sample is traced by a sharp stylus that is scanned line by line over the sample. For most setups, the stylus is sitting at the free end of a cantilever spring. Every elevation on the sample causes the stylus to move up and bend the cantilever upward and every depression makes the lever move down. Stylus and cantilever are usually microfabricated from silicon or silicon nitrite. The cantilever is typically a fraction of a millimeter long and a few micrometers thick. The softer the specimen, the softer the cantilever spring should be for it to trace the sample surface rather than deform it. The shape of the stylus is crucial too. It may be tetrahedral or extended, with a high aspect ratio, depending on need. The radius of curvature at the apex of the stylus may be as small as 2 nm (the apex of the stylus is also referred to as the *tip*) (Figure 16–5).

Accurate measurement of the deflection of the cantilever is the basis for accurate measurement of the sample topology. It also allows proper control of the loading force of the stylus onto the sample. It turns out that this latter aspect is particularly important in life science applications of AFM (see below). The first AFMs used an STM behind the cantilever to measure the deflection. Deflection has also been measured by the change in electrical capacitance between the cantilever and a reference electrode, or by means of a piezoresistor integrated with the cantilever (Tortonese et al., 1993). Most AFMs now use an

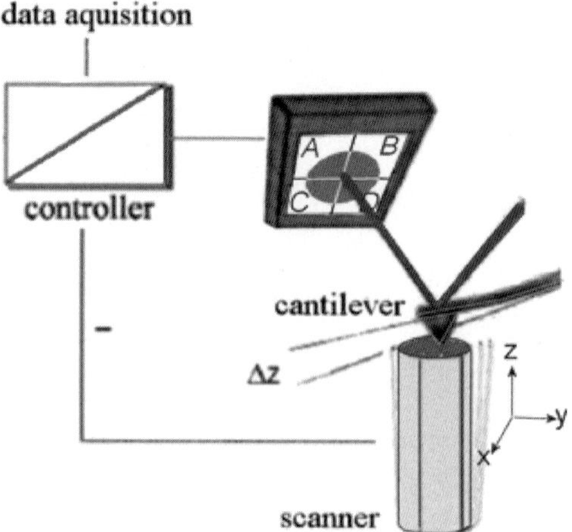

Figure 16–5. The elements of an AFM.

optical pointer to measure cantilever deflection. This detection system is fully adequate as it poses no limitation to AFM resolution.

In the optical pointer detection, a laser beam is focused onto the back of the free end of the cantilever (Figure 16–5). The laser beam is then reflected off the cantilever onto a four-segment photodiode. Prior to imaging, the four-segmented laser diode is moved until all four segments are equally illuminated. For imaging, the stylus is then loaded onto the sample. This causes the free end of the cantilever to bend upward and the laser beam now illuminates the two upper segments more strongly. The signals from the two upper segments of the diode are compared to the two lower segments $[(A + B) - (C + D)]$ to derive the amount of deflection of the lever in the z direction. The load is preset by the user, depending on the application and is related to the deflection of the cantilever:

$$F = c \cdot \Delta S$$

ΔS is the deflection in the z direction and c is the spring constant of the cantilever. The force is typically selected within the range of less than 100 pN to a few nanonewtons, depending on the application.

In operation, the cantilever is deflected from the preset value by the sample topology and the reflected laser beam is moved up or down. The original deflection is then restored via a feedback loop by a motion of the scanner perpendicular to the sample plane (referred to as the z direction). The position of the scanner with respect to the tip is recorded and used as the AFM topographical image.

Torsion of the cantilever may also occur during scanning, when the tip is experiencing friction with the sample. When the cantilever is becoming twisted, the laser beam is moved sideways. The amount of torsion and, hence, friction is then derived from comparison of the signals from the two right and two left segments of the photodiode $[(A + C) - (B + D)]$. Maps of local friction are used to reveal materials contrast in addition to the topographical image.

In dynamic AFM modes, the cantilever is oscillated and the amplitude and phase of the oscillation are monitored using the laser signal on the photodiode rather than a static deflection.

An AFM does not necessarily need to be based on a cantilever at all. In an instrument combining AFM topographical imaging with near-field optical imaging (the scanning near-field optical microscope, SNOM or NSOM), a tapered optical fiber is used as the stylus in most current instruments. It is oscillated parallel to the sample. Dampening of this oscillation is used as the feedback signal. In another alternative setup to the cantilever-based AFM, the sample is mounted on the membrane of an electret microphone and the output of this microphone is used for feedback. This setup performs equally well as the more traditional cantilever setup (Figure 16–6).

In addition to a highly sensitive probe, AFM depends on a precise scanner. The scanner is attached either to the probe or the sample. It allows the sample to be scanned with respect to the stylus in the plane of the sample (referred to as the x,y plane) and adjusting the relative height of the sample and the probe (referred to as the z direction) with

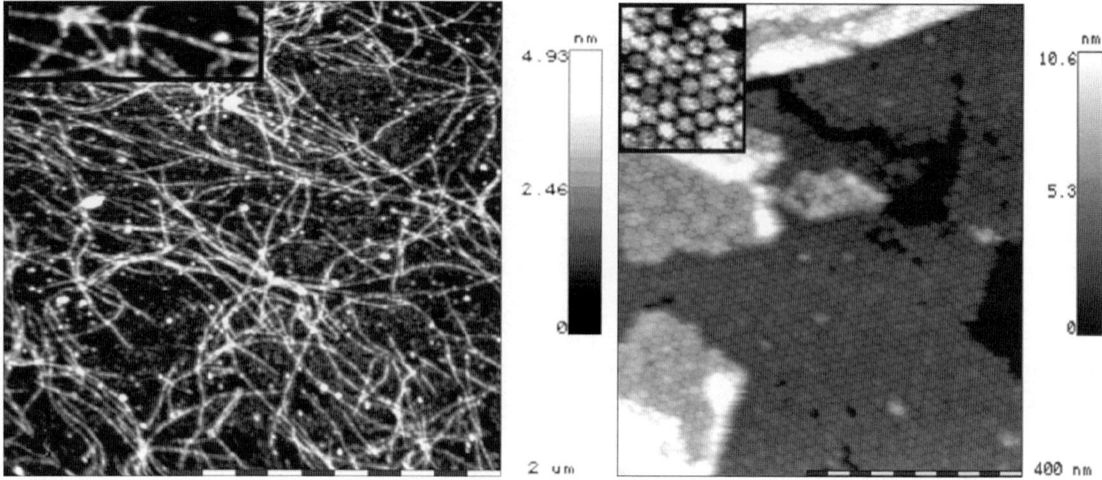

Figure 16–6. Actin filaments (left) and an archebacterial S-layer (HPI-layer, right) imaged with an AFM based on an electret sensor, rather than a cantilever sensor (Schenk et al., 1994, 1996).

subatomic precision (Amrein et al., 1997). AFM scanners are made of voltage-driven piezoceramic elements.

AFM imaging is a mechanical process, usually operated in a closed feedback loop. Even an apparently crisp, high-resolution topographical image need not necessarily reflect the true sample topography. The faithfulness of the topographical images depends on the properties of the feedback loop, the tip and sample geometries, and on how much the sample deforms upon imaging. Eventually, the accuracy of the topographical images is limited by noise (Figure 16–7). Each of these aspects is discussed below. Also discussed is how development of AFM technology might lead to improved instrument performance.

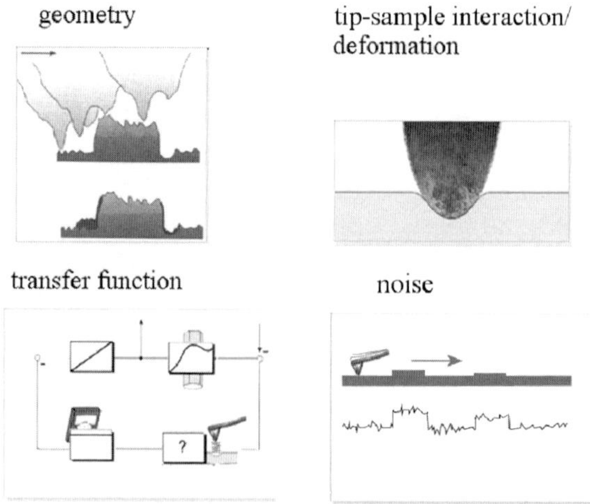

Figure 16–7. The faithfulness of an AFM topographical image depends on the sample and tip geometries, the sample deformation, the quality of the mechanical feedback loop of the instrument, and noise.

2.2 Geometry of the Stylus

When the stylus moves over a sample, the effective point of contact of the tip with the sample also changes. The topographical image is, colloquially speaking, convoluted with the tip geometry. The early days of AFM (and STM) were plagued by ill-characterized tips. Multiple whiskers created images that contained the same object multiple times (Figure 16–8). For a blunt stylus, each prominent object of the sample resulted in a local image of the tip itself rather than the local sample topology, and the overall appearance of such images has been cloudy. Because both the tip geometry and the sample topology matter, high-resolution images were sometimes obtained with an apparent blunt tip for very flat samples. This is because even the bluntest of tips has a rough surface, being covered with fine asperities.

Commercially available cantilevers now come with a well-characterized stylus and the apex may have a very small radius of curvature. The shape of the stylus needs to be selected with the sample in mind. Biological membranes or two-dimensional arrays of proteins with little overall height variations, for example, are well imaged by a pyramid-shaped stylus that ends in an apex of small radius of curvature, whereas a sample with prominent topology with steep flanks needs to be scanned by an elongated, needle-like stylus of high aspect ratio. During imaging, even a well-characterized, sharp stylus may become mechanically damaged or may pick up contaminate that renders it blunt. In these cases, the probe needs to be cleaned (e.g., by washing in ultrapure water, containing a detergent). If not successful, it has to be replaced.

When selecting an appropriate stylus, not only the expected sample topology must be considered. The sample compliance is an equally important aspect as a sharp tip may strongly deform a soft sample. This aspect is described below.

2.3 Tip–Sample Interaction

The tip–sample interactions need to be considered carefully, because they influence critically the success of the experiments. In AFM, most of the time the stylus is loaded onto the sample either intermittently (referred to as intermittent contact mode or tapping mode) or con-

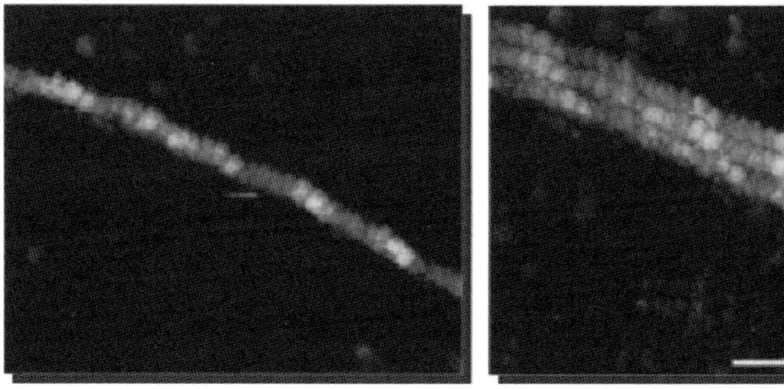

Figure 16–8. A single active filament, imaged with a single tip (left) and after the tip has become a triple tip (right).

stantly (contact mode; see below). This is achieved by approaching the probe and sample and bending the cantilever until the desired loading force is achieved.

In addition to the loading force, exerted by the cantilever spring, there are additional forces F_{StSa} acting between the sample and the stylus (see below). They may be repulsive or attractive. An attractive force makes the effective load of the tip onto the sample greater than what would be assumed from the bending of the cantilever. A repulsive force that acts prior to physical contact reduces the effective load of the tip onto the sample. The total loading force of the tip onto the sample becomes

$$F = c \cdot \Delta S + F_{StSa}$$

where ΔS is the deflection of the cantilever and c is the spring constant of the cantilever.

Evaluating these interactions may be pursued by acquiring force-versus-distance curves (referred to also as "force spectroscopy"). Thereby, the tip is approached to the sample and the deflection of the cantilever recorded as a function of the vertical position of the scanner with respect to the sample. The deflection of the cantilever can then be converted into a force using the spring constant of the lever (Figure 16–9).

2.3.1 Sample Deformation

When physical contact of the tip and the sample is established, the sample will deform until the contact area has sufficiently increased such that the load is accommodated (Figure 16–10).

The deformation strongly determines the resolution and trustworthiness of AFM imaging and must therefore be considered carefully.

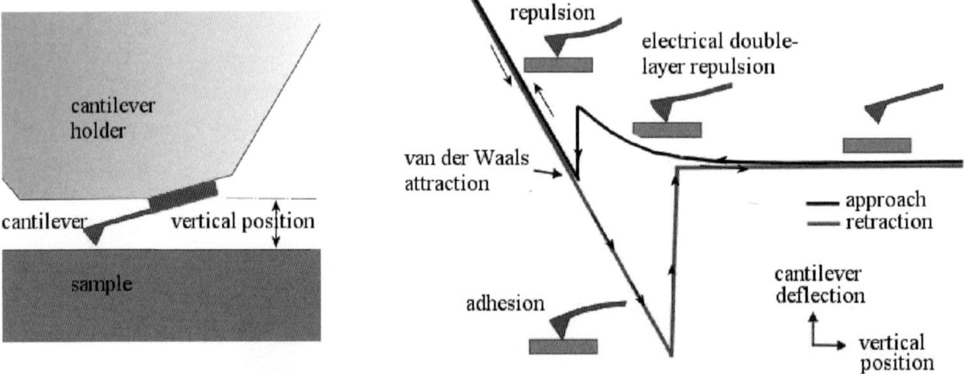

Figure 16–9. Force versus distance curve. For the example shown here, the tip first experiences a long-range repulsive force upon approaching the sample, even before the tip and sample are in physical contact. Close to the sample, the tip becomes strongly attracted by the van der Waals force. In this instance, the attractive force gradient becomes greater than the force gradient by the cantilever spring. This causes the tip to snap into physical contact with the sample (the perpendicular part of the approach curve). Once physical contact has been made, the cantilever is deflected linearly by the approaching scanner. On the way back, the tip may stick to the sample by adhesion until the pull by the cantilever forces it out of contact.

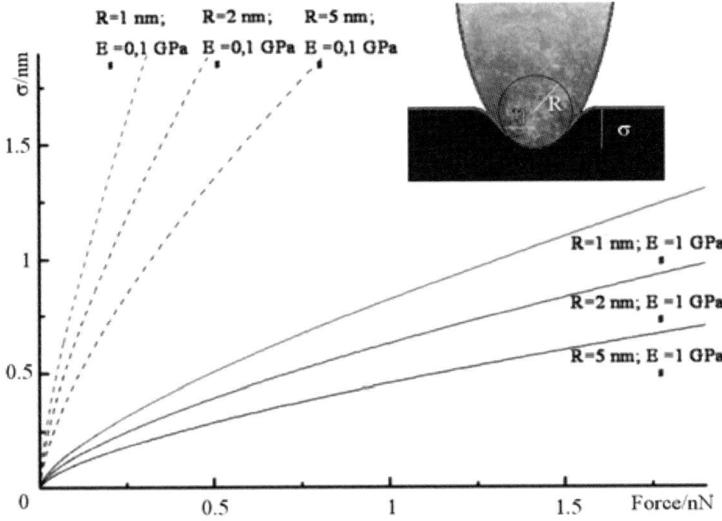

Figure 16–10. Parameter plot of the loading force F versus the penetration depth σ for three different radii of curvature of the apex R and for two types of differently stiff samples. A Young's modulus of 0.1 GPa might reflect a living cell; 1 GPa could be ascribed to a protein structure. The plot shows that even at a very low force, a sharp stylus dives right into the softer sample. It is therefore crucial to select the right kind of an apex radius for each application.

There are a number of models that relate the deformation of a solid body to the loading force of a stylus. According to Sneddon (Heuberger et al., 1996), for example, the repulsive force F for a stylus being loaded onto a solid, homogeneous body is

$$F = \frac{E_s}{2\left(1 - v_s^2\right)}\left[\left(\eta^2 - R^2\right)\ln\left(\frac{R+\eta}{R-\eta}\right) - 2\eta R\right]$$

where E_s is the Young's modulus, v_s is the Poisson's ratio of the sample, and R is the apex radius of the stylus (the deformation of the stylus is neglected). With increasing loading force, the radius η of the contact area between the tip and the sample increases. The penetration depth σ of the tip and the radius of the contact area η are related as (Heuberger et al., 1996):

$$\sigma = \frac{1}{2}\eta\ln\left(\frac{R+\eta}{R-\eta}\right)$$

Hence, for highest resolution, macromolecular samples need to be imaged at minimal load. Under optimal conditions, subnanometer scale resolution has been obtained on protein samples (Schabert and Engel, 1994); for a review, see Engel and Müller (2000). Müller et al. found on a two-dimensional regular array of the protein bacteriorhodopsin that at a load exceeding about 100 pN, the resolution dropped and the molecules became deformed vertically and laterally (Müller and Büldt, 1995). Because the atomic structure of bacteriorhodopsin is known, the change in topography could be assigned to a distinct conformational change of the protein upon the higher load (Figure 16–11).

A convenient way to detect sample deformation upon too high a load comes from comparison of images from trace and retrace (Hoh et al., 1993; Müller et al., 1996; Schabert and Engel, 1994). Trace and retrace refer to the line-by-line motion of the scanner. All commercial AFMs allow two individual images to be acquired, one using the line traces when the stylus moves from left to right and the other one using the traces on the way back.

The actual load leading to high-resolution images might be even smaller than expected from the force setting of the microscope. Yang and co-workers (1996) proposed that short-range interactions with a local asperity give rise to high-resolution contrast while longer-range interactions with blunter parts of the tip help support the load of the tip in contact with the sample. Muller et al. (1997) have argued that in solution, the long-range force required for repulsion of the body of the tip is electrostatic. They adjust the supporting electrolyte so that the asperity just touches the sample lightly.

To obtain details of living cells and tissues by AFM, a meaningful image is often obtained only after a loading force of a few nanonewtons has been applied. Such high loads lead to deformation of the cell up to several hundred nanometers (Hoh and Schoenenberger, 1994). The cell membrane is then pressed onto intracellular structures such

Figure 16–11. Force-dependent surface topography of bacteriorhodopsin (scale bar represents 10 nm) demonstrating the effect of force variations on the topography of the cytoplasmic purple membrane surface. The initial force of 300 pN (bottom of image) was decreased during the scan to 100 pN (top of image). A conformational change is distinct: donut-shaped bacteriorhodopsin trimers transform into units with three pronounced protrusions at their periphery. Inset: noise reduced image at higher magnification (scale bar represents 4 nm). (Courtesy of Müller and Büldt, 1995).

as the nucleus, cytoskeletal elements, and vesicles, which in consequence become visible (Chang et al., 1993; Fritz et al., 1994; Hoh and
Schoenenberger, 1994). In these cases, the AFM images reflect the local
plasticity of the living cells more than their true surface topography.

2.3.2 Forces between the Apex of the Stylus and the Sample

A van der Waals force F_{vdW} is always present between the tip and
the sample. The main contribution to F_{vdW} is the dispersion force,
caused by the dipole-induced dipole interaction and is present
between all kinds of materials. The Lifshitz theory, a combination of
quantum electrodynamics theory and spectroscopic data allows
the forces between two geometrically shaped surfaces to be calculated.
In the case of a flat surface (representing the sample) and a sphere
(being used as an approximation for the apex of the stylus) F_{vdW} is
(Israelachvili, 1991)

$$F_{vdW} = \frac{-H_a R}{6d^2}$$

where R is the radius of the tip (radius of the tip apex) and d is the
distance between apex and sample. H_a represents the Hamaker constant, which characterizes the interaction of the two surfaces (media)
across a third medium. For example, for two mica surfaces in water H_a
is 2.2×10^{-20} J and for two silicon oxide surfaces in water H_a is 8.3×10^{-21} J
(Israelachvili, 1991). For hydrocarbons in water, H_a lies between $(0.2 - 1)$
$\times 10^{-20}$ J (Butt, 1992). The van der Waals force between particles is
always attractive in air and attractive for most situations in an aqueous
solution.

When imaging in aqueous solution, additional interactions between
the apex of the stylus and the sample need to be taken into account.
Many of the commonly used supports and probes as well as most biological samples are charged in an aqueous environment. This is because
they usually carry weak acidic and basic functional groups. They dis-
sociate in an aqueous solution, according to their equilibrium constants. The net charge density of a surface in water depends on the
density of the functional groups, their pK values, and the pH of the
buffer solution.

The DLVO (Derjaguin, Landau, Verwey, Overbeek) theory quantitatively describes the total force between charged interfaces in aqueous
solution. It considers the electrostatic double-layer interaction caused
by surface charges and the van der Waals forces and neglects entropic
or steric contributions. Unlike the van der Waals interaction, the electrical double-layer repulsion depends on the sign and magnitude of the
surface charge density, the ion concentration, and the pH. A charged
surface attracts counterions in the water. At the solid–liquid interface,
a charge cloud on the order of molecular dimensions is created as a
transition region. In this so-called electrical double layer (EDL) the
counterions balance the charge of the surface. The density of the
charges surrounding the surface falls off exponentially with distance
z from the surface (Debye–Hückel approximation):

$$\psi = \psi_0 e^{-z/\lambda_D}$$

ψ_0 represents the potential at the surface. The Debye length λ_D is the thickness of the EDL:

$$\lambda_D = \sqrt{\frac{\varepsilon_0 \varepsilon_e kT}{e^2 \sum_i c_i q_i^2}}$$

where ε_0 represents the vacuum permittivity, ε_e the dielectric permittivity of the electrolyte, k the Boltzmann constant, T the absolute temperature, e the unit charge, and c_i the concentration and q_i the ionic charge of the ith component of the liquid. Note the strong dependence of the double layer thickness on the valence of the ions.

If the tip and the sample are approaching each other, the electrical double layers of the two interfaces become perturbed when they begin to overlap (Figure 16–12). This results in a force that is known as the double layer force F_{el}. F_{el} decreases exponentially with distance d between the two surfaces. For a stylus with an apex radius of curvature of R and a planar sample (at a surface potential <50 mV) (Butt, 1992):

$$F_{el} = \frac{4\pi R \sigma_1 \sigma_2 \lambda_D}{\varepsilon \varepsilon_0} e^{-d/\lambda_D}$$

where σ_1 and σ_2 represent the surface charge density of the stylus and the specimen, respectively, ε_0 is the vacuum permittivity, ε_e the dielectric permittivity of the electrolyte, and d the distance between the two surfaces. This equation is a simplification of any real situation. At a distance much below λ_D, it is necessary to resort to numerical solutions for which there are no simple expressions.

In addition, when the two charged surfaces approach each other, the local ion concentration also changes. This means a shift in the equilibrium conditions of the charged groups with the ions in solution. Hence, the ionizable functional groups of the surface may become neutralized for finite dissociation constants, according to the new equilibrium conditions, and the surface charge density is decreased. This phenomenon is called charge regulation and causes a less strong repulsive force for surfaces charged with similar sign than would occur without charge regulation. F_{el} can even become attractive at very small distances (Israelachvili, 1991).

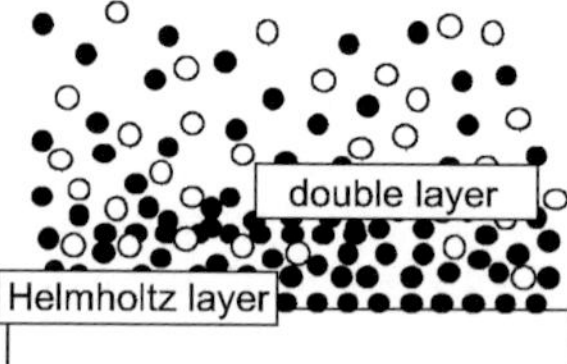

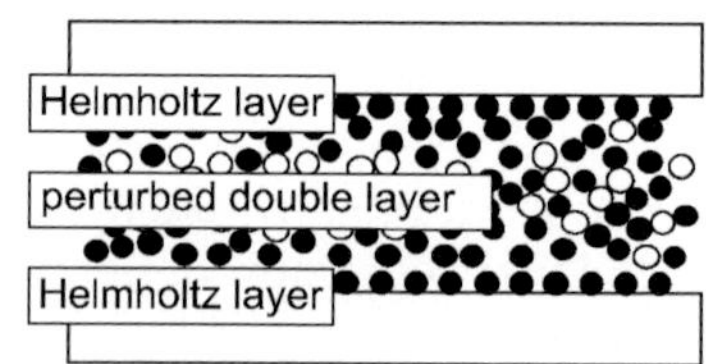

Figure 16–12. Specific interaction of ions at solid–liquid interfaces. The surfaces are negatively charged. (Left) The ions produce an interfacial region of excess solute concentration, the electrical double layer, which consists of the Debye layer and the counterions bound at the surface (Helmholtz layer). (Right) Two negatively charged surfaces at very small separation. The ion clouds of the electrical double layer overlap and cause a repulsive force (Amrein and Müller, 1999). (From Mueller et al., 1997a, reprinted with permission.)

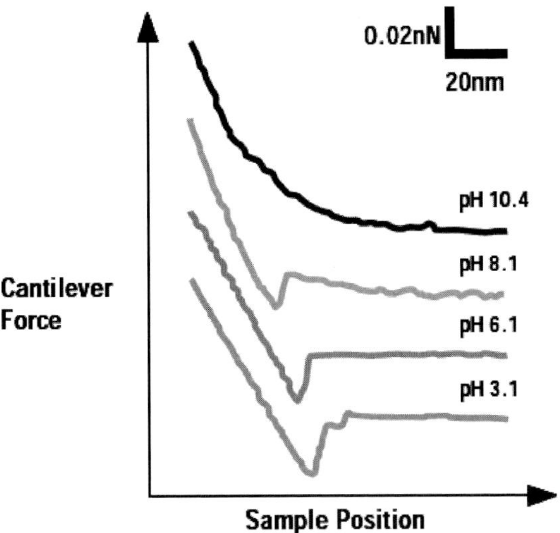

Figure 16–13. Force versus distance curves showing the DVLO force between a silicon nitride cantilever and a mica surface under varying pH conditions and at fixed ion concentration. Mica is naturally negatively charged at neutral pH, less charged at low pH, and increasingly charged at higher pH. At low surface charge, the van der Waals attraction dominates and causes the tip to snap onto the sample. At higher charge, the van der Waals attraction becomes increasingly screened by the repulsive electrical double layer repulsion. (From Butt, 1992, reprinted with permission).

To evaluate the overall force that is relevant for the interaction between the stylus and the sample, the electrical double-layer force and the van der Waals force are summed up to the total DLVO force (F_{DLVO}) (Figure 16–13).

$$F_{DLVO} = \frac{4\pi R\sigma_1\sigma_2\lambda_D}{\varepsilon\varepsilon_0}e^{-d/\lambda_D} + \frac{-H_a R}{6d^2}$$

For high-resolution imaging, the electrical double-layer repulsion may have to be reduced such that the stylus can effectively come in physical contact with the sample rather than "riding" on the electrical double layer. This can be achieved by increasing the (bivalent) ion concentration (Butt 1991, 1992; Butt et al., 1995; Ducker et al., 1991).

There are a number of interactions that need to be considered in addition to the above-mentioned DLVO force, including steric, hydrophobic, and hydrophilic interactions.

When the microscopy is performed in air, there is often a water bridge occurring between stylus and sample. This is because under ambient conditions, most surfaces are covered by a thin water layer. The resulting meniscus force may be quite strong (e.g., in the order of 10^{-7} N) and may pull the tip effectively onto the sample. This will result in poor resolution and may even cause damage to the sample and stylus. It is notable that the van der Waals interaction in air is usually about an order of magnitude stronger than in an aqueous environment. The detrimental effects occurring in air through these forces may partially be overcome by choosing an appropriate imaging mode (intermittent contact mode, see below). The properties of the stylus may also be changed to avoid formation of a water bridge between sample and tip. Knapp et al. (1995) have established a method for rendering the Si_3N_4 probes hydrophobic. The probes became coated with a Teflon-like polymer through glow discharge in a hexafluoropropene (HFP, Hoechst AG, Frankfurt, Germany) atmosphere (Guckenberger et al., 1994; Knapp et al., 1995).

2.4 The Feedback Loop of the AFM

During scanning, the sample topology causes the stylus to move up or down and the cantilever is deflected accordingly. The deflection is measured by means of the optical pointer and compared to the set point in the regulator. The set point represents the loading force chosen by the user. Deviation from the set point is relayed to the driver of the scanner and the height of the sample with respect to the stylus is readjusted, etc. Hence, the scanning process translates the sample topology into a time-dependent signal. The signal may be decomposed (by the Fourier transform) into a spectrum of sine waves. Each sine wave of the spectrum is distinct based on its amplitude and frequency. The smaller the features are in the plane of the sample and the faster the probe is scanned over the sample, the higher the frequencies. The taller the features of the sample, the higher the amplitude of the respective frequencies. Each frequency is also distinct by its phase. The phase describes how much one sine wave is shifted with respect to all other sine waves. The quality of the feedback loop refers to how accurate phase and amplitude is relayed through feedback loop as a function of the frequency.

The feedback loop is only able to respond adequately up to a certain speed (frequency). Above that speed, the stylus does no longer trace properly the sample topology or the feedback becomes unstable. When the probe is scanned fast over the sample, for example, the speed of the feed-back loop may be too low to adjust the tip to sample distance in time. This results in variable and at times high interaction forces (Weisenhorn et al., 1993).

The quality of the feedback loop with respect to speed and accuracy depends on the quality of its elements. The elements of the feedback are the regulator, the cantilever spring, the scanner element that moves the sample up and down, and also the interaction between the sample and the apex of the stylus and the sample topology itself (Figure 16–14).

2.4.1 Stability of the Feedback Loop

Each element of the feedback loop has its own transfer function or frequency response. It amplifies the signal in a frequency-dependent manner. Moreover, it relays the signal with a delay. A given delay relates to a larger and larger phase shift with increasing frequency. The phase shifts from all elements add up. At a certain frequency the total phase shift will have become too large and causes the feedback loop to react in the wrong direction, i.e., instead of moving up, the scanner

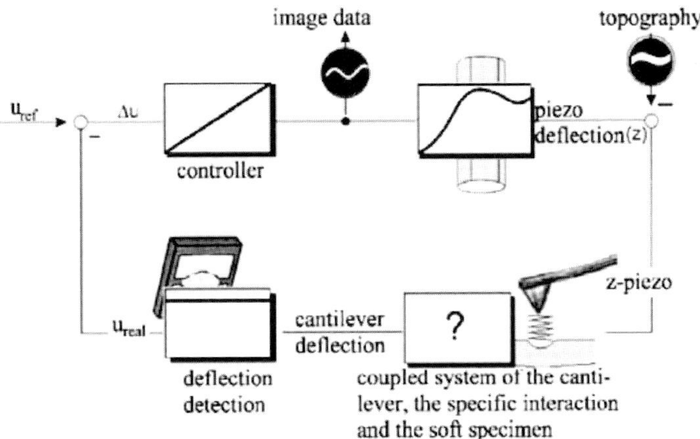

Figure 16–14. The feedback loop of the AFM. The boxes symbolize the frequency response of the elements of the loop.

would move down, etc. Only the frequency range below this limiting frequency must be used for imaging. The higher frequencies must be suppressed by an appropriate setting of the regulator. Otherwise, the feedback loop will oscillate. Because the frequency response of the feedback loop may be different for each new sample and probe and may even change during imaging, the regulator needs to be regularly adjusted by the user. Most of the time, the regulator allows the amplitude for all frequencies to be adjusted proportionally (P-regulator) and in a frequency-dependent way (integral-regulator).

The bandwidth (the usable frequency range) of an AFM is usually limited either by the scanner or the probe in contact with the sample. The optical pointer and the regulator should not pose a limitation to speed or accuracy of the feedback loop.

2.4.2 Elements of the Feedback Loop

2.4.2.1 The Scanner

Subatomic-precision scanners are the key to the success of STM, AFM, and nanotechnology in general. Such scanners are based on piezo-ceramic elements. They deflect in an electrical field. Within a certain range, the deflection is approximately linear to the applied voltage. A scanner is characterized by its scan range (x,y), the z range, and the frequency response. The scan range of the actuator limits the size an image can have. The z range puts a limit on the maximum specimen corrugation the scanner can trace. The first scanners used were tripods, with each leg representing one axis of space (Binnig and Rohrer, 1982). Actuators used today in AFM are often made of single piezoceramic tubes (Binnig and Smith, 1986). The bending of the tube allows the probe or the sample to be moved in two dimensions (on approximately a spherical surface). The length adjusts the sample-to-probe distance. Piezoceramic tube scanners are superior in many respects to piezoceramic tripods that were used in the beginning of SPMs. The tube scanners are symmetrical with respect to the z axis and therefore are less susceptible to thermal drift. They can be made much smaller to obtain a scan range similar to the tripods. Most recent microscopes also have scanners with an open frame for access from above and below the sample. In these cases, the x,y scanner is separate from the elements moving the scanner up and down.

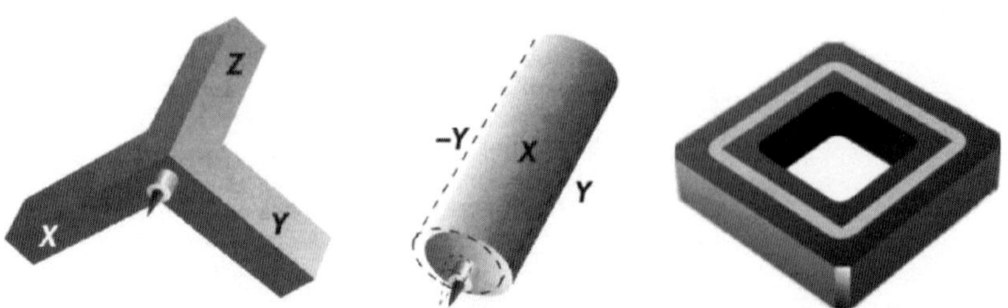

The frequency response determines how fast the actuator can respond to an applied voltage. Today's scan units are a compromise with respect to these properties. On the one hand, a high range and sensitivity are accomplished by increasing the size of the actuators; on the other hand, larger scanners are slow to react. The bandwidth (i.e., the usable frequency range) of the scanner is limited by its first mechanical resonant frequency (also *natural frequency*) in the z direction and this depends on the size and shape of the element as well as material properties (Table 16–1). An almost linear relation between an applied voltage and the deflection is achieved only when the actuator is used well below its resonant frequency. Close to the resonance, the sensitivity rises sharply and, even more important, the deflection becomes increasingly delayed with respect to the driving voltage (i.e., there is a negative phase shift between the driving voltage and the deflection). As discussed above, too much of a delay is not acceptable, because the probe could then no longer trace the relief of a specimen in real time.

To date the scanner is usually the slowest element in the feedback loop of the AFM. It thus determines the bandwidth of the whole instrument. From this condition a slow scan speed results and, hence, a long image acquisition time that not only can make the SPM a tedious task, but also puts a corresponding limit on the time domain for the observation of dynamic events. This is surprising, since there are ways to strongly increase the bandwidth of AFM by a faster scanner without limiting the scan range in any direction (Figure 16–15).

2.4.2.2 The Probe and the Sample

The dynamic response of the probe is more difficult to understand than the frequency response of the scanner. Prior to establishing contact with the sample, the probe (cantilever) may be treated as a beam with one end fixed and one end free. Establishing the frequency response is straightforward, as described in Table 16–1. The smaller and the stiffer the cantilever, the higher its first resonant frequencylies. Note that the frequency response of a free cantilever is always established prior to imaging in the intermittent contact mode (tapping mode). However, after establishing contact, the beam is in an intermediate state between a beam with both ends fixed and one end fixed. Moreover, contact to the

Table 16–1. Natural frequencies for the elements of a scanner.[a]

	Natural frequency (Hz)		
	Longitudinal	Lateral	Torsional
Beam with quadratic cross section: one end fixed (E = Young's modulus, ρ = density, L = length, and a = edge length	$f_z = \dfrac{1}{4L}\sqrt{\dfrac{E}{\rho}}$	$f_{xy} = 0.1615\,\dfrac{a}{L^2}\sqrt{\dfrac{E}{\rho}}$	
Tube with attached mass: one end fixed A = cross-sectional area, m = mass, M = attached mass, l = area-moment of inertia, d = outer diameter, w = wall thickness, l_z = moment of inertia, l_M = moment of inertia of attached mass, and $k = \dfrac{\pi}{32}\dfrac{E}{2.6\,L}\left[d^4 - (d - 2w)^4\right]$	$f_z = \dfrac{1}{2\pi}\sqrt{\dfrac{EA}{L(M + \frac{m}{3})}}$	$f_w = \dfrac{1}{2\pi}\sqrt{\dfrac{3El}{(M + 0.23m)\,L^3}}$ $I = \dfrac{\pi}{4}\left[\left(\dfrac{d}{2}\right)^4 - \left(\dfrac{d}{2} - w\right)^4\right]$	$f_t = \dfrac{1}{2\pi}\sqrt{\dfrac{k}{I_M + \frac{I_z}{3}}}$ $I_z = \dfrac{m}{4}\left[\left(\dfrac{d}{2}\right)^2 - \left(\dfrac{d}{2} - w\right)^2\right]$

[a] From Harris (1987).

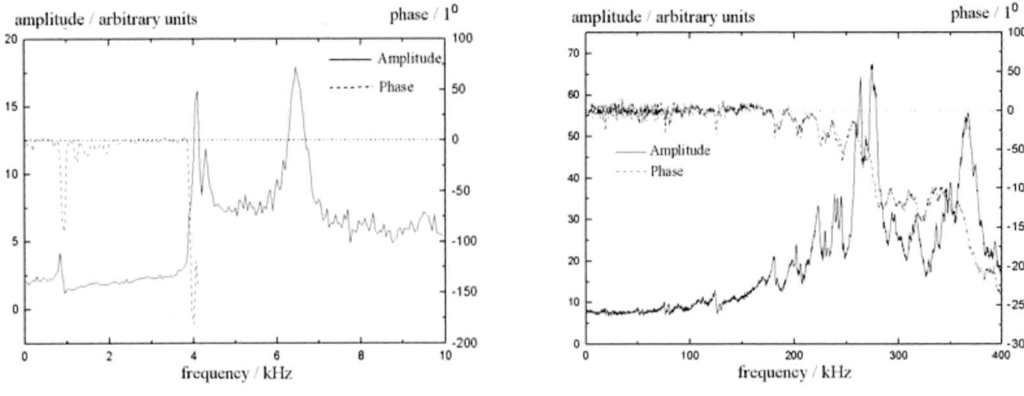

Figure 16–15. (Left) The frequency response of a typical commercial scanner (the J scanner from NanoScope III, digital instruments). The scanner may be used up to approximately 4 kHz. Above, the phase shift is too large for stable imaging. (Right) A scanner developed in the group of the author. The scan range and range in the z direction are comparable to the J-scanner. However, the usable frequency range is one to two orders of magnitude wider (Knebel et al., 1997).

sample is complicated. The cantilever and the sample are coupled via the force interaction between the tip and the sample and the compliant sample itself. Mechanically, the situation resembles a series of three dampened springs, the cantilever, the tip sample interaction, and the viscoelastic sample. There is no analytical solution for this situation (Budó, 1976) and the frequency response is best determined experimentally (Figure 16–16). Smaller and stiffer cantilevers will still lead to a broader bandwidth of this element of the feedback loop.

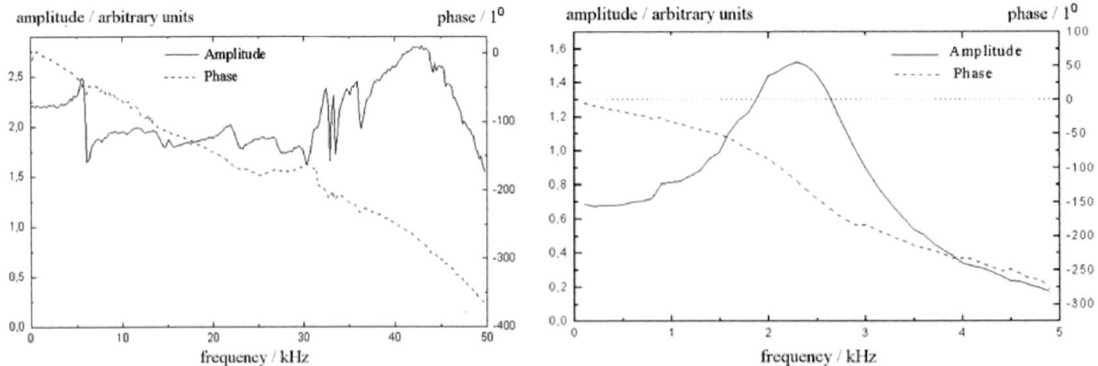

Figure 16–16. (Left) Frequency response of a silicon nitride cantilever (100 µm long; $c = 0.1\,\text{N/m}$; Olympus Ltd., Tokyo, Japan) in contact with a mica sample. The loading force was 1 nN. The frequency limit posed by this element of the feedback loop lies at about 10 kHz, above which the phase shift has become too great for a stable feedback loop. (Right) Frequency response of the closed feedback loop of a NanoScope III (digital imaging, Santa Barbara, CA) equipped with a J-scanner. The same cantilever has been used. The measurement was carried out in water. Instead of scanning across the mica sample, the sample was oscillated, mimicking a corrugated sample. The oscillation was continuously changed from low frequency (mimicking broad objects) to a higher frequency (representing smaller objects). The amplitude of the oscillation was constant over the total measuring range. The regulator of the microscope was adjusted as for imaging to optimally track the oscillation of the sample. The amplitude shown corresponds to the output of the feedback for image acquisition.

2.4.3 Frequency Response of the Closed Feedback Loop

For imaging, all elements work together in the closed feedback loop (Figure 16–16A). The frequency response of the closed feedback loop is therefore a direct measure of the quality and trustworthiness of AFM imaging. In the example shown in Figure 16–16B, lower frequencies (larger object details) up to about 1.5 kHz are all represented with about the same amplitude. Higher frequencies up to about 2.5 kHz (smaller object details) are amplified increasingly more strongly by the system. Hence, in an image these details would be more prominent than they should be. Very small details would no longer be traced. For proper interpretation of the sample structure at high resolution, images should be taken at various scan speeds.

2.5 Noise

The instrument's noise degrades the accuracy with which the topology is traced. For a well-designed instrument, from all the elements of the feedback, the thermal motion of the cantilever is the major source of noise. The cantilever "produces" one $k_B T$ per resonance interval (k_B is Bolzmann's constant and T is the absolute temperature). This will cause the cantilever typically to oscillate by a few Angstroms. When the stylus rests in firm contact on the sample, the thermal motion of the cantilever is strongly reduced and will not affect resolution directly. However, the thermal energy now creates a fluctuation of the loading force. In the case of a soft, susceptible sample, this in turn will create a marked fluctuation of the sample deformation. The thermal force noise within the context of the simple harmonic oscillator model is given by (Sarid, 1991)

$$\Delta F = 23 \sqrt{\frac{4k_B B k}{\omega_0 Q}}$$

where ΔF is the rms force noise, B the bandwidth at which the microscope operates, T the absolute temperature, k the spring constant of the cantilever, ω_0 the radial resonant frequency, and Q the mechanical quality factor. Q parameterizes the sharpness of the resonance, being the ratio of the resonant frequency to the full-width at half-height of the peak. In air or vacuum, very high Q (on the order of thousands) can be achieved. In water, Q is much lower. As an example, for a typical cantilever used to scan a molecular sample in an aqueous solution ($k = 0.1$ N/m, $\omega_0 = 20$ kHz, $Q = 1$, $B = 10$ kHz) the thermal force noise is 5–20 pN. From the equation above it follows that an ideal cantilever for low-noise operation should have a low spring constant, but still a high resonant frequency. For a given size of a cantilever, however, the stiffer the lever, the higher the resonant frequency. This dilemma can be overcome by employing smaller (shorter) cantilevers (Hansma, 1996). The effect of the thermal noise on the image quality will also be reduced by operating the microscope at a lower bandwidth. But this comes at the cost of reduced scan speed.

It is notable that noise is not evenly spread over the entire frequency spectrum. For large Q, there is a large reduction of the thermal noise away from the resonant peak. In water, the thermal noise is quite evenly distributed over the whole frequency range. At low frequencies, other sources of mechanical motion become progressively more important. This is referred to as $1/f$ noise and is notable foremost in contact mode imaging (see below).

2.6 Imaging Modes

2.6.1 Contact Mode

In contact mode, the tip is permanently touching the sample. On a relatively solid sample, the position of the cantilever in the z direction is well defined and is affected only by force noise. It has therefore provided the best resolution to date on molecular samples. Contact mode imaging works best with densely packed samples, to provide stability against the sideways push by the tip. On soft samples, small cantilevers with a low spring constant are desirable. For operation at minimal force, the force during scanning is reduced until the cantilever takes off from the sample and then increases it just enough to touch back down and trace the topology properly.

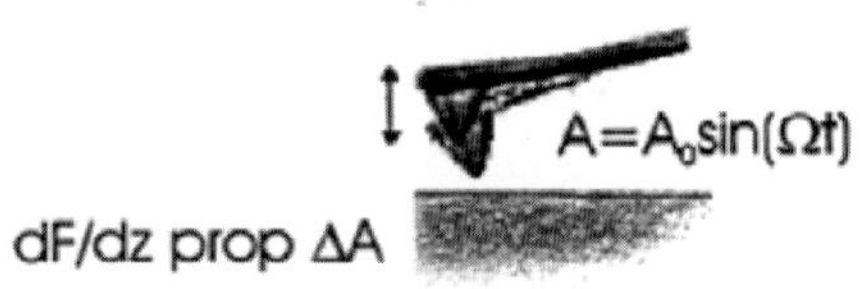

2.6.2 Noncontact Mode

In noncontact mode, the probe tip is excited to oscillate near the resonant frequency. To chose the excitation frequency, it is necessary to perform a sweep from low to high frequencies and observe the amplitude and phase of the cantilever oscillation. The fundamental bending frequency of the cantilever will stand out as a prominent peak and the excitation frequency chosen to be close by (on the lower-frequency flank of the peak). In close vicinity to the sample but prior to contact, the interaction between the tip and the sample will change the resonance conditions for the lever (as the cantilever now enters an intermediate state between a lever free on one side and a lever fixed on both sides). This in turn will cause the amplitude to decrease, because the driving frequency is now further away from the resonant frequency. The shift in resonant frequency is a function of the force gradient (dF/dz) between the tip and the sample. The microscope is operated to keep the amplitude reduction at a defined level and, hence, the tip at a defined distance to the sample. Although noncontact imaging promises to be the most subtle mode for highly deformable, weakly immobilized objects, it has not become popular so far in life science applications because it does not usually allow for stable high-resolution tracing of an interface of a biological sample. This might be because the tip is kept at a distance to the sample where it still can oscillate considerably by thermal noise and instrument instabilities. Very stable microscopes, very small cantilevers, and very sharp tips with a high aspect ratio may, in the future, bring about a strong improvement for this potentially attractive imaging mode.

2.6.3 Intermittent Contact Mode (Tapping Mode, Dynamic Force Microscopy)

Intermittent contact mode AFM in a certain sense combines the advantages of noncontact and contact AFM. The probe oscillates as in the noncontact operation. But the set point for the amplitude is now chosen so that the tip can make brief contact with the sample in each cycle of its oscillation. As in the noncontact mode, there are no lateral forces excerpted on the sample. As in contact mode AFM, the local topological height is determined accurately by the well-defined contact point. This mode has therefore had success in imaging well-separated molecules bound to an underlying support rather weakly that did not withstand the lateral forces of contact mode imaging. It is notable, however, that the loading force during contact is not smaller than with contact mode AFM. Intermittent contact mode has resulted in resolution of molecular samples on the order of 1 nm (Figure 16–17).

As in noncontact mode, intermittent contact mode AFM in air is sensitive to a shift in resonance frequency that then brings about a change in amplitude of the oscillation. In buffer, resonance effects are small an and the microscope is largely sensitive to the restriction of the oscillation brought about by the physical barrier posed by the sample. Its sensitivity is thus limited by thermal fluctuations and is not significantly enhanced by operating at resonance. In water, it is desirable to choose as small an amplitude as possible to minimize disturbance of the sample. In practice, the overall oscillation amplitude is set to the smallest value that provides stable imaging. Then the set point amplitude change is increased (increasing the resolution) to the largest value that does not damage the sample. However, stable operation requires amplitude sufficient to pull the tip out of attractive interactions.

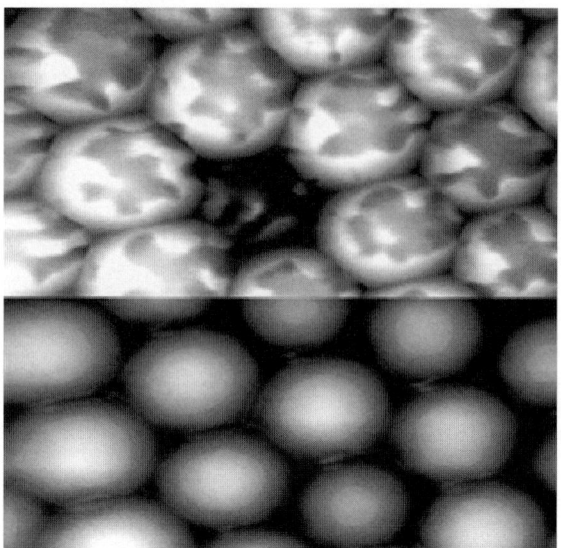

Figure 16–17. Latex-beads imaged in tapping mode. In the top portion of the image, the set change in amplitude was chosen large and the beads were deformed. In the lower portion, the set amplitude change was reduced, resulting in proper topographical imaging (1 µm × 1 µm).

Two methods for exciting the cantilever oscillator are common, acoustic drive (Hansma et al., 1994; Putman et al., 1994) and magnetic drive. In acoustic drive an oscillator supplies a voltage drive to a piezo-electric actuator that generates sound waves in the cantilever holder. The frequencies are typically from tens of kilohertz to megahertz. When the driving frequency is near a bending-mode resonance of the cantilever, the cantilever is driven into an oscillation. A frequency scan with acoustic drive results in many peaks, only some of which yield a usable signal (Putman et al., 1994; Schäffer et al., 1996). This is because the transmission of the acoustic wave is not uniform over the frequency range. In magnetic drive, the cantilever is coated with a magnetic film, or a magnetic particle is glued onto the end of the cantilever (Lantz et al., 1994; Han et al., 1996). The oscillating magnetic field of a solenoid in the vicinity of the cantilever causes the cantilever to oscillate.

In contrast to contact mode, which senses a force, intermittent contact mode AFM is a technique that senses an interfacial stiffness. As frequency is increased, intermittent AFM also becomes sensitive to increased interfacial damping.

2.7 Optimizing AFM Design for Life Science Applications

The range of applications for AFM is very large, from atomic resolution imaging on some crystal surfaces to imaging or manipulation of whole cells. An AFM designed for working under high vacuum to look at defects in atomic lattices has different basic requirements from one that is designed predominantly for samples in liquid, with full optical microscope integration as a priority. On biological samples, atomic resolution is generally not possible; for samples of macromolecules and larger structures, the resolution is mainly defined by issues such as the tip–sample interaction and sample deformation. Then technical issues such as integrated optics, flexibility of handling, and sample environment carry more weight.

2.7.1 Combining AFM and Optical Microscopy

Optical microscopy still dominates life science research, and techniques for enhancing contrast, labeling, or staining to recognize and study biological samples have been developed. It makes no sense to throw away this vast reserve of information about samples that are to be studied using the AFM, and indeed it is not necessary to compromise the quality of optical images for light microscopy to be combined with AFM. For a well-designed AFM it is possible to simultaneously obtain high-contrast and high-resolution optical images, including techniques such as laser scanning microscopy (LSM), total internal reflection fluorescent (TIRF) microscopy, and fluorescence resonance energy transfer (FRET) microscopy (Figure 16–18).

In one sense it is relatively straightforward to build a combined optical and atomic force microscope, since the AFM can be built around a standard inverted optical microscope. This allows the use of the optics through the objective lens under the sample (if the sample is transparent, which is often the case), while at the same time the AFM can approach from above. In practice, care must be taken that the entire system has the required mechanical stability, that sources of drift or vibration are minimized, and that the "open" nature required of such an AFM does not compromise its stability and reduce the resolution achievable. There are also two potential problems that must be considered in the instrument design that relate directly to the optical imaging: the optical quality of the support for the sample, through which the optical image must be formed, and the light path to illuminate the sample from above.

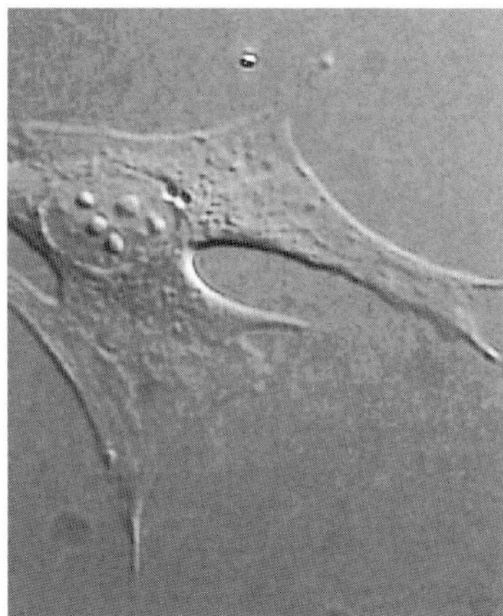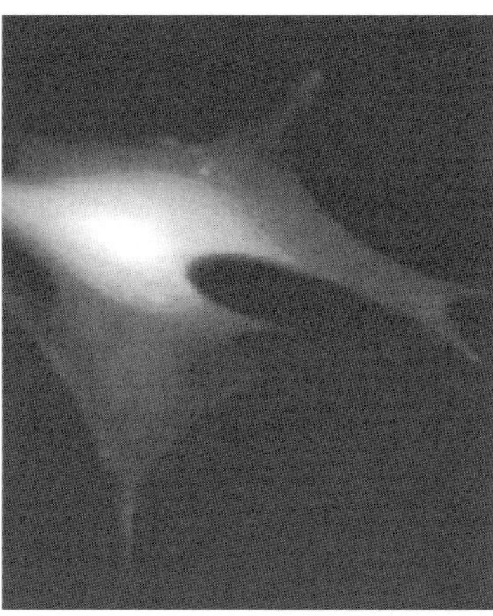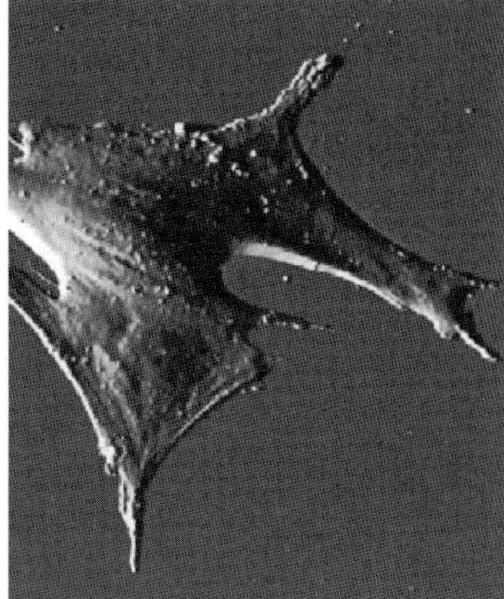

Figure 16–18. Images of a fixed 3T3 cell in buffer, on a coverslip. Left to right: Optical (differential interference contrast, DIC) and simultaneous AFM (JPK Instruments, contact mode) topography and feedback signal. (See color plate.)

For high-quality optical images, the objective lenses are optimized for thin glass coverslips, which are mechanically unstable sample supports for AFM imaging. The working distance of high-magnification objectives is very small, so the objective lens must be positioned close beneath the sample, often with an oil or water droplet between the lens and the coverslip. These considerations must be addressed by appropriate design of sample holders (which must generally also function as a liquid cell) that are able to provide enough mechanical support for the sample without interfering with access for the microscope objective beneath.

For common biological samples such as cells or vesicles, there is very low optical contrast in brightfield illumination, so some enhanced optical contrast mechanism is required. The most common choices are phase contrast (which provides contrast based on the sample refractive index) or differential interference contrast (DIC, which is sensitive to local differences in the refractive index from point to point). In both cases, the path length and optical quality of any components in the illumination path are critical. To use the optical microscope in its standard configuration, and hence have the optimal optical system, the microscope requires illumination from above, using the condensor optics. This means that the illumination light must pass "through" the AFM in some way. The ideal optical solution is to be able to use the condensor optics provided by the optical microscope manufacturer, and build the AFM on an open frame around the optical path that does not interfere with the sample illumination.

Another issue for experiments involving simultaneous AFM and optical microscopy is the laser illumination used to measure the cantilever deflection. Although the majority of this is reflected by the cantilever, a significant proportion spills over the cantilever edges and can pass through the sample into the optical microscope image. For brightfield, phase contrast, or DIC this can be removed using a simple filter in the optical path to cut the required wavelength, but for fluorescence this can severely limit the wavelength regions available and hence dyes that can be used for labeling. This is particularly a problem when the sample should be labeled with more than one dye to identify different components, or in more advanced optical techniques such as FRET, where two fluorescent dyes are used simultaneously. Typically most AFMs used a detection laser in the red region of the optical spectrum, because of the availability of small, low-power red laser diodes. More recently, AFMs are becoming available with infrared laser illumination, which allows the full visible spectrum to be used for optical imaging. Using an infrared detection laser also results in better performance of the AFM itself, in that it removes artifacts caused by optical interference between the light reflected from the cantilever and from the sample in the path to the photodiode detector.

2.7.2 Scanners

Over time, the scan range available from commercial AFMs has increased, particularly the z-range or distance the tip is able to move up and down over the surface. For cell imaging, a z-range of at least

10–15 µm is required, otherwise the tip will be unable to move high enough to scan over the cell nucleus. Larger x and y scan ranges of 100 µm or more are also generally required for cell imaging, and sometimes also for imaging much smaller samples, which may be distributed inhomogeneously over the surface. The piezoceramic material used for all AFM scanners suffers from various problems of nonlinearity, hysteresis, and creep. Although it is possible to move the tip very precisely, this is against a large background of position changes due to longer term effects in the piezo material, as it continues moving slowly for a long time after a voltage jump is applied (creep) or moves variable distances depending on its history (hysteresis). These problems have been addressed by adding position sensors (such as capacitive or strain-based sensors) along the movement axes, so that the tip movement is no longer set merely by converting the desired position into a simple voltage. The current position of the tip is constantly read by the position sensors and the nonlinearity of the piezo material and its changes over time can be constantly corrected, using another feedback loop. These "linearized" piezo systems are also becoming available in commercial AFM systems and are likely to become standard.

There are two possibilites generally used for the lateral scanning for an AFM—tip scanning or sample scanning. For general AFM, these are equivalent, and both have their advantages and disadvantages, but in terms of optimizing an AFM for biological samples, tip scanning generally has clear advantages. On a basic practical level, much of AFM imaging for life science research has to take place in aqueous solutions, usually containing a reasonably high concentration of salt, and there is always a greater risk of expensive damage when the high-voltage piezoelectronics sit at a lower level than the sample. Even using a tip-scanner, the AFM must be carefully designed so that the electronics and piezo elements above the sample are protected from accidental spillage or water vapor. This is particularly important if the sample is to be heated to 37°C, when the evaporation is significant, and condensed water vapor could easily collect in an unsealed AFM head. There is also a more fundamental problem with sample scanning, however, as simultaneous AFM and optical imaging cannot generally be performed. When the sample is moved in the z direction, which usually lies along the optical axis, the objective lens can be tracked with the sample using a piezoactuated lens holder, but this is not possible for lateral scanning movements across the optical axis.

2.7.3 Sample Environment

In addition to the considerations already discussed about the need for using coverslips as a sample support for many optical imaging applications, there are other practical issues raised by life science samples. Temperature control of the sample is often important for studying molecular reactions or whole cells under physiological conditions, and, for example, many lipid bilayers used as model cell membranes undergo phase transitions over the range from around 10 to 37°C and above. Perfusion is also important, particularly for *in situ* experiments and to introduce molecules for reactions, blocking, or to change the properties of the solution. For live cell work, it may also be necessary to allow a gas exchange to equilibrate the cell medium with 5% CO_2.

The priorities for the design of the cantilever and sample holders should also include the easy disassembly and cleaning of all the components that are in contact with the sample, so that, for example, ultrasound or autoclaving is possible. When working in liquid, all parts of the instrument that come in contact with the fluid are sources of possible sample contamination. The issues are somewhat different depending on the applications. For instance, if the aim is to image or stimulate living cells, then molecular-sized contamination is not so important for AFM imaging itself, but the sample must remain sterile and the cells must not be exposed to any chemical contaminants that will produce a biological response. For single molecule imaging, the molecules may not be affected by traces of certain ions leaching from metal surfaces, but every macromolecular contaminant that could adhere to the surface is a problem for AFM imaging.

3 Sample Preparation

To study their native structures and probe their functions, most cellular or macromolecular samples need to be kept in an aqueous environment. Hydrophilic and hydrophobic interactions promote correct folding of the polypeptide chains into a protein and are responsible for the formation of micelles, bilayers, and membranes from lipids and proteins. The conformation of membrane proteins is determined by hydrophobic interactions with lipidic tails and hydrophilic interactions with their heads and the surrounding water (Haltia and Freire, 1995; Engel et al., 1992; Jap et al., 1992). The pH, electrolyte type, and its concentration and temperature also influence the structure and function. The function of macromolecular structures depends not only on their native conformation but often requires even more exacting environmental conditions with respect to pH and temperature and may depend on the presence of coenzymes or ATP, for example. Living cells are sensitive not only to pH, ionic strength, temperature, and CO_2 levels; they usually need a specific and often highly complex medium in which to grow. When biological structures are allowed to air dry, they are subjected to a high force caused by the change in surface tension as the water evaporates. The energy involved is considerable. As a result, even macromolecules become severely flattened and collapsed (Baumeister et al., 1986; Kellenberger et al., 1982; Kellenberger and Kistler, 1979; Wildhaber et al., 1985). For these reasons, most sample preparation techniques described below are for samples in buffer. Most of the time, immobilizing cells and macromolecular structures is all that is needed for AFM analysis. AFM analysis of macromolecular samples is demanding with respect to the cleanliness of the support and the purity of the buffer. All surfaces become immediately covered with hydrocarbons when exposed to ambient air. Even double-distilled water can be a source of organic contaminants. A layer of these hydrocarbons on the sample or the probe can be most disturbing for AFM.

As a result, the sample supports should be prepared or activated immediately before use. Ultrapure water (fresh milli-Q water; $\leq 18\,M\Omega\,cm^{-1}$) should be used to prepare all buffer and rinsing solutions because it contains fewer hydrocarbons and macroscopic contaminants than conventional bidistilled water.

In the following sections, we will describe suitable supports and immobilization techniques for both macromolecular and cellular samples.

3.1 Macromolecular Samples

Immobilizing macromolecular structures aims for a homogeneous distribution of the specimens in a close-to-native conformation. Tight binding of the biological specimens to the support surface will prevent them from clustering. They may also better withstand the forces that arise between the probe and the sample for most AFM. On the other hand, the structure can be substantially distorted and proteins may even denature by strong binding. This is well known from transmission electron microscopy (TEM) as well as from STM of biological specimens (Baumeister et al., 1986; Wang et al., 1990). The specimens may adsorb with preferential orientations depending on the binding conditions (Fisher et al., 1978; Hayward et al., 1978; Karrasch et al., 1993; Müller et al., 1996). In addition to suitable binding properties, the support should be as smooth as possible so that it does not interfere with the structure of the biological specimen in the final image. Furthermore, it should be relatively chemically inert to prevent contamination due to the solution or nonspecific reactions with the biological system.

3.1.1 Specimen Supports

Glass coverslips are widely used as an amorphous specimen support and can be used either unmodified or altered to change their physisorption or chemisorption properties. The surface can be almost featureless on the scale of macromolecular specimens. They are best suited for all experiments in which visible light is transmitted across the sample, as in SNOM or in the combined light and atomic force microscopy. Before use, organic contaminants, dust, and other particles are removed by washing one time with concentrated HCl/HNO_3 (3:1) and five times for 1 min with Millipore water in an ultrasonic bath (50 kHz). This process makes the coverslips clean and smooth (rms roughness ~0.5 nm).

The most commonly used support for imaging biological specimens in the AFM is mica. Mica minerals are characterized by their layered crystal structure. Mica can be readily cleaved for a clean, atomically flat surface. Muscovite mica (Mica New York Corporation, New York, NY) is the most commonly used form. The average surface charge density of muscovite mica in water is $\sigma_m \approx -0.0025\,C/m^2$ (0.015 electron per surface unit cell).

Gold surfaces can be easily prepared by vapor deposition. They are chemically inert against O_2 and stable against radicals. They bind organic thiols or bifunctional disulfides with high affinity, which can be used to covalently attach biological macromolecules (see below). Hegner et al. (1993) developed a relatively simple and reliable method for preparing ultraflat gold (Knebel et al., 1997) surfaces. They consist of atomically flat terraces many microns in diameter. Thin carbon films commonly used in TEM are smooth on molecular dimensions and adsorb macromolecules well when freshly prepared. High-vacuum carbon evaporators are common in electron microscopy laboratories.

3.1.2 Physiorption, DLVO Force

The most common technique for immobilizing biomolecules onto a support is by physisorption. Usually, the objects are immobilized out of an aqueous solution. They become attached to a support when there is an overall attractive force that pulls the surfaces into contact. The relevant force for adsorption is the DLVO force. Hydrophobic and hydrophilic interactions may also play a role. The DLVO force for two planar surfaces is (Israelachvili, 1991)

$$F_{\mathrm{DLVO}} = F_{\mathrm{el}}(z) + F_{\mathrm{vdW}}(z) = \frac{2\sigma_{\mathrm{su}}\sigma_{\mathrm{sp}}}{\varepsilon_0\varepsilon_{\mathrm{e}}}e^{-z/\lambda_{\mathrm{D}}} + \frac{-H_{\mathrm{a}}}{6\pi z^3} ; \qquad \text{per unit area}$$

The DLVO force between charged surfaces is highly susceptible to ion concentration and conditions can thus be adjusted to achieve good adsorption (Müller et al., 1997) (Figure 16–19).

When the electrical double layer repulsion between the two surfaces has "vanished," they rapidly coalesce to minimize the interaction energy (Israelachvili, 1991). For the example shown in Figure 16–19, at electrolyte concentrations above 50 mM KCl, the amount of adsorbed membranes rapidly increased and reached its maximum at about 150 mM KCl. Note that adsorption occurred even though both mica and the sample carried a net negative charge.

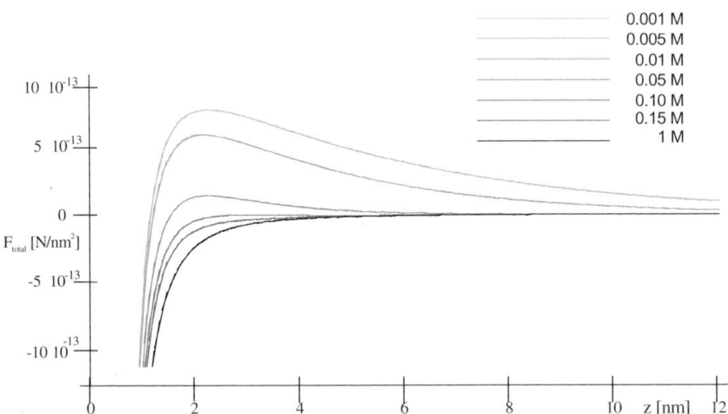

Figure 16–19. Dependence of the DLVO force on ion concentration (1 : 1 monovalent electrolyte) and distance between a macromolecular sample (purple membrane) and a mica support. (From Mueller et al., 1997a, reprinted with permission.) Whereas the attractive van der Waals force is mainly unaffected by the electrolyte, the double layer repulsion decreases with increasing salt concentration. The surface charge densities were −0.0025 C/m² for mica (Israelachvili, 1991) and −0.05 C/m² for purple membrane (Butt, 1992), respectively. The Hamaker constant was 3 × 10⁻¹⁹ J.

3.1.3 Physisorption, Hydrophobic and Hydrophilic Interaction

There is an attractive interaction between hydrophobic surfaces in water. The attractive interaction potential is larger than the van der Waals potential and can be very long range. The nature of these long-range forces is not yet fully elucidated. Hydrophilic molecules, on the other hand, tend to disorder the surrounding water molecules and prefer contact with water molecules. Hence, the molecules repel each other. These repulsive, hydrophilic forces are also referred to as hydration, structural, or solvation forces. They may cause the DLVO theory to fail at small distances between two hydrophilic surfaces. With respect to adsorption, hydrophobic molecules do not attach to a hydrophilic surface and vice versa. For example, hydrophilic purple membranes did not adsorb to highly hydrophobic supports such as derivatized glass. The hydrophilic and hydrophobic interaction can cause an oriented adsorption of molecular structures.

3.1.4 Physisorption, Preparation of the Support

With mica, an active surface is conveniently obtained by cleaving the layered mica crystals prior to specimen adsorption. For most other supports, the active surface cannot be produced so easily. These supports are usually covered by hydrocarbon contaminants and behave more or less hydrophobicly. Glass, silicon wafers, and many thin films can be rendered hydrophilic by exposure to glow discharge (for example, in a Harrick Plasma cleaner, 1 min, $p = 0.1$ m bar, with air as the residual gas) right before use. Thin carbon films become negatively charged. For those specimens that adsorb better to hydrophobic surfaces, glow discharge must be omitted.

Coating is another way to improve physisorption on many specimen supports and it has been used for a long time by electron microscopists (Jacobson and Branton, 1977; Mazia et al., 1975). For example, poly-L-lysine can be used for coating glass and mica and render the coated surfaces positively charged. This allows cells, tissues, and plasma membranes that are usually negatively charged to be readily adsorbed. Objects that carry charge in an uneven distribution can be adsorbed in a defined orientation on a poly-L-lysine-coated surface. For example, purple membrane mainly consists of a light-driven proton pump that builds up an electrochemical potential across the membrane. Illuminated by light, purple membranes show an asymmetric charge distribution and adsorb to polylysine-coated surfaces in an oriented fashion (Fisher et al., 1977, 1978; Hayward et al., 1978). More than 90% of the membranes attach with their cytoplasmic surface toward the poly-L-lysine under specific conditions (pH 9). At a pH below 4, the majority of the membranes (>94%) were directed with their extracellular surface toward the coated surface.

3.1.5 Chemical Bonding

Covalent bonding can be a very reliable technique to allow firm binding of biological specimens to a support. Some of the first high-resolution AFM topographies of protein structures in buffer solution have been obtained using this technique (Karrasch et al., 1993). It appears that covalent binding does not interfere with the macromolecular structure anymore than physisorption. Bonding of the macromolecular specimens can be accomplished using chemically modified supports. Karrasch et al. (1993) developed a protocol to cross-link biological systems to a silanized glass coverslip. The silane (APTES, Fluka Chemie AG, Buchs, Switzerland) contained a free amino group that allowed it to react with the succinimide ester group of the photocrosslinker ANB-NOS (Fluka Chemie; $\lambda = 312\,nm$). Proteins were then bound to the interface by activating the photocrosslinker with UV radiation. This method resulted in the first high-resolution images of protein structures by AFM in buffer (Figure 16–20).

Epitaxial gold surfaces can effectively be functionalized by alkanethiols. They form ordered, self-assembled monolayers that are tightly bound to the gold surface via chemisorption of the sulfur atoms. The monolayers are further stabilized by the lateral hydrophobic interactions of the alkyl chains (Hegner et al., 1993; Sellers et al., 1993; Wagner et al., 1994; Wolf et al., 1995). The latter can carry head groups at the free end that allow oriented covalent anchoring of macromolecular structures (Allison and Thundat, 1993; Hegner et al., 1993, 1996; Wagner et al., 1994). Wagner et al. (1995, 1996) bound protein structures via their amino groups with an N-hydroxysuccinimide-terminated monolayer on gold.

3.1.6 Langmuir–Blodgett Films

There are amphiphilic substances that naturally form insoluble monomolecular films on an air–water interface. They exhibit a water-soluble polar or charged head group and a highly apolar tail. This causes them to attach to an air–water interface with the head group immersed in the water and the tail toward the air. The most prominent example is the pulmonary surfactant that forms at the interface of the respiratory

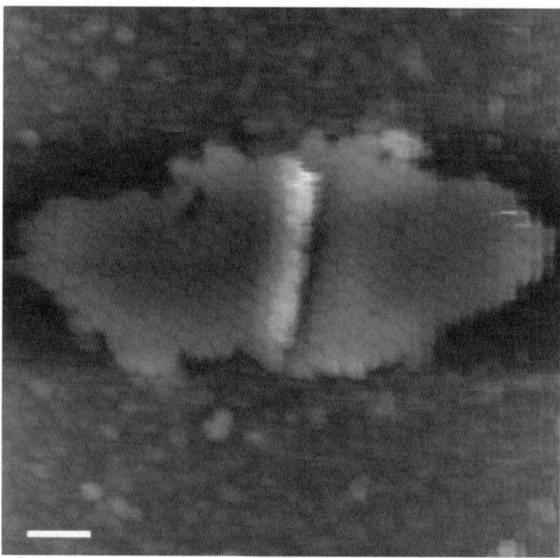

Figure 16–20. Hexagonally packed intermediate layer. Scale bar = 70 nm. (From Karrasch et al., 1993, reprinted with permission.)

gas lumen and the solvation layer that covers the alveolar epithelium of lungs. Surfactant layers can be formed *ex vivo* in a Langmuir trough to study their biophysical properties under defined conditions or for the purpose of microscopic examination. Langmuir films of lipids have also been used to mimic biological membranes (for references see Bader et al., 1984), or they served as a substrate to bind and crystallize proteins in two dimensions for TEM and AFM investigations (e.g., Brisson et al., 1994). AFM proved to be outstandingly well suited to study the structure and mechanical properties of these thin layers.

To prepare films for microscopy, the amphiphilic substances are spread at the air–water interface of a Langmuir trough. They are then compressed by a movable barrier by a desired amount. To perform AFM on the air side of the film, the monolayers may be transferred from the air–water interface onto a solid support by slowly pulling a hydrophilic support out of the aqueous phase across the interface (Langmuir–Blodgett transfer; Blodgett and Langmuir, 1937). The film is deposited as the support is moved vertically across the air–water interface. It is then inspected by AFM in air. To do microscopy on the aqueous side of the film, the monolayer may also be deposited by dipping a hydrophobic substrate from the air side across the interface into the water. If a first lipid layer is deposited on the upstroke onto a hydrophilic substrate and then another layer added on the down stroke of the sample, a complete lipid bilayer has formed. This bilayer may contain membrane proteins. It is interesting to note that deposition of a bilayer onto a mica substrate arrests the lipid of the first lipid layer. These lipids are no longer free to diffuse in the plane of the membrane. If the support is glass, both the lipids bound to the support and those within the second layer facing the aqueous phase are free to diffuse.

Finally, Langmuir–Blodgett transfer may not be necessary and films of pulmonary surfactant have been studied directly at the air–water interface (Knebel et al., 2002).

3.2 Cells

Successful immobilization of living cells largely depends on the cell type. There are cells with adherent growth (for example, epithelial cells, fibroblasts, or glial cells), and cells that grow in suspension without contact to a substrate (for example, bacterial cells or erythrocytes). Adhesive cells are more readily imaged with the AFM, whereas cells that grow in suspension have to be immobilized to be imaged. It is notable that cells may change their shape, physiology, and even their life cycle once bound to a substrate. A variety of techniques have been developed to immobilize living cells. Cells are best imaged with an AFM that is combined with a light microscope.

3.2.1 Adsorbing Cells on Glass Coverslips

Cells that naturally adhere to a substrate can either be cultured on an appropriate support and subsequently imaged, or plated on the support and monitored shortly after they have established cell–substrate contact. For both procedures, the glass coverslip must be thoroughly cleaned. If cleaned with water, the glass support has to be dried in air or a stream of N_2 to prevent plated cells from possible osmotic shock. Coverslips have been coated with poly-L-lysine, collagen (Henderson et al., 1992), proteoglycans, laminin, or fibronectin to improve adhesion.

For imaging individual, adherent cells with the SFM, the density of the cell suspension has to be chosen such that enough space remains for the cells to spread out. The time required for the cells to attach and spread depends on the cell type. Before imaging, the samples have to be rinsed with buffer solution to remove cells that are not firmly attached and, if feasible, examined by conventional light microscopy. Specific cells that were cultured on a solid support spread out to a thickness of less than 100 nm over large areas in the periphery (Fritz et al., 1994; Henderson et al., 1992; Hoh and Schoenenberger, 1994; Kasas and Ikai, 1995). In these thin regions it is possible to monitor the organization of the intracellular cytoskeleton (Figure 16–13).

3.2.2 Immobilizing Nonadhering Cells

A stable immobilization of cells that grow in suspension and do not establish substrate interactions in their natural environment is difficult to obtain. Hörber et al. (1992) have a method to trap single cells by a micropipette and image the exposed part with the AFM. The setup makes it possible to use the advantages of the micropipette technique and to enhance the inner pressure of the cell. This is an advantage, because the "spring constant" of a cell surface may be very low [for example, ~0.002 N/m (Hoh and Schoenenberger, 1994)]. Hence, the cell is extensively deformed by any reasonable interaction force with an AFM probe.

Permeable supports provide the possibility of measuring additional properties of cells (for example, permeability, diffusion, and voltage characteristics of the plasma membrane) while they are imaged by AFM. The cells may attach onto substrates with a much smaller pore size than the average diameter of the cell, or individual cells may be trapped in pores that are only slightly smaller than the average cell diameter. Kasas and Ikai (1995) have used Millipore filters (Millipore PCF, Millipore Corp., Bedford, MA) with pore sizes similar to that of the cell diameter for trapping yeast cells. Hoh and Schoenenberger (1994) have cultured MDCK epithelial cells (average lateral diameter ~10 µm) on polycarbonate filter supports (Millipore PCF, 12 mm diameter) with a much smaller pore size (0.4 µm) than the average cell diameter.

4 Imaging and Locally Probing Macromolecular and Cellular Samples: Examples

4.1 Imaging

By controlling the ionic strength and composition of the deposition buffer, individual collagen molecules can be assembled and adsorbed onto a mica surface in various different conformations (Jiang et al., 2004). Around five collagen molecules associate form microfibrils, which have a lateral size of around 3–5 nm. These microfibrils are also likely to be an intermediate stage in the formation of the larger collagen fibers seen in natural tissue. A monolayer of these microfibrils can then be adsorbed to the surface to form a nanostructured, biologically active surface. The fibrils can be aligned through adsorbing under conditions of hydrodynamic flow (Figure 16–21); Figure 16–21 shows a case in which the fibrils are adsorbed at very low coverage, and the 67-nm banding repeat can be seen along the filament length. For samples in which a complete monolayer is formed, the composition of the adsorption buffer can be used to control whether the filaments organize themselves such that the bands on adjacent filaments are aligned or not (Jiang et al., 2004).

The nanometer-scale topography of these surfaces can be measured using AFM, and then used for cultivating cells *in situ*. The orientation and growth direction of fibroblast cells grown on the aligned collagen supports depend critically on the alignment of the D-banding between adjacent collagen fibrils. The overall alignment of the direction of the collagen fibers is not sufficient to produce a response in the cells, but when the banding of the fibers is aligned, the cells show a strong response (Poole et al., 2005). This is one case in which the ability to combine AFM and optical microscopy allows the study of a biological structure/function question over the size scale from the molecular structure and organization to the response of whole cells.

4.2 Beyond Imaging

One strength of the AFM is that it is able to combine imaging modes sensitive to different properties of the sample with direct measurements of forces and interactions. These measurements can be carried out at particular points (selected, for instance, from a sample that has just been imaged), or built up over a grid to "map" the surface properties.

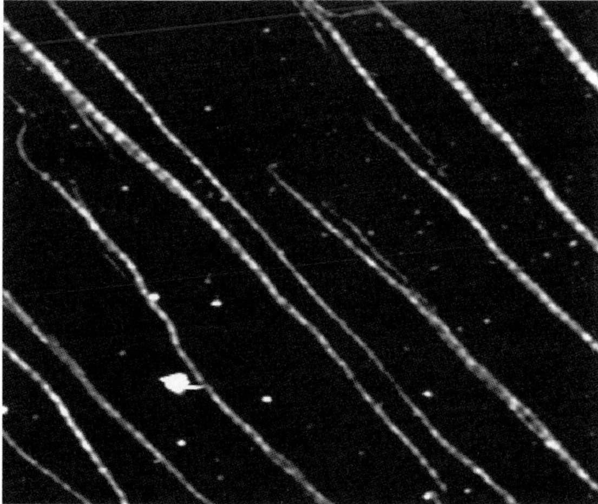

Figure 16–21. Collagen fibrils adsorbed on mica; sample courtesey of Müller; imaging JPK Instruments, intermittent contact mode in buffer. Scan area = 2.6 × 2.3 μm, z range = 2.5 nm. The 67-nm banding along the axis of the individual collagen fibrils can be seen.

Force spectroscopy in AFM refers to the measurement of tip–sample interaction forces at a point as the height of the cantilever base is varied. This allows measurement of forces either pushing into the sample (elasticity, rheology, etc.), or pulling away from the sample (adhesion, including recognition of specific molecular binding, unfolding, or stretching of molecules bound between the tip and the sample surface). Alternatively, manipulation of the surface, such as dissection or alignment, is possible using the tip to apply forces and modify the sample.

The ability to locally probe and manipulate a sample in addition to imaging is a unique strength of AFM. The local experiments are as diverse as the research questions and address both isolated macromolecular structures and cells. This is demonstrated for a few examples.

4.2.1 Pulmonary Surfactant

A mixed lipid–protein film of pulmonary surfactant covers the hydrated lung epithelia to the air. This highly cohesive and mechanically stable film reduces the otherwise high surface tension of the air–aqueous interface to almost zero as is required for the structural stability of the alveolar lung and ease of breathing. The mechanical stability of the film depends on a pattern of monolayer domains of disaturated phospholipids intercepted by multilayer areas, containing the unsaturated lipids also present in surfactant. This molecular architecture is conveyed to the film by the surfactant-specific proteins SP-B and/or SP-C. The proteins cross-link the multilayer areas to the monolayer.

This molecular arrangement has been discovered by AFM (von Nahmen et al., 1997). First, a topographical image was obtained (Figure 16–22). Then, the loading force was increased such as to physically remove the lipid stacks on top of the molecular monolayer. After removal of the multilayers, the monolayer showed crevices on the circumference of the earlier multilayer patch and holes, evenly

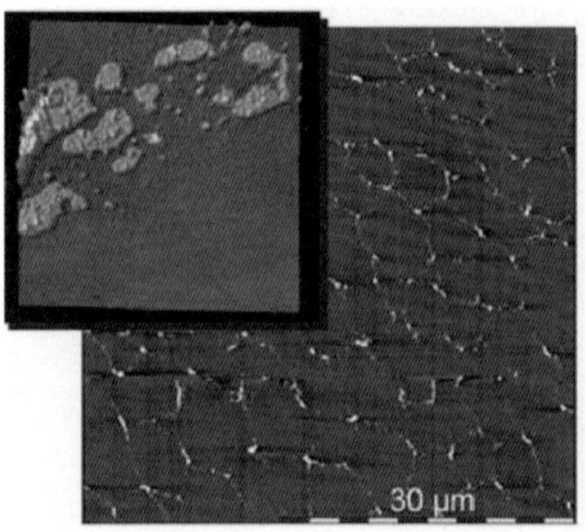
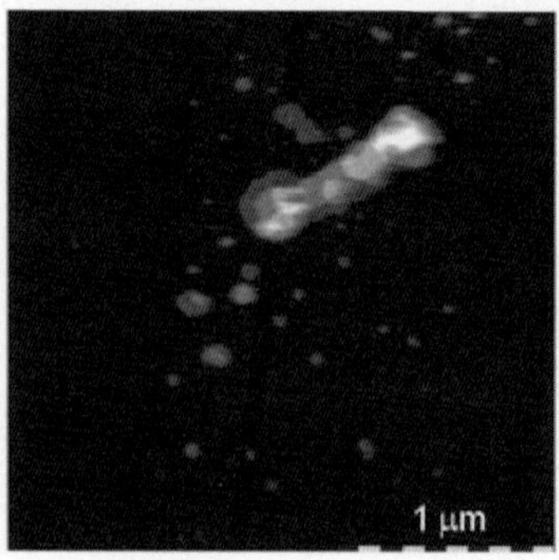
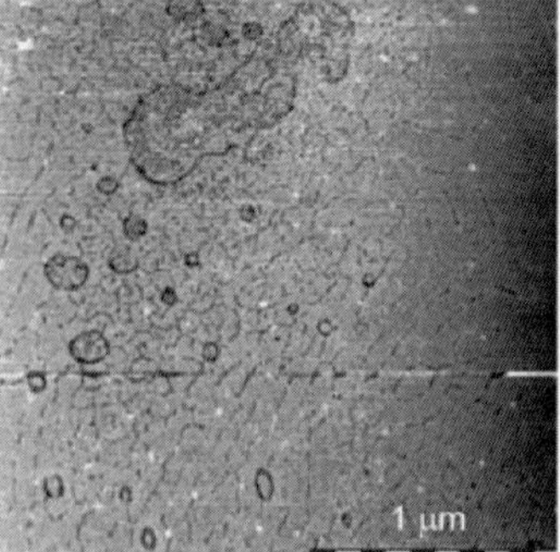

Figure 16–22. Functional pulmonary surfactant forms molecular films of molecular monolayer regions and areas of monolayers with stacks of lipid bilayers cross-linked to it (left). Cross-linking is revealed by physically removing the bilayers by an intentionally high load of the AFM probe (middle and right).

distributed over the area earlier covered. This was indicative of a cross-linking structure that had also been removed by the stylus. Mere lipid layers, formed on top of a lipid monolayer in the absence of surfactant proteins, left no traces behind after being removed by the stylus.

Multilayers, cross-linked to the monolayer as opposed to multilayers, merely adsorbed to the monolayer appear to be crucial for surfactant function. At low surface tension, the interfacial film is subjected to high lateral pressure. Cross-linked multilayered areas take up the lateral load together with the monolayer. Otherwise, the load resides on the monolayer alone. But lipid monolayer films containing unsaturated lipids are not able to withstand a high film pressure without collapsing. They therefore cannot reduce tension by a degree required for proper function of the lung. This condition is likely responsible for lung failure after acute lung injury where the surfactant is distinct by an elevated level of cholesterol (Karagiorga et al., 2006). Excess cholesterol prevents formation of cross-linked multilayer stacks and surface tension is high (Gunasekara et al., 2005).

4.2.2 Collagen Matrices

Müller and his associates (Jang et al., 2003) used the atomic force microscope to mechanically direct collagen into two-dimensional, 3-nm-thick layers of defined structure. During a critical, initial period of time this structure was found to be highly malleable and the collagen molecules could be reoriented by the stylus of the AFM. After this time, the collagen coating hardened and remained stable for several months without loss of fiber orientation or mechanical strength. It was proposed that collagen fiber formation *in vivo* may also have a critical period during which cellular forces remodel a collagen network that acts as a matrix for later tissue growth. In tissues, cells are indeed thought to organize the collagen fibrils by exerting tension on the matrix, and the realignment forces required in the experiment (300 pN) were within the range of forces that fibroblasts can generate. Besides explaining an important phenomenon on how tissue might organize *in vivo*, this discovery might also find technical applications. Engineered collagen matrices might find use as a platform for cell biological and tissue engineering applications (Figure 16–23).

4.2.3 Imaging and Unfolding a Bacterial Cell Wall Protein

Unfolding proteins and DNA and separating molecular binding partners have become important and active offspring of AFM, referred to as force spectroscopy. It provides insight into protein and DNA folding and conformation about the nature of binding pockets. Müller and his associates have used this technique in conjunction with high-resolution imaging to gain insight into the interaction forces between the individual protomers of a regular bacterial surface layer, the hexagonally packed intermediate (HPI) layer of *Deinococcus radiodurans*. After imaging the HPI layer, the AFM stylus was attached to individual protomers by enforced stylus–sample contact to allow force spectroscopy experiments. Imaging of the HPI layer after recording force–extension curves allowed unfolding forces to be correlated with the structures being unfolded. By using this approach, individual protomers of the HPI layer were found to be removed at pulling forces of ≈300 pN. Furthermore, it was possible to sequentially unzip entire bacterial pores formed by six HPI protomers (Figure 16–24).

4.2.4 Sphingolipid–Cholesterol Rafts

In the plasma membrane, nanoscopic, condensed sphingolipid–cholesterol rafts "float" in a fluid matrix and likely act as prefabricated subunits of functional complexes. The functions these platforms perform involve essential cellular processes such as the initiation and down-regulation of signaling cascades that regulate cell growth, survival, and death, or endo- and exocytosis. A rapidly growing list of diseases has been identified as being related to rafts, including Alzheimer's

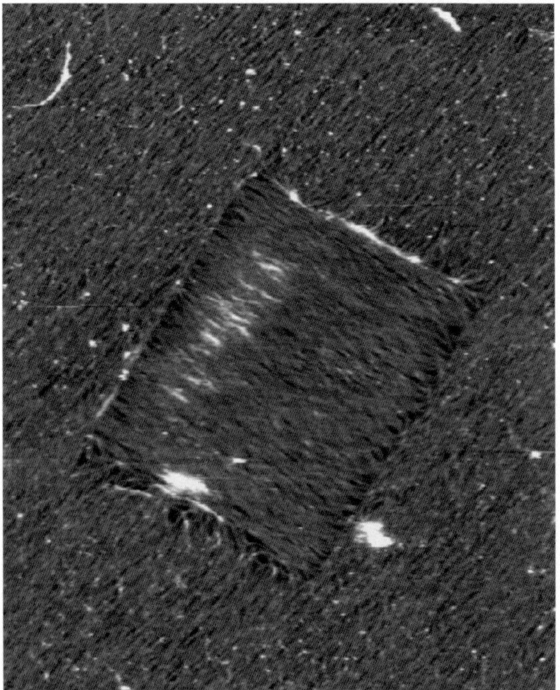

Figure 16–23. After collagen microfibrils were assembled out of solution on the supporting surface, their orientation has been changed in a controlled manner using the stylus of an AFM in the central region of the image. For this, the force was increased to 300 pN while the stylus was scanned over this operator-defined region. The collagen microfibrils in the region were now aligned with the fast-scanning direction of the AFM stylus and were approximately perpendicular to the original orientation. The surface was then reimaged at low force to create the image. (From Jiang et al., 2004b, reproduced with permission.)

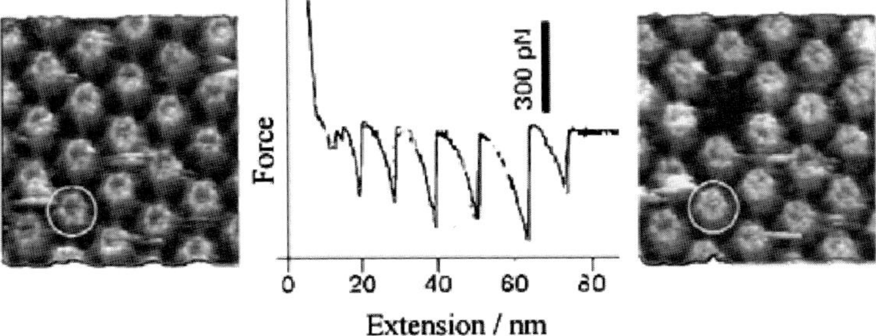

Figure 16–24. The six protomers of an individual pore can be sequentially pulled out of the HPI layer. Left: AFM topograph of the inner surface of the HPI layer prior to (left) and after (right) the pulling experiment. Note that one entire pore is missing. (Middle) The force-extension curve shows a sawtooth pattern with six force peaks of about 300 pN each corresponding to the extraction of one protomer. The height of the force peaks corresponds to the binding force for a protomer to its neighbors; the stretching distances between protomer disruption events (7.3 ± 1.6 nm) corresponds to the length of the molecular linker connecting a protomer to its neighbors. (From Engel et al., 2000, reproduced with permission.)

and Parkinson's diseases, muscular dystrophy, asthma and allergic responses, hypertension, arteriosclerosis, and prion diseases. Bacteria (e.g., *Vibrio cholarae*), viruses such as HIV-1, measles, and Ebola virus, as well as parasites (e.g., malaria) depend on lipid rafts for their purposes. Despite the outstanding importance of lipid rafts, most questions about how rafts are formed, composed, and behave in the plasma membrane of living cells are not well understood. Although several chemical groups that anchor proteins to rafts are known (e.g., the GPI, dipalmitoyl group), it is not known why, and to what extent, certain proteins become associated with one type and others to another kind of raft. But even the most basic questions regarding the size of rafts and whether they are transient structures are controversially discussed (Figure 16–25).

Strong experimental evidence of the existence of rafts as entities of defined size and with an extended lifetime has come from measuring the diffusion and stability over time of individual rafts in the plasma membrane of BHK-21 cells by photonic force microscopy, a variety of AFM (Hörber et al., 1992; Pralle et al., 2000). A bead in a laser trap was scanned over a cell sample. Like with the tip of a cantilever-based AFM, the height of the laser spot was readjusted by feedback when the bead was forced out of the focal spot of the laser trap by the sample. A topographical image of the sample was thus recorded. In a set of experiments, beads were coated with antibodies against a raft-associated protein. After binding to the cell, the laser trap was kept still and the diffusion of the bead attached to a raft was observed within the narrow confinement of the laser focal spot. The diffusion reflected the diffusion of the raft within the membrane and allowed the size of

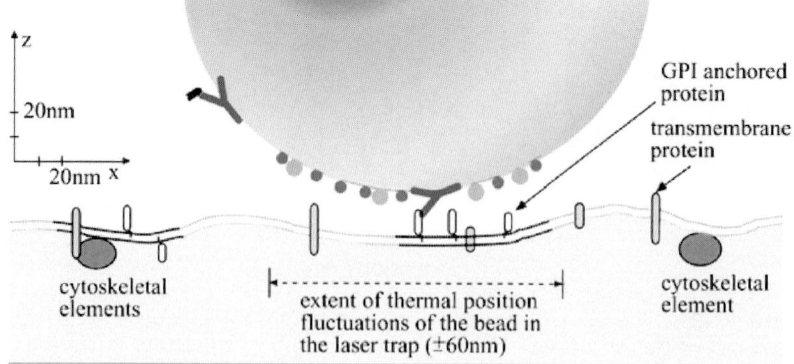

Figure 16–25. The size and temporal stability of a lipid raft may be derived by observing its diffusion within the plasma membrane. Observing the free diffusion of a raft is problematic, because rafts collide frequently with the elements of the cytoskeleton, anchored to the membrane. A more accurate picture is derived from observation of the constraint diffusion of a raft offered by photonic force microscopy. (From Pnalle et al., 2000, reproduced with permission.)

the raft to be derived. For those rafts addressed by the selected antibodies, these entities turned out to be about 50 nm in diameter. The experiment also showed that these rafts were not transient structures within the timeframe of the measurement (Figure 16–26).

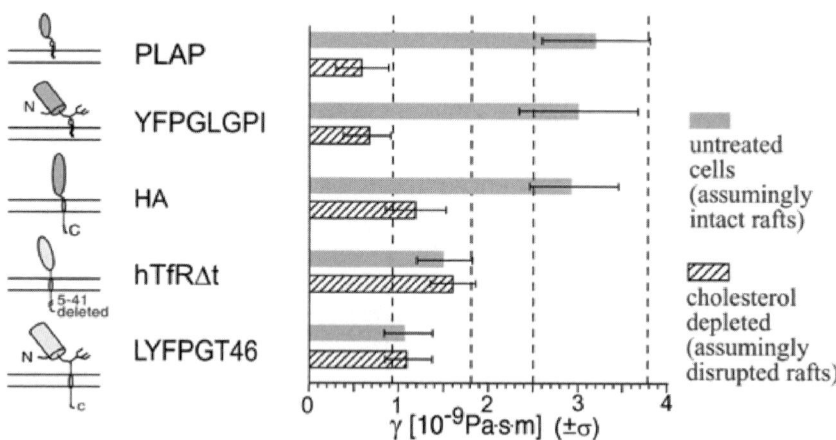

Figure 16–26. Diffusion coefficients for raft proteins (PLAP, YFPGLGPI, HA) and nonraft membrane proteins (hTfRΔt, LYFPGT46). The raft proteins diffuse together with the lipid raft; the nonraft proteins diffuse freely in the fluid phase of the plasma membrane. Depriving cells in culture of one of the mandatory raft elements, cholesterol, abolishes rafts. As a consequence, the former raft proteins diffuse individually in the membrane. (From Pnalle et al., 2000, reproduced with permission.)

References

Allison, D. and Thundat, T. (1993). Imaging entire genetically functional DNA molecules with the scanning tunneling microscope. *J. Vac. Sci. Technol. A* **11**, 816.

Amrein, M., Stasiak, A., Gross, H. and Travaglini, G. (1988). Scanning tunneling microscopy of recA-DNA complexes coated with a conducting film. *Science* **240**, 514–516.

Amrein, M., Dürr, R., Stasiak, S., Gross, H. and Travaglini, G. (1989b). Scanning tunneling microscopy of uncoated recA-DNA complexes. *Science* **243**, 1708–1711.

Amrein, M., Dürr, R., Winkler, H., Travaglini, G., Wepf, R. and Gross, H. (1989b). STM of freeze-dried and Pt-Ir-C-coated bacteriophage T4 polyheads. *J. Ultrastruc. Mol. Struct. Res.* **102**, 170–177.

Amrein, M., Gross, H. and Guckenberger, R. (1993). STM of proteins and membranes. In *STM and AFM in biology* (O. Marti and M. Amrein, Eds.), 127–175 (Academic Press, London).

Amrein, M., von Nahmen, A. and Sieber, M. (1997). A scanning force- and fluorescence light microscopy study of the structure and function of a model pulmonary surfactant. *Eur. Biophys. J.* **26**(5), 349–357.

Arscott, P.G. and Bloomfield, V.A. (1993). *STM of DNA and RNA* (Academic Press, San Diego).

Bader, H., Dorn, K., Hupfer, B. and Ringsdorf, H. (1985). Polymeric monolayers and liposomes as models for biomembranes—how to bridge the gap between polymer science and membrane biology. *Adv. Polymer Sci.* **64**, 1–62.

Baró, A.M., Miranda, R., Alaman, J., Garcia, N., Binnig G., Rohrer, H., Gerber, C. and Carracosa, J.L. (1985). Determination of surface topography of biological specimens at high resolution by scanning tunnelling microscopy. *Nature* **315**, 253–254.

Baumeister, W., Barth, M., Hegerl, R., Guckenberger, R., Hahn, M. and Saxton, W.O. (1986). Three-dimensional structure of the regular surface layer (HPI layer) of *Deinococcus radioudurans*. *J. Mol. Biol.* **187**, 241–253.

Binnig, G. and Rohrer, H. (1982). Scanning tunneling microscopy. *Helv. Phys. Acta* **55**, 726–735.

Binnig, G. and Smith, D.P.E. (1986). Single-tube three-dimensional scanner for scanning tunneling microscopy. *Rev. Sci. Instrum.* **57**(8), 1688–1689.

Blackford, B.L. and Jericho, M.H. (1991). A metallic replica/anchoring technique for scanning tunneling microscope or atomic force microscope imaging of large biological structures. *J. Vac. Sci. Technol. B* **9**(2), 1253–1258.

Blodgett, K.A. and Langmuir, I. (1937). *Phys. Rev.* **51**, 2492.

Brisson, A., Olofsson, A. and Ringler, P. (1994). Two-dimensional crystallization of proteins on planar lipid films and structure determination by electron crystallography. *Biol. Cell* **80**, 221–228.

Budó, A. (1976). *Theoretische Mechanik* (VEB Deutscher Verlag der Wissenschaften, Berlin).

Butt, H.-J. (1991). Measuring electrostatic, van der Waals, and hydration forces in electrolyte solutions with an atomic force microscope. *Biophys. J.* **60**, 1438–1444.

Butt, H.-J. (1992). Measuring local surface charge densities in electrolyte solutions with a scanning force microscope. *Biophys. J.* **63**, 578–582.

Butt, H.-J., Jaschke, M. et al. (1995). Measuring surface forces in aqueous solution with the atomic force microscope. *Bioelect. Bioenerg.* **38**, 191–201.

Chang, L., Kious, T., Yorgancioglu, M., Keller, D. and Pfeiffer, J. (1993). Cytoskeleton of living, unstained cells imaged by scanning force microscopy. *Biophys. J.* **64**, 1282–1286.

Ducker, W.A., Senden, T.J. and Pashley, R.M. (1991). Direct measurements of colloidal forces using an atomic force microscope. *Nature* **353**, 239–241.

Engel, A. and Müller, D.J. (2000). Observing single biomolecules at work with the atomic force microscope. *Nature Struct. Biol.* **7**(9), 715–718.

Engel, A., Hoenger, A., Hefti, A., Henn, C., Ford, R.C., Kistler, F. and Zulauf, M. (1992). Assembly of 2-D membrane protein crystals: dynamics, crystal order, and fidelity of structure analysis by electron microscopy. *J. Struct. Biol.* **109**, 219–234.

Fisher, K.A., Yanagimoto, K. and Stoeckenius, W. (1977). Purple membrane bound to polylysine glass: effects of pH and light. *J. Cell Biol.* **75**, 200a.

Fisher, K.A., Yanagimoto, K. and Stoeckenius, W. (1978). Oriented adsorption of purple membrane to cationic surfaces. *J. Cell Biol.* **77**, 611–621.

Fritz, M., Radmacher, M. and Gaub, H.E. (1994). Granula motion and membrane spreading during activation of human platelets imaged by atomic force microscopy. *Biophys. J.* **66**(5), 1328–1334.

Guckenberger, R., Wiegräbe, W., Hillebrand, A., Hartmann, T., Wang, Z. and Baumeister, W. (1989). Scanning tunneling microscopy of a hydrated bacterial surface protein. *Ultramicroscopy* **31**, 327–332.

Guckenberger, R., Heim, M., Cevc, G., Knass, H.F., Weigräbbe, W. and Hillebrand, A. (1994). Scanning tunneling microscopy of insulators and biological specimens based on lateral conductivity of ultrathin water films. *Science* **266**, 1538–1540.

Gunasekara, L., Schürch, S., Schoel, W.M., Nag, K., Leonenko, Z., Haufs, M. and Amrein, M. (2005). Pulmonary surfactant function is abolished by an elevated proportion of cholesterol. *Biochim. Biophys. Acta* **1737**, S27–35.

Haltia, T. and Freire, E. (1995). Forces and factors that contribute to the structural stability of membrane proteins. *Bba-Bioenergetics* **1228**(1), 1–27.

Han, W., Lindsay, S.M. and Jing, T.W. (1996). A magnetically driven oscillating probe microscope for operation in liquids. *Appl. Phys. Lett.* **69**(26), 4111–4113.

Hansma, P. (1996). New AFM technology for better biological imaging. *Am. Phys. Soc.* **41**, 515.

Hansma, P.K., Cleveland, J.P., Radmacher, M., Walters, D.A., Hillner, P.E., Bezanilla, M., Fritz, M., Vie, D., Hansma, H.G. and Prater, C.B. (1994). Tapping mode atomic force microscopy in liquids. *Appl. Phys. Lett.* **64**(13), 1738–1740.

Harris, C.M. (1987). *Shock and Vibration Handbook* (McGraw-Hill, New York).

Hayward, S.B., Grano, D.A., Glaeser, R.M. and Fisher, K.A. (1978). Molecular orientation of bacteriorhodopsin within the purple membrane of *Halobacterium halobium*. *Proc. Natl. Acad. Sci. USA* **75**(9), 4320–4324.

Hegner, M., Wagner, P. and Semenza, G. (1993). Immobilizing DNA on gold via thiol modification for atomic force microscopy imaging in buffer solutions. *FEBS Lett* **336**(3), 452–456.

Hegner, M., Dreier, M., Wagner, P., Semenza, G. and Güntherodt, H.J. (1996). Modified DNA immobilized on bioreactive self-assembled monolayer on gold for dynamic force microscopy imaging in aqueous buffer solution. *J. Vac. Sci. Technol.* **14**, 1418–1421.

Henderson, E., Haydon, P.G. and Sakaguchi, D.S. (1992). Actin filament dynamics in living glial cells imaged by atomic force microscopy. *Science* **257**, 1944–1945.

Heuberger, M., Dietler, G. and Schlapbach, L. (1996). Elastic deformations of tip and sample during atomic force microscope measurements. *J. Vac. Sci. Technol. B* **14**(2), 1250–1254.

Hoh, J.H. and Schoenenberger, C.-A. (1994). Surface morphology and mechanical properties of MDCK monolayers by atomic force microscopy. *J. Cell Sci.* **107**(5), 1105–1114.

Hoh, J.H., Sosinski, G.E., Revel, J.-P. and Hansma, P.K. (1993). Structure of the extracellular surface of the junction by atomic force microscopy. *Biophys. J.* **65**, 149–163.

Hörber, J.K.H., Häberle, W., Ohnesorge, F., Binnig, G., Liebich, H.G., Czerny, C.P., Bahnel, H. and Mayr, A. (1992). Investigation of living cells in the nanometer regime with the scanning force microscope. *Scan. Microsc.* **6**(4), 919–930.

Israelachvili, J. (1991). *Intermolecular & Surface Forces* (Academic Press, London).

Jacobson, B.S. and Branton, D. (1977). Plasma membrane: rapid isolation and exposure of the cytoplasmic surface by use of positively charged beads. *Science* **195**, 302–304.

Jap, B.K., Zulauf, M., Scheybani, T., Hefti, A., Baumeister, W., Aebi, U. and Engel, A. (1992). 2D crystallization: from art to science. *Ultramicroscopy* **46**, 45–84.

Jiang, F., Hörber, H., Howard, J. and Muller, D.J. (2004a). Assembly of collagen into microribbons: effects of pH and electrolytes. *J. Struct. Biol.* **148**(3), 268–278.

Jiang, F.Z., Khairy, K., Poole, K., Howard, J. and Muller, D.J. (2004b). Creating nanoscopic collagen matrices using atomic force microscopy. *Microsc. Res. Techni.* **64**, 435–440.

Karagiorga, G., Nakos, G., Galiatson, E. and Lekka, M.E. (2006). Biochemical parameters of bronchoalveolar lavage fluid in fat embolism. *Intensive Care med.* **32**, 116–123.

Karrasch, S., Dolder, M., Schabert, F., Ramsden, J. and Engel, A. (1993). Covalent binding of biological samples to solid supports for scanning probe microscopy in buffer solution. *Biophys. J.* **65**, 2437–2446.

Kasas, S. and Ikai, A. (1995). A method for anchoring round shaped cells for atomic force microscope imaging. *Biophys. J.* **68**(5), 1678–1680.

Kellenberger, E. and Kistler, J. (1979). The physics of specimen preparation. In *Advances in Structure Research by Diffraction Methods* (W.H.R. Mason, Ed.), III, 49–79 (Friedrich Vieweg, Wiesbaden).

Kellenberger, E., Häner, M. and Wurtz, M. (1982). The wrapping phenomenon in airdried and negatively stained preparations. *Ultramicroscopy* **9**, 139–150.

Knapp, H., Wiegräbe, W., Heim, M., Eschrich, R. and Guckenberger, R. (1995). Atomic force microscope measurements and manipulation of Langmuir-Blodgett films with modified tips. *Biophys. J.* **69**, 708–715.

Knebel, D., Amrein, M. and Reichelt, R. (1997). A fast and versatile scan-unit for scanning probe microscopy. *Scanning* **19**(4), 264–268.

Knebel, D., Sieber, M., Reichelt, R., Galla, H.J. and Amrein, M. (2002). Scanning force microscopy at the air-water interface of an air bubble coated with pulmonary surfactant. *Biophys. J.* **82**(1 Pt 1), 474–480.

Lantz, M.A., O'Shea, S.J. and Welland, M.E. (1994). Force microscopy imaging in liquids using ac techniques. *Appl. Phys. Lett.* **65**(4), 409–411.

Lindsay, S.M. and Tao, N.J. (1993). Potentiostatic deposition of molecules for SXM. In *STM and AFM in Biology* (O. Marti and M. Amrein, Eds.), 229–258 (Academic Press, London).

Matsko, N. and Mueller, M. (2004). AFM of biological material embedded in epoxy resin. *J. Struct. Biol.* **146**(3), 334–343.

Mazia, D., Schatten, G. and Sale, W. (1975). Adhesion of cells to surfaces coated with polylysin: applications to electron microscopy. *J. Cell. Biol.* **66**, 198–200.

Miles, M.J. and McMaster, T.J. (1993). *Protein Assemblies and Single Molecules Imaged by STM* (Academic Press, San Diego).

Miles, M.J., McMaster, T., Carr, H.J., Tatham, A.S., Shewry, P.R., Field, J.M., Belton, B.S., Jeenes, D., Hanley, B. and Whittam, M. (1990). Scanning tunneling microscopy of biomacromolecules. *J. Vac. Sci. Technol. A* **8**(1), 698–702.

Müller, D.J. and Büldt, G. (1995). Force-induced conformational change of bacteriorhodopsin. *J. Mol. Biol.* **249**, 239–243.

Müller, D.J., Baumeister, W. and Engel, A. (1996). Conformational change of the hexagonally packed intermediate layer imaged by atomic force microscopy. *J. Bacteriol.* **78**, 3025–3030.

Müller, D.J., Amrein, M. and Engel, A. (1997a). Adsorption of biological molecules to a solid support for scanning probe microscopy. *J. Struct. Biol.* **119**(2), 172–188.

Müller, D.J., Engel, A. and Amrein, M. (1997b). Preparation techniques for the observation of native biological systems with the atomic force mocroscope. *Biosensors Bioelectron.* **12**(8), 867–877.

Nahmen von, A., Schenk, M., Sieber, M. and Amrein, M. (1997). The structure of a model pulmonary surfactant as revealed by scanning force microscopy. *Biophys. J.* **72**, 463–469.

Poole, K., Khairy, K., Friedrichs, J., Franz, C., Cisneros, D.A. Howard, J. and Mueller, D. (2005). Molecular-scale topographic cues induce the orientation and directional movement of fibroblasts on two-dimensional collagen surfaces. *J. Mol. Biol.* **349**, 380–386.

Pralle, A., Keller, P., Florin, E.-L., Simons, K. and Horber, J.K.H. (2000) Sphingolipod-cholesterol rafts diffuse as small entities in the plasma membrane of mammalian cells. *J. Cell Biol.* **148**, 997–1008.

Putman, C.A.J., Van der Werf, K.O., Degrooth, B.G., Vanhulst, N.F. and Greve, J. (1994). Tapping mode atomic force microscopy in liquid. *Appl. Phys. Lett.* **64**(18), 2454–2456.

Sarid, D., Ed. (1991). *Scanning Force Microscopy with Application to Electric, Magnetic and Atomic Forces* (Oxford University Press, New York).

Schabert, F.A. and Engel, A. (1994). Reproducible acquisition of *Escherichia coli* porin surface topographs by atomic force microscopy. *Biophys. J.* **67**, 2394–2403.

Schäffer, T.E., Cleveland, J.P., Ohnesorge, F., Walters, D.A. and Hansma, P.K. (1996). Studies of vibrating atomic force microscope cantilevers in liquid. *J. Appl. Phys.* **80**(7), 3622–3627.

Schenk, M., Amrein, M., Dietz, P. and Reichelt, R. (1994). *Simultaneous STM/ AFM Imaging of Biological Macromolecular Assemblies: Experimental Set-up and First Results.* 13th International Congress on Electron Microscopy (Les Editions de Physique Les Ulis, Paris).

Schenk, M., Amrein, M. and Reichelt, R. (1996). An electret microphone as a force sensor for combined scanning probe microscopies. *Ultramicroscopy* **65**, 109–118.

Seelert, H., Poetsch, A., Dencher, N.A., Engel, A., Stahlberg, H. and Muller, D.J. (2000). Structural biology: proton-powered turbine of a plant motor. *Nature* **405**, 418–419.

Sellers, H., Ulman, A., Schnidman, Y. and Eilers, J.E. (1993). Structure and binding of alkanethiolates on gold and silver surfaces—implications for self-assembled monolayers. *J. Am. Chem. Soc.* **115**(21), 9389–9401.

Stemmer, A., Hefti, A., Aebi, U. and Engel, A. (1989). Scanning tunneling and transmission electron microscopy on identical areas of biological specimens. *Ultramicroscopy* **30**, 263–280.

Tortonese, M., Barret, R.C. and Quate, C.F. (1993). Atomic resolution with an atomic force microscope using piezoresistive detection. *Appl. Phys. Lett.* **62**, 834–836.

Travaglini, G., Rohrer, H., Amrein, M. and Gross, H. (1987). Scanning tunneling microscopy on biological matter. *Surf. Sci.* **181**, 380–390.

Voelker, M.A., Hameroff, S.R., He, J.D., Dereniak, E.L., MeCuskey, R.S., Schneiker, C.W., Chvapil, T.A., Bell, T.S. and Weiss, L.B. (1988). STM imaging of molecular collagen and phospholipid membranes. *J. Microsc.* **152**, 557–573.

Wagner, D., Moy, V.T., Hofman, U.G., Benoit, M., Ludwig, M. and Gaub, H.E. (1994). *AFM—The Art of Touching Molecules (Invited).* 13th International Congress on Electron Microscopy (Les Editions de Physique Les Ulis, Paris).

Wagner, P., Hegner, M., Guntherodt, H.J. and Semenza, G. (1995a). Formation and in-situ modification of monolayers chemisorbed on ultraflat template-stripped gold surfaces. *Langmuir* **11**(10), 3867–3875.

Wagner, P., Hegner, M., Kernen, P., Zauga, F. and Semenza, G. (1995b). Covalent immobilization of native biomolecules onto Au (111) via N-hydroxysuccinimide ester functionalized self-assembled monolayers for scanning probe microscopy. *Biophys. J.* **70**, 2052–2066.

Wang, Z., Hartmann, T., Baumeister, W. and Guckenberger, R. (1990). Thickness determination of biological samples with a z-calibrated scanning tunneling microscope. *Proc. Natl. Acad. Sci. USA* **87**, 9343–9347.

Weisenhorn, A.L., Maivald, P., Butt, H.-J. and Hansma, P.K. (1992). Measuring adhesion, attraction, and repulsion between surfaces in liquids with an atomic-force microscope. *Phys. Rev. B* **45**(19), 11226–11232.

Weisenhorn, A.L., Khorsandi, M., Kasas, S., Gotzos, V. and Butt, H.-J. (1993). Deformation and height anomaly of soft surfaces studied with an AFM. *Nanotechnology* **4**, 106–113.

Welland, M.E., Miles, M.J., Lambert, N., Morris, V.J., Loombs, J.H. and Pethia, J.B. (1989). Structure of the globular protein vicilin revealed by scanning tunnelling microscopy. *Int. J. Biol. Macromol.* **11**(1), 29–32.

Wildhaber, I., Gross, H., Engel, A. and Baumeister, W. (1985). The effects of air-drying and freeze-drying of the structure of a regular protein layer. *Ultramicroscopy* **16**, 411–422.

Wolf, H., Ringsdorf, H., Delamarche, E., Takami, T., Kang, H., Michel, B., Gerber, C., Jaschke, M. Butt, H.J. and Bamberg, E. (1995). End-group-dominated molecular order in self-assembled monolayers. *J. Phys. Chem.* **99**(18), 7102–7107.

Yang, J., Mou, J. and Shao, Z. (1996). The effect of deformation on the lateral resolution of   force microscopy. *J. Microsc.* **182**(2), 106–113.

Zasadzinski, J.A.N., Schneir, J., Gurley, J., Elings, V. and Hansma, P.K. (1988). Scanning tunneling microscopy of freeze-fracture replicas of biomembranes. *Science* **239**, 1013–1015.

17

Low-Temperature Scanning Tunneling Microscopy

Uwe Weierstall

1 Introduction

The scanning tunneling microscope (STM) has revolutionized surface science since its invention in 1982 (Binnig and Rohrer, 1982) by providing a means to directly image atomic scale spatial and electronic structure. Using the combination of a coarse approach and piezoelectric transducers, a sharp, metallic tip is brought into close proximity with the sample. The distance between tip and sample is less than 1 nm, which means that the electron wave functions of tip and sample start to overlap. A bias voltage is applied between tip and sample that causes electrons to tunnel through the barrier. The tunneling current is a quantum mechanical effect: tunneling of electrons can occur between two electrodes separated by a thin insulator or a vacuum gap and the tunneling current decays on the length scale of one atomic radius. The tunneling current is in the range of picoamperes to nanoamperes and is measured with a preamplifier. In an STM, the tip is scanned over the surface and electrons tunnel from the very last atom of the tip apex to single atoms on the surface, providing atomic resolution. The exponential dependence of the tunneling current on the tip–sample distance can be exploited to control the tip–sample distance with high precision. There are four basic operation modes for any STM: constant current imaging, constant height imaging, spectroscopic imaging, and local spectroscopy. Their interpretation and realization will be briefly discussed below. For details about other modes and a comprehensive introduction to electron tunneling and STM see Wiesendanger (1994).

To acquire constant current images, a feedback loop adjusts the height of the tip during scanning so that the tunneling current flowing between tip and sample is kept constant. The height z is adjusted by applying an appropriate voltage V_z to the z-piezoelectric drive while the lateral tip position (x,y) is determined by the corresponding voltages applied to the x and y piezoelectric drives. The recorded signal V_z can be translated into a topography $z(x,y)$ if the sensitivity of the piezoelectric drives is known. The word topography should be used with caution: since the local density of states at the Fermi level is measured, a molecule adsorbed on a metal surface that reduces the local density of states and may actually be imaged as a depression.

To acquire constant height images, the feedback loop is switched off, i.e., the tip is scanned at constant height above the surface, and variations in the current are measured. This mode has the advantage that the finite response time of the feedback loop does not limit the scan speed. It can be used to collect images at video rates, offering the opportunity to observe dynamic processes at surfaces. However, thermal drift limits the time of the experiment and there is an increased risk of crashing the tip.

To measure differential conductance (dI/dV) maps with the STM, a high-frequency sinusoidal modulation voltage is superimposed on the constant dc bias voltage V_{bias} between tip and sample. The modulation frequency is chosen higher than the cutoff frequency of the feedback loop, which keeps the tunneling current constant. By recording the tunneling current modulation, which is in phase with the applied bias voltage modulation, with a lock in amplifier, a spatially resolved spectroscopic signal $dI/dV|_{V_{\text{bias}}}$ can be obtained simultaneously with the constant current image (Binnig et al., 1985a,b).

By measuring the differential conductance dI/dV at a fixed tip position with open feedback loop (constant tip–sample distance z) while sweeping the applied bias voltage, an energy-resolved spectrum can be obtained. This is useful for probing, e.g., band-gap states in semiconductors or the onset of surface states on metals.

The tunneling current I at a given tip position is approximately equal to the integrated local density of states (ILDOS), integrated over the energy range between the Fermi energy E_F of the sample and eV, where V is the applied bias voltage. Therefore the differential conductance dI/dV is approximately proportional to the local density of states (LDOS) of the sample at the energy eV, and a constant current image should represent a contour of constant ILDOS. For measurements close to E_F, i.e., at low bias voltages, the LDOS and ILDOS are essentially the same and a constant current image at low bias (a few millivolts) is therefore approximately proportional to the sample LDOS at the Fermi energy E_F (assuming the tip has a uniform density of states and the temperature is low). To illustrate how to arrive at the picture presented above, the theoretical treatment of electron tunneling is briefly outlined.

A one-dimensional WKB approximation predicts that the tunneling current at low temperatures (where the Fermi distribution is a step function) is given by

$$I = \int_0^{eV} \rho_s\left(E, x\right)\rho_t\left(-eV + E, x\right)T\left(E, eV, x\right) dE \qquad (1)$$

where $\rho_s(E)$ and $\rho_t(E)$ are the density of states of the sample and the tip at the location x and energy E, measured with respect to their individual Fermi levels, and V is the applied bias voltage (Hamers, 1989). The tunneling transmission probability $T(E,eV,x)$ for electrons with energy E and applied voltage V is given by

$$T(E, eV, x) = \exp\left(-z(x)\sqrt{\frac{4m}{\hbar^2}(\phi_t + \phi_s + eV - 2E)}\right) \qquad (2)$$

where ϕ_s and ϕ_t are the work functions of sample and tip and z is the tip–sample distance. Therefore, assuming that the tip electronic structure is featureless, Eq. (1) shows that the tunneling current at position x is approximately equal to the ILDOS of the sample integrated between E_F and eV, weighted by the transmission probability T. Examination of Eq. (2) shows that if $eV < 0$ (i.e., negative sample bias), the transmission probability is largest for $E = 0$ (corresponding to electrons at the Fermi level of the sample). If $eV > 0$ (positive sample bias) the probability is largest for $E = eV$ (corresponding to electrons at the Fermi level of the tip). Therefore the tunneling probability is always largest for electrons at the Fermi level of whichever electrode is negatively biased. This is shown schematically in Figure 17–1.

Differentiating Eq. (1) gives the differential conductance

$$\frac{dI}{dV} \propto \rho_s(eV, x)\rho_t(0, x)T(eV, eV, x)$$

$$+ \int_0^{eV} \rho_s(E, x)\rho_t(E - eV, x)\frac{dT(E, eV, x)}{dV}dE \qquad (3)$$

The first term is the product of the density of states of the sample, the density of states of the tip, and the tunneling transmission probability. The second term contains the voltage dependence of the tunneling transmission probability. Since T is a smooth monotonically

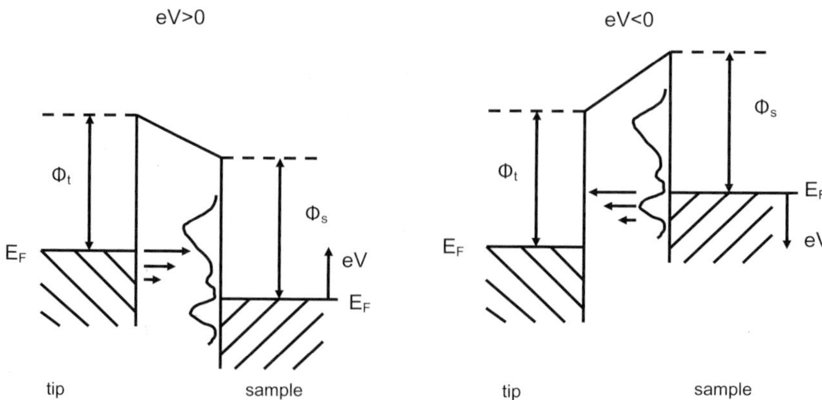

Figure 17–1. Energy level diagram of sample and tip with V being the sample voltage relative to the tip. Left: electrons tunneling from tip to sample with positive sample bias voltage. Right: electrons tunneling from sample to tip with negative sample bias voltage. The curve represents the density of states of the sample; the tip density of states is assumed featureless. The different lengths of the arrows illustrate the fact that the tunneling probability is largest for electrons at the Fermi level of the negatively biased electrode. Φ_t and Φ_s are the tip and sample work functions.

increasing function of V, structure in dI/dV can be assigned to changes in the sample LDOS at eV, assuming a tip with featureless density of states.

In general, however, the tip electronic structure is unknown and may not be featureless. The small size of the STM tip is expected to significantly modify its electronic structure from that of a bulk material. Therefore, it is usually necessary to compare tunneling spectra acquired at different locations on the surface to distinguish the spatially invariant contribution of the tip and the spatially varying contribution from the sample. Comparing spectra taken with different tips also helps to eliminate tip contributions.

At finite temperatures Eq. (1) contains an additional Fermi distribution factor, which imposes a limit on spectroscopic resolution. At room temperature, with $k_B T \approx 0.026\,\text{eV}$, the spread of the tip and sample energy distribution are each $2k_B T \approx 0.052\,\text{eV}$. Therefore the total spread is $\Delta E \approx 4k_B T \approx 0.1\,\text{eV}$. To make high-resolution spectroscopic measurements with ΔE in the millielectron volt range, experiments must be conducted at cryogenic temperatures.

To work with clean metal and semiconductor surfaces, STM measurements have to be done in ultrahigh vacuum (UHV). Operating a UHV-STM at cryogenic temperature has several major advantages:

1. Close coupling of the microscope to a large temperature bath that keeps constant temperature over hours or days ensures a reduction of thermal drift and allows long-term measurements. The reduction of thermal drift during low-temperature operation is further improved by the fact that the thermal expansion coefficients at liquid helium temperature are two or more orders of magnitude smaller than at room temperature.

2. Superior vacuum conditions: If the microscope is incorporated in a cryostat that acts as an effective cryopump, surfaces are kept free from contamination over days.

3. Thermal diffusion of adsorbates and defects is suppressed—stable imaging becomes possible even for weakly bound species. Moreover, low temperatures might also stabilize the atomic configuration at the tip end by preventing sudden jumps of the most loosely bound foremost atoms due to thermal activation.

4. Small thermal broadening at the Fermi energy is a necessary condition for spectroscopic investigations with high-energy resolution.

5. Individual adsorbates can be manipulated with the STM to qualitatively probe their interaction with the substrate.

6. Physical properties can be studied as a function of temperature or physical effects can be examined that occur only at low temperatures (e.g., superconductivity, Kondo effect, nanoscale magnetism).

7. Piezo nonlinearities and hysteresis (creep) affecting the piezoelectric scanners of the STM decrease substantially at low temperatures.

In an effort to improve the stability of the microscope and the atoms or molecules under investigation, a variety of low-temperature UHV STM designs have been developed.

2 Design Principals

All low-temperature STMs (LT-STM) operate in UHV, which is a prerequisite for obtaining a clean surface. To reach a low final temperature and a short cooling time, the thermal conductivity from the microscope to the cryogen has to be maximized, and the thermal load from room temperature has to be minimized. Heat transfer through the electrical connections and contacts has to be considered as well as thermal radiation. The discussion here is restricted to liquid helium (LHe)-cooled instruments. In designing an LT-STM, there is a choice between a flow cryostat and a bath cryostat to cool the STM. Then there are two basic designs: one in which only the sample is cooled and one in which the whole STM is cooled and surrounded by a radiation shield.

2.1 STM with LHe Continuous-Flow Cryostat

If the ability to change temperature on a relatively short time scale is of importance, the heat reservoir should be small. Therefore in variable-temperature STMs that can work from room temperature down to liquid helium temperature, a flow cryostat is usually used and only the sample is cooled to ensure a small thermal mass. The sample is thermally connected to the cryostat via a flexible Cu or Au braid. Cryostat vibrations and instabilities caused by boiling of the coolant as well as different thermal expansion coefficients of various materials during temperature cycling require special attention. To reduce the amplitude of vibrations introduced by the cryostat, it is important to mechanically decouple the braid to a heavy mass (Bott et al., 1995) (see Figure 17–2). Since only the sample is cooled, the scanner has to be thermally isolated from the sample to avoid heat transfer to the sample. One advantage of this approach for variable-temperature operation is that the repeated recalibrations of the STM made necessary by temperature-dependent changes in the piezo coefficients are avoided. But large thermal gradients can create image drift problems and possible disturbances due to heat transfer from the "hot" tip scanning the cold sample (Xu et al., 1994).

An example of a variable-temperature STM with flow cryostat cooling the sample only is shown in Figure 17–3 (Behler et al., 1997). Other examples are given in Bott et al. (1995), Horch et al. (1994), and Petersen et al. (2001).

LHe flow cryostats can also be used to cool the entire STM. This results in improved thermal stability due to thermal equilibrium between all STM parts. If fast cooldown times are required, the STM can be connected rigidly to the cold end of the flow cryostat (Mugele et al., 1998; Zhang et al., 2001). This mandates a high mechanical stability (high resonance frequency) of the STM so that the vibration frequencies caused by the He flow are well below the eigenfrequency of the STM. The microscope is surrounded by one or two radiation shields. Higher mechanical stability can be obtained if the STM is suspended with springs inside a radiation shield to provide vibration isolation

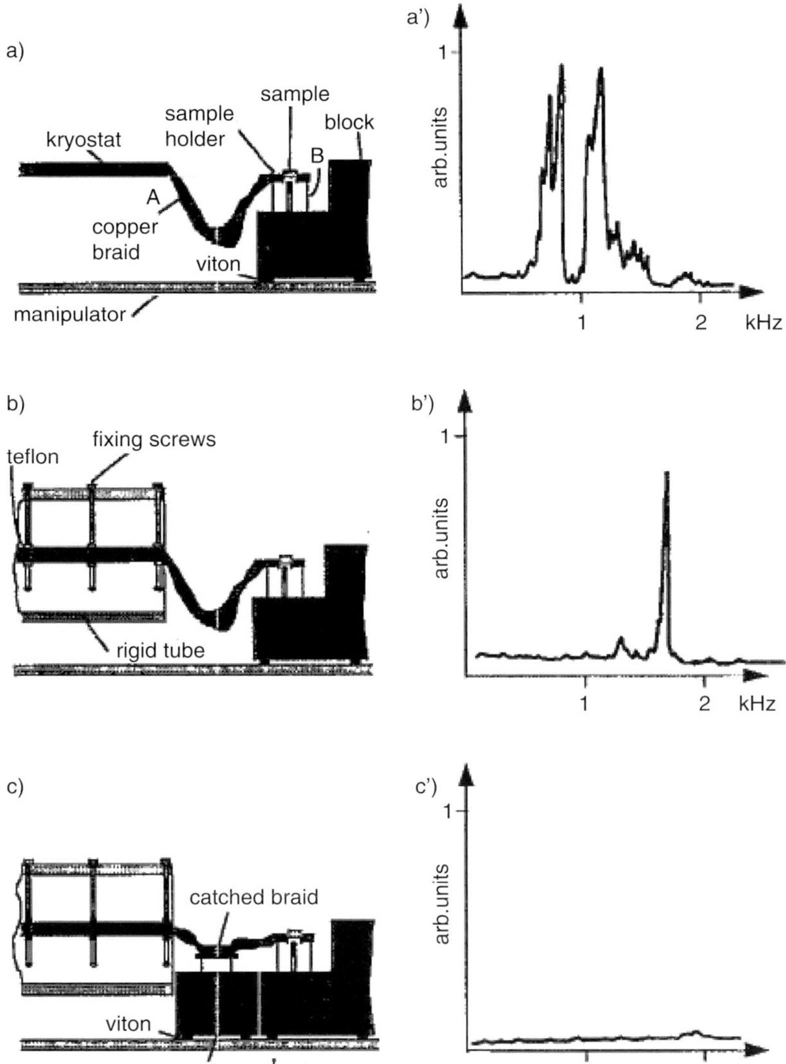

Figure 17–2. Mechanical decoupling of a copper braid connecting a flow cryostat and a sample holder. Shown are modifications and their effect on the noise spectrum. (a) Initial arrangement of the sample holder connected to the cryostat by a copper braid. (a′) Fourier spectrum of the tunneling current with noise showing up as peaks. (b) Resonance frequencies of cryostat are increased by defining nodes with a rigid tube mounted on the cryostat with three screws separated by 120° at each node. (b′) Fourier spectrum after modification; noise frequencies are shifted to higher frequencies. (c) Noise from the copper braid coupling is damped out by fixing the copper braid thermally isolated to a massive block. (c′) Noise is greatly reduced. (From Bott et al., 1995.)

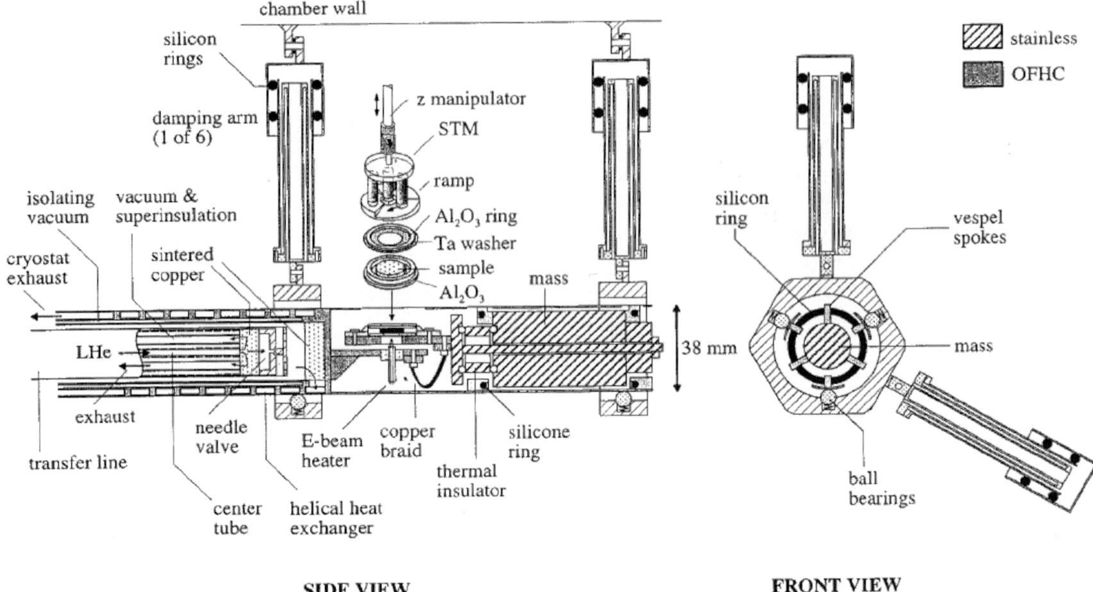

Figure 17–3. Example of a variable-temperature STM for operation in the temperature range of 20–300 K. Only the sample is cooled. Continuous flow cryostat on the left; arrows show helium flow. The sample is thermally connected to the cryostat by a copper braid and mechanically decoupled with a heavy stainless steel mass supported by O-rings. (From Behler et al., 1997.)

from the cryostat (Wolkow, 1995; Stipe et al., 1999b) (see Figures 17–4 and 17–5).

2.2 STM with LHe Bath Cryostat

A bath cryostat is usually used if the whole STM is being cooled. The bath cryostat serves as a large thermal reservoir that is being cooled down prior to the measurements. This approach has been successfully applied on a number of 4 K microscopes (Eigler and Schweizer, 1990; Gaisch et al., 1992; Stranick et al., 1994b; Rust et al., 1997; Meyer, 1996; Becker et al., 1998; Ferris et al., 1998; Harrell and First, 1999; Stroscio, 2000). If all parts of the system are allowed to reach thermal equilibrium, thermal drift may be virtually eliminated. Bath cryostat low-temperature STMs operating in a rotatable magnetic field have also been built (Wittneven et al., 1997); some of them are ^{3}He refrigerated and operate at about 250 mK (Pan et al., 1999; Kugler et al., 2000; Matsui et al., 2003). Sample turnaround times for the very-low-temperature variants are quite long since it can take 36 h to reach thermal equilibrium (Pan et al., 1999). LT-STMs with bath cryostats are difficult to use for variable temperature measurements, e.g., study of diffusion and phase transitions, since they use a large cold reservoir to cool the STM and the temperature cannot be changed easily. Temperature control by altering the exchange gas pressure in combination with a PID-controlled heater has been achieved (Rust et al., 2001). Another way to achieve measurements at different temperatures is to simply remove

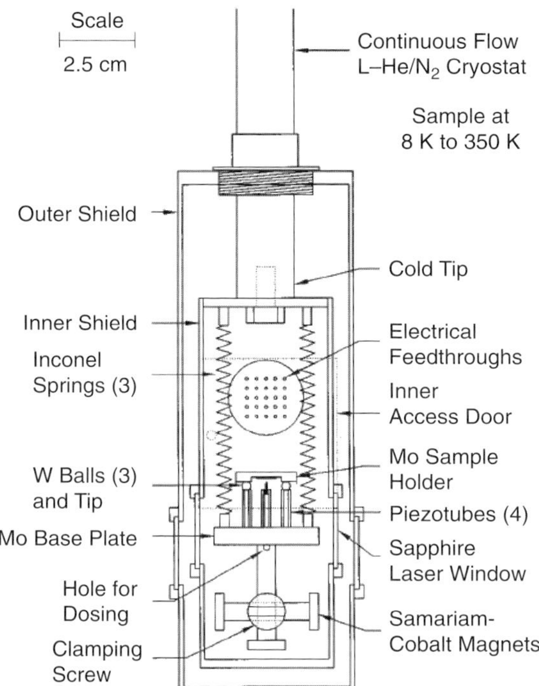

Figure 17–4. A variable-temperature scanning tunneling microscope cooled with a continuous-flow cryostat. The STM and the sample are suspended from three springs that provide the second stage of vibrational isolation. The design allows for *in situ* dosing and irradiation of the sample as well as the exchange of samples and tips. (From Stipe et al., 1999b.)

Figure 17–5. Low and variable temperature STM in the author's laboratory, which is very similar in design to the one by Stipe et al. (1999b). Refer to Figure 17–4 for identification of different parts. The front of the outer radiation shield and the sides of both radiation shields are missing to enable the inside to be viewed.

liquid helium from the cryostat, giving rise to a temperature increase of the microscope over days, thus enabling drift-free STM measurements at well-defined temperatures up to 300 K. (Jeandupeux et al., 1999)

Since the tunneling current has an exponential dependence on the distance between tip and sample, vibrations can cause strong noise. Therefore the construction of the STM scanner, consisting of tip holder, sample holder, and actuators, should be as rigid as possible to increase the mechanical eigenfrequency of the scanner. For typical tube scanner setups this eigenfrequency lies in the range of 1–10 kHz. A two-stage damping system consisting of an external damper (e.g., air damped feet suspension of the whole UHV chamber) and an internal damper (e.g., scanner suspended on springs and damped by eddy current dampers) effectively isolates the tunneling gap from building vibrations and acoustic noise. The two damping stages should have resonance frequencies well below building and acoustic frequencies and should be strongly damped (low Q-factor).

With a low-temperature STM the boiling cryogenic liquid produces additional vibrations after the first damping stage. Even worse, the boiling liquid has to be in close proximity to the vibration-sensitive scanner to enable good thermal contact to the scanner. Therefore there are two contradictory demands for the design of a low-temperature STM: mechanical decoupling of the STM head and thermal coupling to the cryostat.

This problem has been solved by different groups in different ways and in the following we will discuss three design examples. They all have in common a helium bath cryostat surrounded by a liquid nitrogen cryostat, which acts as a radiation shield. They all have a two-stage damping system; in the first stage the whole UHV chamber is mechanically decoupled from the floor by pneumatic dampers. The designs differ in the way the second damping stage is realized (see Figure 17–6):

The IBM-Rueschlikon/University Lausanne LT-STM designed by R. Gaisch (Gaisch et al., 1992) uses a bellow to mechanically decouple the liquid helium cryostat from the liquid nitrogen cryostat and the UHV chamber. The STM, however, is rigidly connected to the liquid helium cryostat, which results in very effective cooling. However, vibrations from the cryostat can reach the STM unattenuated.

The LT-STM designed by G. Meyer and K.H. Rieder (Meyer, 1996) has no damping between the liquid helium and the liquid nitrogen cryostat; instead the STM is suspended from the He cryostat by small extension springs and thereby is mechanically decoupled during measurements. During cooldown the STM is lowered onto a copper plate connected to the He cryostat to achieve good thermal contact. Since the thin springs are not very good thermal conductors, the STM has to be effectively shielded against incoming thermal radiation. This is achieved by a two-stage radiation shield with the inner shield connected to the He bath and the outer shield connected to the liquid nitrogen bath. The STM is then almost thermally isolated during measurements (see Figure 17–7).

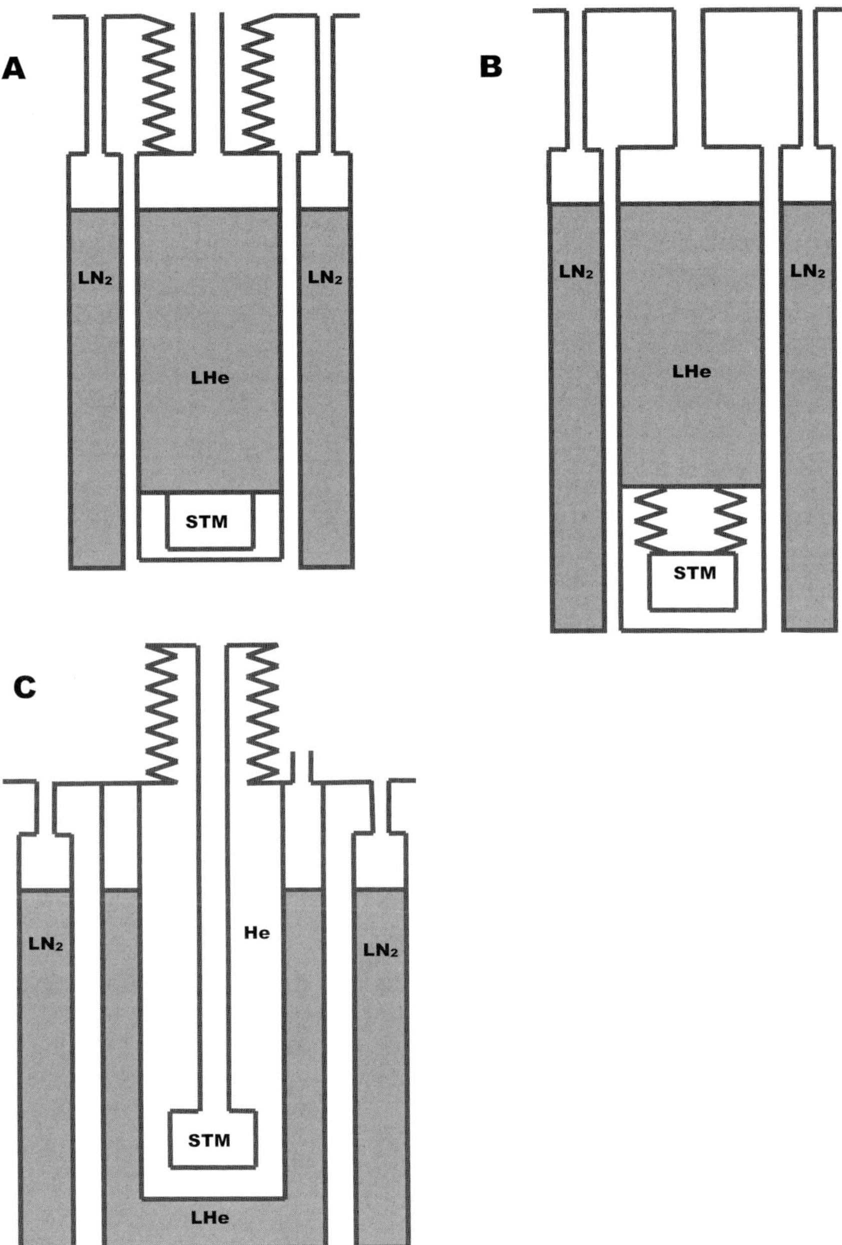

Figure 17–6. Schematic comparison of different LT-STMs with bath cryostat. (A) R. Gaisch/IBM Rueschlikon, (B) G. Meyer/FU Berlin, (C) D. Eigler/IBM Almaden. The connection to the LHe cryostat and the second damping stage of the STM is shown. (A) and (B) have a cryoshield surrounding the STM. In (A) the LHe cryostat is decoupled from the LN₂ cryostat by bellows. In (B) the STM is decoupled from the LHe cryostat by springs. In (C) the STM is decoupled from the LHe cryostat by a bellows-supported pendulum. Thermal contact is made by He exchange gas surrounding the evacuated pendulum.

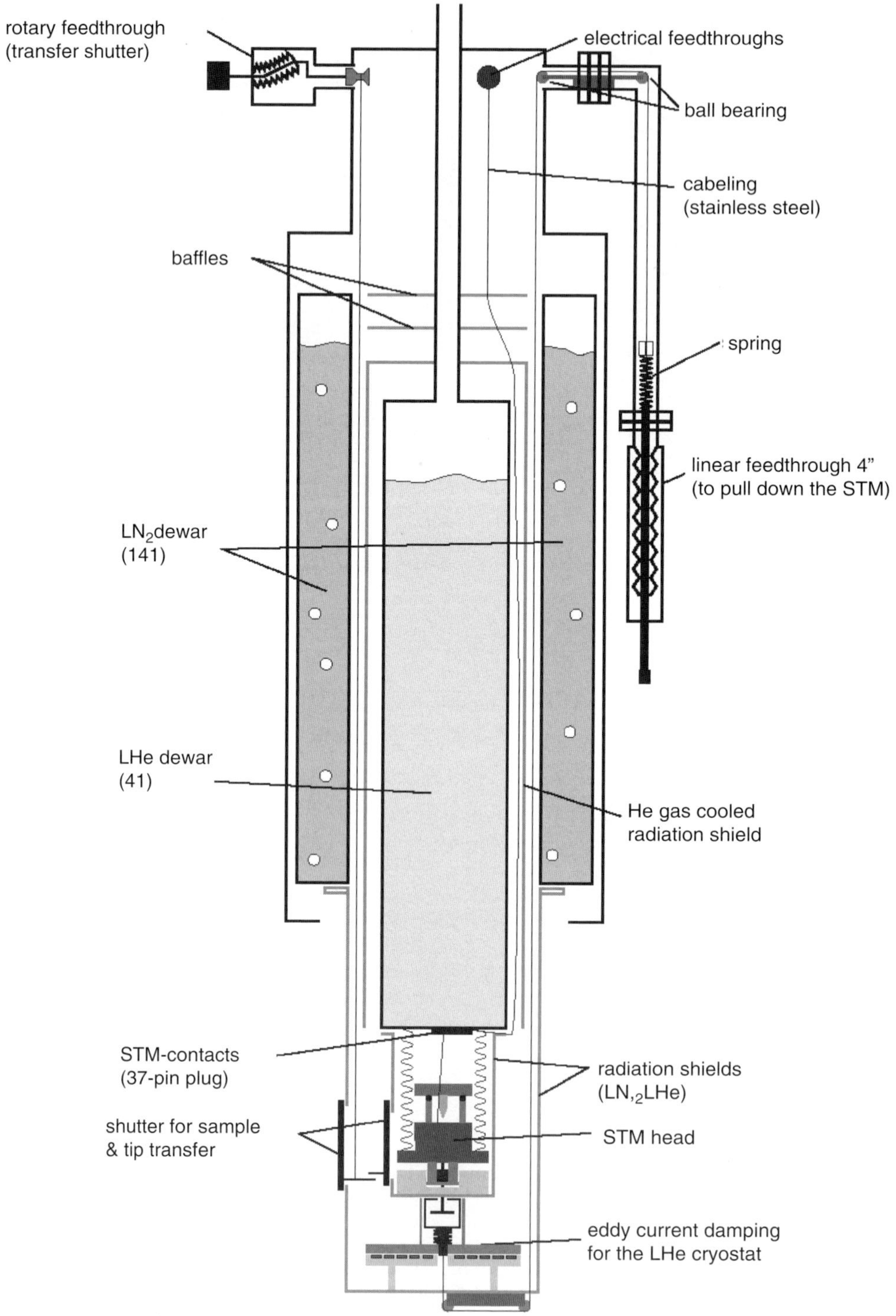

Figure 17–7. Design of the LT-STM by Prof. Rieder and Dr. Meyer (Free University Berlin), courtesy of CreaTec GmbH, distributed by SPECS GmbH.

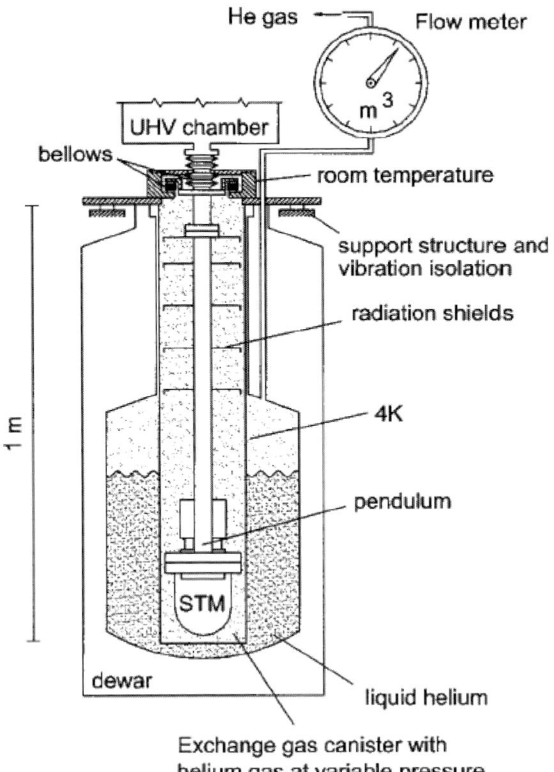

Figure 17–8. Illustration of the basic design principle for the Eigler-style low-temperature STM. The STM is mounted at the end of a bellows-supported pendulum, which is encapsulated in an exchange gas canister inside of a liquid helium dewar. Limited variable temperature operation is possible by controlling the exchange gas pressure. (From Rust et al., 2001.)

The LT-STM build by D. Eigler (Eigler et al., 1990) at IBM-Almaden and a similar instrument at Berlin are different from the previous two. In both previous instruments, the STM vacuum system and the isolation vacuum of the He and nitrogen reservoirs are connected. In the Eigler design the cryogenic vacuum system is not connected to the STM vacuum system. Instead the STM is mounted at the end of a bellows-supported evacuated pendulum with a resonance frequency of about 0.6 Hz. Helium exchange gas provides thermal coupling as well as acoustic isolation between the surrounding liquid helium dewar and the STM flange. The temperature of the STM can be set by controlling the helium gas pressure in the exchange gas chamber (Rust et al., 2001) (see Figure 17–8). In all designs, mechanical disturbances from bubbling liquid nitrogen in the surrounding liquid nitrogen cryostat can be prevented by solidifying the nitrogen by pumping it with a rotary pump (which is located far from the microscope) (Jeandupeux et al., 1999).

3 Applications

There is a large volume of literature on experiments using a LT-STM covering electronic structure and lifetime measurements, measurements on superconductors, atomic and molecular manipulations, molecular vibrational spectroscopy, photon-emission spectroscopy, and measurements of magnetic properties with spin polarized electrons. We can discuss only a small subset of the work done and refer to the literature for further reading. For the applications shown here, low-temperature operation of the STM has proven to be essential.

3.1 Electronic Structure

Understanding the distribution and response of electrons in solids, the *electronic structure*, is key to explaining and exploiting fundamental and technologically important properties of materials. The method of choice to obtain information about the electronic structure of solids and surfaces in the past decades has been angle-resolved photoemission spectroscopy (ARPES) and inverse photoemission spectroscopy (IPES). The development of low-temperature scanning tunneling microscopes operating under ultrahigh vacuum conditions has provided new opportunities for investigating electronic states at metal surfaces. At low temperatures, due to reduced broadening of the Fermi level of the STM tip and sample, rather high-energy resolution is achievable. Moreover, the absence of diffusion together with the spatial resolution of the STM enables detailed studies of the interaction of electronic states with single atoms and other nanoscale structures.

3.1.1 LDOS Oscillations

Electrons occupying surface states on the close-packed surfaces of noble metals are bound in the direction perpendicular to the surface, but have free-electron-like characteristics parallel to the surface (Gartland and Slagsvold, 1975; Heimann et al., 1977; Zangwill, 1988). On the (111) face of noble metals these surface states arise as a result of the gap that exists along the Γ–L line in their bulk Brillouin zone (Shockley, 1939). Surface state electrons play an important role in a variety of physical processes, including epitaxial growth (Memmel and Bertel, 1995), in determining equilibrium crystal shapes (Garcia and Serena, 1995), molecular ordering (Stranick et al., 1994a), surface catalysis (Bertel et al., 1995), and physisorption (Bertel, 1997).

Surface state electrons scattered form defects and step edges give rise to quantum interference patterns in the electron density, which can be probed using the scanning tunneling microscope (Davis et al., 1991). Standing wave patterns in the LDOS on the Cu(111) surface have been observed with an LT-STM for the first time by Crommie et al. (1993b). LDOS oscillations at surfaces are the analog to the Friedel oscillations of the total charge density (Friedel, 1958). The experiments where done at 4 K in an LHe bath cryostat STM (Eigler and Schweizer, 1990). Figure 17–9 shows a constant current image of the Cu(111) surface with static spatial oscillations in the ILDOS. The oscillations decay away from step edges and point defects and have a characteristic period of ~15 Å. The amplitude of the corrugations is ~0.02 Å and is greater near the top of step edges than near the bottom. Spatial variations in the LDOS of the surface at the energy $E = E_{\mathrm{F}} + eV$ can be approximately mapped out by measuring dI/dV at bias voltage V (Crommie et al., 1993b). Figure 17–10 shows dI/dV linescans as a function of distance to a monoatomic step on Cu(111). The wavelength of

Figure 17–9. Standing wave patterns on copper: constant current image of the Cu(111) surface ($V = 0.1\,$V, $I = 1.0\,$nA). Three monoatomic steps and several point defects are visible. Spatial oscillations of the tunneling current originating from step edges and defects are evident. Image size: $500\,\text{Å} \times 500\,\text{Å}$. (From Crommie et al., 1993b.)

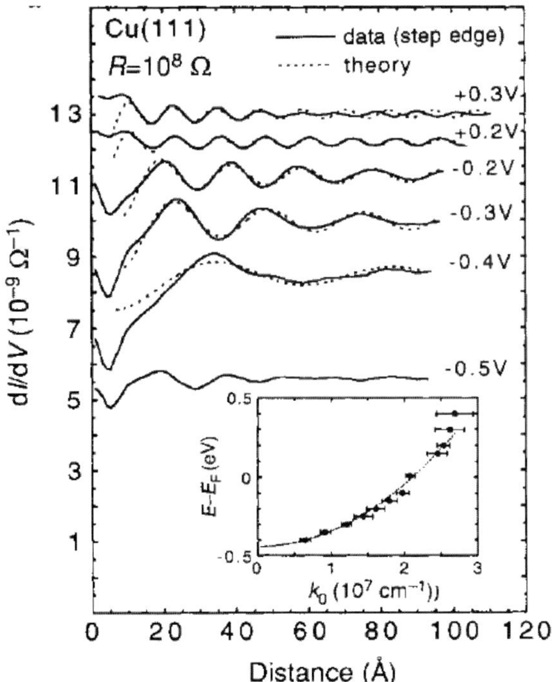

Figure 17–10. Spatial dependence of dI/dV, measured as a function of distance (along upper terrace) from step edge on Cu(111) at different bias voltages. Zero distance corresponds to the lower edge of the step. The measured wavelength of the surface LDOS oscillations changes as a function of energy. Solid lines, experiment; dashed lines, theoretical fit modeling the surface as a two-dimensional electron gas in the presence of a single hard wall barrier. Inset shows the extracted values of k plotted against energy defining an experimental dispersion curve of the surface state. (From Crommie et al., 1993b.)

the oscillation in the surface LDOS increases with decreasing energy. At ~0.45 eV below E_F the LDOS sharply decreases in magnitude. Measuring a full dI/dV spectrum at a fixed point on the surface revealed the origin of this transition: Figure 17–11 shows dI/dV spectra recorded with a stationary tip away from steps and on a step while ramping the bias from 1.0 to –1.0 V. The spectra taken on a terrace show a sharp drop in dI/dV at $V \approx -0.45$ V. This corresponds to a sudden decrease in the surface LDOS at energies more than 0.45 eV below E_F and marks the bottom of the Shockley surface state band on Cu(111).

By fitting the oscillations of the dI/dV linescan data to a simple free particle model with hard-wall confinement at surface steps, the major features of the spectra could be explained and a value of the surface state electron wavevector $k_\parallel$ for each electron energy could be extracted. A plot of these electron wavevectors against electron energy resulted in an experimental dispersion curve, as shown in the inset of Figure 17–10. The $k_\parallel$ data for different energies $E - E_F$ were fitted with the dispersion relation for electrons confined to two dimensions (dotted curve in inset of Figure 17–10):

$$E(k_\parallel) = E_0 + \frac{\hbar^2 k_\parallel^2}{2m^*} \tag{4}$$

where E_0 is the surface state band edge, m^* is the effective mass of the surface state electron, and $k_\parallel = \sqrt{k_x^2 + k_y^2}$ is the electron wavevector parallel to the surface. This procedure yielded the surface state effective mass $m^* = 0.38 \pm 0.02 m_e$ and the surface state band edge $E_0 = -0.44 \pm 0.01$ eV below E_F. The extracted value of the surface state band edge matched the location of the peak in the dI/dV spectrum of Figure 17–11. The measured values where roughly in agreement with previous photoemission results.

In the previous example, scattering of surface state electrons has been modeled as scattering at infinite potential walls. However, absorp-

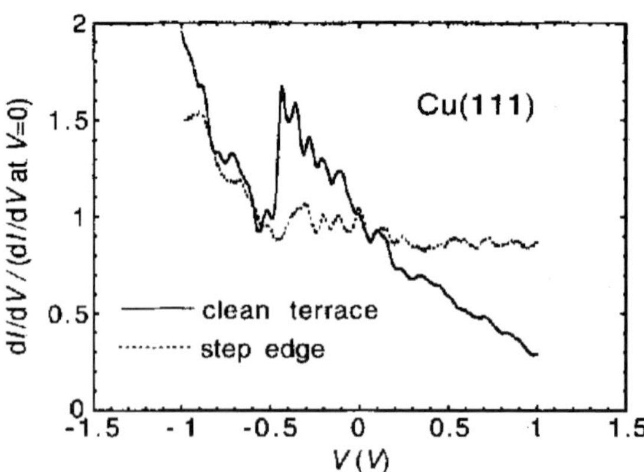

Figure 17–11. Differential conductance dI/dV spectra taken with tip over clean terrace (solid line) and over the center of a step edge (dashed line). The increase at ~0.45 V is due to the onset of the Cu(111) surface state. (From Crommie et al., 1993b.)

tion processes at step edges due to bulk coupling are disregarded with this model.

Therefore, a Fabry–Perot resonator model with a step reflection amplitude and scattering phase has been introduced to analyze the measured LDOS patterns of electrons confined between two parallel steps on Ag(111) at 4.9 K. (Burgi et al., 1998). The electron reflectivity has been found to be independent of crystallographic step structure, but depends on the step morphology (ascending, descending). Reflectivity and scattering phase could be quantified for any two parallel steps, providing insight into the scattering mechanism.

It has been assumed so far that for measurements close to E_F (small bias voltage), the LDOS and ILDOS are essentially the same, i.e., a constant-current image (topograph) and a dI/dV differential conductance image should be identical. However, with increasing bias voltage the constant-current image should systematically diverge from the LDOS (dI/dV) as states from a range of energies contribute to the integrated state density.

One-dimensional electron confinement has been used to measure the bias-dependent difference between the ILDOS and LDOS (Pivetta et al., 2003) with a low-temperature STM. The measurements where performed in a home-built STM, operating in UHV at 4.8 K (Gaisch et al., 1992). The system studied was a quantum box consisting of two-dimensional surface states on Ag(111) confined by two parallel surface steps (see Figure 17–12). Constant current cross sections (measuring the ILDOS) and dI/dV cross sections (measuring the LDOS) from the quantum box perpendicular to the steps were compared at different bias voltages. Oscillations in both, constant current and dI/dV curves were observed (Figure 17–13). These oscillations are due to the interference of the incident and scattered surface state electrons at the box boundaries. The measured data were compared with a model calculation of the LDOS and ILDOS assuming a free particle model with hard wall confinement at surface steps.

The surface state electrons are free in the direction parallel to the steps, but due to the boundary conditions given by the one-dimensional confinement in a box of width L, the component of the electron wavevector perpendicular to the step edges is quantized: $k_{\perp n} = \pi n/L$. Each corresponding eigenenergy $E_n(k_{\perp n})$ defines the onset of a one-dimensional subband $E_n(k_\parallel)$, where $E(k_\parallel)$ is the dispersion relation (4) for the Ag(111) surface state electrons. Since the tunneling current I is calculated by integrating over all energies between the bias eV and E_F, it contains contributions from several subbands $E_n(k_\parallel)$. Calculating the differential conductance dI/dV also requires integration of the DOS over a small range of energies given by the modulation amplitude used for lock-in detection (here 1–10 mV). Comparison of the model calculations and experimental measurements could explain the observed oscillation patterns in I and dI/dV in terms of the way in which different subbands contribute to the signals. The measured data of Pivetta et al. (2003) indicate that the difference between differential conductance and constant-current measurements can already be pronounced for bias voltages as small as 19 mV. Thus, even close to E_F, the interpre-

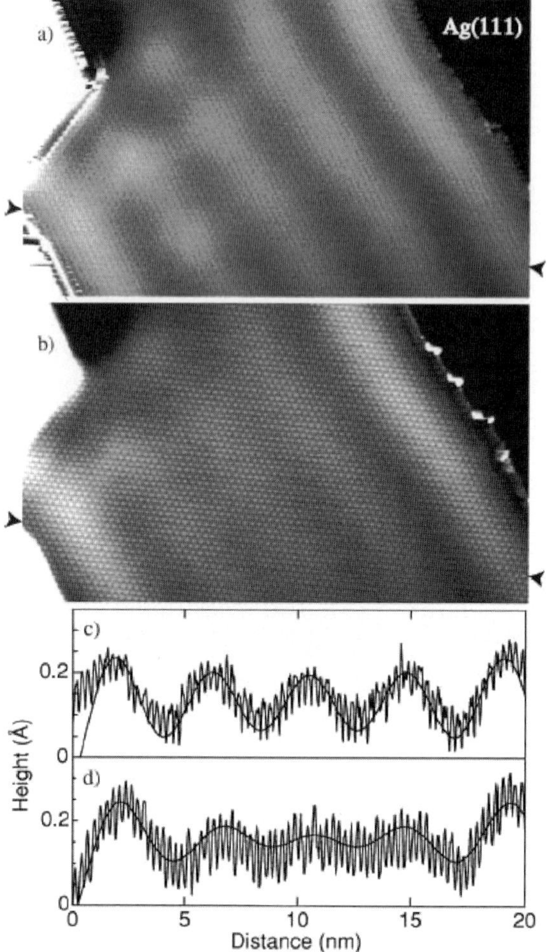

Figure 17–12. Surface states confined by two parallel steps on Ag(111) Atomically resolved STM constant current images at 4.8 K, 20 nm × 12 nm, I = 1.8 nA at (a) +21 mV and (b) +6 mV showing standing-wave patterns. (c and d) Cross sections along the black arrows in (a) and (b), respectively, showing the atomic corrugation superimposed on the standing waves. The solid curves are the calculated ILDOS. (From Pivetta et al., 2003.)

tation of constant current images in terms of the LDOS may not be fully correct.

3.1.2 Energy Dispersion Measurements

Energy dispersion measurements as a function of the electron wavevector $E(k)$ of surface states on metals is usually performed by techniques such as ARPES for occupied states or IPES for empty states, which are selective in both energy and wavevector. As show in Section 3.1.1, the energy dispersion relation of the surface state can also be measured with scanning tunneling spectroscopy (STS) by measuring the LDOS on single line scans perpendicular to a straight substrate step. Measurement of the LDOS ρ_s at the surface by means of measuring the differential conductance dI/dV is affected by local variations of the tip–sample distance z and therefore does not exactly represent ρ_s (Crommie et al., 1993b; Heller et al., 1994; Hormandinger, 1994). As a consequence, the surface state dispersion $E(k)$ determined from dI/dV maps deviates somewhat from photoemission spectroscopy results (Hormandinger, 1994). Simultaneous measurements of z and dI/dV

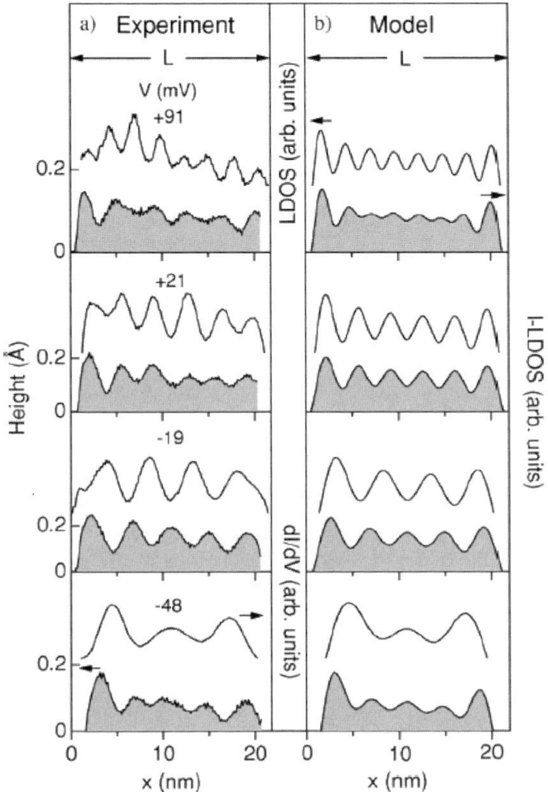

Figure 17–13. Bias-dependent evolution of the LDOS and ILDOS oscillations of surface states confined by parallel surface steps. (a) Simultaneously recorded constant current (gray shaded) and dI/dV (curves above) cross sections from a quantum box of width $L = 21.8$ nm formed by two parallel steps on Ag(111) Bias voltages (mV) are indicated (bias applied to the sample). (b) Calculated electron density (gray shaded) and dI/dV. Note the difference between dI/dV (LDOS) (three maxima) and topography (ILDOS) (five maxima) at -48 mV, a bias voltage that might be considered to be close to E_F. (From Pivetta et al., 2003.)

have been used to recover the oscillating LDOS on a close-packed Ag(111) surface at $T = 50$ K (Li et al., 1997).

A dependence of dI/dV on local z variations follows from Eq. (3). If the second term in Eq. (3) is neglected, the differential conductance is proportional to the product of ρ_s and T. The exponential z-dependence of T causes an error in the measurement of ρ_s in terms of dI/dV, if the tip–sample distance is not constant, but instead is controlled in constant current mode at the same bias voltage where dI/dV is measured.

At low bias, the tunneling current results from an integration over a narrow energy window and consequently the standing wave oscillations in the ILDOS get more pronounced (less k values contribute to the pattern). This causes an oscillation of z, since at every position x, the tip height $z(x)$ is adjusted by the feedback loop to follow the ILDOS oscillations at the preset constant-current value. Then the simultaneously measured dI/dV signal does not reflect ρ_s anymore, because the z adjustments influence the barrier transmission T, which depends exponentially on z. Since dI/dV is proportional to the product of ρ_s and T, ρ_s at $E = eV$ can be recovered from dI/dV by division with T, which is measured by measuring $z(x)$, i.e., a constant-current image. With these corrections, good agreement between LT-STM-derived dispersion curves $E(k)$ and PES-derived dispersion curves has been achieved (Li et al., 1997).

Another solution to avoid convolution between standing waves in the tip height z of the constant-current line scan and those in the simultaneously recorded dI/dV spectra is to control z at a large negative bias voltage (relative to the sample). Under these conditions, the current I contains contributions from electronic states with many different oscillation periods ($k_\parallel$ values), which minimizes the standing waves in the z signal (ILDOS). Thus the tip moves to a good approximation parallel to the surface plane, unaffected by interference patterns in the LDOS. The differential conductance is then roughly proportional to the LDOS of the sample (Hormandinger, 1994). Such a measurement has been published by Jeandupeux et al. (1999), and the result is shown in Figure 17–14. The upper graph displays the constant-current line scan on which the tip was moved while taking differential conductance maps. It can be seen that at a bias voltage of $V = 0.3\,\text{V}$ the tip–surface distance is almost unaffected by standing waves and follows the real topography. The differential conductance data are represented by gray levels as a function of the distance x from the step edge and the energy E. This plot already illustrates the dispersion of the Ag(111) surface state: from top to bottom the wavelength of the LDOS oscillations increases until it diverges at the band edge at $E_0 = -65\,\text{meV}$. Analyzing constant energy cuts of the differential conductance plot in Figure 17–14 in quantitative terms by modeling the reflection of electrons at a potential barrier with a reflected amplitude and a phase shift leads to values for the wave number k for each energy, and thus the energy dispersion relation $E(k)$ of the Ag(111) surface state, which is shown in Figure 17–15. The data were in excellent agreement with other STS-derived data (Li et al., 1997) and PES data (Paniago et al., 1995).

The previously shown measurements of the dispersion relation on noble metal surfaces were limited to $k_\parallel$ vectors around the center of the surface Brillouin zone (SBZ) and it has been found that in this limit the surface state is free electron like, i.e., has an isotropic parabolic dispersion. LT-STS measurements on Ag(111) and Cu(111) over an extended energy range have shown a significant deviation from free electron behavior for large $k_\parallel$ vectors approaching the symmetry points at the SBZ boundary (Burgi et al., 2000b).

Direct visualization of the surface state dispersion on Ag(110) by means of Fourier transformation of differential conductance data taken with an LT-STM at $4\,\text{K}$ at energies up to the vacuum level has also been shown (Pascual et al., 2001b). Low temperatures ($4\,\text{K}$) enabled the necessary high ($2\,\text{meV}$) energy resolution and high stability to allow long recording times for the measurement of the differential conductance (dI/dV) using lock-in amplification techniques.

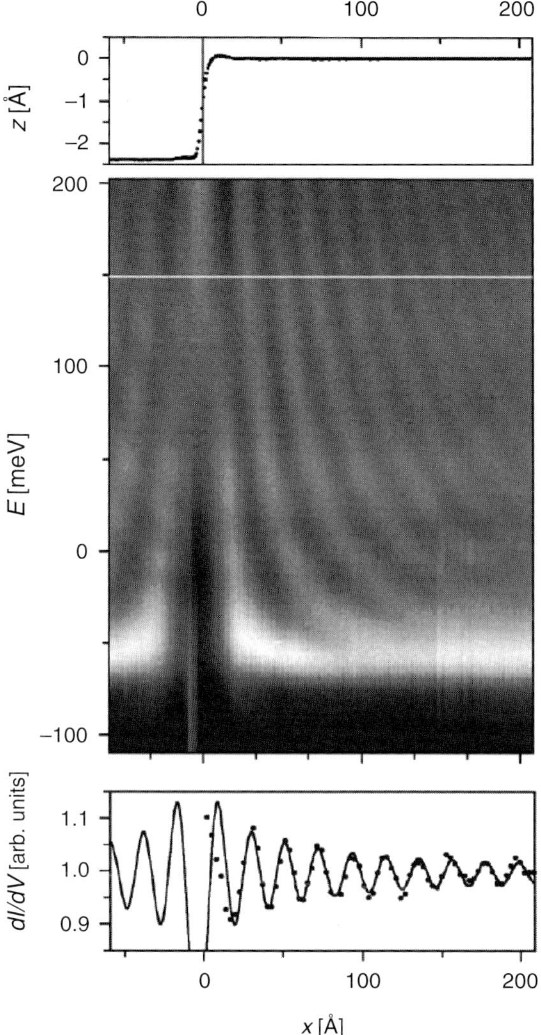

Figure 17–14. LDOS oscillations on Ag(111) at 5 K represented as gray levels as a function of the lateral distance from the step x and of bias energy $E = eV$ with respect to E_F. The upper graph shows the constant-current line scan (V = 0.3 V, I = 2.0 nA) on which the STM tip was moved while taking the dI/dV spectra. The lower graph is a cut of the dI/dV plot taken along the white line at E = 148 meV (dots), and the line is a fit to theory. (From Jeandupeux et al., 1999.)

Figure 17–16a shows the LDOS oscillations in front of a step on Ag(110) represented by a gray scale, plotted as a function of the distance form the step edge and the energy. Figure 17–16b shows the one-dimensional Fourier transform of the spatial one-dimensional conductance profiles in Figure 17–16a and visualizes the dispersion of the surface state in reciprocal space. A calculation shown in Figure 17–16c and d with a Bloch wave function limited to the first-order terms ($G = -1,0,1$) could explain the features of Figure 17–16a and b. It could

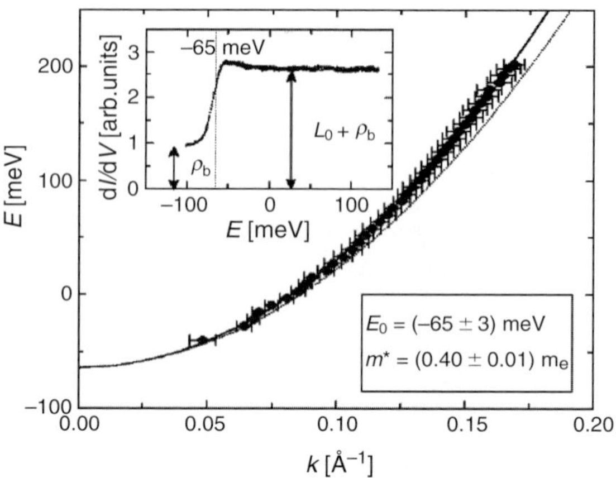

Figure 17–15. Energy dispersion relation of the Ag(111) surface state obtained from the measurement shown in Figure 17–14. The solid line is a quadratic fit to the STM data and the dotted line shows results from photoemission data at 65 K (Paniago et al., 1995). The inset shows a dI/dV spectrum taken on a perfect terrace ($V = 497$ mV, $I = 5.0$ nA) showing the onset of the surface state at 65 mV below the Fermi energy. (From Jeandupeux et al., 1999.)

be shown that the underlying atomic lattice gives rise to additional features in reciprocal space arising from coherent interference with surface state-derived Bloch waves.

3.1.3 Electron Confinement

When electrons are confined to length scales close to the de Broglie wavelength, their behavior is dominated by quantum mechanical

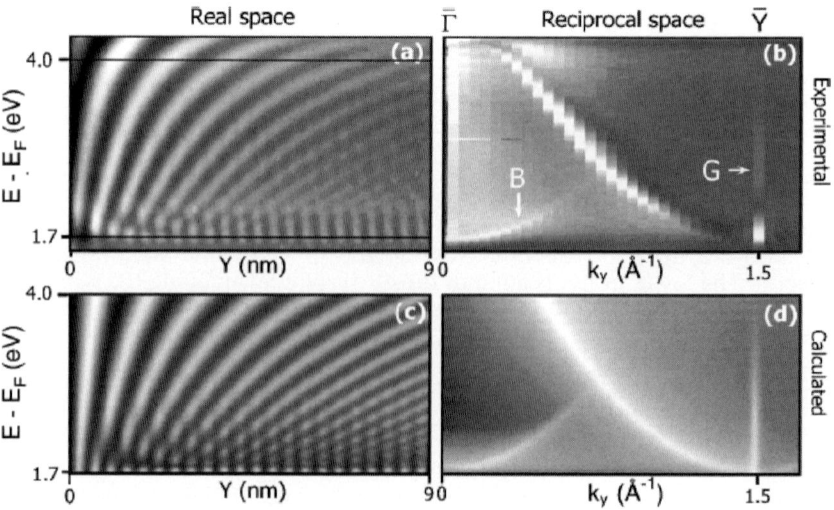

Figure 17–16. Direct visualization of the surface state dispersion $E(k)$. (a) Experimental dI/dV data from Ag(110) at 4 K as a function of energy and distance from a monoatomic step. (b) One-dimensional Fourier transform of the dI/dV vs. Y data, which is a direct visualization of the surface state disperison. (c and d) Results from calculations. (From Pascual et al., 2001b.)

effects, i.e., discrete energy levels are formed. The locality of the probe
in STM has an advantage over ARPES for the measurement of the
effects of lateral confinement of surface-state electrons on the surface-
state energies. Theoretical calculations have suggested that lateral confinement raises the energies of the surface-state electrons, resulting in
a depopulation of the surface-state band and a modification of surface
properties associated with the surface-state electrons (Bertel et al., 1995;
Memmel and Bertel, 1995; Garcia and Serena, 1995). The energy levels
of surface-state electrons confined within artificial nanostructures
have been measured with an LT-STM for the first time by Crommie
et al. (1993a). By manipulation with the STM tip, they positioned 48 Fe
atoms on Cu(111) in the form of a circle, forming a so-called quantum
corral, which confines surface-state electrons in the inner area (Figure
17–17). A standing wave pattern was observed within the corral. dI/dV
spectra taken in the center showed that the energy levels of the confined electrons are discrete. Good agreement with quantum mechanical calculations of the energy levels for a particle in a box was found.
For ideal confinement of electrons (i.e., infinitely high potential walls)

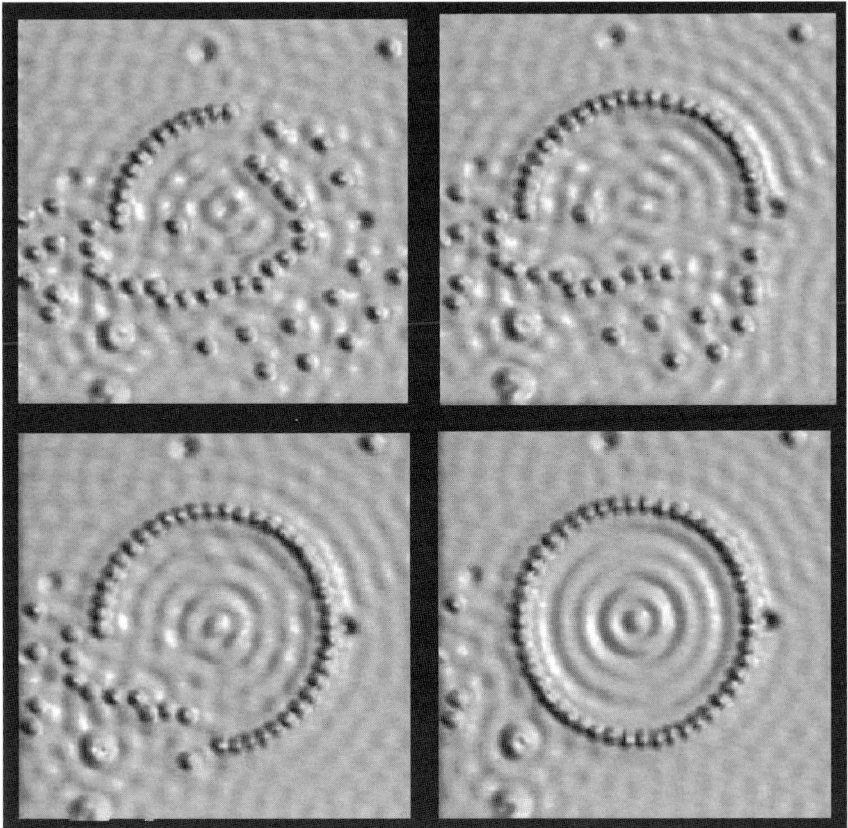

Figure 17–17. Various stages during the construction of a circular quantum
corral consisting of 48 Fe atoms on Cu(111) (measured at 4 K). (From Crommie
et al., 1993a. Reproduced by permission of IBM Research, Almaden Research
Center.)

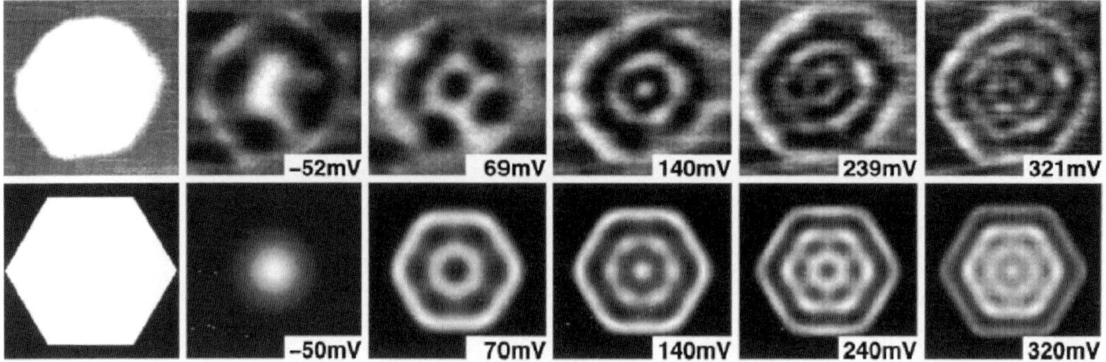

Figure 17–18. Electron confinement on nanoscale hexagonal Ag islands. Upper row: constant-current image of a hexagonal Ag island on Ag(111) (area ~94 nm²) and a series of dI/dV maps recorded at 50 K at different bias voltages as indicated. Lower row: calculated local density of states for the hexagonal box shown at the left. (From Li et al., 1998b.)

the energy levels E_n should be infinitely sharp apart from instrumental contributions. The measured peak width was interpreted as a lifetime effect via $\Delta t_n \approx \hbar/\Delta E_n$, dominated by scattering into bulk states (see Section 3.1.4).

Another experiment with a 76-atom "stadium" by the same group (Heller et al., 1994) gave further information about the scattering mechanism of surface-state electrons at the Fe adatoms. Comparison with s-wave multiple scattering theory showed that the Fe atoms reflect only about 25% of the incident electron wave, while 25% are transmitted and 50% are absorbed. Absorption is related to scattering into bulk states.

More recently a systematic study of electron confinement on nanoscale hexagonal Ag islands on Ag(111) was performed (Li et al., 1998b). In contrast to adatom corrals, these are stable structures even at elevated bias voltages in the STM. Figure 17–18 shows a series of dI/dV maps taken on a hexagonal Ag island on Ag(111). For voltages around −65 mV and lower (tunneling form of the sample) the image of the interior of the island is featureless, since this is below the onset of the surface state at −67 mV (Li et al., 1997). At higher voltages standing wave patterns are observed, which originate from oscillations in the LDOS of surface-state electrons, confined by the rapidly rising potential at the edges of the island. By recording local dI/dV spectra above the center of an island with an open feedback loop and varying tunnel voltage, peaks at different energy where recorded, which correspond to the energy levels of the confined surface-state electrons. The validity of the particle in a box model for these confined surface states was confirmed. Analyzing the measured energy levels in terms of a particle in a box model showed that the energies conform to the expected scaling behavior down to the smallest island sizes. The experiments confirmed that lateral confinement is a mechanism for surface-state depopulation since it causes a discretization of the formerly continuous SS band and an upward shift of the surface-state levels. The fact that ARPES on vicinal Cu(111) surfaces (Sanchez et al., 1995) failed to observe the shifts in the surface state energies expected on the basis of ideal confinement is most likely due to the presence of a sizable distribution of terrace widths within the spot area of the incident radiation.

Electron confinement effects have also been observed in metallic single-walled carbon nanotubes (SWCNTs) (Maltezopoulos et al., 2003). With an LT-STM operating at 14 K (Pietzsch et al., 2000b), dI/dV spectra and maps on SWCNTs have been recorded (Figure 17–19). Peaks close to the Fermi energy are restricted to a certain area of the nanotube and have been assigned to confined states within the SWCNT created by defect scattering (see Figure 17–20). This shows that defects can lead to significant backscattering within individual metallic SWCNTs. These findings are important since SWCNTs are considered promising candidates for molecular electronics.

LT-STM measurements of the two-dimensional structure of individual electron wavefunctions in SWCNTs have been reported (Lemay et al., 2001). An SWCNT was deposited on an Au(111) surface. To increase the electronic energy-level spacing, the nanotube was cut to less than 40 nm length by applying short bias voltage pulses between tip and sample. Local spectroscopy with the tip above the nanotube showed an energy spectrum with a series of sharp peaks, representing confinement-induced energy levels. Differential conductance maps measured at the energies of the peaks revealed spatial patterns that could be understood from the electronic structure of a single graphite sheet. The patterns observed are a representation of individual molecular wavefunctions $|\psi_j(r)|^2$. The wavefunctions exhibited "beating" between electron waves with slightly different wavevectors, and this intereference effect was exploited to directly probe the electronic dispersion relation of an individual nanotube.

3.1.4 Lifetime Measurements

Improved energy resolution (~3 meV) in photoemission spectroscopy has made it possible to study the lifetimes of photoholes in a surface

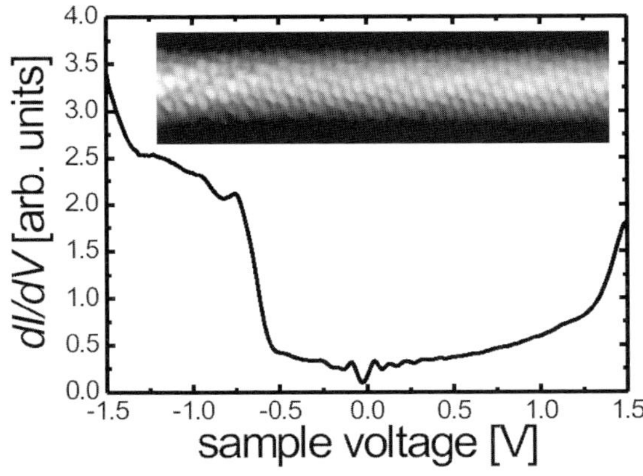

Figure 17–19. Electron confinement in single wall carbon nanotubes: dI/dV spectrum of a metallic SWCNT. The oscillations around the Fermi energy (0 V) are shown in more detail in Figure 17–20. Inset shows atomically resolved STM image of SWCNT at 14 K. (From Maltezopoulos et al., 2003.)

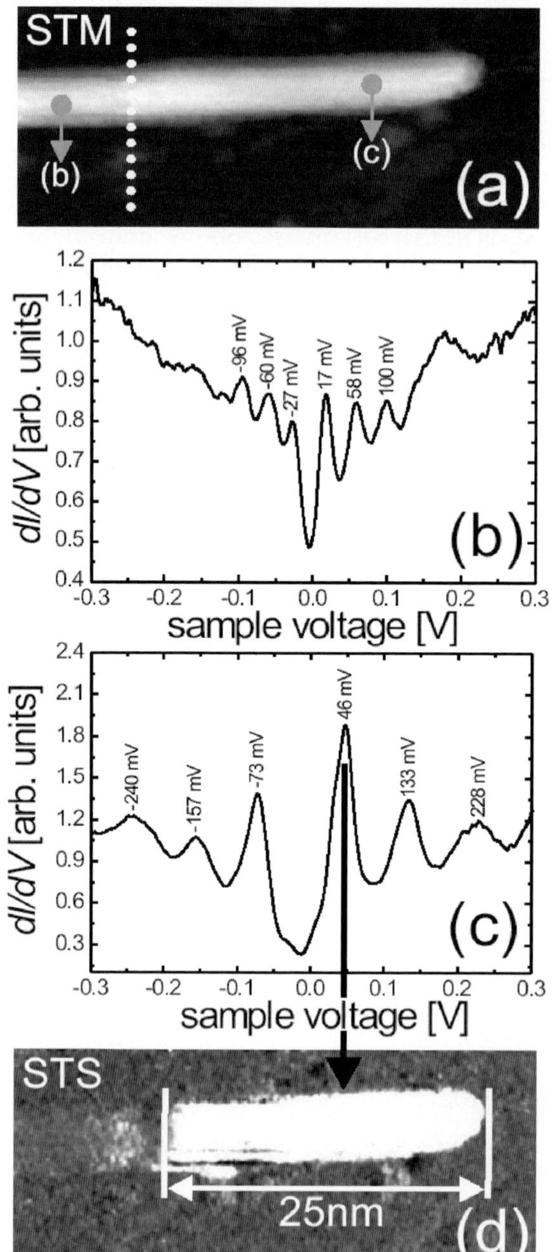

Figure 17–20. (a) STM image of an SWCNT end. (b) STS data (~LDOS) taken on the left-hand side of the dotted line in (a). (c) STS data taken on the right-hand side of the dotted line in (a). (d) Spatially resolved dI/dV map taken at the energy of the 46 mV peak shows confinement of these states to the area on the left. (From Maltezopoulos et al., 2003.)

state. Furthermore, femtosecond time-resolved two photon photoemission (2PPE) experiments opened up a new path to measure lifetimes of hot electrons in metals. A hole in an energy band or a hot electron are quasiparticles, elementary excitations of an interacting Fermi liquid. A quasiparticle has a lifetime that is the duration of the excitation. In combination with its velocity, the lifetime determines the mean free path of the quasiparticle and thereby the range of influence of the excitation. Lifetimes are determined by inelastic scattering processes, i.e., electron–electron (e–e) scattering and electron–phonon (e–ph) scattering. Fermi liquid theory predicts an inverse second power law for the energy dependence of the e–e lifetime. It also predicts that e–e scattering dominates e–ph scattering at low temperatures and large excess energies ($\geq$0.5 eV), whereas at energies very close to the Fermi level inelastic scattering is dominated by e–ph processes at all temperatures (Pines, 1966).

In metallic systems, quasiparticle states in two-dimensional systems are important in surface science (Memmel, 1998) and nanoscale technology (Himpsel et al., 1998). One example for a two-dimensional systems is a surface state on a noble metal, and here the L-gap Shockley-type surface state on the (111) surface of noble metals has played a special role as model systems for the investigation of two-dimensional electron systems. The lifetime of a hole in such a surface state band has been extensively studied both theoretically and experimentally (Matzdorf, 1998), but the discrepancy between experiment and theory remained large. This lifetime is of interest as it determines the effective range of surface-mediated interactions and is important for epitaxial growth and chemical reactions at surfaces (Memmel and Bertel, 1995; Bertel et al., 1995). Theoretical models predicted significantly longer lifetimes than were observed with photoemission spectroscopy. Using

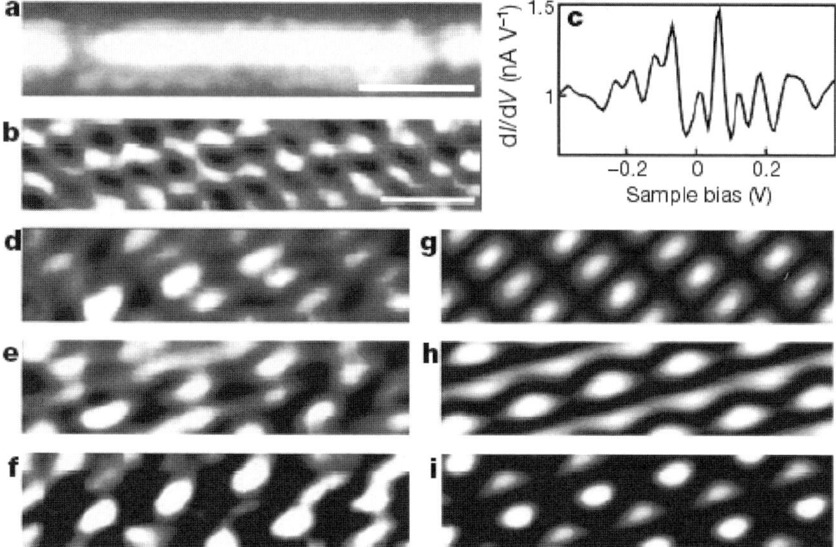

Figure 17–21. Imaging individual molecular wavefunctions in a carbon nanotube at 4.6 K. (a) Constant-current image of metallic SWNT cut to a length of 34 nm by bias voltage pulses (scale bar 10 nm). (b) High-resolution constant-current image of the shortened nanotube (scale bar 0.5 nm). (c) Differential conductance spectrum (LDOS) averaged over the area shown in (b) showing particle in a box states. (d–f) Differential conductance maps showing individual states at −95, 30, and 96 meV. (g–i) Calculated spatial maps of individual molecular wavefunctions. The characteristic features of each image are well reproduced. (From Lemay et al., 2001.)

ARPES from a surface state, the lifetime of a photohole state can be determined by analyzing the spectral linewidth Γ of a photoemission excitation. The spectral linewidth of a photoemission line gives access to the lifetime τ via $\Gamma = \hbar/\tau$. ARPES integrates over a large surface area (typically $\geq 1\,\mathrm{mm}^2$) and contains contributions from surface imperfections over the whole investigated surface range (Theilmann et al., 1997). Therefore peak widths in ARPES are influenced by surface conditions (defects, roughness) as well as finite energy and angular resolution (so-called nonlifetime effects). Defects on the surface are known to strongly couple the surface-state electrons to bulk states and thus reduce lifetimes (Crampin et al., 1994). It would be desirable to apply the concept of measuring lifetimes from the linewidth of spectral features directly to STS. The nonlifetime effects could then be avoided by measuring the lifetime locally with an STM tip on a spot bare of impurities. The ability of the STM to detect surface topography and to identify minute amounts of contamination well below the limits of conventional surface analytical techniques ensures that an effectively defect-free surface is studied. Since STS does not offer momentum resolution, measuring the lifetime of a specific state defined by its momentum $k_\parallel$ and energy E may seem impossible. This problem can be circumvented to some extend as shown below.

The surface-state (excited hole) lifetime on Ag(111) has been determined at the dispersion minimum ($k_\parallel = 0$) by evaluating the linewidth of the surface state onset in dI/dV tunneling spectra (Li et al., 1998a; Pivetta et al., 2003; Kliewer et al., 2000). Calculated dI/dV spectra show (Li et al., 1998a) that this linewidth depends on the imaginary part Σ of the electron self-energy. As Σ increases, the onset of the surface-state contribution is seen to broaden. Therefore the measured linewidth can be used to estimate Σ or the lifetime $\tau = \hbar/2\Sigma$ (Li et al., 1998a) of a hole in the surface state at the band minimum. The first lifetime measurements (Li et al., 1998a) were performed at a temperature of $5\,\mathrm{K}$ using W tips on Ag(111). Differential conductance spectra were recorded under open feedback loop conditions, adding a modulation voltage (1–$10\,\mathrm{mV}$ at $\omega = 230\,\mathrm{Hz}$) and using a lock-in amplifier to record the signal at ω. Figure 17–22 shows a dI/dV spectrum measured on a large defect-free terrace showing the characteristic surface state band onset of width Δ, used to estimate the spectral linewidth (also called lifetime width or inverse lifetime) $\Gamma = \hbar/\tau$. The geometric width Δ is a combination of the spectral linewidth Γ of the surface state, thermal broadening, and additional broadening due to the modulation technique used to measure the spectrum. The electron self-energy Σ and the corresponding lifetime $\tau = 67 \pm 8\,\mathrm{fs}$ were determined from a line-shape analysis. This measured spectral linewidth was considerably smaller than previous PES measurements. A more recent lifetime measurement (Pivetta et al., 2003) with the same method resulted in even smaller spectral linewidths, which were about $1\,\mathrm{meV}$ smaller than recent photoemission (Nicolay et al., 2000) results and theoretical predictions (Eiguren et al., 2002). Since STS is measured on locally defect-free surface regions, a smaller spectral linewidth (i.e., a larger lifetime of surface hole states) than in photoemission is expected. Lifetimes resulting form STS lineshape analysis where $\tau = 120$, 35, and 27 fs for excited holes at the surface-state band edge on Ag, Au, and Cu, respectively (Kliewer et al., 2000). A comparison of linewidths obtained from STM measurements with previous PES data (Kliewer et al., 2000) showed that for Ag(111) and Au(111), STM measures linewidths smaller by a factor of 3 compared to PES (i.e., longer lifetimes). This illustrates how large the contribution from electron scattering at surface imperfections is. Cu(111) extrapolation of the PES linewidth to zero defect density (Theilmann et al., 1997) resulted in good agreement with the STM value, which provides evidence that STS measurements avoid defect-induced broadening of the linewidth.

In the absence of defect scattering, two processes limit the lifetime of the hole: inelastic e–e scattering and e–ph coupling. Inelastic e–e scattering results in the hole being filled by interband transitions involving bulk electrons near the surface or by intraband transitions involving surface-state electrons (Figure 17–23). These contributions have been estimated theoretically and good agreement of STS measurements of the lifetime width with theoretical values has been achieved (Kliewer et al., 2000).

With the method described, lifetimes can be studied only at a single energy, namely the energy $E_{\bar{\Gamma}}$ at the surface band minimum. Therefore, in a different approach, the electron lifetimes have been measured with an LT-STM (Burgi et al., 1999; Jeandupeux et al., 1999; Burgi et al., 2000a; Vitali et al., 2003) by studying the decay of quantum mechanical interference patterns from surface-state electrons scattering off step edges. The decay length is influenced by the loss of coherence and hence the phase relaxation length. The phase-relaxation length, i.e., the distance an electron can travel without losing its phase information, is directly related to the lifetime. If the lifetime is shorter than the time necessary

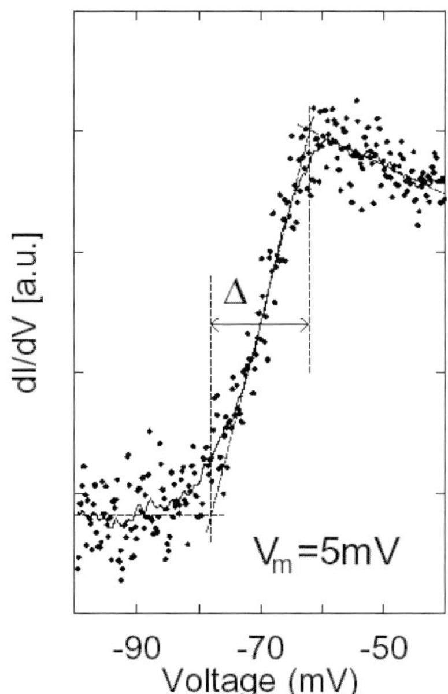

Figure 17–22. Surface state lifetime measurement. dI/dV spectrum measured on a large defect-free terrace on Ag(111), showing steplike onset of the surface state near −70 mV. The surface state lifetime can be determined from the width Δ by fitting it to calculated spectra. (From Li et al., 1998a.)

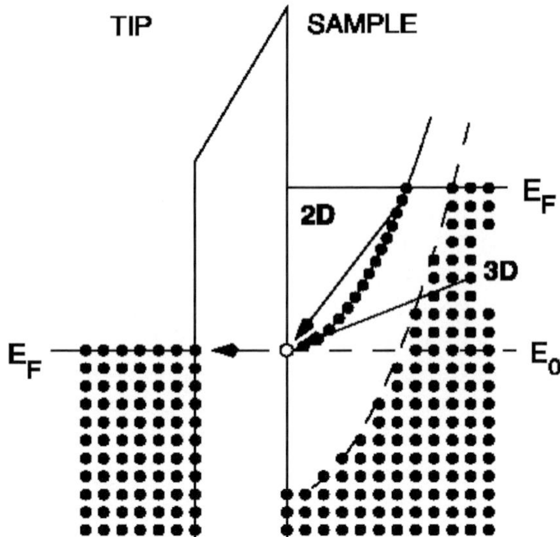

Figure 17–23. Schematic energy diagram of electron (solid circles) tunneling from the bottom of the occupied surface state band of the sample to a metallic tip (sample bias negative with respect to the tip). Sample states are shown as a projected band structure. The dashed parabola indicates the maximum of the bulk bands. The intrinsic lifetime of the hole left behind by an electron tunneling from sample to tip is determined by inelastic electron–electron scattering, which results in either an intraband transition (two-dimensional) or an interband transition (three-dimensional). (From Kliewer et al., 2000.)

to reflect the electron wave back from the step, the phase information is lost and the electron wave cannot interfere with itself anymore. The electron lifetime τ is related to the phase relaxation length by $\tau = L_\phi/v_g$, where v_g is the group velocity (Datta, 1995). The LDOS oscillation induced by scattering at a surface step is intrinsically damped and its intensity falls off like $1/\sqrt{kx}$, where k is the electron wave vector and x the distance from the step. The phase relaxation length leads to an additional damping e^{-x/L_ϕ} of the interference pattern and can be determined by fitting a model function to the data. $L_\phi(E)$ and thus $\tau(E)$ could be extracted from dI/dV scans across step edges for the Shockley-type surface states on Ag(111) and Cu(111). This method has the advantage that quasiparticle lifetimes can be measured as a function of excess energy and not only at the surface-state band edge $\overline{\Gamma}$ as with the STS peak width method described by Li et al. (1998a). Lifetimes of hot electrons injected by the STM tip have been measured for energies from the Fermi level to 3 eV excess energy (Vitali et al., 2003; Burgi et al., 1999). Figure 17–24 shows a schematic energy diagram of the tunneling process: an electron tunnels from the Fermi level of the tip into an unoccupied level of the metal, from which, after a short time, it will decay to the ground state. Experimental lifetimes of hot electrons as a function of excess energy are shown in Figure 17–25. These energy-dependent lifetime measurements of hot electrons have been interpreted in terms of e–e scattering, since e–ph lifetimes are independent of energy and exceed the measured lifetimes considerably.

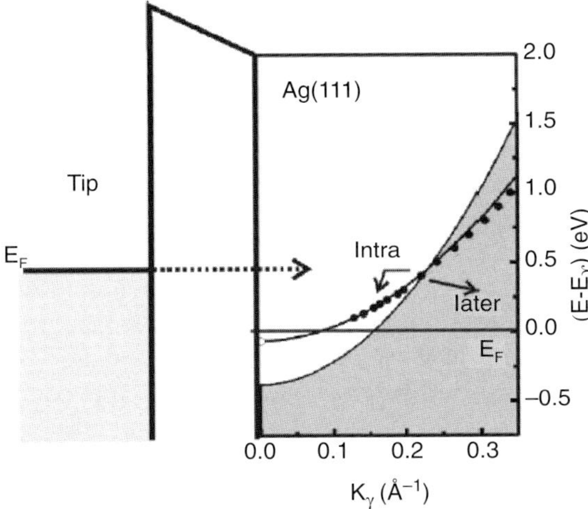

Figure 17–24. Schematic energy diagram of the tunneling process of hot electrons into the surface state band (sample bias positive with respect to the tip). Dots indicate experimental measurements in Vitali et al. (2003). Electrons injected at an energy E scatter inelastically with other electrons into unoccupied states with an energy smaller than E. (From Vitali et al., 2003.)

The experimental data are found to scale with $(E - E_F)^{-2}$ above $0.5\,\text{eV}$ as predicted by Fermi-liquid theory, but deviate from the quadratic scaling law at low energies. Calculations showed that the main scattering process at low energies are inelastic intraband e–e scattering and e–ph scattering, whereas at higher energies e–e interband scattering with bulk electrons dominates.

To determine the e–ph scattering contribution to the lifetime, temperature-dependent lifetime measurements have been done with low-energy quasiparticles (Jeandupeux et al., 1999). Constant-current

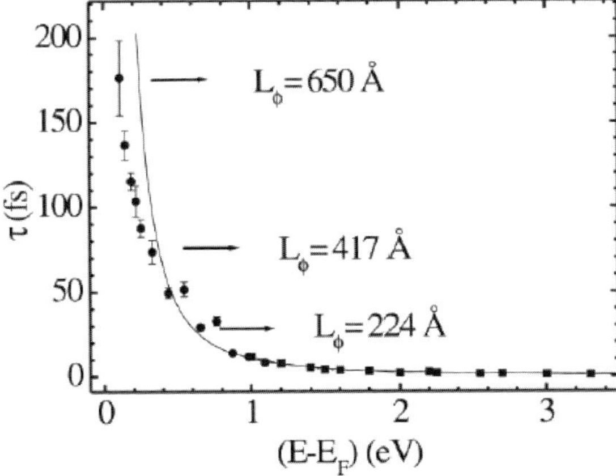

Figure 17–25. Experimental lifetime of hot surface state electrons at the Ag(111) surface as a function of energy. The solid line is a fit of the high energy data (solid squares) to the expected $(E - E_F)^{-2}$ behavior in the Fermi liquid model. (From Vitali et al., 2003.)

line scans across the oscillating LDOS at very low bias voltage perpendicular to step edges on Ag(111) have been measured in the temperature range of 3.5–178 K and compared with theoretical calculations. By measuring at the Fermi energy, the lifetimes are not reduced by e–e scattering (since the e–e lifetime is large at E_F). The measurements showed that the temperature-dependent damping of the standing waves at step edges could be very well described by Fermi–Dirac broadening alone. Therefore only a lower limit for the e–ph lifetime could be given, pointing to significantly longer lifetimes than all previous data.

Below 1 eV the phase relaxation length L_ϕ becomes longer and longer and the damping effect for LDOS oscillations at step edges becomes small. Measuring the additional damping of the oscillating LDOS caused by L_ϕ beyond the inherent $1/\sqrt{kx}$ decay at energies close to the Fermi level is very demanding. It has been proposed that by choosing a closed scattering geometry such as a quantum corral assembled with the low-temperature STM, reflections add up and contribute to a high total intensity, which allows the measurement of small effects. This idea was realized and the lifetime of surface-state electrons confined in an artificial atomic structure has been measured at various energies on an Ag(111) surface at 6 K (Braun and Rieder, 2002). Ag atoms were taken out of the surface by applying short voltage pulses to the tip. By means of lateral manipulation, single Ag atoms were then assembled along closed-packed row directions to form a triangle with a base length of 24 nm as shown in Figure 17–26. The surface-state electrons are scattered by the positioned Ag atoms, resulting in a standing wave pattern as can be seen in Figure 17–26. For the lifetime determination, dI/dV maps were recorded in constant height mode inside the triangle. Lifetimes have been determined by fitting theoretically calculated dI/dV maps to the experimental maps that have been recorded in the energy range of −55 to +796 meV with respect to the Fermi level. The model to calculate the dI/dV maps contained three fit parameters: the phase shift and absorption of the electron wave during scattering events and the phase-relaxation length. The extracted $\tau(E)$ data also reproduced the $\tau \propto E^{-2}$ law predicted by Fermi liquid theory for a 2DEG (Pines, 1966). Electrons injected at an energy E scatter inelastically with other electrons into unoccupied states with an energy smaller than E. This results in a singularity of the lifetime at the Fermi energy due to the reduction of the number of available states into which to scatter. In accordance with theory, all STM lifetime measurements showed a sharp maximum at the Fermi energy.

3.1.5 Stark Effect

The influence of the high electric field between the STM tip and the sample on STS spectra has been measured in an LT-STM study (Kroger et al., 2004). In metals, the electric field is efficiently screened by the conduction electrons. Nevertheless, the wavefunction of surface state electrons is evanescent into vacuum and can thus be affected by the electric field. By varying the electric field of the microscope through the resistance of the junction from 1 GΩ (low field) down to 60 kΩ (high field), it could be demonstrated that the STS spectra of the Au(111) and

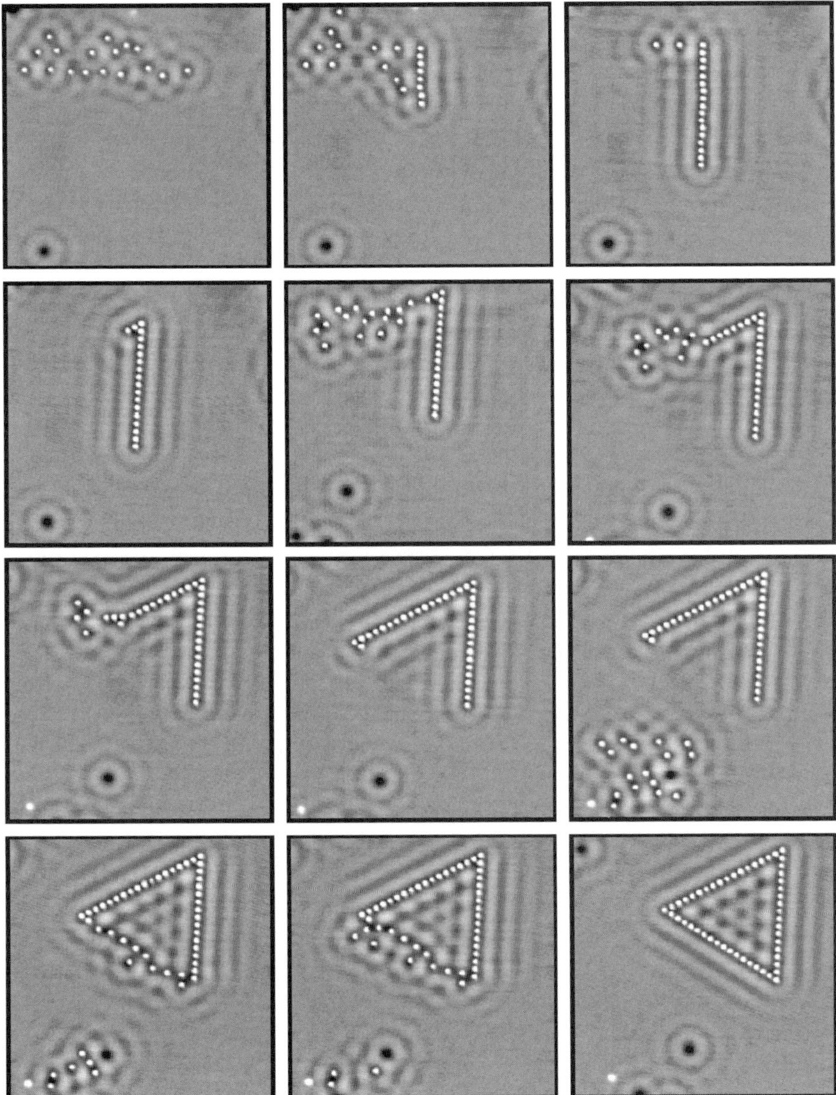

Figure 17–26. Series of images showing the construction of the triangle consisting of 51 Ag atoms on an Ag(111) surface (scan area 49.3 nm × 49.3 nm). (From Braun, 2001.)

Cu(111) surface states undergo a downward shift that was attributed to the Stark effect. The presence of a Stark effect in the noble metal surface states even at usual tunneling parameters suggests that this effect is quite common for STS.

3.1.6 Long-Range Interactions between Adatoms
The previous studies on standing wave patterns in the electron density of surface electron states concentrated on the effects caused by the adatom scatterers on the surface-state electrons. On the other hand, surface-state electrons give rise to an interaction between the scatter-

ers. This surface-state-mediated interaction is long ranged and oscillatory in nature. Quantitative measurements of the long-range interaction energy between single Cu adatoms on the Cu(111) surface have been performed with a low-temperature STM operated at 9–21 K (Repp et al., 2000). The main criteria of the theory of surface-state-mediated interaction energy (Lau and Kohn, 1978) could be reproduced in the experimental data: the interaction energy is oscillatory with a period of $\lambda_F/2$, where λ_F is the Fermi wavelength, and the envelope of the magnitude decays as $1/d^2$, where d is the adatom separation.

3.1.7 Magnetic Atoms at Surfaces: Kondo Effect

The smallest magnetic structure in condensed matter physics is a single magnetic atom in a nonmagnetic host. The localized spin of the magnetic atom interacts with the spin of the surrounding delocalized conduction electrons. For temperatures below a characteristic Kondo temperature T_K, this interaction causes the conduction electrons of the host metal to condense into a many-body ground state that collectively screens the local spin of the impurity. This screening cloud exhibits a set of low-energy excitations called the Kondo resonance. Due to the formation of the screening cloud, the LDOS near the Fermi energy is enhanced at the site of the impurity. The Kondo resonance disappears at temperatures above T_K, and it energetically splits by the Zeeman energy in an applied magnetic field (Hewson, 1993). The Kondo resonance shows up as a peak at the Fermi level in high-resolution photoemission spectroscopy.

This resonance can be probed locally by STS and manifests itself as a sharp (~10 meV wide) depression in the differential conductivity at the Fermi energy. The feature is localized to within ~1 nm of the Kondo impurity, and is observed only at low temperatures. Co impurities on Au(111) have been studied (Madhavan et al., 1998) at 4 K. The Kondo resonance line shape has been interpreted in terms of quantum interference resulting of two possible tunneling channels, one direct channel into the Co d-orbital and another channel into the surrounding continuum of conduction band states. A similar observation has been made for Ce adatoms on Ag(111) (Li et al., 1998c). A later experiment with Co adatoms on Cu(100) and Cu(111) (Knorr et al., 2002) showed that while at the Cu(111) surface both tunneling into the localized impurity state and into the substrate conduction band contribute to the Kondo resonance, tunneling into the conduction band dominates for Cu(111) (Figure 17–27).

Another landmark STM experiment showed that when a magnetic atom is placed at one focus of a properly sized empty elliptical corral built from magnetic atoms, a "mirage" of the Kondo signature in the LDOS is cast to the opposite empty focus (Manoharan et al., 2000) more than 7 nm away. The experiments were performed with Co atoms on Cu(111) with a low-temperature STM at 4 K well below the Kondo temperature of the system. Co atoms were evaporated onto the clean Cu surface at 4 K. The ellipses where constructed by use of the adatom sliding process (Eigler and Schweizer, 1990; Stroscio and Eigler, 1991). Figure 17–28 shows the results of these measurements, simultaneously acquired constant current images, and dI/dV maps of the corrals. dI/dV maps measured on the corrals without an interiour atom have been subtracted from the maps measured with an interior atom to remove the contribution of the LDOS oscillation and emphasize the Kondo signal. The constant current images (Figure 17–28a and b) show the location of the Co atoms and oscillations in the LDOS due to two-dimensional confinement of the surface-state electrons whereas the differential conductance spectra show the Kondo signature of Co inside the corral. The striking results reveal two interior positions that show a Kondo resonance: one signal is centered on the real Co atom at the left focus, while the other signal is centered on the empty right focus. The localized electronic structure has been projected from the occupied focus to the unoccupied focus. Tunneling spectra taken at both foci showed that the mirage at the right focus is a faithful spectroscopic replica of the real atom at the left focus attenuated by a factor of eight.

An energy-dependent phase shift of electrons scattering off magnetic impurities has been shown to explain the observed quantum

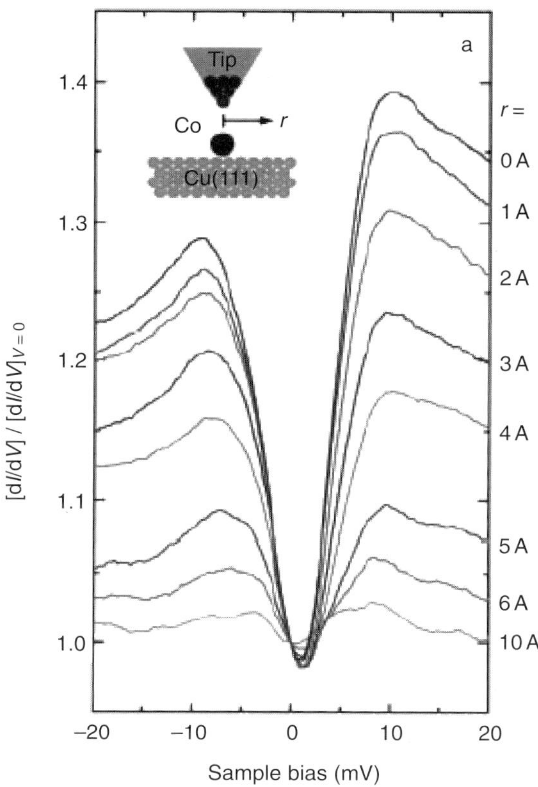

Figure 17–27. Spectroscopic signature of the Kondo resonance around a single Co atom on Cu(111) Tunneling spectra shown have been acquired over the Co atom with increasing lateral displacement r. The spectroscopic feature is narrow (9 mV FWHM) and dies off over a lateral length scale of 1 nm. Measurements with different tips showed nearly identical features. (From Manoharan et al., 2000. Reproduced by permission of IBM Research, Almaden Research Center.)

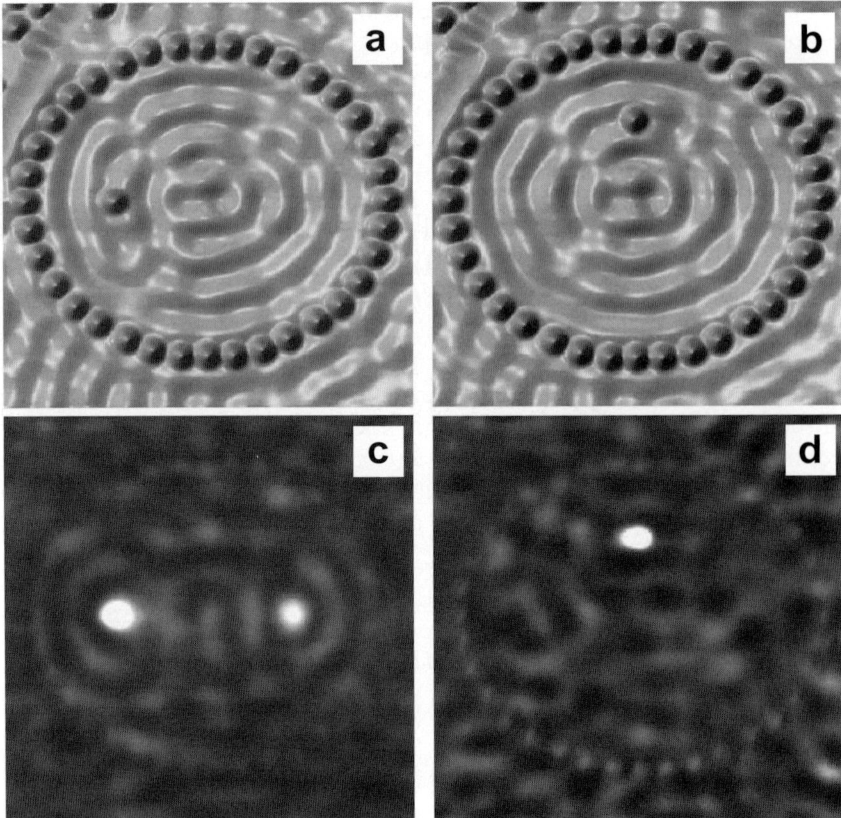

Figure 17–28. Quantum mirage. STM topographs of two elliptical corrals with (a) a Co atom at one focus and (b) a Co atom off focus. Associated dI/dV maps showing in (c) the Kondo effect projected to the empty right focus, resulting in a Co atom mirage; in (d) the mirage vanishes. (From Manoharan et al., 2000. Reproduced by permission of IBM Research, Almaden Research Center.)

mirage (Fiete et al., 2001) and Schneider et al. (2002) have shown experimentally that this phase shift can also be accurately determined with low-temperature STM when the Co atoms are simply adsorbed on Ag(111) and are not placed in artificial resonators.

3.1.8 Fermi Contour Imaging

It has been demonstrated that with a low-temperature STM an image of the surface Fermi contour is directly obtainable by Fourier transforming a low-bias constant-current image of the standing waves in the LDOS (Petersen et al., 1998, 2000ab; Sprunger et al., 1997). The wavevectors of electrons with an energy at the Fermi level are confined to the surface Fermi contour, and consequently standing wave patterns observed in low-bias STM images contain information about the entire Fermi contour. This information can be extracted by performing a simple two-dimensional Fourier transform of the STM image. Thus the power spectrum of a low-bias STM image contains an image of the surface Fermi contour. For an isotropic free electron like surface state a ring is observed in the Fourier transform of the oscillating LDOS image, as shown in Figure 17–29. The radius of this ring is twice the Fermi wave vector $2k_F$. This is due to the fact that waves in the charge density are observed rather than in the actual wavefunction. Application of this method to anisotropic surfaces was undertaken by Hofmann et al. (1997), who studied the Be(10$\bar{1}$0) surface. An image of the standing waves on Be(10$\bar{1}$0) recorded at 4 K is shown in Figure 17–30a. In contrast to isotropic surfaces, standing wave patterns on Be(10$\bar{1}$0) are visible only when originating from steps in the $\overline{\Gamma M}$ direction, whereas no waves from the steps in the $\overline{\Gamma A}$ direction are present. The two-dimensional Fourier transform of the standing wave pattern shows the Fermi contour at the SBZ boundary as half ellipses. Obviously the screening of defects at the surface is highly anisotropic due to the orientation of the Fermi contour. This is a beautiful example of the intimate coupling between the Fermi contour and the screening at the surface.

3.2 Single Atom and Molecule Manipulation

The tip of an STM always exerts a force on an adsorbate atom or molecule. This force contains both van der Waals and electrostatic contributions. By adjusting the position and the voltage of the tip, the magnitude and direction of this force may be tuned, thereby enabling atomic scale manipulation processes. Two basic modes of transport of atoms or molecules have been identified: lateral manipulation, i.e., moving particles along the surface by which they maintain contact with the substrate, and vertical manipulation, by which the particles are picked up by the tip and released back to the surface at a desired place. To realize a lateral manipulation experiment, the tip is positioned on the adsorbate and its height reduced to 0.2–0.4 nm by reducing the tunneling resistance from ~1 MΩ to 100 kΩ. The manipulation is then performed in constant-current mode and the particle is dragged

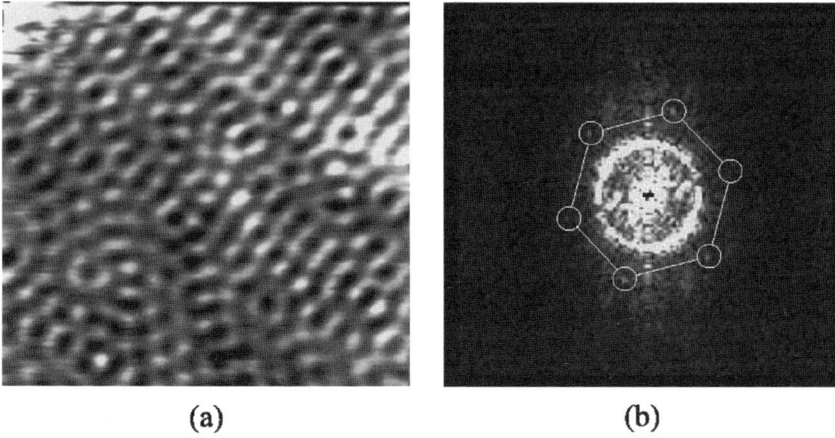

(a)   (b)

Figure 17–29. Imaging the Fermi contour. (a) Constant-current image of the Be(0001) surface at 150 K showing LDOS oscillations. (b) Two-dimensional Fourier transform of (a) showing a ring-shaped Fermi contour as expected for an isotropic free electron-like surface state. The spots are reciprocal lattice vectors and scale reciprocal space. (From Petersen et al., 2000a.)

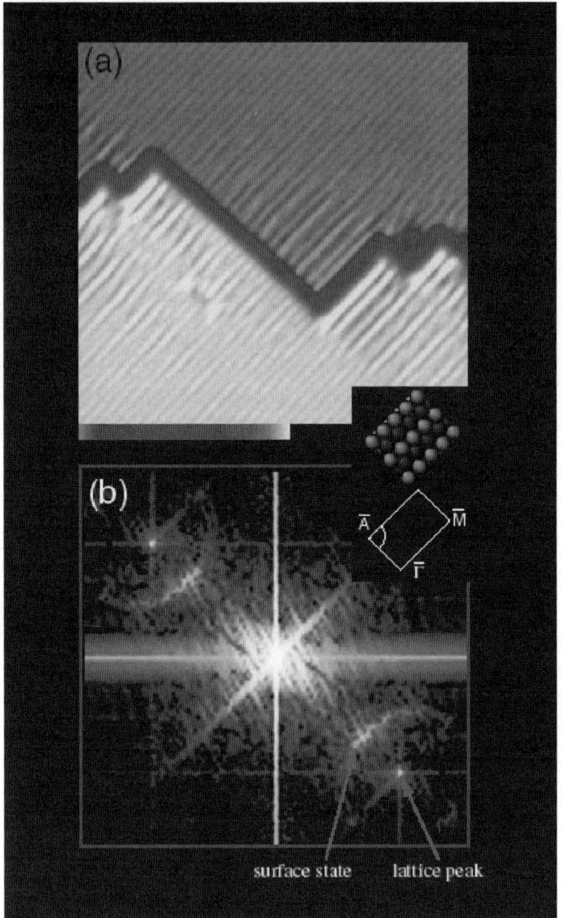

Figure 17–30. Imaging the Fermi contour. (a) Constant current image of the Be(10$\bar{1}$0)surface at 4 K with a surface step. Anisotropic scattering at step edges results in LDOS oscillations mainly in one direction. (b) Logarithm of the power spectrum of the image. The elliptically shaped features originate from the Fermi contour of the anisotropic surface states. Inset shows a real space top view of the surface and the surface Brillouin zone with the Fermi contour. (From Hofmann et al., 1997.)

by lateral movement of the tip to the final position. As single metal atoms or small molecules are mobile at ambient temperatures, experiments on small adsorbates located on low index metal surfaces have been conducted at low temperatures. Using an LT-STM reduces or eliminates thermally activated diffusion of adatoms on the surface and enables the experimenter to build stable nanostructures or synthesize molecules atom by atom. The electronic properties of these artificial structures can then be studied by STS.

Reliable lateral manipulation and build-up of extended nanostructures on an atomic basis with adsorbed atoms and small molecules at low temperatures have been demonstrated by several groups. Don Eigler's group at IBM-Almaden assembled arrays of Xe atoms on Ni(110) at 4 K (Eigler and Schweizer, 1990) (Figure 17–31). An atomic switch has

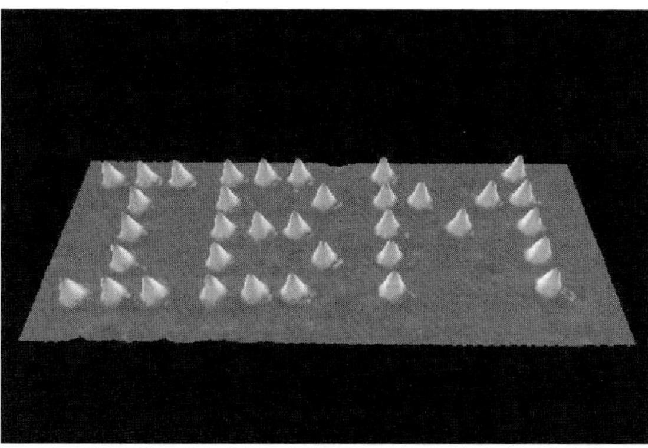

Figure 17–31. Patterned array of xenon atoms on an Ni(110) surface. The atomic structure of the nickel surface is not resolved. Each letter is 5 nm from top to bottom. (From Eigler and Schweizer, 1990. Reproduced by permission of IBM Research, Almaden Research Center.)

been realized where a xenon atom is moved reversibly between two stable positions on the STM tip and a nickel surface (Eigler et al., 1991). Heat-assisted electromigration (Ralls et al., 1989) has been identified as a possible explanation for the vertical manipulation of a xenon atom between a nickel surface and the tip (Eigler et al., 1991). Several other possible mechanisms responsible for lateral and vertical manipulation processes have been investigated (Stroscio and Eigler, 1991). Field-assisted diffusion, which relies on the static and induced dipole moment of the adatom in the strong electric field under the tip, has been identified as being responsible for lateral manipulation of Cs atoms on GaAs(110) at room temperature (Whitman et al., 1991). The lateral manipulation of Xe on Ni(110) has been attributed to the chemical bonding force between the adatom and the tip, since it is not sensitive to the sign or the magnitude of the electric field, the voltage, or the current. It depends only on the separation of the tip from the substrate (Stroscio and Eigler, 1991). A possible mechanism for vertical manipulation of adatoms is transfer on contact, whereby the tip is moved close to the adsorbate until the energy barrier separating them is lowered enough that thermal activation is sufficient for atom transfer. This process does not need an electric field, whereas field evaporation, another possible mechanism, relies on the strong field to create ions that are evaporated over the Schottky barrier. Inelastic tunneling, which leads to multiple vibrational excitations of the atom–surface bond until the bond is broken, has been used to explain the desorption of hydrogen atoms along single dimer rows of the Si(100)-2 × 1 surface. Linewidths as small as 1 nm have been achieved (Foley et al., 1998) in these experiments. Inelastic tunneling was also used to selectively cut single molecular bonds in O_2 on Pt(111) (Stipe et al., 1997b).

Lateral manipulation of Cu atoms and Co and C_2H_4 molecules on Cu(211) in constant current mode has been demonstrated by Meyer et al. (2001) at 15 K. The measured tip height curves during pulling of a Cu atom along a close-packed row on Cu(111) showed characteristic jumps due to the atom hopping from one fcc adsite to the next. By recording the tip height during adatom manipulations, it is possible to distinguish three different types of lateral manipulation: pushing, pulling, and sliding (Bartels et al., 1997a). During a pulling process the adsorbate is situated behind the tip apex with respect to the manipulation direction (attractive tip–adatom interaction). By applying larger forces than for pulling, the adsorbate remains under the tip apex without escaping sideways from the tip trajectory (sliding mode, attractive tip–adatom interaction). Small molecules can be manipulated by the pushing mode, where the adsorbate is in front of the tip (repulsive tip–molecule interaction).

Bias polarity-independent STM-induced desorption of individual NH_3 molecules from Cu(111) has been demonstrated at 15 K, which sometimes leads to transfer of the molecule to the tip (Bartels et al., 1999). Vertical manipulation of C_3H_6 molecules at low temperatures has also been shown (Meyer et al., 1996). Pb monomers and dimmers on Cu(211) have been moved laterally at 20 K. Vertical manipulation of CO molecules on Cu(111) has been shown (Bartels et al., 1997b), whereby a CO molecule could be reliably transferred between the surface and the tip. This ability adds chemical sensitivity to STM: with a tip having a CO molecule at its apex, chemical contrast between otherwise similar appearing adsorbates has been achieved: CO molecules on Cu(111) always appear as depressions independent of the bias polarity when imaged with a clean tip. However, when imaged with a tip having a CO molecule at its apex, they appear independently of bias polarity as protrusions. This inversion of shape allows chemical-sensitive imaging, as shown in Figure 17–32. Here, O_2 and CO molecules were adsorbed on the surface. Both adsorbates where imaged as depressions in the STM images with a clean tip. Picking up the dark spot indicated with a white arrow yielded a contrast reversal for most dark spots, while some of them (black arrow) stayed the same. The conclusion was that such spots originate from oxygen, while the others are CO molecules.

Not only can adsorbates be manipulated, but single native substrate atoms can be removed in a controlled manner from differently coordinated sites of the substrate by using lateral manipulation techniques (Meyer et al., 1997). This ability may be of importance in gathering information about subsurface defects or to identify surface atoms with a time-of-flight analyzer (Weierstall and Spence, 1998) after transferring the atoms to the tip. Reversible lateral displacement of specific Si adatoms on the Si(111)-7 × 7 surface has been reported at low temperatures (30–175 K) (Stipe et al., 1997a). A single adatom could be reversibly displaced as an atomic switch and its position monitored with the tunneling current.

The continued miniaturization of electronic devices is leading to an increasing interest in the application of single molecules in nanoelectronics (Joachim et al., 2000). In this context, low-temperature STM is a fundamental technique to study different molecular conformations and to manipulate single molecules, bringing them in electronic contact with atomically ordered nanoelectrodes. Extension of the lateral manipulation mode from small to large molecules has been demonstrated at low temperatures (Moresco et al., 2001). Although big molecules have been positioned in a controlled way by room temperature STM (Jung et al., 1996, 1997; Gimzewski and Joachim, 1999), the stability and low noise level of the LT-STM are necessary to obtain detailed and quantitative information about the manipulation process. Complex organic molecules show a different translational movement under the influence of the STM tip than atoms or simple molecules. It has been shown that Cu-TBPP, a specially designed porphyrin-based molecule, cannot be moved at low temperatures by lateral manipulation in constant-current mode. Therefore a constant-height mode for lateral manipulation of these molecules was used, where the current signal can be used to extract information about the internal movement of the molecule under the action of the tip (Moresco et al., 2001). The central group of a $C_{90}H_{98}$ molecule (known as Lander) has been used as a model system for a molecular wire. The Lander molecule has the same molecular legs as Cu-TBTT, but it has a central polyaromatic molecular wire instead of a central porphyrin ring (Figure 17–33). The molecule was designed as a central molecular board lifted above the substrate by four legs and should act as a short (1.7-nm-long) molecular wire when contacted to an atomic step (Langlais et al., 1999). In an LT-STM study the conformational changes induced in the Lander upon adsorption on Cu(001) have been investigated (Kuntze et al., 2002). It has been found that the legs of adsorbed Lander molecules are rotated and deformed. This leg rotation results in a lowering of the central wire to

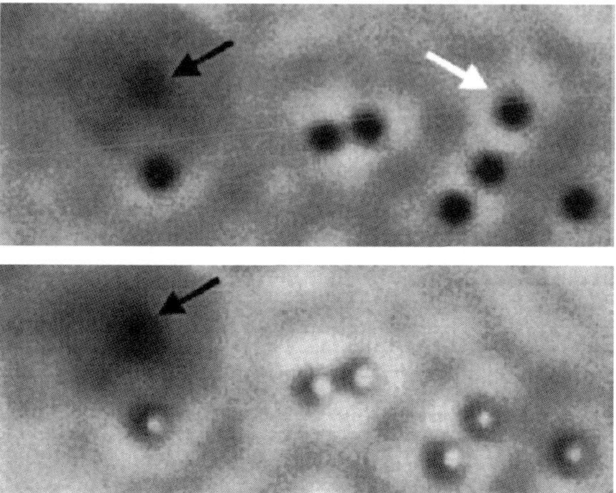

Figure 17–32. Chemical contrast with functionalized tip at 15 K. (Top) STM image of oxygen and CO on Cu(111) obtained with a clean metal tip. All adsorbates appear as indentations (dark spots) in the image. (Bottom) Same part of the surface imaged after picking up the adsorbate marked with a white arrow. All adsorbates imaged as protrusions correspond now to CO molecules, whereas the adsorbate marked by the black arrow still appears as an indentation and is therefore identified as oxygen. The fact that the appearance of CO changes when imaged with a CO molecule at the tip has been verified on a surface where only CO molecules where adsorbed. (Form Bartels et al., 1997b.)

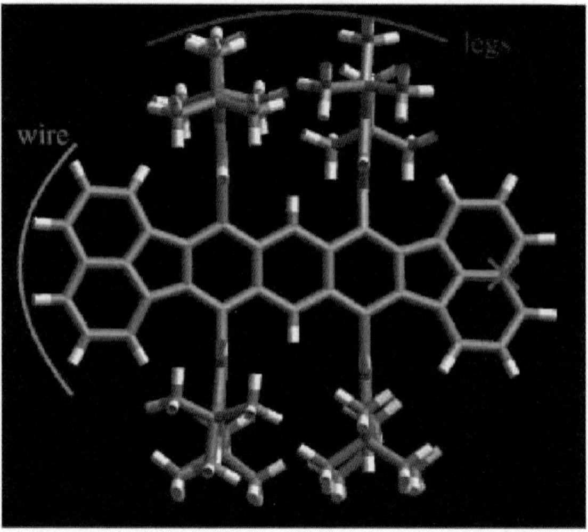

Figure 17–33. Top view of the chemical structure of the Lander molecule, which was designed to be a model system for a molecular wire. The molecular wire and the four legs that support the wire are shown. (From Moresco et al., 2003.)

0.37 nm above the surface compared to 0.7 nm for the gas phase molecule. Upon adsorption at room temperature the molecules diffuse to step edges and are stabilized with their wire parallel to step edges, which prevents good electronic contact (Kuntze et al., 2002).

The problem of controlling the electronic contact between the molecule and its electrodes has been addressed in an investigation with the Lander molecule. Reproducible contact formation was obtained by lateral manipulation of the Lander on Cu(111) with an LT-STM (Moresco et al., 2003). The molecules where adsorbed at 70 K instead of room temperature to avoid postdeposit thermal diffusion. Figure 17–34 (A3) shows an STM image of a single Lander on a terrace where the four bumps are attributed to the four legs of the molecule. When a molecule is pushed to the step with its central wire parallel to it, it reaches a final conformation imaged in Figure 17–34 (B3). Separated by the legs, the central wire is not interacting with the step edge and the standing wave pattern (LDOS oscillations) (B4) on the upper terrace has not changed. If the molecule is repositioned with its wire oriented perpendicular to the step edge, a notable modification of the standing wave pattern on the upper terrace is observed. The amplitude of the standing wave pattern is reduced at the contact point compared to the clean step edge case (4C). Simulations of the standing wave pattern indicated that the perturbation is caused by the terminal part of the molecular wire on the upper terrace. This contact area is visible in the STM image as an additional small bump (C3) and its location was confirmed by elastic-scattering quantum chemistry STM image calculations (Sautet and Joachim, 1991) (C2). The molecule could also be decontacted by reverse lateral manipulation, and the original step edge and molecule image were recovered.

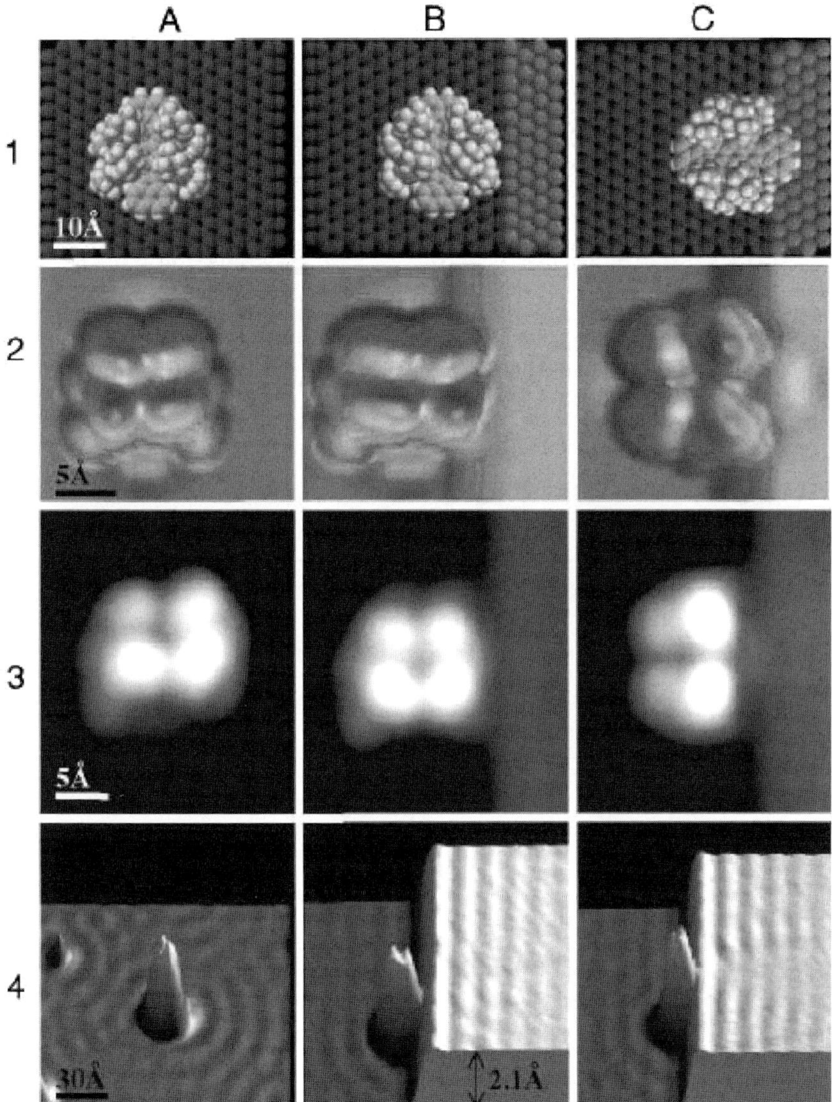

Figure 17–34. Contacting a molecular wire to a step: Lander molecules on a step-free Cu(111) surface (column A) and contacted to a (100) step. Molecular wire parallel (column B) and orthogonal (column C) to the step. Row 1: Sphere models of molecular structures. Row 2: Calculated STM images corresponding to the sphere models above. Row 3: STM images at 8 K. Row 4: STM measurements showing LDOS oscillations. In (C2) and (C3), an additional bump appears corresponding to the contact point of the wire to the step. Modification of the standing wave pattern on the upper terrace is observed only when the wire is orthogonal to the step (C4). (From Moresco et al., 2003.)

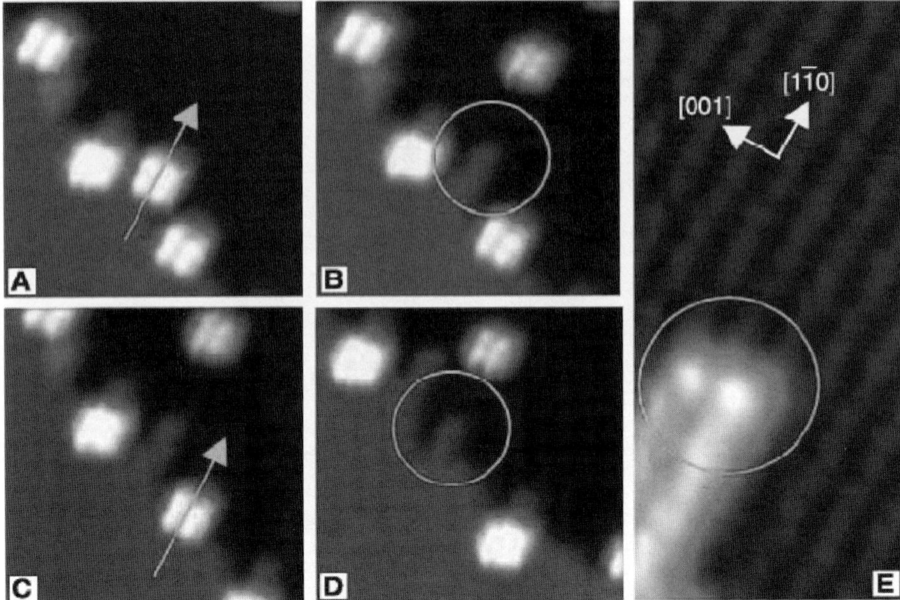

Figure 17–35. Self-assembly of nanowires at step edges initiated by Lander molecule adsorption. (A–D) Low-temperature STM images of a manipulation sequence of the Lander molecules from a step edge on Cu(110) The arrows show which molecule is being pushed aside; the circles mark the tooth-like structures that are visible on the step where the molecule was docked. (E) Zoom-in STM image showing the characteristic two-row width of the tooth-like structure after removal of a single Lander molecule. (From Rosei et al., 2002.)

In a variable-temperature STM experiment it has been found that the Lander molecule can act as a template, self-fabricating short metallic nano-structures at step edges. Lander molecules where adsorbed at room temperature on the Cu(110) surface and their conformation and anchoring at step edges were studied at 100 K. At room temperature the Cu kink atoms are highly mobile and the Lander molecule reshapes the fluctuating Cu step adatoms into tooth-like nanostructures perpendicular to the step edges. The dimension and shape of the Lander molecule form a perfect template for a double row of Cu atoms. Moving the molecule away form the step edge by lateral manipulation at low temperature revealed the underlying restructuring of the step edges (Figure 17–35). Upon adsorption of the Lander molecules at 150 K, no restructuring of the Cu step edges was observed, since the mobility of the Cu kink atoms at this temperature is not high enough for the molecular template to be effective.

At low temperatures, the short Cu nanowire acts as a sliding "rail" for the Lander molecule. By moving the molecule to the end of such a nanostructure, a model geometry can be obtained where one end of the central molecular wire is electronically connected to the metallic wire. A detailed study of the lateral manipulation of the Lander molecule along such an atomic wire has been performed with an LT-STM at 8 K (Grill et al., 2004). Lateral displacements of the molecule have been separated into monoatomic steps and it has been shown that single molecular legs can be rotated reversibly while keeping the central wire fixed. Comparison with theory confirmed that the central wire is contacted to the metallic Cu atomic wire.

In a recent LT-STM study of the Lander molecule, it has been shown that despite the fact that the legs elevate the molecular wire away from the surface, there is still an electronic interaction between the central wire and the surface states of the substrate (Gross et al., 2004). This was shown by comparing the standing wave patterns of surface-state electrons scattered off the molecule with calculated patterns taking into account scattering from different areas of the molecule.

Chemical bond formation was studied with an LT-STM (Lee and Ho, 1999). Individual iron atoms were evaporated and coadsorbed with CO molecules on an Ag(110) surface at 13 K. A CO molecule was transferred from the surface to the STM tip and bonded with an Fe atom on the surface to form Fe(CO). A second CO molecule could then be added to form $Fe(Co)_2$. This is shown in Figure 17–36. The adsorption sites of

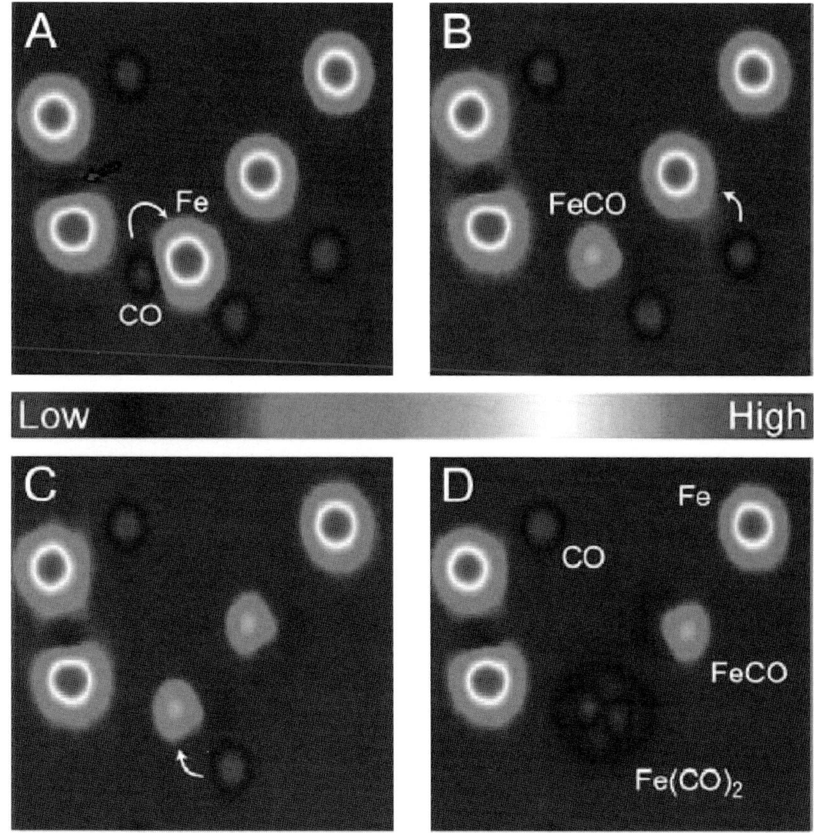

Figure 17–36. Bond formation induced with STM tip. A sequence of STM constant-current images at 13 K showing the formation of Fe–CO bonds by vertical manipulation. Fe atoms are imaged as protrusions and CO molecules as depressions. The white arrows indicate the pair of adsorbates involved in each bond formation step. In (B) and (C) a CO molecule has been picked up and bonded to an Fe atom to form Fe(CO). In (D) a second CO molecule has been bonded to Fe(CO) to form $Fe(CO)_2$. (From Lee and Ho, 1999.) (See color plate.)

the reactants could be determined by resolving the underlying Ag lattice with a CO molecule attached to the tip, which leads to increased resolution in the constant-current image. This increase in spatial resolution can be attributed to the more localized wavefunction of the molecule-terminated tip. Analysis with inelastic tunneling spectroscopy provided spectroscopic support for the identification of the created single molecule products with $Fe(CO)$ and $Fe(CO)_2$.

Assembly of an artificial nanostructure composed of a copper(II) phthalocyanine (CuPc) molecule bonded to two gold atomic chains on NiAl(110) has been realized with an LT-STM (Nazin et al., 2003b). The electronic structure of this model metal–molecule–metal junction was studied by spatially resolved STS and systematically tuned by varying the number of gold atoms in the chains. Splitting and shifting of molecular orbital energies and modification of the local electronic structure of the electrodes were observed. These effects determine the alignment of the molecular orbital energies with respect to the Fermi energy of the metal and affect the conductivity of the junction.

3.3 Local Inelastic Electron Tunneling Spectroscopy

Besides the dominant elastic electron tunneling process, for which the electron energy is equal in the initial and final state, inelastic tunneling can occur if the tunneling electrons couple to some modes ω in the tunneling junction. Figure 17–37 shows an energy diagram for $T = 0$, illustrating elastic and inelastic tunneling processes. In the case of inelastic tunneling the electron loses energy $\hbar\omega$ to a mode in the tunneling barrier. According to the Pauli exclusion principle, tunneling is possible only if the final state after the inelastic tunneling event is initially unoccupied. The bias voltage dependence of the tunneling current with inelastic tunneling is shown schematically in Figure 17–38. The elastic tunneling current increases linearly, proportional to V. As long as the bias voltage is smaller than the lowest energy mode that can be excited in the gap, inelastic tunneling processes cannot occur. At the threshold bias $V = \hbar\omega/e$, the inelastic channel opens up, and the number

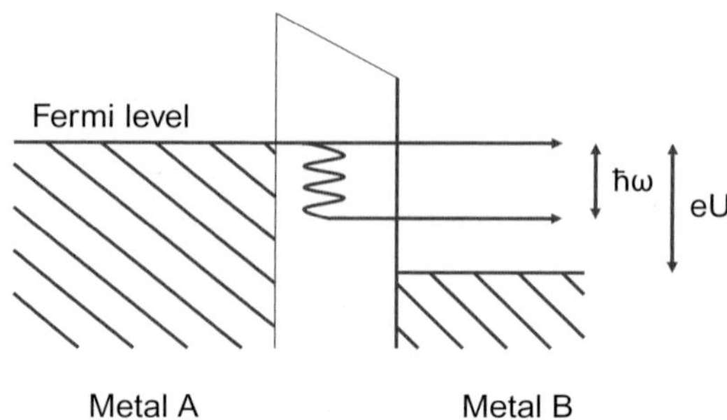

Figure 17–37. Elastic and inelastic tunneling channels. Tunneling electrons can excite a molecular vibration of energy $\hbar\omega$ only if $eU > \hbar\omega$. For smaller energies, there is no final state into which the electron can tunnel. Therefore the inelastic current has a threshold at $\hbar\omega/e$. The increase in conductance at the threshold is typically 1–10% in an STM experiment.

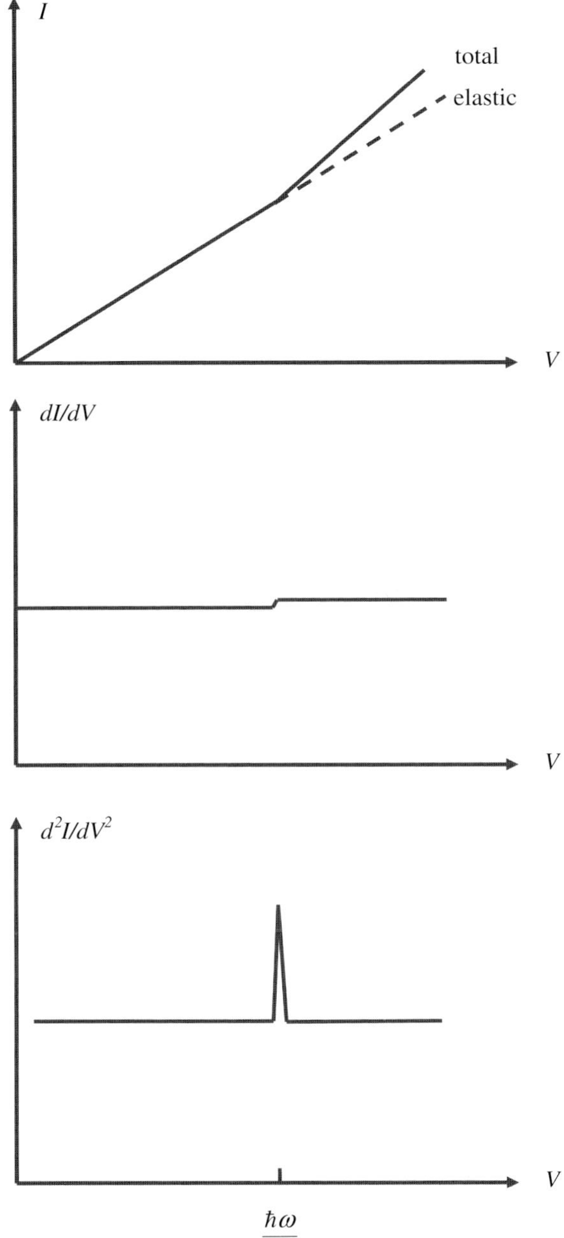

Figure 17–38. Schematic current versus voltage curves with elastic and inelastic tunneling. A kink is observed when the inelastic tunneling channel opens up. The kink becomes a step in the first derivative and a peak in the second derivative.

of electrons using this channel increases linearly with V. Therefore the total current has a kink at the threshold bias voltage. In the differential conductance curve dI/dV, the kink becomes a step and the second derivative d^2I/dV^2 exhibits a peak at the threshold. If several modes ω_i can be excited in the tunneling process, each mode leads to a peak in the differential conductance at the corresponding voltage $V_i = \hbar\omega_i/e$. Inelastic electron tunneling spectroscopy (IETS) can therefore be regarded as a special form of electron energy loss spectroscopy.

IETS has been shown to be a powerful technique for measuring the vibrational spectra of molecules that have been intentionally incorporated into a metal–oxide–metal tunneling junction (Jaklevic and Lambe, 1966). Vibrational spectroscopy can be performed with a variety of other techniques including electron energy loss spectroscopy, infrared absorption spectroscopy, Raman spectroscopy, inelastic neutron scattering, and helium atom scattering. All of these techniques have in common with IETS that they rely on macroscopic numbers of molecules to achieve detectable signal levels. The signal is therefore an average over molecules whose local environment can vary. The major drawback of traditional IETS with planar metal–oxide–metal junctions is that the molecules are buried within the junction, which is difficult to characterize microscopically.

Replacing the oxide layer by vacuum and the top planar electrode by a sharp STM metal tip has made it possible to extend IETS to single adsorbed molecules. One great advantage of performing vibrational spectroscopy with the STM is that the high spatial resolution of STM images permits changes in molecular spectra to be correlated with variations in the local environment on an atomic scale. STM-IETS was proposed as early as 1985 (Binnig et al., 1985b). Since the changes in tunneling conductance resulting from opening of additional inelastic tunneling channels are typically 0.1–1% for planar junctions and 1–10% for STM junctions, the relative stability of the tunneling current has to be better than 1% to obtain reasonable IET spectra with the STM. The physics of tunneling then dictates a tunneling gap stability of better than ~0.005 Å over the time it takes to complete one scan of the spectrum (Lauhon and Ho, 2001). Because the vibrational features are very sharp, liquid helium temperatures are required to avoid thermal broadening of the Fermi levels. Hansma (1982) estimated an effective resolution of $5.4k_BT$ (~140 mV at room temperature) for inelastic tunneling, while vibrational features are typically only a few millivolts wide. For those reasons, vibrational spectroscopy with the STM has proved difficult. First experiments probing a cluster of sorbic acid molecules adsorbed on graphite at 4 K reported large jumps in the *first* derivative spectrum instead of the expected second derivative spectrum (Smith et al., 1987). The peaks where attributed to characteristic vibrations of molecules. However, due to molecular diffusion events during the measurements, the spectra were not very reproducible and the energies of the peaks were different form those measured in bulk tunnel junctions. Reproducible single-molecule vibrational spectroscopy has been achieved only recently with an LT-STM (Stipe et al., 1998). In these landmark experiments, a Cu(100) surface was dosed with acetylene (C_2H_2) and deuterated acetylene (C_2D_2). Vibrational spectra where aquired at 8 K above single molecules with the use of a tracking scheme to position the tip at the center of the molecule, with lateral and vertical resolution of better than 0.1 and 0.01 Å, respectively. Contributions from the electronic spectrum of the tip and the substrate could be minimized by subtracting spectra taken over a clean area of the surface from the molecular spectra. The I–V curves from a single molecule and the clean surface (Figure 17–39A) show the expected linear dependence for metallic junctions. The differential conductance dI/dV shows an increase of 4.2% at 358 mV, resulting from the excitation of the C—H stretch mode (Figure 17–39B). The second derivative d^2I/dV^2 reveals a distinct peak at 358 mV (Figure 17–39C, compare with the idealized view of Figure 17–38). An isotopic shift to 266 mV was observed for deuterated acetylene and the C—D stretch mode. These values are in close agreement with those obtained by EELS for the same molecules on Cu(100). The ability to spectroscopically identify molecules with the STM makes it possible to implement chemical-sensitive STM imaging. This has been demonstrated by recording a d^2I/dV^2 map above both acetylene isotopes. When the dc voltage was fixed at 358 mV, only one of the two molecules (C_2H_2) was imaged, whereas at 266 mV, the other molecule (C_2D_2) was imaged (Figure 17–40). Thus, individual adsorbed molecules can be identified by their vibrational spectra and inelastic images. In contrast, identification and characterization of adatoms and

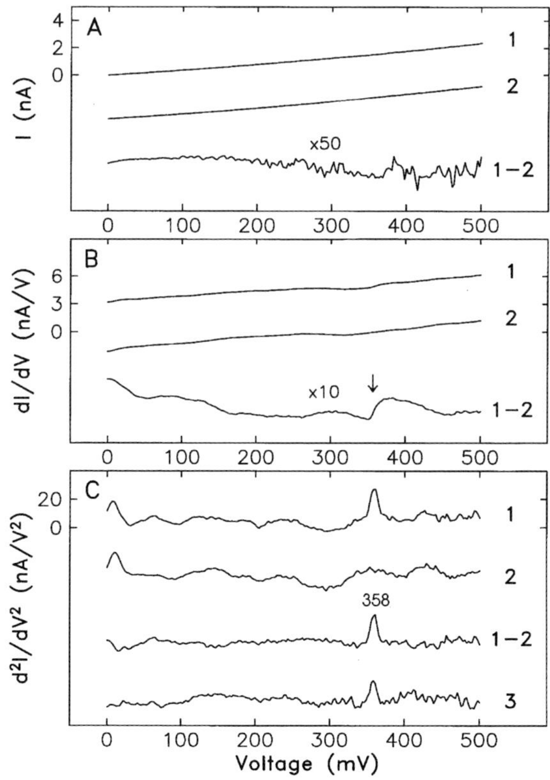

Figure 17–39. Molecular vibrational spectra observed with the STM at 8 K. (A) I–V curves recorded with the STM tip directly over the center of an acetylene molecule (1) and over the bar Cu(100) surface (2). (8) dI/dV on the molecule (1) and on the substrate (2). (C) d^2I/dV^2 on the molecule (1) and on the substrate (2). A peak at 358 mV is visible in the difference spectrum. (3) An average over 279 scans of 2 min each (10 h total data acquisition time directly above the molecule) with a different tip. The conductance change due to inelastic tunneling was 3–4% with different tips. (From Stipe et al., 1998.)

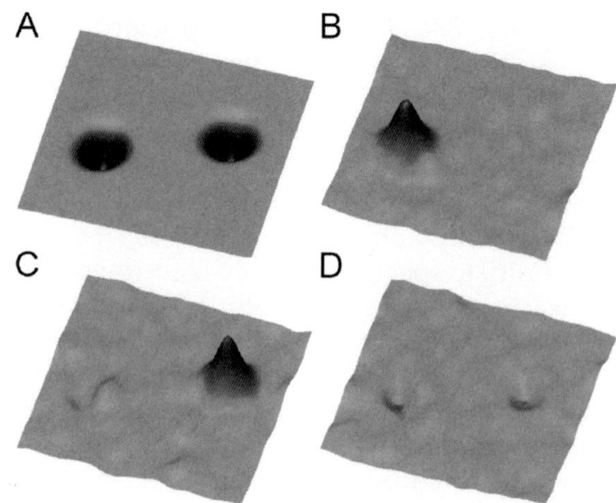

Figure 17–40. Chemical sensitive imaging. Spectroscopic spatial images of the inelastic channels for C_2H_2 and C_2D_2. (A) Constant-current image of a C_2H_2 (left) and a C_2D_2 (right) molecule. This image is an average of STM images recorded simultaneously with the vibrational images. The molecules appear identical in this normal imaging mode. d^2I/dV^2 maps (vibrational images) of the same area were recorded at (B) 358 mV, (C) 266 mV, and (D) 311 mV. In (B) only C_2H_2 is visible, whereas in (C) C_2D_2 is visible. The symmetric, round appearance of the molecules is attributed to the rotation of the molecule between two equivalent orientations during the experiment. (From Stipe et al., 1998.)

molecules by electronic spectroscopy with the STM are problematic because the electronic energy levels are broadened and shifted upon adsorption and the adsorbed molecule's spectrum becomes convolved with the STM tip's electronic spectrum (Crommie et al., 1993c).

STM-IETS has been used to determine the orientation of individual C_2HD molecules (deuterated acetylene) adsorbed on the Cu(100) surface at 8 K (Stipe et al., 1999a). By setting the bias voltage to the vibrational energy of the C—D stretch mode in C_2HD, and simultaneous recording the constant-current image and the vibrational image (d^2I/dV^2 map), the spatial distribution of the C—D stretch signal within the molecule could be determined. Since the inelastic image has its maximum near the midpoint of the C—D bond, it locates the position of the bond in this case.

The previous two examples show the ability of the LT-STM to resolve internal vibrations of molecules adsorbed on surfaces. The internal modes can be used in surface chemical analysis for the identification of adsorbed species. External vibrations, i.e., vibrations of adsorbed molecules with respect to the substrate, are more sensitive to the inter-action of the adsorbed molecule with the substrate. These external modes generally have lower energy than the internal ones, and are not easily accessed by some of the averaging vibrational spectroscopy methods mentioned above. External vibrational modes of benzene molecules on an Ag(110) surface have been detected with an LT-STM at 4 K (Pascual et al., 2001a). These measurements confirmed that the external vibrations are strongly sensitive to the nature of the molecule–substrate bond. For internal vibration modes, the inelastic signal has been shown to be very localized at the position of the particular bond excited (Stipe et al., 1999a). In contrast the spatial distribution of the inelastic tunneling, signal for external modes extends over the whole area of the molecule (Pascual et al., 2001a).

In a detailed study of temperature effects, electronic structure contributions, and tip effects on STM-IETS, Lauhon and Ho (2001) suggested that functionalization of the tip by transfer of a known molecule to the tip offers a means of accessing different vibrational modes. This has been confirmed in later experiments (Moresco et al., 1999; Hahn and Ho, 2001) where CO and C_2H_4 molecules have been transferred to the tip and single CO and O_2 molecules were probed with this tip.

There are a few caveats regarding the application and interpretation of STM-IETS spectra. In STM-IETS as in traditional IETS, there are no strict selection rules. Modes involving motion parallel and perpendicular to the surface can be excited. Modes are not observed for all molecules and not all modes are necessarily observed for any particular molecule. The symmetry of vibrational modes and electronic resonances of an adsorbate seem to give rise to selection rules for vibrational mode detection in STM-IETS (Lorente et al., 2001). A dependence of the STM-IETS signal on the molecular orientation on the surface has been shown for C_{60} on Ag(110) (Pascual et al., 2002). The spectroscopic maps showed a correlation between the enhanced vibrational signal and orientational symmetry of the adsorbed molecule observed in constant-current mode.

To complicate matters more, it has been shown that vibrational excitation can lead to suppression of elastic tunneling and produce dips instead of peaks in the differential conductance spectrum (Hahn et al., 2000). This has been recently explained theoretically by Lorente (2004). Despite this complexity, there are many advantages of STM-IETS, i.e., that the adsorbate geometry is well defined and the effect of adsorbate orientation can be studied systematically. Recent progress in the theoretical analysis of STM-IETS may greatly enhance its ability to probe chemistry at the spatial limit (Mingo and Makoshi, 2000; Makoshi and Mingo, 2002; Lorente and Persson, 2000; Lorente, 2004).

3.4 STM-Induced Photon Emission

Injection of electrons or holes form the tip of an STM into the surface leads to the emission of light for many materials. The first observation of light emitted from the tunneling junction of an STM in the low-bias tunneling regime ($eV < \Phi$ where Φ is the work function) goes back to Coombs et al. (1988). The highly localized tunneling current allows high spatial resolution that enables experiments with single nanostructures and molecules. Photon emission from the tunneling junction of an STM can be used to measure the optical properties of the sample surface in the nanometer regime. Photon emission from metals involves surface plasmons, which are inelastically excited by the tunneling electrons and then decay by the emission of photons (Persson and Baratoff, 1992). Besides these inelastic excitations, photons can also be emitted due to electron-hole recombination in semiconductors (Abraham et al., 1990). Given that the tunneling currents used for excitation are rather small—typically in the nanoampere range—the photon count rates tend to be small as well. The photon creation efficiency is typically of the order of 10^{-4}–10^{-3} photons per tunneling electron (Persson and Baratoff, 1992). Increasing the tunneling current in many cases leads to spontaneous surface modifications, which prevents reproducible experiments. Given the low photon count rates, the photon detector has to be optimized for maximum collection efficiency. A number of designs have been reported, which make use of lenses (Berndt et al., 1991b; Hoffmann et al., 2002a), mirrors (Berndt et al., 1991b; Nilius et al., 2000), transparent tips (Smolyaninov et al., 1990; Murashita, 1999), and optical fibers (Arafune et al., 2001). The emitted light can be experimentally investigated using a variety of methods (Reihl et al., 1989):

1. Isochromat spectroscopy: The photon energy is kept fixed while the bias voltage is being scanned (at constant tunneling current). This allows study of the influence of the excitation energy on the intensity of particular emission features. During the acquisition of isochromat spectra the tip–sample distance is varied by the STM feedback loop to maintain constant current. The detectors used are avalanche photodiodes or photomultipliers, where the spectral response is limited by an optical bandpass filter.

2. Fluorescence spectroscopy: The electron energy (i.e., the bias voltage) is kept fixed while the wavelength-resolved distribution of the emitted photons is measured. This involves a grating spectrometer and a cooled CCD detector. Direct information on light-emitting transitions is obtained. This mode allows the observed photon emission to be assigned to elementary processes such as interface plasmons in the tip–sample cavity or interband transitions in the sample.

3. Luminescence spectroscopy: A special case of (2) for semiconductors, where the injected electron thermalizes to the bottom of the conduction band emitting a characteristic photon.

4. Spatial mapping: A selected feature in the spectra is used to generate a high-resolution picture of the intensity of the emitted photons. Spatial maps are usually acquired simultaneously with a constant-current image to relate the spectral map to topographic features. In the case of luminescence this allows the spatial identification of defects, grain boundaries, and dopant concentrations on semiconductor surfaces similar to cathodoluminescence in electron microscopy.

STM-induced photon emission from semiconductors has been compared to cathodoluminescence (CL) (Gustafsson et al., 1998) in scanning electron microscopy, photoluminescence (PL) (Montelius et al., 1992) and inverse photoemission spectroscopy (IPS) (Reihl et al., 1989). Measuring luminescence from semiconductors locally by inducing it with an STM has distinct advantages: Electron tunneling from the tip provides a bright and extremely localized source of electrons. In addition and unique to STM measurements, holes can be injected by biasing the tip positive with respect to the sample. Comparative measurements of STM-induced photon emission in the field emission regime ($eV > \Phi$) on Si(111) with STS and normal incidence IPS demonstrated a correlation between those methods, but showed also differences due to the local nature and the strong electrostatic field of the STM measurement (Reihl et al., 1989).

STM-induced photon emission from metal surfaces involves excitation of a plasmon localized within the region near the end of the tip (Aizpurua et al., 2000). The plasmon then decays into photons, which are detected in the far field. These plasmons are excited either by inelastic tunneling, or by elastic tunneling into the substrate and subsequent thermalization via plasmon generation. Model calculations favor inelastic tunneling, which occurs in the tunneling gap as the excitation mechanism (Berndt et al., 1991a). This is in contrast to STM-induced luminescence on semiconductors, where hot electron decay within the semiconductor was found to dominate (Abraham et al., 1990). The resulting photon spectrum resulting from plasmon decay is quite broad and has a characteristic energy cutoff determined by the sample bias voltage at eV_{bias}. The spectrum is also sensitive to the geometry of the tunnel junction and the tip material (Berndt et al., 1993). Fluorescence due to hot electron thermalization, on the other hand, is expected to be insensitive to the tunneling gap conditions and produce a distinct peak (or a series of peaks) in the photon spectrum. Measurements on Cu(111) in the field emission regime found a close correlation between oscillations in the conductance, which arise from standing waves in the tip–sample gap, and oscillations in isochromat photon spectra (Berndt and Gimzewski, 1993). These findings supported the view that for metals inelastic tunneling and coupling to a tip-induced plasmon mode occurs in the tunneling gap.

Low temperature in CL (Murashita et al., 1993), PL, and IPS is known to increase the intensities of spectral features and/or reduce thermal broadening, which can obscure important details in the spectrum. This is one reason to perform STM photoemission measurements at low temperatures; the other main reason is to improve the long-term drift stability necessary to record spatial maps (Hoffmann et al., 2004) and photon spectra from single molecules (Qiu et al., 2003).

Low-temperature luminescence experiments on GaAs/AlAs using an LT-STM with a transparent tip as detector have been reported (Murashita, 1997). Isochromat spectra from individual several-atom silver chains assembled with an LT-STM on the NiAl(110) surface have been measured showing sensitivity to the number of atoms in the silver chain (Nazin et al., 2003a). The changes in photon emission with chain size were explained as a quantum size effect. Recently experiments with superconducting tips on a superconducting sample were reported that showed an energy cutoff at $2eV_{\text{bias}}$ in the isochromat photon spectrum. This could not be explained by single electron tunneling and was attributed to tunneling of Cooper pairs (Uehara et al., 2001).

The modes of measurement described above (1–4) either provide lateral or spectral resolution. Spectral and spatial resolution can be obtained simultaneously by recording a spatial map where at each point in the image a complete fluorescence spectrum is recorded. The tip motion during constant-current imaging is delayed at each pixel for a photon detection time of 0.05–1 s (depending on detector sensitivity). This mode was named spectroscopic imaging (Hoffmann et al., 2002a). From the resulting three-dimensional data set isochromat spatial maps at various wavelengths can be extracted. Moreover, changes of the sample or tip are easily recognized. The first measurements of this kind were conducted by Nishitani et al. (1998) at room temperature. The recording time for a complete spectroscopic image was 1 h. Subsequent improvements in the photon detection system enabled faster recording times and the use of an LT-STM provided less thermal drift (Hoffmann et al., 2002a). Atomically resolved isochromat spatial maps have been recorded of the Au(110) surface at 4 K where the atomic rows appear darker then the troughs (see Figure 17–41). The finding that atomic scale structures cause clear variations of the photon emission characteristics was surprising when it was first observed (Berndt et al., 1995). Theoretically the local photon emission on metals was expected to extend laterally over approximately 50 Å, which is the lateral extent of the tip-induced plasmons. The first finding of atomic scale features in STM-induced photoemission on Au(110) was explained in terms of the tip–sample distance dependence of the photon emission, which is given by the coupling to localized plasmon modes (Berndt et al., 1995). This model could not explain later measurements that showed that photon emission is reduced by a factor of 5 at Ag(111) steps, whereas it is

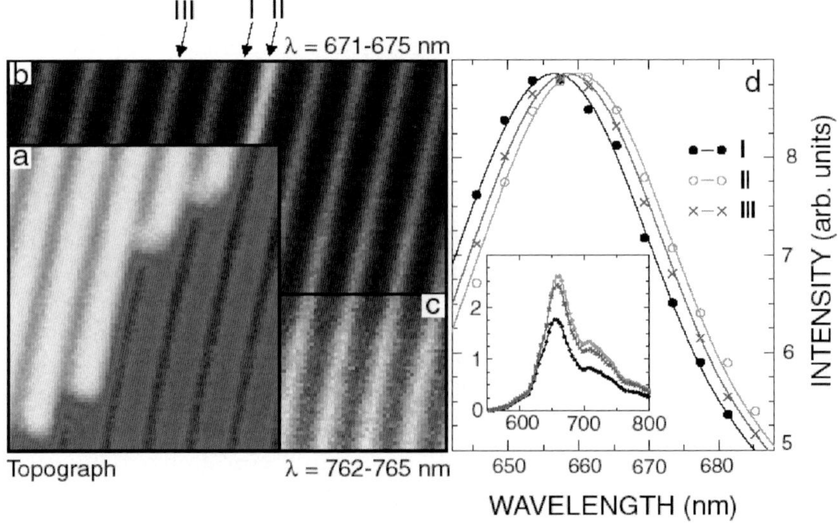

Figure 17–41. Atomic scale detail in photon maps: (a) Constant current image of a reconstructed Au(110) surface showing two terraces separated by a monoatomic step. Marker I: on top of an atomic row. Marker II: below a monoatomic step. Marker III: between atomic rows. (b and c) Simultaneously acquired photon map is in registry with the topographic data. (b) Photon wavelength λ = 671–675 nm. (c) Photon wavelength λ = 762–765 nm. (d) Representative spectra normalized to the same peak intensity. Inset: Unnormalized spectra. (From Hoffmann et al., 2004.)

increased by 50% at Au(110) steps. In the new study of atomic resolution spectroscopic imaging on Au(110) mentioned above (Hoffmann et al., 2004), a new model was presented that is consistent with the existing data. Atomic resolution on Au(110) and the contrast at steps was attributed to the local electronic structure and its effect on the elastic and inelastic tunneling channels (Hoffmann et al., 2004).

The sensitivity of STM-induced photon emission from metals to the conditions of the tunneling gap has been discussed above. New phenomena are observed when molecules are placed in this cavity. In spatial photon maps of close packed C_{60} monolayers on Au(110), individual molecules have been resolved as distinct maxima (Berndt et al., 1994). This could not be explained by photon emission from tip-induced plasmon modes alone, since on metals the emission intensity would be reduced if the tip–metal distance is increased. Therefore if the emission from the C_{60}-covered metal surface was directly due to plasmon modes, the increase in tip–metal distance due to the molecules should weaken the plasmon modes. But actually the opposite is observed. The role of the molecules in the emission process is unclear, although molecular fluorescence has been suggested.

Spatial photon maps and fluorescence spectra from single hexa-*tert*-butyl-decacyclene (HBDC) molecules on Au(111), Ag(111), and Cu(111) have been recorded at 4 K (Hoffmann et al., 2002b). Low temperature was essential for this measurement, since these molecules are mobile on the surface at room temperature. Results similar to C_{60} are reported, i.e., increased emission on the molecule, but no new spectral features attributable to molecular fluorescence were observed. Only a 4-nm shift of the HBDC photon spectrum toward shorter wavelengths was observed compared to the substrate spectrum.

On a metal surface, the electronic levels of a molecule are considerably broadened whereas light emission is strongly quenched (Barnes, 1998), making it difficult to detect molecule-specific emission. In an LT-STM study it has been shown that molecular fluorescence could be observed when the molecule (ZnEtiol) was supported on a thin aluminum oxide (Al_2O_3) film grown on an NiAl(110) surface (Qiu et al., 2003). The oxide spacer reduces the interaction between the molecule and the metal. The photon emission spectra and intensities varied with different tips and increased photon emission efficiency was observed with Ag tips compared to W tips. This has been attributed to plasmon excitation in the tunneling gap. To reduce the plasmon signature in the spectra the tips were voltage pulsed. After that, photon emission spectra with sharp features could be seen when the tip was positioned directly above a molecule. Furthermore, the spectra were very sensitive to the tip position above the molecule. Light emission has been found to be almost identical for molecules of the same conformation. This is shown in photon spectra form three different molecules (in the same conformation) measured with three different STM tips (Figure 17–42A). The spectra could be explained as a superposition of tip-induced plasmon emission (which depend on tunneling gap conditions) and molecular fluorescence (which should be nearly independent of gap conditions). Molecular fluorescence is due to electrons tunneling elasti-

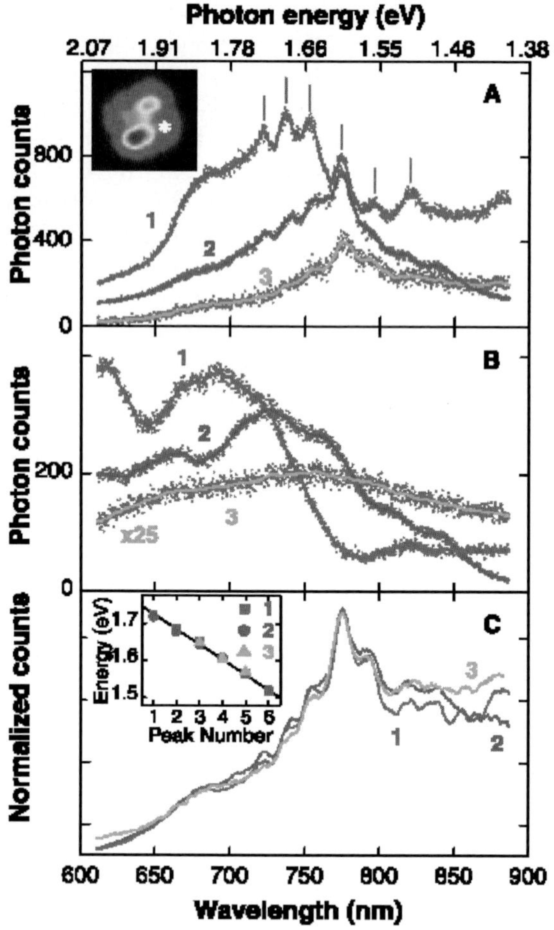

Figure 17–42. Molecular fluorescence from the excitation of vibrational modes in a single molecule. (A) Photon emission spectra for three different experiments with three different tips. The spectra where obtained over a single ZnEtiol molecule shown in the inset (constant current image) at the marked spot. The raw data are plotted together with the smoothed curves. Spectra 1 and 2 were taken with two different Ag tips and have been offset for clarity. Spectrum 3 was obtained with a W tip. The differences in the spectra are due to different tip plasmon properties. (B) NiAl light emission spectra measured with the same tips and voltages as in (A). (C) Molecular spectra from (A) divided by the corresponding NiAl spectra from (B) show remarkable similarity. The inset shows the photon energy of each peak determined for the three spectra. (From Qiu et al., 2003.)

cally into the molecule and creating excited vibrational states that make a radiative transition. To eliminate the influence of plasmons from the spectra, they have been divided by the spectra taken from the NiAl surface (Figure 17–42B) with the same tips. The resulting curves (Figure 17–42C) show remarkable similarity, indicating that the molecules have nearly identical light emission properties independent of the tip. The vibrational features were equidistant in energy with spacing of 40 ± 2 meV (Figure 17–42, inset).

A similar picture for the emission process has been used to explain the photon emission measured with an LT-STM from quantum well structures formed by Na layers on a Cu(111) surface (Hoffmann et al., 2001). The photon emission in this experiment was assigned to transitions between the electronic levels of the quantum structure and was observed together with tip-induced plasmon emission.

Recently another experiment with a room temperature STM has shown molecular fluorescence from organic molecules decoupled from an Au(100) surface by several adsorbed layers of the same molecule (Dong et al., 2004). Molecular fluorescence peaks in the spectra became sharper with increased coverage while the plasmon-related band was suppressed. Due to the higher thermal drift at room temperature, the spectra in this observation were averaged over several molecules.

Weakness of interaction between a molecule and the substrate turns out to be essential for the observation of STM-induced molecular fluorescence. STM-induced fluorescence combined with imaging and STS can be used to probe the interdependence between conformational structure, energy levels, and optical properties of single molecules. The inherent stability of the LT-STM together with the suppressed molecular diffusion at low temperatures should enable the study of electron dynamics in organic molecules with submolecular resolution.

3.5 Spin-Polarized Scanning Tunneling Microscopy

STM can yield information about magnetic properties by use of spin-polarized tunneling between a magnetic tip and substrate. The spin valve effect predicts that the tunneling current depends on the relative orientation of the magnetic moments of the tunneling electrodes (Julliere, 1975; Slonczewski, 1988). A magnetic tip acts as a source of spin-polarized electrons, probing the spin-split density of states of the magnetic sample. This technique allows imaging with atomic spatial resolution and, like conventional STM, is mostly sensitive to the topmost atomic layer. The ability to probe topography, crystallography, and magnetism at the same time renders spin-polarized scanning tunneling microscopy (SP-STM) a very powerful tool for the investigation of magnetic surfaces and monolayers. Since many magnetic phenomena exist only below a certain critical temperature, it is advantageous to use a low- or variable-temperature STM to observe magnetic materials. In addition, measurements on low-dimensional magnetic systems like ultrathin films or superparamagnetic particles demand low temperatures, since the Curie temperature T_c in general scales with dimension. In addition to those advantages of low temperatures specific to SP-STM, all other advantages mentioned in the introduction apply.

Wiesendanger et al. (1990) explored SP-STM using CrO_2 tips on an antiferromagnetic Cr(001) surface. These early experiments represented a mixture between topographic and spin-dependent contrast. Discriminating between magnetically caused contrast and contrast from other features of the electronic density of states near the Fermi level requires images taken with different tips (magnetic and nonmagnetic). An experimental setup for an LT-STM, which uses an external magnetic field combined with the ability to rotate the sample without changing the tip position, has been described (Wittneven et al., 1997). This setup enables the determination of the relative magnetic orientation between the end of the tip and the sample.

Other magnetic imaging techniques and their respective lateral resolution are magnetooptical Kerr effect (MOKE) microscopy (~300 nm), MOKE with scanning near-field optical microscopy (SNOM) (~50 nm), scanning electron microscopy with polarization analysis (SEMPA) (~40 nm), magnetic force microscopy (MFM) (~20 nm), X-ray magnetic linear dichroism photoelectron emission microscopy (XMLD-PEEM) (~20 nm), off-axis electron holography, and Lorentz microscopy for thin films (~20 nm). SP-STM has proven to be capable of atomic resolution, which has been shown for the first time on a manganese monolayer on W(110) at 16 K (Heinze et al., 2000). A two-dimensional antiferromagnetic structure as predicted by theory with a periodicity of 4.5 Å has been observed as shown in Figure 17–43. This is a considerable advance, considering that previous characterizations of antiferromagnetic domains could not go beyond micrometer resolution. Spectroscopic studies with SP-STM allow the correlation of structural, local electronic, and local magnetic properties down to the atomic level. This has been demonstrated in low-temperature studies of ferromagnetic rare earth (Bode et al., 1998) and transition metal (Pietzsch et al., 2000a) systems as well as antiferromagnets (Kleiber et al., 2000). The tips used in these experiments are nonmagnetic tips coated with a thin layer of ferromagnetic (Kleiber et al., 2000) or antiferromagnetic (Kubetzka et al., 2002) material. Fe-coated tips are magnetized perpendicular to the tip axis at the apex and are therefore sensitive to the in-plane component of the sample's magnetization (Bode et al., 1998, Kleiber et al., 2000). FeGd-coated tips exhibit a perpendicular magnetic anisotropy and are therefore sensitive to the out-of-plane component of the sample's magnetization (Kubetzka et al., 2002).

If the LT-STM is built completely from nonmagnetic materials, SP-STM can be performed in the presence of high magnetic fields, which enables observation of hysteresis on a nanometer scale (Pietzsch et al., 2001). The internal spin structure of magnetic vortex states on three-dimensional Fe islands on W(110) has been resolved recently with an LT-STM, as shown in Figure 17–44. A Cr-coated tip was used to map out both the curling in plane magnetization around the vortex core as well as the perpendicular magnetization within the vortex core (see Figure 17–45). The width of the core was determined as 9 ± 1 nm in agreement with theory.

The extreme surface sensitivity of the SP-STM (as of all types of STM) may be its only weakness with regard to the investigation of coupling phenomena. Possible changes in the magnetic configuration of buried layers therefore are hidden from the measurement. These can be analyzed with XMLD-PEEM (Scholl, 2003), which, because of its elemental specificity and relatively long probing depth (3–5 nm), is able to investigate layered systems at more modest spatial resolution. However, SP-STM is unrivaled for the investigation of the magnetic structure of magnetic monolayers, surfaces, or surface alloys at atomic resolution.

Figure 17–43. Atomic resolution magnetic imaging with SP-STM. (A) Constant-current image of one monolayer of Mn on W(110) recorded with a non-magnetic W tip at 16 K. (B) Image recorded with a magnetic Fe tip showing an antiferromagnetic configuration as predicted by theory. The colored insets show calculated STM images. (C) Experimental and theoretical line sections from (A) and (B). The image size is 2.7 nm by 2.2 nm. (From Heinze et al., 2000.) (See color plate.)

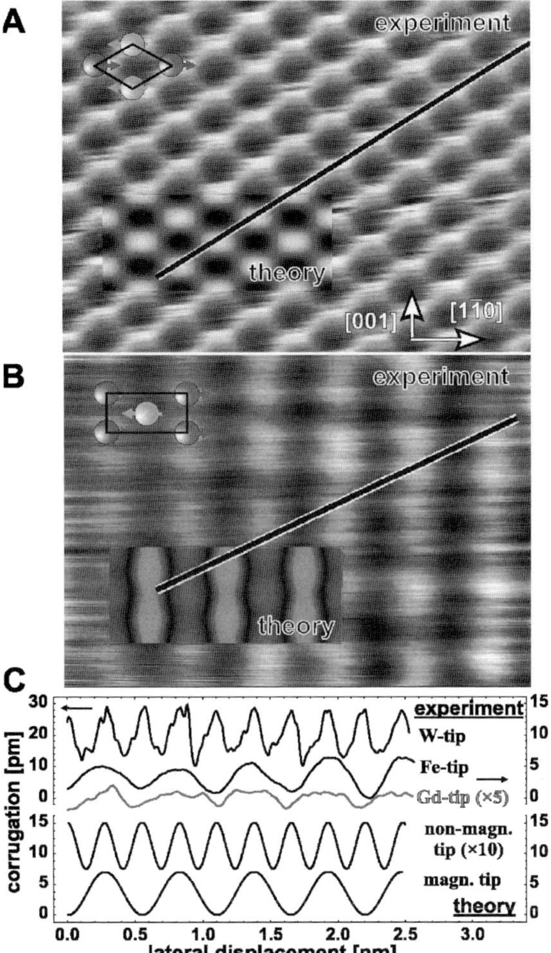

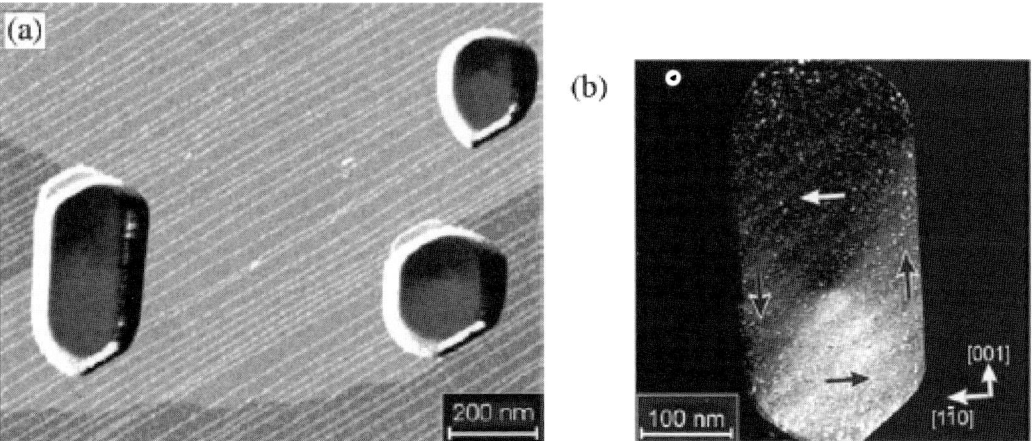

Figure 17–44. Magnetic vortex states on Fe islands imaged with SP-STM. (a) Spin-resolved dI/dV map of seven monolayers Fe on W(110) measured at 16 K with a Cr-coated tip being sensitive to the in-plane component of the sample magnetization. The islands exhibit a magnetic vortex state, as visible in (b). (From Wiesendanger et al., 2004.)

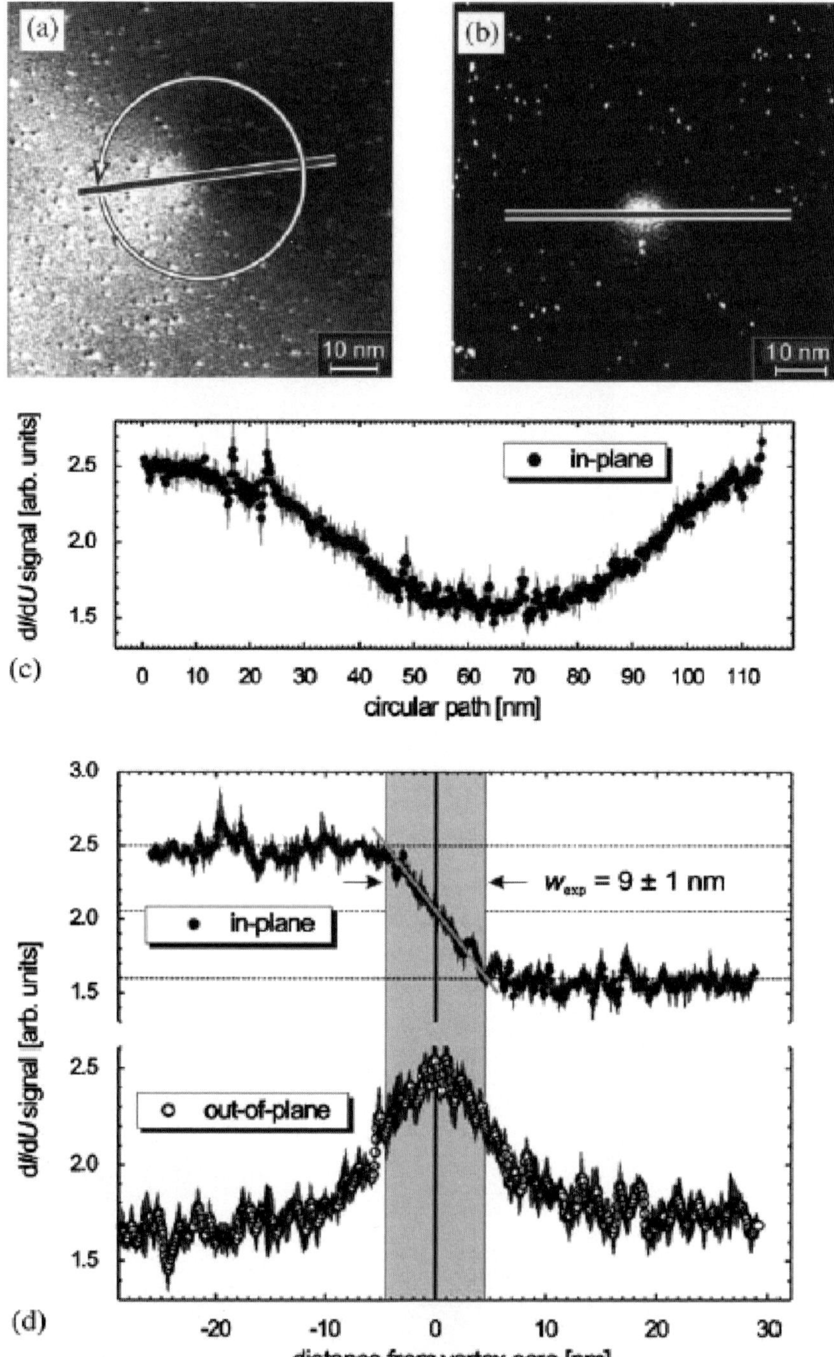

Figure 17–45. Magnetic vortex states on Fe islands at imaged higher magnification. Spin resolved dI/dV maps measured with an (a) in-plane and (b) out-of-plane sensitive Cr-coated tip. The curling in plane magnetization around the vortex core is visible in (a), whereas the magnetization component in the direction of the surface normal is seen in (b) as a bright area. (c) Spin-resolved dI/dV signal measured around the circle in (a) indicates that the in-plane component continuously curls around the core. (d) Spin-resolved dI/dV signal measured along the lines in (a) and (b) showing a vortex core width of 9 ± 1 nm. (From Wiesendanger et al., 2004.)

3.6 Superconductivity

The technique of tunneling was initially used as a tool to understand the bulk properties of superconductors. Electron tunneling through thin insulating films into superconductors is used to study how the density of states in a superconductor is modified. This technique averages over a macroscopic area and requires a high-quality insulating tunneling barrier. By using an STM for these measurements, a controllable vacuum barrier is used instead and superconductors can be studied, which do not support the formation of good, high-quality insulating films. Moreover, tunneling takes place over only a small area and variations in the density of states can be studied on an atomic scale.

According to BCS theory, the superconducting state is a new ground state where electrons of opposite momentum and spin form pairs. Any excitation of this ground state requires a Cooper pair to be broken into two single particle excitations. A minimum energy of Δ, the energy gap, is required to create one of these quasiparticles. If an electron tunnels from a metallic STM tip into the superconductor, it will enter into one of these quasiparticle states. Therefore the tunneling current provides a direct measure of the density of these states. Superconductors expel a magnetic field up to a certain critical field (Meissner effect). For larger fields, superconductivity is destroyed. In type II superconductors a mixed phase exists for certain magnetic field strengths, consisting of hexagonal lattices of normal conducting vortices (flux lines) in the superconductor, the so-called Abrikosov vortex lattice. Many experiments have confirmed the existence of these flux lines, e.g., electron microscopy (Bosch et al., 1985) and electron holography (Matsuda et al., 1989; Harada et al., 1992; Bonevich et al., 1993). These techniques are sensitive to the magnetic field variations. STM, in contrast, is sensitive to the electronic structure. The first LT-STM observation of flux lines and the flux lattice has been reported by Hess (Hess, 1991; Hess et al., 1989, 1990a,b, 1991). Surfaces of $NbSe_2$ have been studied in a ultralow temperature STM in a temperature range from 50 mK to 7 K. Differential conductance spectra in zero applied field showed the development of the superconducting energy gap below 7.2 K, which opens up to about 1.1 meV at the lowest temperatures. A second discontinuity was observed in the density of states at ± 34 mV. This feature resulted from the charge density wave (CDW) gap, since this material also supports a CDW state. Differential conductance spectra taken at 0 mV from a single vortex core at an applied field of 500 G showed the normal conducting vortex as a star pattern, which was explained as resulting from the 6-fold anisotropy of the atomic crystalline band structure. At 0 mV no state should exist in the superconductor, since this is the middle of the superconducting gap. Therefore the normal conducting vortex core was imaged bright on a dark background. At higher fields, the vortices organize into a hexagonal lattice. Figure 17–46 shows this lattice at a field of 10 kG. The dI/dV map at 1.3 mV was recorded under constant current feedback conditions with 1.3 mV bias. The question arises whether the localized current from the STM tip is sufficiently large to cause local perturbations such as heating, exceeding the local critical

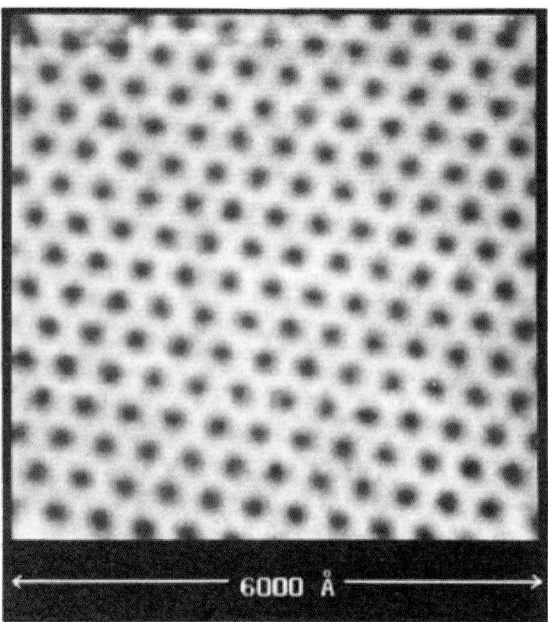

Figure 17–46. Abrikosov flux lattice imaged by LT-STM. Differential conductance map at 1.3 mV (above the superconducting gap) recorded on $NbSe_2$ at 1.8 K and 10 kG magnetic field. An enhanced density of states is observed away from the core and the density of states decreases at the vortex center. (From Hess et al., 1989. Reproduced by permission of Bell Labs, Lucent Technologies Inc.)

current or creating other nonequilibrium effects in the superconductor. No such current-dependent effects could be observed.

LT-STM provides a novel probe of inhomogeneous superconducting structures, of which the normal metal-to-superconductor planar interface is the most basic configuration. If the electrical contact between the two metals is good, superconductivity is weakened in the superconductor and induced in the normal metal. The phenomenon is known as the proximity effect. LT-STM has been used as a local probe of the superconducting proximity effect across a normal metal–superconductor interface of a short coherence length superconductor (Tessmer et al., 1996). Both the topography and the local electronic density of states were measured at 1.6 K on a superconducting $NbSe_2$ crystal decorated with nanometer-sized Au islands. A quasiparticle bound state was observed not only in the Au islands, but even when tunneling directly into the $NbSe_2$, which is evidence for a significant reduction of the superconductivity inside the $NbSe_2$ induced by the proximity of the Au overlayer. Local tunneling spectroscopy for an Nb/In/As/Nb superconducting proximity system was demonstrated with a low-temperature scanning tunneling microscope at 4.2 K (Inoue and Takayanagi, 1991). It was found that the local electron density of states in the InAs region is spatially modulated by the neighboring superconductor.

The local effects of individual Zn impurity atoms in the high-temperature superconductor $Bi_2Sr_2CaCu_2O_{8+\delta}$ (BSCCO) have been studied with an LT-STM at 4.2 K (Figure 17–47) (Pan et al., 2000). Recording a dI/dV map at zero bias, i.e., in the superconducting gap, revealed the Zn impurity scattering centers as bright dots. This is an image of the quasiparticle density of states in the vicinity of the impurity atoms. Imaging of the spatial dependence of the quasiparticle density of states in the vicinity of the impurity atoms reveals a four-fold symmetric quasiparticle "cloud" aligned with the nodes of the d-wave supercon-

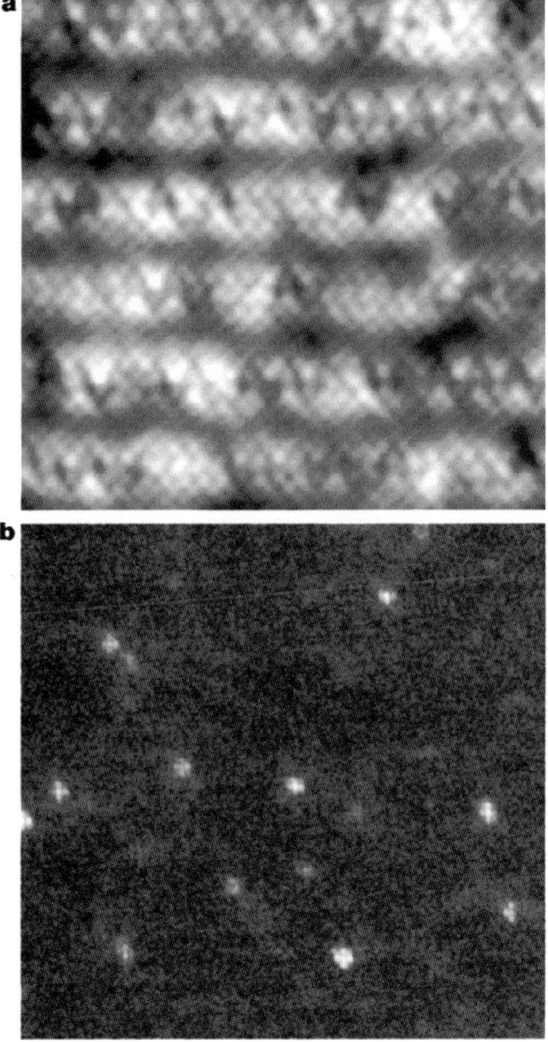

Figure 17–47. Constant current image (a) and zero-bias differential conductance map (b) of a cleaved single crystal of BSCCO at 4 K. (a) The surface BiO plane is exposed after cleavage. The atoms form a supermodulation along the crystal b-axis. (b) A map of the DOS at the Fermi energy ($V = 0$) Most of the image is dark, which is due to a low quasiparticle DOS at the Fermi level as expected for a superconductor far below T_c. The Zn impurities show up as bright features ~15 Å in extent, each exhibiting a cross-shaped structure. (From Pan et al., 2000.)

Figure 17–48. Quasiparticle density of states around a Zn impurity on the high T_c superconductor BSCCO: Relationship between the position of the Bi atoms on the crystal surface (lower layer), and the resonant DOS structure at the Zn atom (upper layer). Both layers where simultaneously acquired; the lower layer is a 6 × 6-nm high-spatial-resolution image of the topography of the exposed BiO plane and the upper layer is the differential conductance at –1.5 mV bias voltage. The bright center of the scattering resonance coincides with the position of a Bi atom on the exposed surface BiO plane. The Zn scattering center is sitting on the CuO_2 plane two layers below. The inner bright cross in the quasiparticle DOS around the impurity (green bumps around the center) is aligned with the nodes of the d-wave superconducting gap. The weaker outer features, oriented at 45° to the inner cross, are aligned with the gap maxima. (From Pan et al., 2000.)

ducting gap, which is believed to characterize superconductivity in these materials (Figure 17–48).

References

Abraham, D.L., Veider, A., Schonenberger, C., Meier, H.P., Arent, D.J. and Alvarado, S.F. (1990). *App. Phys. Lett.* **56** 1564–1566.

Aizpurua, J., Apell, S.P. and Berndt, R. (2000). *Phys. Rev. B*, **62**, 2065–2073.

Arafune, R., Sakamoto, K., Meguro, K., Satoh, M., Arai, A. and Ushioda, S. (2001). *JPN J. App. Phys. Part 1* **40**, 5450–5453.

Barnes, W.L. (1998). *J. Mod. Opt.* **45**, 661–699.

Bartels, L., Meyer, G. and Rieder, K.H. (1997a). *Phys. Rev. Lett.* **79**, 697–700.

Bartels, L., Meyer, G. and Rieder, K.H. (1997b). *App. Phys. Lett.* **71**, 213–215.

Bartels, L., Wolf, M., Klamroth, T., Saalfrank, P., Kuhnle, A., Meyer, G. and Rieder, K.H. (1999). *Chem. Phys. Lett.* **313**, 544–552.

Becker, T., Hovel, H., Tschudy, M. and Reihl, B. (1998). *App. Phys. a—Mat. Sci. Proc.* **66**, S27–S30.

Behler, S., Rose, M.K., Dunphy, J.C., Ogletree, D.F., Salmeron, M. and Chapelier, C. (1997). *Rev. Sci. Inst.* **68**, 2479–2485.

Berndt, R. and Gimzewski, J.K. (1993). *Ann. Phys.* **2**, 133–140.

Berndt, R., Gimzewski, J.K. and Johansson, P. (1991a). *Phys. Rev. Lett.* **67**, 3796–3799.

Berndt, R., Schlittler, R.R. and Gimzewski, J.K. (1991b). *J. Va. Sci. Tech. B* **9**, 573–577.

Berndt, R., Gimzewski, J.K. and Johansson, P. (1993). *Phys. Rev. Lett.* **71**, 3493–3496.

Berndt, R., Gaisch, R., Schneider, W.D., Gimzewski, J.K., Reihl, B., Schlittler, R.R. and Tschudy, M. (1994). *Surface Sci.* **309**, 1033–1037.

Berndt, R., Gaisch, R., Schneider, W.D., Gimzewski, J.K., Reihl, B., Schlittler, R.R. and Tschudy, M. (1995). *Phys. Rev. Lett.* **74**, 102–105.

Bertel, E. (1997). *Phys. Status Sol. a—App. Res.* **159**, 235–242.

Bertel, E., Roos, P. and Lehmann, J. (1995). *Phys. Rev. B* **52**, 14384–14387.

Binnig, G. and Rohrer, H. (1982). *Helvetica Phys. Acta* **55**, 726–735.

Binnig, G., Frank, K.H., Fuchs, H., Garcia, N., Reihl, B., Rohrer, H., Salvan, F. and Williams, A.R. (1985a). *Phys. Rev. Lett.* **55**, 991–994.

Binnig, G., Garcia, N. and Rohrer, H. (1985b). *Phys. Rev. B* **32**, 1336–1338.

Bode, M., Getzlaff, M. and Wiesendanger, R. (1998). *Phys. Rev. Lett.* **81**, 4256–4259.

Bonevich, J.E., Harada, K., Matsuda, T., Kasai, H., Yoshida, T., Pozzi, G. and Tonomura, A. (1993). *Phys. Rev. Lett.* **70**, 2952–2955.

Bosch, J., Gross, R., Koyanagi, M. and Huebener, R.P. (1985). *Phys. Rev. Lett.* **54**, 1448–1451.

Bott, M., Michely, T. and Comsa, G. (1995). *Rev. Sci. Instr.* **66**, 4135–4139.

Braun, K.-F. (2000). PhD Thesis. Freie Universitaet, Berlin.

Braun, K.F. and Rieder, K.H. (2002). *Phys. Rev. Lett.* **88**, 096801.

Burgi, L., Jeandupeux, O., Hirstein, A., Brune, H. and Kern, K. (1998). *Phys. Rev. Lett.* **81**, 5370–5373.

Burgi, L., Jeandupeux, O., Brune, H. and Kern, K. (1999). *Phys. Rev. Lett.* **82**, 4516–4519.

Burgi, L., Brune, H., Jeandupeux, O. and Kern, K. (2000a). *J. Electron Spectrosc. Rel. Phenomena* **109**, 33–49.

Burgi, L., Petersen, L., Brune, H. and Kern, K. (2000b). *Surface Sci.* **447**, L157–L161.

Coombs, J.H., Gimzewski, J.K., Reihl, B., Sass, J.K. and Schlittler, R.R. (1988). *J. Microsc.—Oxf.* **152**, 325–336.

Crampin, S., Boon, M.H. and Inglesfield, J.E. (1994). *Phys. Rev. Lett.* **73**, 1015–1018.

Crommie, M.F., Lutz, C.P. and Eigler, D.M. (1993a). *Science* **262**, 218–220.

Crommie, M.F., Lutz, C.P. and Eigler, D.M. (1993b). *Nature* **363**, 524–527.

Crommie, M.F., Lutz, C.P. and Eigler, D.M. (1993c). *Phys. Rev. B* **48**, 2851–2854.

Datta, S. (1995). *Electronic Transport in Mesoscopic Systems* (Cambridge University Press, Cambridge).

Davis, L.C., Everson, M.P., Jaklevic, R.C. and Shen, W.D. (1991). *Phys. Rev. B* **43**, 3821–3830.

Dong, Z.C., Guo, X.L., Trifonov, A.S., Dorozhkin, P.S., Miki, K., Kimura, K., Yokoyama, S. and Mashiko, S. (2004). *Phys. Rev. Lett.* **92**, 086801.

Eigler, D.M. and Schweizer, E.K. (1990). *Nature* **344**, 524–526.

Eigler, D.M., Weiss, P.S., Schweizer, E.K. and Lang, N.D. (1990). *Abstr. Pap. Amer. Chem. Soc.* **200**, 85–PHYS.

Eigler, D.M., Lutz, C.P. and Rudge, W.E. (1991). *Nature* **352**, 600–603.

Eiguren, A., Hellsing, B., Reinert, F., Nicolay, G., Chulkov, E.V., Silkin, V.M., Hufner, S. and Echenique, P.M. (2002). *Phys. Rev. Lett.* **88**, 066805.

Ferris, J.H., Kushmerick, J.G., Johnson, J.A., Youngquist, M.G.Y., Kessinger, R.B., Kingsbury, H.F. and Weiss, P.S. (1998). *Rev. Sci. Instr.* **69**, 2691–2695.

Fiete, G.A., Hersch, J.S., Heller, E.J., Manoharan, H.C., Lutz, C.P. and Eigler, D.M. (2001). *Phys. Rev. Lett.* **86**, 2392–2395.

Foley, E.T., Kam, A.F., Lyding, J.W. and Avouris, P. (1998). *Phys. Rev. Lett.* **80**, 1336–1339.

Friedel, J. (1958). *Nuovo Cimento* **7**, 287.

Gaisch, R., Gimzewski, J.K., Reihl, B., Schlittler, R.R., Tschudy, M. and Schneider, W.D. (1992). *Ultramicroscopy* **42**, 1621–1626.

Garcia, N. and Serena, P.A. (1995). *Surface Sci.* **330**, L665–L667.

Gartland, P.O. and Slagsvold, B.J. (1975). *Phys. Rev. B* **12**, 4047–4058.

Gimzewski, J.K. and Joachim, C. (1999). *Science* **283**, 1683–1688.

Grill, L., Moresco, F., Jiang, P., Joachim, C., Gourdon, A. and Rieder, K.H. (2004). *Phys. Rev. B* **69**, 035416

Gross, L., Moresco, F., Savio, L., Gourdon, A., Joachim, C. and Rieder, K.H. (2004). *Phys. Rev. Lett.* **93**.

Gustafsson, A., Pistol, M.E., Montelius, L. and Samuelson, L. (1998). *J. App. Phys.* **84**, 1715–1775.

Hahn, J.R. and Ho, W. (2001). *Phys. Rev. Lett.* **8719**, art. no. 196102.

Hahn, J.R., Lee, H.J. and Ho, W. (2000). *Phys. Rev. Lett.* **85**, 1914–1917.

Hamers, R.J. (1989). *Ann. Rev. Phys. Chem.* **40**, 531–559.

Hansma, P. (1982). *Tunneling Spectroscopy: Capabilities, Applications and New Techniques* (Plenum Press, New York).

Harada, K., Matsuda, T., Bonevich, J., Igarashi, M., Kondo, S., Pozzi, G., Kawabe, U. and Tonomura, A. (1992). *Nature* **360**, 51–53.

Harrell, L.E. and First, P.N. (1999). *Rev. Sci. Instr.* **70**, 125–132.

Heimann, P., Neddermeyer, H. and Roloff, H.F. (1977). *J. Phys. C-Sol. State Phys.* **10**, L17–L22.

Heinze, S., Bode, M., Kubetzka, A., Pietzsch, O., Nie, X., Blugel, S. and Wiesendanger, R. (2000). *Science* **288**, 1805–1808.

Heller, E.J., Crommie, M.F., Lutz, C.P. and Eigler, D.M. (1994). *Nature* **369**, 464–466.

Hess, H.F. (1991). *Physica C* **185**, 259–263.

Hess, H.F., Robinson, R.B., Dynes, R.C., Valles, J.M. and Waszczak, J.V. (1989). *Phys. Rev. Lett.* **62**, 214–216.

Hess, H.F., Robinson, R.B., Dynes, R.C., Valles, J.M. and Waszczak, J.V. (1990a). *J. Vac. Sci. Tech. a—Vac. Surfaces Films* **8**, 450–454.

Hess, H.F., Robinson, R.B. and Waszczak, J.V. (1990b). *Phys. Rev. Lett.* **64**, 2711–2714.

Hess, H.F., Robinson, R.B. and Waszczak, J.V. (1991). *Physica B* **169**, 422–431.

Hewson, A.C. (1993). *The Kondo Problem to Heavy Fermions* (Cambridge University Press, Cambridge).

Himpsel, F.J., Ortega, J.E., Mankey, G.J. and Willis, R.F. (1998). *Adv. Phys.* **47**, 511–597.

Hoffmann, G., Kliewer, J. and Berndt, R. (2007). *Phys. Rev. Lett.* **87**, 176803.

Hoffmann, G., Kroger, J. and Berndt, R. (2002a). *Rev. Sci. Instr.* **73**, 305–309.

Hoffmann, G., Libioulle, L. and Berndt, R. (2002b). *Phys. Rev. B* **65**, 212107

Hoffmann, G., Maroutian, T. and Berndt, R. (2004). *Phys. Rev. Lett.* **93**, 076102.

Hofmann, P., Briner, B.G., Doering, M., Rust, H.P., Plummer, E.W. and Bradshaw, A.M. (1997). *Phys. Rev. Lett.* **79**, 265–268.

Horch, S., Zeppenfeld, P., David, R. and Comsa, G. (1994). *Rev. Sci. Inst.* **65**, 3204–3210.

Hormandinger, G. (1994). *Phys. Rev. Lett.* **73**, 910.

Inoue, K. and Takayanagi, H. (1991). *Phys. Rev. B* **43**, 6214–6215.

Jaklevic, R.C. and Lambe, J. (1966). *Phys. Rev. Lett.* **17**, 1139.

Jeandupeux, O., Burgi, L., Hirstein, A., Brune, H. and Kern, K. (1999). *Phys. Rev. B* **59**, 15926–15934.

Joachim, C., Gimzewski, J.K. and Aviram, A. (2000). *Nature* **408**, 541–548.

Julliere, M. (1975). *Phys. Lett. A* **54**, 225–226.

Jung, T.A., Schlittler, R.R., Gimzewski, J.K., Tang, H. and Joachim, C. (1996). *Science* **271**, 181–184.

Jung, T.A., Schlittler, R.R. and Gimzewski, J.K. (1997). *Nature* **386**, 696–698.

Kleiber, M., Bode, M., Ravlic, R. and Wiesendanger, R. (2000). *Phys. Rev. Lett.* **85**, 4606–4609.

Kliewer, I., Berndt, R., Chulkov, E.V., Silkin, V.M., Echenique, P.M. and Crampin, S. (2000). *Science* **288**, 1399–1402.

Knorr, N., Schneider, M.A., Diekhoner, L., Wahl, P. and Kern, K. (2002). *Phys. Rev. Lett.* **88**, 096804.

Kroger, J., Limot, L., Jensen, H., Berndt, R. and Johansson, P. (2004). *Phys. Rev. B* **70**, 033401.

Kubetzka, A., Bode, M., Pietzsch, O. and Wiesendanger, R. (2002). *Phys. Rev. Lett.* **88**, 057201

Kugler, M., Renner, C., Fischer, O., Mikheev, V. and Batey, G. (2000). *Rev. Sci. Inst.* **71**, 1475–1478.

Kuntze, J., Berndt, R., Jiang, P., Tang, H., Gourdon, A. and Joachim, C. (2002). *Phys. Rev. B* **65**, 233405.

Langlais, V.J., Schlittler, R.R., Tang, H., Gourdon, A., Joachim, C. and Gimzewski, J.K. (1999). *Phys. Rev. Lett.* **83**, 2809–2812.

Lau, K.H. and Kohn, W. (1978). *Surface Sci.* **75**, 69–85.

Lauhon, L.J. and Ho, W. (2001). *Rev. Sci. Inst.* **72**, 216–223.

Lee, H.J. and Ho, W. (1999). *Science* **286**, 1719–1722.

Lemay, S.G., Janssen, J.W., van den Hout, M., Mooij, M., Bronikowski, M.J., Willis, P.A., Smalley, R.E., Kouwenhoven, L.P. and Dekker, C. (2001). *Nature* **412**, 617–620.

Li, J.T., Schneider, W.D. and Berndt, R. (1997). *Phys. Rev. B* **56**, 7656–7659.

Li, J.T., Schneider, W.D., Berndt, R., Bryant, O.R. and Crampin, S. (1998a). *Phys. Rev. Lett.* **81**, 4464–4467.

Li, J.T., Schneider, W.D., Berndt, R. and Crampin, S. (1998b). *Phys. Rev. Lett.* **80**, 3332–3335.

Li, J.T., Schneider, W.D., Berndt, R. and Delley, B. (1998c). *Phys. Rev. Lett.* **80**, 2893–2896.

Lorente, N. (2004). *App. Phys. a—Mat. Sci. Proc.* **78**, 799–806.

Lorente, N. and Persson, M. (2000). *Phys. Rev. Lett.* **85**, 2997–3000.

Lorente, N., Persson, M., Lauhon, L.J. and Ho, W. (2001). *Phys. Rev. Lett.* **86**, 2593–2596.

Madhavan, V., Chen, W., Jamneala, T., Crommie, M.F. and Wingreen, N.S. (1998). *Science* **280**, 567–569.

Makoshi, K. and Mingo, N. (2002). *Surface Sci.* **502**, 34–40.

Maltezopoulos, T., Kubetzka, A., Morgenstern, M., Wiesendanger, R., Lemay, S.G. and Dekker, C. (2003). *App. Phys. Lett.* **83**, 1011–1013.

Manoharan, H.C., Lutz, C.P. and Eigler, D.M. (2000). *Nature* **403**, 512–515.

Matsuda, T., Hasegawa, S., Igarashi, M., Kobayashi, T., Naito, M., Kajiyama, H., Endo, J., Osakabe, N., Tonomura, A. and Aoki, R. (1989). *Phys. Rev. Lett.* **62**, 2519–2522.

Matsui, T., Kambara, H., Ueda, I., Shishido, T., Miyatake, Y. and Fukuyama, H. (2003). *Phys. B—Cond. Matt.* **329**, 1653–1655.

Matzdorf, R. (1998). *Surface Sci. Rep.* **30**, 153–206.

Memmel, N. (1998). *Surface Sci. Rep.* **32**, 93–163.

Memmel, N. and Bertel, E. (1995). *Phys. Rev. Lett.* **75**, 485–488.

Meyer, G. (1996). *Rev. Sci. Inst.* **67**, 2960–2965.

Meyer, G., Bartels, L., Zophel, S., Henze, E. and Rieder, K.H. (1997). *Phys. Rev. Lett.* **78**, 1512–1515.

Meyer, G., Zophel, S. and Rieder, K.H. (1996). *App. Phy. Lett.* **69**, 3185–3187.

Meyer, G., Moresco, F., Hla, S.W., Repp, J., Braun, K.F., Folsch, S. and Rieder, K.H. (2001). *J. J. App. Phy. Part 1* **40**, 4409–4413.

Mingo, N. and Makoshi, K. (2000). *Phys. Rev. Lett.* **84**, 3694–3697.

Montelius, L., Pistol, M.E. and Samuelson, L. (1992). *Ultramicroscopy* **42**, 210–214.

Moresco, F., Meyer, G. and Rieder, K.H. (1999). *Mod. Phys. Lett. B* **13**, 709–715.

Moresco, F., Meyer, G., Rieder, K.H., Tang, H., Gourdon, A. and Joachim, C. (2001). *App. Phys. Lett.* **78**, 306–308.

Moresco, F., Gross, L., Alemani, M., Rieder, K.H., Tang, H., Gourdon, A. and Joachim, C. (2003). *Phys. Rev. Lett.* **91**, 036601.

Mugele, F., Rettenberger, A., Boneberg, J. and Leiderer, P. (1998). *Rev. Sci. Inst.* **69**, 1765–1769.

Murashita, T. (1997). *J. Vac. Sci. Tech. B* **15**, 32–37.

Murashita, T. (1999). *J. Vac. Sci. Tech. B* **17**, 22–28.

Murashita, T., Wada, K. and Inoue, N. (1993). *App. Surface Sci.* **68**, 223–226.

Nazin, G.V., Qiu, X.H. and Ho, W. (2003a). *Phys. Rev. Lett.* **90**, 216110.

Nazin, G.V., Qiu, X.H. and Ho, W. (2003b). *Science* **302**, 77–81.

Nicolay, G., Reinert, F., Schmidt, S., Ehm, D., Steiner, P. and Hufner, S. (2000). *Phys. Rev. B* **62**, 1631–1634.

Nilius, N., Ernst, N. and Freund, H.J. (2000). *Phys. Rev. Lett.* **84**, 3994–3997.

Nishitani, R., Umeno, T. and Kasuya, A. (1998). *App. Phys. a—Mat. Sci. Proc.* **66**, S139–S143.

Pan, S.H., Hudson, E.W. and Davis, J.C. (1999). *Rev. Sci. Inst.* **70**, 1459–1463.

Pan, S.H., Hudson, E.W., Lang, K.M., Eisaki, H., Uchida, S. and Davis, J.C. (2000). *Nature* **403**, 746–750.

Paniago, R., Matzdorf, R., Meister, G. and Goldmann, A. (1995). *Surface Sci.* **336**, 113–122.

Pascual, J.I., Jackiw, J.J., Song, Z., Weiss, P.S., Conrad, H. and Rust, H.P. (2001a). *Phys. Rev. Lett.* **86**, 1050–1053.

Pascual, J.I., Song, Z., Jackiw, J.J., Horn, K. and Rust, H.P. (2001b). *Phys. Rev. B*, **6324**, art. no. 241103.

Pascual, J.I., Gomez-Herrero, J., Sanchez-Portal, D. and Rust, H.P. (2002). *J. Chem. Phys.* **117**, 9531–9534.

Persson, B N.J. and Baratoff, A. (1992). *Phys. Rev. Lett.* **68**, 3224–3227.

Petersen, L., Sprunger, P.T., Hofmann, P., Laegsgaard, E., Briner, B.G., Doering, M., Rust, H.P., Bradshaw, A.M., Besenbacher, F. and Plummer, E.W. (1998). *Phys. Rev. B* **57**, R6858–R6861.

Petersen, L., Hofmann, P., Plummer, E.W. and Besenbacher, F. (2000a). *J. Electron Spectrosc. Rel. Phenomena* **109**, 97–115.

Petersen, L., Schaefer, B., Laegsgaard, E., Stensgaard, I. and Besenbacher, F. (2000b). *Surface Sci.* **457**, 319–325.

Petersen, L., Schunack, M., Schaefer, B., Linderoth, T.R., Rasmussen, P.B., Sprunger, P.T., Laegsgaard, E., Stensgaard, I. and Besenbacher, F. (2001). *Rev. Sci. Inst.* **72**, 1438–1444.

Pietzsch, O., Kubetzka, A., Bode, M. and Wiesendanger, R. (2000a). *Phys. Rev. Lett.* **84**, 5212–5215.

Pietzsch, O., Kubetzka, A., Haude, D., Bode, M. and Wiesendanger, R. (2000b). *Rev. Sci. Inst.* **71**, 424–430.

Pietzsch, O., Kubetzka, A., Bode, M. and Wiesendanger, R. (2001). *Science* **292**, 2053–2056.

Pines, D. (1966). *The Theory of Quantum Liquids* (Benjamin, New York).

Pivetta, M., Silly, F., Patthey, F., Pelz, J.P. and Schneider, W.D. (2003). *Phys. Rev. B* **67**, 193402.

Qiu, X.H., Nazin, G.V. and Ho, W. (2003). *Science* **299**, 542–546.

Ralls, K.S., Ralph, D.C. and Buhrman, R.A. (1989). *Phys. Rev. B* **40**, 11561–11570.

Reihl, B., Coombs, J.H. and Gimzewski, J.K. (1989). *Surface Sci.* **211**, 156–164.

Repp, J., Moresco, F., Meyer, G., Rieder, K.H., Hyldgaard, P. and Persson, M. (2000). *Phys. Rev. Lett.* **85**, 2981–2984.

Rosei, F., Schunack, M., Jiang, P., Gourdon, A., Laegsgaard, E., Stensgaard, I., Joachim, C. and Besenbacher, F. (2002). *Science* **296**, 328–331.

Rust, H.P., Buisset, J., Schweizer, E.K. and Cramer, L. (1997). *Rev. Sci. Inst.* **68**, 129–132.

Rust, H.P., Doering, M., Pascual, J.I., Pearl, T.P. and Weiss, P.S. (2001). *Rev. Sci. Inst.* **72**, 4393–4397.

Sanchez, O., Garcia, J.M., Segovia, P., Alvarez, J., Deparga, A.L.V., Ortega, J.E., Prietsch, M. and Miranda, R. (1995). *Phys. Rev. B* **52**, 7894–7897.

Sautet, P. and Joachim, C. (1991). *Chem. Phys. Lett.* **185**, 23–30.

Schneider, M.A., Vitali, L., Knorr, N. and Kern, K. (2002). *Phys. Rev. B* **65**, 121406.

Scholl, A. (2003). *Curr. Opinion Sol. State Mat. Sci.* **7**, 59–66.

Shockley, W. (1939). *Phys. Rev.* **56**, 317–323.

Slonczewski, J.C. (1988). *J. Phys.* **49**, 1629–1630.

Smith, D.P.E., Kirk, M.D. and Quate, C.F. (1987). *J. Chem. Phys.* **86**, 6034–6038.

Smolyaninov, II, Khaikin, M.S. and Edelman, V.S. (1990). *Phys. Lett. A* **149**, 410–412.

Sprunger, P.T., Petersen, L., Plummer, E.W., Laegsgaard, E. and Besenbacher, F. (1997). *Science* **275**, 1764–1767.

Stipe, B.C., Rezaei, M.A. and Ho, W. (1997a). *Phys. Rev. Lett.* **79**, 4397–4400.

Stipe, B.C., Rezaei, M.A., Ho, W., Gao, S., Persson, M. and Lundqvist, B.I. (1997b). *Phys. Rev. Lett.* **78**, 4410–4413.

Stipe, B.C., Rezaei, M.A. and Ho, W. (1998). *Science* **280**, 1732–1735.

Stipe, B.C., Rezaei, H.A. and Ho, W. (1999a). *Phys. Rev. Lett.* **82**, 1724–1727.

Stipe, B.C., Rezaei, M.A. and Ho, W. (1999b). *Rev. Sci. Inst.* **70**, 137–143.

Stranick, S.J., Kamna, M.M. and Weiss, P.S. (1994a). *Science* **266**, 99–102.

Stranick, S.J., Kamna, M.M. and Weiss, P.S. (1994b). *Rev. Sci. Inst.* **65**, 3211–3215.

Stroscio, J.A. and Eigler, D.M. (1991). *Science* **254**, 1319–1326.

Stroscio, J.A., Celotta, R.J., Blankenship, S., Hudson, E. and Fein, A.P. (2000). Proceedings of the 4th International Workshop on Quantum Functional Devices, 145–146 (Kanazawa, Japan).

Tersoff, J. and Hamann, D.R. (1985). *Phys. Rev. B* **31**, 805–813.

Tessmer, S.H., Tarlie, M.B., Van Harlingen, D.J., Maslov, D.L. and Goldbart, P.M. (1996). *Phys. Rev. Lett.* **77**, 924–927.

Theilmann, F., Matzdorf, R., Meister, G. and Goldmann, A. (1997). *Phys. Rev. B* **56**, 3632–3635.

Uehara, Y., Fujita, T., Iwami, M. and Ushioda, S. (2001). *Rev. Sci. Inst.* **72**, 2097–2099.

Vitali, L., Wahl, P., Schneider, M.A., Kern, K., Silkin, V.M., Chulkov, E.V. and Echenique, P.M. (2003). *Surface Sci.* **523**, L47–L52.

Weierstall, U. and Spence, J.C.H. (1998). *Surface Sci.* **398**, 267–279.

Whitman, L.J., Stroscio, J.A., Dragoset, R.A. and Celotta, R.J. (1991). *Science* **251**, 1206–1210.

Wiesendanger, R. (1994). *Scanning Probe Microscopy and Spectroscopy.* (Cambridge University Press, Cambridge).

Wiesendanger, R., Guntherodt, H.J., Guntherodt, G., Gambino, R.J. and Ruf, R. (1990). *Phys. Rev. Lett.* **65**, 247–250.

Wiesendanger, R., Bode, M., Kubetzka, A., Pietzsch, O., Morgenstern, M., Wachowiak, A. and Wiebe, J. (2004). *J. Magn. Magn. Mat.* **272–76**, 2115–2120.

Wittneven, C., Dombrowski, R., Pan, S.H. and Wiesendanger, R. (1997). *Rev. Sci. Inst.* **68**, 3806–3810.

Wolkow, R.A. (1995). *Abstr. Papers Amer. Chem. Soc.* **209**, 381–PHYS.

Xu, J.B., Lauger, K., Moller, R., Dransfeld, K. and Wilson, I.H. (1994). *J. App. Phys.* **76**, 7209–7216.

Zangwill, A. (1988). *Physics at Surfaces* (Cambridge University Press, Cambridge).

Zhang, H., Memmert, U., Houbertz, R. and Hartmann, U. (2001). *Rev. Sci. Inst.* **72**, 2613–2617.

Part IV

HOLOGRAPHIC AND LENSLESS MODES

18

Electron Holography

Rafal E. Dunin-Borkowski, Takeshi Kasama,
Martha R. McCartney, and David J. Smith

1 Introduction

Electron holography, as originally described by Gabor (1949), is based on the formation of an interference pattern or "hologram" in the transmission electron microscope (TEM). In contrast to most conventional TEM techniques, which only record spatial distributions of image intensity, electron holography allows the phase shift of the high-energy electron wave that has passed through the specimen to be measured directly. The phase shift can then be used to provide information about local variations in magnetic induction and electrostatic potential. This chapter provides an overview of the technique of electron holography. It begins with an outline of the experimental procedures and theoretical background that are needed to obtain phase information from electron holograms. Medium-resolution applications of electron holography to the characterization of magnetic domain structures and electrostatic fields are then described, followed by a description of high-resolution electron holography and alternative modes of electron holography. The majority of the experimental results described below are obtained using the off-axis, or "sideband," TEM mode, which is the most widely used mode of electron holography at present. For further details about electron holography, the interested reader is referred to several recent books (e.g., Tonomura et al., 1995; Tonomura, 1998; Völkl et al., 1998) and review papers (e.g., Tonomura, 1992; Midgley, 2001; Lichte, 2002; Matteucci et al., 2002).

1.1 Basis of Off-Axis Electron Holography

The off-axis mode of electron holography involves the examination of an electron-transparent specimen using defocused illumination from a highly coherent field-emission gun (FEG) electron source. To acquire an off-axis electron hologram, the region of interest on the specimen should be positioned so that it covers approximately half the field of view. An electron biprism, which takes the form of a fine (<1 μm diameter) wire (Möllenstedt and Düker, 1954), is located below the sample, usually in place of one of the conventional selected-area apertures. The application of a voltage to the biprism results in overlap of a "reference" electron wave that has passed through vacuum (or through a thin region of support film) with the electron wave that has passed through the specimen, as shown schematically in Figure 18–1a. If the illumination is sufficiently coherent, then holographic interference fringes are

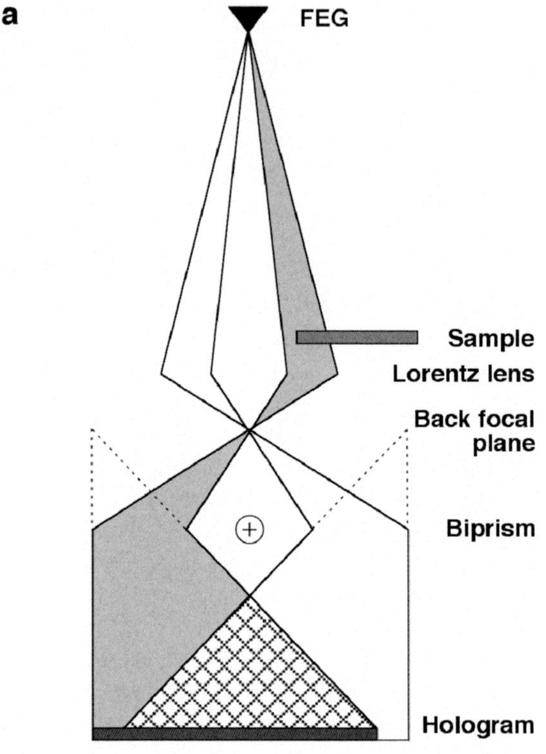

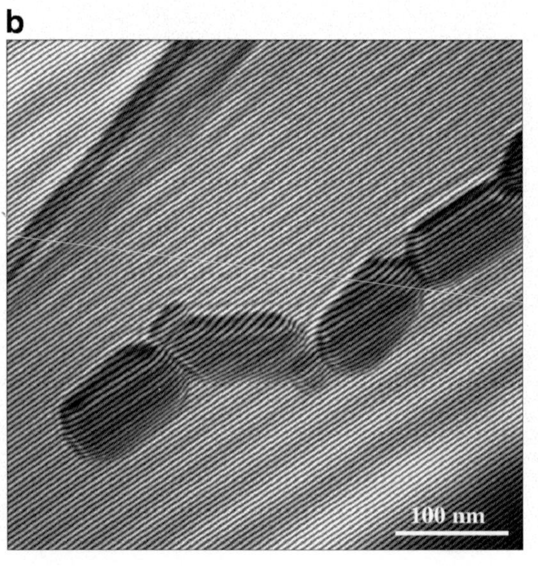

Figure 18–1. (a) Schematic illustration of the set-up used for generating off-axis electron holograms. The specimen occupies approximately half the field of view. Essential components are the field-emission gun (FEG) electron source, which provides coherent illumination, and the electron biprism, which causes overlap of the object and (vacuum) reference waves. The Lorentz lens allows imaging of magnetic materials in close to field-free conditions. (b) Off-axis electron hologram of a chain of magnetite (Fe_3O_4) crystals, recorded at 200 kV using a Philips CM200 FEGTEM. The crystals are supported on a holey carbon film. Phase changes can be seen in the form of bending of the holographic interference fringes as they pass through the crystals. Fresnel fringes from the edges of the biprism wire are also visible. (Reprinted from Dunin-Borkowski et al., 2004a.)

formed in the overlap region, with a spacing that is inversely proportional to the biprism voltage (Matteucci et al., 1998). The amplitude and the phase shift of the electron wave from the specimen are recorded in the intensity and the position of the holographic fringes, respectively. A representative off-axis electron hologram of a chain of magnetite nanocrystals is shown in Figure 18–1b.

For coherent TEM imaging, the electron wavefunction in the image plane can be written in the form

$$\psi_i(r) = A_i(r)\,\exp[i\phi_i(r)] \tag{1}$$

where r is a two-dimensional vector in the plane of the sample, A and ϕ refer to amplitude and phase, and the subscript i refers to the image plane. The recorded intensity distribution is then given by the expression

$$I_i(r) = |A_i(r)|^2 \tag{2}$$

Thus, the image intensity can be described as the modulus squared of an electron wavefunction that has been modified by the specimen and the objective lens. The intensity distribution in an off-axis electron hologram can be represented by the addition of a tilted plane reference wave to the complex specimen wave, in the form

$$I_{\mathrm{hol}}(r) = |\psi_i(r) + \exp[2\pi i q_c \cdot r]|^2 \tag{3}$$
$$= 1 + A_i^2(r) + 2A_i(r)\cos[2\pi i q_c \cdot r + \phi_i(r)] \tag{4}$$

where the tilt of the reference wave is specified by the two-dimensional reciprocal space vector $q = q_c$. It can be seen from Eq. (4) that there are three separate contributions to the intensity distribution in a hologram: the reference wave, the image wave, and an additional set of cosinusoidal fringes with local phase shifts and amplitudes that are exactly equivalent to the phase and amplitude of the electron wavefunction in the image plane, ϕ_i and A_i, respectively.

1.2 Hologram Reconstruction

To obtain amplitude and phase information, the off-axis electron hologram is first Fourier transformed. From Eq. (4), the complex Fourier transform of the hologram is given by

$$\begin{aligned} FT[I_{\mathrm{hol}}(r)] = {} &\delta(q) + FT[A_i^2(r)] \\ &+ \delta(q + q_c) \otimes FT\{A_i(r)\,\exp[i\phi_i(r)]\} \\ &+ \delta(q - q_c) \otimes FT\{A_i(r)\,\exp[-i\phi_i(r)]\} \end{aligned} \tag{5}$$

Equation (5) describes a peak at the reciprocal space origin corresponding to the Fourier transform of the reference image, a second peak centered on the origin corresponding to the Fourier transform of a bright-field TEM image of the sample, a peak centered at $q = -q_c$ corresponding to the Fourier transform of the desired image wavefunction, and a peak centered at $q = +q_c$ corresponding to the Fourier transform of the complex conjugate of the wavefunction.

The reconstruction of a hologram to obtain amplitude and phase information is illustrated in Figure 18–2. Figure 18–2a–c shows a

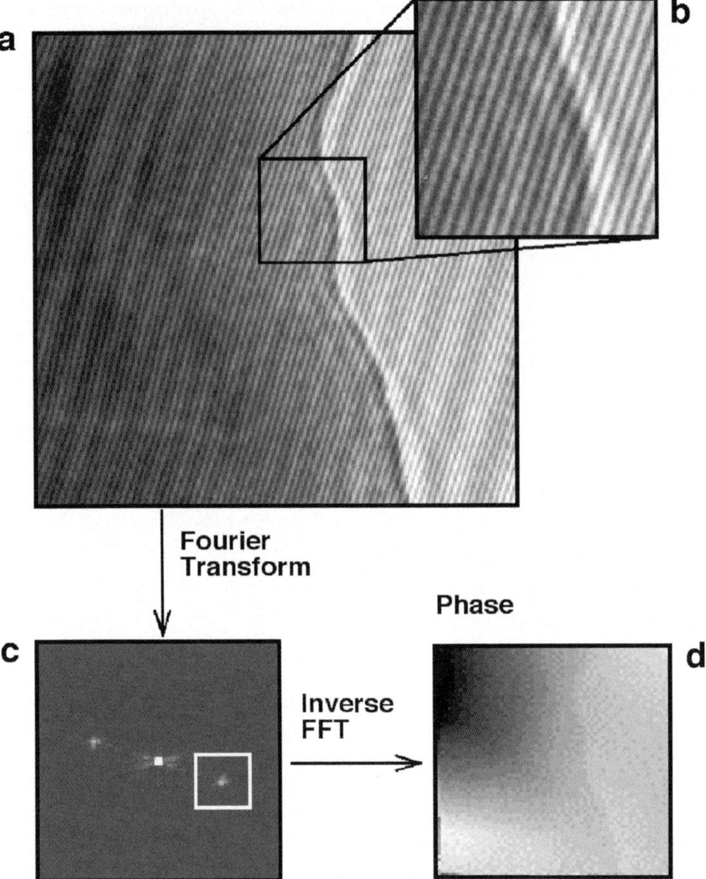

Figure 18–2. (a) Off-axis electron hologram recorded from a thin crystal. (b) Enlargement showing interference fringes within the specimen. (c) Fourier transform of the electron hologram. (d) Phase image obtained after inverse fast Fourier transformation (FFT) of the sideband marked with a box in (c). (Reprinted from Dunin-Borkowski et al., 2000.)

hologram of a thin crystal, an enlargement of part of the hologram, and a Fourier transform of the entire hologram, respectively. To recover the amplitude and the relative phase shift of the electron wavefunction, one of the two sidebands is selected digitally and inverse Fourier transformed, as shown in Figure 18–2d. The amplitude and phase of this complex wavefunction are then easily calculated.

The impact of electron holography results from the dependence of the phase shift on the electrostatic potential and the in-plane component of the magnetic induction in the specimen. Neglecting dynamical diffraction (i.e., assuming that the specimen is thin and weakly diffracting), the phase shift can be expressed in the form

$$\phi(x) = C_E \int V(x, z)\,dz - \left(\frac{e}{\hbar}\right) \iint B_\perp(x, z)\,dx\,dz \tag{6}$$

where

$$C_E = \left(\frac{2\pi}{\lambda}\right)\left[\frac{E+E_0}{E(E+2E_0)}\right] \tag{7}$$

z is in the incident electron beam direction, x is in the plane of the specimen, $B_\perp$ is the component of the magnetic induction within and outside the specimen perpendicular to both x and z, V is the electrostatic potential, λ is the (relativistic) electron wavelength, and E and E_0 are, respectively, the kinetic and rest mass energies of the incident electron (Reimer, 1991). C_E has values of 7.29×10^6, 6.53×10^6, and $5.39 \times 10^6 \, \mathrm{rad} \, \mathrm{V}^{-1} \mathrm{m}^{-1}$ at 200 kV, 300 kV, and 1 MV, respectively. If neither V nor $B_\perp$ varies along the electron beam direction within a sample of thickness t, then Eq. (6) can be simplified to

$$\phi(x) = C_E V(x)t(x) - \left(\frac{e}{\hbar}\right)\int B_\perp(x)t(x)dx \tag{8}$$

By making use of Eqs. (6) and (8), information about V and $B_\perp$ can be recovered from a measurement of the phase shift ϕ, as described below.

1.3 Experimental Considerations

In practice, several issues must be addressed to record and analyze an electron hologram successfully. A key experimental requirement is the availability of a vacuum reference wave that can be overlapped onto the region of interest on the specimen, which usually implies that the hologram must be recorded from a region close to the specimen edge. This restriction can be relaxed if a thin, clean, and weakly diffracting region of electron-transparent support film, rather than vacuum, can be overlapped onto the region of interest.

As phase information is stored in the lateral displacement of the holographic interference fringes, long-range phase modulations arising from inhomogeneities in the charge and thickness of the biprism wire, as well as from lens distortions and charging effects (e.g., at apertures), must be taken into account by using a reference hologram obtained by removing the specimen from the field of view without changing the optical parameters of the microscope. Correction is then possible by performing a complex division of the sample and reference waves in real space to obtain the distortion-free phase of the image wave (de Ruijter and Weiss, 1993).

The need for this procedure is illustrated in Figure 18–3. Figure 18–3a shows a reconstructed phase image of a wedge-shaped crystal of indium phosphide (InP) obtained before distortion correction, with the vacuum region outside the sample edge on the left side of the image. Figure 18–3b shows the corresponding vacuum reference phase image, which was acquired with the sample removed from the field of view but with all other imaging parameters unchanged. The equiphase contour lines now correspond to distortions that must be removed from the phase image of the sample. Figure 18–3c shows the distortion-corrected phase image of the sample, which was obtained by dividing the two complex image waves. The vacuum region in Figure 18–3c is

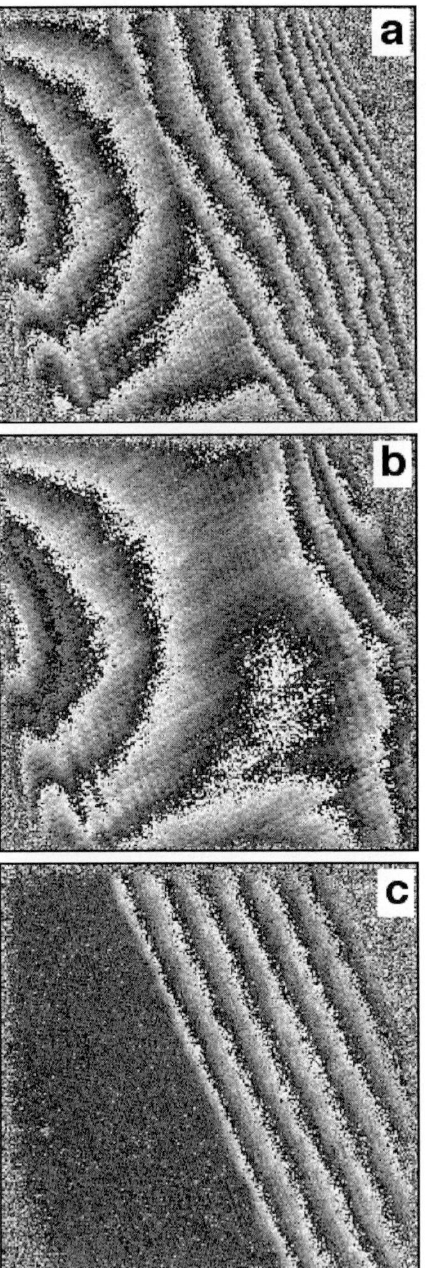

Figure 18–3. Illustration of distortion correction procedure. (a) Initial (six times amplified) phase image of a wedge-shaped InP crystal recorded at 100 kV using a Philips 400ST FEGTEM equipped with a Gatan 679 slow-scan CCD (charge coupled device) camera. (b) Phase image obtained from vacuum, with the specimen removed from the field of view. (c) Corrected phase image, obtained by subtracting image (b) from image (a). This procedure is carried out by dividing the complex image waves obtained by inverse Fourier transforming the sideband obtained from each hologram, and by calculating the phase of the resulting wavefunction. (Reprinted from Smith et al., 1998.)

flattened substantially by this procedure, which allows relative phase changes within the sample to be interpreted much more reliably. The acquisition of a reference hologram has the additional advantage that it allows the center of the sideband in Fourier space to be determined accurately. The use of the same location for the sideband in the Fourier transforms of the sample and reference waves removes any tilt of the recorded wave that might be introduced by an inability to locate the exact (subpixel) position of the sideband frequency in Fourier space.

Electron holograms have traditionally been recorded on photographic film, but digital acquisition using charge coupled device (CCD) cameras is now widely used due to their linear response, dynamic range, and high detection quantum efficiency, as well as the immediate accessibility to the recorded information (de Ruijter, 1995; Meyer and Kirkland, 2000). Whether a hologram is recorded on film or digitally, the field of view is typically limited to approximately $5\,\mu\text{m}$ by the dimensions of the recording medium and the sampling of the holographic fringes. A phase image that is calculated digitally is usually evaluated modulo 2π, meaning that 2π phase discontinuities that are unrelated to specimen features will appear at positions where the phase shift exceeds this amount. The phase image must often then be "unwrapped" using suitable algorithms (Ghiglia and Pritt, 1998).

The high electron beam coherence that is required for electron holography requires the use of an FEG electron source, a small spot size, a small condenser aperture, and a low gun extraction voltage. The coherence may be improved further by adjusting the condenser lens stigmators in the microscope to provide elliptical illumination that is wide in the direction perpendicular to the biprism when the condenser lens is overfocused (Smith and McCartney, 1998). The contrast of the holographic interference fringes is determined primarily by the lateral coherence of the electron wave at the specimen level, the mechanical stability of the biprism wire, and the point spread function of the recording medium. The fringe contrast

$$\mu = \left(\frac{I_{\max} - I_{\min}}{I_{\max} + I_{\min}}\right) \tag{9}$$

can be determined from a holographic interference fringe pattern that has been recorded in the absence of a sample, where $I_{\max}$ and $I_{\min}$ are the maximum and minimum intensities of the interference fringes, respectively (Völkl et al., 1995). Should the fringe contrast decrease too much, reliable reconstruction of the image wavefunction will no longer be possible.

The phase detection limit for electron holography (Harscher and Lichte, 1996) can be determined from the effect on the recorded hologram of Poisson-distributed shot noise, the detection quantum efficiency and point spread function of the CCD camera, and the fringe contrast. The minimum phase difference between two pixels that can be detected is given by the expression

$$\Delta\phi_{\min} = \frac{SNR}{\mu}\sqrt{\frac{2}{N_{\text{el}}}} \tag{10}$$

where SNR is the signal-to-noise ratio, μ is defined in Eq. (9), and N_{el} is the number of electrons collected per pixel (Lichte, 1995). In practice, some averaging of the measured phase is often implemented, particularly if the features of interest vary slowly across the image or only in one direction. A final artifact results from the presence of Fresnel diffraction at the biprism wire, which is visible in Figure 18–1b and causes phase and amplitude modulations of both the image and the reference wave (Yamamoto et al., 2000). These effects can be removed to some extent by using a reference hologram, and by Fourier filtering the sideband before reconstruction of the image wave. More advanced approaches for removing Fresnel fringes from electron holograms based on image analysis (Fujita and McCartney, 2005) and double-biprism electron holography (Harada et al., 2004; Yamamoto et al., 2004) have recently been introduced. Great care should also be taken to assess the effect on the reference wave of long-range electromagnetic fields that may extend outside the sample and affect both the object wave and the reference wave (Matteucci et al., 1994).

2 Measurement of Mean Inner Potential and Sample Thickness

Before describing the application of electron holography to the characterization of magnetic and electrostatic fields, the use of the technique to measure local variations in specimen morphology and composition is considered. Such measurements are possible from a phase image that is associated solely with variations in mean inner potential and specimen thickness. When a specimen has uniform structure and composition in the electron beam direction, and in the absence of magnetic and long-range electrostatic fields (such as those at depletion regions in semiconductors), Equation (8) can be rewritten in the form

$$\phi(x) = C_E V_0(x) t(x) \tag{11}$$

where the mean inner potential of the specimen, V_0, is the volume average of the electrostatic potential (Spence, 1993). Values of V_0 can in principle be calculated from the equation

$$V_0 = \left(\frac{h^2}{2\pi m e \Omega} \right) \sum_\Omega f_{el}(0) \tag{12}$$

by treating the specimen as an array of neutral atoms. In Eq. (12), $f_{el}(0)$ are electron scattering factors at zero scattering angle for each atom, which have been calculated by, for example, Doyle and Turner (1968) and Rez et al. (1994), and Ω is typically the volume of the unit cell in a crystalline material. However, values of V_0 that are calculated using Eq. (12) are invariably overestimated as a result of bonding in the specimen (Radi, 1970; Gajdardziska-Josifovska and Carim, 1998). It is therefore important to obtain experimental measurements of V_0. According to Eq. (11), an independent measure of the specimen thickness profile is required to determine V_0, for example, by examining a specimen in which the thickness changes in a well-defined manner, as shown in Figure 18–4a for a phase profile obtained from a 90° wedge of GaAs tilted to a weakly diffracting orientation. If the specimen thickness profile is known, then V_0 can be determined by measuring the gradient of the phase $d\phi/dx$, and making use of the relation

$$V_0 = \left(\frac{1}{C_E} \right) \left(\frac{d\phi/dx}{dt/dx} \right) \tag{13}$$

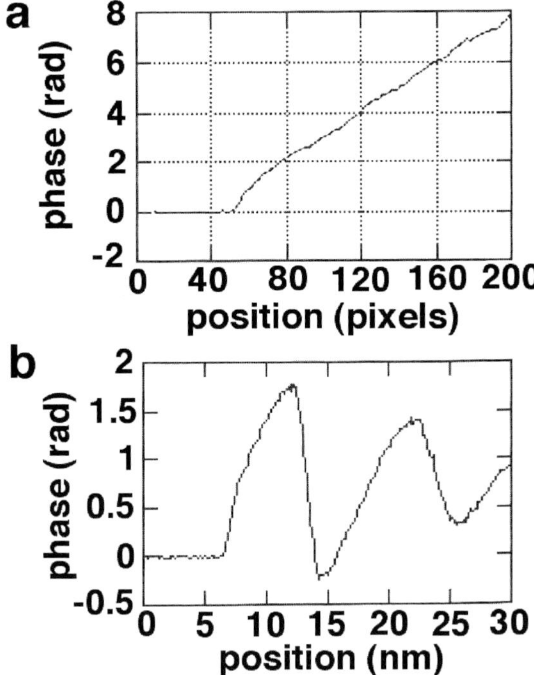

Figure 18–4. (a) Phase profile plotted as a function of distance into a 90° GaAs cleaved wedge specimen tilted to a weakly diffracting orientation. The phase change increases approximately linearly with specimen thickness. (b) Phase profile obtained from a GaAs wedge tilted close to a [100] zone axis, showing strong dynamical effects. (Reprinted from Gajdardziska-Josifovska and Carim, 1998.)

This approach has been used successfully to measure the mean inner potential of cleaved wedges and cubes of Si, MgO, GaAs, PbS (Gajdardziska-Josifovska et al., 1993; de Ruijter et al., 1992), and Ge (Li et al., 1999). The resulting values of V_0 that were determined for MgO, GaAs, PbS, and Ge using this approach are 13.0 ± 0.1, 14.5 ± 0.2, 17.2 ± 0.1, and $14.3 \pm 0.2\,V$, respectively. In a similar study, wedge-shaped Si samples with stacked Si oxide layers on their surfaces were used to measure the mean inner potentials of the oxide layers (Rau et al., 1996). Experimental measurements of V_0 have been obtained from 20- to 40-nm-diameter Si nanospheres coated in layers of amorphous SiO_2 (Wang et al., 1997). The mean inner potential of crystalline Si was found to be $12.1 \pm 1.3\,V$, that of amorphous Si $11.9 \pm 0.9\,V$, and that of amorphous SiO_2 $10.1 \pm 0.6\,V$. Similar measurements obtained from spherical latex particles embedded in vitrified ice have provided values for V_0 of 8.5 ± 0.7 and $3.5 \pm 1.2\,V$ for the two materials, respectively (Harscher and Lichte, 1998).

Dynamical contributions to the phase shift complicate the determination of V_0 from crystalline samples. These corrections can be taken into account by using either multislice or Bloch wave algorithms. The fact that Eq. (11) is no longer valid when the sample is tilted to a strongly diffracting orientation is demonstrated in Figure 18–4b for a 90° cleaved wedge sample of GaAs that has been tilted to a [100] zone axis. The phase shift varies nonlinearly with sample thickness, and is also very sensitive to small changes in sample orientation. Additional experimental factors that may affect measurements of V_0 include the chemical and physical state and the crystallographic orientation of the specimen surface (O'Keeffe and Spence, 1994), and specimen charging (Lloyd et al., 1997).

When V_0 is already known, then measurements of the phase shift can be used to determine the local specimen thickness t. Alternatively, the specimen thickness can be inferred from a holographic amplitude image in units of λ_{in}, the mean free path for inelastic scattering, by making use of the relation

$$\frac{t(x)}{\lambda_{in}} = -2\ln\left[\frac{A_i(x)}{A_r(x)}\right] \tag{14}$$

where $A_i(x)$ and $A_r(x)$ are the measured amplitudes of the sample and reference holograms, respectively (McCartney and Gajdardziska-Josifovska, 1994). If desired, the thickness dependence of both the phase and the amplitude image can be removed by combining Eqs. (11) and (14) in the form

$$\frac{\phi(x)}{-2C_E\ln\left[\dfrac{A_i(x)}{A_r(x)}\right]} = V_0(x)\lambda_{in}(x) \tag{15}$$

Equation (15) can be used to generate an image where the contrast is the product of the local values of the mean inner potential and the inelastic mean free path. These parameters depend only on the local composition of the sample, and thus can be useful for interpreting images obtained from samples with varying composition and thickness (Weiss et al., 1991).

When the mean inner potential in a specimen is constant or if its variation across the specimen is known, then the morphologies of nanoscale particles (for which dynamical contributions to the phase shift are likely to be small) can be measured using electron holography by making use of Eq. (11). Examples of the measurement of particle shapes using this approach include the characterization of faceted ZrO_2 crystals (Allard et al., 1996), carbon nanotubes (Lin and Dravid, 1996), bacterial flagellae (Aoyama and Ru, 1996), and atomic-height steps on clean surfaces of MoS_2 (Tonomura et al., 1985). Such measurements can, in principle, be extended to three dimensions by combining electron holography with electron tomography, as demonstrated by the acquisition and analysis of tilt series of electron holograms of latex particles (Lai et al., 1994). The demanding nature of these latter measurements results from the fact that specimen tilt angles of at least ±60°, as well as small tilt steps and accurate alignment of the resulting phase images, are all essential to avoid reconstruction artifacts.

3 Measurement of Magnetic Fields

The most successful and widespread applications of electron holography have involved the characterization of magnetic fields within and
surrounding materials at medium spatial resolution. When examining
magnetic materials, the normal objective lens is usually switched off,
as its strong magnetic field will likely saturate the magnetization in
the sample along the electron beam direction. A high-strength mini-
lens located below the objective lens can instead be used to provide
reasonably high magnification (~50–75 kx) with the sample still in a
magnetic field-free environment.

3.1 Early Experiments

Early examples of the examination of magnetic materials using
electron holography involved the reconstruction of holograms using a
laser bench, and included the characterization of horseshoe magnets
(Matsuda et al., 1982), magnetic recording media (Osakabe et al., 1983),
and vortices in superconductors (Matsuda et al., 1989, 1991; Bonevich
et al., 1993). The most elegant series of experiments involved the confirmation of the Aharonov–Bohm effect (Aharonov and Bohm, 1959),
which states that when an electron wave from a point source passes
on either side of an infinitely long solenoid then the relative phase shift
that occurs between the two parts of the wave should result from the
presence of a vector potential. In this way, the Aharonov–Bohm effect
provides the only observable confirmation of the physical reality of
gauge theory. Electron holography experiments were carried out on
20-nm-thick permalloy toroidal magnets that were covered with 300-
nm-thick layers of superconducting Nb, which prevented electrons
from penetrating the magnetic material and confined the magnetic flux
by exploiting the Meissner effect (Tonomura et al., 1982, 1983). The
observations showed that the phase difference between the center of
the toroid and the region outside was quantized to a value of 0 or π
when the temperature was below the Nb superconducting critical temperature (5 K), i.e., when a supercurrent was induced to circulate in the
magnet. The observed quantization of magnetic flux, and the measured phase differences with the magnetic field entirely screened
by the superconductor, provided unequivocal confirmation of the
Aharonov–Bohm effect.

3.2 Digital Acquisition and Analysis

Recent applications of electron holography to the characterization of
magnetic fields in nanostructured materials have been based on digital
recording and processing. The examples chosen here highlight the different approaches that can be used to separate a weak magnetic signal
from a recorded phase image, as well as illustrating the magnetic
properties of the materials. The off-axis mode of electron holography
is ideally suited to the characterization of such materials because
unwanted contributions to the contrast from local variations in

composition and specimen thickness can usually be removed from a phase image more easily than from images recorded using other TEM phase contrast techniques. For example, the Fresnel and Foucault modes of Lorentz microscopy and differential phase contrast (DPC) imaging provide signals that are approximately proportional to either the first or the second differential of the phase shift. These techniques inherently enhance contributions to the contrast from rapid variations in specimen thickness and composition, as compared to the weak and slowly varying magnetic signal.

The *digital* acquisition, reconstruction, and analysis of electron holograms have allowed magnetic fields within samples with small feature sizes and rapid variations in thickness or composition to be examined. The key advantage of digital analysis is that the magnetic and mean inner potential contributions to the measured holographic phase shift can be separated, particularly at the edges of nanostructured particles, where rapid changes in specimen thickness often dominate both the measured phase and the phase gradient. The approaches that can be used to achieve this separation are described below. Digital analysis also facilitates the construction of line profiles from phase images, which can provide quantitative information such as the widths of domain walls.

Determination of the phase gradient is particularly useful for a magnetic material because of the following relationship, obtained by differentiating Eq. (8):

$$\frac{d\phi(x)}{dx} = C_E \frac{d}{dx}[V(x)t(x)] - \left(\frac{e}{\hbar}\right)B_\perp(x)t(x) \tag{16}$$

According to Eq. (16), for a specimen of uniform thickness and composition the phase gradient is proportional to the in-plane component of the magnetic induction in the specimen

$$\frac{d\phi(x)}{dx} = -\left(\frac{et}{\hbar}\right)B_\perp(x) \tag{17}$$

A direct graphic representation of the magnetic induction can therefore be obtained by adding contours to a magnetic phase image, where a phase difference of 2π corresponds to an enclosed magnetic flux of 4×10^{-15} Wb. Significantly, an experimental phase image does not need to be unwrapped to evaluate its first differential digitally. Instead, if the reconstructed image wave is designated ψ, then the phase differential can be determined directly from the expression

$$\frac{d\phi(x, y)}{dx} = \text{Im}\left(\frac{\dfrac{d\psi(x, y)}{dx}}{\psi(x, y)}\right) \tag{18}$$

Most of the results that are described below were acquired using Philips CM200ST and CM300ST FEGTEMs equipped with rotatable electron biprisms, and with Lorentz minilenses located in the bores of their objective lens pole-pieces. The Lorentz lenses allow electron holograms to be recorded at magnifications of up to ~75 kx with the specimens located in a magnetic field-free environment.

3.2.1 NdFeB Hard Magnets

Figure 18–5a shows a Lorentz (Fresnel defocus) image of a $Nd_2Fe_{14}B$ specimen, in which magnetic domains can be seen (McCartney and Zhu, 1998). Such images provide little information about the direction of the local magnetic induction in the specimen. An electron holographic phase image acquired from the same area using an interference fringe spacing of 2.5 nm is shown in Figure 18–5b. Gradients of the phase image were calculated along the $+x$ and $-y$ directions, as shown in Figure 18–5c and d, respectively. These images were combined to form a vector map of the magnetic induction, as shown in Figure 18–5e. The map is divided into 20-nm squares, and has a low contrast phase gradient image superimposed on it for reference purposes. The minimum vector length is zero (corresponding to out-of-plane induction), while the maximum vector length is consistent with a measured induction $B = 1.0$ T. A vector map of the region marked in Figure 18–5e is shown at higher magnification in Figure 18–5f. In this map, magnetic vortices show Bloch-like character, with vanishingly small vector lengths. Care is needed when interpreting the fine details in such induction maps due to the undetermined effects of magnetic fringing fields immediately above and below the sample, as well as possible contributions from variations in specimen thickness. A single pixel line scan across a 90° domain wall, which appears as the bright ridge near the central part of Figure 18–5b, is shown in Figure 18–5g. This line profile places an upper limit of 10 nm on the domain wall width, which agrees well with theoretical estimates.

3.2.2 Co Nanoparticle Chains

The dominant nature of the mean inner potential contribution to the phase shift recorded from a nanoscale magnetic particle is illustrated in Figure 18–6. Figure 18–6a and b shows a hologram and a reconstructed phase image of a chain of Co particles suspended over a hole in a carbon support film (de Graef et al., 1999). Figure 18–6c and d shows corresponding line traces determined from the phase image across the centers of two particles. Each trace is obtained in a direction perpendicular to the chain axis. The magnetic induction and mean inner potential of each particle can be determined by fitting simulations to the experimental line traces. Analytical expressions for the expected phase shifts can be derived for a uniformly magnetized sphere of radius a, magnetic induction $B_\perp$ (along y), and mean inner potential V_0 in the form

$$\phi(x,y)\Big|_{(x^2+y^2)\leq a^2} = 2C_E V_0 \sqrt{a^2 - (x^2 + y^2)}$$
$$+ \left(\frac{e}{\hbar}\right) B_\perp a^3 \left(\frac{x}{x^2 + y^2}\right)$$
$$\left\{ 1 - \left[1 - \left(\frac{x^2 + y^2}{a^2}\right) \right]^{\frac{3}{2}} \right\} \tag{19}$$

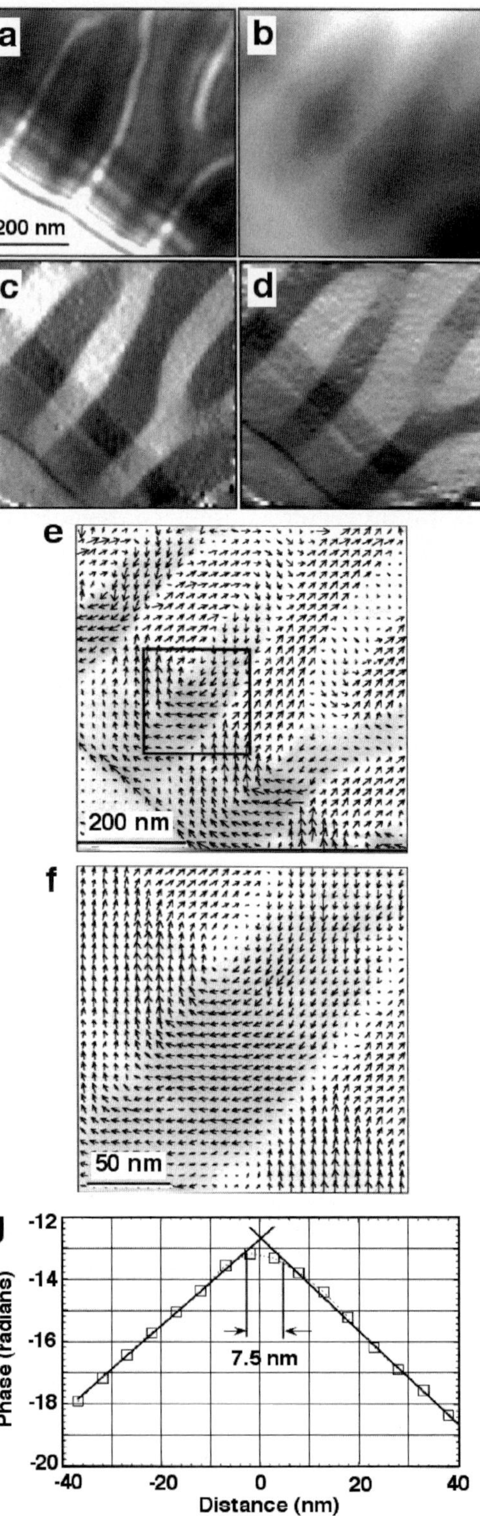

Figure 18–5. (a) Room temperature Lorentz (Fresnel underfocus) image of an $Nd_2Fe_{14}B$ hard magnet, recorded at 200 kV using a Philips CM200 FEGTEM operated in Lorentz mode. (b) Phase image of the same region of the specimen reconstructed from an electron hologram obtained using an interference fringe spacing of 2.5 nm. (c and d) Gradients of the phase image shown in (b), calculated parallel to the $+x$ and $-y$ directions, respectively. (e) Induction map derived from the phase gradients shown in (c) and (d). (f) Enlargement of the area indicated in (e). (g) Linescan obtained across a 90° domain wall that appears as a bright ridge near the center of image (b). The line profile provides an upper limit for the domain wall width of 10 nm. (Reprinted from McCartney and Zhu, 1998.)

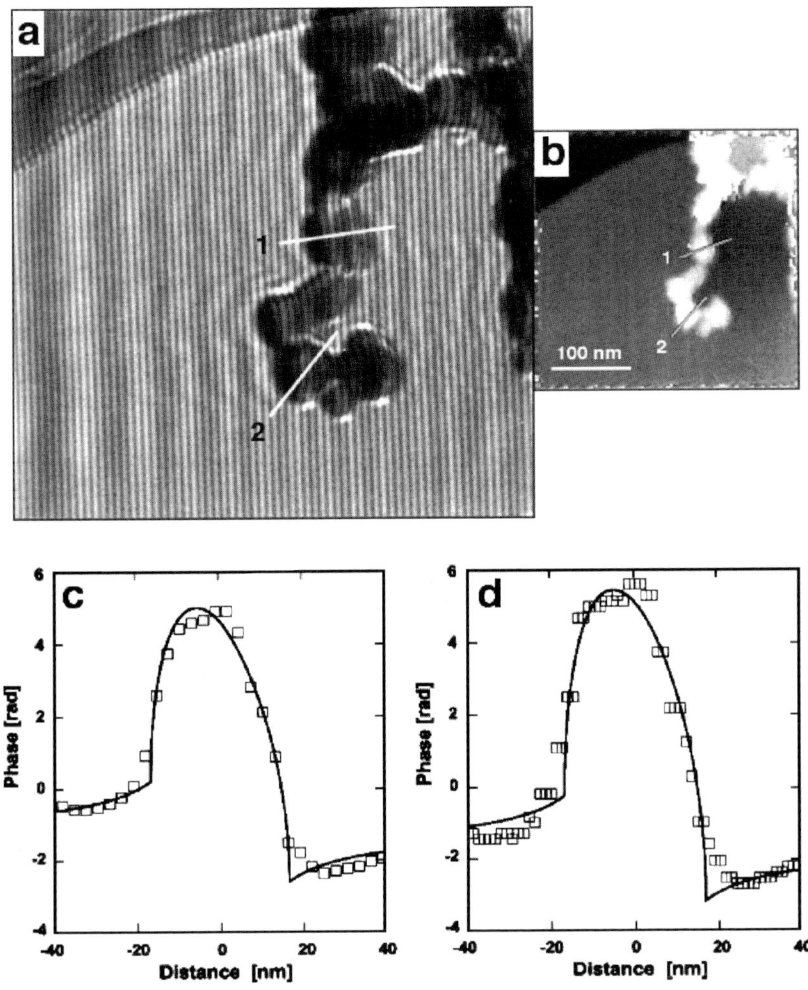

Figure 18–6. (a) Off-axis electron hologram of a chain of Co particles suspended over a hole in a carbon support film, acquired at 200 kV using a Philips CM200 FEGTEM and a biprism voltage of 90 V. (b) Corresponding unwrapped phase image. (c and d) Experimental line profiles formed from lines 1 and 2 in (b), and fitted phase profiles generated for spherical Co nanoparticles. (Reprinted from de Graef et al., 1999.)

$$\phi(x, y)\Big|_{(x^2+y^2)>a^2} = \left(\frac{e}{\hbar}\right)B_\perp a^3\left(\frac{x}{x^2+y^2}\right) \tag{20}$$

For line profiles obtained through the centers of the particles in a direction perpendicular to that of $B_\perp$, these expressions reduce to

$$\phi(x)\big|_{x\le a} = 2C_E V_0 \sqrt{a^2-x^2}$$
$$+\left(\frac{e}{\hbar}\right)B_\perp\left[\frac{a^3-\left(a^2-x^2\right)^{\frac{3}{2}}}{x}\right] \tag{21}$$

$$\phi(x)|_{x>a} = \left(\frac{e}{\hbar}\right)B_\perp\left(\frac{a^3}{x}\right) \tag{22}$$

Least squares fits of Eqs. (21) and (22) to the experimental data points, which are also shown in Figure 18–6c and d, provide best-fitting values for a, $B_\perp$, and V_0 of 17 nm, 1.7 T, and 26 V, respectively.

3.3 Separation of Magnetic and Mean Inner Potential Contributions

When characterizing magnetic fields inside nanostructured materials, the mean inner potential contribution to the measured phase shift must in general be removed to interpret the magnetic contribution of primary interest. Several approaches can be used. The sample may be inverted to change the sign of the magnetic contribution to the signal and a second hologram recorded. The sum and the difference of the two phase images can then be used to provide twice the magnetic contribution, and twice the mean inner potential contribution, respectively (Wohlleben, 1971; Tonomura et al., 1986). Alternatively, two holograms may be acquired from the same area of the specimen at two different microscope accelerating voltages. In this case, the magnetic signal is independent of accelerating voltage, and subtraction of the two phase images can be used to provide the mean inner potential contribution. A more practical method of removing the mean inner potential contribution involves performing magnetization reversal *in situ* in the electron microscope, and subsequently selecting pairs of holograms that differ only in the (opposite) directions of the magnetization. The magnetic and mean inner potential contributions to the phase can be calculated by taking half the difference, and half the sum, of the phases. The mean inner potential contribution can then be subtracted from all other phase images acquired from the same specimen region (Dunin-Borkowski et al., 1998). *In situ* magnetization reversal, which is required both for this purpose and for performing magnetization reversal experiments in the TEM, can be achieved by exciting the conventional microscope objective lens slightly and tilting the specimen to apply known magnetic fields, as shown schematically in Figure 18–7. Subsequently, electron holograms can be recorded with the conventional microscope objective lens switched off and the Lorentz lens

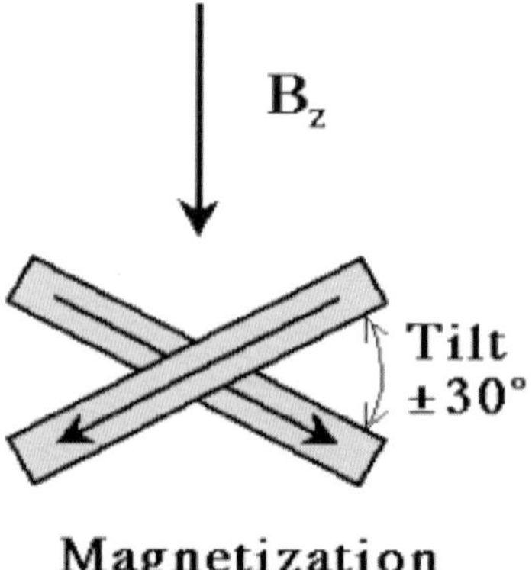

Figure 18–7. Schematic diagram illustrating the use of specimen tilt to provide an in-plane component of the external field for *in situ* magnetization reversal experiments. (Reprinted from Dunin-Borkowski et al., 2000.)

switched on, while the magnetic specimen is located in a magnetic field-free environment.

3.3.1 Magnetite Nanoparticle Chains

The chain of magnetite nanoparticles shown in Figure 18–8 illustrates the fact that both the mean inner potential and the magnetic contribution to the phase shift can provide useful information. In particular, the mean inner potential contribution can be used to interpret the morphologies and orientations of nanoparticles, as discussed above. Figure 18–8a and b shows phase contours generated from, respectively, the mean inner potential and magnetic contributions to the phase shift at the end of a chain of magnetite (Fe_3O_4) crystals from a magnetotactic bacterium collected from a brackish lagoon at Itaipu in Brazil. The magnetic moment that the crystals impart to the bacterial cell results in its alignment and subsequent migration along the Earth's magnetic field lines (Blakemore, 1975; Bazylinski and Moskowitz, 1997; Dunlop and Özdemir, 1997). Separation of the mean inner potential and magnetic contributions to the phase shift was achieved *in situ* by using the field of the conventional microscope objective lens to magnetize each chain parallel and then antiparallel to its length, as illustrated in Figure 18–7. The contours in Figure 18–8a and b have been overlaid onto the mean inner potential contribution to the phase. In Figure 18–8a, they are associated with variations in specimen thickness and are confined primarily to the crystals, while in Figure 18–8b they correspond to magnetic lines of force, which extend smoothly from within the crystals to the surrounding region. Figure 18–8c shows line profiles measured across the large and small magnetite crystals visible close to the centers of Figure 18–8a and b, in a direction perpendicular to the chain axis. Individual experimental data points are shown as open circles. Corresponding simulations based on Eq. (8) are shown on the same axes. The darker solid line shows the best-fitting simulation to the data for the larger crystal, on the assumption that the external shape is formed from a combination of {111}, {110}, and {100} faces. The simulation corresponds to a distorted hexagonal shape in cross section (shown as an inset above the figure). The lighter line shows a worse fit, provided by assuming a diamond shape in cross section.

3.3.2 Co Nanoparticle Rings

An illustration of the characterization of magnetostatic interactions between particles that each contains a single magnetic domain is provided by the examination of rings of 20-nm-diameter crystalline Co particles, as shown in Figure 18–9. Such rings are appealing candidates for high density information storage applications because they are expected to form chiral domain states that exhibit flux closure (FC). Nanoparticle rings are also of interest for the development of electron holography because their magnetization directions cannot be reversed by applying an in-plane external field. As a result, phase images were obtained both before and after inverting the specimen. The resulting pairs of phase images were aligned in position and angle, and their sum and difference calculated as described above. Figure 18–9a shows a low magnification bright-field image of the Co rings (Tripp et al.,

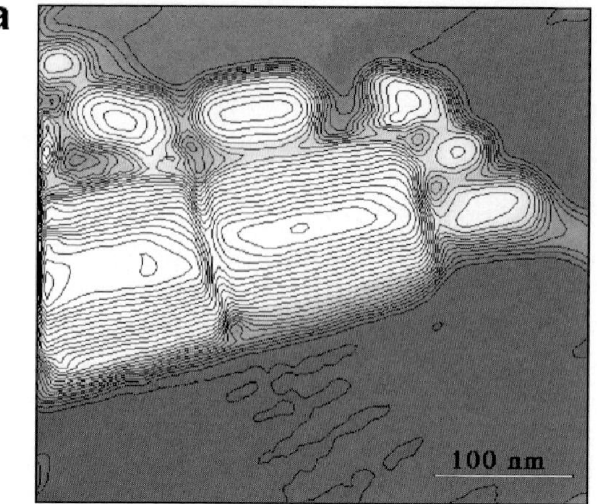

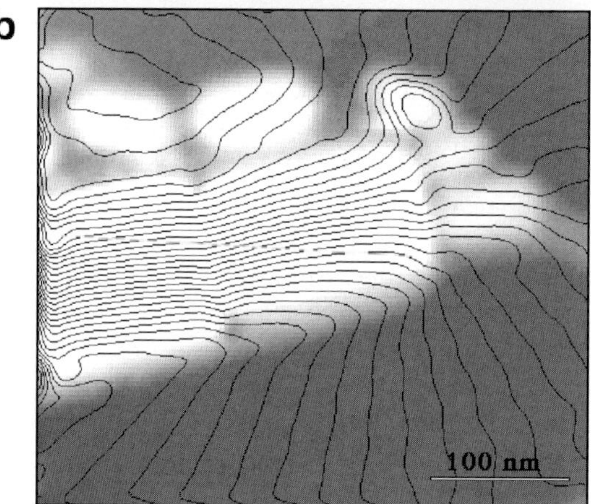

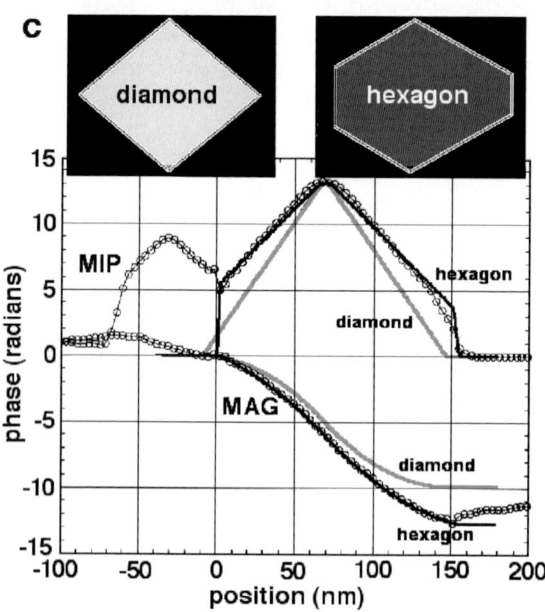

Figure 18–8. Phase contours showing (a) the mean inner potential and (b) the magnetic contribution to the phase shift at the end of a chain of magnetite crystals from magnetotactic bacteria collected from a brackish lagoon at Itaipu in Brazil. The contours have been overlaid onto the mean inner potential contribution to the phase. (c) Line profiles obtained from images (a) and (b) across the large and small magnetite crystals close to the center of each image. The experimental data are shown as open circles. The darker line shows the best-fitting simulation to the data for the larger crystal, corresponding to a distorted hexagonal cross section (shown above the figure). The lighter line shows the worse fit that results from assuming a diamond shape in cross section (also shown above the figure). (Reprinted from Dunin-Borkowski et al., 2004a.)

Figure 18–9. (a) Low magnification bright-field image of self-assembled Co nano-particle rings and chains deposited onto an amorphous carbon support film. Each Co particle has a diameter of between 20 and 30 nm. (b–e) Magnetic phase contours (128× amplification; 0.049 radian spacing), formed from the magnetic contribution to the measured phase shift, in four different nanoparticle rings. The outlines of the nanoparticles are marked in white, while the direction of the measured magnetic induction is indicated both using arrows and according to the color wheel shown in (f) (red = right, yellow = down, green = left, blue = up). (Reprinted from Dunin-Borkowski et al., 2004b.) (See color plate.)

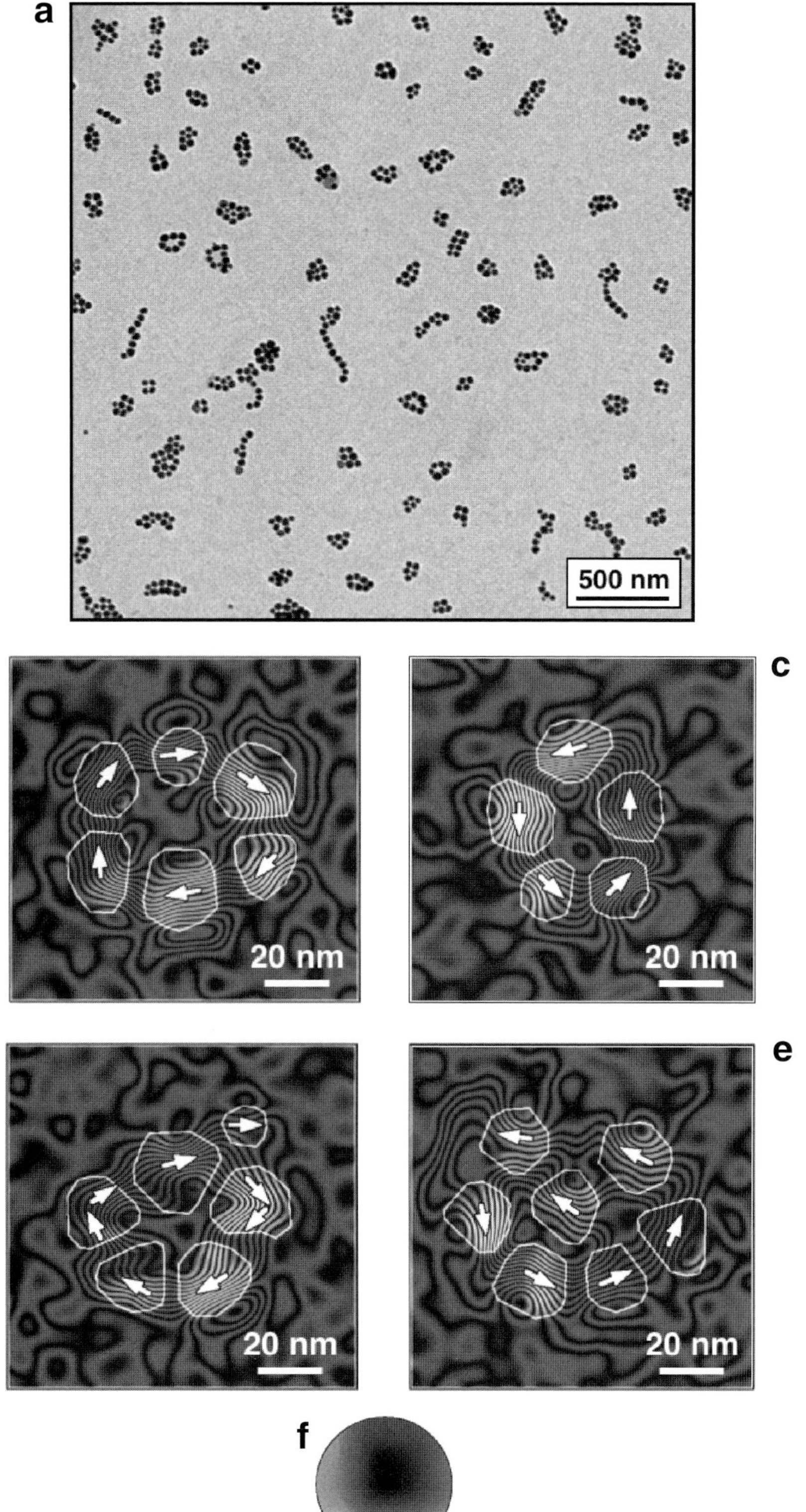

2002). A variety of self-assembled structures is visible, including five- and six-particle rings, chains, and closely packed aggregates. The particles are each encapsulated in a 3- to 4-nm oxide shell. Figure 18–9b–d shows magnetic FC states in four different Co particle rings, measured using electron holography at room temperature in zero-field conditions (Tripp et al., 2003). The magnetic flux lines, which are formed from the cosine of 128 times the magnetic contribution to the measured phase shift, reveal the in-plane induction within each ring ensemble. Further electron holography experiments show that the chirality of the FC states can be switched *in situ* in the TEM by using an out-of-plane magnetic field.

3.3.3 FeNi Nanoparticle Chains

The magnetic properties of nanoparticle chains have been studied for many years (e.g., Jacobs and Bean, 1955). However, there are few experimental measurements of the critical sizes at which individual particles that are arranged in chains are large enough to support magnetic vortices rather than single domains. Previous electron holography studies of magnetic nanoparticle chains (e.g., Seraphin et al., 1999; Signoretti et al., 2003) have never provided direct images of such vortex states. Here, we illustrate the use of electron holography to characterize chains of ferromagnetic FeNi crystals, whose average diameter of 50 nm is expected to be close to the critical size for vortex formation (Hÿtch et al., 2003). Figure 18–10a shows a chemical map of a chain of $Fe_{0.56}Ni_{0.44}$ nanoparticles, acquired using a Gatan imaging filter. The particles are each coated in a 3-nm oxide shell. A defocused bright-field image and a corresponding electron hologram from part of a chain are shown in Figure 18–10b and c, respectively. The mean inner potential contribution to the phase shift was again determined by using the field of the microscope objective lens to magnetize each chain parallel and then antiparallel to its length. The external field was removed before finally recording holograms in field-free conditions. Figure 18–11a and b shows the remanent magnetic states of two chains of $Fe_{0.56}Ni_{0.44}$ particles, measured using electron holography. For a 75-nm $Fe_{0.56}Ni_{0.44}$ particle sandwiched between two smaller particles (Figure 18–11a), closely spaced contours run along the chain in a channel of width 22 ± 4 nm. A comparison of the result with micromagnetic simulations (Hÿtch et al., 2003) indicates that the particle contains a vortex with its axis parallel to the chain axis, as shown schematically in Figure 18–11c. In Figure 18–11b, a vortex can be seen end-on in a 71-nm particle at the end of a chain. The positions of the particle's neighbors determine the handedness of the vortex, with the flux channel from the rest of the chain sweeping around the core to form concentric circles (Figure 18–11d). The vortex core, which is now perpendicular to the chain axis, is only 9 ± 2 nm in diameter. The larger value of 22 nm observed in Figure 18–11a results from magnetostatic interactions along the chain. Vortices were never observed in particles below 30 nm in size, while intermediate states were observed in 30- to 70-nm particles. Similar particles with an alloy concentration of $Fe_{0.10}Ni_{0.90}$ contain wider flux channels of diameter ~70 nm, and single domain states when the par-

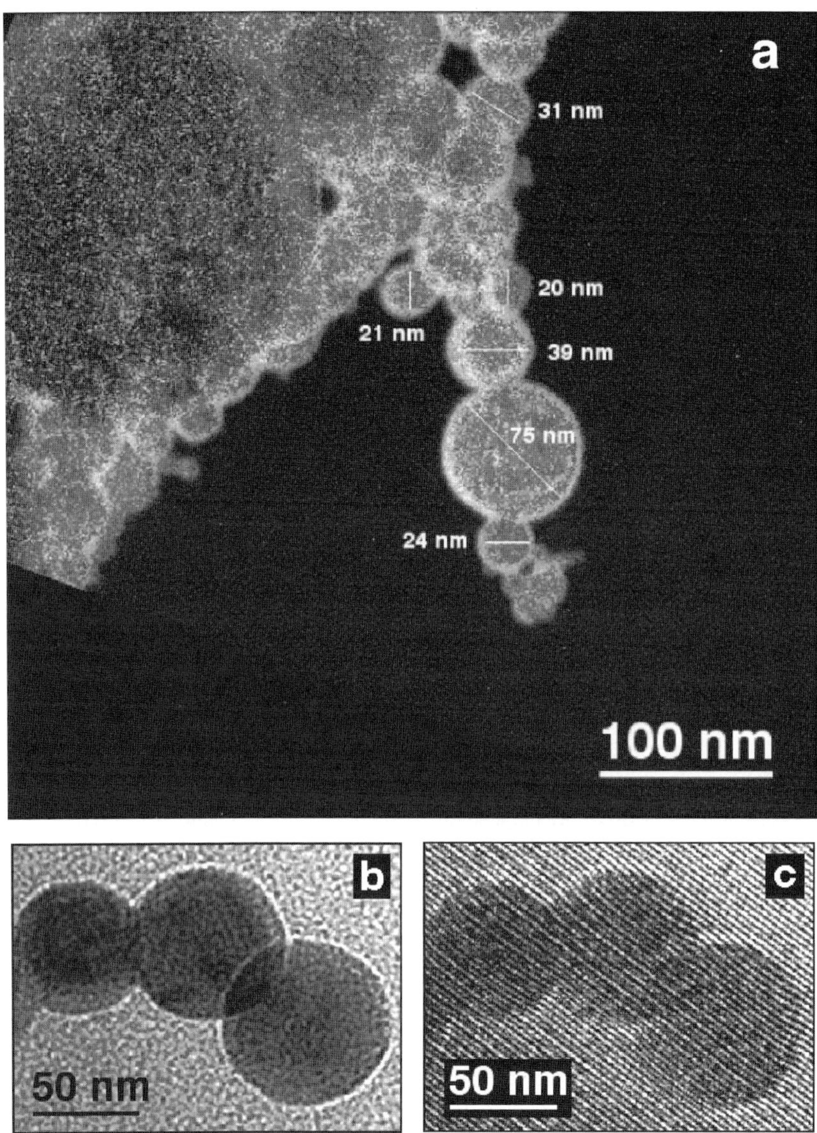

Figure 18–10. (a) Chemical map of $Fe_{0.56}Ni_{0.44}$ nanoparticles, obtained using three-window background-subtracted elemental mapping with a Gatan imaging filter, showing Fe (red), Ni (blue), and O (green). (b) Bright-field image and (c) electron hologram of the end of a chain of $Fe_{0.56}Ni_{0.44}$ particles. The hologram was recorded using an interference fringe spacing of 2.6nm. (Reprinted from Dunin-Borkowski et al., 2004b.) (See color plate.)

ticles are above ~100nm in size. The complexity of such vortex states highlights the importance of controlling the shapes, sizes, and positions of closely spaced magnetic nanocrystals for applications in magnetic storage devices.

3.3.4 Planar Arrays of Magnetite Nanoparticles

The magnetic behavior of the chains and rings of nanomagnets described above contrasts with that of a regular two-dimensional array

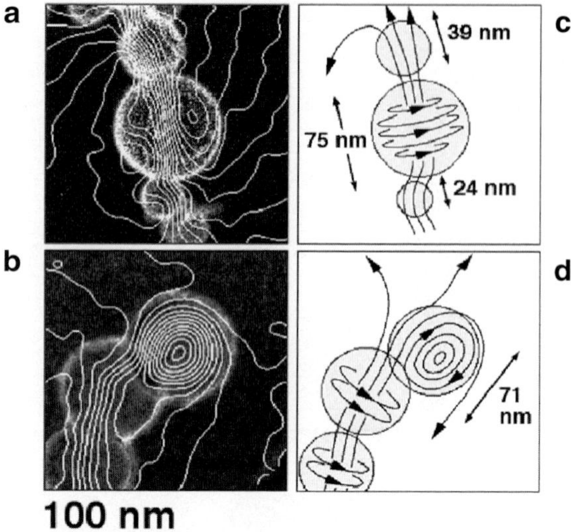

Figure 18–11. (a and b) Experimental phase contours showing the strength of the local magnetic induction (integrated in the electron beam direction) in two different chains of $Fe_{0.56}Ni_{0.44}$ particles, recorded with the electron microscope objective lens switched off. The particle diameters are (a) 75 nm between two smaller particles and (b) 71 nm at the end of a chain. Contours, whose spacings are 0.083 and 0.2 radians for images (a) and (b), respectively, have been overlaid onto oxygen maps of the particles recorded using a Gatan imaging filter. The mean inner potential contribution to the measured phase shift has been removed from each image. (c and d) Schematic representations of the magnetic microstructure in the chains. Magnetic vortices spinning about the chain axis are visible in (c) and (d). A vortex spinning perpendicular to the chain axis is also visible in (d). (Reprinted from Dunin-Borkowski et al., 2004b.)

of closely spaced crystals. Figure 18–12 shows chemical maps of a crystalline region of a naturally occurring magnetite–ulvöspinel $(Fe_3O_4–Fe_2TiO_4)$ mineral specimen, which has exsolved during slow cooling to yield an intergrowth of magnetite-rich blocks separated by nonmagnetic ulvöspinel-rich lamellae (Price, 1981). The Fe and Ti chemical maps shown in Figure 18–12 were obtained using three-window background-subtracted elemental mapping with a Gatan imaging filter. Exsolution lamellae subdivide the grain into a fairly regular array of magnetite-rich blocks. The specimen thickness increases from 70 nm at the top of the region to 195 nm at the bottom. The magnetite blocks are, therefore, roughly equidimensional. Remanent magnetic states were recorded by tilting the specimen in zero field and then turning the objective lens on fully to saturate the sample, to provide a known starting point from which further fields could be applied. The objective lens was then turned off, the specimen tilted in zero field in the opposite direction, and the objective lens was excited partially to apply a known in-plane field component to the specimen in the opposite direction. The objective lens was switched off and the sample tilted back to 0° in zero field to record each hologram. This procedure was repeated for a number of different applied fields (Harrison et al., 2002). Mean inner potential contributions to the measured phase shifts were removed using a procedure different from that used for the chains and rings of nanoparticles described above. Although both thickness and composition vary in the magnetite–ulvöspinel specimen, the different compositions of magnetite and ulvöspinel are compensated by their densities in such a way that their mean inner potentials are exactly equal. As a result, only a thickness correction is required. The local specimen thickness across the region of interest was determined in units of inelastic mean free path by using energy-filtered imaging. This thickness measurement was then used to determine the mean inner potential contribution to the phase shift, which was in turn used to establish the magnetic contribution to the phase. Figure 18–13 shows eight of the resulting remanent magnetic states recorded after applying the in-plane fields indicated. The black contour lines provide the direction and magnitude of the magnetic induction in the plane of the sample, which can be correlated with the positions of the magnetite blocks (outlined in white). The direction of the induction is also indicated using colors and arrows, according to the color wheel shown at the bottom of the figure. Figure 18–13 shows that the magnetic domain structure in this sample is extremely complex. In Figure 18–13, the smallest block observed to form a vortex is larger than the predicted minimum size of 70 nm for vortices to form in *isolated* cubes of magnetite. The abundance of single domain states implies that they have lower energy than vortex states in the presence of strong interactions. The demagnetizing energy, which normally destabilizes the single domain state with respect to the vortex state in isolated particles, is greatly reduced in an array of strongly interacting particles.

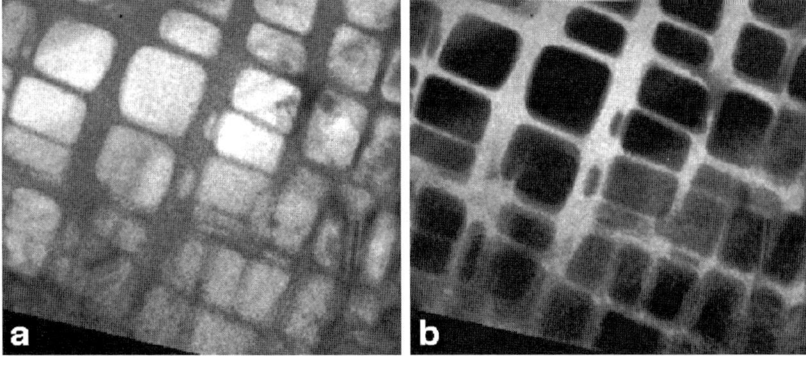

Figure 18–12. Three-window background-subtracted elemental maps acquired from a naturally occurring titanomagnetite sample with a Gatan imaging filter using (a) the Fe L edge and (b) the Ti L edge. Brighter contrast indicates a higher concentration of Fe and Ti in (a) and (b), respectively. (Reprinted from Dunin-Borkowski et al., 2004b.)

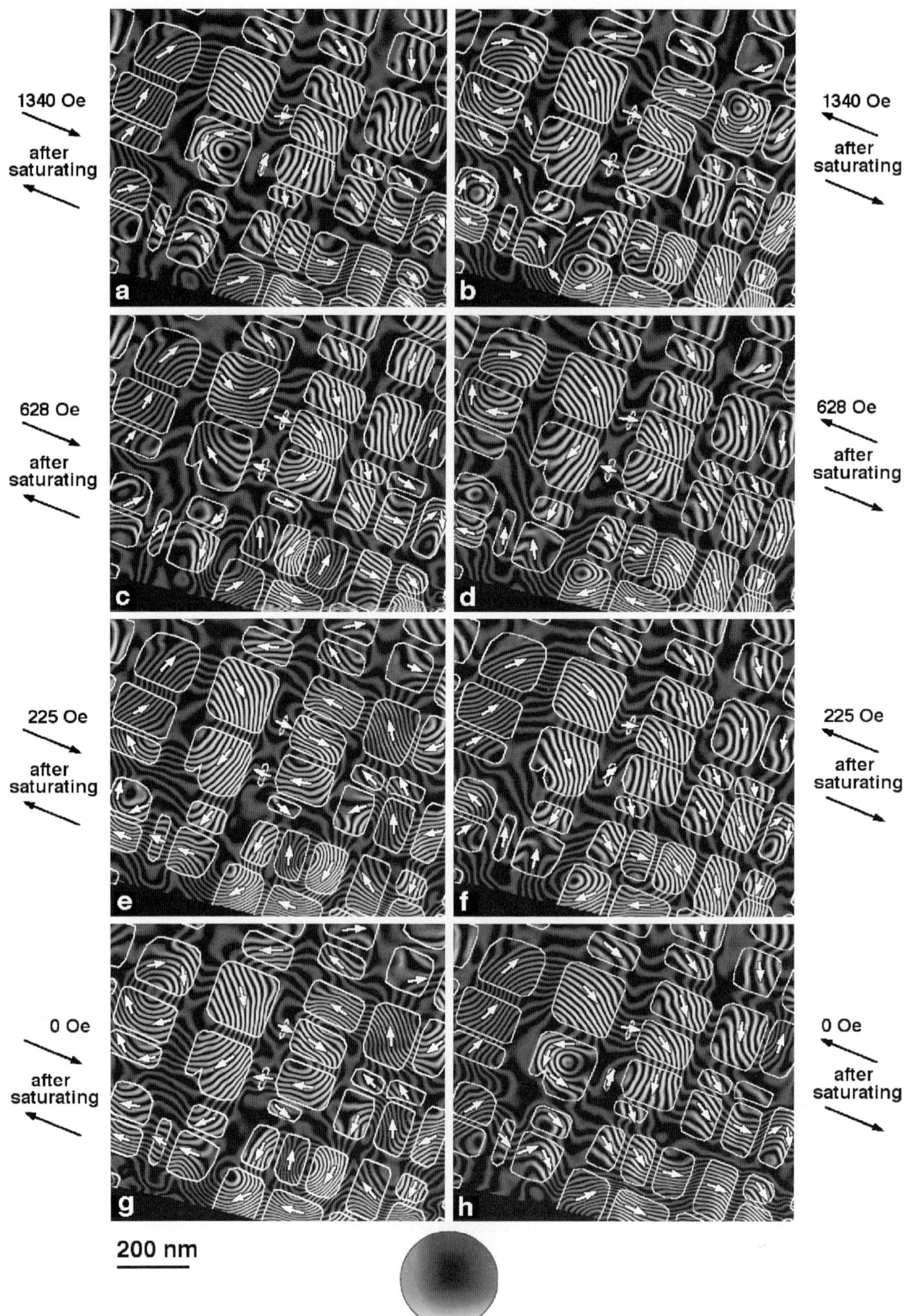

1340 Oe
after saturating
1340 Oe
after saturating
628 Oe
after saturating
628 Oe
after saturating
225 Oe
after saturating
225 Oe
after saturating
0 Oe
after saturating
0 Oe
after saturating
a
b
c
d
e
f
g
h
200 nm

3.3.5 *Lithographically Patterned Magnetic Nanostructures*

Specimen preparation presents a challenge for many samples of interest that contain nanostructured magnetic materials. One example is provided by nanomagnet arrays that have been fabricated directly onto an Si substrate using interferometric lithography (Ross, 2001). Figure 18–14a shows a scanning electron microscope image of nominally 100-nm-diameter 20-nm-thick Co dots fabricated on Si in a square array of side 200 nm. The dots were prepared for TEM examination using focused ion beam milling in plan-view geometry, by micromachining a trench from the substrate side of the specimen to leave a free-standing 10×12-μm membrane of crystalline Si, which was approximately 100 nm in thickness and contained over 3000 Co dots. Figure 18–14b shows an off-axis electron hologram recorded from part of the electron-transparent membrane containing the dots. The specimen was tilted slightly away from zone axis orientations of the underlying Si substrate to minimize diffraction contrast. The specimen edge is toward the bottom left of the figure (Dunin-Borkowski et al., 2001). Figure 18–14c and d shows contours of spacing $0.033 \approx \pi/94$ radians that have been added to the (slightly smoothed) magnetic contribution to the holographic phase, for two different remanent magnetic states of the Co dots. In Figure 18–14c, which was recorded after saturating the dots upward and then removing the external field, the dots are oriented magnetically in the direction of the applied field. In contrast, in Figure 18–14d, which was formed by saturating the dots upward, applying a 382 Oe downward field and then removing the external field, the dots are magnetized in a range of directions. The experiments show that the dots are sometimes magnetized out of the plane (e.g., at the bottom left of Figure 18–14d). The measured saturation magnetizations are smaller than expected for pure Co, possibly because of oxidation or damage sustained during specimen preparation. Similar electrodeposited 57-nm-diameter 200-nm-high Ni pillars arranged in square arrays of side 100 nm, which were prepared for TEM examination using focused ion beam milling in a cross-sectional geometry, have also been examined. Despite their shape, not all of the Ni pillars were magnetized parallel to their long axes. Instead, they interacted with each other strongly, with two, three, or more adjacent pillars combining to form vortices.

◀──

Figure 18–13. Magnetic phase contours from the region shown in Figure 8–12, measured using electron holography. Each image was acquired with the specimen in magnetic field-free conditions. The outlines of the magnetite-rich regions are marked in white, while the direction of the measured magnetic induction is indicated both using arrows and according to the color wheel shown at the bottom of the figure (red = right, yellow = down, green = left, blue = up). Images (a), (c), (e), and (g) were obtained after applying a large (>10,000 Oe) field toward the top left, then the indicated field toward the bottom right, after which the external magnetic field was removed for hologram acquisition. Images (b), (d), (f), and (h) were obtained after applying identical fields in the opposite directions. (Reprinted from Harrison et al., 2002.) (See color plate.)

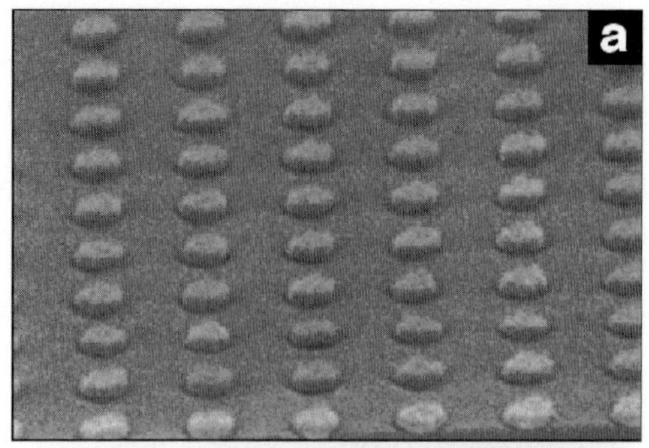

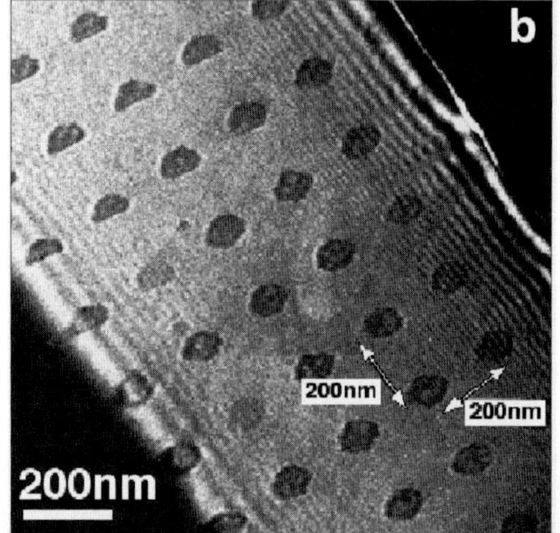

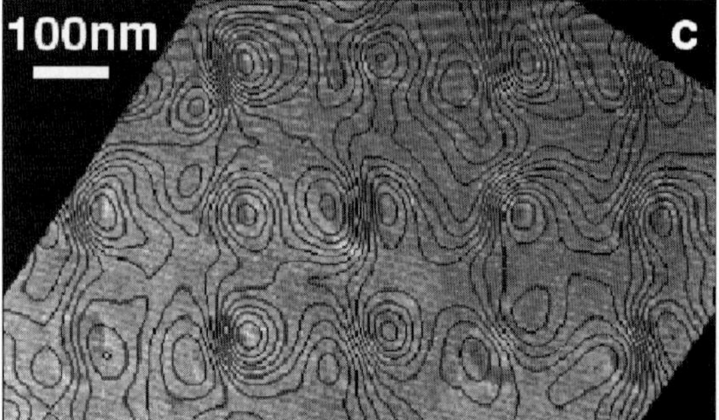

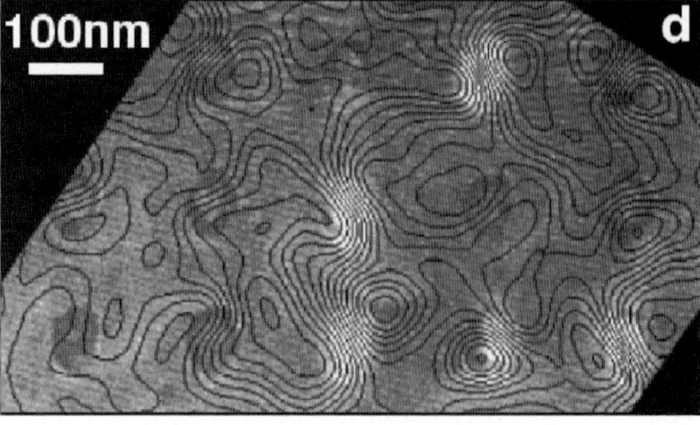

Figure 18–14. (a) Scanning electron microscope image of 100-nm-diameter 20-nm-thick Co dots fabricated on Si in a square array of side 200 nm using interferometric lithography. (b) Off-axis electron hologram of part of an electron-transparent membrane containing the dots, prepared using focused ion beam milling. The hologram was acquired at 200 kV using a Philips CM200 FEGTEM, a biprism voltage of 160 V, a holographic interference fringe spacing of 3.05 nm, and an overlap width of 1.04 μm. (c and d) Magnetic contributions to the measured electron holographic phase shift for two remanent magnetic states. The contour spacing is 0.033 radians: (c) was formed by saturating the dots upward and then removing the external field; (d) was formed by saturating the dots upward, applying a 382 Oe downward field, and then removing the external field. (Reprinted from Dunin-Borkowski et al., 2001.)

Results similar to those shown in Figure 18–14 have been obtained from a wide range of larger lithographically patterned structures, many of which show multidomain behavior (Dunin-Borkowski et al., 2000; Hu et al., 2005; Heumann et al., 2005). Few phase contours are visible outside such elements when they support magnetic flux closure states. Electron holography has also been used to provide information about magnetic interactions between closely separated ferromagnetic layers within individual Co/Au/Ni spin-valve elements (Smith et al., 2000). The presence of two different contour spacings at different applied fields in such elements is associated with the reversal of the magnetization direction of the Ni layer in each element before the external field is reduced to zero as a result of flux closure associated with the strong fringing field of the magnetically more massive and closely adjacent Co layer.

3.3.6 Co Nanowires

An important question relates to the minimum size of a nanostructure in which magnetic fields can be characterized successfully using electron holography. This point is now addressed through the characterization of 4-nm-diameter single crystalline Co nanowires (Snoeck et al., 2003). The difficulty of this measurement results from the fact that the mean inner potential contribution to the phase shift at the center of a 4-nm wire relative to that in vacuum is 0.57 radians (assuming a value for V_0 of 22 V), whereas the step in the magnetic contribution to the phase shift across the wire is only 0.032 radians (assuming a value for B of 1.6 T). Figure 18–15a shows a bright-field TEM image of a bundle of 4-nm-diameter Co wires, which are each between a few hundred nanometers and several hundred micrometers in length. Magnetic contributions to the phase shift were obtained by recording two holograms from each area of interest, where the wires were magnetized parallel and then antiparallel to their length by tilting the sample by ±30° about an axis perpendicular to the wire axis and using the conventional microscope objective lens to apply a large in-plane field to the specimen. The lens was then switched off and the sample returned to zero tilt to record each electron hologram. This procedure relies on the ability to reverse the magnetization in the sample exactly, which is a good assumption for these narrow and highly anisotropic wires. Figure 18–15b shows the magnetic contribution to the measured phase shift for an isolated wire, in the form of contours that are spaced 0.005 radians apart. The contours have been overlaid onto an image showing the mean inner potential contribution to the phase shift, so that they can be correlated with the position of the wire. The magnetic signal is weak and noisy, and was smoothed before forming the contours. The closely spaced contours along the length of the wire confirm that it is magnetized along its axis. The fact that they are not straight is intriguing. However, this effect may result from smoothing of the signal, which is noisy and weak. Figure 18–16a shows a montage of three holograms obtained close to the end of a bundle of Co wires, which was magnetized approximately parallel to its length. The magnetic contribution to the measured phase shift is shown in Figure

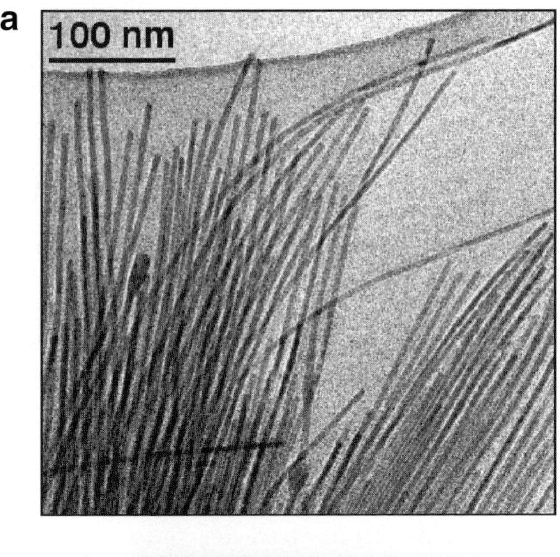

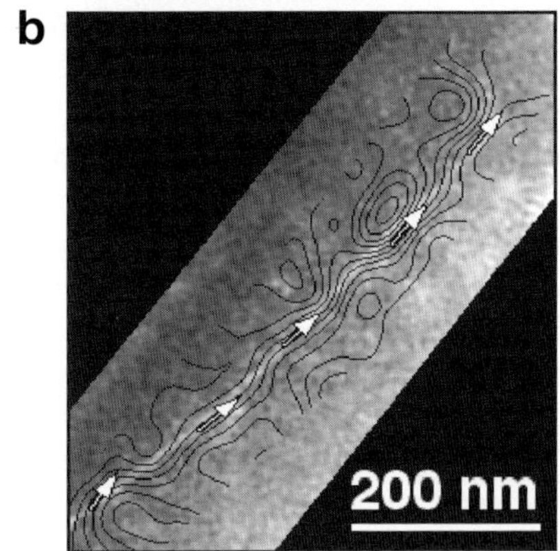

Figure 18–15. (a) Bright-field image of the end of a bundle of Co nanowires adjacent to a hole in a carbon support film. (b) Contours (0.005 radian spacing) generated from the magnetic contribution to the phase shift for a single isolated Co nanowire, superimposed onto the mean inner potential contribution to the measured phase shift. (Reprinted from Snoeck et al., 2003.)

18–16b in the form of contours, which are spaced 0.25 radians apart. The wires channel the magnetic flux efficiently along their length, and they fan out as the field decreases in strength at the end of the bundle. Although the signal from the bundle appears overall to obscure that from individual wires and junctions, these details can be recovered by increasing the density of the contours (Snoeck et al., 2003). The slight asymmetry between the contours on either side of the bundle in Figure 18–16b may result from the fact that the reference wave is affected by the magnetic leakage field of the bundle, which acts collectively as though it were a single wire of larger diameter. The step in phase across the bundle is (9.0 ± 0.2) radians, which is consistent with the presence of (280 ± 7) ferromagnetically coupled wires.

3.3.7 Cross-Sectional Specimens

One of the most challenging problems for electron holography of magnetic materials is the quantitative measurement of the magnetic properties of nanometer-scale magnetic layers when examined in cross section. The primary difficulty is the presence of rapid and unknown variations in both the composition and the thickness of the specimen, from which the weak magnetic signal must be separated. In a cross-sectional sample, the effects of variations in specimen thickness on the measurements cannot be eliminated by using the normalized amplitude of the hologram [Eq. (14)], both because the mean free path in each material in such a cross-sectional specimen is usually unknown

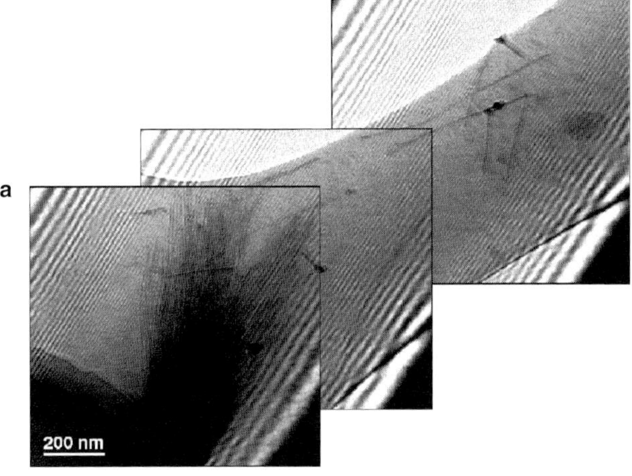

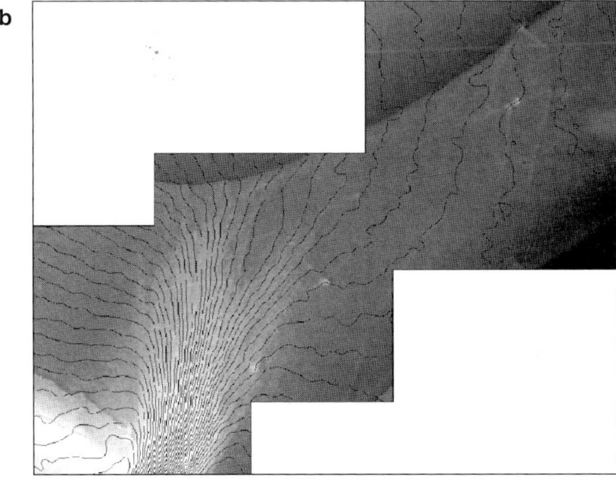

Figure 18–16. (a) Montage of three electron holograms acquired from the end of a bundle of Co nanowires. The biprism voltage is 210 V, the acquisition time for each hologram 16 s, the holographic interference fringe spacing is 3.9 nm, and the holographic overlap width is 1160 nm. No objective aperture was used. (b) Magnetic remanent state, displayed in the form of contours (0.25 radian spacing), generated from the measured magnetic contribution to the electron holographic phase shift after saturating the wires in the direction of the axis of the bundle. The contours are superimposed onto the mean inner potential contribution to the phase shift. (Reprinted from Snoeck et al., 2003.)

and because the amplitude image is in general noisy and may contain strong contributions from diffraction and Fresnel contrast. However, by rearranging Eqs. (8) and (16), it can be shown that specimen thickness effects may be removed by plotting the difference in the phase gradient between images in which the magnetization has reversed divided by the average of their phases, multiplied by a constant and by the value of the mean inner potential of each magnetic layer separately. Formally, this procedure can be written

$$\left(\frac{C_E \hbar V_0 (x, y)}{e}\right)\left\{\frac{\Delta[d\phi(x, y)/dx]}{\langle\phi(x, y)\rangle}\right\}$$
$$= \frac{\Delta B_\perp (x, y)}{\{1-[e/C_E \hbar V_0 (x, y)]\}\left[\langle\int B_\perp (x, y)t(x, y)dx\rangle/t(x, y)\right]} \quad (23)$$

According to Eq. (23), by combining phase profiles and their gradients (evaluated in a direction perpendicular to the layers) from successive holograms with the magnetization direction reversed, the specimen thickness profile can be eliminated and the magnetic induction in each layer determined quantitatively. Both the magnitude and the sign of $\Delta B_\perp(x,y) = 2B_\perp(x,y)$ are obtained exactly using Eq. (23) if the magnetization reverses exactly everywhere in the sample. (The denominator on the right-hand side of the equation is then unity.) Furthermore, nonzero values are returned only in regions where the magnetization has changed. Figure 18–17 illustrates the application of Eq. (23) to a cross-sectional magnetic tunnel junction that contains a layer sequence of 22nm Co/4nm HfO_2/36nm CoFe on an Si substrate (McCartney and

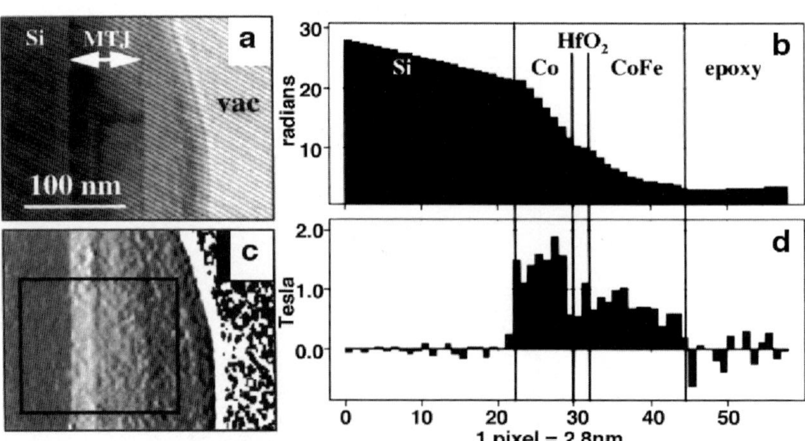

Figure 18–17. (a) Off-axis electron hologram obtained from a magnetic tunnel junction containing a 4-nm HfO_2 tunnel barrier. (b) Measured phase profile across the layers in the tunnel junction structure. (c) Image formed by recording two holograms with opposite directions of magnetization in the specimen, and subsequently taking the difference between the recorded phase gradients (calculated in a direction perpendicular to the layers) and dividing by the average of the two phases. (d) Measured magnetic induction in the tunnel junction sample, generated by multiplying a line profile obtained from image (c) by a constant (see text for details), with the vertical scale now plotted in units of Tesla. (Reprinted from McCartney and Dunin-Borkowski, 1998.)

Dunin-Borkowski, 1998). Two holograms were obtained, similar to that shown in Figure 18–17a, between which the magnetization directions of the Co and CoFe layers in the specimen were reversed *in situ* in the electron microscope. Figure 18–17b shows an unwrapped phase profile obtained from the hologram in Figure 18–17a by taking a line profile in the direction perpendicular to the layers. Phase profiles from the two holograms appeared almost identical irrespective of the direction of magnetization. The application of Eq. (23) to the two phase images results in the image shown in Figure 18–17c. The line profile in Figure 18–17d was obtained by averaging Figure 18–17c parallel to the direction of the layers. As predicted, Figure 18–17d, which should by now be independent of variations in composition and specimen thickness, is nonzero only in the magnetic layers and yields a value for the magnetic induction in the Co layer of 1.5 T (assuming a mean inner potential of 25 V). In a similar experiment, holograms of $La_{0.5}Ca_{0.5}MnO_3$ have recently been acquired both above and below the Curie temperature of the material to remove specimen thickness and mean inner potential contributions from the measured phase (Loudon et al., 2003).

3.4 Quantitative Measurements, Micromagnetic Simulations, and Resolution

A particular strength of electron holography is its ability to provide quantitative information about magnetic properties. For example, the magnetic moment of a nanoparticle can be obtained from the relation

$$m_x = \left(\frac{\hbar}{e}\right) \int_{y=-\infty}^{y=+\infty} \int_{x=-\infty}^{x=+\infty} \frac{\partial}{\partial y} \phi_{mag}(x, y)dxdy \tag{24}$$

where ϕ_{mag} is the magnetic contribution to the phase shift and y is a direction perpendicular to x in the plane of the specimen. According to Eq. (24), the magnetic moment in a given direction can be obtained by measuring the area under the first differential of ϕ_{mag} evaluated in the perpendicular direction. The contribution of stray magnetic fields to the moment is included in this calculation if the integration is carried out over a large enough distance from the particle.

The need to compare electron holographic measurements with micromagnetic simulations results from the sensitivity of the magnetic domain structure in nanoscale materials and devices to their detailed magnetic history. Differences in the starting magnetic states on a scale that is too small to be distinguished visually, as well as interelement coupling and the presence of out-of-plane magnetic fields, are all important for the formation of subsequent domain states, and, in particular, to the sense (the handedness) with which magnetic vortices unroll (Dunin-Borkowski et al., 1999). The sensitivity of the domain structure to such effects emphasizes the need to correlate high quality experimental holographic measurements with micromagnetic simulations.

The spatial resolution that can be achieved in phase images is determined primarily by the spacing of the holographic interference fringes.

However, the contrast of these fringes decreases as their spacing is reduced, and the recording process is also dominated by Poisson-distributed shot noise (Lichte et al., 1987). These parameters are affected by the illumination diameter, exposure time, and biprism voltage. The final "phase resolution" (Harscher and Lichte, 1996) and "spatial resolution" are always inherently linked, in the sense that a small phase shift can be measured with high precision and poor spatial resolution, or with low precision but high spatial resolution. In each of the examples described above, the recorded phase images were always smoothed slightly to remove noise, and the spatial resolution of the magnetic information was estimated typically to be between 10 and 20 nm. This procedure is necessarily subjective, and great care is required to ensure that artifacts are not introduced. Higher spatial and phase resolution might possibly also be achieved by recording several holograms of each area of interest and subsequently averaging the resulting phase images.

4 Measurement of Electrostatic Fields

In this section, the application of electron holography to the characterization of electrostatic fields is reviewed. Initial examples are taken from the characterization of electrostatic fringing fields outside electrically biased nanowires. The challenges that are associated with imaging dopant contrast at depletion layers in semiconductors are then described, before discussing the characterization of interfaces at which both charge redistribution and changes in chemistry are possible.

4.1 Field-Emitting Carbon Nanotubes

Early experiments on tungsten microtips demonstrated that electron holography could be used to measure electrostatic fringing fields in biased samples (Matteucci et al., 1992). Further studies were made on pairs of parallel 1-μm-diameter Pt wires held at different potentials (Matteucci et al., 1988) and on single conducting wires (Kawasaki et al., 1993), and simulations were presented for electrostatic phase plates (Matsumoto and Tonomura, 1996). A more recent example involves the use of electron holography to map the electrostatic potential around the end of an electrically biased multiwalled carbon nanotube (Cumings et al., 2002). Nanotubes were mounted on a three-axis manipulation electrode using conducting epoxy and positioned approximately 6 μm from a gold electrode, as shown in Figure 18–18a. Depending on the applied bias, electrons were emitted from the nanotube. The left hand column of Figure 18–18b shows contoured phase images recorded before a bias V_b was applied to the specimen, and for a bias above the threshold for field emission (approximately 70 V). The upper phase shift map ($V_b = 0$) is featureless around the nanotube, whereas the lower map ($V_b = 120$ V) shows closely spaced 2π phase contours. The right hand column in Figure 18–18b shows the corresponding phase gradient for each image. When $V_b = 0$, the phase gradient is featureless around the nanotube, whereas it is concentrated around the nanotube tip when V_b = 120 V. The images shown in Figure 18–18b were interpreted by comparison with simulations, calculated on the assumption that the nanotube could be approximated by a line charge, where the charge distribution was varied until a close fit to the data was found. The fit to the 120 V phase data in Figure 18–18b provided a value of 1.22 V/nm for the electric field at the nanotube tip. This field was stable over time, even when the emission current varied.

4.2 Dopant Potentials in Semiconductors

One of the most elusive yet tantalizing challenges for electron holography has been the quest for a reliable, quantitative approach to the characterization of electrostatic potentials associated with charge redistribution at depletion regions in doped semiconductors. Attempts to tackle this problem have been made since the 1960s using many forms of electron interferometry, both experimentally (e.g., Titchmarsh et al., 1969; Frabboni et al., 1987) and theoretically (e.g., Pozzi and Vanzi, 1982; Beleggia et al., 2000). It is now recognized that TEM specimen preparation can have a profound effect on the contrast visible in holographic phase images of doped semiconductors, either because of physical damage to the specimen surface or because of the implantation of dopant ions such as Ar or Ga during ion milling. An electrically inactive surface layer and/or a doped layer, with a thickness depending on the specimen preparation method, may form at the sample surface. In addition, the specimen may charge up during observation, to such an extent that all dopant contrast is lost. The effects of specimen

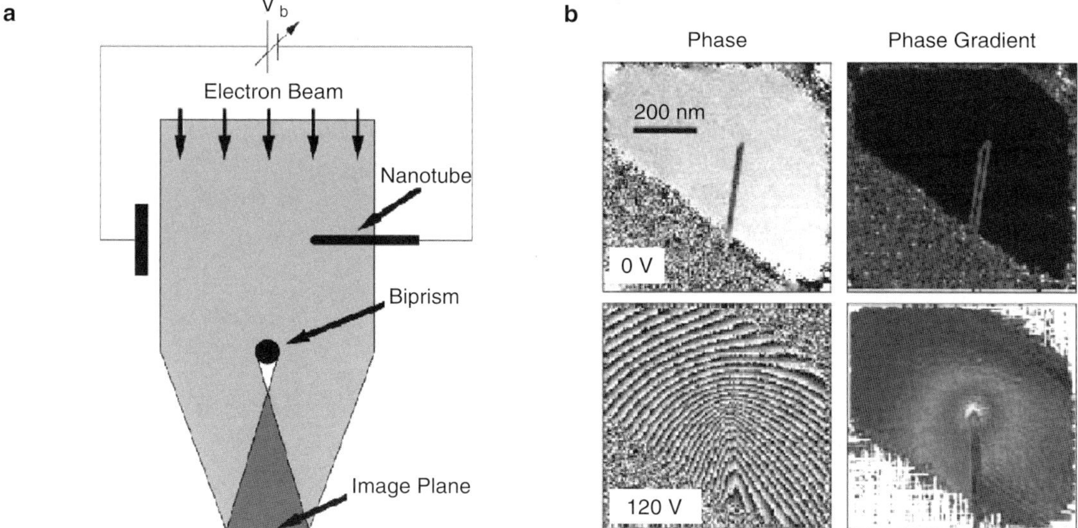

Figure 18–18. (a) Schematic diagram of the experimental set-up used to record electron holograms of field-emitting carbon nanotubes. (b) Phase shift and phase gradient maps determined from electron holograms of a single multiwalled carbon nanotube at bias voltages of 0 and 120 V. The phase gradient indicates where the electric field is strongest. Note the concentration of the electric field at the nanotube tip when the bias voltage is 120 V. (Reprinted from Cumings et al., 2002.)

preparation, and in particular the electrical state of near-surface regions, most likely account for many of the anomalous results in early experiments. Recent studies indicate that it may be possible to resolve these problems.

The first unequivocal demonstration of two-dimensional mapping of the electrostatic potential in an unbiased doped semiconductor using electron holography was achieved for metal–oxide–semiconductor (MOS) Si transistors (Rau et al., 1999). The source and drain regions were visible in phase images with a spatial resolution of 10 nm and an energy resolution of close to 0.10 eV. Differential thinning was discounted as a cause of the observed phase shifts, and an optimal specimen thickness of 200–400 nm was identified for such experiments. The transistors were prepared for TEM examination using conventional mechanical polishing and Ar ion milling. A 25-nm-thick electrically altered layer was identified on each surface of the specimen, which resulted in measured built-in voltages of 0.9 ± 0.1 V across each $p–n$ junction, which was lower than the value of 1.0 V predicted for the specified dopant concentration.

More recently, electron holography studies of transistors have been compared with process simulations (Gribelyuk et al., 2002). Figure 18–19a shows a contoured image of the electrostatic potential associated with a 0.35-µm Si device inferred from an electron hologram, where the contours correspond to potential steps of 0.1 V. The B-doped source and drain regions are delineated clearly. In this study, the specimen was prepared primarily using tripod wedge-polishing, followed by limited low-angle Ar ion milling at 3.5 kV. Significantly, no electrically dead surface layer had to be taken into account to quantify the results. Figure 18–19b and c shows a comparison between line profiles obtained from Figure 18–19a and simulations, both laterally across the junction and with depth from the Si surface. Simulations for "scaled loss" and "empirical loss" models, which account for B-implant segregation into the adjacent oxide and nitride layers, are shown. The scaled loss model, which leads to stronger B diffusion, assumes uniform B loss across the device structure, whereas the empirical loss model assumes segregation of the implanted B at the surfaces of the source and drain regions. In both Figure 18–19b and Figure 18–19c, the empirical loss model provides a closer match to the experimental results. Figure 18–19d shows a simulated electrostatic potential map for the same device based on the "empirical loss" model, which matches closely with the experimental image in Figure 18–19a. Overall, this study demonstrated successful mapping of the electrostatic potential in 0.13-µm and 0.35-µm device structures with a spatial resolution of 6 nm and a sensitivity of 0.17 eV.

In early applications of electron holography to dopant delineation, which were carried out on chemically thinned Si samples under conditions of reverse bias (e.g., Frabboni et al., 1985), differences between phase images recorded at different bias voltages were used to visualize external electrostatic fringing fields close to the positions of $p–n$ junctions. Electrostatic potential profiles have recently been measured for reverse-biased Si $p–n$ junctions that were prepared for TEM

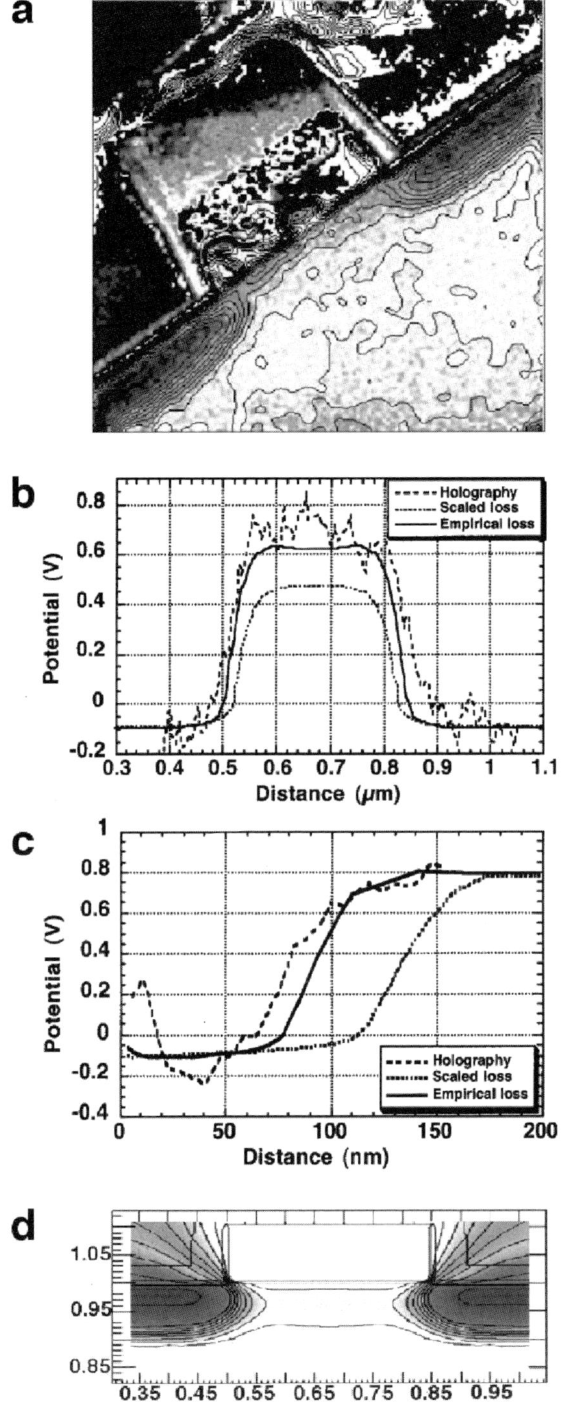

Figure 18–19. (a) Reconstructed maps of the electrostatic potential distribution in a 0.35-μm semiconductor device structure, with a contour step of 0.1 V, recorded at an accelerating voltage of 200 kV using a Philips CM200 FEGTEM. (b) Lateral and (c) depth profiles obtained from the image shown in (a). Predictions from process simulations for "scaled loss" and "empirical loss" models are also shown. (d) Two-dimensional simulated map of the potential based on the "empirical loss" model, with a contour step of 0.1 V. The dimensions are in micrometers. (Reprinted from Gribelyuk et al., 2002.)

examination using focused ion beam milling (Twitchett et al., 2002, 2004, 2005). It is significant to note here that focused ion beam milling is currently the technique of choice for preparing TEM specimens from site-specific regions of integrated circuits. It is therefore important to establish whether holography results obtained from unbiased specimens prepared by focused ion beam milling are reliable. It is also useful to develop a specimen geometry that allows electrical currents to be passed through TEM specimens prepared using this technique.

Specimens for *in situ* electrical biasing were prepared by using a 30-kV FEI 200 focused ion beam workstation to machine parallel-sided electron-transparent membranes at the corners of 1×1-mm $90°$ cleaved squares of wafer, as shown schematically in Figure 18–20a. This geometry allowed electrical contacts to be made to the front and back surfaces of each specimen using a modified single-tilt holder, as shown in Figure 18–20b. Care was taken to expose the region of interest to the focused beam of Ga ions only at a glancing angle to its surface. Figure 18–20c shows a representative holographic phase image recorded from an unbiased Si p–n junction sample prepared by focused ion beam milling. The crystalline thickness was measured independently to be 550 nm using convergent beam electron diffraction. The p-type and n-type regions are delineated clearly as areas of darker and lighter contrast, respectively. The additional "gray" band at the specimen edge is likely to be associated with the presence of an electrically altered layer, which is visible in cross section but is thought to extend around the entire specimen surface. No electrostatic fringing field is visible outside the specimen, indicating that its surface must be an equipotential. Line profiles across the junction were obtained from phase images acquired with different reverse bias voltages applied to a specimen of 390 nm crystalline thickness (Figure 18–20d), as well as from several unbiased specimens. Each profile in Figure 18–20d is qualitatively consistent with the expected potential profile for a p–n junction in a specimen of uniform thickness. The height of the potential step across the junction, $\Delta\phi$, increases linearly with reverse bias voltage V_{appl}, as shown in Figure 18–20e. This behavior is described by the equation

$$\Delta\phi = C_E(V_{bi} + V_{appl})t_{active} \qquad (25)$$

where C_E is defined in Eq. (7) and the p–n junction is contained in an electrically active layer of thickness t_{active} in a specimen of total thickness t. Measurement of the gradient of Figure 18–20e, which is equal to $C_E t_{active}$, provides a value for t_{active} of 340 ± 10 nm, indicating that 25 ± 5 nm of the crystalline thickness on each surface of the TEM specimen is electrically inactive. The intercept with the vertical axis is $C_E V_{bi} t_{active}$, which provides the expected value for the built-in voltage across the junction of 0.9 ± 0.1 V. Depletion widths across the junction measured from the line profiles are higher than expected, suggesting that the electrically active dopant concentration in the specimen is lower than the nominal value. These experiments also show that electrical biasing reactivates some of the dopant that has been passivated by specimen preparation (Dunin-Borkowski et al., 2002). Figure 18–20f shows a four-times-amplified phase image obtained from a $90°$ cleaved

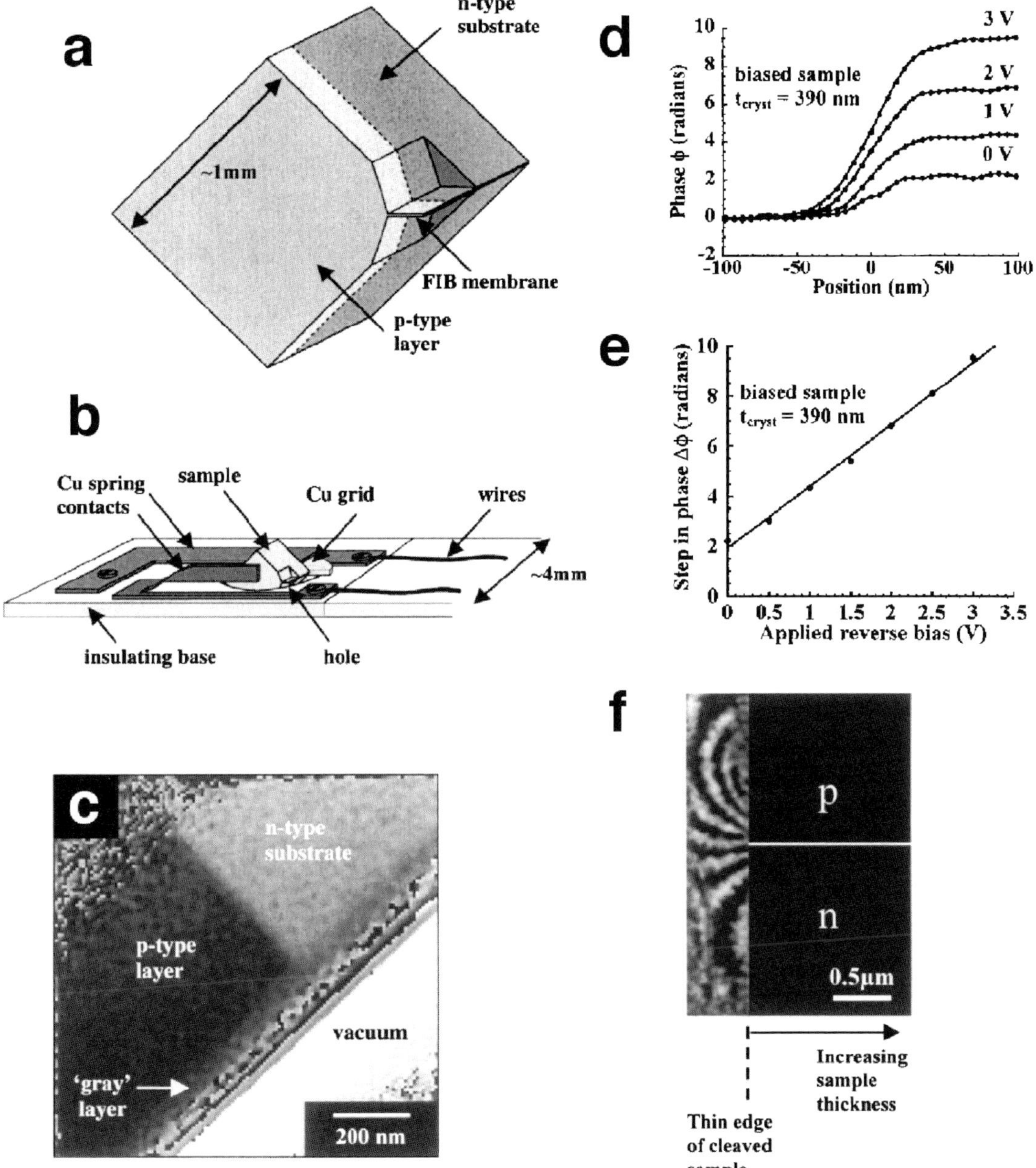

Figure 18–20. (a) Schematic diagram showing the specimen geometry used for applying external voltages to focused ion beam milled semiconductor device specimens containing $p\text{–}n$ junctions *in situ* in the TEM. In the diagram, focused ion beam (FIB) milling has been used to machine a membrane of uniform thickness that contains a $p\text{–}n$ junction at one corner of a $90°$ cleaved wedge. (b) Schematic diagram showing the specimen position in a single tilt electrical biasing holder. The specimen is glued to the edge of a Cu grid using conducting epoxy and then clamped between two spring contacts on an insulating base. (c) Reconstructed phase image acquired from an unbiased Si sample containing a $p\text{–}n$ junction. Note the "gray" layer running along the edge of the specimen, which is discussed in the text. No attempt has been made to remove the 2π phase "wraps" at the edge of the specimen. (d) Phase shift measured across a $p\text{–}n$ junction as a function of reverse bias for a single sample of 390 nm crystalline thickness (measured using convergent beam electron diffraction). (e) The height of the measured step in phase across the junction is shown as a function of reverse bias. (f) Four times-amplified reconstructed phase image, showing the vacuum region outside a $p\text{–}n$ junction in a 2-V reverse-biased cleaved wedge sample that had not been focused ion beam milled. (Reprinted from Twitchett et al., 2002.)

wedge that had *not* been prepared by focused ion beam milling, for an applied reverse bias of 2 V, where an external electrostatic fringing field is visible. Such fringing fields were never observed outside *unbiased* cleaved wedges or any focused ion beam milled specimens, indicating that the surfaces of the present TEM specimens prepared by focused ion beam milling are equipotentials under applied bias.

The importance of minimizing and assessing damage, implantation, and specimen thickness variations when examining focused ion beam milled TEM specimens that contain $p–n$ junctions has been highlighted by results from unbiased samples (Wang et al., 2002a–c, 2005). The most elegant of these experiments involved the use of focused ion beam milling to form a 45° specimen thickness profile, from which both the phase change across the junction and the absolute phase shift relative to vacuum on each side of the junction could be plotted as a function of specimen thickness. The slopes of the phase profiles were then used to determine the built-in voltage across the junction, the mean inner potentials on the p and n sides of the junction, and the electrically altered layer thickness. Using this approach, the built-in voltage across a junction with a dopant concentration of approximately 10^{15} cm^{-3} was measured to be 0.71 ± 0.05 V, while the mean inner potentials of the p and n sides of the junction were measured to be 11.50 ± 0.27 and 12.1 ± 0.40 V, respectively. The electrically altered layer thickness was measured to be approximately 25 nm on each surface of the specimen.

The electrical nature of the surface of a TEM specimen that contains a doped semiconductor can be assessed by comparing experimental holography results with simulations. Such a comparison, performed using commercial semiconductor process simulation software (Beleggia et al., 2001), suggests that electron beam-induced positive charging of the surface of a TEM specimen, at a level of $10^{13}–10^{14}$ cm^{-2}, creates an inversion layer on the p-side of the junction. This layer may explain the absence of electrostatic fringing fields outside the specimen surface, which would otherwise dominate the observed phase contrast (Dunin-Borkowski and Saxton, 1996). Figure 18–21 shows the results of an alternative set of numerical simulations, in which semiclassical equations are used to determine the charge density and potential in a parallel-sided Si sample that contains a $p–n$ junction. The Fermi level on the surface of the specimen is set to a single value to ensure that it is an equipotential (Somodi et al., 2005). The simulations in Figure 18–21 are for symmetrical junctions with dopant concentrations of 10^{18}, 10^{17}, and 10^{16} cm^{-3}. Contours of spacing 0.05 V are shown in each figure. As either the dopant concentration or the specimen thickness decreases, a correspondingly smaller fraction of the specimen retains electrical properties that are close to those of the bulk device. In the simulations, the average step in potential across the junction through the thickness of the specimen, which is insensitive to the surface state energy, is reduced from that in the bulk device. This reduction is greatest for low sample thicknesses and low dopant concentrations. In practice, as a result of additional complications from oxidation, physical damage, and implantation, the simulations shown in Figure 18–21 are likely to

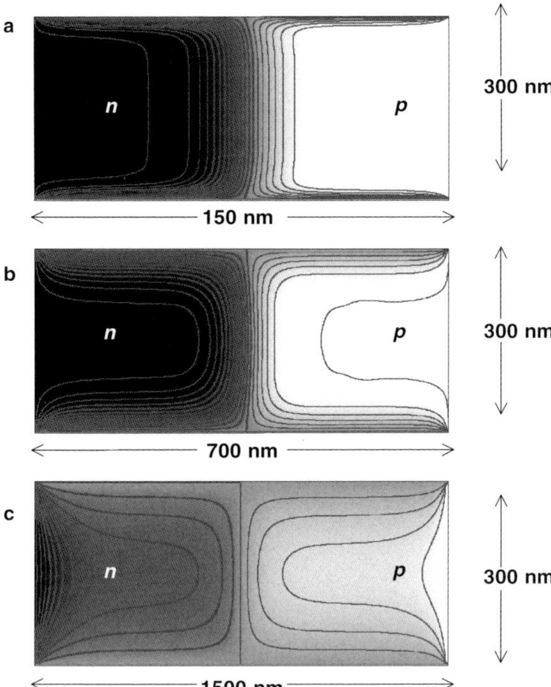

Figure 18–21. Simulations of electrostatic potential distributions in parallel-sided slabs of thickness 300 nm containing abrupt, symmetrical Si p–n junctions formed from (a) 10^{18}, (b) 10^{17}, and (c) 10^{16} cm^{-3} of Sb (n-type) and B (p-type) dopants. The potential at the specimen surfaces is 0.7 eV above the Fermi level, and contours of spacing 0.05 V are shown. The horizontal scale is different in each figure to show the variation in potential close to the position of the junction. The simulations were generated using a two-dimensional rectangular grid. (Reprinted from Somodi et al., 2005.)

be an underestimate of the full modification of the potential from that in the original device.

The ways in which the sample preparation technique of "wedge-polishing" affects both the dead layer thickness and specimen charging have been explored experimentally for a one-dimensional p–n junction in Si by McCartney et al. (2002). A specimen was prepared from a p-type wafer that had been subjected to a shallow B implant and a deeper P implant, resulting in the formation of an n-type well and a p-doped surface region. Phase images were obtained before and after coating one side of the specimen with approximately 40 nm of carbon. Profiles obtained from the uncoated sample showed an initial increase in the measured phase going from vacuum into the specimen, then dropping steeply and becoming negative at large thicknesses. This behavior was not observed after carbon coating, suggesting that it is associated with sample charging that results from the electron beam-induced emission of secondary electrons.

Similar charging effects can be seen directly in two dimensions in Figure 18–22. Figure 18–22a shows a bright-field image of a linear array

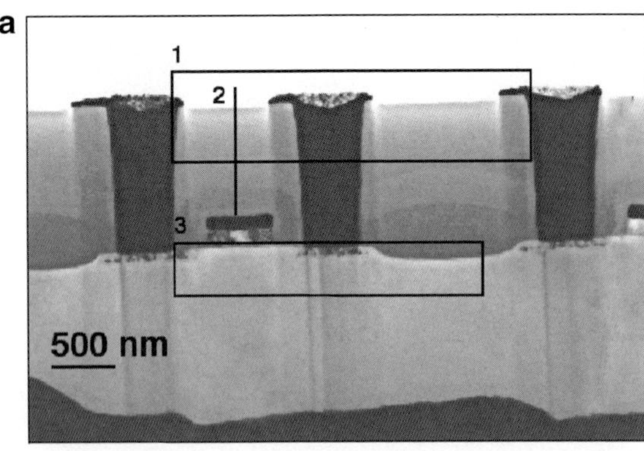

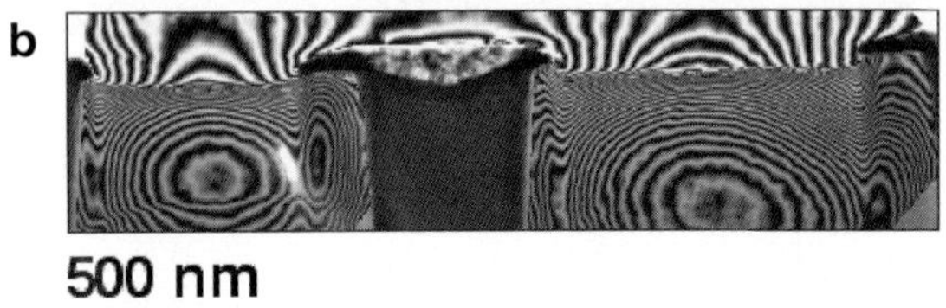

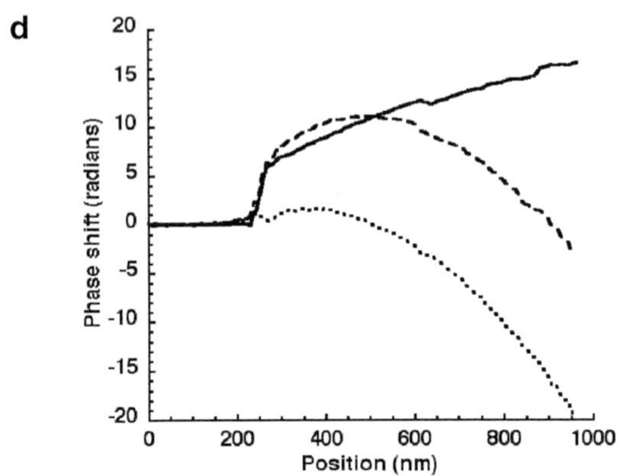

Figure 18–22. Results obtained from a cross-sectional semiconductor device specimen of nominal thickness 400 nm prepared using conventional "trench" focused ion beam milling. (a) Bright-field TEM image of a PMOS (0.5-μm gate) transistor, which forms part of a linear array of similar transistors, indicating the locations of the regions analyzed in more detail in the subsequent figures. The dark bands of contrast above the transistors are W contacts. Thickness corrugations are visible in the Si substrate in each image. The gates are formed from W silicide, while the amorphous layers above the gates and between the W plugs are formed from Si oxides that have different densities. (b) Eight times-amplified phase contours were calculated by combining phase images from several holograms obtained across the region marked "1" in (a) using a microscope accelerating voltage of 200 kV and a biprism voltage of 160 V. Specimen charging results in the presence of electrostatic fringing fields in the vacuum region outside the specimen edge, as well as elliptical phase contours within the Si oxide layers between the W contacts. (c) An equivalent phase image obtained after coating the specimen on one side with approximately 20 nm of carbon to remove the effects of charging. The phase contours now follow the expected mean inner potential contribution to the phase shift in the oxide layers, and there is no electrostatic fringing field outside the specimen edge. (d) One-dimensional line profiles obtained from the phase images in (b) and (c) along the line marked "2" in (a). The dashed and solid lines were obtained before and after coating the specimen with carbon, respectively. The dotted line shows the difference between the solid and dashed lines. (Reprinted from Dunin-Borkowski et al., 2005.)

of transistors, which were originally located ~5 μm below the surface of a wafer and separated from its surface by metallization layers. Such transistors present a significant but representative challenge for TEM specimen preparation for electron holography both because the metallization layers are substantial and can result in thickness corrugations in the doped regions of interest and because these overlayers must, at least in part, be removed to provide a vacuum reference wave for electron holography. An additional difficulty results from the possibility that the overlayers, which contain silicon oxides, may charge during examination in the electron microscope. Conventional "trench" focused ion beam milling (Park, 1990; Szot et al., 1992) was used to prepare the specimen, which has a nominal thickness of 400 nm. Figure 18–22b shows eight-times-amplified phase contours obtained from the region marked "1" in Figure 18–22a. Instead of the expected phase distribution, which should be proportional to the mean inner potential multiplied by the specimen thickness, elliptical contours are visible in each oxide region, and an electrostatic fringing field is present outside the specimen (at the top of Figure 18–22b). Both the elliptical contours and the fringing field are associated with the build-up of positive charge in the oxide layer. The elliptical contours are centered is several hundreds of nanometers from the specimen edge. Figure 18–22c shows a similar phase image obtained after coating the specimen on one side with approximately 20 nm of carbon. The effects of charging are now absent, there is no fringing field outside the specimen edge, and the phase contours follow the change in specimen thickness. One-dimensional phase profiles were generated from the phase images used to form Figure 18–22b and c along the line marked "2" in Figure 18–22a, and are shown in Figure 18–22d. The dashed and solid lines correspond to results obtained before and after coating the specimen with carbon, respectively, while the dotted line shows the difference between the solid and dashed lines. If the charge is assumed to be distributed through the thickness of the specimen, then the electric field in the oxide is approximately 2×10^7 V/m. This value is just below the breakdown electric field for thermal SiO_2 of 10^8 V/m (Sze, 2002). Equivalent results obtained from a specimen of 150 nm nominal thickness show that the elliptical contours are closer to the specimen edge. The effect of specimen charging on the dopant potential (in the source and drain regions of the transistors) is just as significant. The phase gradient continues into the substrate, and the dopant potential is undetectable before carbon coating, whether or not a phase ramp is subtracted from the images. If focused ion beam milling from the substrate side of the wafer (Schwarz et al., 2003) is used, then specimen charging no longer occurs, presumably as a result of Si redeposition onto the specimen surface. McCartney et al. (2003) provide an overview of this and other techniques for the preparation of semiconductor devices for electron holography.

Although questions still remain about phase contrast observed at simple p–n junctions, electron holographic data have been interpreted from more complicated semiconductor device structures, in which changes in composition as well as doping concentration are present. One example is a strained n-$Al_{0.1}Ga_{0.9}N/In_{0.1}Ga_{0.9}N/p$-$Al_{0.1}Ga_{0.9}N$ heterojunction diode, in which strong piezoelectric and polarization fields are used to induce high two-dimensional electron gas concentrations (McCartney et al., 2000). To interpret experimental measurements of the potential profile across the heterojunction, after corrections for specimen thickness changes (assuming a linear thickness profile and neglecting contributions to the measured phase from variations in mean inner potential), additional charge had to be added to simulations. In particular, a sheet of negative charge was included at the bottom of the InGaN well. The sheet charge density at this position was $2.1 \times 10^{13} cm^{-2}$. More recently, electron holography has been used to measure internal electrostatic potentials across InGaN quantum wells with thicknesses ranging from 2 to 10 nm (Stevens et al., 2004). In this study, the electric field strengths across the wells were observed to decrease in strength above a well thickness of 6 nm.

4.3 Space Charge Layers at Grain Boundaries

Electron holography has been used to characterize space charge layers at doped and undoped grain boundaries in electroceramics, although several contributions to the electron holographic phase shift can complicate interpretation. The space charge distribution that is predicted to form at such a grain boundary (Frenkel, 1946) is often described as a double (back-to-back) Schottky barrier. For Mn-doped and undoped grain boundaries in $SrTiO_3$, a decrease in the measured phase shift at the boundary relative to that in the specimen was observed (Ravikumar et al., 1995). The changes in phase measured at the doped boundaries were larger in magnitude and spatial extent than at similar undoped boundaries. Possible contributions to the contrast from changes in density, composition, specimen thickness, dynamic diffraction, and electrostatic fringing fields (Pozzi, 1996; Dunin-Borkowski and Saxton, 1997) were considered, and the remaining contributions to the measured phase shifts at the doped boundaries were attributed to space charge. The sign of the space charge contribution to the specimen potential was consistent with the presence of Mn^{2+} and Mn^{3+} ions on Ti sites at the boundaries. The results were finally interpreted in terms of a narrow (1–2 nm) region of negative grain boundary charge and a wider (3–5 nm) distribution of positive space charge.

A similar approach has recently been applied to the characterization of grain boundaries in ZnO, at which a space charge layer width of approximately 150 nm has been measured (Elfwing and Olsson, 2002). In an earlier study, defocus contrast recorded from delta-doped layers in Si and GaAs was also attributed to the presence of space charge (Dunin-Borkowski et al., 1994). Defocus contrast has been used to assess possible space charge contributions to electrostatic potential profiles across grain boundaries in doped and undoped $SrTiO_3$ (Mao et al., 1998). The contrast observed in these experiments was not consistent with a dominant contribution to the signal from space charge. Related experiments have been performed to measure polarization distributions across domain boundaries in ferroelectric materials such as $BaTiO_3$ and $PbTiO_3$ (Lichte, 2000; Lichte et al., 2003). There are many opportunities for further work on this topic.

5 High-Resolution Electron Holography

Aberrations of the objective lens, which result in modifications to the amplitude and phase shift of the electron wave, rarely need to be taken into account when characterizing magnetic and electrostatic fields at medium spatial resolution, as described in Sections 3 and 4. However, these aberrations must be considered when interpreting electron holograms that have been acquired at atomic resolution, in which lattice fringes are visible.

The back focal plane of the objective lens contains the Fraunhofer diffraction pattern, i.e., the Fourier transform, of the specimen wave $\psi_s(r) = A_s(r)\exp[i\phi_s(r)]$, denoted $\psi(q) = FT[\psi_s(r)]$. Transfer from the back focal plane to the image plane is then represented by an inverse Fourier transform. For a perfect thin lens, neglecting magnification and rotation of the image, the complex image wave would be equivalent to the object wave $\psi_s(r)$. Modifications to the electron wave that result from objective lens aberrations can be represented by multiplication of the electron wavefunction in the back focal plane by a transfer function of the form

$$T(q) = B(q)\exp[i\chi(q)] \tag{26}$$

In Eq. (26), $B(q)$ is an aperture function that takes a value of unity for q within the objective aperture and zero beyond the edge of the aperture. The effects of two objective lens aberrations, defocus and spherical aberration, can be included in the phase factor in the form

$$\chi(q) = \pi \Delta z \lambda q^2 + \frac{\pi}{2}C_S\lambda^3 q^4 \tag{27}$$

where Δz is the defocus of the lens and C_S is the spherical aberration coefficient. The complexity of Eq. (27) increases rapidly as further aberrations are considered. The complex wave in the image plane can then be written in the form

$$\psi_i(r) = FT^{-1}[FT[\psi_s(r)] \times T(q)] \tag{28}$$

$$= \psi_s(r) \otimes t(r) \tag{29}$$

where $t(r)$ is the inverse Fourier transform of $T(q)$, and the convolution $\otimes$ of the specimen wave $\psi_s(r)$ with $t(r)$ represents the smearing of information that results from lens imperfections. Since both $\psi_s(r)$ and $t(r)$ are in general complex, the intensity of a conventional bright-field image, which can be expressed in the form

$$I(r) = \left|\psi_s(r)\otimes t(r)\right|^2 \tag{30}$$

is no longer related simply to the structure of the specimen.

The effects of lens aberrations can be removed by multiplying the complex image wave by a suitable phase plate corresponding to $T^*(q)$ to provide the amplitude and the phase shift of the *specimen* wave $\psi_s(r)$ rather than the *image* wave $\psi_i(r)$. Hence, the interpretable resolution of the image can be improved beyond the point resolution of the electron microscope. The optimal defocus that maximizes the resolution of the reconstructed specimen wave after correction of aberrations (Lichte, 1991, 1992; Lichte and Rau, 1994; Ishizuka et al., 1994) is given by the expression

$$\Delta z_{opt} = -\frac{3}{4}C_S\left(\lambda q_{max}\right)^2 \tag{31}$$

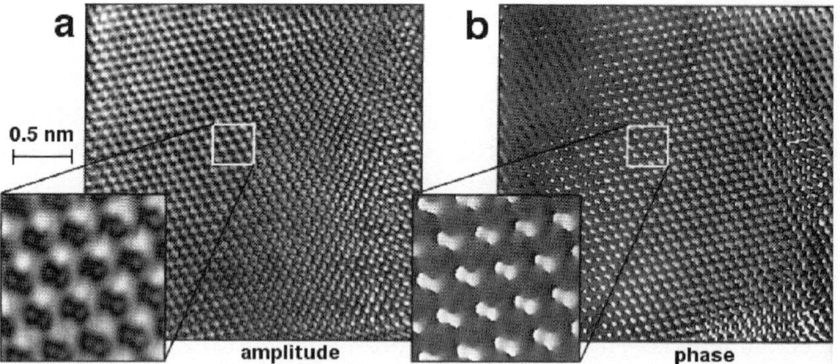

Figure 18–23. High-resolution (a) amplitude and (b) phase images of the aberration-corrected specimen wave reconstructed from an electron hologram of [110] Si, obtained at 300 kV on a CM30 FEGTEM. The spacing of the hologram fringes was 0.05 nm. The sideband contained {111}, {220}, {113}, and {004} reflections, corresponding to lateral information of 0.136 nm. The characteristic Si dumbbell structure is visible only after aberration correction. (Reprinted from Orchowski et al., 1995.)

where q_{max} is the maximum desired spatial frequency.

Figure 18–23 illustrates the application of aberration correction to a high-resolution electron hologram of crystalline Si imaged at the ⟨110⟩ zone axis, at which characteristic "dumbbell" contrast, of spacing 0.136 nm, is expected (Orchowski et al., 1995). The original hologram was acquired using an interference fringe spacing of 0.05 nm on a CM30 FEGTEM, which has a point resolution of 0.198 nm and an information limit of 0.1 nm at 300 kV. Figures 18–23a and b show, respectively, the reconstructed amplitude and phase shift of the hologram after aberration correction using a phase plate. The phase image reveals the expected white "dumbbell" contrast, at a spatial resolution that is considerably better than the point resolution of the microscope, after lens aberrations, including residual astigmatism and off-axis coma, have been measured and removed. Note also that the projected atom column positions are visible as black contrast in the amplitude image.

More recent developments in high-resolution electron holography, including the application of the technique to a wide range of materials problems, have been reviewed by Lehmann et al. (1999) and Lehmann and Lichte (2005).

6 Alternative Forms of Electron Holography

Many different forms of electron holography can be envisaged and implemented both in the TEM and in the scanning TEM (STEM) (Cowley, 1992). Equally, there are several ways in which the off-axis mode of TEM electron holography can be implemented. A full discussion of these various schemes, which include interferometry in the diffraction plane of the microscope (Herring et al., 1995) and reflection electron holography (Banzhof and Herrmann, 1993), is beyond the scope of this chapter. Here, some of the more important developments are reviewed.

The need for a vacuum reference wave is a major drawback of the standard off-axis mode of TEM holography since this requirement restricts the region that can be examined to near the specimen edge. In many applications, the feature of interest is not so conveniently located. The implementation of a DPC mode of electron holography in the TEM enables this restriction to be overcome. DPC imaging is well established as a technique in the STEM, involving the use of various combinations of detectors to obtain magnetic contrast (Dekkers and de Lang, 1974; Rose, 1977; Chapman et al., 1978). It has also been shown (Mankos et al., 1994) that DPC contrast can be obtained using far-out-of-focus STEM electron holography (see below). An equivalent TEM configuration can be achieved by using an electron biprism located in the condenser aperture plane of the microscope (McCartney et al., 1996). Figure 18–24a shows a schematic ray diagram that illustrates the electron-optical configuration for this differential mode of off-axis TEM holography. The application of a positive voltage to the biprism results in the formation of two closely spaced, overlapping plane waves, which appear to originate from sources S_1 and S_2 to create an interference fringe pattern at the specimen level. When the observation plane is defocused by a distance Δz with respect to the specimen plane, the two coherent beams produced by the beam splitter, which are labeled k_1 and k_2 in Figure 18–24a, impinge upon different parts of the specimen. For a magnetic material, the difference in the component of the magnetic induction parallel to the biprism wire between these two points in the specimen plane determines the relative phase shift of the holographic fringes, thus giving differential phase contrast. Since the hologram is acquired under out-of-focus conditions, it is in effect the superposition of a pair of Fresnel images. The biprism voltage must be adjusted so that the feature of interest or the desired spatial resolution is sampled by at least three interference fringes. An appropriate postspecimen magnification should be chosen to ensure that the interference fringes are properly sampled by the recording medium. Figure 18–24b shows a composite phase image formed from a series of eight DPC holograms of a 30-nm-thick Co film. The fringe system was shifted progressively across the specimen plane between exposures. In addition to the holographic interference fringes, the image shows black and white lines that delineate walls between magnetic domains, with magnetization ripple visible within the domains. All of the image features are doubled due to the split incident beam. Figure 18–24c shows the final reconstructed DPC image obtained from Figure 18–24b, in which the contrast is proportional to the component of the magnetic induction parallel to the holographic fringes. The arrow below the image indicates the direction of the component of the induction analyzed in this experiment. Several magnetic vortices, at which the measured field direction circles an imperfection in the film, are visible. One such vortex is indicated by an arrow in the lower right corner of the image. For characterization of both components of the in-plane induction

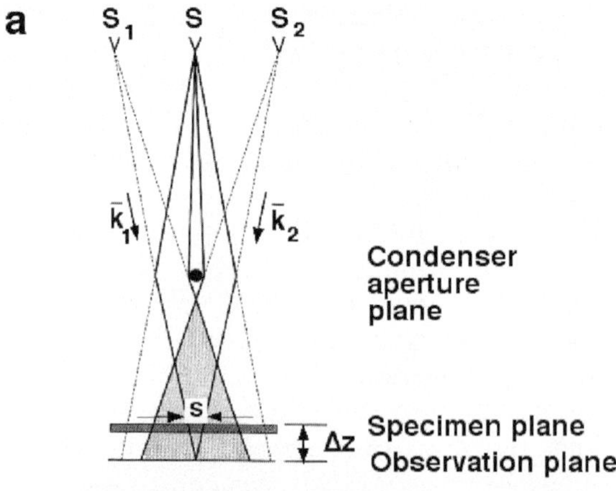

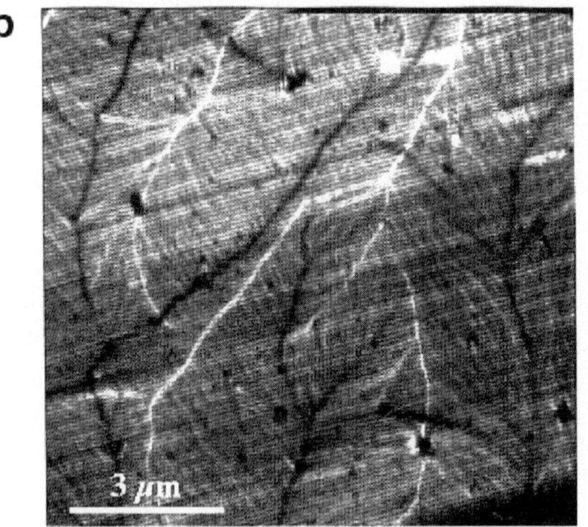

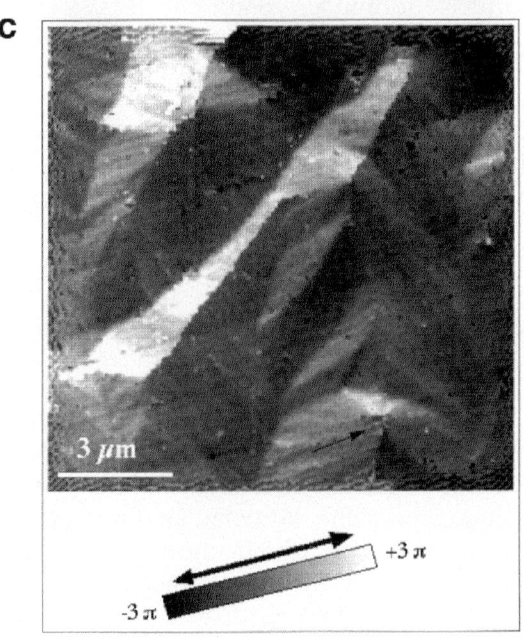

Figure 18–24. (a) Schematic ray diagram for the differential mode of off-axis TEM electron holography. The symbols are defined in the text. (b) Composite differential mode electron hologram formed from a series of eight holograms of a 30-nm-thick Co film. (c) Differential phase contrast image obtained from the hologram shown in (b). The arrow below the image indicates the direction of the magnetic signal analyzed. The arrow close to the bottom of the image indicates the position of a magnetic vortex. (Reprinted from McCartney et al., 1996.)

without removing the sample from the microscope, a rotating biprism or a rotating sample holder is required.

An alternative scheme that is conceptually similar to the differential mode of electron holography in the TEM, but which does not require the use of an electron biprism or a field emission electron gun, is termed amplitude division electron holography. Whereas conventional modes of electron holography involve splitting the wavefront of the incident illumination and thus require high spatial coherence to form interference fringes, this coherence requirement can be removed by dividing the amplitude of the electron beam instead of the wavefront. Division of the amplitude of the electron wave can be achieved by using a crystal film located before the specimen. The lattice fringes of the crystal film are then used as carrier fringes. The original configuration for this scheme involved placing the specimen in the selected-area-aperture plane of the microscope (Pozzi, 1983; Matteucci et al., 1982). The specimen can also be inserted into the normal object plane by placing a single-crystal thin film and the sample of interest on top of each other, in close proximity (Ru, 1995). The single crystal film is then tilted to a strong Bragg condition and used as an electron beam splitter. As a result of the separation of the crystal and the specimen, the hologram plane contains two defocused images of the specimen that are shifted laterally with respect to one another. One of these images is carried by the direct beam and the other by the Bragg-reflected beam. When the distance between the two images is greater than the size of the object, the images separate perfectly and interfere with adjacent plane waves to form an off-axis electron hologram. Because the single crystal is in focus and the object is out of focus, a Fresnel electron hologram of the object is obtained. The defocus of the object can be corrected at the reconstruction stage by using a phase plate, although high coherence of the incident illumination is then required. The coherence used when forming the image therefore determines the spatial resolution of the final reconstructed image. Although amplitude-division electron holography has several disadvantages over wavefront-division holography, the final phase image is not affected by effects such as Fresnel diffraction from the edges of the biprism.

An approach that can be used to increase the phase sensitivity of electron holography is termed phase-shifting electron holography. This approach is based on the acquisition of several off-axis holograms while the phase offset (the initial phase) of the image is changed, either by tilting the incident electron beam or by shifting the biprism (Ru et al., 1992). Electron holograms are recorded at successive values of the incident beam tilt, such that the phase is shifted by at least 2π over the image series. The fringe shift can be monitored in the complex Fourier spectra of the holograms. Although three holograms can in principle be used to reconstruct the object wave, in practice as many holograms as possible should be used to reduce noise. The advantages of the phase-shifting approach are greatly improved phase sensitivity and spatial resolution. Moreover, objects that are smaller than one fringe width can be reconstructed. Care is required if the object is out of focus, as tilting the beam will also induce an image shift between successive images. Very small phase shifts have been observed from individual unstained ferritin molecules using this approach (Kawasaki et al., 1986).

Electron holograms can be acquired at video rate and subsequently digitized and processed individually to record dynamic events, but this procedure is time consuming. An alternative real-time approach for acquiring and processing holograms has been demonstrated by using a liquid-crystal panel to reconstruct holograms (Chen et al., 1994). The holograms are recorded at TV rate and transferred to a liquid-crystal spatial light modulator, which is located at the output of a Mach–Zender interferometer. The liquid crystal panel is illuminated using an He–Ne laser, and interference micrographs are observed at video rate on the monitor beside the microscope as the specimen is examined. In an alternative configuration, a liquid-crystal panel can also be used as a computer-controlled phase plate to correct for aberrations.

Whereas an off-axis electron hologram is formed by the interference of an object and a reference wave that propagate in different directions in the electron microscope, the simplest way of recording an electron hologram without using an electron biprism involves using the transmitted wave as the reference wave to form an in-line hologram. Gabor's original paper described the reconstruction of an image by illuminating an in-line hologram with a parallel beam of light and using a spherical aberration correcting plate and an astigmatism corrector. The reconstructed image is, however, disturbed by the presence of a "ghost" or "conjugate" twin image. If the hologram is recorded and subsequently illuminated by a plane wave, then the reconstructed image and a defocused conjugate image of the object are superimposed on each other. The most effective method of separating the twin images is to use Fraunhofer in-line holography. Here, in-line holograms are recorded in the Fraunhofer diffraction plane of the object (Thomson et al., 1966; Tonomura et al., 1968). Under this condition, the conjugate image is so blurred that its effect on the reconstructed image is negligible (Matsumoto et al., 1994).

The STEM holographic mode used for DPC imaging, which has similarities with the TEM differential mode of electron holography described above, is a point projection technique in which a stationary beam in an STEM is split by a biprism preceding the sample so that two mutually coherent electron point sources are formed just above the specimen. In this operating mode, the objective lens is excited weakly so that the hologram is formed in the diffraction plane rather than the image plane (Cowley, 1990). By greatly defocusing the objective lens, a shadow image of the object is formed, which has the appearance of a TEM hologram, although it is distorted by spherical aberration and defocus. The image magnification and the separation of the sources relative to the specimen are flexible in this configuration, and can be adjusted by changing the biprism voltage and/or the objective or post-specimen lens settings. The far-out-of-focus mode of STEM holography has been applied to the characterization of a range of magnetic materials (Mankos et al., 1995; Cowley et al., 1996).

A rapid approach that can be used to visualize equiphase contours involves superimposing a hologram of the specimen onto a reference hologram acquired under identical conditions, with the specimen removed from the field of view (Matteucci and Muccini, 1994). Interference effects between the holographic fringes in the two images then provide widely spaced, low-contrast bright and dark bands that reveal phase contours directly. By defocusing the combined image slightly, the unwanted finely spaced holographic interference fringes can be removed. The technique has been applied to image both electrostatic and magnetic fields.

A related approach involves the use of two parallel or perpendicular electron biprisms to generate an interference pattern between either three or four electron waves, respectively. Equiphase contours are then displayed in the recorded hologram. This method has been used to form images of electric fields outside charged latex and alumina particles and magnetic fields outside ferrite particles (Hirayama, 1999), and to expose a resist to fabricate a 100-nm-period two-dimensional grating lithographically (Ogai et al., 1995). Two parallel biprisms have also been used to form a "trapezoidal" biprism to record a double-exposure hologram with the biprism voltage changed (Tanji et al., 1996) and with the reference wave unaffected by the biprism voltage (Tanji et al., 1999).

7 Discussion and Conclusion

In this chapter, the technique of off-axis electron holography has been described, and its recent application to a wide variety of materials has been reviewed. Results have been presented from the characterization of magnetic fields in arrangements of closely spaced nanocrystals, patterned elements and nanowires, and electrostatic fields in field emitters and doped semiconductors. *In situ* experiments, which allow magnetization reversal processes to be followed and electrostatic fields in working semiconductor devices to be characterized, have been described, and the advantages of using digital approaches to record and analyze electron holograms have been highlighted. High-resolution electron holography and alternative modes of electron holography have also been described. Although the results that have been presented are specific to the dimensions and morphologies of the particular examples chosen, they illustrate the ways in which electron holography can be adapted to tackle different materials problems.

Future developments in electron holography are likely to include the development and application of new forms of electron holography and instrumentation, the introduction of new approaches for enhancing weak magnetic and electrostatic signals, the formulation of a better understanding of the effect of different TEM sample preparation techniques on phase images recorded from semiconductors and ferroelectrics, and the combination of electron holography with electron tomography to record both electrostatic and magnetic fields *inside* nanostructured materials in *three* dimensions rather than simply in projection. If the application of electron holographic tomography to the characterization of magnetic vector fields inside materials in three dimensions, which requires two high-tilt series of holograms to be recorded about orthogonal axes, is ultimately successful, the primary difficulty may lie in the measurement and subtraction of the unwanted mean inner potential contribution to the measured phase shift at every one of the tilt angles required.

The unique capability of electron holography to provide quantitative information about magnetic and electrostatic fields in materials at a resolution approaching the nanometer scale, coupled with the increasing availability of field-emission-gun transmission electron microscopes and quantitative digital recording, ensures that the technique has a very promising future.

Acknowledgments. Thanks are due to the Royal Society, the EPSRC, FEI, and the Isaac Newton Trust for support, and to C. B. Boothroyd, P. R. Buseck, R. B. Frankel, K. Harada, R. J. Harrison, M. J. Hÿtch, B. Kardynal, J. Li, J. C. Loudon, P. A. Midgley, S. B. Newcomb, M. Pósfai, G. Pozzi, A. Putnis, C. A. Ross, M. R. Scheinfein, E. Snoeck, A. Tonomura, S. L. Tripp, A. C. Twitchett, A. Wei, and Y. Zhu for discussions and for ongoing collaborations. R.D.B. acknowledges the C.N.R.S. and the European Community for support through a European research network (GDR-E) entitled "Quantification and Measurement in Transmission Electron Microscopy," grouping laboratories in France, the United Kingdom, Germany, and Switzerland. We acknowledge the use of electron microscopy facilities at the John M. Cowley Center for High Resolution Electron Microscopy at Arizona State University.

References

Aharonov, Y. and Bohm, D. (1959). *Phys. Rev.* **115**, 485.

Allard, L.F., Völkl, E., Carim, A., Datye, A.K. and Ruoff, R. (1996). *Nano. Mater.* **7**, 137.

Aoyama, K. and Ru, Q. (1996). *J. Microsc.* **182**, 177.

Banzhof, H. and Herrmann, K.-H. (1993). *Ultramicroscopy* **48**, 475.

Bazylinski, D.A. and Moskowitz, B.M. (1997). *Mineralog. Soc. Am. Rev. Mineral.* **35**, 181.

Beleggia, M., Cristofori, D., Merli, P.G. and Pozzi, G. (2000). *Micron* **31**, 231.

Beleggia, M., Cardinali, G.C., Fazzini, P.F., Merli, P.G. and Pozzi, G. (2001). *Inst. Phys. Conf. Ser.* **169**, 427.

Beleggia, M., Fazzini, P.F., Merli, P.G. and Pozzi, G. (2003). *Phys. Rev. B* **67**, 045328.

Blakemore, R.P. (1975). *Science* **190**, 377.

Bonevich, J.E., Harada, K., Matsuda, T., Kasai, H., Yoshida, T., Pozzi, G. and Tonomura, A. (1993). *Phys. Rev. Lett.* **70**, 2952.

Chapman, J.N., Batson, P.E., Waddell, E.M. and Ferrier, R.P. (1978). *Ultramicroscopy* **3**, 203.

Chen, J., Hirayama, T., Tanji, T., Ishizuka, K. and Tonomura, A. (1994). *Opt. Commun.* **110**, 33.

Cowley, J.M. (1990). *Ultramicroscopy* **34**, 293.

Cowley, J.M. (1992). *Ultramicroscopy* **41**, 335.

Cowley, J.M., Mankos, M. and Scheinfein, M.R. (1996). *Ultramicroscopy* **63**, 133.

Cumings, J., Zettl, A., McCartney, M.R. and Spence, J.C.H. (2002). *Phys. Rev. Lett.* **88**, 056804.

de Graef, M., Nuhfer, T. and McCartney, M.R. (1999). *J. Microsc.* **194**, 84.

Dekkers, N.H. and de Lang, H. (1974). *Optik* **41**, 452.

de Ruijter, W.J. (1995). *Micron* **26**, 247.

de Ruijter, W.J. and Weiss, J.K. (1993). *Ultramicroscopy* **50**, 269.

de Ruijter, W.J., Gajdardziska-Josifovska, M., McCartney, M.R., Sharma, R., Smith, D.J. and Weiss, J.K. (1992). *Scanning Microsc. Suppl.* **6**, 347.

Doyle, P.A. and Turner, P.S. (1968). *Acta Crystallogr. A* **24**, 390.

Dunin-Borkowski, R.E. and Saxton, W.O. (1996). In *Atomic Resolution Microscopy of Surfaces and Interfaces* (D.J. Smith and R.J. Hamers, Eds.), Materials Research Society Symposium Proceedings **466**, 73. (Materials Research Society, Warrendale, PA.)

Dunin-Borkowski, R.E. and Saxton, W.O. (1997). *Acta Crystallogr. A* **53**, 242.

Dunin-Borkowski, R.E., Stobbs, W.M., Perovic, D.D. and Wasilewski, Z.R. (1994). In *Proceedings of ICEM13* (B. Jouffrey and C. Colliex, Eds.), **1**, 411. Paper presented at the 13th International Conference on Electron Microscopy, Paris, France, July 17–22, 1994.

Dunin-Borkowski, R.E., McCartney, M.R., Smith, D.J. and Parkin, S.S.P. (1998). *Ultramicroscopy* **74**, 61.

Dunin-Borkowski, R.E., McCartney, M.R., Kardynal, B., Smith, D.J. and Scheinfein, M.R. (1999). *Appl. Phys. Lett.* **75**, 2641.

Dunin-Borkowski, R.E., McCartney, M.R., Kardynal, B., Parkin, S.S.P., Scheinfein, M.R. and Smith, D.J. (2000). *J. Microsc.* **200**, 187.

Dunin-Borkowski, R.E., Newcomb, S.B., McCartney, M.R., Ross, C.A. and Farhoud, M. (2001). *Inst. Phys. Conf. Ser.* **168**, 485.

Dunin-Borkowski, R.E., Twitchett, A.C. and Midgley, P.A. (2002). *Microsc. Microanal.* **8**(Suppl. 2), 42.

Dunin-Borkowski, R.E., McCartney, M.R. and Smith, D.J. (2004a). *Encyclopaedia of Nanoscience and Nanotechnology* (H.S. Nalwa, Ed.), **3**, 41. (American Scientific Publishers, Stevenson Ranch, CA.)

Dunin-Borkowski, R.E., Kasama, T., Wei, A., Tripp, S.L., Hÿtch, M.J., Snoeck, E., Harrison, R.J. and Putnis, A. (2004b). *Microsc. Res. Techn.* **64**, 390.

Dunin-Borkowski, R.E., Newcomb, S.B., Kasama, T., McCartney, M.R., Weyland, M. and Midgley, P.A. (2005). *Ultramicroscopy* **103**, 67.

Dunlop, D.J. and Özdemir, Ö. (1997). *Rock Magnetism.* (Cambridge University Press, Cambridge, UK.)

Elfwing, M. and Olsson, E. (2002). *J. Appl. Phys.* **92**, 5272.

Frabboni, S., Matteucci, G., Pozzi, G. and Vanzi, M. (1985). *Phys. Rev. Lett.* **55**, 2196.

Frabboni, S., Matteucci, G. and Pozzi, G. (1987). *Ultramicroscopy* **23**, 29.

Frenkel, J. (1946). *Kinetic Theory of Liquids* (Oxford University Press, Oxford).

Fujita, T. and McCartney, M.R. (2005). *Ultramicroscopy* **102**, 279.

Gabor, D. (1949). *Proc. R. Soc. London A* **197**, 454.

Gajdardziska-Josifovska, M. and Carim, A. (1998). In *Introduction to Electron Holography* (E. Völkl, L.F. Allard and D.C. Joy, Eds.). (Kluwer Academic/ Plenum Publishers, New York.)

Gajdardziska-Josifovska, M., McCartney, M.R., de Ruijter, W.J., Smith, D.J., Weiss, J.K. and Zuo, J.M. (1993). *Ultramicroscopy* **50**, 285.

Ghiglia, D.C. and Pritt, M.D. (1998). *Two-Dimensional Phase Unwrapping. Theory, Algorithms and Software* (Wiley, New York).

Gribelyuk, M.A., McCartney, M.R., Li, J., Murthy, C.S., Ronsheim, P., Doris, B., McMurray, J.S., Hegde, S. and Smith, D.J. (2002). *Phys. Rev. Lett.* **89**, 025502.

Harada, H., Tonomura, A., Togawa, Y., Akashi, T. and Matsuda, T. (2004). *Appl. Phys. Lett.* **84**, 3229.

Harrison, R.J., Dunin-Borkowski, R.E., and Putnis, A. (2002). *Proc. Natl. Acad. Sci. USA* **99**, 16556.

Harscher, A. and Lichte, H. (1996). *Ultramicroscopy* **64**, 57.

Harscher, A. and Lichte, H. (1998). In *Electron Microscopy 98* (H.A. Calderón Benavides and M.J. Yacamán, Eds.), **1**, 553. Paper presented at the 14th International Conference on Electron Microscopy, Cancun, Mexico, August 31 to September 4, 1998 (Institute of Physics Publishing, Bristol, UK).

Herring, R.A., Pozzi, G., Tanji, T. and Tonomura, A. (1995). *Ultramicroscopy* **60**, 153.

Heumann, M., Uhlig, T. and Zweck, J. (2005). *Phys. Rev. Lett.* **94**, 077202.

Hirayama, T. (1999). *Mater. Characterization* **42**, 193.

Hu, H., Wang, H., McCartney, M.R. and Smith, D.J. (2005). *J. Appl. Phys.* **97**, 054305.

Hÿtch, M.J., Dunin-Borkowski, R.E., Scheinfein, M.R., Moulin, J., Duhamel, C., Mazelayrat, F. and Champion, Y. (2003). *Phys. Rev. Lett.* **91**, 257207.

Ishizuka, K., Tanji, T., Tonomura, A., Ohno, T. and Murayama Y. (1994). *Ultramicroscopy* **55**, 197.

Jacobs, I.S. and Bean, C.P. (1955). *Phys. Rev.* **100**, 1060.

Kawasaki, T., Endo, J., Matsuda, T., Osakabe, N. and Tonomura, A. (1986). *J. Electron Microsc.* **35**, 211.

Kawasaki, T., Missiroli, G.F., Pozzi, G. and Tonomura, A. (1993). *Optik* **92**, 168.

Lai, G., Hirayama, K., Ishizuka, T., Tanji, T. and Tonomura, A. (1994). *Appl. Opt.* **33**, 829.

Lehmann, M. and Lichte, H. (2005). *Cryst. Res. Technol.* **40**, 149.

Lehmann, M., Lichte, H., Geiger, D., Lang, G. and Schweda, E. (1999). *Mater. Characterization* **42**, 249.

Li, J., McCartney, M.R., Dunin-Borkowski, R.E. and Smith, D.J. (1999). *Acta Crystallogr. A* **55**, 652.

Lichte, H. (1991). *Ultramicroscopy* **38**, 13.

Lichte, H. (1992). *Ultramicroscopy* **47**, 223.

Lichte, H. (1995). In *Electron Holography* (A. Tonomura, L.F. Allard, G. Pozzi, D.C. Joy and Y.A. Ono, Eds.), pp. 11–31. (Elsevier, Amsterdam.)

Lichte, H. (2000). *Crystal Res. Technol.* **35**, 887.

Lichte, H. (2002). *Phil. Trans. R. Soc. Lond. A* **360**, 897.

Lichte, H. and Rau, W.D. (1994). *Ultramicroscopy* **54**, 310.

Lichte, H., Herrmann, K.H. and Lenz, F. (1987). *Optik* **77**, 135.

Lichte, H., Reibold, M., Brand, K. and Lehmann, M. (2003). *Ultramicroscopy* **93**, 199.

Lin, X. and Dravid, V.P. (1996). *Appl. Phys. Lett.* **69**, 1014.

Lloyd, S.J., Dunin-Borkowski, R.E. and Boothroyd, C.B. (1997). *Inst. Phys. Conf. Ser.* **153**, 113.

Loudon, J.C., Mathur, N.D., and Midgley, P.A. (2002). *Nature* **420**, 797.

Mankos, M., Scheinfein, M.R. and Cowley, J.M. (1994). *J. Appl. Phys.* **75**, 7418.

Mankos, M., Cowley, J.M. and Scheinfein, M.R. (1995). *Mat. Res. Soc. Bull.* **20**, 45.

Mao, Z., Dunin-Borkowski, R.E., Boothroyd, C.B. and Knowles, K.M. (1998). *J. Am. Ceram. Soc.* **81**, 2917.

Matsuda, T., Tonomura, A., Suzuki, R., Endo, J., Osakabe, N., Umezaki, H., Tanabe, H., Sugita, Y. and Fujiwara, H. (1982). *J. Appl. Phys.* **53**, 5444.

Matsuda, T., Hasegawa, S., Igarashi, M., Kobayashi, T., Naito, M., Kajiyama, H., Endo, J., Osakabe, N. and Tonomura, A. (1989). *Phys. Rev. Lett.* **62**, 2519.

Matsuda, T., Fukuhara, A., Yoshida, T., Hasegawa, S., Tonomura, A. and Ru, Q. (1991). *Phys. Rev. Lett.* **66**, 457.

Matsumoto, T. and Tonomura, A. (1996). *Ultramicroscopy* **63**, 5.

Matsumoto, T., Tanji, T. and Tonomura, A. (1994). *Ultramicroscopy* **54**, 317.

Matteucci, G. and Muccini, M. (1994). *Ultramicroscopy* **53**, 19.

Matteucci, G., Missiroli, G.F. and Pozzi, G. (1982). *Ultramicroscopy* **8**, 403.

Matteucci, G., Missiroli, G.F. and Pozzi, G. (1988). *Physica B* **151**, 223.

Matteucci, G., Missiroli, G.F. Muccini, M. and Pozzi, G. (1992). *Ultramicroscopy* **45**, 77.

Matteucci, G., Muccini, M. and Hartmann, U. (1994). *Phys. Rev. B* **50**, 6823.

Matteucci, G., Missiroli, G.F. and Pozzi, G. (1998). *Adv. Imaging Electron Phys.* **99**, 171.

Matteucci, G., Missiroli, G.F. and Pozzi, G. (2002). *Adv. Imaging Electron Phys.* **122**, 173.

McCartney, M.R. and Gajdardziska-Josifovska, M. (1994). *Ultramicroscopy* **53**, 283.

McCartney, M.R. and Zhu, Y. (1998). *Appl. Phys. Lett.* **72**, 1380.

McCartney, M.R., Kruit, P., Buist, A.H. and Scheinfein, M.R. (1996). *Ultramicroscopy* **65**, 179.

McCartney, M.R. and Dunin-Borkowski, R.E. (1998). In *Electron Microscopy 98* (H.A. Calderón Benavides and M. José Yacamán, Eds.), **2**, 497. Paper presented at the 14th International Conference on Electron Microscopy, Cancun, Mexico, August 31 to September 4, 1998. (Institute of Physics Publishing, Bristol, UK.)

McCartney, M.R., Ponce, F.A., Cai, J. and Bour, D.P. (2000). *Appl. Phys. Lett.* **76**, 3055.

McCartney, M.R., Gribelyuk, M.A., Li, J., Ronsheim, P., McMurray, J.S. and Smith, D.J. (2002). *Appl. Phys. Lett.* **80**, 3213.

McCartney, M.R., Li, J., Chakraborty, P., Giannuzzi, L.A. and Schwarz, S.M. (2003). *Microsc. Microanal.* **9**(Suppl. 2), 776.

Meyer, R.R. and Kirkland, A.I. (2000). *Micr. Res. Techn.* **49**, 269.

Midgley, P.A. (2001). *Micron* **32**, 167.

Möllenstedt, G. and Düker, H. (1954). *Naturwissenschaften* **42**, 41.

Ogai, K., Kimura, Y., Shimizu, R., Fujita, J. and Matsui, S. (1995). *Appl. Phys. Lett.* **66**, 1560.

O'Keeffe, M. and Spence, J.C.H. (1994). *Acta Crystallogr. A* **50**, 33.

Orchowski, A., Rau, W.D. and Lichte, H. (1995). *Phys. Rev. Lett.* **74**, 399.

Osakabe, N., Yoshida, K., Horiuchi, Y., Matsuda, T., Tanabe, H., Okuwaki, T., Endo, J., Fujiwara, H. and Tonomura, A. (1983). *Appl. Phys. Lett.* **42**, 746.

Park, K.H. (1990). *Mater. Res. Soc. Symp. Proc.* **199**, 271.

Pozzi, G. (1983). *Optik* **66**, 91.

Pozzi, G. (1996). *J. Phys. D: Appl. Phys.* **29**, 1807.

Pozzi, G. and Vanzi, M. (1982). *Optik* **60**, 175.

Price, G.D. (1981). *Am. Mineral.* **66**, 751.

Radi, G. (1970). *Acta Crystallogr. A* **26**, 41.

Rau, W.D., Baumann, F.H., Rentschler, J.A., Roy, P.K. and Ourmazd, A. (1996). *Appl. Phys. Lett.* **68**, 3410.

Rau, W.D., Schwander, P., Baumann, F.H., Höppner, W. and Ourmazd, A. (1999). *Phys. Rev. Lett.* **82**, 2614.

Ravikumar, V., Rodrigues, R.P. and Dravid, V.P. (1995). *Phys. Rev. Lett.* **75**, 4063.

Reimer, L. (1991). *Transmission Electron Microscopy* (Springer-Verlag, Berlin).

Rez, D., Rez, P. and Grant, I. (1994). *Acta Crystallogr., Sect. A* **50**, 481.

Rose, H. (1977). *Utramicroscopy* **2**, 251.

Ross, C.A. (2001). *Ann. Rev. Mater. Res.* **31**, 203.

Ru, Q. (1995). *J. Appl. Phys.* **77**, 1421.

Ru, Q., Endo, J., Tanji, T. and Tonomura, A. (1992). *Appl. Phys. Lett.* **59**, 2372.

Schwarz, S.M., Kempshall, B.W., Giannuzzi, L.A. and McCartney, M.R. (2003). *Microsc. Microanal.* **9**(Suppl. 2), 116.

Seraphin, S., Beeli, C., Bonard, J.M., Jiao, J., Stadelmann, P.A. and Chatelain, A. (1999). *J. Mat. Res.* **14**, 2861.

Signoretti, S., Del Blanco, L., Pasquini, L., Matteucci, G., Beeli, C. and Bonetti, E. (2003). *J. Magn. Magn. Mat.* **262**, 142.

Smith, D.J. and McCartney, M.R. (1998). In *Introduction to Electron Holography* (E. Völkl, L.F. Allard and D.C. Joy, Eds.), 87–106 (Kluwer Academic/Plenum Publishers, New York).

Smith, D.J., de Ruijter, W.J., Weiss, J.K. and McCartney, M.R. (1998). In *Introduction to Electron Holography* (E. Völkl, L.F. Allard and D.C. Joy, Eds.), 116 (Kluwer Academic/Plenum Publishers, New York).

Smith, D.J., Dunin-Borkowski, R.E., McCartney, M.R., Kardynal, B. and Scheinfein, M.R. (2000). *J. Appl. Phys.* **87**, 7400.

Snoeck, E., Dunin-Borkowski, R.E., Dumestre, F., Renaud, P., Amiens, C., Chaudret, B. and Zurcher, P. (2003). *Appl. Phys. Lett.* **82**, 88.

Somodi, P.K., Dunin-Borkowski, R.E., Twitchett, A.C., Barnes, C.H.W. and Midgley, P.A. (2005). In *Electron Microscopy of Molecular and Atom-Scale Mechanical Behavior, Chemistry and Structure* (D. Martin, D.A. Muller, E. Stach and P.A. Midgley, Eds.), *Mater. Res. Soc. Proc.* **839**, P3.2 (Materials Research Society, Pittsburgh, PA).

Spence, J.C.H. (1993). *Acta Crystallogr. A* **49**, 231.

Stevens, M., Bell, A., McCartney, M.R., Ponce, F.A., Marui, H. and Tanaka, S. (2004). *Appl. Phys. Lett.* **85**, 4651.

Sze, S.M. (2002). *Physics of Semiconductor Devices* (Wiley, New York).

Szot, J., Hornsey, R., Ohnishi, T. and Minagawa, J. (1992). *J. Vac. Sci. Technol. B* **10**, 575.

Tanji, T., Ru, Q. and Tonomura, A. (1996). *Appl. Phys. Lett.* **69**, 2623.

Tanji, T., Manabe, S., Yamamoto, K. and Hirayama, T. (1999). *Ultramicroscopy* **75**, 197.

Thomson, B.J., Parrent, G.B., Ward, J.H. and Justh, B. (1966). *J. Appl. Meteorol.* **5**, 343.

Titchmarsh, J.M., Lapworth, A.J. and Booker, G.R. (1969). *Phys. Stat. Sol.* **34**, K83.

Tonomura, A. (1992). *Adv. Phys.* **41**, 59.

Tonomura, A. (1998). *The Quantum World Unveiled by Electron Waves* (World Scientific, Singapore).

Tonomura, A., Fukuhara, A., Watanabe, H. and Komoda, T. (1968). *Jpn. J. Appl. Phys.* **7**, 295.

Tonomura, A., Matsuda, T., Suzuki, R., Fukuhara, A., Osakabe, N., Umezaki, H., Endo, J., Shinagawa, K., Sugita, Y. and Fujiwara, H. (1982). *Phys. Rev. Lett.* **48**, 1443.

Tonomura, A., Umezaki, H., Matsuda, T., Osakabe, N., Endo, J. and Sugita, Y. (1983). *Phys. Rev. Lett.* **51**, 331.

Tonomura, A., Matsuda, T., Kawasaki, T., Endo, J. and Osakabe, N. (1985). *Phys. Rev. Lett.* **54**, 60.

Tonomura, A., Matsuda, T., Endo, J., Arii, T. and Mihama, K. (1986). *Phys. Rev. B: Solid State* **34**, 3397.

Tonomura, A., Allard, L.F., Pozzi, G., Joy, D.C. and Ono, Y.A., Eds. (1995). *Electron Holography* (Elsevier, Amsterdam).

Tripp, S.L., Pusztay, S.V., Ribbe, A.E. and Wei, A. (2002). *J. Am. Chem. Soc.* **124**, 7914.

Tripp, S.L., Dunin-Borkowski, R.E. and Wei, A. (2003). *Angew. Chem.* **42**, 5591.

Twitchett, A.C., Dunin-Borkowski, R.E. and Midgley, P.A. (2002). *Phys. Rev. Lett.* **88**, 238302.

Twitchett, A.C., Dunin-Borkowski, R.E., Hallifax, R.J., Broom, R.F. and Midgley, P.A. (2004). *J. Microsc.* **214**, 287.

Twitchett, A.C., Dunin-Borkowski, R.E., Hallifax, R.J., Broom, R.F. and Midgley, P.A. (2005). *Microsc. Microanal.* **11**, 66.

Völkl, E., Allard, L.F., Datye, A. and Frost, B. (1995). *Ultramicroscopy* **58**, 97.

Völkl, E., Allard, L.F. and Joy, D.C., Eds. (1998). *Introduction to Electron Holography* (Plenum, New York).

Wang, Y.C., Chou, T.M., Libera, M. and Kelly, T.F. (1997). *Appl. Phys. Lett.* **70**, 1296.

Wang, Z., Hirayama, T., Kato, T., Sasaki, K., Saka, H. and Kato, N. (2002a). *Appl. Phys. Lett.* **80**, 246.

Wang, Z., Sasaki, K., Kato, N., Urata, K., Hirayama, T. and Saka, H. (2002b). *J. Electron Microsc.* **50**, 479.

Wang, Z., Kato, T., Shibata, N., Hirayama, T., Kato, N., Sasaki, K. and Saka, H. (2002c). *Appl. Phys. Lett.* **81**, 478.

Wang, Z., Kato, T., Hirayama, T., Kato, N., Sasaki, K. and Saka, H. (2005). *Surf. Int. Anal.* **37**, 221.

Weiss, J.K., de Ruijter, W.J., Gajdardziska-Josifovska, M., Smith, D.J., Völkl, E. and Lichte, H. (1991). In *Proceedings of the 49th Annual EMSA Meeeting* (G.W. Bailey, Ed.), 674 (San Francisco Press, San Francisco).

Wohlleben, D.J. (1971). *Electron Microscopy in Materials Science*, Vol. 2, 712 (Academic Press, New York).

Yamamoto, K., Tanji, T. and Hibino, M. (2000). *Ultramicroscopy* **85**, 35.

Yamamoto, K., Hirayama, T. and Tanji, T. (2004). *Ultramicroscopy* **101**, 265.

19

Diffractive (Lensless) Imaging

John C.H. Spence

1 Introduction

Diffractive (or lensless) imaging refers to the use of theoretical methods and computer algorithms to solve the phase problem for scattering by a nonperiodic object. The name coherent X-ray diffractive imaging (CXDI) is used in the X-ray community, which we could generalize to CDI. Additional information about the object, such as the sign of the scattering potential and the approximate boundary of the object, may be combined with the measured scattered intensity to solve for the phases of the scattered amplitudes. In this way, under conditions of single scattering (and other approximations that often apply in optics and in electron, X-ray, and neutron diffraction) it may be possible to reconstruct a real-space image of an object by Fourier transform of the complex scattering distribution, or Fraunhoffer diffraction pattern. (Applications to Fresnel near-field imaging are also possible. In this geometry, resolution is, however, limited by detector pixel size, since the magnification is unity if lenses are not used.) In this review we will not discuss the recently developed and powerful transport of intensity method, which is also applicable to the near field (Paganin and Nugent, 1998). By avoiding the need for a lens, the aberrations and resolution limits introduced by lenses are thus avoided. Within the past decade this process has been demonstrated experimentally for neutron, X-ray, and electron scattering, so that the field has reached an exciting point. The electron work has produced atomic-resolution images, while experiments with soft X-rays have finally produced three-dimensional (tomographic) reconstructions. It now offers the real possibility of diffraction-limited imaging with any radiation for which lenses do not exist. Since each radiation interacts differently with matter, the method can be expected to provide us with new information on matter in fields as diverse as biology, materials science, and astronomy. Much current interest focuses on tomographic imaging of whole cells, nanoparticles and mesoporous materials, and on the imaging of proteins at near-atomic resolution. At present, the inversion of even a two-dimensional diffraction pattern is a lengthy process, and while three-dimensional inversion may involve weeks of work for a $1024 \times 1024 \times 1024$ voxel data set (with much time devoted to data merging and preparation issues), it is clear that automated processing will soon be possible. While it can be shown rigorously that some of the algorithms described here do not diverge, because of the nonconvex nature of the constraints involved no formal proof of convergence or of the uniqueness of solutions exists. Yet computational trials for simulated data with little noise practically always converge rapidly to the known solution, while the inevitable stagnation occurs in the presence of excessive noise. [The ability of these algorithms to "climb out" of local minimum and find a global minimum in a data set with perhaps a million unknown parameters (phases) is one of their most remarkable features.] The challenge now is to perform experiments under conditions in which the experimental parameters are sufficiently well known that inversion of experimental data performs as well as the simulations. In summary, over the past decade many algorithms for lensless imaging have been published that work well on simulated noisy data; the more difficult problem is experimental data collection under the conditions for which the algorithms converge. (This may require, for example, that the experimental scattered intensity distribution possesses inversion symmetry, consistent with the theoretical assumption of a two-dimensional hermitian potential or "real object.")

2 History

The phase problem in optics has early origins: in a letter to Michelson, Rayleigh comments that "the phase problem in interferometry is insoluble without a-priori information on the symmetry of the data" (Strutt, 1892). Several fields have contributed to the development of our current working solution to this problem, including signal processing, wavefront sensing, astronomy, electron microscopy, image processing, and X-ray crystallography, in addition to applied mathematics. Thus a wonderfully rich set of ideas has contributed, from communications theory to diffraction physics and set theory. Since my background is in diffraction physics, I must apologize to those workers in other areas (such as acoustics) whose contributions to the vast literature on this subject I have overlooked. In this review we will not discuss the crystallographic phase problem, except to note parallels (such as solvent flattening) with our noncrystallographic or "single particle" problem (Carrozzini et al., 2004). [By the obvious method of periodic continuation, it has been shown that the highly successful direct methods of crystallography can be applied immediately to the continuous scattering from an isolated object, if the correct choice of sampling interval is made (Spence et al., 2003b), however this may not be the most efficient solution.] In retrospect, we can now see that the key to a successful solution was given in the crystallography literature as early as 1952. D. Sayre in "Some implications of a theorem due to Shannon" (Sayre, 1952) appears to have been the first to consider the relationship between Shannon's sampling theorem and Bragg's law. He pointed out that for centrosymmetric crystals, where only the sign of structure factors F_g is unknown, these could be determined if intensities $|F|^2_g$ were observable at half-integral values of g, since a change of sign requires a zero crossing of the molecular scattering factor there. By Shannon's theorem, these fractional and integral order intensities are needed and sufficient to completely define the autocorrelation function of the molecule. Sayre's fractional orders could thus be generated only by a molecule that filled half the cell, leaving the remaining "half" empty. [The Patterson function or autocorrelation function of an inorganic, continuously bonded crystal will differ from that of the basis (one "molecule"), since the autocorrelation of the molecule is twice as large as the unit cell. These concepts will be clarified in the next section.] Sayre concludes that a solution to the phase problem for centric crystals could as well be obtained from measurements of the *intensity* of half-order reflections as from study of the Bragg intensities. [The paper was prompted by the observation of nonintegral reflections (following hydration) in experimental patterns from hemoglobin by Perutz and others, and a talk by Gay.] Since diffraction from an isolated nonperiodic object is a continuous function of scattering angle it provides immediate access to these "fractional orders." Thus a solution in principle to the phase problem for real, centric single particles has existed in the literature since 1952.

Sayre's paper had no impact in the image processing or signal processing communities, and we will not trace here its important influence on the development of density modification techniques in protein crystallography. (Here the molecule is often surrounded by a water jacket of comparable volume to the molecule, effectively doubling the cell size and possibly giving rise to "half-integral" reflections.) For imaging, the next important development was work by Gerchberg and Saxton (G–S) (Gerchberg and Saxton, 1971), which posed the following question: if the intensity of a general complex two-dimensional image and its corresponding diffraction pattern intensity is known, can the complex image (and diffraction pattern) be reconstructed? (By a complex image we refer here to the sample exit-face wavefunction, as discussed in the next section.) These data are available in a modern electron microscope. By solving the resulting quadratic equations (equal in number, as they comment, to the number of unknown phases) using an iterative algorithm, they showed successful inversion for one-dimensional data. It has often been noted that these Fourier equations, one for each pixel in the detector, may not be independent, are nonlinear if a sign constraint is used, and are corrupted by noise. Dependence in the equations may be introduced by choice of support shape. Other approaches have been explored for solving these nonlinear Fourier equations, but the G–S paper established the use of iterations between real and reciprocal space, with known information imposed repeatedly in each domain. G–S do not establish uniqueness or convergence in general.

The connection with Sayre's paper was not made; however, in retrospect we see that the sampling interval used by G–S for their analysis of a nonperiodic object necessarily corresponded to the Shannon half-integral orders of Sayre's paper. Modern work is focused on three-dimensional data in which only the diffraction pattern intensity from a real or complex object is measured. However, for a pure phase object, the modulus of the exit face wavefunction is known a priori to be unity everywhere, greatly reducing the number of unknowns in the G–S analysis. (This unit modulus constraint is, unfortunately, nonconvex.) For a real object (or, equivalently, a weak phase object) the diffraction pattern has Hermitian symmetry (Friedel's law), resulting in a further reduction in unknowns. Following the Gerchberg and Saxton paper (and a second paper, Gerchberg and Saxton, 1972), it was still necessary to determine the minimum information needed about a real object to solve the phase problem if the diffracted intensity was given. The role of the object boundary (defining a "support") and the need to sample the continuous scattering finely enough to satisfy Shannon's theorem (treating the autocorrelation function as the "bandlimit") were all recognized at this time.

The autocorrelation of a two-dimensional object is twice as large as the object in any direction, so Shannon's theorem requires sampling at half the Bragg angle for a periodically continued object. Bates referred to this as "oversampling," a term still in use. (Since it is in fact *optimum* sampling of the diffracted *intensity*, we avoid that term here.) The problem was vigorously attacked by authors such as Bates, Fiddy, Fienup, Gonslaves, and Papoulis in the early nineteen eighties, and the use of lensless imaging based on these ideas using X-ray scattering was advocated (Sayre, 1980). The early series of papers by Bates and co-workers are especially significant, and can be traced through Bates and McDonnell (1989), while Fiddy provided a new approach based on analysis of zeros in the diffraction pattern (e.g., Liao et al., 1997, and earlier work). It was shown, for example, that the set of all bandlimited functions having a given set of real zero crossings is convex, and that, in the absence of noise, the data might be factorizable. By 1982 a useful working solution had been obtained (Fienup, 1982) for real two-dimensional images by the addition of a feedback feature to the G–S error-reduction (ER) algorithm. An important realization at about this time was that the "landscape" (a map of error metric in the N-dimensional search space) for the phase problem was not rugged, and usually consisted of a single shallow global minimum with a few small bumps; the essential difficulty is not caused by the number of local false minima, but rather by the large number of directions in which search is possible, in a space containing one dimension for each image pixel. This Fienup algorithm has come to be known as the hybrid input–output or HIO algorithm. Feedback greatly speeds up convergence and improves the ability of the algorithm to escape from local minima, however, most modern work is based on a combination of the ER and HIO algorithms, since only the ER algorithm supplies a reliable error metric. Fienup showed that the ER algorithm converges, in the sense that the error monotonically decreases (Fienup, 1997). Real images could then be reconstructed from Fraunhoffer diffraction pattern intensity data, provided that the sign of the scattering potential was known together with an approximate estimate of the object boundary (support). Important work on the uniqueness problem for the HIO algorithm was published by Bruck and Sodin (1979) and Baraket and Newsam (1984) who showed that ambiguous solutions are obtained with these constraints only on "pathologically rare" occasions. The aim of this work is to show that only one image function (or images equivalent to this) satisfies the sign, support, and Fourier modulus constraints.

The HIO algorithm, which we give below, is the basis for most successful modern work, and has led to the development of many variants and much detailed mathematical analysis. The book by Stark (1987) provided an excellent overview of the subject up to 1987, and can be strongly recommended to students, together with an important review article by Millane (1990), which unified the crystallographic and single-particle approaches and so brought the field to the attention of a much larger audience. [The striking resemblance of the HIO algorithm to the independently developed solvent flattening methods of X-ray crystallography can be traced through Wang (1985).] The water-jacket around a protein in a crystal plays the same role as the "zero-density" region outside the support of our single particles.

The HIO algorithm and its variants have been highly successful for real objects, however, rather less attention has been paid to the more difficult problem of complex image reconstruction. Except for special cases, it is found that a very precise knowledge of the support is needed, and that in one and two dimensions this support should be disjoint (separated into two parts) (Fienup, 1987). Iterative inversion schemes fail for one-dimensional real data that are not disjoint, and work better in higher dimensions.

The third more recent period of work since about 1990, during which the field has grown rapidly, is discussed in the next sections. During that time a major effort was made to apply these methods to soft X-ray transmission data at the Brookhaven synchrotron (Sayre et al., 1998; Miao et al., 1998). Our understanding of the success of the HIO algorithm has increased considerably in that period due to the contribution of mathematicians using powerful methods based on convex set theory; Bregman projections and constraint theory were first described as early as 1982 (Youla and Webb, 1982) but not taken up until recently (for an excellent review, see Bauschke et al., 2002). Questions of uniqueness in dimensions greater than two have been considered, where the problem is overdetermined (Millane, 1995), and it is generally found that the iterative algorithms work better. In addition, a number of developments of the HIO algorithm have been published (Bauschke et al., 2003; Elser, 2003); the standard symbols used by mathematicians for the theory of projections has been adopted to describe the problem, and a new algorithm, which dispenses with any need for knowledge of the object support [the "shrinkwrap" algorithm (Marchesini et al., 2003b)], has been published. The problem of lost information within the synchrotron beam-stop has been addressed by reconstructing the diffuse scattering around a Bragg reflection from a nanocrystal (Robinson et al., 2001). In 2001 the first of a biannual series of conferences on the noncrystallographic phase problem was held at Lawrence Berkeley Laboratory (Spence et al., 2001), with a second in Cairns, Australia. The latest, held at Porquelles in France in 2005, can be found at http://www.esrf.fr/NewsAndEvents/Events/Non-Crys20-06-05/. An entirely different approach to the solution of the phase problem, based on the transport of intensity equations, has also been fully developed during this period and applied to optical, X-ray, and electron imaging (Paganin and Nugent, 1998).

These theoretical developments, often based on computer simulations, have been supported by far fewer demonstrations of inversions from experimental data. Only experimental results can truly give confidence in theory, and early results were somewhat disappointing. An early application of the HIO algorithm to experimental optical speckle data can be found in Cederquist et al. (1988), and applications to images formed with laser light can be found in Weierstall et al. (2001). In the X-ray imaging community the work at Brookhaven finally paid off in 1999, when images of lithographed characters were reconstructed from soft X-ray diffraction patterns (Miao et al., 1999). More recently, an atomic-resolution image of a single double-walled carbon nanotube has been reconstructed from its electron microdiffraction pattern (Zuo et al., 2003). These successes have led to rapid growth of the subject in recent years.

3 Objects, Images, and Diffraction Patterns: Validity Domains of Approximations

The terms *object, image, diffraction pattern, exit-face wavefunction, real and complex object, and transmission function* have sometimes been confused in the literature and are defined here for clarity. We define the *object* by its ground-state charge density $\rho(\mathbf{R})$ (which diffracts X-rays) and by the corresponding electrostatic potential $V(\mathbf{R})$ (which diffracts electrons). $V(\mathbf{R})$ is related to the density $\rho(\mathbf{R})$ by Poisson's equation. As a result of inelastic processes, both may be complex, and could be referred to generally as the "complex optical potential," but in this chapter unless otherwise stated we take them to be real. Here $\mathbf{R}$ is a three-dimensional vector while $\mathbf{r}$ will be two dimensional. We define the *exit-face wavefunction* $\Psi(\mathbf{r})$ across the downstream face of a thin slab of sample in the transmission diffraction geometry with *transmission function* $T(\mathbf{r})$ by

$$\Psi(\mathbf{r}) = T(\mathbf{r})\,\Psi_0(\mathbf{r})$$

where $\Psi_0(\mathbf{r})$ is the wavefield incident on the sample. [For perfectly coherent radiation from a point source, collimated by a lens, $\Psi_0(\mathbf{r})$ is approximately a plane-wave.]

Transmission functions for X-ray and electron diffraction are derived below in terms of the wanted object properties $\rho(\mathbf{R})$ or $V(\mathbf{R})$. We define the *image* as any magnified, resolution-limited or aberrated copy of $\Psi_0(\mathbf{r})$, formed by a lens. Unfortunately this term is now widely used in the CDI literature as a synonym for either *object* or *exit-face wavefunction*. This use of "image" to refer to an exit-face wavefunction is now firmly established and will be continued here; however, the more important distinction between object and exit-face wavefunction is likely to become important in future work, and should be preserved. The terms "real object" and "complex object" have also been widely adopted in optics, but should strictly refer to the nature of the exit-face wavefunction for two-dimensional imaging. [For a strong phase object described by Eq. (1), the object Δn may be real, but the exit-face wavefunction complex. Most authors would refer to this as a "complex object," however, the object property we wish to recover is $\Delta n(\mathbf{r})$.] It is also useful to reserve the word *image* for real-space functions, and *diffraction* for reciprocal space, so that the term "diffraction image," when referring to a diffraction pattern, should be avoided. The measured diffraction pattern intensity is $I(\mathbf{u}) = |\Phi(\mathbf{u})|^2$, with $\Phi(\mathbf{u})$ the Fourier transform of $\Psi(\mathbf{r})$ and scattering vector $|\mathbf{u}| = \Theta/\lambda$ for small scattering angles Θ. (Here λ is the X-ray, or relativistically corrected de Broglie wavelength for electrons.) $I(\mathbf{u}) = I(-\mathbf{u})$ if $\Psi(\mathbf{r})$ is real, or, for complex objects, if $\Psi(\mathbf{r}) = \Psi(-\mathbf{r})$. $\Phi(\mathbf{u}) = \Phi^*(-\mathbf{u})$ for real $\Psi(\mathbf{r})$, while $\Phi(\mathbf{u})$ is a signed real quantity if $\Psi(\mathbf{r}) = \Psi(-\mathbf{r})$ and $\Psi(\mathbf{r})$ is real. The aim of diffractive imaging is to reconstruct the *object* from the scattered intensity $I(\mathbf{u})$; however, as a first step the exit-face wavefunction $\Psi(\mathbf{r})$, which is simply related to $\Phi(\mathbf{u})$ by a Fourier transform, is obtained. [This reduces to determination of the sign of $\Phi(\mathbf{u})$ if $\Psi(\mathbf{r}) = \Psi(-\mathbf{r})$ and $\Psi(\mathbf{r})$ is real.] The further recovery of object properties $\rho(\mathbf{R})$ or $V(\mathbf{R})$ from $\Psi(\mathbf{r})$ may be possible only in the absence of multiple scattering or inelastic scattering. For electron diffraction, the weak-phase approximation generates a "real object" in the language of optics and diffractive imaging, as described below. For X-ray diffraction, where single-scattering conditions are common, the absence of spatially dependent absorption (due to the photoelectric effect) provides such a "real object," so that imaging should be performed at energies that avoid absorption edges for any elements present if phase contrast is expected. A single, spatially uniform absorptive process, however, may allow the phase contrast formulation to be used. The introduction of the transmission function allows a simple extension to the case of coherent convergent-beam illumination and related methods for phasing (Spence and Zuo, 1992).

We now relate $\Psi(\mathbf{r})$ to the wanted object properties $\rho(\mathbf{R})$ or $V(\mathbf{R})$ for the case of visible light, electron beams, and X-rays in the projection, iconal, or "flat Ewald sphere" approximation. Both refractive and dissipative (inelastic) processes may occur. In each case it is necessary to consider whether an iconal or projection approximation may be made, and the question of whether three-dimensional (tomographic) information (discussed later) may be extracted.

The simplest case for each radiation is that in which the projection approximation holds, in which case the image $\Psi(\mathbf{r})$ may be treated as a simple projection of some property of the sample, taken in the beam direction. Then, if a transmission sample in the form of a thin plate of thickness t is illuminated by a plane-wave,

$$\psi(\mathbf{r}) = T(\mathbf{r})\psi_0(\mathbf{r}) = \exp\left[-2\pi i\Delta n_p(\mathbf{r})/\lambda\right]\exp(-2\pi i\mathbf{u}_0 \cdot \mathbf{r}) \tag{1}$$

For normal-incidence plane-wave illumination $\mathbf{u}_o = 0$, and we may set $\Psi_0(\mathbf{r}) = 1$, and $\Psi(\mathbf{r}) = \exp[-i\theta(\mathbf{r})]$. Here Δn_p is proportional to the complex refractive index of the sample for the radiation concerned. We see that a real ("mask-like") object can be obtained only if the real part of Δn_p is independent of $\mathbf{r}$, and all structural information is contained in the imaginary part. These experimental conditions (pure "absorption" contrast) can be obtained only at relatively low spatial resolution (under incoherent conditions) for both electrons and X-rays.

For X-rays,

$$\Delta n_p(\mathbf{r}) = \int_0^t [\delta(\mathbf{R}) - i\beta(\mathbf{R})]\,dz \tag{2}$$

where d is a positive quantity (Kirz et al., 1995). In terms of mean values, the complex index of refraction for X-rays is $n = 1 - \Delta n, = (1 - \delta) + i\beta$, where δ describes refraction and β absorption (mainly the photoelectric effect, arising from absorption edges). The linear absorption coefficient is $\mu = 4\pi\beta/\lambda$. The dependence of $\Delta n(\mathbf{r})$ on the real and imaginary parts of the atomic scattering factors f and f' is given by $\delta = (r_e\lambda^2/2\pi)n_a f$, and $\beta = (r_e\lambda^2/2\pi)n_a f'$, with n_a atoms per unit volume and r_e the classical electron radius. Away from absorption edges, the electronic charge density (excluding the nuclear contribution) is

$$\rho(\mathbf{R}) = \left(\frac{2\pi}{r_e\lambda^2}\right)\delta(\mathbf{R}) \tag{3}$$

If small bonding effects are ignored, $\rho(\mathbf{R})$ is obtainable from tabulated X-ray scattering factors for neutral atoms.

Since δ is about 10^{-3} at 6 kV for light materials, a thickness of about $0.3\,\mu\text{m}$ of sapphire is needed to obtain a phase shift of $\pi/2$, allowing a first-order expansion of Eq.(1). Then the diffracted amplitudes are simply proportional to the Fourier transform of the projected charge density of the object.

For an electron beam of kinetic energy eV_o,

$$\Delta n_p(\mathbf{r}) = \int_0^t \frac{V_c(\mathbf{R})}{2V_0}\,dz \tag{4}$$

where $V_c(\mathbf{R})$ is the complex "optical" potential for high energy electrons (Radi, 1970; Howie and Stern, 1972), and the mean refractive index for electrons is $n = 1 + \Delta n$. The real part of this, for high energies, is the positive electrostatic or Coloumb potential, also obtainable from X-ray scattering factors using Poisson's equation (Spence and Zuo, 1992). The imaginary part (typically about a tenth of the real part) accounts for depletion of the elastic wavefield by inelastic scattering events such as plasmon, inner-shell, and phonon scattering. The average value V_0 of $\text{Re}(Vc)$ is about 12 eV for light materials, and is positive, so this may be used as a constraint. Both real and imaginary parts of the potential $V_c(\mathbf{R})$ are positive if the mean inner potential is included, since electron beams are attracted predominantly to the positive nuclei. Electron diffraction in the transmission geometry is not, however, sensitive to the mean potential V_0, which produces only a constant phase shift. The mean inner potential can be meaningfully defined only for a finite object with zero total charge (O'Keeffe and Spence, 1993). Although the potential due to an (unphysical) isolated negative ion may have small negative excursions, the mean value depends on the volume assumed, and, in a real crystal, any long range ionic potential is also screened. In summary, in the absence of its mean value, the real part of the optical potential used to describe electron diffraction (when synthesized from diffraction data, with resolution limited by a temperature factor) is positive, except for possible very small negative excursions around negative ions. The positivity of the imaginary part is guaranteed by the requirement that energy gain is forbidden in inelastic processes if very small virtual processes are ignored. Fourier coefficients of the total optical potential may have either sign, and those of the real (elastic) and imaginary (inelastic) potential become mixed in objects without inversion symmetry. Hence a sign condition on the optical potential may be used as a convex constraint unless atomic resolution reconstructions are attempted of sufficiently high accuracy to detect the very weak bonding effects.

By estimating the maximum value of $[\text{Re}(Vc) - V_o]$, the maximum thickness can be estimated for which a first-order expansion of Eq.(1) may be made, the weak-phase approximation, in which $\theta(\mathbf{r}) \ll \pi/2$. This is also a limited case of the single scattering approximation. For X-rays, the maximum thickness allowable in the weak-phase object approximation can be estimated using the CXRO web calculator page supported by the Lawrence Berkeley Laboratory at http://www-cxro. lbl.gov/. For electrons the Fourier coefficients of electrostatic potential published for many materials by Radi may be useful (Radi, 1970), since these also provide an estimate of the imaginary part of the optical potential for electron diffraction due to inelastic scattering.

For visible light, $\Delta n_p(r)$ has a similar interpretation as for X-rays (with n given by the square root of the complex dielectric constant), however, $n > 1$. Thus visible light and electrons ($n > 1$) are bent toward the normal on entering a denser medium, while X-rays undergo total external reflection, with $n < 1$.

In summary, if inelastic processes are neglected (away from absorption edges), at high energies, transmission samples of thickness t are phase objects for electrons and X-rays, for which the refractive index is proportional to the electron density for X-rays and to the total electrostatic potential, including the nuclear contribution, for electrons. Poisson's equation relates these. The magnitudes of these quantities are such that a first-order expansion of Eq.(1) (weak phase object approximation) is justified $[2\pi i n_p(\mathbf{r})/\lambda < \pi/2]$ if $t < 20\,\text{nm}$ for electrons (light elements, $V_0 = 200\,\text{kV}$) but $t < 0.3\,\mu\text{m}$ for X-rays (light inorganic material, $6\,\text{kV}$). Higher order terms in the expansion of Eq.(1) correspond to the multiple-scattering terms of the Born series in a "flat Ewald sphere" approximation.

This "flat sphere" or projection approximation, on which Eq.(1) is based, depends on the ratio of wavelength to smallest detail d of interest (with spatial frequency $u = 1/d$), and on the thickness of the sample. Scattering kinematics restrict elastic scattering to regions of reciprocal space near the energy and momentum-conserving Ewald sphere of radius $1/\lambda$. The projection approximation holds if the "excitation error" distance $S_u \approx \lambda u^2/2$ from this sphere onto a plane in reciprocal space normal to the beam (passing through the origin) is less than either $1/t$ or $1/\xi_u$, whichever is the smallest. (ξ_u is a multiple-scattering extinction distance for spatial frequency u. Thus samples never look thicker than ξ_u, to diffracting radiation.) Hence we require

$$\frac{\lambda u^2}{2} < \frac{1}{t} \; or \; \frac{1}{\xi_u} \tag{5}$$

or

$$\left(\frac{\lambda t}{2}\right)^{1/2} < d \tag{6}$$

for the validity of Eq.(1). Since the width of the first Fresnel fringe due to propagation over distance t is approximately $w = (\lambda t/2)^{1/2}$, this condition requires that the spreading of the wavefield due to free-space propagation over a distance equal to the thickness of the sample be small compared to the resolution required.

The failure of Eq.(1) may require either single or multiple scattering treatments, depending on the strength of the interaction and sample thickness. We note that Eq.(1) is an exact solution that sums the Born series, including all multiple scattering effects, in the limit of vanishing wavelength. For X-ray tomography, use of Eq.(1) greatly simplifies the collection of data for three-dimensional reconstruction.

The preceding discussion has concerned two-dimensional imaging. For tomography there is a different analysis. Again, a single-scattering approximation must be used. But the introduction of a curved Ewald sphere does not prevent three-dimensional reconstruction (Miao et al., 2002), since data may be collected for various object orientations, and in each case assigned to points on the sphere until all of the reciprocal space volume is filled. (Experimentally this is done by rotating the sample about an axis normal to the beam and recording a diffraction pattern at each orientation.) Use of the Fienup algorithm with three-dimensional Fourier transform iterations can then take advantage of the improved convergence in three dimensions. A sign constraint may or may not be applied to both real and imaginary parts of the scattering potential, and a general spherical support, enclosing the object, has been found useful.

4 The HIO Algorithm and Its Variants

We now assume the weak-phase approximation, in which the exit-face wavefunction is computed along a single optical path along the beam direction, and no multiple scattering or inelastic processes are permitted other than an overall, spatially independent exponential attenuation with thickness. The recorded intensity of the diffraction pattern, excluding the central portion, is then given as

$$I(\mathbf{u}) = \Phi(\mathbf{u})\Phi(\mathbf{u})^* = |FT(\Psi(\mathbf{r})|^2 \tag{7}$$

If the phase of the Fourier transform $FT[\Psi(\mathbf{r})]$ could be recovered, than the aberration-free complex exit wave $\Psi(\mathbf{r})$ could be reconstructed. Aberrations of any diffraction lenses in the electron case that magnify the diffraction pattern (astigmatism, distortion, etc.) could alter the intensity distribution $I(\mathbf{u})$ and so complicate recovery.

For nonperiodic objects the phase problem can now be solved if the object has a finite support, i.e., the image is nonzero only within a finite region of image space and if the image is real and nonnegative. For a weak phase object the exit-face wavefunction is $\Psi(\mathbf{r}) \sim 1 - i\theta(\mathbf{r})$ where $\theta(\mathbf{r})$ may be treated as real and positive. The diffracted amplitude is $\Psi(\mathbf{u}) = \delta(u) - i\Phi(u)$, for which the scattered part differs in phase by $90°$ from the direct beam given by the first term. This direct "unscattered" beam is absorbed by the beam-stop; difficulties arising from the finite size of this are discussed later, and we note that the average value of $\theta(\mathbf{r})$ is also lost within the beam stop. Including a larger area around the object known a priori to have zero or unit transmission function is equivalent to "oversampling" in Fourier space, and the spatial coherence of the illumination must span this larger area, at least equal to that of the autocorrelation function of the object. Hence diffractive imaging requires spatial coherence twice that of coherent imaging with a lens (Spence et al., 2004), increasing exposure time by a factor of four for a given source and resolution limit.

We assume that $|FT[\Psi(x, y)]|$ has been measured, and that the support $S(x, y)$ of $\theta(\mathbf{r})$ (which is the same as that of the object) is known. $S(x, y)$ is the region outside which the object is known to be zero. The iterations start with an initial estimate $\tilde{G}_1(u, v) = |FT[\Psi(x, y)]|\exp[i\theta_1(u, v)]$ of the spectrum. $\theta_1(u, v)$ is chosen to be an array of independent pseudorandom real numbers distributed between 0 and 2π. The iterative Fourier transform algorithm consists of the following steps (with subscript k labeling quantities at the kth iteration):

1. Inverse Fourier transform $\tilde{G}_k(u, v)$ to obtain the image $\tilde{g}_k(x, y)$
2. Define $g_{k+1}(x, y)$ as

$$g_{k+1}(x, y) = \begin{cases} \tilde{g}_k(x, y) & \text{if } (x, y) \in S(x, y) \\ g_k(x, y) - \beta\tilde{g}_k(x, y) & \text{if } (x, y) \notin S(x, y) \end{cases} \tag{8}$$

This constitutes the HIO version of the algorithm. β is a constant chosen between 0.5 and 1. In the ER version of the algorithm this step is replaced by

$$g_{k+1}(x, y) = \begin{cases} \tilde{g}_k(x, y) & \text{if } (x, y) \in S(x, y) \\ 0 & \text{if } (x, y) \notin S(x, y) \end{cases} \tag{9}$$

3. Fourier transform $g_{k+1}(x, y)$ to obtain $G_{k+1}(u, v)$
4. Define new Fourier domain function $\tilde{G}_{k+1}(u, v)$ using the known Fourier modulus $|FT[\Psi(x, y)]|$ with the computed phase:

$$\tilde{G}_{k+1}(u, v) = |FT[\Psi(x, y)]|\exp[i\theta_{k+1}(u, v)]$$

5. Go to step 1 with k replaced by $(k + 1)$.

To monitor the progress of the algorithm, the object space error metric ε_k is calculated during each iteration:

$$\varepsilon_k = \frac{\sum\limits_{(x,y)\notin S} |\tilde{g}_k(x,y)|^2}{\sum\limits_{(x,y)} |\tilde{g}_k(x,y)|^2} \tag{10}$$

ε_k is the amount by which the reconstructed image violates the image-space constraints. Physically, in the X-ray case, it is the normalized amount of charge that remains outside the boundary of the object, which should be zero. In all our calculations we have used $\beta = 0.7$ and a combination of the ER and HIO algorithms, with 20 ER iterations followed by 50 HIO cycles, all repeated until the error ε_k drops below a certain level. Simulations based on the above procedure with small noise levels invariably converge to the correct solution (Weierstall et al., 2001; Spence et al., 2002). Since $g(x, y)$, $g^*(-x - a_1, -y - a_2)\exp(i\theta)$ and $g(x - a_1, y - a_2)\exp(i\theta)$ all have the same Fourier modulus they cannot be distinguished by the algorithm. Each run of the algorithm started with different random phases may thus produce images centered on different origins or related by inversion symmetry. We call these equivalent images. (Similarly, in three dimensions, enantiomorphs cannot be distinguished.)

Some understanding of the inversion can be obtained by considering the Fourier equations relating the object to its diffraction pattern. We reverse the domains originally considered by Shannon, and treat the object space as if it applies a "bandlimit" to the diffraction pattern (the object is assumed compact). Shannon's theorem then specifies either the sampling interval on the diffraction pattern *intensity* needed to fully reconstruct the autocorrelation function of the object, or the sampling of the complex scattered amplitude needed to reconstruct the object. Consider a simple one-dimensional complex exit-face wavefunction $f(x)$, which is nonzero only for $0 < x < W$, so the support has width W. We first treat this function as the "bandlimit" on the diffraction pattern $F(u)$, which must therefore be sampled at intervals $u_n = n/W$ by the detector to satisfy Shannon's theorem. We then have the N equations (one for each pixel in the detector)

$$|F(u_n)| = \left| \sum_{i=1}^{N} f(x_i)\exp(2\pi inx_i/W) \right| \quad n = 1 \ldots N$$

relating measured intensities $|F|$ to the $2N$ unknown complex values of $f(x_i)$ we seek. Since there are more unknowns than equations, the complex values of $f(x)$ cannot be found from these measurements.

However, now consider the same function placed within a domain of width $2W$, so that the values of $f(x)$ in $W < x < 2W$ are known a priori to be zero. The sampling interval on $F(u)$ is now $1/(2W)$ and we thus have $2N$ equations, but the number of unknown values of $f(x)$ remains at $2N$, so that the system of equations now becomes solvable in principle. Loosely speaking, we compensate for the missing half of the data in the diffraction domain (the phases) by requiring that half of the object values be known (they are zero outside the support of width W). Since the HIO algorithm assigns a random set of phases initially, the equations are also linear within the algorithm unless a sign constraint is applied. Experimentally, real data from a CCD may not provide independent equations (symmetrical support shapes also lead to dependent equations) and the effects of noise must be considered. Large systems of nonlinear equations are normally extremely difficult or impossible to solve, so it is remarkable that the HIO algorithm does this so effectively, for reasons that are not fully understood, since few methods exist for analyzing nonconvex optimization. However, although the Fourier modulus constraint is nonconvex, it is known that the error surface in hyperspace is not a rugged landscape, but consists of a single rather smooth minimum with a few small bumps. Use of a boundary estimate (support) without symmetry also drives the algorithm toward one particular enantiomorph $\rho(\mathbf{r})$ rather than a spurious mixture involving $\rho(-\mathbf{r})$.

This example may be extended to other cases: for complex, two-dimensional objects there are $2N^2$ unknowns but N^2 equations (if the CCD has linear pixel dimension N). But by placing empty space around our compact object, which increases the diffracting volume by $2^{1/2}$ in both dimensions, we sample the diffraction pattern more finely and recover the $2N^2$ equations needed to solve the phase problem. It is the knowledge of the support (the object boundary) that ensures that known object pixel values may be inserted in the algorithm correctly outside the support. Experimentally, this creates the greatest difficulty of the diffractive imaging method; it is necessary to know a priori that the diffraction pattern comes from *an isolated object whose size is approximately known.* The existence of any material outside the assumed support that unwittingly contributes to the diffraction pattern will result in inconsistent constraints being applied by the algorithm, which will then not converge.

When experimental data are analyzed, the physical support in the object may not be known accurately. To avoid confusion we call the support actually present in the experiment the "physical support" and the support estimate used in the data analysis the "computational support." The term "loose support" describes the situation in which the computational support is larger than the physical support. "Tight support" means the physical support is the same as the experimental support. For two-dimensional simulations with real objects, it is found that a "default" triangular-shaped support may invariably be used, if it is chosen to lack any symmetry and enclose the object. With experimental data, convergence may be slower (or nonexistent) due to high noise levels, the absence of data around the origin at the beam-stop, and excessively loose support. Then the simplest initial choice of support is the boundary of the autocorrelation function (obtained by Fourier transform of the diffracted intensity). This estimate is rapidly improved upon by the shrinkwrap algorithm, which uses a specified intensity threshold to find an improved smaller boundary at every iteration of the HIO algorithm (Marchesini et al., 2003b). This appears to be the most useful practical algorithm at present. Depending on noise levels, the support estimate has been found to be sufficiently accurate to deal with complex objects in many cases.

We conclude this section with some comments on additional constraints, the beam stop problem, support determination, and recent algorithm developments. In the general case of strong multiple scattering no simple closed-form expression relates the object to the exit-face wavefunction, and the two are related only by symmetry constraints. For a general ("strong") phase object [Eq. (1)], a unit modulus constraint may be applied, so that reconstructed pixels are forced to lie on a unit circle on an Argand diagram; however, this constraint is nonconvex and so has not been found very useful in practice. For such an object, to which a spatially independent absorption term is added [so that $\rho(\mathbf{R})$ or $V_C(\mathbf{R})$ has a constant known imaginary part], the reconstructed image pixels may be constrained to lie on a given spiral on an Argand diagram. A summary of the many constraints that have been tried and experimental results is given in Weierstall et al. (2001). Other constraints include atomicity, symmetry (most useful for reducing computing time), imposition of a known histogram of gray-levels for the object density, and low-resolution imaging by a different technique to provide a support estimate. In the following section the use of prepared objects is demonstrated to allow the method of Fourier transform holography to be used to provide a support estimate. The reference object need not be a simple point scatterer (He et al., 2004), and may have an extended complicated known shape (Szoke, 1997). The shrink-wrap algorithm described below rapidly improves iteratively on any initial support estimate and is found to converge for both real and complex objects under a wide range of conditions. For arrays of semiconductor devices the support will often be known, so that tomographic imaging of defects within an array element might be based on a support provided by the lithography pattern used to make the array.

A remarkable new "flipping" algorithm has recently appeared (Oszlanyi and Suto, 2004) for the crystallographic phase problem, which may also be adapted for nonperiodic objects, in which case it reduces to Fienup's output–output algorithm with feedback parameter $\beta = 2$ and a dynamic support defined by an adjustable threshold (Wu et al., 2004b). This algorithm operates as follows: First, random phases are assigned to the structure factors, which must extend to atomic resolution. These are transformed to yield a real charge density. All density values below a certain threshold have their sign reversed. The result is transformed, and Fourier magnitudes are replaced with measured values. The process is continued to convergence. This algorithm (far simpler than the direct methods normally used in crystallography) has been used to solve new crystal structures from X-ray data (Wu et al., 2004a), and is found to perform very well. It does not require a support estimate or knowledge of atomic scattering factors, but does require atomic-resolution data. The "atomicity" constraint in crystallography assumes that the solution density consists of a set of smooth peaks, and requires that diffraction data extend to atomic resolution. The support then consists of spheres around each atom; most of the density within a crystal consists of empty space between atoms, akin to the zero-density band generated by oversampling around an object in the HIO algorithm. Since all matter in the cold universe consists of atoms (whose scattering factors are known), this constraint provides an extremely powerful *ab initio* assumption if one has atomic-resolution data, and is the basis of the direct methods algorithms of X-ray crystallography. Other known "building blocks" (such as gold balls or lithographed dots) have been used to reduce the number of unknown parameters in image definition (Spence et al., 2003b). (For proteins, for example, both the sequence and the atomic structure of the 20 amino acids of which they consist are usually known a priori, together with a typical gray-level histogram for the density maps.) An algorithm that has used the atomicity idea for nonperiodic data is Speden (Hau-Riege et al., 2004), which has been applied to the data of Figure 19–3. Finally, we note the use of the HIO algorithm for two-dimensional protein crystals in electron microscopy. This technique uses Fourier transforms of conventional weak-phase-object images (formed with a magnetic lens) to provide the phases of structure factors, and diffraction patterns for their magnitudes. The subnanometer resolution images must be obtained over as large a range of tilts as possible, creating serious experimental difficulty. These two-dimensional monolayer crystals are nonperiodic in the direction normal to the plane of the crystal, so that lines of diffraction are generated in reciprocal space. By oversampling along these lines it is possible to phase these reciprocal lattice rods individually. The phase relation linking them can then be obtained form a few high-resolution images recorded at small tilts. In this way the number of electron microscope images needed for three-dimensional imaging of two-dimensional organic crystals can be greatly reduced, with most of the reconstruction based on easy-to-obtain diffraction data (Spence et al., 2003a).

We note in passing that the nonuniqueness of the crystallographic phase problem has been well studied; the so called "homometric" crystal structures studied by Pauling, Burger, and others have the same diffraction patterns, but different structures. (They are not enantiomorphs.) Fortunately these are very rare.

Several approaches have been made to the problem of data lost behind a synchrotron beam-stop, which is essential to protect a sensitive area detector. Any additional blooming can mean much loss of low-frequency data. In several HIO applications these missing values have simply been treated as free adjustable parameters, and the algorithm was found to converge. Calibrated absorption filters have been placed in front of the inner portion of the detector. Another solution is to use a sample consisting of an unknown object filling a small hole in an otherwise opaque mask (Eisebett et al., 2004; Weierstall et al., 2001). Stray scattering of X-rays from the edges of the holes in the mask can be a difficulty; however, surface roughness, which is smaller than the wavelength of the X-rays, produces little scattering. The use of very small silicon nitride windows (e.g., $2\,\mu m$ width) greatly reduces the intensity of the direct beam and blooming effects; however, the detailed shape of the partially transparent silicon wedge around the window must then be modeled and used as a support for inversion. (When making samples of small particles deposited from solution, it is most efficient to use a focused ion beam to remove all but a favorably located and isolated particle in the center of the window.) Finally, the "diffuse" X-ray scattering around Bragg peaks (which forms the "shape transform" of the object) from a crystallite has been inverted to an image, thus avoiding the direct-beam scattering (Robinson et al., 2001; Williams et al., 2003).

5 Experimental Results

Figure 19–1 shows the first X-ray images reconstructed by this lensless method in 1999 (Miao et al., 1999). The test object consists of letters formed from gold dots, 100 nm in diameter and 80 nm thick, on a transparent silicon nitride membrane. The transmission X-ray diffraction pattern formed with 1.7-nm monochromatic soft X-rays is shown in Figure 19–1A and the reconstructed image in Figure 19–1B. The object was illuminated through a coherently filled 10-µm-diameter pinhole, and a 25-cm camera length was used. Missing data from the central region within the beamstop were obtained from a lower-resolution optical image. The exposure time was 15 mins at the Brookhaven synchrotron. The Fienup algorithm was used for reconstruction, with a sign constraint applied to both real and imaginary parts of the scattering potential. A square support was used (irregular shapes usually work better) and 1000 iterations were needed for convergence. The resolution is about 75 nm. A detailed description of an improved version of the apparatus used to obtain this and other recent results at the Advanced Light Source is given in Beetz et al. (2005).

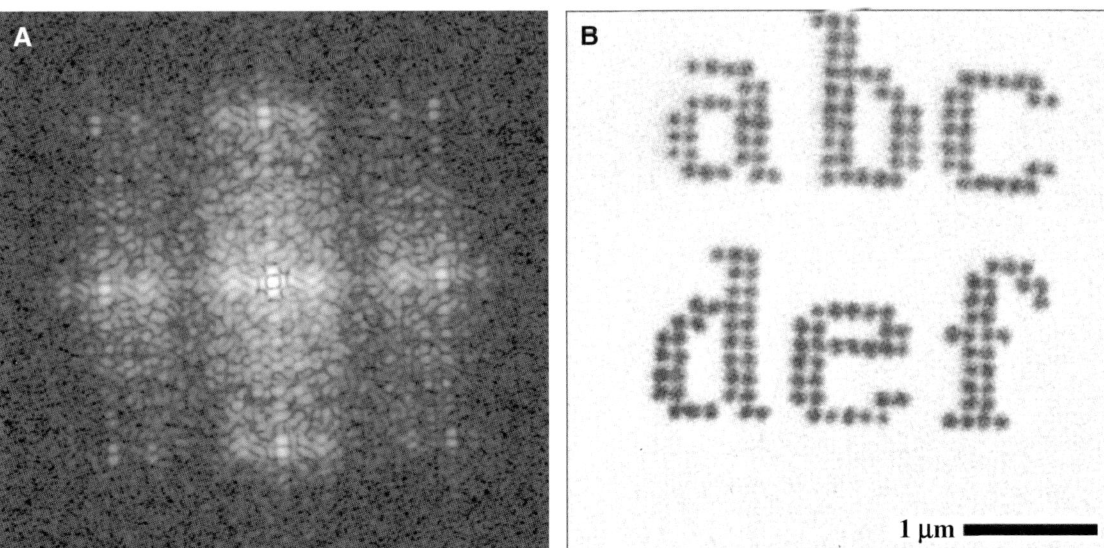

Figure 19–1. (A) Soft X-ray transmission diffraction pattern formed with 1.7-nm X-rays from the set of lithographed letters shown in (B). (B) Image recovered from (A) using a modified form of the HIO algorithm. (From Miao et al., 1999.)

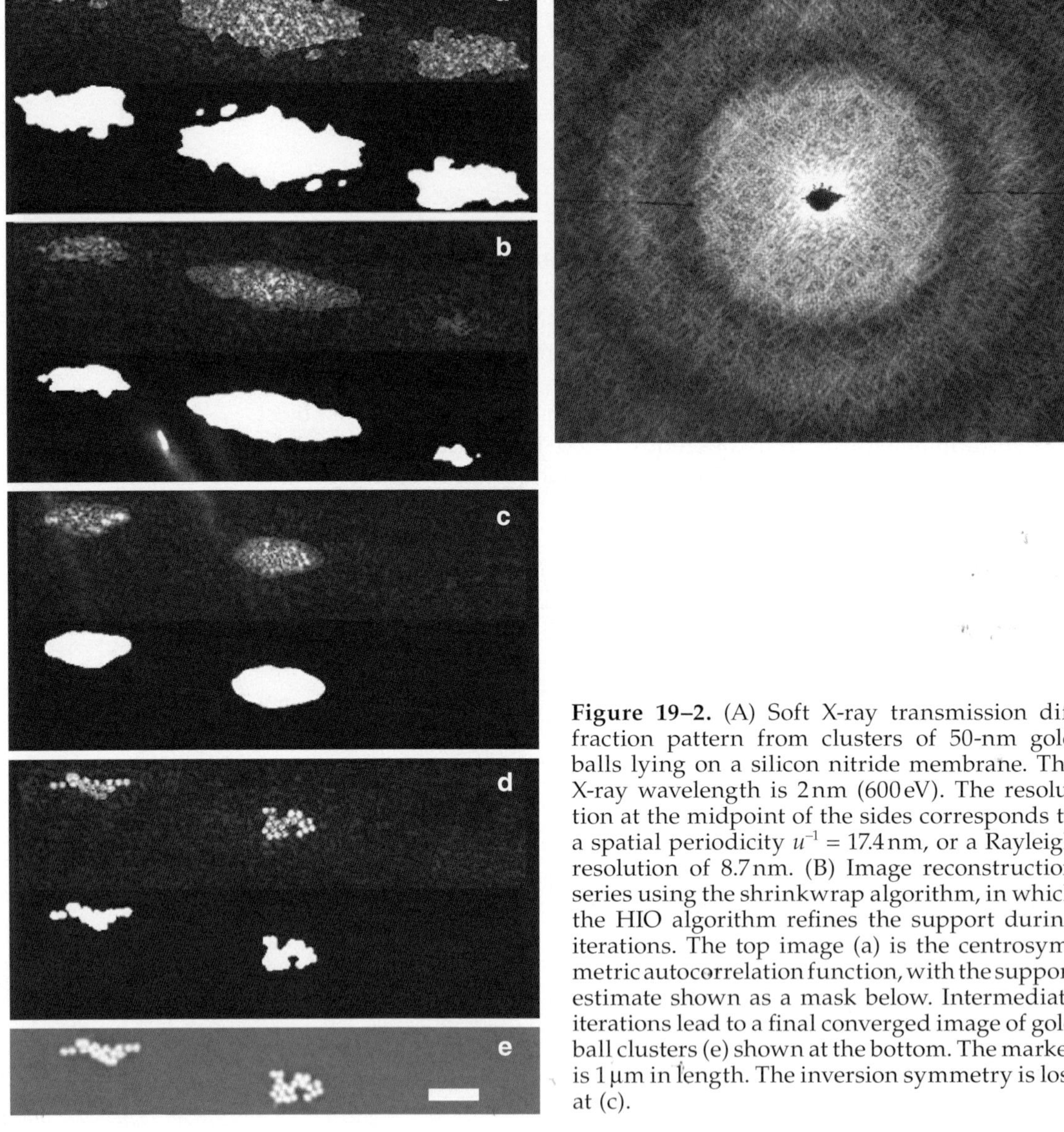

Figure 19–2. (A) Soft X-ray transmission diffraction pattern from clusters of 50-nm gold balls lying on a silicon nitride membrane. The X-ray wavelength is 2 nm (600 eV). The resolution at the midpoint of the sides corresponds to a spatial periodicity $u^{-1} = 17.4$ nm, or a Rayleigh resolution of 8.7 nm. (B) Image reconstruction series using the shrinkwrap algorithm, in which the HIO algorithm refines the support during iterations. The top image (a) is the centrosymmetric autocorrelation function, with the support estimate shown as a mask below. Intermediate iterations lead to a final converged image of gold ball clusters (e) shown at the bottom. The marker is 1 μm in length. The inversion symmetry is lost at (c).

Figure 19–2A shows a transmission diffraction pattern obtained using 600-eV monochromatic soft X-rays from clusters of gold balls, 50 nm in diameter, lying on a silicon nitride membrane. The silicon nitride membrane is almost transparent to the X-rays, so the object provides a useful test object for reconstruction. The pattern resembles the Airey's disk-like pattern from one ball, crossed by "speckle" fringes due to interference between different balls. An image reconstruction series using the shrinkwrap algorithm, in which the HIO algorithm refines the support during iterations, is shown in Figure 19–2B. The top image is the centrosymmetric autocorrelation function, with the support estimate shown as a mask below. This mask was obtained by Fourier transform of the diffraction pattern intensity (to produce the autocorrelation function shown), followed by the selection of a contour corresponding to a certain threshold of intensity. This thresholding operation is repeated after each HIO-ER iteration cycle, to generate a new improved estimate of the object support (see Marchesini et al., 2003b, for details). Intermediate iterations lead to the final converged image of the gold ball clusters (e) shown at the bottom. The marker is 1 µm in length. We note that the inversion symmetry necessarily possessed by the autocorrelation function at (a) is lost at (c) as it changes smoothly into the correctly phased image.

Figure 19–3 shows an instructive case, indicating the way in which "prepared objects" may be used to assist reconstruction. [Full experimental details for CXDI are given in He et al. (2003), from which Figure 19–3 is taken.] Figure 19–3A shows a scanning electron microscope (SEM) image of a set of gold balls lying on a silicon nitride membrane. One ball at A is isolated. The autocorrelation function obtained from an experimental soft X-ray transmission diffraction pattern (not shown) taken from this object is given in Figure 19–3B. This may be interpreted as the self-convolution of the object with its inverse, or, for a collection of point-like objects, as the set of all interpoint vectors. Some interball vectors are shown in Figure 19–3A and indicated again in Figure 19–3B. The convolution of the single isolated ball A with the three balls at B produces the autocorrelation function in Figure 19–3B a faithful image of the three balls, blurred by the image of one ball. This process is

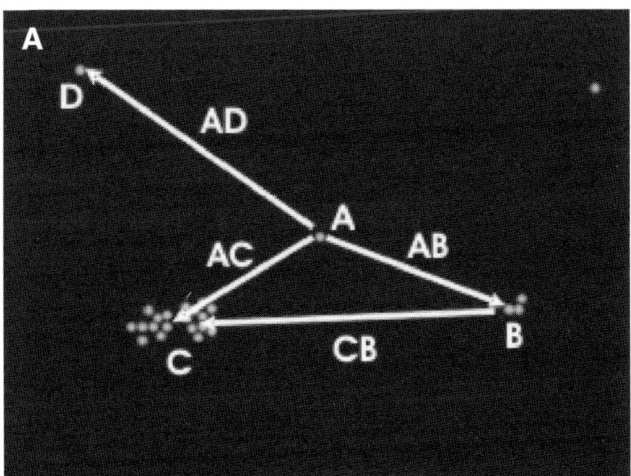

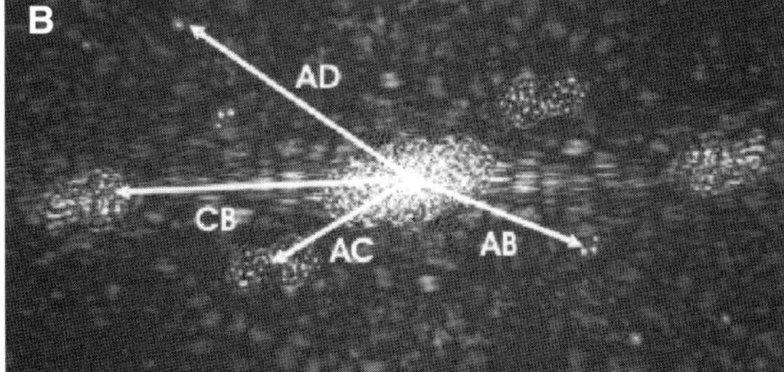

Figure 19–3. (A) SEM image of several clusters of gold balls, each 50 nm in diameter. Some interball vectors are indicated. The balls lie on an X-ray transparent substrate. (B) The Fourier transform of the X-ray diffraction pattern taken from (A). This is the autocorrelation function of the density in (A) and is a map of all interball vectors or the self-convolution of the object with its inverse. Because the object in (A) includes a single isolated ball at A, the vector AB leads to a faithful image of the triple-ball cluster at B. (The convolution of one ball with three gives a blurred image of three.) (From He et al., 2003.)

similar to the heavy-atom method of X-ray crystallography, or the method of Fourier transform holography in optics (Collier et al., 1971). If such a gold ball or strong "point" scatterer can be placed near an unknown object, the autocorrelation function will contain a useful first estimate of the desired image of the unknown, which can also be used to provide a support for further HIO iterations aimed at improving resolution. This process is demonstrated in He et al. (2004), where, following the original suggestion of Stroke (1997), it is found that the resolution in the autocorrelation "image" may be considerably improved beyond the size of the reference ball by simple deconvolution. By using a larger reference object, or one consisting of a cluster of small balls, the intensity of scattering from the reference object can be increased. It has been noted that a randomly placed cluster of point scatterers can provide a high resolution image in Fourier transform holography when used as a reference object (see He et al., 2004; Collier et al., 1971; Eisebett et al., 2004, for more details and references).

The first successful application of CDI to the electron diffraction patterns provided by a transmission electron microscope (TEM) is described in Weierstall et al. (2001), where a complete description of the method can be found. An important asset of the TEM is its ability to provide an image of the same region that contributes to the micro-diffraction pattern, so that this image can be used to supply the support. However, electron scattering is so strong that any scattering contribution from a supporting film, however thin, is found to prevent successful CDI. The resolution of the best TEM instruments in direct phase-contrast imaging mode using lenses is now about 1 Å. Figure 19–4 shows a more recent application of CDI to an electron diffraction pattern using a TEM (Zuo et al., 2003). This remarkable image is the first atomic-resolution CDI image, and possibly the first atomic-resolution image of a nanotube. The image gives us the helicity of the tube, its dimensions, and the number of walls. The double-walled nanotube spans a hole in a thin amorphous carbon film, while the electron beam diameter (about 50 nm) is smaller than the hole, so that there is no background contribution from the carbon film. A conventional TEM image was used to provide the support function for HIO iterations along the edges of the tube, and it is suggested that the boundary of the support across the tube is provided by loss of coherence at the edge of the electron probe due to rapid phase variations arising from the aberrations of the probe-forming lens. (In general a tight support is desirable for CDI.) The image shows higher resolution detail than conventional TEM images of nanotubes. Resolution is limited perhaps only by the temperature factor, or by distortions in electron lenses used to magnify the diffraction pattern. It remains to be seen if tomographic imaging at atomic resolution is simplest by this method or by direct TEM imaging using lenses. If CDI is used, the difficult problem of supporting a nanoparticle for diffraction over a range of orientations will need to be solved. Radiation damage may be reduced in diffraction mode under some conditions.

Three-dimensional (tomographic) CDI of inorganic samples has now been demonstrated (Miao et al., 2002; Williams et al., 2003; Chapman

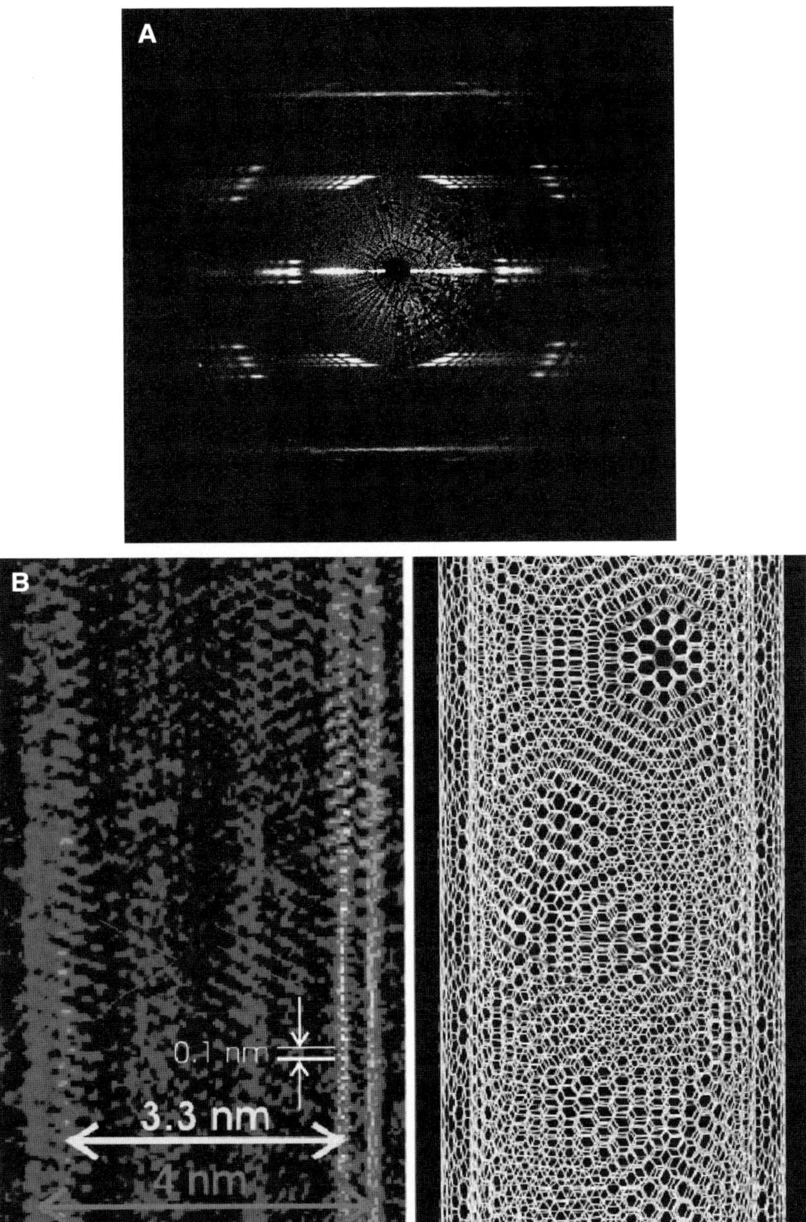

Figure 19–4. (A) Electron microdiffraction pattern from a single double-walled nanotube. This consists of a rolled-up sheet of graphite. Fine details arise from the helical structure (Zuo et al., 2003). (B) At left is the experimental image of the double-walled nanotube reconstructed from the electron diffraction pattern in (A). At right is shown a corresponding model of the structure (Zuo et al., 2003).

et al., 2006) using soft X-rays at a resolution of about 10 nm. This raises hopes of direct imaging, for example, of whole cells by this method if radiation damage considerations allow this at usefully high resolution. The X-ray source is a synchrotron and undulator, providing coherent radiation at about 600 eV. In recent work (Chapman et al., 2006), a simple zone plate was used as a monochromator, following by a beam-defining aperture of about 10 µm in diameter, coherently filled. A nude soft X-ray CCD camera, employing 1024×1024 24-µm pixels was used. The sample is mounted in the center of a silicon nitride window fitted to a TEM single-tilt holder, which provides automated rotation about a single axis normal to the X-ray beam. The window is rectangular, with the long axis normal to both the beam and the holder axis. Diffraction patterns are recorded at 1° rotation increments, with a typical recording time of about 15 min per orientation. The maximum tilt angle is then limited by the thickness of the silicon frame around the window to perhaps 80°, resulting in a missing wedge of data. In addition, data may be missing around the axial beamstop. The development of software for automated tomographic diffraction data collection and merging is a large undertaking (Frank et al., 1996), and much can be learned from the prior experience of tomography in biological electron microscopy, where these techniques have been perfected (Frank, 1992). In that case, however, the registration of successive images at different tilts is greatly facilitated by direct observation of image features. The use of shadow images or X-ray zone-plate images for similar purposes has been suggested. With no direct imaging mode, much time is wasted in X-ray work locating the beam on the sample, which, with current CCD detectors, will typically be smaller than 2 µm in diameter. The final resolution (in one dimension), allowing for an "oversampling" factor of 2, will then be $4000/1024 = 3.9$ nm. The camera length (sample-to-detector distance) of the diffraction camera must then be selected to allow half this spatial frequency to fall at the edge of the CCD camera at $u_{max} = \theta_{max}/\lambda = 0.5/3.9 \, \text{nm}^{-1}$, so that the maximum scattering angle is $\theta_{max} = 0.25$ rad for $\lambda = 2$ nm. Then the finest periodicity in the object (3.9/0.5 nm) is sampled twice in every period, according to Shannon's requirement (two points are required to define the period and amplitude of a sine wave if aliasing is excluded). For a CCD with linear pixel number N, the ratio of the finest detail to largest dimension is $N/2$, so that developments in detector technology limit CDI. The transverse spatial coherence of the beam must exceed 4 µm, as discussed together with monochromator requirements below.

Tomographic or three-dimensional imaging can provide the ability to "see inside" an object, but this requires that the intensity at a point in a projection be proportional to a line integral of some simple property of the object, such as the charge density. Then methods such as filtered back-projection can reassemble these two-dimensional projections into a volume density. Contours of equal density may then be isolated and presented to show the internal structure. For CXDI, a different approach is used, and some simplifications occur. It is no longer necessary to make the resolution-limiting "flat Ewald sphere" approximation, since diffraction data collected at one tilt can be assigned to points lying on the curved Ewald sphere in reciprocal space. (This is the momentum and energy-conserving sphere that describes elastic scattering in reciprocal space.) The sample is then rotated through this sphere around a single axis, until all of the reciprocal space is filled, out to a given resolution. Three-dimensional interpolation of data points near the sphere is needed, and careful intensity scaling may be necessary if several exposures with different times are required to cover the full dynamic range of the data. It is often found that missing data points in the central region can be treated as adjustable parameters in the HIO iterations. Once a roughly spherical volume has been filled in reciprocal space (perhaps with missing wedge and beam-stop region), the three-dimensional iterations of the HIO algorithm may be applied [Eqs. (8), (9), etc., extended to three dimensions]. The computing demands are severe, as outlined below. The converged data will provide a three-dimensional density map, proportional to the local charge density, if the single-scattering approximation of X-ray diffraction theory applies and if the spatial variation in attenuation of the beam due to the photoelectric effect can be neglected. Figure 19–5 shows such a tomographic reconstruction, from which three-dimensional surfaces of constant density may be obtained. These surfaces allow us to "see inside" materials, and may eventually permit maps to be obtained that distinguish regions of different chemical composition.

The usefulness of tomographic CXDI in biology remains to be determined; at present the method appears to have the advantages over electron microscopy by allowing observation of thicker samples under a wider range of environments (for example, in the "water window" around 580 eV for soft X-rays). By comparison with X-ray zone-plate "full-field" imaging, the method allows a much larger numerical aperture to be used, and hence makes more efficient use of scattered

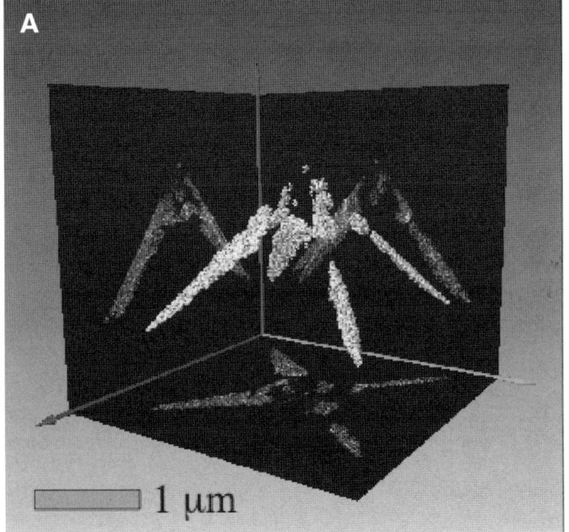
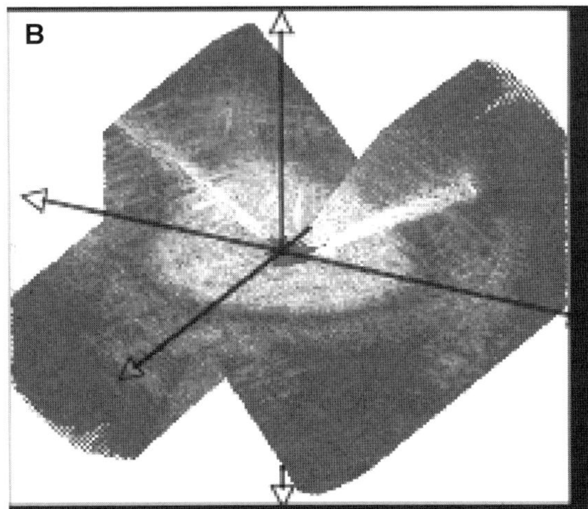

Figure 19–5. (A) Tomographic reconstruction from a soft X-ray diffraction pattern shown in (B). The object consists of gold balls (50 nm diameter) lying along the edges of a pyramidal-shaped silicon nitride structure. This is one image from a rotation series. From the complete series, three-dimensional surfaces of constant density can be constructed. (B) The volume of soft X-ray diffraction data collected to obtain the three-dimensional reconstruction in (A). (See color plate.)

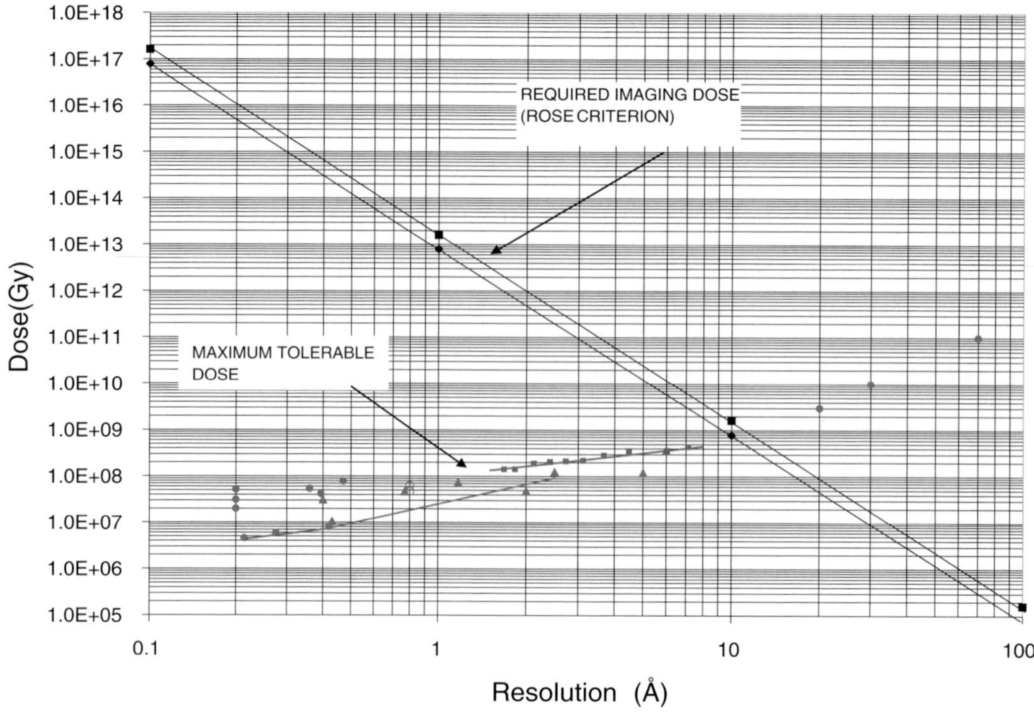

Figure 19–6. Summary of experimental measurement on organics for various microscopies on a plot of dose against resolution (Howells et al., 2005). Literature experimental values (e.g., Reimer, 1989) are shown as follows: Filled circles are from X-ray crystallography, filled triangles from electron crystallography, open circles from single-particle electron cryoelectron microscopy, open triangles from electron tomography, diamonds from soft X-ray microscopy, and filled squares from recent X-ray crystallography work on spot-fading experiments by Holton on the ribosome at the Advanced Light Source. Viable single-molecule microscopies must fall above the Rose equation line (to give a statistically significant image) and below the maximum tolerable damage line. Those below the Rose equation line succeeded by using crystallographic redundancy and form a periodically averaged image of perhaps 10^8 molecules in a crystal. The required imaging dose is calculated for a protein of empirical formula H50C30N9O10S1 and density $1.35\,g/cm^3$ against a background of water, imaged with $10\,keV$ X-rays (upper Rose line) and $1\,keV$ (lower Rose line).

photons, while providing potentially higher resolution. At high resolution the depth of focus λ/θ^2 may become less than the sample thickness, which prevents tomographic reconstruction by back-projection methods based on simple projections. Then tomography may best be undertaken using optical sectioning rather than reconstruction from projections. CDI, based on three-dimensional diffraction data, provides a third alternative. The resolution limit imposed by radiation damage in CXDI, as expressed by the Rose equation (Spence, 2003), remains to be determined experimentally, but is likely to be significantly poorer than 1 nm, which has already been achieved in three-dimensional single-particle electron microscopy of proteins. Howells et al. (2005) and Marchesini et al. (2003a) provide a detailed discussion of this large subject, including a plot of dose against resolution for various microscopies in biology, as discussed in Section 9, shown in

Figure 19–6 (see also Henderson, 1995). The dose-fractionation theorem of Hegel and Hoppe is also relevant (Reimer, 1989). Recently, dramatic images of whole yeast cells have been imaged by CDI (Shapiro et al., 2005) using the apparatus described by Beetz et al. (2005).

6 Iterated Projections

A breakthrough in understanding the remarkable success of the HIO algorithm occurred in 1984, when Levi and Stark (1984) (based on earlier work by Youla and Webb, 1982) showed that the algorithm could be understood as successive Bregman projections between convex and nonconvex sets. Here an image is represented as a single vector $\mathbf{R}$ in an N-dimensional space, with one coordinate for each pixel. The addition of two such vectors adds together two images. Distance between images (vectors) in this space has the form of the familiar χ^2 goodness of fit index, so that similar images are near each other. The set of all images subject to a given constraint (e.g., known symmetry, known Fourier modulus, known sign of density, known support) is considered to occupy a volume in this space. The operation of taking a current estimate of the image, performing a Fourier transform, replacing the magnitudes of the diffracted amplitudes with the measured values, and inverse transforming was shown to be a projection onto the set of images subject to the Fourier modulus constraint. Vectors $\mathbf{R}$ between the boundaries of two constrained sets of images are considered. If it is shown that all the images within the only overlap between two constrained sets are equivalent solutions, then the phase problem reduces to finding this volume, where $\mathbf{R} = \chi^2$ and the ER error metric are a minimum. Constraints may be of two types—convex and nonconvex. For a convex set, all points on any line segment terminating within the set lie inside the set. A set $\mathbf{P}$ is convex if $\alpha\mathbf{R} + (1 - \alpha)\mathbf{R}'$ lies within $\mathbf{P}$ for all $\mathbf{R}$ and $\mathbf{R}'$. Here $0 < \alpha < 1$ is a scalar defining position along the line. In two dimensions, a kidney-shaped set is nonconvex and an ellipse is convex. Bregman has shown that iterative projections between convex sets must lead directly to a unique solution if it exists, without stagnation. In this manner the global optimization problem is solved without an exhaustive search for the case in which a unique solution and convex constraints are known to exist. For our problem the Fourier modulus constraint (the known diffraction intensities) is nonconvex, so that this approach has been of limited value. However, it provides a powerful geometric way of thinking about the algorithm as a trajectory in Hilbert space, which is usually drawn in two dimensions for simplicity. The effects of variations in feedback parameter β can be understood, convergence properties studied, and new algorithms proposed. For the ER algorithm, the path is a zig-zag between the boundaries of sets; for the HIO it is a spiral. Some desirable convex constraints include known support, a knowledge of phase rather than amplitude, the sign constraint, symmetry, a known histogram of density levels (Zhang and Main, 1990) (such as exists for proteins), entropy minimization, and (for nonoverlapping atoms) atomicity. A list of constraints used in protein crystallography can be found in the relevant section of volume F of the international tables on crystallography. Application of these constraints thus avoids the common problem whereby optimization programs become trapped in local minima.

The application of constraints can be viewed as projections in the N-dimensional space. Recall that the support **S** is defined as the set of points for which the density is nonzero. For example, application of the support constraint $\mathbf{P}_s$ corresponds to setting many pixels to zero, that is, to projecting onto a space of lower dimension. For the Fourier modulus constraint, we note that Parseval's theorem ensures that distances in the N-dimensional real space are equal to those in a similar N-dimensional space of Fourier coefficients. Consider an Argand diagram for a particular Fourier component, for which the modulus constraint restricts solutions to a circle, whose radius is given by the measured value of the Fourier modulus. An estimate provided by the algorithm (e.g., outside this circle) must be projected (by operator $\mathbf{P}_m$) onto the nearest point on the circle, along a line that will pass through the origin. (This corresponds to the numerical process during one iteration of retaining the current phase estimate, but replacing the magnitude with the measured magnitude, in the HIO algorithm.) Since the linear addition of two vectors terminating on the circle does not produce a third that terminates on the circle, the modulus constraint is not convex. Note, however, that the addition of two vectors of arbitrary length but equal phase produces a new complex number with the same phase, so that a knowledge of phase is a convex constraint, and thus more powerful than a knowledge of amplitudes. The identity operation $\mathbf{I}$ is also useful, and a reflector operation $\mathbf{R}_s = 2\mathbf{P}_s - \mathbf{I}$ can be defined, which reverses the sign of the density outside the support **S**. Using these operators, all the iterative algorithms can be represented simply and analyzed as alternating projections onto convex (and nonconvex) sets (POCS). In this context, we may define these more limited projections as "projectors," which takes a given vector **R** to the nearest point of a nearby constrained set (usually on its boundary). Then

1. The error-reduction (ER) (Gerchberg–Saxton) algorithm may be written

$$\rho^{(n+1)} = \mathbf{P}_s\mathbf{P}_m\rho^{(n)}$$

2. The charge-flipping (CF) algorithm may be written

$$\rho^{(n+1)} = \mathbf{R}_s\mathbf{P}_m\rho^{(n)}$$

3. The hybrid input–output (HIO) algorithm may be written

$$\rho^{(n+1)}(r) = \mathbf{P}_m\rho^{(n)}(r) \quad \text{if } r \in \mathbf{S}$$

$$\rho^{(n+1)}(r) = (\mathbf{I} - \beta)\mathbf{P}_m\rho^{(n)}(r) \quad \text{if } r \notin \mathbf{S}$$

4. The averaged successive reflections (ASR) algorithm may be written

$$\rho^{(n+1)} = 0.5(\mathbf{R}_s\mathbf{R}_m + \mathbf{I})\rho^{(n)}$$

Similar descriptions of the difference map method (Elser, 2003), the hybrid projection reflection (HPR) (Bauschke et al., 2002), and the relaxed averaged alternating reflectors (RAAR) (Luke, 2005) have been given. For $\beta = 1$, the HIO, HPR, ASR, and RAAR algorithms are identical. A comparison of the performance of all of these, together with the powerful shrinkwrap algorithm (HIO with dynamic support) and simple geometric representations of the trajectory of the error metric for few-dimensional cases, can be found in Marchesini (2006).

Using this approach, it has also been shown that the HIO algorithm is equivalent to the Douglas–Rachford algorithm, and is related to classical convex optimization methods (Bauschke et al., 2002). The text by Stark (1987) is recommended as a tutorial introduction to this large subject.

7 Coherence Requirements for CDI: Resolution

It is readily shown (Spence et al., 2004) that the lateral or spatial coherence requirement for diffractive imaging is, in one dimension, that the coherence width $X_c \sim \lambda/\theta_c$ be at least equal to twice the largest lateral dimension W of the object. (This is similar to the requirement in crystallography that X_c exceed the dimensions of a primitive unit cell to avoid overlap of Bragg beams, with beam divergence θ_c. For phasing by the oversampling method, this cell must be about twice as large as the molecule.) This fixes the incident beam divergence and hence the exposure time for a given object size and source. Since at the unapertured diffraction limit ($\Theta = 90°$) the resolution is approximately equal to the wavelength and about two pixels are required per resolution element, a total of about $(4X_c/\lambda)^2$ image pixels would be needed for a coherence width X_c and oversampling factor 2. Physically, this just means that the coherence patch must include the "known" region of vacuum (zero density) surrounding the object boundary (support). It is necessary to diffract coherently from an area twice as large as the isolated object of interest.

The temporal coherence length L_c is also important. For a field of view W at the object (so that the first oversampling point occurs at scattering angle λ/W) and finest (bandlimited) object spatial frequency d^{-1}, the optical path difference between points on opposite sides of the object and a distant detector point is $W\sin\theta = W\lambda/d$, which should not exceed the longitudinal coherence length for X-rays $L_c = \lambda E/\Delta E$. Hence the fractional energy spread allowable in the beam to record spatial frequency d^{-1} is $E/\Delta E > W/d = N$, where d is the sampling interval in the object and N the linear number of pixels needed to sample the object space in the HIO algorithm. A more detailed calculation, considering the shape of the temporal coherence function, gives the requirement on longitudinal coherence as about $E/\Delta E > N/3$, which improves on the estimate in Spence et al. (2004). This determines the quality of the monochromator needed. In practice values of $E/\Delta E = 500$ have yielded good results in soft X-ray work using CCD detectors with N^2 pixels, where $N = 1024$. Then the in-line arrangement of a simple zone-plate monochromator can be used (Howells et al., 2002). For CDI using electron beams, the coherence requirements are easily met for the nanostructures of most interest. We note that the drive for higher resolution, for a fixed number of object pixels, reduces the demand on coherence. The number of pixels, especially in tomography, is likely to be limited by computer processing power to less than 1024^3 in the medium-term future, as discussed below.

It is important to devise a consistent definition of resolution in CDI. Each spatial frequency $u = d^{-1} = \theta/\lambda$ in the object diffracts energy at scattering angle θ into the far field. Consider a square area detector used for diffractive imaging for which the largest scattering angle into the midpoint of the side of the detector (not the corner) is θ_{max}. Then, in the small angle approximation, this angle defines a cutoff in the "transfer function" (see below) at $u_{max} = \theta_{max}/\lambda$, and the full period of the corresponding finest periodicity in the object that can be reconstructed is $d_{min} = \lambda/\theta_{max} = 1/u_{max}$. This value of d_{min} has frequently been quoted as the resolution limit in CXDI. However, while it is a most important experimental parameter, it fails to consider the accuracy of the phasing process, and is not equal to the corresponding Rayleigh resolution limit. There are other considerations, which we now discuss.

The Rayleigh resolution limit was intended for the imaging of binary stars, which are incoherent point sources, unlike the phase contrast usually important for CXDI. If we do nevertheless wish to apply the Rayleigh condition to this situation, we may consider that the CXDI area detector in the far field plays the role of a square aperture in the back-focal plane of an ideal lens. The numerical aperture imposed on the reconstruction is then θ_{max}, and, if the reconstruction is perfect (no errors in phasing), the image will be given by the ideal object charge density convoluted with a sinc function amplitude (the impulse response for linear imaging), whose full width at half maximum is $d_{min}/2 = 0.5/u_{max}$. (The distance between first minima is d_{min}. For a circular area detector the factor 0.5 becomes 0.61—the square detector does slightly better because of contributions from the corners.) Thus our adapted "Rayleigh resolution" is half the finest spatial periodicity, and is therefore equal to the sampling interval needed in real space for linear phase-contrast imaging, to avoid loss of information. Two samples are needed for every full periodicity in the object. (For an incoherent imaging model the impulse response becomes a sinc squared function, for which a sampling interval of $d_{min}/4$ is needed, since the autocorrelation of the transform of the sinc squared function is a triangular function, doubling the bandwidth and resolution, as pointed out in Rayleigh's original paper.) For a lens-based system with aberrations, the factor 0.5 depends on the aberrations of the lens, but takes its minimum value for the diffraction-limited CXDI case.

There are two further considerations. For phase contrast, the ability to distinguish adjacent small objects will depend on the phase shift each introduces, and thus the resolution becomes a property of the sample, not only of the instrument. It is then impossible to define resolution in a meaningful sample-independent manner. In coherent optics, this problem is partly addressed by introducing the concept of a lens coherent transfer function (CTF). Such a function has been introduced in a way that also tests the reliability of the phasing process in an excellent recent proposal by V. Elser for a resolution definition for CXDI. The algorithm is repeatedly run to convergence, each time starting with a different set of random phases. The results for image contrast are plotted as a function of spatial frequency, showing, if noise is not too severe, a relatively smooth curve that falls to zero at some u_{max}. (If the phasing process fails, the average of these many runs will be zero at each spatial frequency.) The resolution is then $0.5/u_{max}$ if a square detector is used. This appears to be the best current definition of resolution for CXDI; however, it ignores the dependence of resolution on sample properties for phase contrast.

For a known object, the faithfulness of the reconstruction may be indicated by a cross-correlation function between the reconstructed estimate and the known object, or crystallographic R-factor, as discussed in detail elsewhere (Spence et al., 2003a). For an unknown object, the ER error metric ε defined in Eq. (10) has been shown by computational trials against known objects to vary monotonically with a cross-correlation function (Fienup, 1997). The best resolution achieved in CXDI is currently about 8 nm.

8 Computer Processing Demands

While noise-free data converge in a few iterations, several hundred iterations of the HIO algorithm are typically needed for convergence of good quality two-dimensional experimental data, or several thousand if many data are missing in three dimensions. Computer processing power can therefore impose serious limitations on CXI, especially for tomography. Recall that for an area detector of linear pixel dimension N, the ratio of largest to smallest feature size is $N/2$. In many cases we wish to obtain phase-contrast images of a real object, so that sample thickness is limited by the need to avoid spatial variations in absorption, which would lead to the possible complications of complex object restoration. (For CXDI, avoiding a beam energy near an inner-shell ionization edge may therefore be important.) A second limit on thickness may be set by the need to satisfy the weak-phase object condition, or avoid extinction effects, both of which introduce multiple scattering and a complex object. [The inversion of Eq. (1) for Δn_p is also referred to as the phase unwrapping problem.] Fixing this limiting thickness and N for a data cube then fixes the lowest resolution and largest field of view possible.

For $N = 1024$, a double-precision complex N^3 array occupies 16 gigabytes, and three or four of these must be accessible directly in RAM (plus a single-bit array for the support mask) to perform HIO iterations. For the Mac G5, a single fast Fourier transform (FFT) on this array takes about 1 min with the FFTW routine, although it is not possible to store sufficient data in RAM at present to work with a single processor, and hard disk transfers are prohibitively time consuming. Clusters of 16 dual processor G5 machines with TCP gigabit ethernet or Infiniband appear to be ideal, when times of approximately 15 and 7 s per iteration, respectively, are achieved for complex $1K \times 1K \times 1K$ arrays (Barty, 2005). Many procedures may be used to reduce the total data processing time. For example, working with some low-resolution 512^2 projections selected from the three-dimensional data set can provide useful support estimates using the two-dimensional shrinkwrap algorithm, before attempting full three-dimensional Fourier iterations. Thus overnight (or longer) computing times must be expected for tomographic CDI in the near future. Once the three-dimensional data set has been phased and transformed to real space, a variety of commercial tomographic viewing programs can be used to present the data, making it possible to "see inside" a nanostructure, such as a cell or metallic foam.

9 Summary

The past decade has been an exciting time for diffractive imaging. In it, we have seen the first successful applications of the phasing algorithms developed in the 1980s to experimental data, together with the development of many new algorithms, ideas, and stimulating intellectual discussion at workshops and conferences. The interdisciplinary nature of the subject has been striking and exciting. Experimental results have appeared for phase-contrast imaging by neutrons (Nugent, 2003), electrons (Weierstall et al., 2001), electrons at atomic resolution (Zuo et al., 2003), and X-rays in both two and three dimensions (Miao et al., 1999, 2002; Chapman et al., 2006). Applications of the method have been slower to develop, but already both in materials science and biology it is clear that the power of imaging *inside* nanostructures (going beyond the projection approximation) in three dimensions presents an exciting prospect for the future. Preliminary applications of phase-contrast X-ray imaging at lower resolution have included the imaging of nanostructures within composite materials, mesoporous silicates and foams, and the imaging of crack tips (Salvo et al., 2003). Bone is also of interest, since it consists of 20-nm nanocrystals of hydroxyapatite in a labyrinthine structure. There is every reason to suppose that the pursuit by tomographic CXDI imaging of these type of materials to higher resolution will be possible and of great interest to scientists. Using medium energy X-rays, the imaging of much thicker material should be possible than that studied by tomographic electron microscopy in materials science (Midgely et al., 2001). For CDI by electron diffraction in materials science the limiting problem is the method of sample support in the microscope. One obvious solution is to use a thin crystal of known structure (such as graphite) as the physical supporting membrane for the nanostructure. Then the known atom positions of the graphite may be used instead of the zero-density region in the HIO iterations. (This allows a connection with the phasing method of fragment completion in crystallography—the graphite atoms provide a kind of reference structure in projection.) Experiments along these lines are in progress. In biology, the situation will be clarified only when careful studies of radiation damage by spot-fading and other methods have been completed. These are needed to determine the domain of applicability of the CXDI method in comparison with other microscopies. Figure 19–6 shows a plot dose against resolution. The domain of applicability of many microscopies is discussed on this figure (and on plots of resolution against thickness) by Howells et al. (2005), showing the niche for CXDI. Whole cell imaging has been pursued with both the zone-plate X-ray microscope and by cryoelectron microscopy, where, using the latter technique, a resolution of about 2 nm is possible in samples up to 100 nm thick or more. It seems likely that radiation damage will prevent competitive performance by CXDI; however, the method may provide useful images at perhaps 5 nm resolution in much thicker samples using medium energy X-rays in an environment of vitreous ice. The optimum choice of X-ray energy involves many issues, including the variation of synchrotron undulator brightness with beam energy and the variation in phase contrast with beam energy. While the coherent flux B available for a given synchrotron source brightness varies as λ^2, the required fluence (from the X-ray cross section) A scales as λ^{-2}, so that the recording time A/B varies as λ^{-4} in a most unfavorable manner as X-ray beam energy increases. The dose in Grays needed to scatter a given number of photons into a voxel varies inversely as the fourth power of the resolution in tomography. An analysis of the variation of dose against resolution for several microscopies including CDI can be found in Marchesini et al. (2003a). Here the statistical demands of good imaging (based on the Rose equation) are compared with the maximum tolerable dose for a given resolution for single-particle imaging.

In summary, the demand for higher resolution three-dimensional noninvasive imaging with old and new radiation sources continues unabated in both materials science and biology, and it seems clear now that diffractive imaging will soon be able to make a decisive contribution.

Spectacular 25 femtosecond single-shot images have just been obtained at DESY by CXDI using 30 nm X-rays from a free-electron laser with 90 nm spatial resolution (H. Chapman et al., Nature, 2006, in press). It is clear that diffractive imaging will play a major role in future time-resolved imaging efforts.

Acknowledgments. This chapter has summarized the work of many groups and many of my collaborators, as indicated in the references. I am particularly grateful for the help of Malcolm Howells, Uwe Weierstall, and Anton Barty during the preparation of this review. The work was supported by NSF, CBST, and IDBR award.

References

Barakat, R. and Newsam, G. (1984). *J. Math. Phys.* **25**, 3190.

Barty, A. (2005). Personal communication.

Bates, R. and McDonnell, M. (1989). *Image Restoration and Reconstruction.* (Oxford University Press, New York).

Bauschke, H., Combettes, P.I. and Luke, D.R. (2002). *J. Opt. Soc. Am.* **19**, 1334.

Bauschke, H., Combettes, P.I. and Luke, D.R. (2003). *J. Opt. Soc. Am.* **A20**, 1025.

Beetz, T., Howells, M., Jacobsen, C., Koa, C., Kirz, J., Lima, E., Mentes, T., Miao, H., Sanchez-Hanke, C., Sayre, D. and Shapiro, D. (2005). *Nucl. Instr. Methods* **545**, 459.

Bruck, Y. and Sodin, L. (1979). *Opt. Commun.* **30**, 304.

Carrozzini, B., Cascarnao, G., Giacovazzo, C., Chapman, H., Marchesini, S., Howells, M., He, H., Wu, J., Weierstall, H. and Spence, J. (2004). *Acta. Crystallogr.* **A60**, 331.

Cederquist, J.N., Fienup, J.R. et al. (1988). *Opt. Lett.* **13**, 619.

Chapman, H., Barty, A., Beetz, C., Cui, He, H., Howells, M., Marchesini, S., Noy, A., Rosen, R., Spence, J.C.H. and Weierstall, U. (2006). *J. Opt. Soc. Am.* **23**, 1179.

Collier, R., Burkhadt, C.B. Lin. (1971). *Optical Holography.* (Academic Press, New York).

Eisebett, S., Lorgen, M. and Eberhart, W. (2004). *Appl. Phys. Lett.* **84**, 3373.

Elser, V. (2003). *J. Opt. Soc. Am.* **20**, 40.

Fienup, J.R. (1982). *Appl. Opt.* **21**, 2758.

Fienup, R. (1987). *J. Opt. Soc. Am.* **A4**, 118.

Fienup, J.R. (1997). *Appl. Opt.* **36**, 8352.

Frank, J. (1992). *Electron Tomography.* (Plenum, New York).

Frank, J., Radermacher, M. Penczek, P., Zhu, J., Li, Y., Ladiadj, M. and Leith, A. (1996). *J. Struct. Biol.* **116**, 190.

Gerchberg, R.W. and Saxton, W.O. (1971). *Optik.* **34**, 275.

Gerchberg, R. and Saxton, W. (1972). *Optik.* **35**, 237.

Goodman, J.W. (1968). *Introduction to Fourier Osptics.* (McGraw-Hill, New York).

Hau-Riege, S., Szoke, H., Chapman, H., Marchesini, S., Noy, A., He, H., Howells, M., Weierstall, U. and Spence, J. (2004). *Acta. Crystallogr.* **1760**, 294.

He, H., Howells, M., Marchesini, S., Chapman, H., Weierstall, U., Padmore, H. and Spence, J. (2003). *Acta. Crystallogr.* **A59**, 143.

He, H., Marchesini, S., Howells, M., Weierstall, U., Chapman, H., Hau-Riege, S., Noy, A. and Spence, J.C.H. (2003). *Phys. Rev.,* p. 174114.

Henderson, R. (1995). *Q. Rev. Biophys.* **28**, 171.

Howells, M., Charalambous, P., He, H., Marchesini, S. and Spence, J.C.H. (2002). A zone-plate monochromator for soft X-rays. *Design and Microfabrication of Novel X-Ray Optics.* (D. Mancini, Ed.). Bellingham, WA.

Howells, M., Beetz, T., Chapman, H., Cui, C., Holton, J., Jacobsen, C., Kirz, J., Lima, E., Marchesini, S., Miao, H., Sayre, D., Shapiro, D. and Spence, J.C.H. (2006). *J. Spectrosc. Rel. Phenom.* (in press).

Howie, A. and Stern, R.M. (1972). *Z. Naturforsch.* **A27**, 382.

Kirz, J., Jacobsen, C. and Howells, M. (1995). *Q. Rev. Biophys.* **28**.

Levi, A. and Stark, H. (1984). *J. Opt. Soc. Am.* **A1**, 932.

Liao, C., Fiddy, M. and Byrne, C. (1997). *J. Opt. Soc. Am.* **A14**, 3155.

Luke, R. (2005). *Inverse Prob.* **21**, 37.

Marchesini, S., Chapman, H.N., Hau-Riege, S.P., London, R.A., Szoke, A., He, H., Howells, M.R., Padmore, H., Rosen, R., Spence, J.C.H. and Weierstall, U. (2003a). *Opt. Express* **11**, 2344.

Marchesini, S., He, H., Chapman, H., Hau-Riege, S., Noy, M., Howells, M., Weierstall, U. and Spence, J.C.H. (2003b). *Phys. Rev.* **B68**, 140101(R).

Marchesini, S. (2006). *Rev. Sci. Instr.* (arxiv:physics/0603201).

McMahon, P., Allman, B., Jacobsen, D., Arif, M., Werner, S. and Nugent, K. (2003). *Phys. Rev. Lett.* **91**, 145502.

Miao, J., Sayre, D. and Chapman, H. (1998). *J. Opt. Soc. Am.* **A15**, 1662.

Miao, J., Charalambous, C., Kirz, J. and Sayre, D. (1999). *Nature* **400**, 342.

Miao, J., Ishikawa, T., Johnson, E., Lai, B. and Hodgson, K. (2002). *Phys. Rev. Lett.* **89**, 088303.

Midgely, P.A., Weyland, M., Thomas, J.M. and Johnson, F.G. (2001). *Chem. Commun.* **2001**, 907.

Millane, R. (1990). *J. Opt. Soc. Am.* **7**, 394.

Millane, R.P. (1995). *Opt. Soc. Am.* **13**, 725.

O'Keeffe, M.A. and Spence, J.C.H. (1993). *Acta. Crystallogr.* **A50**, 33.

Oszlanyi, G. and Suto, A. (2004). *Acta. Crystallogr.* **A60**, 134.

Paganin, D. and Nugent, K. (1998). *Phys. Rev. Lett.* **80**, 2586.

Radi, G. (1970). *Acta. Crystallogr.* **A26**, 41.

Reimer, L. (1989). *Transmission Electron Microscopy.* (Springer-Verlag, New York).

Robinson, I.K., Vartanyants, I.A., Williams, G., Pfeifer, M. and Pitrey, J. (2001). *Phys. Rev. Lett.* **87**, 195505.

Salvo, L., Cloetens, P., Maire, E., Zabler, S., Blandin, J., Buffiere, J., Ludwig, W., Boller, E., Bellet, D. and Josserond, C. (2003). *Nucl. Instr. Methods* **B200**, 273.

Sayre, D. (1952). *Acta. Crystallogr.* **5**, 843.

Sayre, D. (1980). *Image Processing and Coherence in Physics.* Springer Lecture Notes in Physics (M. Schlenker, Ed.), Vol. 112, 229. (Springer, New York).

Sayre, D., Chapman, H. and Miao, J. (1998). *Acta. Crystallogr.* **A54**, 232.

Shapiro, D., Thibault, P., Beetz, T., Elser, V., Howells, M., Jacobsen, C., Kirz, J., Lima, E., Miao, H., Neiman, A.M. and Sayre, D. (2005). *Proc. Natl. Acad. Sci. USA* **102**, 15343.

Spence, J.C.H. (2003). *High Resolution Electron Microscopy.* (Oxford University Press, New York).

Spence, J.C.H. and Zuo, J.M. (1992). *Electron Microdiffraction.* (Plenum, New York).

Spence, J.C.H., Howells, M., Marks, L.D. and Maio, J. (2001). *Ultramicroscopy* **90**, 1.

Spence, J., Weierstall, U. and Howells, M. (2004). *Ultramicroscopy.* **101**, 149.

Spence, J., Weierstall, U., Fricke, J. Glaeser, R. and Downing, K. (2003a). *J. Struct. Biol.* **144**, 209.

Spence, J.C.H., Wu, J., Giacovazzo, C., Carrozzini, B., Cascarano, G. and Padmore, H. (2003b). *Acta. Crystallogr.* **A59**, 255.

Spence, J.C.H., Weierstall, U. and Howells, M. (2004). *Ultramicroscopy* (in press).

Stark, H. (1987). *Image Recovery: Theory and Applications.* (Academic Press, New York).

Strutt, J.W. (1892). *Phil. Mag.* **34**, 407.

Szoke, A. (1997). *J. Imaging Sci. Technol.* **41**, 332.

Wang, B.-C. (1985). *Methods Enzymol.* **115**, 90.

Weierstall, U., Chen, Q., Spence, J.C.H., Howells, M., Isaacson, M. and Panepucci, R. (2001). *Ultramicroscopy* **90**, 171.

Williams, G.J., Pfeifer, M.A. and Vartanyants, I.A. (2003). *Phys. Rev. Lett.* **90**, 175501.

Wu, J., Spence, J., O'Keeffe, M. and Groy, T. (2004a). *Acta. Crystallogr.* **A60**, 326.

Wu, J., Weierstall, U., Spence, J.C.H. and Koch, C. (2004b). *Opt. Lett.* **29**, 1.

Youla, D. and Webb, H. (1982). *IEEE-Trans. Med. Imaging* **MI-1**, 81.

Zhang, K.Y.J. and Main, P. (1990). *Acta. Crystallogr.* **A46**, 41.

Zuo, J.M., Vartanyants, I.A., Gao, M., Zhang, M. and Nagahara, L.A. (2003). *Science* **300**, 1419.

20

The Notion of Resolution

S. Van Aert, Arnold J. den Dekker, D. Van Dyck, and A. Van den Bos

1 Introduction

In microscopy, resolution has always been, and still is, an important issue. Since it is not an unambiguously defined physical quantity, it is interpreted in many ways (den Dekker and van den Bos, 1997). The purpose of this chapter is, on the one hand, to briefly review past and existing resolution definitions and methods, and, on the other hand, to present alternative quantitative definitions of resolution based on model fitting. Throughout this chapter, emphasis will be placed on electron microscopy.

Using model fitting, the resolution will principally be discussed in terms of the precision with which unknown quantities, atom positions in particular, can be measured. It will be shown that a precision of the order of 0.01 Å is in principle possible even with an electron microscope that is not corrected for spherical and chromatic aberration. Once atom positions can be measured with a precision of 0.01 Å they can be used as input data for theoretical *ab initio* calculations (Muller, 1998, 1999; Spence, 1999; Kisielowski et al., 2001b). Such calculations make it possible to calculate the properties of a material with a given structure. In this way, the combination of precise experimental structure determination and theoretical calculations contributes to the understanding of the properties–structure relation. A complete understanding of this relation, combined with recent progress in building materials atom by atom, will enable materials science to evolve toward materials design, that is, from describing and understanding toward predicting materials with interesting properties (Wada, 1996; Olson, 1997, 2000; Reed and Tour, 2000; Browning et al., 2001).

The detailed outline of the sections in this chapter is as follows. In Section 2, classical two-point resolution criteria are discussed. The most widely known classical resolution criterion is that of Lord Rayleigh. The Rayleigh criterion is derived from the assumption that the human visual system needs a minimal contrast to discriminate two points in its composite intensity distribution. Other classical criteria can be seen as modified versions of Rayleigh's criterion. The classical criteria are expressed in terms of the width of the point spread function of the imaging instrument. Section 3 discusses resolution in the spatial frequency domain, addressing literature relating linear systems theory to resolution. In this section, the diffraction limit to resolution and its relation to the classical resolution criteria are introduced. Furthermore, the notion of superresolution is discussed. It will be shown that to attain superresolution, prior knowledge is required. In Section 4, deterministic model-based resolution is considered. Here, prior knowledge is taken into account in the form of a parametric model. However, the images are supposed to be noise free. The unknown parameters, such as the positions of projected atoms, are measured by means of model fitting. Then, in Section 5, statistical model-based resolution is studied. It is taken into account that the observations fluctuate about their expectations due to the unavoidable presence of electron counting noise in the images. The parameters are measured by means of model fitting using parameter estimation methods. Hence, statistical model-based resolution is discussed in terms of the precision with which the parameters can be estimated. Finally, in Section 6, ultimate model-based resolution is contemplated. It is shown that depending on the observations, the solutions of the position estimates may be exactly coinciding. Section 7 includes a discussion and conclusions.

2 Classical Two-Point Resolution

Two-point resolution is a widely used criterion for the resolving capabilities of an imaging system. It is defined as the system's ability to resolve two point sources of equal brightness. Due to the finite size of the system's optical components, a point source is not imaged as a point but as the diffraction pattern of the system's effective aperture. This diffraction pattern therefore represents the system's point spread function. In astronomical problems, two-point resolution is more than just a resolution measure. It has direct practical significance, since in astronomical problems many objects are effectively point sources.

2.1 Rayleigh Resolution

The most widely used criterion for two-point resolution is that of Lord Rayleigh (Strutt, 1899). Rayleigh estimated the minimal resolvable distance between two points of equal brightness that are imaged by a diffraction-limited imaging system. According to the Rayleigh criterion, two point sources are just resolved if the central maximum of the diffraction pattern generated by one point source coincides with the first zero of the diffraction pattern generated by the second. This means that Rayleigh's resolution limit is given by the distance between the central maximum and the first zero of the point spread function of the imaging system. The criterion can be generalized to include point spread functions that have no zero in the neighborhood of their central maximum, by taking the resolution limit as the distance for which the intensity at the central dip in the composite image is 81% of that at the maxima on either side. This corresponds to the original Rayleigh limit for a rectangular aperture.

For example, let us consider the Rayleigh limit, or alternatively, the *point resolution*, of an imaging system with a (two-dimensional) Gaussian point spread function of the following form:

$$p(\mathbf{r}) = p(r) = \frac{1}{2\pi\rho^2}\exp\left(-\frac{x^2+y^2}{2\rho^2}\right) = \frac{1}{2\pi\rho^2}\exp\left(-\frac{r^2}{2\rho^2}\right) \tag{1}$$

where r is the absolute value of the two-dimensional vector $\mathbf{r} = (x\ y)^{\mathrm{T}}$ and ρ is the width of the Gaussian function. The superscript T denotes transposition. According to Rayleigh, the point resolution ρ_p, which is the smallest distance at which two points can be resolved, is given by the requirement that the value of the cross section of the composite intensity distribution halfway between these two points is about 0.8 times the value at the maxima. Thus

$$2\exp\left(\frac{-\rho_p^2}{8\rho^2}\right) = 0.8 \tag{2}$$

from which it follows that

$$\rho_p \approx 2\sqrt{2}\rho \tag{3}$$

2.2 Sparrow Resolution

Rayleigh's choice of resolution limit is based on presumed resolving abilities of the human visual system. Since Rayleigh's days, several other resolution criteria have been proposed that are similar to Rayleigh's (for a review, refer to den Dekker and van den Bos, 1997). A notable example of such so-called classical criteria for two-point resolution is that of Sparrow (1916). Compared to Rayleigh resolution, which is based on presumed capabilities of the human visual system, the Sparrow resolution is based on a less subjective criterion that is valid for a hypothetical perfect imaging instrument. It states that the smallest resolvable distance between two points is the point at which the minimum in the composite image intensity distribution just disappears. When this definition of resolution is applied to an imaging system with a Gaussian point spread function, it follows from Eqs. (1) and (3) that,

$$\rho_s \approx \frac{\sqrt{2}\rho_p}{2} \tag{4}$$

From the examples given in this section, it can be noted that classical two-point resolution criteria are expressed in terms of the width of the point spread function of the imaging instrument. A narrower point spread function corresponds to an improved Rayleigh or Sparrow resolution.

3 Resolution in the Spatial Frequency Domain: Diffraction Limit and Superresolution

3.1 The Diffraction Limit

In the previous section, classical resolution criteria have been discussed. An alternative resolution criterion is based on linear systems theory. It is assumed that the imaging system is linear and shift invariant. Coherent imaging systems are linear in complex amplitude and incoherent imaging systems are linear in intensity. The characteristics of a shift-invariant linear imaging system are defined by its point spread function, or, equivalently, by its transfer function, which is the Fourier transform of the point spread function. Point spread functions are more directly useful in the assessment of telescopes or spectroscopic instruments. Transfer functions, on the other hand, are more often used in the case of microscopes and cameras. The amplitude (coherent imaging) or intensity (incoherent imaging) distribution in the image produced by a linear and shift-invariant system is the convolution of the amplitude (or intensity) distribution of the object and the amplitude (or intensity) point spread function of the imaging system. For the spatial frequency domain, the imaging system acts as a filter for spatial frequencies. Each spatial frequency is transferred from the object to the image plane independently of all other frequencies present; the corresponding amplitude (coherent imaging) or intensity (incoherent imaging) is multiplied by the transfer function, or frequency response function, of the system. Due to the finite size of the system's aperture, transfer functions of (both coherent and incoherent) imaging systems are band limited, i.e., they are equal to zero for all frequencies above a certain cutoff frequency. Born and Wolf (1999) discuss the more general class of partially coherent imaging systems and show that these systems are also strictly band limited. Therefore, independent of the degree of coherence, spectral components beyond the cutoff frequency are not transferred by the imaging system. For this reason, the cutoff frequency is called the *diffraction limit* to resolution.

3.2 The Diffraction Limit and Its Relation to Rayleigh and Sparrow Resolution

The diffraction limit, or equivalently, the cutoff frequency, is related to the Rayleigh resolution. This can been seen as follows. It is well known that a narrow function has a broad Fourier transform and vice versa. Therefore a point spread function with a narrow main lobe corresponds to a transfer function with a high cutoff frequency. In fact, the product of the Rayleigh limit and the cutoff frequency is a constant that is close or even equal to one. For example, this constant is equal to 1 and 1.22 for incoherent imaging systems with a rectangular and a circular aperture, respectively.

As mentioned above, transfer functions of practical imaging systems are strictly band limited (due to the finite size of the system's aperture). In this work, it will often be assumed that the point spread function of the imaging systems under study can be described by a Gaussian function, such as (1). This assumption is made so as to keep calculations simple. The transfer function of the imaging system of which the point spread function is described by Eq. (1) is given by the two-dimensional Fourier transform of Eq. (1):

$$P(g) = \exp(-2\pi^2\rho^2 g^2) \tag{5}$$

where g is the absolute value of the two-dimensional spatial frequency vector **g**. Although this transfer function tends to zero for increasing frequency values, it is not *strictly* band limited. Nevertheless, the Gaussian approximation is sufficiently accurate for the purpose of this chapter. From the condition that the product of the Rayleigh resolution limit and the cutoff frequency should approximately be equal to one, it follows from Eq. (3) that the cutoff frequency corresponding to Rayleigh resolution is

$$g_{\mathrm{p}} = \frac{1}{\rho_{\mathrm{p}}} \approx \frac{1}{2\sqrt{2}\rho} \tag{6}$$

At that spatial frequency the modulus of the transfer function, which is given by Eq. (5), is reduced to only 8%. Thus, for Gaussian point spread functions as described by Eq. (1), the Rayleigh resolution limit can also be defined as the inverse of the spatial frequency for which the transfer function is reduced to 8% of its peak value.

The diffraction limit and its relation to Rayleigh and Sparrow resolution will now be discussed for conventional transmission electron microscopy (TEM). Thus far, only the diffraction limited point spread function of the imaging instrument has been taken into account. However, for electron microscopy, this should be extended to include the point spread function describing the effect of thermal vibrations of the atom, the effect of the environment, and the detector (de Jong and Dyck, 1993). Moreover, it has to be noted that the atoms are not point scatterers. Hence, an extension from points to objects of finite size has to be made. As shown in Figure 20–1, each effect contributing to the imaging process can be represented by a transfer function, which acts as a low pass filter. The transfer function of the electron microscope consists of a damping function, which is mainly due to chromatic aberration, and a phase shift, which causes the oscillations. Since there are many ways to get rid of the oscillations, such as focal series reconstruction (Schiske, 1973; Saxton, 1978; Van Dyck and Coene, 1987; Van Dyck et al., 1993; Coene et al., 1996; Thust et al., 1996) and correction of the spherical aberration (Rose, 1990), the Rayleigh resolution of the electron microscope can be assumed to be given by the so-called *information limit*, which is proportional to the inverse of the highest spatial frequency that is still transferred with appreciable intensity. For simplicity, it will first be assumed that the imaging process is linear. This requires that the interaction between the electron and the object also is linear, which means that there is a simple linear relation of the *electron exit wave* and the projected electrostatic potential. The electron exit wave is a complex wave function in the plane at the exit face of the object, resulting from the interaction of the electron beam with the object. For example, the imaging process of weak phase objects, for which the so-called weak phase object approximation holds (Buseck et al., 1988), may be considered to be linear. If the object is a crystal, viewed along a zone axis, the electrostatic potential of all the atoms along the atom column is superimposed, which makes the interaction very strong and highly nonlinear. In that particular case, due to the focusing effect of the successive atoms, the scattering is increased to

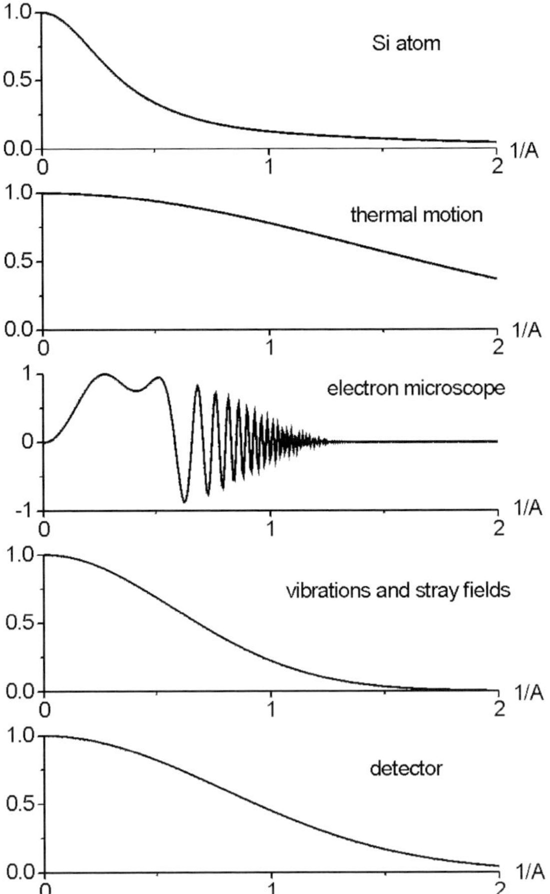

Figure 20–1. Transfer functions of the different subchannels of electron microscopic imaging.

higher angles. This effect is explained by the channeling theory (Howie, 1966; Van Dyck et al., 1989; Pennycook and Jesson, 1991; Van Dyck and Chen, 1999). However, for amorphous objects, the atoms are stacked in a disordered fashion, so that in projection their cores do not overlap, except by coincidence. As a result, the interaction remains linear for much larger object thicknesses and may be described by the weak phase object approximation. If the imaging is linear, all transfer functions have to be multiplied, or, equivalently, the point spread functions have to be convolved. If it is assumed that all constituent point spread functions are Gaussian, such as in Eq. (1), the resulting function is a Gaussian as well, with a Rayleigh resolution ρ_p determined by

$$\rho_P^2 = \rho_A^2 + \rho_T^2 + \rho_{EM}^2 + \rho_v^2 + \rho_D^2 \tag{7}$$

with ρ_A the "width" of the electrostatic potential of the atom, ρ_T the Rayleigh resolution limited by thermal vibrations of the atom, ρ_{EM} the Rayleigh resolution of the electron microscope, ρ_v the Rayleigh resolution limited by the environment (vibrations and stray fields), and ρ_D

the Rayleigh resolution limited by the detector. Today, for the best electron microscopes, ρ_{EM} is somewhat below 1 Å (O'Keefe et al., 2001; Kisielowski et al., 2001a; Batson et al., 2002a). In future instrumental developments the Rayleigh and Sparrow resolution can be improved by improving the resolutions of all different subchannels. However, a factor that cannot be improved is ρ_A, that is, the intrinsic "width" of the atom itself. It is important to note that beyond a certain point, it will be useless to further improve the Rayleigh resolution of the electron microscope since the transfer at high spatial frequencies is limited by the scattering factor. It is already difficult to find suitable objects that can be used to demonstrate the true Rayleigh resolution ρ_{EM} of an electron microscope. Consider, for example, amorphous silicon. From Figure 20–1, it follows that ρ_A is about 1 Å. Therefore, for the best electron microscopes, it follows from Eq. (7) that for amorphous silicon

$$\rho_p \approx 1 \,\text{Å} \tag{8}$$

and from Eq. (4) that

$$\rho_s \approx 0.7 \,\text{Å} \tag{9}$$

From this example, it can be concluded that for the best electron microscopes, the atoms themselves limit classical resolution criteria and hence the diffraction limit. However, it should be noted that the discussion of Rayleigh resolution and the diffraction limit is far more complicated in case of nonlinear electron–object interaction, for example, in the case of atom columns viewed along the column direction. It will then also depend on the assumptions regarding the scattering of the electrons on their way through the object. Furthermore, for coherent imaging, such as in TEM, Goodman (1968) has shown that the Rayleigh resolution will depend on the "phase distribution" associated with the object. Depending on the relative phase associated with two atoms or atom columns, the central dip in the composite image will be absent or present. For particular values of the relative phase shift, the dip will even be greater than the dip corresponding to an incoherent image of these two atoms or atom columns. From this example, it can be concluded that there is no simple generalization as to which type of imaging, coherent or incoherent, is preferred in the sense of Rayleigh resolution. So the assumption that incoherent imaging, for example scanning transmission electron microscopy (STEM), will yield a "better" resolution than coherent imaging, for example, TEM, is in general not valid.

In the remainder of this chapter, it will be shown that by using superresolution algorithms, frequency components lying beyond the diffraction limit of the imaging system may be reconstructed. Then, other definitions of resolution are of interest.

3.3 Superresolution

Superresolution refers to reconstructing frequency components that lie beyond the cutoff frequency of the imaging system. At first sight, superresolution seems impossible. Knowledge of the system's transfer function makes it possible to reconstruct the object spectrum within the passband of the imaging system by means of inverse filtering of the image spectrum, but frequency components beyond the diffraction limit seem irrevocably lost. Indeed, in the absence of any restriction as to the nature of the object, there are an infinite number of objects that can produce the same image. Under certain conditions, however, superresolution is possible. The key to superresolution is prior knowledge. For example, suppose that it is known that the object is of finite size, that is, it is nonzero only in a region of finite extent. This single condition guarantees that the object spectrum is analytic. A well-known property of an analytic function is that if it is known over a specified interval, it can always be reconstructed in its entirety (Castleman, 1979). This process of reconstruction is called analytic continuation. It can be shown that this method is perfect in theory. If the images are noise free, it leads to an exact and complete reconstruction of the object spectrum (Harris, 1964). However, noise limits its practical use (Frieden, 1967). Nevertheless, many effective superresolution algorithms (digital image processing methods) have been proposed in the literature [for a review, see Frieden (1975), Hunt (1994), and Meinel (1986)]. Both empirically and theoretically, it has been shown that there are certain necessary conditions to be satisfied by a superresolution algorithm to be successful (Hunt, 1994). First, the algorithm should explicitly utilize a mathematical description of the image formation process that relates object and image via the point spread function of the imaging system. Second, the images should be sufficiently oversampled to avoid aliasing after reconstruction of spatial frequencies beyond the diffraction limit. For Nyquist sampled images (Gonzalez and Woods, 2002), the algorithm should contain some suitable form of interpolation. Last, but not least, the algorithm should contain prior knowledge of the object. Examples of such prior knowledge used by superresolution algorithms include the following:

- Finite extent of the object (as discussed above) (e.g., Harris, 1964; Gerchberg, 1974, 1989).
- Positivity of the object (e.g., Schell, 1965; Biraud, 1969; Walsh and Nielsen-Delaney, 1994).
- Upper and lower bounds on the object intensity (e.g., Janson et al., 1970).
- Object statistics (e.g., Frieden, 1980; Hunt and Sementilli, 1992).
- Parametric model of the object.

Obviously, the performance of any superresolution algorithm will be limited by noise. In the remainder of this chapter, we will assume that the available prior knowledge of the object to be reconstructed consists of a parametric model. Then, superresolution can be achieved by computing the relatively small number of unknown parameters characterizing the object from the available observations. In electron microscopy these observations may be electron counting results made at the pixels of a CCD camera. The image reconstruction problem thus becomes a parameter estimation problem. For example, in the case of two-point resolution, the object can be described by a two-component model parametric in the locations of the point sources. Hence, if the point spread function is known, a parametric model of the composite image of the two-point sources can be derived. By fitting this model to the image optimizing some criterion of goodness of fit, we obtain estimates of the parameters of the model. For a correct model, in the absence of noise, and apart from potential computational problems that will be discussed in Section 4, this would result in a perfect fit. That is, the object parameters can be estimated with unlimited precision so that the object can be reconstructed perfectly. This means that there would be no limit to resolution no matter how closely the point sources are spaced. However, in practice it is noise that limits the accuracy and precision with which parameters can be measured and therefore limits the resolution. This will be the subject of Sections 5 and 6.

4 Deterministic Model-Based Resolution

Classical resolution criteria disregard the possibility of using prior knowledge to extract analytic results from observations by means of model fitting (den Dekker and van den Bos, 1997). In this section, prior knowledge is taken into account in the form of a model describing the observations. Thus far, the observations are assumed to be noise free. Compared to Section 2, the model will be extended from two-peak models to one or more-peak models. Instead of classical resolution, we will speak of *deterministic* model-based resolution. It will be shown that the relevant limits to deterministic model-based resolution are in any case computational. However, if the model is inaccurate, which means that it systematically deviates from the exact noise-free observations, it will be shown that the relevant limits are both computational and fundamental.

Imagine that Lord Rayleigh would image stars today. First, the image of one star, which can be treated as a point object, will be considered. Now, there exists a model for the object, namely that it consists of a point. The point spread function of the telescope is also exactly known. So, it is known how an image of one star should look like. Thus, there is no interest in the detailed form of this image, but only in the position of the star. The only objective of the experiment is to determine this position. Obviously, in the absence of noise, numerically fitting the known one-peak model to the image with respect to the position parameter would result in a perfect fit. The resulting solution for this location would be exact, and despite the blurring effect of the point spread function, it imposes no limit to location resolution.

This line of reasoning can be extended to position measurements of atoms or atom columns from noise-free electron microscopic observations. Suppose that these observations λ_{kl} are made at the pixels (k, l) at the position $(x_k \ y_l)^{\mathrm{T}}$. The model that describes these observations is called $f_{kl}(\tau)$ with τ the vector of unknown parameters, among which are the locations of the atoms or atom columns. An example of such a model is the following:

$$\lambda_{kl} = f_{kl}(\tau) = \zeta + \sum_{n=1}^{n_c} \frac{\eta_n}{2\pi\rho^2} \exp\left[\frac{-(x_k - \beta_{xn})^2 - (y_l - \beta_{yn})^2}{2\rho^2}\right] \tag{10}$$

where ζ is the constant background, η_n is the column-dependent height of the Gaussian peak, ρ is the width of the Gaussian peak, n_c is the total number of atom columns, and β_{xn} and β_{yn} are the x- and y-coordinate of the nth atom or atom column, respectively. The width ρ is supposed to be identical for different atom columns. The parameter vector τ is equal to $(\beta_{x1} \ldots \beta_{xnc}\ \beta_{y1} \ldots \beta_{ync}\ \eta_1 \ldots \eta_{nc}\ \rho\ \zeta)^T$ and contains $R = 3n_c + 2$ elements. The unknown parameters can be measured by fitting the model to the observations. In a sense, we are then looking for the optimum value of a criterion in a parameter space whose dimension is equal to R, that is, the number of parameters to be measured. Each possible combination of the R parameters can be represented by a point in an R-dimensional space. The search for the global optimum of the criterion of goodness of fit in this space is an iterative numerical optimization procedure.

The problem may be of a computational kind. The existing optimization methods fail if the dimension of the parameter space is so high that it is not possible to avoid ending up at a local optimum instead of at the global optimum of the criterion of goodness of fit, so that the wrong structure is derived. To solve this dimensionality problem, that is, to find a pathway to the global optimum, a good starting structure is required, that is, initial conditions should be available for the parameters. For example, neighboring atoms or atom columns should be discriminated in an image. In other words, the structure has to be resolved. This corresponds to X-ray crystallography, where it is first necessary to resolve a structure by using, for example, direct methods, and afterward to refine the structure. Moreover, the computing time needed to reach convergence of the iterative procedure increases with the dimension of the parameter space. In the following example, the problems related to the study of the amorphous object with atomic resolution TEM will be discussed. More details can be found in Van Dyck et al. (2003).

Example 1 (*Amorphous Object*) *For an amorphous object, the number of parameters increases with thickness. Therefore, from a certain thickness on, it will be difficult to resolve the structure in projection. For example, consider Figures 20–2 and 20–3. In Figure 20–2, the amorphous foil is thin, whereas in Figure 20–3, it is thicker. Therefore, the number of projected atoms is larger in Figure 20–3 than in Figure 20–2. It is clear that it will be more likely to resolve the structure for the example given in Figure 20–2 than for that corresponding to Figure 20–3. To resolve the structure, it will be assumed that the distances between neighboring projected atom positions should be larger than or equal to the Sparrow resolution ρ_s. The reason for choosing this criterion is that the computer will then be able to distinguish the individual atoms, since the observations are assumed to be noise free. However, it should be noted that this criterion is not exact and, therefore, it will give only rough guidelines. Suppose that the mean concentration of atoms per cubic ångstrom is equal to*

REAL SPACE

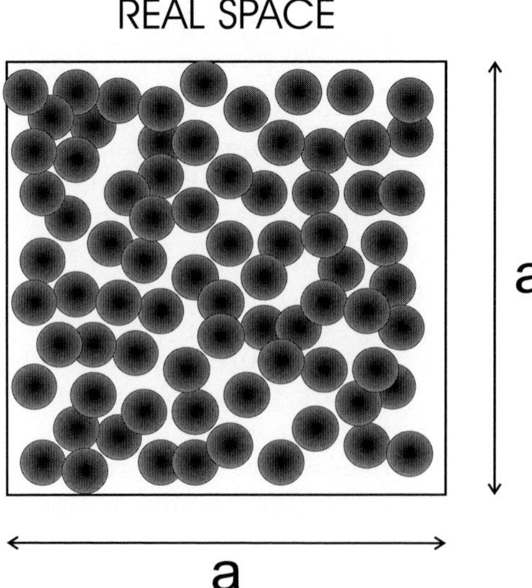

Figure 20–2. Amorphous structure containing clearly separable projected atoms.

V. Then, the mean concentration A of projected atoms per square ångstrom is given by

$$A = Vz \tag{11}$$

where z is the thickness of the amorphous foil. On the other hand, if it is assumed that each projected atom occupies a circle with a diameter equal to the average distance d, averaged over distances between nearest-neighbor projected atoms, then

$$A \approx \frac{1}{\pi(d/2)^2} \tag{12}$$

From Eqs. (11) and (12), it follows that the thickness of the amorphous foil is approximately given by

REAL SPACE

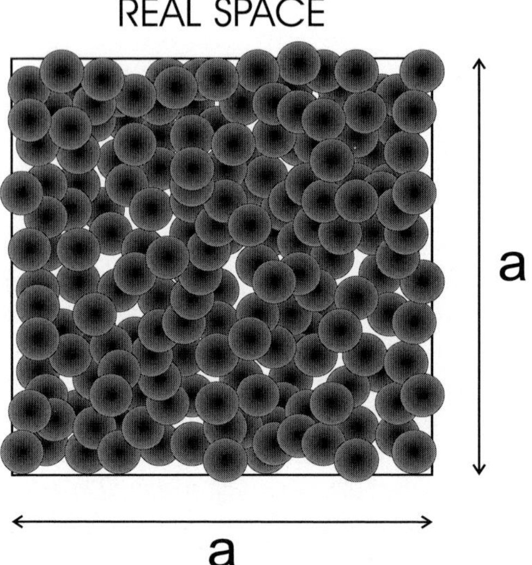

Figure 20–3. Amorphous structure containing severely overlapping projected atoms.

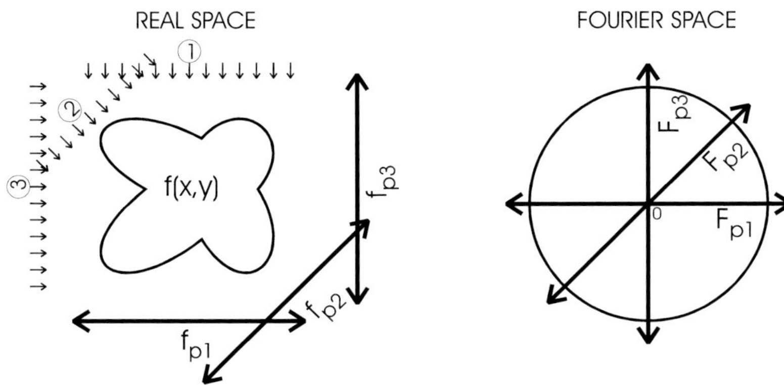

Figure 20–4. The principle of tomography.

$$z \approx \frac{4}{\pi d^2 V} \tag{13}$$

To resolve the structure and therefore to avoid dimensionality problems, it will be assumed that the following condition is met:

$$d \geq 2\rho_s \tag{14}$$

The factor 2 is arbitrarily chosen. Requiring that d is larger than or equal to ρ_s would not be sufficient. In that case, a substantial part of distances between neighboring atoms would be smaller than ρ_s and hence it is not possible to resolve the structure. In principle, this can still occur if inequality (14) is fulfilled, but the probability that it occurs is lower. Then, it follows from Eqs. (13) and (14), that

$$z \leq \frac{1}{\pi \rho_s^2 V} \tag{15}$$

For amorphous silicon, it follows from Eq. (9) that ρ_s is approximately equal to 0.7 Å and, furthermore, V is approximately equal to 0.05 atoms/Å^3. Hence, it follows from Eq. (15) that the amorphous silicon foil should not exceed thicknesses of the order of 13 Å so as to avoid dimensionality problems. This thickness is rather small, which means that it is unrealistic to expect that atomic resolution TEM is able to resolve amorphous silicon samples with realistic foil thicknesses from only one projection (Cowley, 2001). It can thus be stated that the structure of a realistic amorphous object cannot be determined from one image alone. However, the situation can be improved drastically by using a tomographic technique in which the sample is tilted and many projections from different viewing directions are combined as shown in Figure 20–4. The Fourier transform of a projection yields a section through the origin of the three-dimensional Fourier space. By combining many different projections, it is possible to reconstruct the whole Fourier space (Frank, 1992). In this way an ideal microscope can resolve about one atom per cubic ångstrom, which is sufficient to resolve amorphous structures.

In the foregoing, it has been assumed that the model is accurate. However, if the model is inaccurate, the estimated position parameters may deviate from the true positions, even if the observations would be noise free. This is called systematic error. Obviously, systematic errors would place fundamental limits on deterministic model-based resolution.

In this section, it has been shown that deterministic model-based resolution is computationally limited. Computational problems can be overcome only if the structure can be resolved. Moreover, if the model is inaccurate, the resolution is also fundamentally limited, since this results in a systematic error.

5 Statistical Model-Based Resolution

In Section 4, it was assumed that the observations are noise free. However, in any real-life experiment, the observations will "contain errors." Then, the resolution depends fundamentally on the signal-to-noise ratio in the detected image. In this section, the resolution will be considered in the framework of statistical parameter estimation theory and will be called *statistical* model-based resolution. It will be shown that the relevant limits are both computational and fundamental.

Suppose there is a CCD camera that is able to count the individual photons forming the image of a single point object or of two point objects. The images as measured by this camera appear as in Figure 20–5 or as in Figure 20–6 for single or two-point objects, respectively. The noise on these images stems from the counting statistics. The position parameters can be estimated by numerically fitting the known parameterized mathematical model to the images with respect to the component positions, in the same way as expressed in Section 4. However, if one repeated this experiment several times, one would, due to the statistical nature of the observations, find different values

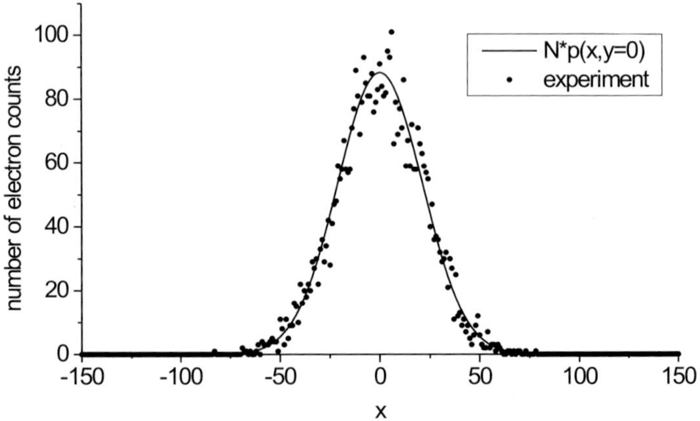

Figure 20–5. Simulation experiment of the image of a point as measured by a CCD camera (300 × 300 pixels), with pixel size $\Delta x = \Delta y = 1$. The point spread function used is a two-dimensional normalized Gaussian function $p(x, y)$ as in Eq. (1) with $\rho = 21$. The experimental as well as the expectation values $N \times p(x, y)$ are shown within the section $y = 0$. The number of imaging particles N is equal to 250,000.

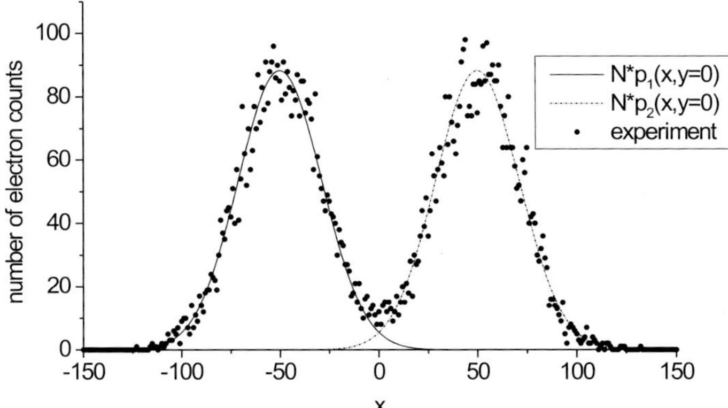

Figure 20–6. The results of a computer-simulated image of two neighboring points. The experimental as well as the expectation values of the individual peaks $[N \times p_{1,2}(x, y)]$ are plotted within the section $y = 0$, where $p_{1,2}(x, y)$ are two-dimensional normalized Gaussian point spread functions, as in Eq. (1). The number of imaging particles $2N$ is equal to 500,000. The peaks are clearly separable.

for the position or the distance estimates for single or two-point objects, respectively. The position and distance estimates are statistically distributed about their mean values. Then, the obvious criterion to quantify statistical model-based resolution is the precision of the estimate, which is given by the variance of this distribution, or, by its square root, the standard deviation. In a sense, the standard deviation is the "error bar" on the position or on the distance. Applying statistical parameter estimation theory, the *attainable* precision can be adequately quantified in the form of the so-called *Cramér–Rao lower bound* (CRLB) (van den Bos and den Dekker, 2001). This is a lower bound on the variance of any unbiased estimator of a parameter. It means that the variance of different estimators, such as, the least squares or the maximum likelihood (ML) estimator, can never be lower than the theoretical CRLB on the variance. Fortunately, the ML estimator attains the CRLB asymptotically, that is, if the number of observations is sufficiently large. Note that in accordance with the available literature, the CRLB on the variance of the position estimate of the image of a single point object and the CRLB on the variance of the distance estimate of the image of two point objects are measures of what is called *single-source* (Falconi, 1964) and *differential* (*two-source*) resolution (Falconi, 1967), respectively. Single-source resolution is defined as the instrument's capacity to determine the position of a point object that is observed in a background of noise. Differential resolution is defined as the instrument's ability to determine the separation of two point objects.

In this section the fluctuating behavior of the observations will be described in Section 5.1 using parametric statistical models of observations. Next, in Section 5.2, it will be shown how an adequate expression for the attainable statistical precision of the parameter estimates, that is, the CRLB, can be derived from such a parametric statistical model. Then, in Section 5.3, the ML estimator of the parameters will be derived from the parametric model. This estimator is important since it achieves the CRLB asymptotically. In Section 5.4, an important purpose for which the expressions for the CRLB can be used will be introduced, namely, *statistical experimental design*.

5.1 Parametric Statistical Models of Observations

Generally, due to the inevitable presence of noise, sets of observations made under the same conditions nevertheless differ from experiment to experiment. The usual way to describe this behavior is to model the observations as *stochastic variables*. The reason is that there is no viable alternative and that it has been found to work (van den Bos, 1999; van den Bos and den Dekker, 2001). By definition, a stochastic variable is characterized by its probability density function, while a set of stochastic variables has a joint probability density function.

Consider a set of stochastic observations w_m, $m = 1, \ldots, M$ made at the measurement points $x_1, \ldots, x_M$. These measurement points are assumed to be exactly known. In electron microscopy, the observations are, for example, electron counting results made at the pixels of a CCD camera, where M represents the total number of pixels. The reader should not be misled by the fact that the observations and measurement points are here represented in a one-dimensional way. It is intended as a general representation. Also if the observations and measurement points would be two- or higher-dimensional, they can easily be transformed to a one-dimensional representation. For example, if the observations and measurement points are two-dimensional, say, w_{kl}, $k = 1, \ldots, K$, $l = 1, \ldots, L$ and x_{kl}, $k = 1, \ldots, K$, $l = 1, \ldots, L$, respectively, they can also be represented as w_m, $m = 1, \ldots, M$ and x_m, $m = 1, \ldots, M$, respectively, with $M = K \times L$. The $M \times 1$ vector w defined as

$$w = (w_1 \ldots w_M)^{\mathrm{T}} \tag{16}$$

is the column vector of these observations. It represents a point in the Euclidean M space having $w_1, \ldots, w_M$ as coordinates. This will be called *space of observations* (van den Bos and den Dekker, 2001). The expectations of the observations, that is, the mean values of the observations, are defined by their probability density function. The vector of expectations

$$E[w] = (E[w_1] \ldots E[w_M])^{\mathrm{T}} \tag{17}$$

is also a point in the space of observations and the observations are distributed about this point. The symbol $E[\cdot]$ denotes the expectation operator. The expectations of the observations are described by the *expectation model*, that is, a physical model, that contains the unknown parameters to be estimated, such as the position coordinates of the projected atoms or atom columns. In a sense, this model has first been introduced in Section 4 on the understanding that it now describes the expectations of the observations whereas in Section 4, the model describes noise-free observations. The unknown parameters are represented by the $R \times 1$ parameter vector $\tau = (\tau_1 \ldots \tau_R)^{\mathrm{T}}$. Thus, it is supposed that the expectation of the mth observation is described by

$$E[w_m] = f_m(\tau) = f(x_m; \tau) \tag{18}$$

where $f_m(\tau)$ represents the expectation model, which is evaluated at the measurement point x_m and which depends on the parameter vector τ. Apart from the unknown parameters τ, the expectation model may contain known parameters and experimental settings as well. An example of an expectation model is given by Eq. (10).

Electron microscopic observations are electron counting results detected, for example, with a CCD camera. Under the assumption that the quantum efficiency of this detector is sufficiently large to detect single electrons, these observations may be assumed to be Poisson distributed. This means that the probability that the observation w_m is equal to ω_m is given by (Papoulis, 1965)

$$\frac{\lambda_m^{\omega_m}}{\omega_m!}\exp(-\lambda_m) \tag{19}$$

where the parameter λ_m is equal to the expectation of the observation w_m, which, in its turn, is described by the expectation model. Therefore,

$$E[w_m] = \lambda_m = f_m(\tau) \tag{20}$$

with $f_m(\tau)$ given by Eq. (18). A property of the Poisson distribution is that the variance of the observation w_m is equal to λ_m:

$$\mathrm{var}(w_m) = \lambda_m \tag{21}$$

Moreover, electron microscopic observations may be assumed to be statistically independent. The probability $P(\omega; \tau)$ that a set of observations $w = (w_1 \ldots w_M)^{\mathrm{T}}$ is equal to $\omega = (\omega_1 \ldots \omega_M)^{\mathrm{T}}$ is thus equal to the product of all probabilities described by Eq. (19):

$$P(\omega; \tau) = \prod_{m=1}^{M} \frac{\lambda_m^{\omega_m}}{\omega_m!}\exp(-\lambda_m) \tag{22}$$

This function is called the *joint probability density function* of the observations. It represents the parametric statistical model of the observations. The parameters τ to be estimated enter $P(\omega; \tau)$ via λ_m.

If the expectation $E[w_m] = \lambda_m$ increases, the Poisson distribution tends to a normal distribution with both expectation $E[w_m]$ and variance $\mathrm{var}(w_m)$ equal to λ_m of the Poisson distribution (Mood et al., 1974). This normal approximation is justified if the magnitude of the observations w_m is large with respect to the square root of this number (Koster et al., 1987). Moreover, if the contrast in the images is low, the deviations of the observations from their expectations may be supposed to be identically distributed, that is, $\mathrm{var}(w_m) = \sigma^2$ (Miedema et al., 1994). Under these conditions the joint probability density function of the observations is given by

$$P(\omega; \tau) = \prod_{m=1}^{M} \frac{1}{\sqrt{2\pi}\sigma}\exp\left[-\frac{1}{2}\left(\frac{\omega_m - \lambda_m}{\sigma}\right)^2\right] \tag{23}$$

In Section 5.2, the parameterized joint probability density function will be used to derive the CRLB, that is, an expression for the attainable precision with which the unknown parameters can be estimated unbiasedly from the observations. In Section 5.3, from the joint probability density function, the ML estimator of the parameters will be derived. This estimator actually achieves the CRLB asymptotically, that is, for the number of observations going to infinity.

5.2 The Cramér–Rao Lower Bound

In this section, the parameterized probability density function of the observations, which is derived in Section 5.1, will be used to define the *Fisher information matrix* and to compute the CRLB on the variance of unbiased estimators of the parameters of the expectation model. The CRLB will also be extended to include unbiased estimators of vectors of *functions of these parameters*. As discussed in the introduction of Section 5, the CRLB is used as a criterion for statistical model-based resolution. The reader is referred to van den Bos (1982), Frieden (1998), and van den Bos and den Dekker (2001) for the details of the CRLB.

First, the *Fisher information matrix F* with respect to the elements of the $R \times 1$ parameter vector $\tau = (\tau_1 \ldots \tau_R)^{\mathrm{T}}$ is introduced. It is defined as the $R \times R$ matrix

$$F = -E\left[\frac{\partial^2 \ln P(\mathbf{w}; \tau)}{\partial \tau \partial \tau^{\mathrm{T}}}\right] \tag{24}$$

where $P(\omega; \tau)$ is the joint probability density function of the observations $w = (w_1 \ldots w_M)^{\mathrm{T}}$. The expression between square brackets represents the Hessian matrix of $\ln P$, for which the (r, s)th element is defined by $\partial^2 \ln P(\omega; \tau)/\partial \tau_r \partial \tau_s$. For electron microscopic observations, where $P(\omega; \tau)$ is given by Eq. (22), it follows from Eqs. (20), (22), and (24) that the (r, s)th element of F is equal to

$$F_{rs} = \sum_{m=1}^{M} \frac{1}{\lambda_m} \frac{\partial \lambda_m}{\partial \tau_r} \frac{\partial \lambda_m}{\partial \tau_s} \tag{25}$$

Furthermore, if the approximation of Eq. (22) by Eq. (23) is justified, the (r, s)th element of F is equal to

$$F_{rs} = \sum_{m=1}^{M} \frac{1}{\sigma^2} \frac{\partial \lambda_m}{\partial \tau_r} \frac{\partial \lambda_m}{\partial \tau_s} \tag{26}$$

Next, it can be shown that the covariance matrix $\mathrm{cov}(\hat{\tau})$ of any unbiased estimator $\hat{\tau}$ of τ satisfies

$$\mathrm{cov}(\hat{\tau}) \geq F^{-1} \tag{27}$$

This inequality indicates that the difference of the matrices $\mathrm{cov}(\hat{\tau})$ and F^{-1} is positive semidefinite. Since the diagonal elements of $\mathrm{cov}(\hat{\tau})$ represent the variances of $\hat{\tau}_1, \ldots, \hat{\tau}_R$ and since the diagonal elements of a positive semidefinite matrix are nonnegative, these variances are larger than or equal to the corresponding diagonal elements of F^{-1}:

$$\text{var}(\hat{\tau}_r) \geq [F^{-1}]_{rr} \tag{28}$$

where $r = 1, \ldots, R$ and $[F^{-1}]_{rr}$ is the (r, r)th element of the inverse of the Fisher information matrix. In this sense, F^{-1} represents a lower bound to the variances of all unbiased $\hat{\tau}$. The matrix F^{-1} is called the CRLB on the variance of $\hat{\tau}$.

The CRLB can be extended to include unbiased estimators of vectors of functions of the parameters instead of the parameters proper. Let $\gamma(\tau) = [\gamma_1(\tau) \ldots \gamma_C(\tau)]^T$ be such a vector and let $\hat{\gamma}$ be an unbiased estimator of $\gamma(\tau)$. Then, it can be shown that

$$\text{cov}(\hat{\gamma}) \geq \frac{\partial \gamma}{\partial \tau^T} F^{-1} \frac{\partial \gamma^T}{\partial \tau} \tag{29}$$

where $\partial \gamma / \partial \tau^T$ is the $C \times R$ Jacobian matrix defined by its (r, s)th element $\partial \gamma_r / \partial \tau_s$ (van den Bos, 1982). The right-hand member of this inequality is the CRLB on the variance of $\hat{\gamma}$.

It should be noted that the CRLB may be computed only if the probability density function of the observations is known. At first sight, this seems to be a problem since the true parameters of the probability density function are unknown. Nevertheless, even if the CRLB is a function of the unknown parameters, it remains an extremely useful tool. For nominal values of the unknown parameters it enables one to quantify variances that might be achieved, to detect possibly strong covariances between parameter estimates, and, as will be shown in Section 5.4, to optimize the experimental design (van den Bos, 1982). Moreover, the estimates obtained using an estimator that achieves the CRLB may be substituted for the true parameters in the expression for the CRLB so as to obtain a level of confidence to be attached to these estimates (den Dekker and Van Aert, 2002; den Dekker et al., 2005; Van Aert et al., 2005). This will briefly be discussed in Section 5.3.

The *attainable* precision with which position and distance parameters of one or two point objects can be measured has been investigated. The observations consist of counting results in a one- or two-dimensional pixel array. The model describing the expectations of these observations has been assumed to consist of Gaussian peaks with unknown position. Under this assumption, the CRLB, which usually has to be calculated numerically, may be approximated by a simple rule in closed analytical form. Although the expectation model of images obtained in practice are generally of higher complexity than Gaussian peaks, the rules are suitable to provide insight in the attainable precision of position and distance parameters using quantitative atomic resolution TEM. In Bettens et al. (1999) and Van Aert et al. (2002a), the details of this study can be found. In the following examples, only the main results will be presented.

Example 2 (*Position Measurement*) *In this example, the attainable precision of the position measurement of one isolated point object will be considered. If the point spread function is assumed to be Gaussian and defined by Eq. (1) and if the total number of imaging particles is N, the lower bound on the* *standard deviation s_{LB} of the coordinates estimates of the position, that is, the square root of the CRLB, is given by*

$$s_{LB} \approx \frac{\rho}{\sqrt{N}} \tag{30}$$

or, from Eq. (3)

$$s_{LB} \approx \frac{\rho_P}{2\sqrt{2N}} \tag{31}$$

Thus, the precision with which the position can be determined is a function of both the Rayleigh resolution ρ_P and the number of imaging photons N. If N is large, the precision can be orders of magnitude higher than the point resolution ρ_P.

Example 3 (*Distance Measurement*) *Here, the attainable precision of the distance measurement of two neighboring point objects will be considered. If the imaging is supposed to be linear, the image consists of the superposition of the two corresponding point spread functions as simulated in Figures 20–6 and 20–7. The total number of imaging photons used in the simulation is equal to 2 N. This means that in agreement with Figure 20–5, the number of photons per peak is equal to N. The results are different for distances smaller than or larger than $\rho_P/2$ and may be summarized as follows.*

- *$d > \rho_P/2$*
  This is shown in Figure 20–6. Now, the distance d between the atoms is larger than the half of the Rayleigh resolution. Then, the lower bound on the standard deviation s_{MIN} on the distance d is minimal, independent of d, and given by

$$s_{MIN} \approx \frac{\rho_P}{2\sqrt{N}} \tag{32}$$

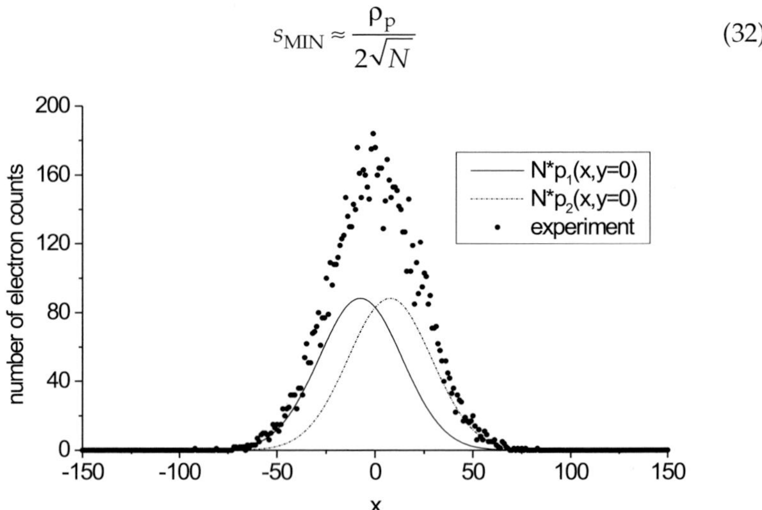

Figure 20–7. The results of a computer-simulated image of two neighboring points with a Gaussian point spread function, where $p_{1,2}(x, y)$ are two-dimensional normalized Gaussian point spread functions, as in Eq. (1). The experimental as well as the expectation values of the individual peaks $[N \times p_{1,2}(x, y)]$ are plotted within the section $y = 0$. The number of imaging particles $2N$ is equal to 500,000. The peaks overlap severely.

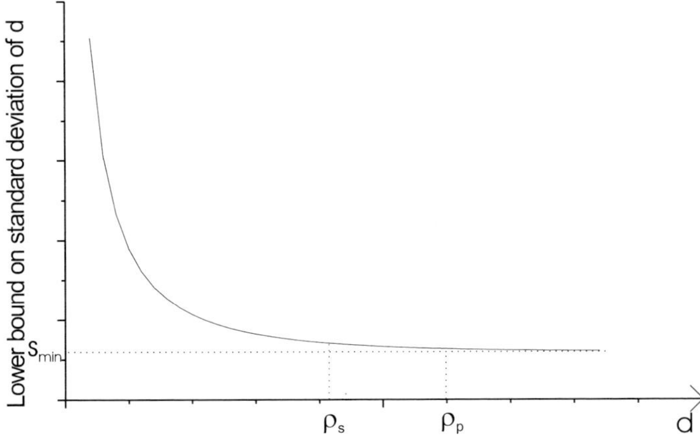

Figure 20–8. The lower bound on the standard deviation of the distance d as a function of the distance.

Thus, from Eq. (31) it follows that the CRLB on the variance of the distance is twice the CRLB on the variance of the position of an isolated point object:

$$s^2_{\mathrm{MIN}} = 2s^2_{\mathrm{LB}} \tag{33}$$

The reason for this is that the coordinate estimates of neighboring atoms are uncorrelated for distances larger than $\rho_p/2$.

- *$d \le \rho_p/2$*
  This is shown in Figure 20–7. Now, the distance d between the atoms is smaller than the half of the Rayleigh resolution. Then, the lower bound on the standard deviation s_{LB} increases inversely proportionally to the distance d, following approximately the relation

$$s_{\mathrm{LB}}(d) \approx \frac{s_{\mathrm{MIN}}\rho_p}{2d} \tag{34}$$

where s_{MIN} is given by Eq. (32).

In Figure 20–8, the lower bound on the standard deviation of the distance d is shown as a function of the distance.

In this section, statistical model-based resolution has been discussed in terms of the CRLB. In this discussion, the parametric statistical model of the observations, that is, the joint probability density function, has been assumed to be exact. Then, it is clear that statistical model-based resolution is limited by the lower bound on the variance or on the standard deviation. This limit is fundamental. Moreover, if the model is inaccurate, the introduced systematic error determines another fundamental limit to statistical model-based resolution. Apart from these fundamental limits, computational limits exist as well, just as for deterministic model-based resolution. The computing time needed to fit the model to the images with respect to the unknown parameters increases with the total number of parameters to be measured. Furthermore, the optimization methods fail if the structure is not resolved.

5.3 Maximum Likelihood Estimation

In this section, the derivation of the ML estimator of the parameters from the parameterized probability density function, which is discussed in Section 5.1, will be discussed. This estimator is very important since it achieves the CRLB asymptotically, that is, for the number of observations going to infinity. Thus, it is asymptotically most precise and is therefore often used in practice. The ML estimator is clearly discussed in van den Bos and den Dekker (2001) and den Dekker et al. (2005). A summary is given here.

The ML method for estimation of the parameters consists of three steps:

1. The available observations $w = (w_1 \ldots w_M)^{\mathrm{T}}$ are substituted for the corresponding independent variables $\omega = (\omega_1 \ldots \omega_M)^{\mathrm{T}}$ in the probability density function, for example, in Eq. (22) or in Eq. (23). Since the observations are numbers, the resulting expression depends only on the elements of the parameter vector $\tau = (\tau_1 \ldots \tau_R)^{\mathrm{T}}$.

2. The elements of $\tau = (\tau_1 \ldots \tau_R)^{\mathrm{T}}$, which are the hypothetical true parameters, are considered to be variables. To express this, they are replaced by $t = (t_1 \ldots t_R)^{\mathrm{T}}$. The logarithm of the resulting function, $\ln P(w; t)$, is called the *log-likelihood function* of the parameters t for the observations w, which is denoted as $q(w; t)$.

3. The ML estimates $\hat{\tau}_{\mathrm{ML}}$ of the parameters τ are defined by the values of the elements of t that maximize $q(w; t)$, or

$$\hat{\tau}_{\mathrm{ML}} = \arg \max_t q(w; t) \tag{35}$$

For independent normally distributed observations with equal variance, for which the joint probability density function is given by Eq. (23), it can easily be shown that the ML estimator is equal to the well-known *uniformly weighted least-squares estimator* (van den Bos and den Dekker, 2001; den Dekker et al., 2005). The uniformly weighted least-squares estimates $\hat{\tau}_{\mathrm{LS}}$ of the unknown parameters τ are given by the values of t that minimize the *uniformly weighted least-squares criterion*:

$$\hat{\tau}_{\mathrm{LS}} = \arg \min_t \sum_{m=1}^{M} [w_m - f_m(t)]^2 \tag{36}$$

where the function f_m is defined by Eq. (18).

The most important properties of the ML estimator are the following:

- *Consistency.* Generally, an estimator is said to be consistent if the probability that an estimate deviates more than a specified amount from the true value of the parameter can be made arbitrarily small by increasing the number of observations used.
- *Asymptotic normality.* If the number of observations increases, the probability density function of an ML estimator tends to a normal distribution.

- *Asymptotic efficiency.* The asymptotic covariance matrix of an ML estimator is equal to the CRLB. In this sense, the ML estimator is most precise.
- *Invariance property.* The ML estimates $\hat{\gamma}_{ML}$ of a vector of functions of the parameters τ, that is, $\gamma(\tau) = [\gamma_1(\tau) \ldots \gamma_C(\tau)]^T$, are equal to $\gamma(\hat{\tau}_{ML}) = [\gamma_1(\hat{\tau}_{ML}) \ldots \gamma_C(\hat{\tau}_{ML})]^T$ (Mood et al., 1974).

To evaluate the level of confidence to be attached to the obtained ML estimates, confidence regions and intervals associated with these estimates are required. A summary of existing methods to compute such regions and intervals is presented in den Dekker et al. (2005). One of these methods is based on the asymptotic normality of the ML estimator. It is preferred, especially if the experiment cannot be replicated. For example, an approximate $100(1 - \alpha)\%$ confidence interval for the rth element of τ, τ_r, is given by

$$[\hat{\tau}_{ML}]_r \pm \lambda_{(1-\alpha/2)} \sqrt{\left[F^{-1}\right]_{rr}} \tag{37}$$

with $[\hat{\tau}_{ML}]_r$ the rth element of the parameter vector $\hat{\tau}_{ML}$, $[F^{-1}]_{rr}$ the (r, r)th element of F^{-1}, and $\lambda_{(1-\alpha/2)}$ the $(1 - \alpha/2)$ quantile of the standard normal distribution. The meaning of a $100(1 - \alpha)\%$ confidence interval is that it covers the true element τ_r of τ with probability $1 - \alpha$. Usually, F^{-1} is a function of the parameters to be estimated. The confidence intervals (37) are then derived by using approximations of F^{-1}. A useful approximation $\hat{F}^{-1}$ of F^{-1} may be obtained by substituting $\hat{\tau}_{ML}$ for τ in the expression for the CRLB yielding

$$\hat{F}^{-1} = F^{-1}\big|_{\tau=\hat{\tau}_{ML}} \tag{38}$$

The effectiveness of the ML method will be shown in the following example, where it will be applied to experimental high-resolution TEM (HRTEM) images of an aluminum crystal. The details of this study can be found in Van Aert et al. (2005).

Example 4 (*Maximum Likelihood Estimation of Structure Parameters from HRTEM Images of an Aluminum Crystal*) *Structure parameters, atom column distances in particular, have been estimated from HRTEM images of an aluminum crystal using the ML method. Therefore, 20 images have been recorded and afterward been corrected for specimen drift using cross-correlation. From these corrected images, a particular region consisting of 51 times 50 pixels has been selected. The individual images resulting from this procedure are shown in Figure 20–9. The observations corresponding to these individual images, that is, the 51 times 50 image pixel values, are supposed to be independent normally distributed with equal variance. For such observations, the joint probability density function is given by Eq. (23). Moreover, the expectation model has been supposed to be given by Eq. (10). The unknown parameters of this model have then been estimated using the ML estimator, which for these types of observations is equal to the uniformly weighted least-squares estimator given by Eq. (36). Furthermore, ML estimates of the atom column distances have been obtained using the invariance property of the ML estimator. Also confidence intervals for the parameters have been computed. For example, for the atom column distances, the*

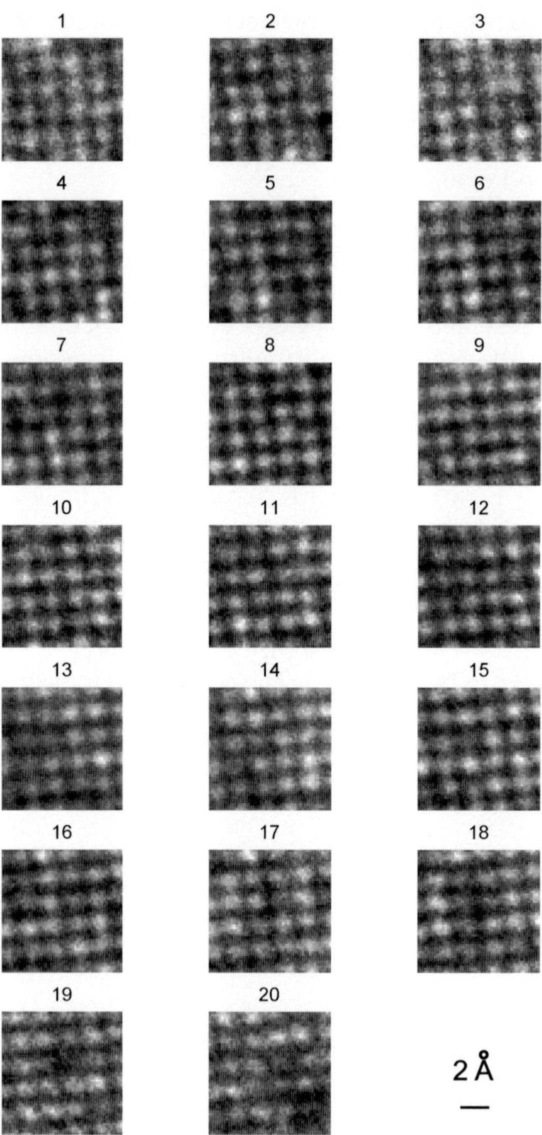

Figure 20–9. HRTEM observations of an aluminum crystal.

precision has been shown to be of the order of 0.03 Å, with the precision represented by the square root of the approximated CRLB on the variance of the distance with the approximated CRLB given by Eq. (38). Obviously, the value to be attached to the obtained estimates and confidence intervals depends on the validity of the model. Therefore, it is important to test the proposed model before attaching confidence to these estimates. This has been done using the model assessment methods described in den Dekker et al. (2005) and Van Aert et al. (2005). From these methods, it could be concluded that the proposed model is a sufficiently adequate description of the experimental observations. The model evaluated at the obtained ML estimates is shown in Figure 20–10. Note that this figure may be regarded as an optimal image reconstruction.

*Finally, the variation of the atom column distance estimates has been com-
pared to the variance that would be expected if the statistical fluctuations of
the observations would be the only source of variation of the distance estimates. The former has been measured by means of the well-known sample
variance (Mood et al., 1974) and the latter by means of the CRLB. If statistical fluctuations would be the only source of contribution to the variation of
the distance estimates, the sample variance and the estimated CRLB should
be about equal to one another. However, from a comparison of both, it followed
that the sample variance was about 10 times larger than the estimated CRLB.
Hence, there must be a further, dominant contribution to the variation of
the distance estimates. Because of this the atom columns of the crystal were
not observed at a perfectly periodic crystal lattice in the experiment and, in*

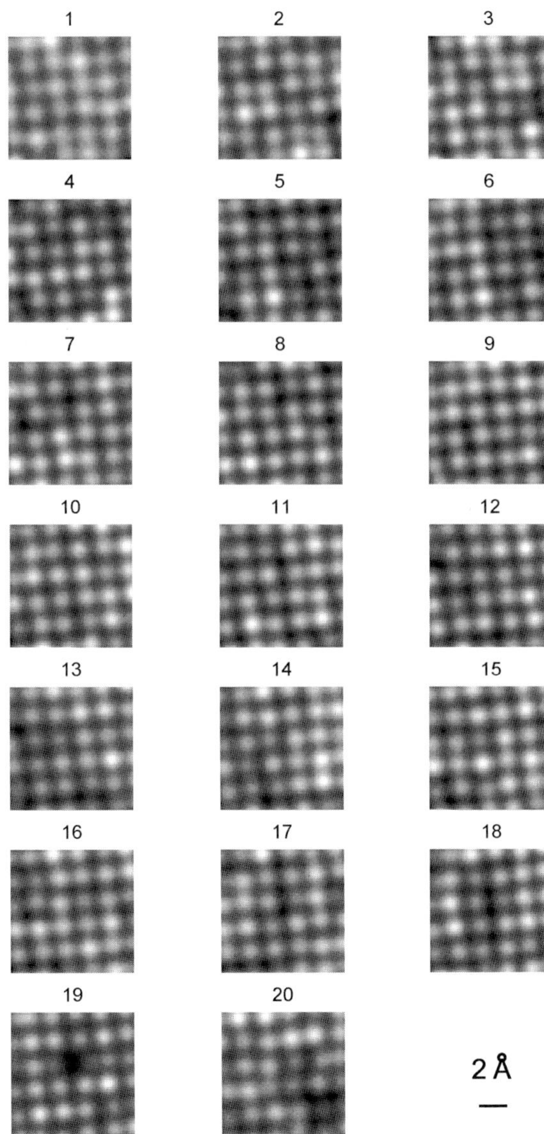

Figure 20–10. The exp-
ection models as descr-
ibed by Eq. (10) evaluated
at the estimated par-
ameters obtained from
the experimental images
shown in Figure 20–9
using the ML method.

addition, the atom columns were spotted at different locations in the course of time.

The ML method is, of course, also applicable to other types of experiments for the estimation of unknown parameters. An interesting application to EELS spectra can be found in Verbeeck and Van Aert (2004).

5.4 Statistical Experimental Design

In this section, it will be shown how the statistical model-based resolution of a quantitative atomic resolution TEM experiment may be improved by optimizing its design.

Usually, instrumental developments in microscopy, telescopy, or even spectrometry aim at improving the Rayleigh resolution or Rayleigh-like resolution criteria. Examples of such developments in electron microscopy are the spherical aberration corrector (Rose, 1990; Haider et al., 1998) and the monochromator (Mook and Kruit, 1999). A resolution (in the sense of Rayleigh) beyond 1 Å allows microscopists to visually distinguish atom columns of solids oriented along a main zone axis. In addition, electron microscopes now have increased versatility of microscope settings, which will preferably be computer controllable in the future. Electron microscopists can choose between TEM or STEM, between imaging or diffraction techniques, focus, accelerating voltage, spherical aberration, energy spread of incoming electrons, beam tilt, and crystal tilt. The main constraints of the experiment will be the radiation sensitivity of the object or the specimen drift. The question then arises as to which microscope and which settings are optimal in keeping the incident electron dose per square ångstrom or the recording time subcritical. In this section this question will be addressed using statistical experimental design. Statistical experimental design can be defined as the selection of free experimental settings in an experiment to improve the precision with which unknown parameters can be estimated. For quantitative atomic resolution TEM experiments, these parameters are the atom or atom column positions in particular. The optimal design is then given by the combination of experimental settings for which the precision of the atom or atom column positions is highest. In the optimization of the experimental design, the CRLB is taken as the optimal criterion. This criterion thus aims at an improvement of statistical model-based resolution rather than an improvement of Rayleigh resolution or Rayleigh-like resolution criteria. Note that the optimal statistical experimental design may be different for different objects under study.

In Section 5.2, it was shown how from the joint probability density function, which was introduced in Section 5.1, the elements of the Fisher information matrix may be calculated explicitly. From the latter, the CRLB on the variance of the parameters of the expectation model and on the variance of functions of these parameters may be computed from the right-hand side of Eqs. (27) and (29), respectively. The diagonal elements of the CRLB give a lower bound on the variance of any unbiased estimator of the parameters. Since the joint probability density function is a function of the experimental settings, the CRLB is a function of these settings as well. Therefore, the CRLB may be used to evaluate and to optimize the experimental design in terms of precision. However, simultaneous minimization of the diagonal elements of the CRLB, that is, the right-hand side of Eq. (28), is usually impossible. Therefore, statistical parameter estimation theory provides different optimality criteria, which are functions of the elements of the CRLB. These are scalar measures. Experimenters may choose one of these criteria or may produce their own criterion, reflecting their purpose. The interested reader may find a selection of such criteria in Fedorov (1972), Pázman (1986), and Van Aert et al. (2004).

Statistical experimental design can be illustrated as follows. Suppose the microscope is able to visualize individual atoms or atom columns so that the structure can be resolved. In other words, neighboring atoms or atom columns can be discriminated in the images. Furthermore, it will be assumed that the expectations of the Poisson distributed image pixel values can be modeled as a summation of Gaussian peaks centered at the atom or atom column positions. The position coordinates of these atoms or atom columns are the unknown parameters in the expectation model. It then follows from Eq. (31) that the lower bound on the standard deviation of these coordinate estimates, that is, the square root of the CRLB, is approximately given by $\rho_p / 2\sqrt{2N}$. In this expression, ρ_p represents the Rayleigh resolution, which depends on the Rayleigh resolution of all different subchannels contributing to the imaging process as follows from Eq. (7) and N represents the number of detected electrons per atom or atom column. It is now clear that it is not only the Rayleigh resolution that matters but the electron dose as well. If the Rayleigh resolution can be improved only at the expense of a decrease in the number of detected electrons, both effects have to be balanced under the existing physical constraints so as to produce the highest precision. Furthermore, it follows from Eq. (7) that below a certain value for ρ_p, it will be useless to further improve the Rayleigh resolution of the electron microscope since scattering is dominant or, equivalently, ρ_p is almost equal to the width of the electrostatic potential of an atom, that is, ρ_A. This result is meaningful in practice. For example, in STEM experiments, further narrowing of the probe, which represents the point spread function of the electron microscope, is not so beneficial in terms of precision since the width of the probe is currently almost equal to the width of an atom (Krivanek et al., 2002).

In the following examples, the main results obtained from the numerical optimization and evaluation of the design of quantitative atomic resolution TEM experiments will be presented. Also the possible benefits of new developments in instruments will be discussed. In these examples, expectation models with a solid physical base, instead of Gaussian peaks, have been assumed and the observations are assumed to be independent and Poisson distributed. The results obtained may intuitively be interpreted using the rules for the CRLB, which have been obtained from Gaussian peaked expectation models.

Comprehensive descriptions of these results can be found in den Dekker et al. (1999, 2001), Van Aert and Van Dyck (2001), and Van Aert et al. (2002a–c, 2004).

Example 5 *(Statistical Experimental Design of Scanning Transmission Electron Microscopy) The expectation model that has been used in the evaluation and optimization of the experimental design of STEM experiments is based on the simplified channelling theory to describe the dynamic, elastic scattering of the electrons with atom columns (Broeckx et al., 1995; Geuens and Van Dyck, 2002; Pennycook and Jesson, 1992; Van Aert et al., 2002b; Van Dyck and Op de Beeck, 1996). Therefore, it has been assumed that the dynamic motion of an electron in a column may be primarily expressed in terms of so-called tightly bound 1s states, which are peaked at the atom column positions. Furthermore, in this model, temporal incoherence due to chromatic aberration has not been taken into account since STEM imaging seems to be robust to chromatic aberration (Batson, et al., 2002; Krivanek et al., 2002; Nellist and Pennycook, 1998, 2000). Thermal diffuse, inelastic scattering has not been taken into account for the following reasons. Thermal diffuse scattered electrons are predominantly collected in the detector at high angles (Treacy, 1982). Therefore, increasing the inner detector angle of an annular detector has the effect of increasing thermal diffuse, inelastic scattering relative to elastic scattering (Wang, 2001). Usually, the inner detector angle is chosen to be relatively large since the detected signal then strongly depends on the atomic number Z, hence the name, Z-contrast imaging. The disadvantage of increasing this detector angle, however, is the accompanied decrease of dose efficiency, which leads to a decrease in signal-to-noise ratio (SNR). It can be shown that as a result of this decrease in SNR, the optimal inner collection angle in terms of precision is small compared to the angles where thermal diffuse scattering is important. This justifies the fact that thermal diffuse scattering has not been taken into account. The details of the expectation model can be found in Van Aert et al. (2002b, 2004).*

It has been shown in Van Aert and Van Dyck (2001) and Van Aert et al. (2002b, 2004) that the optimal aperture radius in terms of precision may be considerably smaller than the aperture radius at the Scherzer conditions for incoherent imaging, which is referred to as being optimal in terms of direct visual interpretability (Scherzer, 1949; Pennycook and Jesson, 1991). This indicates that the optimal width of diffraction-limited probes, for which the width is inversely proportional to the aperture radius, is not as small as possible. The optimal objective aperture radius, which determines the size of a diffraction-limited probe, has been found to be mainly determined by the object under study. More specifically, for isolated atom columns, it is proportional to the depth of the two-dimensional electrostatic potential of the atom column projected along the column direction, that is, the difference between the maximum and minimum potential energy. In terms of the column-specific 1s state, this means that the main lobe of the optimal probe is broader than the 1s state. This is shown in Figure 20–11, where both the 1s state and the amplitude of the optimal probe are shown for a silicon and a gold [100] atom column. However, in the presence of neighboring columns, it has been found that the optimal aperture radius may increase so as to avoid strong overlap of neighboring columns in the image.

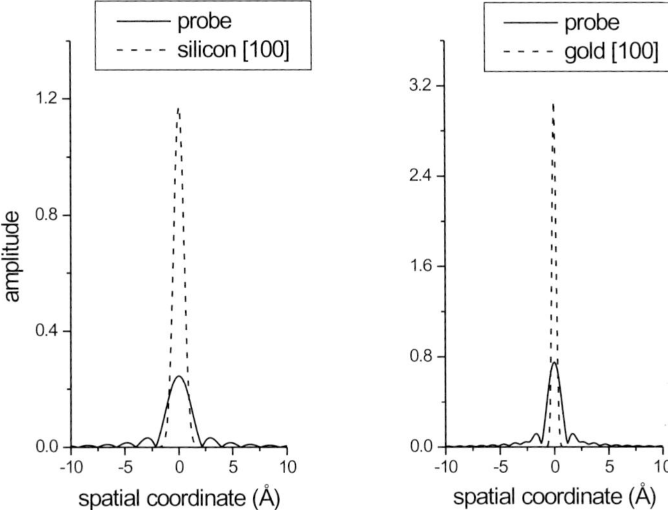

Figure 20–11. The dashed curves represent the 1s state for a silicon [100] (left) and a gold [100] (right) atom column, respectively. The solid curves represent the amplitude of their associated optimal STEM probes for C_s equal to 0.5 mm.

In the evaluation of the experimental design, annular and axial detector types have been compared. Apart from some exceptions, an annular detector usually results in higher precision than an axial one. The optimal inner radius of an annular detector has been found to be equal to the optimal objective aperture radius, whereas in a conventional approach the radius of the hole is usually taken to be much larger than the aperture radius (Pennycook et al., 1995; Hartel et al., 1996). The optimal outer radius of an axial detector is usually slightly smaller than the optimal aperture radius. However, if this detector leads to very low contrast of the image, the optimal detector radius decreases.

Moreover, it has been found that a spherical aberration corrector improves the precision. The accompanied gain depends on the object under study. Correction of the spherical aberration is more useful in terms of precision for heavy than for light atom columns. This is shown in Figures 20–12 and 20–13, where the ratio of the lower bound on the standard deviation of the position coordinates for a given spherical aberration constant to the lower bound for a spherical aberration constant of 0 mm is shown as a function of the spherical aberration constant. This is done for an isolated silicon and gold [100] atom column, respectively. Moreover, the evaluation is done for an annular as well as an axial detector. For each spherical aberration constant considered, the objective aperture radius is set to its optimal value. Furthermore, the detector radius is taken to be equal to this optimal objective aperture radius. For silicon, which is a light atom column, it follows from Figure 20–12 that the precision that is gained by reducing the spherical aberration constant from 0.5 mm to 0 mm is a factor of 1.0009 and 1.0011 for an annular and axial detector, respectively. These gains are negligible. For gold, which is a heavy atom column, it follows from Figure 20–13 that the precision that is gained by reducing the spherical aberration constant from 0.5 mm to 0 mm is a factor of 1.21 and 1.39 for an annular and axial detector, respectively, which is still

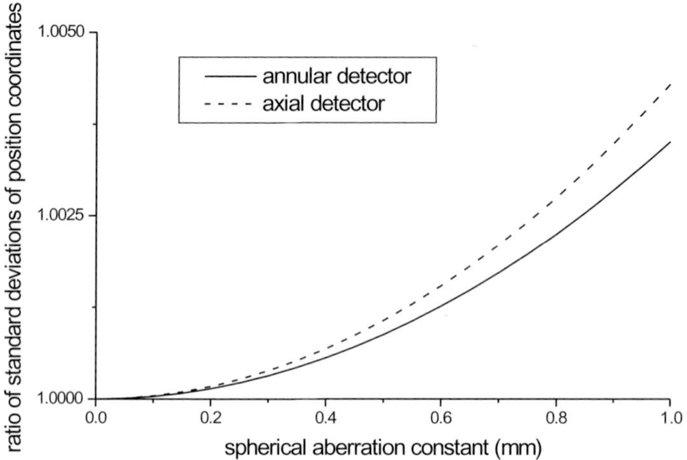

Figure 20–12. Ratio of the lower bound on the standard deviation of the position coordinates for a given spherical aberration constant to the lower bound for a spherical aberration constant of 0 mm for an isolated silicon [100] atom column as a function of the spherical aberration constant for an annular as well as for an axial STEM detector.

relatively small. This makes it questionable whether such a corrector is needed to obtain a prespecified precision of the atom column positions.

Example 6 *(**Statistical Experimental Design of High-Resolution Transmission Electron Microscopy**) The optimal experimental design of HRTEM experiments has been reconsidered in terms of statistical model-based resolution (den Dekker et al., 1999, 2001; Van Aert et al., 2004). The expecta-*

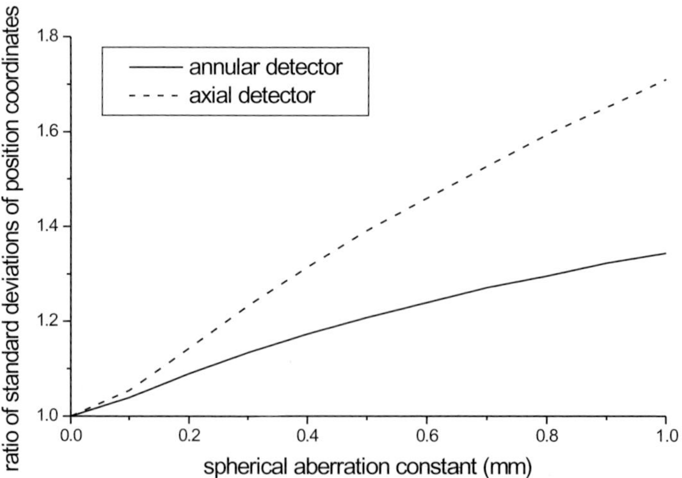

Figure 20–13. Ratio of the lower bound on the standard deviation of the position coordinates for a given spherical aberration constant to the lower bound for a spherical aberration constant of 0 mm for an isolated gold [100] atom column as a function of the spherical aberration constant for an annular as well as for an axial STEM detector.

tion model that has been used in the evaluation and optimization of HRTEM experiments is based on the simplified channeling theory (Geuens and Van Dyck, 2002; Van Dyck and Op de Beeck, 1996). For microscopes operating at intermediate accelerating voltages of the order of 300 kV, it has been shown that use of a spherical aberration corrector or a chromatic aberration corrector is only of limited value and that the use of a monochromator usually is not worthwhile in terms of precision, presuming that specimen drift places a practical constraint on the experiment. These results follow from Figure 20–14, where the lower bound on the standard deviation of the position coordinates of a gold atom column is evaluated as a function of the spherical aberration constant. The solid curve corresponds to a microscope without correction for chromatic aberration, that is, a microscope without a chromatic aberration corrector and monochromator. The dashed curve corresponds to a microscope with a chromatic aberration corrector, for which the chromatic aberration constant is equal to 0 mm. The dotted curve corresponds to a microscope with a monochromator, for which the energy spread corresponds to a typical full width at half maximum height of 200 meV (Batson, 1999). In Figure 20–14 it is assumed that the specimen drift is the relevant physical constraint. Hence, the recording time is kept constant. Consequently, the total number of detected electrons is smaller in the presence of a monochromator. The values for the other microscope settings, that is, for the original, nonoptimized microscope settings, are in accordance with values that are typical for today's electron microscopes. Therefore, it also follows from Figure 20–14 that for today's electron microscopes it is in principle possible to obtain a precision of the order of 0.01 Å, even with a microscope that is not corrected for spherical and chromatic aberration.

However, for amorphous instead of crystalline structures, the conclusions are different since amorphous structures are very sensitive to radiation damage,

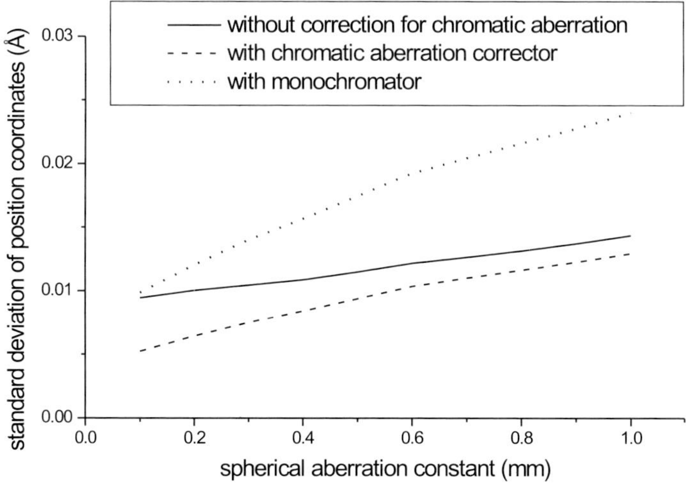

Figure 20–14. The lower bound on the standard deviation of the position coordinates of a gold [100] atom column as a function of the spherical aberration constant for a microscope operating at 300 kV equipped with or without a chromatic aberration corrector or monochromator. In this evaluation, the recording time is kept constant.

so that radiation sensitivity rather than specimen drift will be the limiting factor in the experiment. Although the precision improves by increasing the incident dose, it has to be taken into account that every incident electron has a finite probability of damaging the structure. The structure can be damaged either by displacing an atom from its position or by changing chemical bonds due to ionization. Therefore, a compromise between precision and radiation damage has to be made, which turns out to be optimal when the accelerating voltage is reduced. However, at low accelerating voltages, the instrumental aberrations become important. For this reason, correction of both the spherical aberration and the chromatic aberration by either a chromatic aberration corrector or a monochromator will be essential. By means of statistical experimental design, it has been shown that a substantial improvement of the precision may be obtained if both the spherical and chromatic aberration are corrected (Van Dyck et al., 2003).

6 Ultimate Model-Based Resolution

In the preceding section, the CRLB on the precision of the estimated distance of two peaks was used as a resolution criterion. Such a criterion makes sense only if an estimator exists that actually achieves this bound. Such an estimator exists in the form of the so-called ML estimator. This estimator produces estimates that are normally distributed about the true values of the parameters with a variance equal to the CRLB. However, this is true only if the number of observations used by the ML estimator is sufficiently large. For small numbers of observations, the estimates have statistical properties that are very different. A detailed description is given in van den Bos (1992) and van den Bos and den Dekker (1995, 2001).

6.1 Two Types of Observations

Central in this section is the space of observations, which has been introduced in Section 5.1. This is the Euclidean space in which every coordinate direction corresponds to a particular observation. Therefore, the dimension of the space of observations is equal to the number of observations and a particular set of observations is represented by a single point in the space. For example, the observations shown in Figure 20–7 correspond to a single point in the space of observations. It has been shown in the references mentioned that two types of sets of two-peak observations may occur. If the two-peak model is fitted to the first type of observations, *distinct* values for the locations are obtained. Then the distance of the peaks is different from zero. If, on the other hand, the two-peak model is fitted to observations of the second type, the estimates for the locations *exactly coincide*. Then the distance of the peaks is equal to zero. This means that the peaks exactly coincide. It is clear that from the first type of observations both peaks are resolved in the sense that they occur at distinct locations. On the other hand, for the second type of observations, the peaks occur at the same location. Therefore, the peaks add up to form a single peak and cannot be resolved. From these considerations, it follows that the (hyper)surface separating both types of observations in the space of observations may be used as a resolution criterion. If the point representing a particular set of observations is on the one side of the hypersurface, the peaks are resolved, but if it is on the other side, they are not.

This perhaps unexpected behavior of the solutions for the locations of the peaks has to do with the *structure* of criteria of goodness of fit for the estimation of parameters such as the locations. The structure is the pattern of the stationary points of the criterion, where a stationary point is a point at which the gradient is equal to zero, such as a minimum or maximum. The structure is important since the solution of the model fitting problem is represented by a stationary point: the absolute optimum of the criterion of goodness of fit. It has been found that for observations made on overlapping peaks, two different structures may occur and that these correspond to distinct or coinciding solutions for the locations, respectively. Which of these structures occurs depends on the particular realization of the observations.

6.2 Consequences for Resolution

If the observations are statistical, they are, in the space of the observations, distributed about the point representing their expectations. This *expectation point*, therefore, represents ideal exact observations. If the model fitted is correctly specified, it perfectly fits these exact observations. Then the solutions found for the locations are distinct and equal the hypothetical true locations, no matter how closely located the peaks. Therefore, the expectation point will be located on the "two-peak side" of the separating hypersurface and not on the "one-peak side," since the solutions are distinct. Observations distributed about the expectation point will, with a certain probability, only be on the two-peak side of the separating hyperplane. It is concluded that resolution should be described in terms of *probability of resolution*. The distance of the expectation point and the separating hypersurface becomes smaller and smaller as the hypothetical true locations are closer. This means that the probability that a set of observations is located on the one-peak side of the separating hyperplane increases and the probability of resolution decreases correspondingly.

The separating hypersurface is computed from and, therefore, specific to the peak model fitted. In this chapter, this is often the Gaussian peak. The expectation point, however, is specific to the model actually present in the observations. If the model fitted and the model present differ, the expectation point may even be located on the one-peak side of the separating hyperplane. This implies that the peaks are not resolved even if the statistical errors in the observations are very small.

It is clear that, ultimately, resolution is limited by errors, both systematic (modeling) errors and nonsystematic (statistical) errors. This limit is fundamental: neither an increase in computer power nor improved software can remove it. It would be present even if initial conditions would be available for the locations of all peaks and sufficient computer power would be available for fitting them. Removing these limits requires more and preferably different observations of the same peaks.

7 Discussion and Conclusions

In this chapter, resolution has been investigated in detail in successive steps. Throughout this chapter, emphasis has been put on electron microscopy.

First, classical resolution criteria, such as Rayleigh's and Sparrow's, have been discussed. These criteria are expressed in terms of the width of the point spread function. The narrower the point spread function, the higher the resolution.

Next, the diffraction limit to resolution and its relation to Rayleigh and Sparrow resolution have been considered. The diffraction limit is given by the cutoff frequency, which is the highest spatial frequency that is transferred by the imaging system. It has been shown that the diffraction limit is inversely proportional to the Rayleigh resolution. For electron microscopy applications, it has been shown that the "width" of the electrostatic potential of the atoms as well as the effect of thermal vibrations of the atoms, the environment, and the detection have to be taken into account. As a result of these extensions, it should be concluded that the atoms themselves limit the diffraction limit and classical resolution criteria. Also the notion of superresolution has been introduced. Using superresolution algorithms, frequency components lying beyond the diffraction limit can be reconstructed. Such algorithms use prior knowledge of the object imaged by the imaging system.

Such prior knowledge has been incorporated in the form of a parametric model. Indeed, classical resolution criteria are in fact concerned with calculated images, that is, noise-free images exactly describable by a known parameterized mathematical model. Then, the unknown parameters, such as the positions of projected atoms, can be measured by means of model fitting. The practical limits to this so-called deterministic model-based resolution are of a computational kind. As an example, amorphous structures studied by atomic resolution TEM have been discussed. For these structures, computational problems can be overcome only if the thickness is very small. An upper bound to the thickness is derived, which for amorphous silicon is of the order of 10 Å. For realistic thicknesses, electron tomography is needed. Moreover, fundamental limits to deterministic model-based resolution exist if the model is inaccurate, since then, systematic errors are introduced.

Then, unavoidable noise in the observations has been incorporated by considering detected images instead of calculated images. The fundamental limit to this so-called statistical model-based resolution is determined by the CRLB, which is a lower bound on the variance with which the positions of, or the distance between, projected atoms or atom columns can be measured using parameter estimation methods. In this discussion, the ML estimator is very important since it achieves the CRLB asymptotically. The effectiveness of the ML method has been shown in an example, in which it has been applied to experimental HRTEM images of an aluminium crystal. Moreover, it has been shown that if the model is inaccurate, systematic errors determine fundamental limits to statistical model-based resolution as well. Apart from these fundamental limits, computational limits exist, just as for deterministic model-based resolution. It has also been shown how to improve statistical model-based resolution using statistical experimental design.

Finally, ultimate model-based resolution is considered. Depending on the observations, the solutions of the position estimates may coincide exactly. It is shown that ultimately, resolution is limited by errors, both systematic and nonsystematic. This limit is fundamental: no increase in computer power or improved software can remove it. It would be present even if the structure is resolved and sufficient computer power would be available for fitting them. Removing these limits requires more and preferably different observations of the same structure.

Acknowledgments. Dr. S. Van Aert gratefully acknowledges the financial support of the Fund for Scientific Research—Flanders (FWO). The research of Dr. A.J. den Dekker has been made possible by a fellowship of the Royal Netherlands Academy of Arts and Sciences.

References

Batson, P.E. (1999). Advanced spatially resolved EELS in the STEM. *Ultramicroscopy* **78**, 33–42.

Batson, P.E., Dellby, N. and Krivanek, O.L. (2002). Sub-ångstrom resolution using aberration corrected electron optics. *Nature* **418**, 617–620.

Bettens, E., Van Dyck, D., den Dekker, A.J., Sijbers, J. and van den Bos, A. (1999). Model-based two-object resolution from observations having counting statistics. *Ultramicroscopy* **77**, 37–48.

Biraud, Y. (1969). A new approach for increasing the resolving power by data processing. *Astron. Astrophys.* **1**, 124–127.

Born, M. and Wolf, E. (1999). *Principles of Optics*, 7th (expanded) ed. (Cambridge University Press, Cambridge).

Broeckx, J., Op de Beeck, M. and Van Dyck, D. (1995). A useful approximation of the exit wave function in coherent STEM. *Ultramicroscopy* **60**, 71–80.

Browning, N.D., Arslan, I., Moeck, P. and Topuria, T. (2001). Atomic resolution scanning transmission electron microscopy. *Phys. Stat. Sol. B* **227**, 229–245.

Buseck, P.R., Cowley, J.M. and Eyring, L. (1988). *High-Resolution Transmission Electron Microscopy and Associated Techniques.* (Oxford University Press, Oxford).

Castleman, K.R. (1979). *Digital Image Processing.* (Prentice-Hall International, London).

Coene, W.M.J., Thust, A., Op de Beeck, M. and Van Dyck, D. (1996). Maximum-likelihood method for focus-variation image reconstruction in high resolution transmission electron microscopy. *Ultramicroscopy* **64**, 109–135.

Cowley, J.M. (2001). The quest for ultra-high resolution. In *Progress in Transmission Electron Microscopy—Concepts and Techniques* (X.F. Zhang and Z. Zhang, Eds.), Vol. 1, 35–79, Berlin (Springer, 1).

de Jong, A.F. and Van Dyck, D. (1993). Ultimate resolution and information in electron microscopy II. The information limit of transmission electron microscopes. *Ultramicroscopy* **49**, 66–80.

den Dekker, A.J. and Van Aert, S. (2002). Quantitative high resolution electron microscopy and Fisher information. In *Proceedings of the 15th International Congress on Electron Microscopy, Interdisciplinary and Technical Forum Abstracts*

2002 in Durban, South Africa (R. Cross, Ed.), Vol. 3, 185–186 (Microscopy Society of Southern Africa, Onderstepoort).

den Dekker, A.J. and van den Bos, A. (1997). Resolution: A survey. *J. Opt. Soc. Amer. A* **14**, 547–557.

den Dekker, A.J., Sijbers, J. and Van Dyck, D. (1999). How to optimize the design of a quantitative HREM experiment so as to attain the highest precision. *J. Microsc.* **194**, 95–104.

den Dekker, A.J., Van Aert, S., Van Dyck, D., van den Bos, A. and Geuens, P. (2001). Does a monochromator improve the precision in quantitative HR-TEM? *Ultramicroscopy* **89**, 275–290.

den Dekker, A.J., Van Aert, S., van den Bos, A. and Van Dyck, D. (2005). Maximum likelihood estimation of structure parameters from high resolution electron microscopy images. Part I: a theoretical framework. *Ultramicroscopy*, **104**, 83–106.

Falconi, O. (1964). Maximum sensitivities of optical direction and twist measuring instruments. *J. Opt. Soc. Amer.* **54**, 1315–1320.

Falconi, O. (1967). Limits to which double lines, double stars, and disks can be resolved and measured. *J. Opt. Soc. Amer.* **57**, 987–993.

Fedorov, V.V. (1972). *Theory of Optimal Experiments.* (Academic Press, New York).

Frank, J. (1992). *Electron Tomography—Three-Dimensional Imaging with the Transmission Electron Microscope.* (Plenum Press, New York).

Frieden, B.R. (1967). Band-unlimited reconstruction of optical objects and spectra. *J. Opt. Soc. Amer.* **57**, 1013–1019.

Frieden, B.R. (1975). *Image Enhancement and Restoration* Vol. 6. (Springer-Verlag, New York).

Frieden, B.R. (1980). Statistical models for the image restoration problem. *Comp. Graph. Image. Proc.* **12**, 40–59.

Frieden, B.R. (1998). *Physics from Fisher Information—A Unification.* (Cambridge University Press, Cambridge).

Gerchberg, R.W. (1974). Super-resolution through error energy reduction. *Opt. Acta.* **21**, 709–720.

Gerchberg, R.W. (1989). Superresolution through error function extrapolation. *IEEE Transact. Acoustics, Speech, Sign Proc.* **37**, 1603–1606.

Geuens, P. and Van Dyck, D. (2002). The S-state model: A work horse for HRTEM. *Ultramicroscopy* **93**, 179–198.

Gonzalez, R.G. and Woods, R.E. (2002). *Digital Image Processing*, 2nd Ed. (Prentice Hall, Upper Saddle River, NJ).

Goodman, J.W. (1968). *Introduction to Fourier Optics.* (McGraw-Hill, San Francisco).

Haider, M., Uhlemann, S., Schwan, E., Rose, H., Kabius, B. and Urban, K. (1998). Electron microscopy image enhanced. *Nature* **392**, 768–769.

Harris, J.L. (1964). Diffraction and resolving power. *J. Opt. Soc. Amer.* **54**, 931–936.

Hartel, P., Rose, H. and Dinges, C. (1996). Conditions and reasons for incoherent imaging in STEM. *Ultramicroscopy* **63**, 93–114.

Howie, A. (1966). Diffraction channelling of fast electrons and positrons in crystals. *Philos. Mag.* **14**, 223–237.

Hunt, B.R. (1994). Prospects for image restoration. *Internat. J. Mod. Phys. C* **5**, 151–178.

Hunt, B.R. and Sementilli, P. (1992). Description of a poisson imagery super resolution algorithm. In *Astronomical Data Analysis Software and Systems* (Worrall et al., Ed.), Vol. 25, 196–199 (Astronomical Society of the Pacific, San Francisco).

Janson, P.A., Hunt, R.H. and Plyler, E.K. (1970). Resolution enhancement of spectra. *J. Opt. Soc. Amer. A* **60**, 596.

Kisielowski, C., Hetherington, C.J.D., Wang, Y.C., Kilaas, R., O'Keefe, M.A. and Thust, A. (2001a). Imaging columns of the light elements carbon, nitrogen and oxygen with sub Ångstrom resolution. *Ultramicroscopy* **89**, 243–263.

Kisielowski, C., Principe, E., Freitag, B. and Hubert, D. (2001b). Benefits of microscopy with super resolution. *Physica. B* **308–310**, 1090–1096.

Koster, A.J., van den Bos, A. and van der Mast, K.D. (1987). An autofocus method for a TEM. *Ultramicroscopy* **21**, 209–222.

Krivanek, O.L., Dellby, N. and Nellist, P.D. (2002). Aberration correction in the STEM. In *Proceedings of the 15th International Congress on Electron Microscopy, Interdisciplinary and Technical Forum Abstracts 2002 in Durban, South Africa* (R. Cross, Ed.), Vol. 3, 29–30 (Microscopy Society of Southern Africa, Onderstepoort).

Meinel, E.S. (1986). Origins of linear and nonlinear recursive restoration algorithms. *J. Opt. Soc. Amer. A* **3**, 787–799.

Miedema, M.A.O., van den Bos, A. and Buist, A.H. (1994). Experimental design of exit wave reconstruction from a transmission electron microscope defocus series. *IEEE Transact. Instr. Meas.* **43**, 181–186.

Mood, A.M., Graybill, F.A. and Boes, D.C. (1974). *Introduction to the Theory of Statistics*, 3rd ed. (McGraw-Hill, Singapore).

Mook, H.W. and Kruit, P. (1999). Optics and design of the fringe field monochromator for a Schottky field emission gun. *Nucl. Instr. Methods. Phys. Res. A* **427**, 109–120.

Muller, D.A. (1998). Core level shifts and grain boundary cohesion. In *Proceedings Microscopy and Microanalysis 1998 in Atlanta, Georgia* (G.W. Bailey, Ed.), Vol. 4, 766–767. (Springer, New York).

Muller, D.A. (1999). Why changes in bond lengths and cohesion lead to core-level shifts in metals, and consequences for the spatial difference method. *Ultramicroscopy* **78**, 163–174.

Nellist, P.D. and Pennycook, S.J. (1998). Subangstrom resolution by underfocused incoherent transmission electron microscopy. *Phys. Rev. Lett.* **81**, 4156–4159.

Nellist, P.D. and Pennycook, S.J. (2000). The principles and interpretation of annular dark-field Z-contrast imaging. In *Advances in Imaging and Electron Physics* (P.W. Hawkes, Ed.), Vol. 113, 147–199 (Academic Press, San Diego).

O'Keefe, M.A., Hetherington, C.J.D., Wang, Y.C., Nelson, E.C., Turner, J.H., Kisielowski, C., Malm, J.-O., Mueller, R., Ringnalda, J., Pan, M. and Thust, A. (2001). Sub-Ångstrom high-resolution transmission electron microscopy at 300 keV. *Ultramicroscopy* **89**, 215–241.

Olson, G.B. (1997). Computational design of hierarchically structured materials. *Science* **277**, 1237–1242.

Olson, G.B. (2000). Designing a new material world. *Science* **288**, 993–998.

Papoulis, A. (1965). *Probability, Random Variables, and Stochastic Processes.* (McGraw-Hill, New York).

Pázman, A. (1986). *Foundations of Optimum Experimental Design.* (D. Reidel Publishing Company, Dordrecht).

Pennycook, S.J. and Jesson, D.E. (1991). High-resolution Z-contrast imaging of crystals. *Ultramicroscopy* **37**, 14–38.

Pennycook, S.J. and Jesson, D.E. (1992). Atomic resolution Z-contrast imaging of interfaces. *Acta. Met. Mat.* Supplement **40**, 149–159.

Pennycook, S.J., Jesson, D.E., Chisholm, M.F., Browning, N.D., McGibbon, A.J. and McGibbon, M.M. (1995). Z-contrast imaging in the scanning transmission electron microscope. *J. Microsc. Soc. Amer.* **1**, 231–251.

Reed, M.A. and Tour, J.M. (2000). Computing with molecules. *Sci. Amer.* **282**, 68–75.

Rose, H. (1990). Outline of a spherically corrected semiaplanatic medium-voltage transmission electron microscope. *Optik.* **85**(1), 19–24.

Saxton, W.O. (1978). *Computer Techniques for Image Processing in Electron Microscopy*, 236–248 (Academic Press, New York).

Schell, A.C. (1965). Enhancing the angular resolution of incoherent sources. *The Radio and Electronic Engineer* **29**, 21–26.

Scherzer, O. (1949). The theoretical resolution limit of the electron microscope. *J. App. Phy.* **20**, 20–28.

Schiske, P. (1973). Image processing using additional statistical information about the object. In *Image Processing and Computer-aided Design in Electron Optics* (P.W. Hawkes, Ed.), 82–90 (Academic Press, London).

Sparrow, C.M. (1916). On spectroscopic resolving power. *Astrophys. J.* **44**, 76–86.

Spence, J.C.H. (1999). The future of atomic resolution electron microscopy for materials science. *Mat. Sci. Eng. R* **26**, 1–49.

Strutt, J.W. (1899). Investigations in optics, with special reference to the spectroscope. In *Scientific papers by John William Strutt, Baron Rayleigh*, Vol. 1, 415–459 (Cambridge University Press, Cambridge).

Thust, A., Coene, W.M.J., Op de Beeck, M. and Van Dyck, D. (1996). Focal-series reconstruction in HRTEM: Simulation studies of non-periodic objects. *Ultramicroscopy* **64**, 211–230.

Treacy, M.M.J. (1982). Optimising atomic number contrast in annular dark field images of thin films in the scanning transmission electron microscope. *J. Microsc. Spectrosc. Elect.* **7**, 511–523.

Van Aert, S. and Van Dyck, D. (2001). Do smaller probes in a scanning tranmission electron microscope result in more precise measurement of the distances between atom columns? *Philos. Mag. B* **81**, 1833–1846.

Van Aert, S., den Dekker, A.J., Van Dyck, D. and van den Bos, A. (2002a). High-resolution electron microscopy and electron tomography: Resolution versus precision. *J. Struct. Biol.* **138**, 21–33.

Van Aert, S., den Dekker, A.J., Van Dyck, D. and van den Bos, A. (2002b). Optimal experimental design of STEM measurement of atom column positions. *Ultramicroscopy* **90**, 273–289.

Van Aert, S., den Dekker, A.J., van den Bos, A. and Van Dyck, D. (2002c). High-resolution electron microscopy: From imaging toward measuring. *IEEE Transact. Instr. Meas.* **51**, 611–615.

Van Aert, S., den Dekker, A.J., van den Bos, A. and Van Dyck, D. (2004). Statistical experimental design for quantitative atomic resolution transmission electron microscopy. In *Advances in Imaging and Electron Physics* (P.W. Hawkes, Ed.), Vol. 130, 1–164 (Academic Press, San Diego).

Van Aert, S., den Dekker, A.J., van den Bos, A., Van Dyck, D. and Chen, J.H. (2005). Maximum likelihood estimation of structure parameters from high resolution electron microscopy images. Part II: A practical example. *Ultramicroscopy*, **104**, 107–125.

van den Bos, A. (1982). Parameter estimation. In *Handbook of Measurement Science* (P.H. Sydenham, Ed.), Vol. 1, 331–337 (Wiley, Chicester).

van den Bos, A. (1992). Ultimate resolution—A mathematical framework. *Ultramicroscopy* **47**, 298–306.

van den Bos, A. (1999). Measurement errors. In *Encyclopedia of Electrical and Electronics Engineering* (J.G. Webster, Ed.), 448–459 (Wiley, New York).

van den Bos, A. and den Dekker, A.J. (1995). Ultimate resolution in the presence of coherence. *Ultramicroscopy* **60**, 345–348.

van den Bos, A. and den Dekker, A.J. (2001). Resolution reconsidered—Conventional approaches and an alternative. In *Advances in Imaging and Electron Physics*, (P.W. Hawkes, Ed.), Vol, **117**, 241–360 (Academic Press, San Diego).

Van Dyck, D. and Chen, J.H. (1999). A simple theory for dynamical electron diffraction in crystals. *Sol. State Comm.* **109**, 501–505.

Van Dyck, D. and Coene, W. (1987). A new procedure for wave function restoration in high resolution electron microscopy. *Optik* **77**(3), 125–128.

Van Dyck, D. and Op de Beeck, M. (1996). A simple intuitive theory for electron diffraction. *Ultramicroscopy* **64**, 99–107.

Van Dyck, D., Danckaert, J., Coene, W., Selderslaghs, E., Broddin, D., Van Landuyt, J. and Amelinckx, S. (1989). The atom column approximation in dynamical electron diffraction calculations. In *Computer Simulation of Electron Microscope Diffraction and Images*. (W. Krakow and M. O'Keefe, Eds.), 107–134. (The Minerals, Metals & Materials Society, Warrendale).

Van Dyck, D., Op de Beeck, M. and Coene, W. (1993). A new approach to object wave function reconstruction in electron microscopy. *Optik* **93**(3), 103–107.

Van Dyck, D., Van Aert, S., den Dekker, A.J. and van den Bos, A. (2003). Is atomic resolution transmission electron microscopy able to resolve and refine amorphous structures? *Ultramicroscopy* **98**, 27–42.

Verbeeck, J. and Van Aert, S. (2004). Model based quantification of EELS spectra. *Ultramicroscopy* **101**, 207–224.

Wada, Y. (1996). Atom electronics: A proposal of nano-scale devices based on atom/molecule switching. *Microelectr. Eng.* **30**, 375–382.

Walsh, D.O. and Nielsen-Delaney, P.A. (1994). Direct method for superresolution. *J. Opt. Soc. Amer. A* **11**, 572–579.

Wang, Z.L. (2001). Inelastic scattering in electron microscopy—Effects, spectrometry and imaging. In *Progress in Transmission Electron Microscopy 1—Concepts and Techniques* (X.-F. Zhang and Z. Zhang, Eds.), 113–159. (Springer-Verlag, Berlin).

Index

Printed in the United States of America